Handbook of
Experimental Pharmacology

Volume 131

Editorial Board

G.V.R. Born, London
P. Cuatrecasas, Ann Arbor, MI
D. Ganten, Berlin
H. Herken, Berlin
K. Starke, Freiburg i. Br.

Springer
*Berlin
Heidelberg
New York
Barcelona
Budapest
Hong Kong
London
Milan
Paris
Santa Clara
Singapore
Tokyo*

Antisense Research and Application

Contributors

S. Agrawal, T. Akiyama, C.F. Bennett, M. Butler,
B.J. Chiasson, T.P. Condon, P.D. Cook, S.J. Craig,
R.M. Crooke, S.T. Crooke, G. Davidkova, N.M. Dean,
F.A. Dorr, D. Fabbro, R.S. Geary, T. Geiger, A.M. Gewirtz,
M.O. Hebb, S.P. Henry, M. Hogan, S.L. Hutcherson,
F. Kalkbrenner, D.L. Kisner, A.M. Krieg, J.M. Leeds,
A.A. Levin, R.R. Martin, B.P. Monia, D.K. Monteith,
M. Müller, P.L. Nicklin, P.E. Nielsen, J.A. Phillips,
H.A. Robertson, P.J. Schechter, G. Schultz, W.R. Shanahan,
Jr., M.V. Templin, B. Weiss, B. Wittig, R. Zhang

Editor
Stanley T. Crooke

 Springer

Stanley T. Crooke, M.D., Ph.D.
ISIS Pharmaceuticals, Inc.
Carlsbad Research Center
2292 Faraday Avenue
Carlsbad, CA 92008
USA

With 118 Figures and 42 Tables

ISBN 3-540-63833-4 Springer-Verlag Berlin Heidelberg New York

Library of Congress Cataloging-in-Publication Data

Antisense research and application/contributors, S. Agrawel ... [et al.]; editor, Stanley T. Crooke. p. cm. — (Handbook of experimental pharmacology; v. 131) Includes bibliographical references and index. ISBN 3-540-63833-4 (hardcover: alk. paper) 1. Antisense nucleic acids—Therapeutic use. I. Agrawal, Sudhir. II. Crooke, Stanley T. III. Series. [DNLM: 1. Antisense Elements (Genetics)—phermacology. 2. Antisense Elements (Genetics)—therapeutic use. W1 HA51L v. 131 1998/QV 185 A833 1998] QP905.H3 vol. 131 [RM666.A564] 815'.1 s—dc21 [815'.31] DNLM/DLC for Library of Congress
97-48588
CIP

This work is subject to copyright. All rights are reserved, whether the whole or part of the material is concerned, specifically the rights of translation, reprinting, reuse of illustrations, recitation, broadcasting, reproduction on microfilm or in any other way, and storage in data banks. Duplication of this publication or parts thereof is permitted only under the provisions of the German Copyright Law of September 9, 1965, in its current version, and permission for use must always be obtained from Springer-Verlag. Violations are liable for prosecution under the German Copyright Law.

© Springer-Verlag Berlin Heidelberg 1998
Printed in Germany

The use of general descriptive names, registered names, trademarks, etc. in this publication does not imply, even in the absence of a specific statement, that such names are exempt from the relevant protective laws and regulations and therefore free for general use.

Product liability: The publishers cannot guarantee the accuracy of any information about dosage and application contained in this book. In every individual case the user must check such information by consulting the relevant literature.

Cover design: *design & production* GmbH, Heidelberg

Typesetting: Best-set Typesetter Ltd., Hong Kong

Production Editor: Angélique Gcouta

SPIN: 10559768   27/3020 – 5 4 3 2 1 0 – Printed on acid-free paper

Preface

For nearly two decades advances in molecular biology have resulted in extraordinary insights into normal and pathophysiological processes. Indeed, I believe we are in the midst of a process that will redefine diseases in molecular pathological terms. This process has already shown the central role that a relatively few complex multigene families play in physiological processes and in pathophysiological events. In the next few years the rate at which new molecular targets will be identified will increase exponentially as the sequencing of the human and other genomes proceeds.

As exciting as the progress in the elucidation of molecular physiological process is, the identification of potential molecular targets is only the very first step in the discovery and development of improved therapeutic modalities for the vast array of diseases that remain undertreated. In fact, the identification of multigene families, homologies between individual members of these families, and the likely roles of individual isotypes in disease states raises an important question: How will we create drugs of sufficient specificity to selectively interact with individual isotypes of multigene families?

That antisense technology has arguably the best potential to rationally design isotype selective inhibitors was obvious from its conception and led to the excitement about the technology, both as a tool to dissect the roles of various genes and to create new therapeutics. The question about antisense was: Would it work?

The chapters in this volume, recent publications, and recent symposia – such as the meeting sponsored recently by *Nature Biotechnology* – provide reasonably compelling answers to the questions about the technology. All in all, the data provide ample justification for cautious optimism. This technology is clearly a remarkably valuable tool for dissecting pharmacological processes and confirming the roles of various genes. Perhaps more importantly, it appears that even the first-generation compounds, the phosphorothioates, may have properties that support their use as drugs for selected indications, and new generations of antisense drugs may broaden the therapeutic utility of drugs based on this technology.

Nevertheless, it is important to remember that we are less than a decade into the aggressive creation and evaluation of antisense technology. Moreover, we are attempting to create an entirely new branch of pharmacology: a new chemical class, oligonucleotides; a new receptor, RNA; a new drug receptor binding motif, hybridization; and new post-receptor binding mechanisms. Thus there are still many more questions than answers. So arguably we are at the end of the beginning of this technology. There is a great deal more to do before we understand the true value and limits of antisense, but we are buoyed by the progress to date and look forward to the challenges ahead.

STANLEY T. CROOKE

List of Contributors

AGRAWAL, S., Hybridon, Inc., 620 Memorial Drive, Cambridge, MA 02139, USA

AKIYAMA, T., Department of Oncology, The Institute of Medical Science, The University of Tokyo, 4-6-1 Shirokanedai Minato-ku, Tokyo 108, Japan

BENNETT, C.F., Isis Pharmaceuticals, Inc., Carlsbad Research Center, 2292 Faraday Ave., Carlsbad, CA 92008, USA

BUTLER, M., Isis Pharmaceuticals, Inc., Carlsbad Research Center, 2292 Faraday Ave., Carlsbad, CA 92008, USA

CHIASSON, B.J., Neurobiology Research Group, Department of Anatomy and Cell Biology, Medical Sciences Bldg., 1 King's College Circle, University of Toronto, Toronto, Ontario, Canada M5S 1A8
Current address: Novartis Pharmaceuticals Canada Inc.
525 University Avenue, Suite 1025, Toronto,
Ontario, M5G 2L3 Canada

CONDON, T.P., Isis Pharmaceuticals, Inc., Carlsbad Research Center, 2292 Faraday Ave., Carlsbad, CA 92008, USA

COOK, P.D., Isis Pharmaceuticals, Inc., Carlsbad Research Center, 2292 Faraday Ave., Carlsbad, CA 92008, USA

CRAIG, S.J., Novartis Horsham Research Centre, Wimblehurst Rd., Horsham, West Sussex RH12 4AB, Great Britain

CROOKE, R.M., Isis Pharmaceuticals, Inc., Carlsbad Research Center, 2292 Faraday Ave., Carlsbad, CA 92008, USA

CROOKE, S.T., Isis Pharmaceuticals, Inc., Carlsbad Research Center, 2292 Faraday Ave., Carlsbad, CA 92008, USA

DAVIDKOVA, G., Division of Neuropharmacology, Department of Pharmacology, MCP Hahnemann School of Medicine, Allegheny University of Health Sciences, 3200 Henry Avenue, Philadelphia, PA 19129, USA

DEAN, N.M., Department of Pharmacology, Isis Pharmaceuticals, Inc.,
  Carlsbad Research Center, 2292 Faraday Ave.,
  Carlsbad, CA 92008, USA

DORR, F.A., Isis Pharmaceuticals, Inc., Carlsbad Research Center,
  2292 Faraday Ave., Carlsbad, CA 92008, USA

FABBRO, D., Pharmaceuticals Research, Oncology, K-125.4.10,
  Unit: Growth Control, Novartis, Ltd., CH-4002 Basel, Switzerland

GEARY, R.S., Isis Pharmaceuticals, Inc., Carlsbad Research Center,
  2292 Faraday Ave., Carlsbad, CA 92008, USA

GEIGER, T., Pharmaceuticals Research, Oncology, K-125.2.12,
  Unit: Growth Control, Novartis, Ltd., CH-4002 Basel, Switzerland

GEWIRTZ, A.M., Departments of Pathology and Laboratory Medicine and
  Internal Medicine, University of Pennsylvania School of Medicine,
  Room 513b Stellar Chance Building, 422 Curie Boulevard, Philadelphia,
  PA 19104, USA

HEBB, M.O., Laboratory of Molecular, Neurobiology, Department of
  Pharmacology, Dalhousie University, Halifax, Nova Scotia,
  Canada B3H 4H7

HENRY, S.P., Isis Pharmaceuticals, Inc., Carlsbad Researh Center,
  2292 Faraday Ave., Carlsbad, CA 92008, USA

HOGAN, M., Department of Molecular Physiology and Biophysics,
  Baylor College of Medicine, T405, One Baylor Plaza, Houston,
  TX 77030, USA

HUTCHERSON, S.L., Isis Pharmaceuticals, Inc., Carlsbad Research Center,
  2292 Faraday Ave., Carlsbad, CA 92008, USA

KALKBRENNER, F., Institut für Pharmakologie, Freie Universität Berlin,
  Thielallee 69/73, D-14195 Berlin, FRG

KISNER, D.L., Isis Pharmaceuticals, Inc., Carlsbad Research Center,
  2292 Faraday Ave., Carlsbad, CA 92008, USA

KRIEG, A.M., University of Iowa, Department of Internal Medicine,
  540 EMRB, Iowa City, IA 52242, USA

LEEDS, J.M., Isis Pharmaceuticals, Inc., Carlsbad Research Center,
  2292 Faraday Ave., Carlsbad, CA 92008, USA

LEVIN, A.A., Isis Pharmaceuticals, Inc., Carlsbad Research Center,
  2292 Faraday Ave., Carlsbad, CA 92008, USA

MARTIN, R.R., Drug Development, Hybridon, Inc., 620 Memorial Drive,
  Cambridge, MA 02139, USA

List of Contributors

MONIA, B.P., Department of Molecular Pharmacology, Isis Pharmaceuticals, Inc., Carlsbad Research Center, 2292 Faraday Ave., Carlsbad, CA 92008, USA

MONTEITH, D.K., Isis Pharmaceuticals, Inc., Carlsbad Research Center, 2292 Faraday Ave., Carlsbad, CA 92008, USA

MÜLLER, M., Pharmaceuticals Research, Oncology, K-125.4.43, Unit: Growth Control, Novartis, Ltd., CH-4002 Basel, Switzerland

NICKLIN, P.L., Novartis Horsham Research Centre, Wimblehurst Rd., Horsham, West Sussex RH12 4AB, Great Britain

NIELSEN, P.E., Center for Biomolecular Recognition, Department of Medical Biochemistry and Genetics, Biochemical Laboratory B, The Panum Institute, Blegdamsvej 3c, DK-2200 Copenhagen N, Denmark

PHILLIPS, J.A., Novartis Horsham Research Centre, Wimblehurst Rd., Horsham, West Sussex RH12 4AB, Great Britain

ROBERTSON, H.A., Laboratory of Molecular, Neurobiology, Department of Pharmacology, Dalhousie University, Halifax, Nova Scotia, Canada B3H 4H7

SCHECHTER, P.J., Drug Development, Hybridon, Inc., 620 Memorial Drive, Cambridge, MA 02139, USA

SCHULTZ, G., Institut für Pharmakologie, Freie Universität Berlin, Thielallee 69/73, D-14195 Berlin, FRG

SHANAHAN, W.R., Jr., Isis Pharmaceuticals, Inc., Carlsbad Research Center, 2292 Faraday Ave., Carlsbad, CA 92008, USA

TEMPLIN, M.V., Isis Pharmaceuticals, Inc., Carlsbad Research Center, 2292 Faraday Ave., Carlsbad, CA 92008, USA

WEISS, B., Division of Neuropharmacology, Department of Pharmacology, MCP Hahnemann School of Medicine, Allegheny University of Health Sciences, 3200 Henry Avenue, Philadelphia, PA 19129, USA

WITTIG, B., Institut für Molekularbiologie und Biochemie, Freie Universität Berlin, Arminallee 22–24, D-14195 Berlin, FRG

ZHANG, R., Division of Clinical Pharmacology and Department of Pharmacology and Toxicology, University of Alabama at Birmingham, Volker Hall 113, Box 600, 1670 University Boulevard, Birmingham, AL 35294-0019, USA

Contents

CHAPTER 1

Basic Principles of Antisense Therapeutics
S.T. CROOKE. With 3 Figures 1

  A. Introduction 1
  B. Proof of Mechanism 1
      I. Factors that May Influence Experimental
         Interpretations 1
         1. Oligonucleotide Purity 1
         2. Oligonucleotide Structure 2
         3. RNA Structure 2
         4. Variations in In Vitro Cellular Uptake and
            Distribution 2
         5. The Binding to and Effects of Binding to
            Nonnucleic Acid Targets 3
         6. Terminating Mechanisms 3
         7. Effects of "Control Oligonucleotides" 4
         8. Kinetics of Effects 4
      II. Recommendations 4
         1. Positive Demonstration of Antisense
            Mechanism and Specificity 4
  C. Molecular Mechanisms of Antisense Drugs 5
      I. Occupancy Only Mediated Mechanisms 5
         1. Inhibition of Splicing 5
         2. Translational Arrest 6
         3. Disruption of Necessary RNA Structure 7
      II. Occupancy Activated Destabilization 8
         1. 5′ Capping 8
         2. Inhibition of 3′ Polyadenylation 9
      III. Other Mechanisms 9
      IV. Activation of RNase H 10
      V. Activation of Double-Strand RNase 12
  D. Characteristics of Phosphorothioate Oligodeoxynucleotides . . . 12
      I. Introduction 12
      II. Hybridization 13

III.	Interactions with Proteins	13
IV.	Pharmacokinetic Properties	15
	1. Nuclease Stability	15
	2. In Vitro Cellular Uptake	16
	3. In Vivo Pharmacokinetics	16
V.	Pharmacological Properties	20
	1. Molecular Pharmacology	20
	2. In Vivo Pharmacological Activities	22
VI.	Toxicological Properties	26
	1. In Vitro	26
	2. Genotoxicity	27
	3. In Vivo	28
VII.	Therapeutic Index	29
VIII.	Conclusions	30
E. The Medicinal Chemistry of Oligonucleotides		31
I.	Introduction	31
II.	Heterocycle Modifications	32
	1. Pyrimidine Modifications	32
	2. Purine Modifications	33
	3. Oligonucleotide Conjugates	33
	a) Nuclease Stability	34
	b) Enhanced Cellular Uptake	34
	c) RNA Cleaving Groups	35
	d) In Vivo Effects	35
	4. Sugar Modifications	36
	5. Backbone Modifications	37
III.	Conclusions	38
References		38

CHAPTER 2

Antisense Medicinal Chemistry
D. Cook. With 11 Figures 51

A.	Introduction	51
B.	Scope of the Review	51
C.	Phosphorothioates as Antisense Drug Candidates	52
D.	Phosphorothioate Limitations	53
	I. Pharmacodynamic Limitations	53
	II. Pharmacokinetic Limitations	54
	III. Toxicological Limitations	54
E.	Types of Oligonucleotide Modifications	55
F.	Drug Properties of Antisense Oligonucleotides that May Be Altered by Chemical Modifications	55

Contents   XIII

   G. Why Is it Important to Continue to Design and Synthesize
      Antisense Oligonucleotides with Increased Affinity for
      Target RNA and with Greater Resistance to Nucleolytic
      Degradation? 56
   H. What Are the Standards for Binding Affinity,
      Nuclease Resistance, and Other Properties Recently Achieved
      by Structure–Property/Activity Relationship Studies
      of Oligonucleotides? 57
   I. Summary of Key Oligonucleotide Modifications 59
      I. Heterocycles (Nucleobases) 59
          1. 5′-Position Pyrimidine- and Tricyclic Cytidine-
             Modified Oligonucleotides 59
          2. N^2-Purine-Modified Oligonucleotides 61
          3. 7-Deaza-Modified Oligonucleotides 63
          4. Additional Advantages of Heterocycle-Modified
             Oligonucleotides 64
          5. Toxicity Liability of Heterocyclic Modifications 65
      II. Carbohydrate-Modified Oligonucleotides 66
          1. Basis for Design 67
          2. 4′-Modified Oligonucleotides 67
          3. Sugar Conformationally Restricted
             Oligonucleotides 67
          4. Morpholino Diamidate-Modified Oligomers 69
          5. 2′-Modified Oligonucleotides 69
             a) Structure–Property/Activity Relationship
                Studies 69
                α) Binding Properties of 2′-O-Modified
                   Oligonucleotides 70
                β) Stability of 2′-O-Modified Oligonucleotides 71
             b) Modes of Action of 2′-Modified
               Oligonucleotides 72
                α) Chimera 2′-O-Modified Oligonucleotides –
                   RNase H-Dependent Mode of Action (Gapmer
                   Technology) 72
                β) 2′-O-Modified Oligonucleotides with an RNase
                   H-Independent Mode of Action 76
             c) Selected, Interesting Antisense Results With 2′-O-
               Modified Oligonucleotides in Gapmer Motifs 77
             d) Pharmacokinetic Properties of 2′-Modified
               Oligonucleotides 78
             e) Toxicity Implications of 2′-Modified
               Oligonucleotides 80
             f) Summary of 2′-Sugar Modified
               Oligonucleotides 81

    III. Backbone- or Linkage-Modified Oligonucleotides 82
        1. MMI- and Amide-3-Modified
           Oligonucleosides/tides 82
        2. 3'-Amidate-Modified Oligonucleotides 84
        3. 3'-Thioformacetal- and Formacetal-Modified
           Oligomers 85
        4. 2'-Modified Rp-Me-Phosphonates 85
        5. Peptide Nucleic Acid 85
        6. General Observations About Backbone
           Research 86
    IV. Oligonucleotide Pendants (Conjugates) 87
J. Conclusions 90
References 92

CHAPTER 3

In Vitro Cellular Uptake, Distribution, and Metabolism of Oligonucleotides
R.M. CROOKE. With 14 Figures 103

A. Introduction 103
B. Stability in the Extracellular Milieu 104
    I. Phosphodiester Oligonucleotides 104
    II. Methylphosphonate Oligonucleotides 105
    III. Phosphorothioate Oligonucleotides 106
C. In Vitro Uptake, Intracellular Distribution, and Efflux 106
    I. Phosphodiester Oligonucleotides 106
        1. Uptake 106
        2. Intracellular Distribution 109
        3. Efflux 110
    II. Methylphosphonate Oligonucleotides 111
        1. Uptake 111
        2. Intracellular Distribution 112
        3. Efflux 112
    III. Phosphorothioate Oligonucleotides 113
        1. Uptake 113
        2. Intracellular Distribution 116
        3. Efflux 119
D. Intracellular Stability 120
    I. Phosphodiester Oligonucleotides 101
    II. Methylphosphonate Oligonucleotides 122
    III. Phosphorothioate Oligonucleotides 122
E. Uptake Enhancement Methods 128
F. Conclusions 130
References 133

CHAPTER 4

Pharmacokinetic Properties of Phosphorothioates in Animals – Absorption, Distribution, Metabolism and Elimination
P.L. NICKLIN, S.J. CRAIG, and J.A. PHILLIPS. With 12 Figures 141

- A. Introduction 141
    - I. Phosphorothioate Modification 143
    - II. Oligonucleotide Detection and Analysis 144
- B. Distribution 145
    - I. Blood Kinetics 145
    - II. Tissue Distribution 147
    - III. Cellular Uptake In Vivo 147
    - IV. Dose Dependence 150
    - V. Sequence Dependence 150
    - VI. Multiple Dosing 153
- C. Metabolism 153
    - I. Plasma 155
    - II. Tissues 155
- D. Elimination 159
- E. Absorption 161
    - I. Parenteral Administration 162
    - II. Local Administration 164
    - III. Non-parenteral Administration 164
- F. Conclusion 165
- References 166

CHAPTER 5

Toxicity of Oligodeoxynucleotide Therapeutic Agents
A.A. LEVIN, D.K. MONTEITH, J.M. LEEDS, P.L. NICKLIN, R.S. GEARY, M. BUTLER, M.V. TEMPLIN, and S.P. HENRY. With 13 Figures 169

- A. Introduction 169
- B. Pharmacokinetics and Metabolism 173
- C. Toxicity of Phosphorothioate Oligodeoxynucleotides 179
    - I. Genetic Toxicity 179
    - II. Acute Toxicities 179
        - 1. Complement Activation 180
            - a) Effects of Oligodeoxynucleotide Chemistries 186
        - 2. Hemostasis 187
            - a) Effects of Oligodeoxynucleotide Chemistries 191
    - III. Toxicologic Effects Associated with Chronic Exposure 192
        - 1. Immune Stimulation 192
            - a) Effects of Oligodeoxynucleotide Chemistries 199

       2. Hepatic Effects of Oligodeoxynucleotide
          Treatment 200
       3. Renal Effects of Oligodeoxynucleotide Treatment ... 202
       4. Hematopoiesis 205
  IV. Reproductive Effects of Antisense
      Oligodeoxynucleotides 205
  V. Ocular Toxicity of Antisense Therapeutic Agents 206
       1. Ocular Pharmacokinetics 207
       2. Ocular Toxicity Profile 208
References 210

CHAPTER 6

Pharmacokinetic Properties of Phosphorothioate Oligonucleotides in Humans
J.M. LEEDS and R.S. GEARY. With 4 Figures 217

  A. Introduction 217
  B. Human Pharmacokinetics 218
    I. Absorption 219
    II. Plasma Pharmacokinetics 220
       1. Peak Plasma Concentrations 220
       2. Calculated Parameters 221
    III. Metabolism 223
    IV. Elimination 228
  C. Pharmodynamics 230
References 231

CHAPTER 7

Safety and Tolerance of Phosphorothioates in Humans
P.J. SCHECHTER and R.R. MARTIN. With 3 Figures 233

  A. Introduction 233
  B. Cardiovascular Safety 233
  C. Hematological Safety 234
    I. Coagulation 234
    II. Platelets 235
    III. Red Blood Cells 236
  D. Complement Activation 237
  E. Specific Organ Toxicity 237
  F. Conclusion 239
References 239

CHAPTER 8

Immune Stimulation by Oligonucleotides
A.M. KRIEG 243

A. Sequence-Independent Immune Effects of the DNA
   Backbone 243
B. Sequence-Specific Immune Effects of CpG
   Motifs in Oligodeoxynucleotides 245
   I. Immune Activation by Bacterial DNA and the
      Identification of Stimulatory Palindromes 245
   II. Immune Activation by Antisense and Control
       Oligodeoxynucleotides 246
   III. Discovery of Immune Activation by CpG Motifs 247
   IV. Immune Effects of CpG Motifs 250
       1. Mitogenic and Anti-apoptotic Effects of
          Oligodeoxynucleotides 250
          a) Virtually All B Cells Can Respond to the Mitogenic
             Effects of CpG Oligodeoxynucleotides 250
          b) CpG DNA Blocks B Cell Apoptosis 251
          c) Induction of B Cell Cytokine and Ig Secretion by
             CpG DNA 251
       2. Induction of Monocyte Cytokine Secretion by
          CpG DNA 252
       3. Induction of Natural Killer Interferon-g Secretion
          and Lytic Activity by CpG DNA 253
   V. Role of the Phosphorothioate Backbone in the Immune
      Effects of CpG Oligodeoxynucleotides 254
   VI. Mechanism of Leukocyte Activation by CpG
       Oligodeoxynucleotides 254
   VII. Teleologic Significance of Immune Activation by CpG
        Motifs in DNA 256
C. Sequence-Specific Immune Effects of Poly(G) Motifs in
   Oligodeoxynucleotides 258
References 258

CHAPTER 9

**Pharmacological Inhibition of Dopaminergic and Other
Neurotransmitter Receptors Using Antisense Oligodeoxynucleotides**
G. DAVIDKOVA and B. WEISS. With 5 Figures 263

A. Introduction 263
B. Advantages of Antisense Oligodeoxynucleotides over
   Traditional Pharmacological Antagonists 264
C. Factors to Consider in the Development of Antisense
   Oligodeoxynucleotides Targeted to Neurotransmitter
   Receptors 266
   I. Structure of Antisense Oligodeoxynucleotides: Optimal
      Recognition Site on the Targeted Transcript and

Oligodeoxynucleotide Length and Chemical
  Structure 266
 II. In Vitro and In Vivo Substrates for Studying Antisense
  Oligodeoxynucleotides 272
 III. Delivery Systems 273
D. Uptake and Distribution of Antisense
 Oligodeoxynucleotides 273
E. Mechanism of Action of Antisense Oligodeoxynucleotides 276
 I. Mechanisms of Inhibition of Gene Expression 276
 II. Appropriate Controls to Determine the Specificity of
  Antisense Effects 277
 III. Functional, Biochemical, and Molecular Consequences
  of Antisense Inhibition 277
 IV. Antisense Oligodeoxynucleotides Inhibit a Pool of
  Functional Neurotransmitter Receptors 278
F. In Vitro and In Vivo Effects of Antisense
 Oligodeoxynucleotides Targeted to the Transcripts Encoding
 Neurotransmitter Receptors 281
 I. Acetylcholine Receptors 281
 II. Adrenergic Receptors 282
 III. Dopamine Receptors 283
  1. In Vivo Models for Studying the Effects of Dopamine
   Receptor Antisense Oligodeoxynucleotides 285
  2. Design and Synthesis of Antisense
   Oligodeoxynucleotides Targeted to the Dopamine
   Receptor Intended for In Vivo Use 286
  3. D_1 Dopamine Receptor Antisense 286
   a) In Vivo Studies 286
   b) In Vitro Studies 289
  4. D_2 Dopamine Receptor Antisense 290
   a) In Vivo Studies 290
   b) In Vitro Studies 293
  5. D_3 Dopamine Receptor Antisense 294
 IV. GABA Receptors 295
 V. NMDA Receptors 298
 VI. Serotonin Receptors 299
G. Future Directions: Comparison Between the Properties of
 Antisense Oligodeoxynucleotides and Antisense RNA
 Produced by Mammalian Expression Vectors 300
H. Conclusions 301
 References 302

CHAPTER 10

Pharmacological Effects of Antisense Oligonucleotide Inhibition of Immediate-Early Response Genes in the CNS
B.J. CHIASSON, M.O. HEBB, and H.A. ROBERTSON. With 3 Figures 309

A. Introduction 309
B. Some Correlative Studies Suggesting a Role of Immediate-
   Early Genes in Brain Function 309
C. Immediate-Early Genes and Stimulus-Transcription
   Coupling 311
   I. A Model for Change 311
   II. The Origin of Immediate-Early Genes 312
   III. The Immediate-Early Gene as an Inducible
        Transcription Factor 312
   IV. Immediate-Early Gene Transcription Factors and Their
        Targets 313
D. Studies Using Antisense Oligodeoxynucleotides in the CNS ... 315
   I. The Induction of c-*fos* in the Striatum 315
   II. Knockdown of Amphetamine-Induced c-*fos* Using
        Antisense Oligodeoxynucleotides: Studies of the Basal
        Ganglia Function 317
   III. The Amygdala and Gene Expression 322
   IV. Exploring Alternative Modifications to
        Oligonucleotides 328
   V. Reconciling Differences 330
   VI. Other Studies Using Antisense Oligonucleotides in
        Nervous Tissue 331
E. Neural Plasticity and the Role of c-*fos* as Demonstrated
   by Antisense Technology 332
References 333

CHAPTER 11

Inhibition of G Proteins by Antisense Drugs
F. KALKBRENNER, B. WITTIG, and G. SCHULTZ. With 4 Figures 341

A. Introduction 341
B. Antisense Approaches to Study Signal Transduction
   Pathways 344
C. Inhibition of G-Protein Functions by Antisense RNA and
   Genomic Knockout 345
D. Inhibition of G-Protein Functions by Antisense
   Oligodeoxynucleotides in Cell Culture 350
   I. Nuclear Microinjection of G-Protein Antisense
        Oligodeoxynucleotides 350
   II. Signal Transduction Pathways Leading to G-Protein-
        Mediated Inhibition of Voltage-Gated Calcium
        Channels in Neuroendocrine Cells 352
   III. Signal Transduction Pathways Leading to a G-Protein-
        Mediated Increase in the Intracellular Calcium
        Concentration 354

　　　　IV. Specificity of Antisense Effects 356
　　　　V. Application of G-Protein Antisense
　　　　　　Oligodeoxynucleotides Through the Patch-
　　　　　　Clamp Pipette 357
　　　　VI. Application of G-Protein Antisense
　　　　　　Oligodeoxynucleotides by Adding to the Cell
　　　　　　Culture Medium 358
　　E. Inhibition of G-Protein Functions by Antisense
　　　　Oligodeoxynucleotides in Animals 360
　　F. Perspectives 363
　　References 364

CHAPTER 12

Use of Antisense Oligonucleotides to Modify Inflammatory Processes
C.F. BENNETT and T.P. CONDON. With 1 Figure 371

　　A. Introduction 371
　　B. Cell Adhesion Molecules 373
　　　　I. Intercellular Adhesion Molecule 1 376
　　　　　　1. Pharmacology of ICAM-1 Antisense
　　　　　　　　Oligonucleotides 377
　　　　　　　　a) Proof of Mechanism 377
　　　　　　　　b) Human Xenografts 378
　　　　　　　　c) Rodent Allografts 379
　　　　　　　　d) Colitis 381
　　　　　　　　e) Renal Ischemia 382
　　　　　　2. Toxicology of ICAM-1 Antisense Oligonucleotides ... 382
　　　　　　3. Clinical Studies with ISIS 2302 383
　　　　　　4. Second and Third Generation Chemistry 383
　　　　II. Other Endothelial Cell–Leukocyte Adhesion
　　　　　　Molecules 384
　　C. Nuclear Factor-κB 385
　　D. Interleukin-1 Receptors 386
　　E. Conclusions 386
　　References 387

CHAPTER 13

Antisense Oligonucleotides and Their Anticancer Activities
D. FABBRO, M. MÜLLER, and T. GEIGER. With 6 Figures 395

A. Potential Cancer Targets for Pharmaceutical Intervention 395
B. Antisense-Based Approaches and Cancer 396
C. Antisense Targets Related to Solid Tumors 396
　　I. *ras* and *raf* 396

>   II. Growth Factors and Growth Factor Receptors 404
>   III. Proteases 405
>   IV. Multidrug Resistance 405
>   V. Miscellaneous 406
> D. Antisense Targets Related to Hematological Malignancies 407
>   I. Chronic Myelogenous Leukemia and bcr-abl 407
>   II. *myc*, *myb*, and *bcl*-2 408
> E. Protein Kinase C-a and *raf* as Cancer Targets for Antisense ... 408
>   I. In Vivo Activity of Phosphorothioate Antisense Oligonucleotide Targeting Protein Kinase C-a and c-*raf* 408
>   II. In Vivo Activity of Modified and Formulated Antisense Oligonucleotides Targeting c-*raf* 410
>   1. 2′-Methoxy-ethoxy-Modified Oligonucleotides 410
>   2. Antitumor Activity of CGP 69846A in Stealth Liposomes 411
>   3. Antitumor Activity of CGP 69846A with Dextran Sulfate or as Cholesterol Conjugates of CGP 69846A 411
> F. Concluding Remarks 414
> References 417

CHAPTER 14

Pharmacological Activity of Antisense Oligonucleotides in Animal Models of Disease
B.P. MONIA and N.M. DEAN 427

> A. Introduction 427
> B. Vascular System 427
>   I. Restenosis 427
>   II. Atherosclerosis 432
>   III. Hypertension 434
>   IV. Ocular Neovascularization 435
> C. Skin 436
> D. Peripheral Organs 437
>   I. Liver 438
>   II. Kidney 439
>   III. Lung 439
> E. Conclusions and Future Directions 440
> References 440

CHAPTER 15

Clincial Antiviral Activities
S.L. HUTCHERSON. With 2 Figures 445

A. Introduction 445
B. Antisense Oligonucleotide Antiviral Pharmacology 447
C. Pharmacokinetic Assessment Strategies 448
D. Toxicology for Antisense Oligonucleotide Antivirals 449
E. Chemical Synthesis Scale-Up and Chemical Development 450
F. Clinical Research of Antisense Oligonucleotides 450
G. Future Antisense Antiviral Development 455
References 456

CHAPTER 16

Antisense Oligonucleotides to Protein Kinase C-α and C-*raf* Kinase: Rationale and Clinical Experience in Patients with Solid Tumors
F.A. DORR and D.L. KISNER 463

A. Introduction 463
B. Rationale for the Development of an Inhibitor of Protein
   Kinase C (ISIS 3521) 464
C. Rationale for the Development of an Inhibitor of C-*raf*
   Kinase (ISIS 5132) 465
D. Preclinical Toxicology and Pharmacokinetics 466
E. Status of Clinical Development 468
   I. ISIS 3521 468
      1. Clinical Study 1 468
      2. Clinical Study 2 469
   II. ISIS 5132 471
      1. Clinical Study 1 471
      2. Clinical Study 2 472
F. Future Clinical Development 473
References 474

CHAPTER 17

Nucleic Acid Therapeutics for Human Leukemia: Development and Early Clinical Experience with Oligodeoxynucleotides Directed at c-*myb*
A.M. GEWIRTZ. With 10 Figures 477

A. Introduction 477
B. c-*myb* Proto-oncogene 477
   I. Targeting the c-*myb* Gene 479
   II. In Vitro Experience in the Hematopoietic
       Cell System 479
       1. Role of c-*myb*-Encoded Protein in Normal Human
          Hematopoiesis 479

2. Myb Protein Requirement for Leukemic
     Hematopoiesis 480
  3. Differential Reliance of Normal and Leukemic
     Progenitor Cells on c-*myb* Function 483
  4. Use of c-*myb* Oligodeoxynucleotides as Bone
     Marrow-Purging Agents 484
  5. Efficacy of c-*myb* Oligodeoxynucleotides In Vivo:
     Development of Animal Models 485
 III. Why Downregulating *myb* Kills Leukemic Cells
      Preferentially: A Hypothesis 487
 IV. Pharmacodynamic Studies with an *myb*-Targeted
     Oligodeoxynucleotide 488
C. Use of Antisense Oligonucleotides in a Clinical Setting 491
D. Future Outlook 492
E. Conclusions 494
References 495

CHAPTER 18

Properties of ISIS 2302, an Inhibitor of Intercellular Adhesion Molecule-1, Humans
W.R. SHANAHAN JR. With 8 Figures 499

A. Introduction 499
B. Phase I Intravenous Trial 501
    I. Methods and Material 501
   II. Results 502
       1. Safety 502
       2. Pharmacokinetics 504
  III. Discussion 510
C. Phase IIa Trials 513
    I. Crohn's Disease 514
   II. Rheumatoid Arthritis 517
  III. Ulcerative Colitis 517
   IV. Psoriasis 517
    V. Renal Transplantation 518
D. Exploration of Subcutaneous Dosing 519
E. Future Plans 523
References 521

CHAPTER 19

Pharmacokinetics and Bioavailability of Antisense Oligonucleotides Following Oral and Colorectal Administrations in Experimental Animals
S. AGRAWAL and R. ZHANG. With 10 Figures 525

A. Introduction 525
B. Oral Administration 526
   I. Experimental Design 526
      1. Synthesis of Unlabeled and [^{35}S]-Labeled
         Oligonucleotides 526
      2. Animals and Drug Administration 526
      3. Absolute Bioavailability 527
      4. Sample Preparation and Total Radioactivity
         Measurements 528
      5. Gel Electrophoresis 528
      6. High-Performance Liquid Chromatography
         Analysis 528
   II. Results and Discussion 529
      1. Phosphorothioate Oligonucleotide 529
      2. Mixed-Backbone Oligonucleotide-1 529
      3. Mixed-Backbone Oligonucleotide-2 532
      4. General Discussion 533
C. Colorectal Administration 534
   I. Special Considerations in Experimental Design 534
   II. Results and Discussion 535
      1. Mixed-Backbone Oligonucleotide-1 535
      2. Phosphorothioate Oligonucleotide and End-Modified
         Mixed-Backbone Oligonucleotide-2 536
      3. General Discussion 538
D. Conclusions 540
References 541

CHAPTER 20

Antisense Properties of Peptide Nucleic Acid
P.E. NIELSEN. With 7 Figures 545

A. Introduction 545
B. Chemistry 545
C. Hybridization Properties 545
D. Structure of Peptide Nucleic Acid Complexes 547
E. Antisense Activity (Mechanism of Action) 548
F. Cellular Biology 549
G. Peptide Nucleic Acid Derivatives (Structure–Activity
   Relationships) 549
H. Peptide Nucleic Acid–DNA Chimeras 553
I. Antigene Activity 555
J. Future Prospects 557
References 557

CHAPTER 21

Triple Helix Strategies and Progress
T. AKIYAMA and M. HOGAN. With 2 Figures 561

- A. Introduction 561
- B. Discovery of the Triple Helix Structure 561
- C. Detailed Structure of the Triple Helix 562
  - I. Nuclear Magnetic Resonance Studies 564
    1. Hydrogen Bonding 564
    2. Base Orientation 564
  - II. Other Methods to Investigate Triple Helix Structure ... 565
    1. Infrared Spectroscopy 565
    2. Phase-Sensitive Electrophoresis 565
    3. Electron Microscopy 565
    4. X-ray Crystallography 566
    5. Molecular Dynamics Simulation 566
- D. Stability of the Triple Helix 566
  - I. Thermodynamic Analysis by Spectroscopic and Calorimetric Methods 566
  - II. Thermodynamic Analysis by Biochemical Methods 568
  - III. Factors Influencing the Stability of the Triple Helix 569
    1. Mismatching at Inversion Sites 569
    2. Ionic Strength 570
    3. Effect of pH 571
  - IV. Kinetics of Triple Helix Formation 571
- E. Nucleoside Modification to Improve Triple Helix Stability 571
  - I. Modified Nucleoside to Overcome pH Dependency of the Parallel Triple Helix 571
    1. Cytidine Derivatives 572
    2. Adenosine Derivatives 572
    3. Guanosine Derivatives 573
  - II. New Nucleosides to Recognize Invasion Sites 573
  - III. Modification to Overcome Alkali Metal Mediated Inhibition of Antiparallel Triple Helix Formation 573
- F. Triple Helix Binding Ligands 574
  - I. The Triple Helix as a Target for Small Molecule Recognition 574
  - II. Ethidium Bromide 574
  - III. Benzopyridoindole Derivatives 575
  - IV. Naphtyl-Quinoline Derivative 575
  - V. Other Intercalators 575
  - VI. Minor Groove Binding Ligands 576
  - VII. Polyamines and Basic Oligopeptides 576
- G. Backbone Modification and Strand Switching 577
  - I. DNA and RNA Backbone 577

	II. Chemical Modification of Backbones	577
	III. Alternate Strand Recognition and Minor Helix Groove Bridging	578
H.	Conjugation of Small Compounds Enhance Triple Helix Stability	579
	I. Conjugation of Intercalators	579
	II. Conjugation of Other Small Compounds	580
I.	Inhibition of Transcription by Triple Helix Technology	580
	I. Prospects for Triple Helix Mediated Gene Regulation	580
	II. Inhibition of Transcription Factor Binding by Triple Helix Formation	581
	III. Inhibition of Transcription by Triple Helix Formation in a Cell-free System	581
	IV. Change of Nucleosome Positioning by Triple Helix Formation	584
	V. Transcription Inhibition by Triple Helix Formation in Cell	584
J.	Other Pharmaceutically Interesting Effects of Triple Helix Formation	587
	I. Inhibition of DNA Replication	587
	II. Inhibition of HIV Integration by Triple Helix	587
K.	Targeting Single-Stranded DNA or RNA by Triple Helix Formation	588
L.	DNA Modification Mediated by TFO Binding	589
	I. Sequence-Specific DNA Cleavage	589
	II. Sequence-Specific Alkylation	589
	III. Single-Site Enzymatic Cleavage	589
	IV. Photo-Induced Cross-Linking	590
	V. Targeted Mutagenesis by a Psoralen-Linked TFO	590
M.	Recent Topics and Prospects	591
	I. Artificial DNA Bending by Triple Helix Formation	591
N.	Conclusion	592
References		593
Subject Index		611

CHAPTER 1
Basic Principles of Antisense Therapeutics

S.T. CROOKE

A. Introduction

During the past few years, interest in developing antisense technology and in exploiting it for therapeutic purposes has been intense. Although progress has been gratifyingly rapid, the technology remains in its infancy and the questions that remain to be answered still outnumber the questions for which there are answers. Appropriately, considerable debate continues about the breadth of the utility of the approach and about the type of data required to "prove that a drug works through an antisense mechanism."

The objectives of this review are to provide a summary of recent progress, to assess the status of the technology, to place the technology in the pharmacological context in which it is best understood, and to deal with some of the controversies with regard to the technology and the interpretation of experiments.

B. Proof of Mechanism

I. Factors that May Influence Experimental Interpretations

Clearly, the ultimate biological effect of an oligonucleotide will be influenced by the local concentration of the oligonucleotide at the target RNA, the concentration of the RNA, the rates of synthesis and degradation of the RNA, the type of terminating mechanism, and the rates of the events that result in termination of the RNA's activity. At present, we understand essentially nothing about the interplay of these factors.

1. Oligonucleotide Purity

Currently, phosphorothioate oligonucleotides can be prepared consistently and with excellent purity (S.T. CROOKE and LEBLEU 1993). However, this has only been the case for the past 3–4 years. Prior to that time, synthetic methods were evolving and analytical methods were inadequate. In fact, our laboratory reported that different synthetic and purification procedures resulted in oligonucleotides that varied in cellular toxicity (R.M. CROOKE 1991) and that potency varied from batch to batch. Though there are no longer synthetic problems with phosphorothioates, they, undoubtedly, complicated earlier studies. More importantly, with each new analog class, new synthetic, purification, and analytical challenges are encountered.

2. Oligonucleotide Structure

Antisense oligonucleotides are designed to be single stranded. We now understand that certain sequences, e.g., stretches of guanosine residues, are prone to adopt more complex structures (WYATT et al. 1994). The potential to form secondary and tertiary structures also varies as a function of the chemical class. For example, higher affinity 2′-modified oligonucleotides have a greater tendency to self hybridize, resulting in more stable oligonucleotide duplexes than would be expected based on rules derived from the behavior of oligodeoxynucleotides (S.M. FREIER, unpublished results).

3. RNA Structure

RNA is structured. The structure of the RNA has a profound influence on the affinity of the oligonucleotide and on the rate of binding of the oligonucleotide to its RNA target (FREIER 1993; ECKER 1993). Moreover, RNA structure produces asymmetrical binding sites that then result in very divergent affinity constants, depending on the position of oligonucleotide in that structure (ECKER 1993; LIMA et al. 1992; ECKER et al. 1992). This in turn influences the optimal length of an oligonucleotide needed to achieve maximal affinity. We understand very little about how RNA structure and RNA protein interactions influence antisense drug action.

4. Variations in In Vitro Cellular Uptake and Distribution

Studies in several laboratories have clearly demonstrated that cells in tissue culture may take up phosphorothioate oligonucleotides via an active process and that the uptake of these oligonucleotides is highly variable depending on many conditions (R.M. CROOKE 1991; S.T. CROOKE et al. 1994). Cell type has a dramatic effect on total uptake, kinetics of uptake, and pattern of subcellular distribution. At present, there is no unifying hypothesis to explain these differences. Tissue culture conditions, such as the type of medium, degree of confluence, and the presence of serum, can all have enormous effects on uptake (S.T. CROOKE et al. 1994). Oligonucleotide chemical class obviously influences the characteristics of uptake as well as the mechanism of uptake. Within the phosphorothioate class of oligonucleotides, uptake varies as a function of length, but not linearly. Uptake varies as a function of sequence, and stability in cells is also influenced by sequence (S.T. CROOKE et al. 1994, 1995a).

Given the foregoing, it is obvious that conclusions about in vitro uptake must be very carefully made and generalizations are virtually impossible. Thus, before an oligonucleotide could be said to be inactive in vitro, it should be studied in several cell lines. Furthermore, while it may be absolutely correct that receptor-mediated endocytosis is a mechanism of uptake of phosphorothioate oligonucleotides (LOKE et al. 1989), it is obviously simply unwarranted to generalize that all phosphorothioates are taken up by all cells in vitro primarily by receptor mediated endocytosis.

Finally, extrapolations from in vitro uptake studies to predictions about in vivo pharmacokinetic behavior are entirely inappropriate and, in fact, there are now several lines of evidence in animals and man demonstrate that, even after careful consideration of all in vitro uptake data, one cannot predict in vivo pharmacokinetics of the compounds based on in vitro studies (S.T. CROOKE et al. 1994; COSSUM et al. 1993, 1994; SANDS et al. 1995).

5. The Binding to and Effects of Binding to Nonnucleic Acid Targets

Phosphorothioate oligonucleotides tend to bind to many proteins and those interactions are influenced by many factors. Protein binding can influence cell uptake, distribution, metabolism and excretion. It may induce nonantisense effects that can be mistakenly interpreted as antisense or complicate the identification of an antisense mechanism. By inhibiting ribonuclease H (RNase H), protein binding may inhibit the antisense activity of some oligonucleotides. Finally, binding to proteins can certainly have toxicological consequences.

In addition to proteins, oligonucleotides may interact with other biological molecules, such as lipids or carbohydrates, and such interactions, like those with proteins, will be influenced by the chemical class of oligonucleotide studied. Unfortunately, essentially no data bearing on such interactions are currently available.

An especially complicated experimental situation is encountered in many in vitro antiviral assays. In these assays, high concentrations of drugs, viruses, and cells are often coincubated. The sensitivity of each virus to nonantisense effects of oligonucleotides varies depending on the nature of the virion proteins and the characteristics of the oligonucleotides (COWSERT 1993; AZAD et al. 1993). This has resulted in considerable confusion. In particular for HIV, herpes simplex viruses, cytomegaloviruses, and influenza virus, the nonantisense effects have been so dominant that identifying oligonucleotides that work via an antisense mechanism has been difficult. Given the artificial character of such assays, it is difficult to know whether nonantisense mechanisms would be as dominant in vivo or result in antiviral activity.

6. Terminating Mechanisms

It has been amply demonstrated that oligonucleotides may employ several terminating mechanisms. The dominant terminating mechanism is influenced by RNA receptor site, oligonucleotide chemical class, cell type, and probably many other factors (S.T. CROOKE 1995b). Obviously, as variations in terminating mechanism may result in significant changes in antisense potency and studies have shown significant variations from cell type to cell type in vitro, it is essential that the terminating mechanism be well understood. Unfortunately, at present, our understanding of terminating mechanisms remains rudimentary.

7. Effects of "Control Oligonucleotides"

A number of types of control oligonucleotides have been used, including randomized oligonucleotides. Unfortunately, we know little to nothing about the potential biological effects of such "controls" and the more complicated a biological system and test the more likely that "control" oligonucleotides may have activities that complicate interpretations. Thus, when a control oligonucleotide displays a surprising activity, the mechanism of that activity should be explored carefully before concluding that the effects of the "control oligonucleotide" prove that the activity of the putative antisense oligonucleotide are not due to an antisense mechanism.

8. Kinetics of Effects

Many rate constants may affect the activities of antisense oligonucleotides, e.g., the rate of synthesis and degradation of the target RNA and its protein, the rates of uptake into cells, the rates of distribution, extrusion, and metabolism of an oligonucleotide in cells, and similar pharmacokinetic considerations in animals. Despite this, relatively few time courses have been reported, and in vitro studies have been reported that range from a few hours to several days. In animals, we have a growing body of information on pharmacokinetics, but in most studies reported to date, the doses and schedules were chosen arbitrarily and, again, little information on duration of effect and onset of action has been presented. Clearly, more careful kinetic studies are required and rational in vitro and in vivo dose schedules must be developed.

II. Recommendations

1. Positive Demonstration of Antisense Mechanism and Specificity

Until more is understood about how antisense drugs work, it is essential to positively demonstrate effects consistent with an antisense mechanism. For RNase H activating oligonucleotides, northern blot analysis showing selective loss of the target RNA is the best choice and many laboratories are publishing reports in vitro and in vivo of such activities (CHIANG et al. 1991; DEAN and MCKAY 1994; SKORSKI et al. 1994; HIJIYA et al. 1994). Ideally, a demonstration that closely related isotypes are unaffected should be included. In brief, then, for proof of mechanism, the following steps are recommended:

1. Perform careful dose response curves in vitro using several cell lines and methods of in vitro delivery
2. Correlate the rank order potency in vivo with that observed in vitro after thorough dose response curves are generated in vivo

3. Perform careful "gene walks" for all RNA species and oligonucleotide chemical classes.
4. Perform careful time courses before drawing conclusions about potency
5. Directly demonstrate proposed mechanism of action by measuring the target RNA and/or protein
6. Evaluate specificity and therapeutic indices via studies on closely related isotypes and with appropriate toxicological studies
7. Perform sufficient pharmacokinetics to define rational dosing schedules for pharmacological studies
8. When control oligonucleotides display surprising activities, determine the mechanisms involved

C. Molecular Mechanisms of Antisense Drugs
I. Occupancy Only Mediated Mechanisms

Classic competitive antagonists are thought to alter biological activities because they bind to receptors, preventing natural agonists from binding and in this way inducing normal biological processes. Binding of oligonucleotides to specific sequences may inhibit the interaction of the RNA with proteins, other nucleic acids, or other factors required for essential steps in the intermediary metabolism of the RNA or its utilization by the cell.

1. Inhibition of Splicing

A key step in the intermediary metabolism of most mRNA molecules is the excision of introns. These "splicing" reactions are sequence specific and require the concerted action of spliceosomes. Consequently, oligonucleotides that bind to sequences required for splicing may prevent binding of necessary factors or physically prevent the required cleavage reactions. This then would result in inhibition of the production of the mature mRNA. Although there are several examples of oligonucleotides directed to splice junctions, none of the studies present data showing inhibition of RNA processing, accumulation of splicing intermediates, or a reduction in mature mRNA. Nor are there published data in which the structure of the RNA at the splice junction was probed and the oligonucleotides demonstrated to hybridize to the sequences for which they were designed (MCMANAWAY et al. 1990; KULKA et al. 1989; ZAMECNIK et al. 1986; SMITH et al. 1986). Activities have been reported for anti-c-*myc* and antiviral oligonucleotides with phosphodiester, methylphosphonate and phosphorothioate backbones. Very recently, an oligonucleotide was reported to induce alternative splicing in a cell-free splicing system and, in that system, RNA analyses confirmed the putative mechanism (DOMINSKI and KOLE 1993).

In our laboratory, we have attempted to characterize the factors that determine whether splicing inhibition is effected by an antisense mechanism

List of Abbreviations

CMV	cytomegalovirus
Ha-*ras*	Harvey *ras*
HIV	human immune deficiency virus
HPV	human papillomavirus
HSV	herpes simplex virus
ICAM	intercellular adhesion molecule
IL	interleukin
mRNA	messenger RNA
NK	natural killer
PKC	protein kinase C
PNA	peptide nucleic acid
PTHrP	parathyroid hormone-related peptide
RT	reverse transcriptase
TAR	transactivator response element
T_m	melting transition

(HODGES and CROOKE 1995). To this end, a number of luciferase-reporter plasmids containing various introns were constructed and transfected into HeLa cells. Then the effects of antisense drugs designed to bind to various sites were characterized. The effects of RNase H-competent oligonucleotides were compared to those of oligonucleotides that do not serve as RNase H substrates. The major conclusions from this study were as follows. First, most of the earlier studies in which splicing inhibition was reported were probably due to nonspecific effects. Second, less effectively spliced introns are better targets than those with strong consensus splicing signals. Third, the 3′-splice site and branchpoint are usually the best sites to which to target to the oligonucleotide to inhibit splicing. Fourth, RNase H-competent oligonucleotides are usually more potent than even higher affinity oligonucleotides that inhibit by occupancy only.

2. Translational Arrest

Many oligonucleotides have been designed to arrest translation of targeted protein by binding to the translation initiation codon. The positioning of the initiation codon within the area of complementarity of the oligonucleotide and the length of oligonucleotide used have varied considerably. Again, unfortunately, only in relatively few studies have the oligonucleotides, in fact, been shown to bind to the sites for which they were designed, and data that directly support translation arrest as the mechanism have been lacking.

Target RNA species that have been reported to be inhibited by a translational arrest mechanism include HIV, vesicular stomatitis virus (VSV), N-*myc* and a number of normal cellular genes (AGRAWAL et al. 1988; LEMAITRE et al.

1987; ROSOLEN et al. 1990; VASANTHAKUMAR and AHMED 1989; SBURLATI et al. 1991; ZHENG et al. 1989; MAIER et al. 1990). In our laboratories, we have shown that a significant number of targets may be inhibited by binding to translation initiation codons. For example, ISIS 1082 hybridizes to the AUG codon for the UL13 gene of herpes virus types 1 and 2. RNase H studies confirmed that it binds selectively in this area. In vitro protein synthesis studies confirmed that it inhibited the synthesis of the UL13 protein and studies in HeLa cells showed that it inhibited the growth of herpes type 1 and type 2 with IC_{50} of 200–400 nM by translation arrest (MIRABELLI et al. 1991). Similarly, ISIS 1753, a 30-mer phosphorothioate complementary to the translation initiation codon and surrounding sequences of the *E2* gene of bovine papilloma virus, was highly effective and its activity was shown to be due to translation arrest. ISIS 2105, a 20-mer phosphorothioate complementary to the same region in human papilloma virus, was shown to be a very potent inhibitor. Compounds complementary to the translation initiation codon of the E2 gene were the most potent of the more than 50 compounds studied complementary to various other regions in the RNA (COWSERT et al. 1993). We have also shown inhibition of translation of a number of other mRNA species by compounds designed to bind to the translation codon.

In conclusion, translation arrest represents an important mechanism of action for antisense drugs. A number of examples purporting to employ this mechanism have been reported, and recent studies on several compounds have provided data that unambiguously demonstrate that this mechanism can result in potent antisense drugs. However, very little is understood about the precise events that lead to translation arrest.

3. Disruption of Necessary RNA Structure

RNA adopts a variety of three-dimensional structures induced by intramolecular hybridization, the most common of which is the stem loop. These structures play crucial roles in a variety of functions. They are used to provide additional stability for RNA and as recognition motifs for a number of proteins, nucleic acids, and ribonucleoproteins that participate in the intermediary metabolism and activities of RNA species. Thus, given the potential general activity of the mechanism, it is surprising that occupancy-based disruption RNA has not been more extensively exploited.

As an example, we designed a series of oligonucleotides that bind to the important stem loop present in all RNA species in HIV, the TAR element. We synthesized a number of oligonucleotides designed to disrupt TAR and showed that several did indeed bind to TAR, disrupt the structure, and inhibit TAR-mediated production of a reporter gene (VICKERS et al. 1991). Furthermore, general rules useful in disrupting stem-loop structures were also developed (ECKER et al. 1992).

Although designed to induce relatively nonspecific cytotoxic effects, two other examples are noteworthy. Oligonucleotides designed to bind to a 17-

nucleotide loop in Xenopus 28 S RNA required for ribosome stability and protein synthesis inhibited protein synthesis when injected into Xenopus oocytes (SAXENA and ACKERMAN 1990). Similarly, oligonucleotides designed to bind to highly conserved sequences in 5.8S RNA inhibited protein synthesis in rabbit reticulocyte and wheat germ systems (WALKER et al. 1990).

II. Occupancy Activated Destabilization

RNA molecules regulate their own metabolism. A number of structural features of RNA are known to influence stability, various processing events, subcellular distribution and transport. It is likely that, as RNA intermediary metabolism is better understood, many other regulatory features and mechanisms will be identified.

1. 5′ Capping

A key early step in RNA processing is 5′ capping (Fig. 1). This stabilizes pre-mRNA and is important for the stability of mature mRNA. It is also important in binding to the nuclear matrix and transport of mRNA out of the nucleus. As the structure of the cap is unique and understood, it presents an interesting target.

Several oligonucleotides that bind near the cap site have been shown to be active, presumably by inhibiting the binding of proteins required to cap the

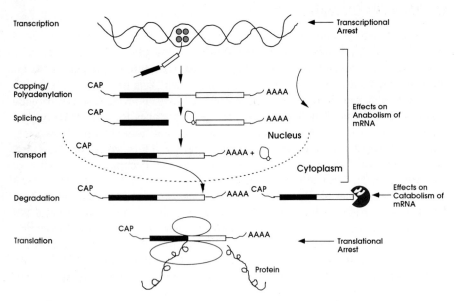

Fig. 1. RNA processing

RNA. For example, the synthesis of SV40 T-antigen was reported to be most sensitive to an oligonucleotide linked to polylysine and targeted to the 5'-cap site of RNA (WESTERMANN et al. 1989). However, once again, in no published study has this putative mechanism been rigorously demonstrated. In fact, in no published study have the oligonucleotides been shown to bind to the sequences for which they were designed.

In our laboratory, we have designed oligonucleotides to bind to 5'-cap structures and reagents to specifically cleave the unique 5'-cap structure (BAKER 1993). These studies demonstrate that 5'-cap targeted oligonucleotides were capable of inhibiting the binding of the eukaryotic translation initiation factor 4a (eIF-4a) (BAKER et al. 1992).

2. Inhibition of 3' Polyadenylation

In the 3'-untranslated region of pre-mRNA molecules, there are sequences that result in the post-transcriptional addition of long (hundreds of nucleotides) tracts of polyadenylate. Polyadenylation stabilizes mRNA and may play other roles in the intermediary metabolism of RNA species. Theoretically, interactions in the 3'-terminal region of pre-mRNA could inhibit polyadenylation and destabilize the RNA species. Although there are a number of oligonucleotides that interact in the 3'-untranslated region and display antisense activities, to date, no study has reported evidence for alterations in polyadenylation (CHIANG et al. 1991).

III. Other Mechanisms

In addition to 5' capping and 3' adenylation, there are clearly other sequences in the 5'- and 3'-untranslated regions of mRNA that affect the stability of the molecules. Again, there are a number of antisense drugs that may work by these mechanisms.

ZAMECNIK and STEPHENSON (1978) reported that a 13 mer targeted to untranslated 3'- and 5'-terminal sequences in Rous sarcoma viruses was active. Oligonucleotides conjugated to an acridine derivative and targeted to a 3'-terminal sequence in type A influenza viruses were reported to be active. Against several RNA targets, studies in our laboratories have shown that sequences in the 3'-untranslated region of RNA molecules are often the most sensitive (ZERIAL et al. 1987; THUONG et al. 1989; HELENE and TOULME 1989). For example, ISIS 1939 is a 20-mer phosphorothioate that binds to and appears to disrupt a predicted stem-loop structure in the 3'-untranslated region of the mRNA for ICAM and is a potent antisense inhibitor. However, inasmuch as a 2'-methoxy analog of ISIS 1939 was much less active, it is likely that, in addition to destabilization to cellular nucleolytic activity, activation of RNase H (see below) is also involved in the activity of ISIS 1939 (CHIANG et al. 1991).

IV. Activation of RNase H

RNase H is an ubiquitous enzyme that degrades the RNA strand of an RNA–DNA duplex. It has been identified in organisms as diverse as viruses and human cells (CROUCH and DIRKSEN 1985). At least two classes of RNase H have been identified in eukaryotic cells. Multiple enzymes with RNase H activity have been observed in prokaryotes (CROUCH and DIRKSEN 1985).

Although RNase H is involved in DNA replication, it may play other roles in the cell and is found in the cytoplasm as well as the nucleus (CRUM et al. 1988). However, the concentration of the enzyme in the nucleus is thought to be greater and some of the enzyme found in cytoplasmic preparations may be due to nuclear leakage.

RNase H activity is quite variable in cells. It is absent or minimal in rabbit reticulocytes but present in wheat germ extracts (CROUCH and DIRKSEN 1985; HAEUPTLE et al. 1986). In HL-60 cells, for example, the level of activity in undifferentiated cells is greatest; it is relatively high in DMSO and Vitamin D differentiated cells and much lower in phorbol myristic acid (PMA) -differentiated cells (G. HOKE, unpublished data).

The precise recognition elements for RNase H are not known. However, it has been shown that oligonucleotides with DNA-like properties as short as tetramers can activate RNase H (DONIS-KELLER 1979). Changes in the sugar influence RNase H activation as sugar modifications that result in RNA-like oligonucleotides, e.g., 2′ fluoro or 2′ methoxy do not appear to serve as substrates for RNase H (KAWASAKI et al. 1993; SPROAT et al. 1989). Alterations in the orientation of the sugar to the base can also affect RNase H activation as α-oligonucleotides are unable to induce RNase H or may require parallel annealing (MORVAN et al. 1991; GAGNOR et al. 1989). Additionally, backbone modifications influence the ability of oligonucleotides to activate RNase H. Methylphosphonates do not activate RNase H (MAHER et al. 1989; MILLER 1989). In contrast, phosphorothioates are excellent substrates (MIRABELLI et al. 1991; STEIN and CHENG 1993; CAZENAVE et al. 1989). In addition, chimeric molecules have been studied as oligonucleotides that bind to RNA and activate RNase H (QUARTIN et al. 1989; FURDON et al. 1989). For example, oligonucleotides comprised of wings of 2′-methoxy phosphorothioates and a five-base gap of deoxyoligonucleotides bind to their target RNA and activate RNase H (QUARTIN et al. 1989; FURDON et al. 1989). Furthermore, a single ribonucleotide in a sequence of deoxyribonucleotides was shown to be sufficient to serve as a substrate for RNase H when bound to its complementary oligodeoxynucleotide (EDER and WALDER 1991).

That it is possible to take advantage of chimeric oligonucleotides designed to activate RNase H that have greater affinity for their RNA receptors and to enhance specificity has also been demonstrated (MONIA et al. 1993; GILES and TIDD 1992). In a recent study, RNase H mediated cleavage of target transcript was much more selective when oligodeoxynucleotides comprised of methylphosphonate oligodeoxynucleotide wings and phosphodiester gaps were compared to full phosphodiester oligonucleotides (GILES and TIDD 1992).

Despite the information about RNase H and the demonstration that many oligonucleotides may activate RNase H in lysate and purified enzyme assays, relatively little is yet known about the role of structural features in RNA targets in activating RNase H (WALDER and WALDER 1988; MINSHULL and HUNT 1986; GAGNOR et al. 1987). In fact, direct proof that RNase H activation is, in fact, the mechanism of action of oligonucleotides in cells is to a large extent lacking.

Recent studies in our laboratories provide additional, albeit indirect, insights into these questions. ISIS 1939 is a 20-mer phosphorothioate complementary to a sequence in the 3'-untranslated region of ICAM-1 RNA (CHIANG et al. 1991). It inhibits ICAM production in human umbilical vein endothelial cells, and northern blots demonstrate that ICAM-1 mRNA is rapidly degraded. A 2'-methoxy analog of ISIS 1939 displays higher affinity for the RNA than the phosphorothioate and is stable in cells, but inhibits ICAM-1 protein production much less potently than ISIS 1939. It is likely that ISIS 1939 destabilizes the RNA and activates RNase H. In contrast, ISIS 1570, an 18-mer phosphorothioate that is complementary to the translation initiation codon of the ICAM-1 message inhibited production of the protein, but caused no degradation of the RNA. Thus, two oligonucleotides that are capable of activating RNase H had different effects, depending on the site in the mRNA at which they bound (CHIANG et al. 1991).

A more direct demonstration that RNase H is likely a key factor in the activity of many antisense oligonucleotides was provided by studies in which reverse-ligation polymerase chain reaction (PCR) was used to identify cleavage products from *bcr-abl* mRNA in cells treated with phosphorothioate oligonucleotides (GILES et al. 1995).

Given the emerging role of chimeric oligonucleotides with modifications in the 3' and 5' wings designed to enhance affinity for the target RNA and nuclease stability and a DNA-type gap to serve as a substrate for RNase H, studies focused on understanding the effects of various modifications on the efficiency of the enzyme(s) are also of considerable importance. In one such study on *Escherichia coli* RNase H, we have recently reported that the enzyme displays minimal sequence specificity and is processive. When a chimeric oligonucleotide with 2'-modified sugars in the wings was hybridized to the RNA, the initial site of cleavage was the nucleotide adjacent to the methoxy-deoxy junction closest to the 3' end of the RNA substrate. The initial rate of cleavage increased as the size of the DNA gap increased, and the efficiency of the enzyme was considerably less against an RNA target duplexed with a chimeric antisense oligonucleotide than a full DNA-type oligonucleotide (S.T. CROOKE et al. 1995).

In subsequent studies, we have evaluated in more detail the interactions of antisense oligonucleotides with structured and unstructured targets and the impacts of these interactions on RNase H (LIMA and CROOKE 1997). Using a series of noncleavable substrates and Michaelis-Menten analyses, we were able to evaluate both binding and cleavage. We showed that, in fact, *E. coli* RNase H1 is a double-strand RNA binding protein. The K_d for an RNA duplex was 1.6 μM; the K_d for a DNA duplex was 176 μM; and the K_d for single-strand DNA was 942 μM. In contrast, the enzyme could only cleave RNA in an RNA–DNA duplex. Any 2′ modification in the antisense drug at the cleavage site inhibited cleavage, but significant charge reduction and 2′ modifications were tolerated at the binding site. Finally, placing a positive charge (e.g., 2′ propoxyamine) in the antisense drug reduced affinity and cleavage.

We have also examined the effects of antisense oligonucleotide-induced RNA structures on the activity of *E. Coli* RNase H1 (LIMA et al., in press). Any structure in the duplex substrate was found to have a significant negative effect on the cleavage rate. Further, cleavage of selected sites was inhibited entirely, and this was explained by the sterric hindrance imposed by the RNA loop traversing either the minor or major grooves or the hetroduplex.

V. Activation of Double-Strand RNase

By using phosphorothioate oligonucleotides with 2′ modified wings and a ribonucleotide center, we have shown that mammalian cells contain enzymes that can cleave double-strand RNAs (WU et al., submitted). This is an important step forward because it adds to the repertoire of intracellular enzymes that may be used to cleave target RNAs and because chimeric oligonucleotides 2′ modified wings and oligoribonucleotide gaps have higher affinity for RNA targets than chimeras with oligodeoxynucleotide gaps.

D. Characteristics of Phosphorothioate Oligodeoxynucleotides

I. Introduction

Of the first generation oligonucleotide analogs, the class that has resulted in the broadest range of activities and about which the most is known is the phosphorothioate class. Phosphorothioate oligonucleotides were first synthesized in 1969 when a poly rI-rC phosphorothioate was synthesized (DE CLERCQ et al. 1969). This modification clearly achieves the objective of increased nuclease stability. In this class of oligonucleotides, one of the oxygen atoms in the phosphate group is replaced with a sulfur. The resulting compound is negatively charged, is chiral at each phosphorothioate, and much more resistant to nucleases than the parent phosphodiester (COHEN 1993).

II. Hybridization

The hybridization of phosphorothioate oligonucleotides to DNA and RNA has been thoroughly characterized (CROOKE and LEBLEU 1993; S.T. CROOKE 1992; R.M. CROOKE 1993a). The melting transition (T_m) of a phosphorothioate oligodeoxynucleotide for RNA is approximately 0.5°C less per nucleotide than for a corresponding phosphodiester oligodeoxynucleotide. This reduction in T_m per nucleotide is virtually independent of the number of phosphorothioate units substituted for phosphodiesters. However, sequence context has some influence as the ΔT_m can vary from –0.3°C to 1.0°C depending on sequence. Compared to RNA and RNA duplex formation, a phosphorothioate oligodeoxynucleotide has a T_m approximately 2.2°C lower per unit (FREIER 1993). This means that to be effective in vitro, phosphorothioate oligodeoxynucleotides must typically be 17–20 mer in length and that invasion of double-stranded regions in RNA is difficult (LIMA et al. 1992; VICKERS et al. 1991; MONIA et al. 1992, 1993).

Association rates of phosphorothioate oligodeoxynucleotide to unstructured RNA targets are typically $10^6-10^7 M^{-1}S^{-1}$, independent of oligonucleotide length or sequence (FREIER 1993; LIMA et al. 1992). Association rates to structured RNA targets can vary from 10^2 to $10^8 M^{-1}S^{-1}$, depending on the structure of the RNA, the site of binding in the structure, and other factors (FREIER 1993). Put in another way, association rates for oligonucleotides that display acceptable affinity constants are sufficient to support biological activity at therapeutically achievable concentrations.

The specificity of hybridization of phosphorothioate oligonucleotides is, in general, slightly greater than that of phosphodiester analogs. For example, a T-C mismatch results in a 7.7°C or 12.8°C reduction in T_m, respectively, for a phosphodiester or phosphorothioate oligodeoxynucleotide 18 nucleotides in length with the mismatch centered (FREIER 1993). Thus, from this perspective, the phosphorothioate modification is quite attractive.

III. Interactions with Proteins

Phosphorothioate oligonucleotides bind to proteins. The interactions with proteins can be divided into nonspecific, sequence specific, and structure-specific binding events, each of which may have different characteristics and effects. Nonspecific binding to a wide variety of proteins has been demonstrated. Exemplary of this type of binding is the interaction of phosphorothioate oligonucleotides with serum albumin. The affinity of such interactions is low. The K_d for albumin is approximately 200 μM, and thus in a similar range with aspirin or penicillin (S.T. CROOKE et al. 1996; JOOS and HALL 1969). Furthermore, in this study, no competition between phosphorothioate oligonucleotides and several drugs that bind to bovine serum albumin was observed. In this study, binding and competition were determined in an assay in which electrospray mass spectrometry was used. In contrast, in a study in which an equilibrium dissociation constant was derived from an assay using albumin loaded on a CH-sephadex column, the K_m ranged from 1 to 5×10^{-5} M for bovine serum albumin and from 2 to 3×10^{-4} M for human serum albumin. Moreover, warfarin and indomethacin were reported to compete for binding to serum albumin (SRINIVASAN et al. 1995). Clearly, much more work is required before definitive conclusions can be drawn.

Phosphorothioate oligonucleotides can interact with nucleic acid binding proteins such as transcription factors and single-strand nucleic acid binding proteins. However, very little is known about these binding events. Additionally, it has been reported that phosphorothioates bind to an 80-KDa membrane protein that was suggested to be involved in cellular uptake processes (LOKE et al. 1989). However, again, little is known about the affinities, sequence or structure specificities of these putative interactions. More recently interactions with 30-KDa and 46-KDa surface proteins in T15 mouse fibroblasts were reported (HAWLEY and GIBSON 1996).

Phosphorothioates interact with nucleases and DNA polymerases. These compounds are slowly metabolized by both endo- and exonucleases and inhibit these enzymes (R.M. CROOKE et al. 1995; S.T. CROOKE 1992). The inhibition of these enzymes appears to be competitive, and this may account for some early data suggesting that phosphorothioates were almost infinitely stable to nucleases. In these studies, the oligonucleotide to enzyme ratio was very high and, thus, the enzyme was inhibited. Phosphorothioates also bind to RNase H when in an RNA–DNA duplex, and the duplex serves as a substrate for RNase H (GAO et al. 1992). At higher concentrations, presumably by binding as a single strand to RNase H, phosphorothioates inhibit the enzyme (S.T. CROOKE et al. 1995; S.T. CROOKE 1992). Again, the oligonucleotides appear to be competitive antagonists for the DNA–RNA substrate.

Phosphorothioates have been shown to be competitive inhibitors of DNA polymerase α and β with respect to the DNA template, and noncompetitive inhibitors of DNA polymerases χ and δ (GAO et al. 1992). Despite this inhibition, several studies have suggested that phosphorothioates might serve as primers for polymerases and be extended (STEIN and CHENG 1993; S.T. CROOKE 1995a; AGRAWAL et al. 1991). In our laboratories, we have shown extensions of 2–3 nucleotides only. At present, a full explanation as to why no longer extensions are observed is not available.

Phosphorothioate oligonucleotides have been reported to be competitive inhibitors for HIV-reverse transcriptase and to inhibit reverse transcriptase (RT)-associated RNase H activity (MAJUMDAR et al. 1989; CHENG et al. 1991). They have been reported to bind to the cell surface protein, CD4, and to protein kinase C (STEIN et al. 1991). Various viral polymerases have also been shown to be inhibited by phosphorothioates (STEIN and CHENG 1993). Additionally, we have shown potent, nonsequence specific inhibition of RNA splicing by phosphorothioates (HODGES and CROOKE 1995).

Like other oligonucleotides, phosphorothioates can adopt a variety of secondary structures. As a general rule, self-complementary oligonucleotides are avoided, if possible, to avoid duplex formation between oligonucleotides. However, other structures that are less well understood can also form. For example, oligonucleotides containing runs of guanosines can form tetrameric structures called G (guanosine) quartets, and these appear to interact with a number of proteins with relatively greater affinity than unstructured oligonucleotides (WYATT et al. 1994).

In conclusion, phosphorothioate oligonucleotides may interact with a wide range of proteins via several types of mechanisms. These interactions may influence the pharmacokinetic, pharmacologic, and toxicologic properties of these molecules. They may also complicate studies on the mechanism of action of these drugs, and may, in fact, obscure an antisense activity. For example, phosphorothioate oligonucleotides were reported to enhance lipopolysaccharide-stimulated synthesis or tumor necrosis factor (HARTMANN et al. 1996). This would obviously obscure antisense effects on this target.

IV. Pharmacokinetic Properties

To study the pharmacokinetics of phosphorothioate oligonucleotides, a variety of labeling techniques have been used. In some cases, 3'- or 5' ^{32}P-end-labeled or fluorescently labeled oligonucleotides have been used in in vitro or in vivo studies. These are probably less satisfactory than internally labeled compounds because terminal phosphates are rapidly removed by phosphatases and fluorescently labeled oligonucleotides have physicochemical properties that differ from the unmodified oligonucleotides. Consequently, either uniformly (COWSERT et al. 1993) S-labeled, or base-labeled phosphorothioates are preferable for pharmacokinetic studies. In our laboratories, a tritium exchange method that labels a slowly exchanging proton at the C8 position in purines was developed and proved to be quite useful (GRAHAM et al. 1993). Very recently, a method that added radioactive methyl groups via S-adenosylmethionine has also been successfully used (SANDS et al. 1994). Finally, advances in extraction, separation and detection methods have resulted in methods that provide excellent pharmacokinetic analyses without radiolabeling (S.T. CROOKE et al. 1996).

1. Nuclease Stability

The principle metabolic pathway for oligonucleotides is cleavage via endo- and exonucleases. Phosphorothioate oligonucleotides, while quite stable to various nucleases are competitive inhibitors of nucleases (S.T. CROOKE 1995b; GAO et al. 1992; HOKE et al. 1991; WICKSTROM 1986; CAMPBELL et al. 1990). Consequently, the stability of phosphorothioate oligonucleotides to nucleases is probably a bit less than initially thought, as high concentrations (that inhibited nucleases) of oligonucleotides were employed in the early studies. Similarly, phosphorothioate oligonucleotides are degraded slowly by cells in tissue culture with a half-life of 12–24h and are slowly metabolized in animals (S.T.

Crooke 1995b; Cossum et al. 1993; Hoke et al. 1991). The pattern of metabolites suggests primarily exonuclease activity with perhaps modest contributions by endonucleases. However, a number of lines of evidence suggest that, in many cells and tissues, endonucleases play an important role in the metabolism of oligonucleotides. For example, 3′- and 5′-modified oligonucleotides with phosphodiester backbones have been shown to be relatively rapidly degraded in cells and after administration to animals (Sands et al. 1995; Miyao et al. 1995). Thus, strategies in which oligonucleotides are modified at only the 3′ and 5′ terminus as a means of enhancing stability have not proven to be successful.

2. In Vitro Cellular Uptake

Phosphorothioate oligonucleotides are taken up by a wide range of cells in vitro (R.M. Crooke 1991, 1993a; R.M. Crooke et al. 1995; Gao et al. 1992; Neckers 1993). In fact, uptake of phosphorothioate oligonucleotides into a prokaryote, *Vibrio parahaemolyticus*, has been reported, as has uptake into *Schistosoma mansoni* (Chrisey et al. 1993; Tao et al. 1995). Uptake is time and temperature dependent. It is also influenced by cell type, cell-culture conditions, media and sequence, and length of the oligonucleotide (R.M. Crooke et al. 1995). No obvious correlation between the lineage of cells, whether the cells are transformed or whether they are virally infected and uptake has been identified (S.T. Crooke 1995a). Nor are the factors that result in differences in uptake of different sequences of oligonucleotide understood. Although several studies have suggested that receptor-mediated endocytosis may be a significant mechanism of cellular uptake, the data are not yet compelling enough to conclude that receptor-mediated endocytosis accounts for a significant portion of the uptake in most cells (Loke et al. 1989).

Numerous studies have shown that phosphorothioate oligonucleotides distribute broadly in most cells once taken up (S.T. Crooke 1995a; R.M. Crooke 1993a). Again, however, significant differences in subcellular distribution between various types of cells have been noted.

Cationic lipids and other approaches have been used to enhance uptake of phosphorothioate oligonucleotides in cells that take up little oligonucleotide in vitro (Bennett et al. 1992, 1993; Quattrone et al. 1994). Again, however, there are substantial variations from cell type to cell type. Other approaches to enhanced intracellular uptake in vitro have included streptolysin D treatment of cells and the use of dextran sulfate and other liposome formulations as well as physical means such as microinjections (S.T. Crooke 1995a; Giles et al. 1995; Wang et al. 1995).

3. In Vivo Pharmacokinetics

Phosphorothioate oligonucleotides bind to serum albumin and α_2-macroglobulin. The apparent affinity for albumin is quite low (200–400 μM) and comparable to the low-affinity binding observed for a number of drugs, e.g., aspirin and penicillin (S.T. CROOKE et al. 1996; JOOS and HALL 1969; SRINIVASAN et al. 1995). Serum protein binding, therefore, provides a repository for these drugs and prevents rapid renal excretion. As serum protein binding is saturable at higher doses, intact oligomer may be found in urine (AGRAWAL et al. 1991; IVERSEN 1991). Studies in our laboratory suggest that in rats, oligonucleotides administered intravenously at doses of 15–20 mg/kg saturate the serum protein binding capacity (J. LEEDS, unpublished data).

Phosphorothioate oligonucleotides are rapidly and extensively absorbed after parenteral administration. For example, in rats, after an intradermal dose of 3.6 mg/kg of ^{14}C-labeled ISIS 2105, a 20-mer phosphorothioate, approximately 70% of the dose was absorbed within 4 h and total systemic bioavailability was in excess of 90% (COSSUM et al. 1994). After intradermal injection in man, absorption of ISIS 2105 was similar to that observed in rats (S.T. CROOKE et al. 1994). Subcutaneous administration to rats and monkeys results in somewhat lower bioavailability and greater distribution to lymph, as would be expected (J. LEEDS, unpublished observations).

Distribution of phosphorothioate oligonucleotides from blood after absorption or intravenous administration is extremely rapid. We have reported distribution half-lives of less than 1 h, and similar data have been reported by others (COSSUM et al. 1993, 1994; AGRAWAL et al. 1991; IVERSEN 1991). Blood and plasma clearance is multiexponential with a terminal elimination half-life from 40 to 60 h in all species except man. In man, the terminal elimination half-life may be somewhat longer (S.T. CROOKE et al. 1994).

Phosphorothioates distribute broadly to all peripheral tissues. Liver, kidney, bone marrow, skeletal muscle, and skin accumulate the highest percentage of a dose, but other tissues display small quantities of drug (COSSUM et al. 1993, 1994). No evidence of significant penetration of the blood–brain barrier has been reported. The rates of incorporation and clearance from tissues vary as a function of the organ studied, with liver accumulating drug most rapidly (20% of a dose within 1–2 h) and other tissues accumulating drug more slowly. Similarly, elimination of drug is more rapid from liver than any other tissue, e.g., terminal half-life from liver: 62 h, from renal medulla: 156 h. The distribution into the kidney has been studied more extensively and drug shown to be present in Bowman's capsule, the proximal convoluted tubule, the brush border membrane, and within renal tubular epithelial cells (RAPPAPORT et al. 1995). The data suggest that the oligonucleotides are filtered by the glomerulus, then reabsorbed by the proximal convoluted tubule epithelial cells. Moreover, the authors suggest that reabsorption might be mediated by interactions with specific proteins in the brush border membranes.

At relatively low doses, clearance of phosphorothioate oligonucleotides is due primarily to metabolism (COSSUM et al. 1993, 1994; IVERSEN 1991). Metabolism is mediated by exo- and endonucleases that result in shorter oligonucleotides and, ultimately, nucleosides that are degraded by normal metabolic pathways. Although no direct evidence of base excision or modification has been reported, these are theoretical possibilities that may occur. In one study, a larger molecular weight radioactive material was observed in urine, but not fully characterized (AGRAWAL et al. 1991). Clearly, the potential for conjugation reactions and extension of oligonucleotides via these drugs serving as primers for polymerases must be explored in more detail. In a very thorough study, 20 nucleotide phosphodiester and phosphorothioate oligonucleotides were administered intravenously at a dose of 6 mg/kg to mice. The oligonucleotides were internally labeled with ^{3}H-CH$_3$ by methylation of an internal deoxycytidine residue using Hha1 methylase and S-(3H) adenosylmethionine (SANDS et al. 1994). The observations for the phosphorothioate oligonucleotide were entirely consistent with those made in our studies. Additionally, in this paper, autoradiographic analyses showed drug in renal cortical cells (SANDS et al. 1994).

One study of prolonged infusions of a phosphorothioate oligonucleotide to human beings has been reported (BAYEVER et al. 1993). In this study, five patients with leukemia were given 10-day intravenous infusions at a dose of 0.05 mg/kg per hour. Elimination half-lives reportedly varied from 5.9 to 14.7 days. Urinary recovery of radioactivity was reported to be 30%–60% of the total dose, with 30% of the radioactivity being intact drug. Metabolites in urine included both higher and lower molecular weight compounds. In contrast, when GEM-91 (a 25-mer phosphorothioate oligodeoxynucleotide) was administered to humans as a 2-h, I.V. infusion at a dose of 0.1 mg/kg, a peak plasma concentration of 295.8 mg/ml was observed at the cessation of the infusion. Plasma clearance of total radioactivity was biexponential, with initial and terminal elimination half-lives of 0.18 and 26.71 h, respectively. However, degradation was extensive and intact drug pharmacokinetic models were not presented. Nearly 50% of the administered radioactivity was recovered in urine, but most of the radioactivity represented degradates. In fact, no intact drug was found in the urine at any time (ZHANG et al. 1995a).

In a more recent study in which the level of intact drug was carefully evaluated using capillary gel electrophoresis, the pharmacokinetics of ISIS 2302, a 20-mer phosphorothioate oligodeoxynucleotide, were determined after a 2-h infusion. Doses from 0.6 to 2.0 mg/kg were studied and the peak plasma concentrations were shown to increase linearly with dose, with the 2 mg/kg dose resulting in peak plasma concentrations of intact drug of approximately 9.5 μg/ml. Clearance from plasma, however, was dose dependent, with the 2 mg/kg dose having a clearance of 1.28 ml min^{-1} kg^{-1}, while that of 0.5 mg/kg was 2.07 ml min^{-1} kg^{-1}. Essentially, no intact drug was found in urine.

Clearly, the two most recent studies differ from the initial report in several facets. Although a number of factors may explain the discrepancies, the most likely explanation is related to the evolution of assay methodology, not to a difference between compounds. Overall, the behavior of phosphorothioates in the plasma of humans appears to be similar to that in other species.

In addition to the pharmacological effects that have been observed with phosphorothioate oligonucleotides, there are a number of lines of evidence supporting the notion that these drugs enter cells in various organs. As an

Basic Principles of Antisense Therapeutics

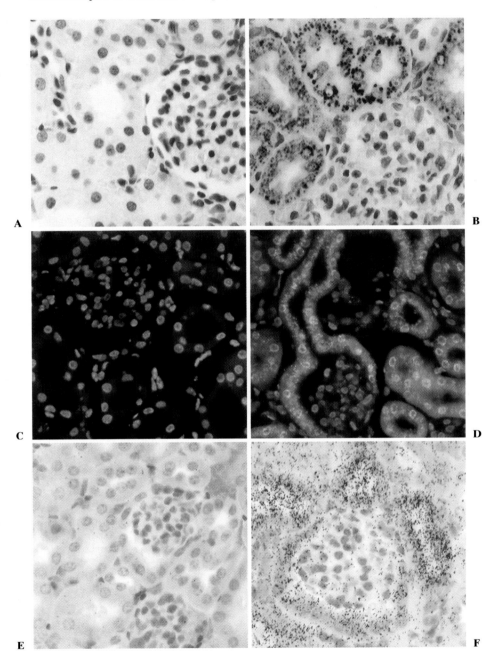

Fig. 2A–F. Autoradiographic, fluorescent, and immunohistochemical data demonstrating the intracellular location of phosphorothioate oligodeoxynucleotides in renal proximal convoluted tubular cells

example, Fig. 2 shows autoradiographic, fluorescent, and immunohistochemical data demonstrating the intracellular location of phosphorothioate oligonucleotides in renal proximal convoluted tubular cells. Similar results have been observed in liver, skin, and bone marrow in similar studies. Using radiolabeled drugs and isolated perfused rat liver cells, uptake into parenchymal and nonparenchymal cells of the liver (TAKAKURA et al. 1996) has been reported.

We have also performed oral bioavailability experiments in rodents treated with an H_2 receptor antagonist to avoid acid-mediated depurination or precipitation. In these studies, very limited (<5%) bioavailability was observed (S. CROOKE, unpublished observations). However, it seems likely that the principal limiting factor in the oral bioavailability of phosphorothioates may be degradation in the gut rather than absorption. Studies using everted rat jejunum sacs demonstrated passive transport across the intestinal epithelium (HUGHES et al. 1995). Further, studies using more stable 2'-methoxy phosphorothioate oligonucleotides showed a significant increase in oral bioavailability that appeared to be associated with the improved stability of the analogs (AGRAWAL et al. 1995).

In summary, pharmacokinetic studies of several phosphorothioates demonstrate that they are well absorbed from parenteral sites, distribute broadly to all peripheral tissues, do not cross the blood–brain barrier and are eliminated primarily by metabolism. In short, systemic dosing should be feasible once a day or every other day. Although the similarities between oligonucleotides of different sequences are far greater than the differences, additional studies are required before determining whether there are subtle effects of sequence on the pharmacokinetic profile of this class of drugs.

V. Pharmacological Properties

1. Molecular Pharmacology

Antisense oligonucleotides are designed to bind to RNA targets via Watson-Crick hybridization. As RNA can adopt a variety of secondary structures via Watson-Crick hybridization, one useful way to think of antisense oligonucleotides is as competitive antagonists for self-complementary regions of the target RNA. Obviously, creating oligonucleotides with the highest affinity per nucleotide unit is pharmacologically important, and a comparison of the affinity of the oligonucleotide to a complementary RNA oligonucleotide is the most sensible comparison. In this context, phosphorothioate oligodeoxynucleotides are relatively competitively disadvantaged as the affinity per nucleotide unit of oligomer is less than RNA (>−2.0°C T_m per unit; COOK 1993). This results in a requirement of at least 15–17 nucleotides in order to have sufficient affinity to produce biological activity (MONIA et al. 1992).

Although multiple mechanisms by which an oligonucleotide may terminate the activity of an RNA species to which it binds are possible, examples of biological activity have been reported for only three of these mechanisms. Antisense oligonucleotides have been reported to inhibit RNA splicing, affect translation of mRNA, and induce degradation of RNA by RNase H (CHIANG et al. 1991; KULKA et al. 1989; AGRAWAL et al. 1988). Without question, the mechanism that has resulted in the most potent compounds and is best understood is RNase H activation. To serve as a substrate for RNase H, a duplex between RNA and a "DNA-like" oligonucleotide is required. Specifically, a sugar moiety in the oligonucleotide that induces a duplex conformation equivalent to that of a DNA–RNA duplex and a charged phosphate are required (MIRABELLI and CROOKE 1993). Thus, phosphorothioate oligodeoxynucleotides are expected to induce RNase H-mediated cleavage of the RNA when bound. As will be discussed later, many chemical approaches that enhance the affinity of an oligonucleotide for RNA result in duplexes that are no longer substrates for RNase H.

Selection of sites at which optimal antisense activity may be induced in a RNA molecule is complex, dependent on terminating mechanism and influenced by the chemical class of the oligonucleotide. Each RNA appears to display unique patterns of sites of sensitivity. Within the phosphorothioate oligodeoxynucleotide chemical class, studies in our laboratory have shown antisense activity can vary from undetectable to 100% by shifting an oligonucleotide by just a few bases in the RNA target (CHIANG et al. 1991; S.T. CROOKE 1992; BENNETT and CROOKE 1996). Although significant progress has been made in developing general rules that help define potentially optimal sites in RNA species, to a large extent, this remains an empirical process that must be performed for each RNA target and every new chemical class of oligonucleotides.

Phosphorothioates have also been shown to have effects inconsistent with the antisense mechanism for which they were designed. Some of these effects are due to sequence or are structure specific. Others are due to nonspecific interactions with proteins. These effects are particularly prominent in in vitro tests for antiviral activity as high concentrations of cells, viruses and oligonucleotides are often coincubated (AZAD et al. 1993; WAGNER et al. 1993). Human immune deficiency virus (HIV) is particularly problematic as many oligonucleotides bind to the gp120 protein (WYATT et al. 1994). However, the potential for confusion arising from the misinterpretation of an activity as being due to an antisense mechanism when, in fact, it is due to nonantisense effects is certainly not limited to antiviral or just in vitro tests (BARTON and LEMOINE 1995; BURGESS et al. 1995; HERTL et al. 1995). Again, these data simply urge caution and argue for careful dose-response curves, direct analyses of target protein or RNA, and inclusion of appropriate controls before drawing conclusions concerning the mechanisms of action of oligonucleotide-based drugs. In addition to protein interactions, other factors, such as overrepresented sequences of RNA and unusual structures that may be adopted by oligonucleotides, can contribute to unexpected results (WYATT et al. 1994).

Given the variability in cellular uptake of oligonucleotides, the variability in potency as a function of binding site in an RNA target, and potential nonantisense activities of oligonucleotides, careful evaluation of dose-response curves and clear demonstration of the antisense mechanism are required before drawing conclusions from in vitro experiments. Nevertheless, numerous well-controlled studies have been reported in which antisense activity was conclusively demonstrated. As many of these studies have been reviewed previously, suffice it to say that antisense effects of phosphorothioate oligodeoxynucleotides against a variety of targets are well documented (CROOKE and LEBLEU 1993; STEIN and CHENG 1993; S.T. CROOKE 1992, 1993, 1995b; NAGEL et al. 1993).

2. In Vivo Pharmacological Activities

A relatively large number of reports of in vivo activities of phosphorothioate oligonucleotides have now appeared documenting activities both after local and systemic administration (Table 1; S.T. CROOKE 1995b). However, for only a few of these reports have sufficient studies been performed to draw relatively firm conclusions concerning the mechanism of action. Consequently, I will review in some detail only a few reports that provide sufficient data to support a relatively firm conclusion with regard to mechanism of action. Local effects have been reported for phosphorothioate and methylphosphonate oligonucleotides. A phosphorothioate oligonucleotide designed to inhibit c-*myb* production and applied locally was shown to inhibit intimal accumulation in the rat carotid artery (SIMONS et al. 1992). In this study, a Northern blot analysis showed a significant reduction in c-*myb* RNA in animals treated with the antisense compound, but no effect when treated with a control oligonucleotide. In a recent study, it was suggested that the effects of the oligonucleotide were due to a nonantisense mechanism (BURGESS et al. 1995). However, only one dose level was studied, so much remains to be done before definitive conclusions are possible. Similar effects were reported for phosphorothioate oligodeoxynucleotides designed to inhibit cyclin-dependent kinases (CDC-2 and CDK-2). Again, the antisense oligonucleotide inhibited intimal thickening and cyclin-dependent kinase activity, while a control oligonucleotide had no effect (ABE et al. 1994). Additionally, local administration of a phosphorothioate oligonucleotide designed to inhibit N-*myc* resulted in reduction in N-*myc* expression and slower growth of a subcutaneously transplanted human tumor in nude mice (WHITESELL et al. 1991).

Antisense oligonucleotides administered intraventricularly have been reported to induce a variety of effects in the central nervous system. Intraventricular injection of antisense oligonucleotides to neuropeptide Y-Y1 receptors reduced the density of the receptors and resulted in behavioral signs of anxiety (WAHLESTEDT et al. 1993). Similarly, an antisense oligonucleotide designed to bind to N-methyl-D-aspartate (NMDA)-R1 receptor channel RNA inhibited the synthesis of these channels and reduced the volume of focal

Table 1. Reported activity of antisense oligonucleotides in animal models

Target	Route	Species	Reference
Cardiovascular models			
c-*myb*	Topically	Rat	(SIMONS et al. 1992)
cdc2 kinase	Topically	Rat	(MORISHITA et al. 1993)
PCNA	Topically	Rat	(MORISHITA et al. 1993)
cdc2 kinase	Topically	Rat	(ABE et al. 1994)
CDK2	Topically	Rat	(ABE et al. 1994)
Cyclin B_1	Topically	Rat	(MORISHITA et al. 1994)
PCNA	Topically	Rat	(SIMONS et al. 1994)
Angiotensin type 1 receptor	Intracerebral	Rat	(GYURKO et al. 1993)
Angiotensinogen	Intracerebral	Rat	(PHILLIPS et al. 1994)
c-*fos*	Intracerebral	Rat	(SUZUKI et al. 1994)
Inflammatory models			
Type 1 IL-1 receptor	Intradermal	Mouse	(BURCH and MAHAN 1991)
ICAM-1	Intravenous	Mouse	(STEPKOWSKI et al. 1994)
ICAM-1	Intravenous	Mouse	(KUMASAKA et al. 1996)
ICAM-1	Intravenous	Mouse	(KATZ et al. 1995)
ICAM-1	Intravenous	Mouse	(STEPKOWSKI et al. 1995)
ICAM-1	Intravenous	Mouse	(BENNETT et al. 1997)
Adenosine type 1 receptor	Aerosol	Rabbit	(NYCE and METZGER 1997)
Cancer models			
N-*myc*	Subcutaneous	Mouse	(WHITESELL et al. 1991)
NF-kB p65	Intraperitoneal	Mouse	(KITAJIMA et al. 1992)
c-*myb*	Subcutaneous	Mouse	(RATAJCZAK et al. 1992)
p120 nucleolar antigen	Intraperitoneal	Mouse	(PERLAKY et al. 1993)
NK-kB p65	Subcutaneous	Mouse	(HIGGINS et al. 1993)
Protein kinase C-a	Intraperitoneal	Mouse	(DEAN and McKAY 1994)
c-*myb*	Subcutaneous	Mouse	(HIJIYA et al. 1994)
Ha-*ras*	Intratumor	Mouse	(SCHWAB et al. 1994)
BCR-ABL	Intravenous	Mouse	(SKORSKI et al. 1994)
PTHrP	Intraventricular	Rat	(AKINO et al. 1996)
c-raf kinase	Intravenous	Mouse	(MONIA et al. 1995)
Protein kinase C-a	Intravenous	Mouse	(DEAN et al. 1996)
Protein kinase C-a	Intravenous	Mouse	(YAZAKI et al. 1996)
Neurological models			
c-*fos*	Intracerebral	Rat	(CHIASSON et al. 1992)
SNAP-25	Intracerebral	Chicken	(OSEN-SAND et al. 1993)
Kinesin heavy chain	Intravitreal	Rabbit	(AMARATUNGA et al. 1993)
Arginine vasopressin	Intracerebral	Rat	(FLANAGAN et al. 1993)
c-*fos*	Intracerebral	Rat	(HEILIG et al. 1993)
Progesterone receptor	Intracerebral	Rat	(POLLIO et al. 1993)
Dopamine type 2 receptor	Intracerebral	Rat	(ZHANG and CREESE 1993)
Y-Y1 receptor	Intracerebral	Rat	(WAHLESTEDT et al. 1993)
Neuropeptide Y	Intracerebral	Rat	(AKABAYASHI et al. 1994)
k-opioid receptor	Intracerebral	Rat	(ADAMS et al. 1994)
IGF-1	Intracerebral	Rat	(CASTRO-ALAMANCOS and TORRES-ALEMAN 1994)
c-*fos*	Intraspinal	Rat	(GILLARDON et al. 1994)
c-*fos*	Intracerebral	Rat	(HOOPER et al. 1994)
c-*fos*	Intraspinal	Rat	(WOODBURN et al. 1994)
NMDA receptor	Intracerebral	Rat	(KINDY 1994)
CREB	Intracerebral	Rat	(KONRADI et al. 1994)

Table 1. *Continued*

Target	Route	Species	Reference
Delta-opioid receptor	Intracerebral	Mice	(LAI et al. 1994)
Progesterone receptor	Intracerebral	Rat	(MANI et al. 1994)
GAD65	Intracerebral	Rat	(MCCARTHY et al. 1994)
GAD67	Intracerebral	Rat	(MCCARTHY et al. 1994)
AT1-angiotensin receptor	Intracerebral	Rat	(SAKAI et al. 1995)
Tryptophan hydroxylase	Intracerebral	Mouse	(MCCARTHY et al. 1995)
AT1-angiotensin receptor	Intracerebral	Rat	(AMBUHL et al. 1995)
CRH_1-corticotropin-releasing Hormone receptor	Intracerebral	Rat	(LIEBSCH et al. 1995)
• Opiod receptor	Intracerebral	Rat	(CHA et al. 1995)
• Opiod receptor	Intracerebral	Mouse	(MIZOGUCHI et al. 1995)
Oxytocin	Intracerebral	Rat	(MORRIS et al. 1995)
Oxytocin	Intracerebral	Rat	(NEUMANN et al. 1994)
Substance P receptor	Intracerebral	Rat	(OGO et al. 1994)
Tyrosine hydroxylase	Intracerebral	Rat	(SKUTELLA et al. 1994)
c-*jun*	Intracerebral	Rat	(TISCHMEYER et al. 1994)
Dopamine type 1 receptor	Intracerebral	Mouse	(ZHANG et al. 1994)
Dopamine type 2 receptor	Intracerebral	Mouse	(ZHOU et al. 1994)
Dopamine type 2 receptor	Intracerebral	Mouse	(WEISS et al. 1993)
Dopamine type 2 receptor	Intracerebral	Mouse	(QIN et al. 1995)
Viral models			
HSV-1		Mouse	(KULKA et al. 1989)
Tick-born encephalitis		Mouse	(VLASSOV 1989)
Duck hepatitis virus	Intravenous	Duck	(OFFENSPERGER et al. 1993)

ischemia produced by occlusion of the middle cerebral artery in rats (WAHLESTEDT et al. 1993).

In a series of well-controlled studies, antisense oligonucleotides administered intraventricularly selectively inhibited dopamine type-2 receptor expression, dopamine type-2 receptor RNA levels, and behavioral effects in animals with chemical lesions. Controls included randomized oligonucleotides and the observation that no effects were observed on dopamine type-1 receptor or RNA levels (WEISS et al. 1993; ZHOU et al. 1994; QIN et al. 1995). This laboratory also reported the selective reduction of dopamine type-1 receptor and RNA levels with the appropriate oligonucleotide (ZHANG et al. 1994).

Similar observations were reported in studies on angiotensin type 1 (AT-1) receptors and tryptophan hydroxylase. In studies in rats, direct observations of AT-1 and AT-2 receptor densities in various sites in the brain after administration of different doses of phosphorothioate antisense, sense, and scrambled oligonucleotides were reported (AMBUHL et al. 1995). Again, in rats, intraventricular administration of an antisense phosphorothioate oligonucleotide resulted in a decrease in tryptophan hydroxylase levels in the brain, while a scrambled control did not (MCCARTHY et al. 1995).

Injection of antisense oligonucleotides to synaptosomal-associated protein-25 into the vitreous body of rat embryos reduced the expression of the protein and inhibited neurite elongation by rat cortical neurons (OSEN-SAND et al. 1993).

Aerosol administration to rabbits of an antisense phosphorothioate oligodeoxynucleotide designed to inhibit the production of antisense A_1 receptor has been reported to reduce receptor numbers in the airway smooth muscle and to inhibit adenosine, house dust mite allergen, and histamine-induced bronchoconstriction (NYCE and METZGER 1997). Neither control nor an oligonucleotide complementary to bradykinin B_2 receptors reduced adenosine A_1 receptors' density, although the oligonucleotides complementary to bradykinin B_2 receptor mRNA reduced the density of these receptors.

In addition to local and regional effects of antisense oligonucleotides, a growing number of well-controlled studies have demonstrated systemic effects of phosphorothioate oligodeoxynucleotides. Expression of interleukin-1 in mice was inhibited by systemic administration of antisense oligonucleotides (BURCH and MAHAN 1991). Oligonucleotides to the NF-kB p65 subunit administered intraperitoneally at 40 mg/kg every 3 days slowed tumor growth in mice transgenic for the human T-cell leukemia viruses (KITAJIMA et al. 1992). Similar results with other antisense oligonucleotides were shown in another in vivo tumor model after either prolonged subcutaneous infusion or intermittent subcutaneous injection (HIGGINS et al. 1993).

Several recent reports further extend the studies of phosphorothioate oligonucleotides as antitumor agents in mice. In one study, a phosphorothioate oligonucleotide directed to inhibition of the *bcr-abl* oncogene was administered at a dose of 1 mg/day for 9 days intravenously to immunodeficient mice injected with human leukemic cells. The drug was shown to inhibit the development of leukemic colonies in the mice and to selectively reduce *bcr-abl* RNA levels in peripheral blood lymphocytes, spleen, bone marrow, liver, lungs, and brain (SKORSKI et al. 1994). However, it is possible that the effects on the RNA levels were secondary to effects on the growth of various cell types. In the second study, a phosphorothioate oligonucleotide antisense to the protooncogene *myb*, inhibited the growth of human melanoma in mice. Again, *myb* mRNA levels appeared to be selectively reduced (HIJIYA et al. 1994).

A number of studies from our laboratories that directly examined target RNA levels, target protein levels, and pharmacological effects using a wide range of control oligonucleotides and examination of the effects on closely-related isotypes have been completed. Single and chronic daily administration of a phosphorothioate oligonucleotide designed to inhibit mouse protein kinase C-α, (PKC-α), selectively inhibited expression of PKC-α RNA in mouse liver without effects on any other isotype. The effects lasted at least 24 h after a dose and a clear dose-response curve was observed, with a dose of 10–15 mg/kg intraperitoneally reducing PKC-α RNA levels in liver by 50% 24 h after a dose (DEAN and McKAY 1994).

A phosphorothioate oligonucleotide designed to inhibit human PKC-α expression selectively inhibited expression of PKC-α RNA and PKC-α

protein in human tumor cell lines implanted subcutaneously in nude mice after intravenous administration (DEAN et al. 1996). In these studies, effects on RNA and protein levels were highly specific and observed at doses lower than 6 mg/kg per day and antitumor effects were detected at doses as low as 0.6 mg/kg per day. A large number of control oligonucleotides failed to show activity.

In a similar series of studies, Monia et al. demonstrated highly specific loss of human c-*raf* kinase RNA in human tumor xenografts and antitumor activity that correlated with the loss of RNA. Moreover, a series of control oligonucleotides with 1–7 mismatches showed decreasing potency in vitro and precisely the same rank order potencies in vivo (MONIA et al. 1995, 1996).

Finally, a single injection of a phosphorothioate oligonucleotide designed to inhibit c-AMP-dependent protein kinase type 1 was reported to selectively reduce RNA and protein levels in human tumor xenografts and to reduce tumor growth (NESTEROVA and CHO-CHUNG 1995).

Thus, there is a growing body of evidence that phosphorothioate oligonucleotides can induce potent systemic and local effects in vivo. More importantly, there are now a number of studies with sufficient controls and direct observation of target RNA and protein levels to suggest highly specific effects that are difficult to explain via any mechanism other than antisense. As would be expected, the potency of these effects varies depending on the target, the organ, and the endpoint measured as well as the route of administration and the time point after administration at which the effect is measured.

In conclusion, although it is of obvious importance to interpret in vivo activity data cautiously, and it is clearly necessary to include a range of controls and to evaluate effects on target RNA and protein levels and control RNA and protein levels directly, it is difficult to argue with the conclusion today that some effects have been observed in animals that are most likely primarily due to an antisense mechanism.

Additionally, in studies on patients with cytomegalovirus-induced retinitis, local injections of ISIS 2922 have resulted in impressive efficacy, though it is obviously impossible to prove the mechanism of action is antisense in these studies (HUTCHERSON et al. 1995). More recently, ISIS 2302, an ICAM-1 inhibitor, was reported to result in statistically significant reductions in steroid doses and prolonged remissions in a small group of steroid-dependent patients with Crohn's Disease. As this study was randomized, double-blinded, and included serial colonoscopies, it may be considered the first study in humans to demonstrate the therapeutic activity of an antisense drug after systemic administration (YACYSHYN et al. 1997).

VI. Toxicological Properties

1. In Vitro

In our laboratory, we have evaluated the toxicities of scores of phosphorothioate oligodeoxynucleotides in a significant number of cell lines in tissue culture. As a general rule, no significant cytotoxicity is induced at concentrations below 100 μM oligonucleotide. Additionally, with a few exceptions, no significant effect on macromolecular synthesis is observed at concentrations below 100 μM (R.M. CROOKE 1993a,b).

Polynucleotides and other polyanions have been shown to cause release of cytokines (COLBY 1971). Also, bacterial DNA species have been reported to be mitogenic for lymphocytes in vitro (MESSINA et al. 1991). Furthermore, oligodeoxynucleotides (30–45 nucleotides in length) were reported to induce interferons and enhance natural killer cell activity (KURAMOTO et al. 1992). In the latter study, the oligonucleotides that displayed natural killer cell (NK)-stimulating activity contained specific palindromic sequences and tended to be guanosine rich. Collectively, these observations indicate that nucleic acids may have broad immunostimulatory activity.

It has been shown that phosphorothioate oligonucleotides stimulate B-lymphocyte proliferation in a mouse splenocyte preparation (analogous to bacterial DNA), and the response may underlie the observations of lymphoid hyperplasia in the spleen and lymph nodes of rodents caused by repeated administration of these compounds (see below; PISETSKY and REICH 1994). We also have evidence of enhanced cytokine release by immunocompetent cells when exposed to phosphorothioates in vitro (R.M. CROOKE et al. 1996). In this study, both human keratinocytes and an in vitro model of human skin released interleukin-1α when treated with 250 μM–1 mm of phosphorothioate oligonucleotides. The effects seemed to be dependent on the phosphorothioate backbone and independent of sequence or $2'$ modification. In a study in which murine B lymphocytes were treated with phosphodiester oligonucleotides, B-cell activation was induced by oligonucleotides with unmethylated CpG dinucleotides (KRIEG et al. 1995). This has been extrapolated to suggest that the CpG motif may be required for immune stimulation of oligonucleotide analogs such as phosphorothioates. This clearly is not the case with regard to release of IL-1α from keratinocytes (R.M. CROOKE et al. 1996). Nor is it the case with regard to in vivo immune stimulation (see below).

2. Genotoxicity

As with any new chemical class of therapeutic agents, concerns about genotoxicity cannot be dismissed as little in vitro testing has been performed and no data from long-term studies of oligonucleotides are available. Clearly, given the limitations in our understanding about the basic mechanisms that might be involved, empirical data must be generated. We have performed mutagenicity studies on two phosphorothioate oligonucleotides, ISIS 2105 and ISIS 2922, and found them to be nonmutagenic at all concentrations studied (S.T. CROOKE et al. 1994).

Two mechanisms of genotoxicity that may be unique to oligonucleotides have been considered. One possibility is that an oligonucleotide analog could be integrated into the genome and produce mutagenic events. Although integration of an oligonucleotide into the genome is conceivable, it is likely to be extremely rare. For most viruses, viral DNA integration is itself a rare event and, of course, viruses have evolved specialized enzyme-mediated mechanisms to achieve integration. Moreover, preliminary studies in our laboratory have shown that phosphorothioate oligodeoxynucleotides are generally poor substrates for DNA polymerases, and it is unlikely that enzymes such as integrases, gyrases, and topoisomerases (that have obligate DNA cleavage as intermediate steps in their enzymatic processes) will accept these compounds as substrates. Consequently, it would seem that the risk of genotoxicity due to genomic integration is no greater and probably less than that of other potential mechanisms, for example, alteration of the activity of growth factors, cytokine release, nonspecific effects on membranes that might trigger arachidonic acid release, or inappropriate intracellular signaling. Presumably, new analogs that deviate more significantly from natural DNA would be even less likely to be integrated.

A second concern that has been raised about possible genotoxicity is the risk that oligonucleotides might be degraded to toxic or carcinogenic metabolites. However, metabolism of phosphorothioate oligodeoxynucleotides by base excision would release normal bases, which presumably would be nongenotoxic. Similarly, oxidation of the phosphorothioate backbone to the natural phosphodiester structure would also yield nonmutagenic (and probably nontoxic) metabolites. Finally, it is possible that phosphorothioate bonds could be hydrolyzed slowly, releasing nucleoside phosphorothioates that presumably would be rapidly oxidized to natural (nontoxic) nucleoside phosphates. However, oligonucleotides with modified bases and/or backbones may pose different risks.

3. In Vivo

The acute LD_{50} in mice of all phosphorothioate oligonucleotides tested to date is in excess of 500mg/kg (D. KORNBRUST, unpublished observations). In rodents, we have had the opportunity to evaluate the acute and chronic toxicities of multiple phosphorothioate oligonucleotides administered by multiple routes (HENRY et al. 1997c,d). The consistent dose limiting toxicity was immune stimulation manifested by lymphoid hyperplasia, splenomegaly, and a multiorgan monocellular infiltrate. These effects occurred only with chronic dosing at doses greater than 20mg/kg and were dose dependent. The liver and kidney were the organs most prominently affected by monocellular infiltrates. All of these effects appeared to be reversible and chronic intradermal administration appeared to be the most toxic route, probably because of high local concentrations of the drugs resulting in local cytokine release and initiation of a cytokine cascade. There were no obvious effects of sequence. At doses of 100mg/kg and greater, minor increases in liver enzyme levels and mild thrombocytopenia were also observed.

In monkeys, however, the toxicological profile of phosphorothioate oligonucleotides is quite different. The most prominent dose-limiting side effect is sporadic reductions in blood pressure associated with bradycardia. When these events are observed, they are often associated with activation of C5 complement and they are dose related and peak plasma concentration related. This appears to be related to the activation of the alternative pathway (HENRY et al. 1997a). All phosphorothioate oligonucleotides tested to date appear to induce these effects, though there may be slight variations in potency as a function of sequence and/or length (CORNISH et al. 1993; GALBRAITH et al. 1994; HENRY et al. 1997c,d).

A second prominent toxicologic effect in the monkey is the prolongation of activated partial thromboplastin time. At higher doses, evidence of clotting abnormalities is observed. Again, these effects are dose and peak plasma concentration dependent (GALBRAITH et al. 1994; HENRY et al. 1997b). Although no evidence of sequence dependence has been observed, there appears to be a linear correlation between number of phosphorothioate linkages and potency between 18–25 nucleotides (P. NICKLIN, unpublished observations). The mechanisms responsible for these effects are likely very complex, but preliminary data suggest that direct interactions with thrombin may be at least partially responsible for the effects observed (HENRY et al. 1997b).

In man, the toxicological profile again differs a bit. When ISIS 2922 is administered intravitreally to patients with cytomegalovirus retinitis, the most common adverse event is anterior chamber inflammation, which is easily managed with steroids. A relatively rare and dose-related adverse event is morphological changes in the retina associated with loss in peripheral vision (HUTCHERSON et al. 1995).

ISIS 2105, a 20-mer phosphorothioate designed to inhibit the replication of human papilloma viruses that cause genital warts, is administered intradermally at doses as high as 3 mg/wart weekly for 3 weeks; essentially, no toxicities have been observed, including remarkably, a complete absence of local inflammation (L. GRILLONE, unpublished results).

Every other day administration of 2-h intravenous infusions of ISIS 2302 at doses as high as 2 mg/kg resulted in no significant toxicities, including no evidence of immune stimulation and no hypotension. A slight subclinical increase in activated partial thromboplastin time was observed at the 2 mg/kg dose (GLOVER et al. 1996).

VII. Therapeutic Index

In Fig. 3, an attempt to put the toxicities and their dose-response relationships in a therapeutic context is shown. This is particularly important as considerable confusion has arisen concerning the potential utility of phosphorothioate oligonucleotides for selected therapeutic purposes, deriving from unsophisticated interpretation of toxicological data. As can readily be seen, the immune stimulation induced by these compounds appears to be particularly prominent in rodents and unlikely to be dose limiting in man. Nor have we, to date, observed hypotensive events in humans. Thus, this toxicity

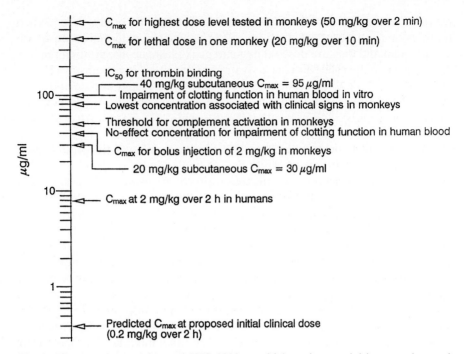

Fig. 3. Plasma concentrations of ISIS 2302 at which various activities are observed. These concentrations are determined by extracting plasma and analyzing by capillary gel electrophoresis and represent intact ISIS 2302

appears to occur at lower doses in monkeys than man and certainly is not dose limiting in man.

Based on our experience to date, we believe that the dose-limiting toxicity in man will be clotting abnormalities, and this will be associated with peak plasma concentrations well in excess of 10 µg/ml. In animals, pharmacological activities have been observed with I.V. bolus doses from 0.006 to 10–15 mg/kg depending on the target, the endpoint, the organ studied and the time after a dose when the effect is measured. Thus, it would appear that phosphorothioate oligonucleotides have a therapeutic index that supports their evaluation for a number of therapeutic indications.

VIII. Conclusions

Phosphorothioate oligonucleotides have perhaps outperformed many expectations. They display attractive parenteral pharmacokinetic properties. They have produced potent systemic effects in a number of animal models and, in many experiments, the antisense mechanism has been directly demonstrated as the hoped-for selectivity. Further, these compounds appear to display satisfactory therapeutic indices for many indications.

Table 2. Phosphorothioate oligonucleotides

Limits

Pharmacodynamic
- Low affinity per nucleotide unit
- Inhibition of RNase H at high concentrations

Pharmacokinetic
- Limited bioavailability
- Limited blood–brain barrier penetration
- Dose-dependent pharmacokinetics
- Possible drug–drug interactions

Toxicologic
- Release of cytokines
- Complement associated effects on blood pressure?
- Clotting effects

Nevertheless, phosphorothioates clearly have significant limits (Table 2). Pharmacodynamically, they have relatively low affinity per nucleotide unit. This means that longer oligonucleotides are required for biological activity and that invasion of many RNA structures may not be possible. At higher concentrations, these compounds inhibit RNase H as well. Thus, the higher end of the pharmacologic dose response curve is lost. Pharmacokinetically, phosphorothioates do not cross the blood–brain barrier, are not significantly orally bioavailable, and may display dose-dependent pharmacokinetics. Toxicologically, clearly the release of cytokines, activation of complement, and interference with clotting will pose dose limits if they are encountered in the clinic.

As several clinical trials are in progress with phosphorothioates and others will be initiated shortly, we shall soon have more definitive information about the activities, toxicities, and value of this class of antisense drugs in human beings.

E. The Medicinal Chemistry of Oligonucleotides

I. Introduction

The core of any rational drug discovery program is medicinal chemistry. Although the synthesis of modified nucleic acids has been a subject of interest for some time, the intense focus on the medicinal chemistry of oligonucleotides arguably predates this chapter by no more than 5 years. Consequently, the scope of medicinal chemistry has recently expanded enormously, but the biological data to support conclusions about synthetic strategies are only beginning to emerge.

Modifications in the base, sugar, and phosphate moieties of oligonucleotides have been reported. The subjects of medicinal chemical programs include approaches to create enhanced affinity and more selective affinity for RNA or duplex structures, the ability to cleave nucleic acid targets, enhanced nuclease stability, cellular uptake and distribution, and in vivo tissue distribution, metabolism and clearance.

II. Heterocycle Modifications

1. Pyrimidine Modifications

A relatively large number of modified pyrimidines have been synthesized and are now incorporated into oligonucleotides and evaluated. The principle sites of modification are C-2, C-4, C-5, and C-6. These and other nucleoside analogs have recently been thoroughly reviewed (SANGHVI 1993). Consequently, a very brief summary of the analogs that displayed interesting properties is incorporated here.

Inasmuch as the C-2 position is involved in Watson-Crick hybridization, C-2 modified pyrimidine containing oligonucleotides have shown unattractive hybridization properties. An oligonucleotide containing 2-thiothymidine was found to hybridize well to DNA and, in fact, even better to RNA (ΔT_m 1.5°C modification) (E. SWAYZE et al., unpublished results).

In contrast, several modifications in the 4 position that have interesting properties have been reported. 4-Thiopyrimidines have been incorporated into oligonucleotides with no significant negative effect on hybridization (NIKIFOROV and CONNOLLY 1991). A bicyclic and an N4-methoxy analog of cytosine were shown to hybridize with both purine bases in DNA with T_m's approximately equal to natural base pairs (LIN and BROWN 1989). Additionally, a fluorescent base has been incorporated into oligonucleotides and shown to enhance DNA–DNA duplex stability (INOUE et al. 1985).

A large number of modifications at the C-5 position have also been reported, including halogenated nucleosides. Although the stability of duplexes may be enhanced by incorporating 5-halogenated nucleosides, the occasional mispairing with G and the potential that the oligonucleotide might degrade and release toxic nucleosides analogs cause concern (SANGHVI 1993).

Furthermore, oligonucleotides containing 5-propynylpyrimidine modifications have been shown to enhance the duplex stability (ΔT_m 1.6°C/modification) and support the RNase H activity. The 5-heteroarylpyrimidines were also shown to influence the stability of duplexes (WAGNER et al. 1993; GUTIERREZ et al. 1994). A more dramatic influence was reported for the tricyclic 2′-deoxycytidine analogs, exhibiting an enhancement of 2°C–5°C/ modification, depending on the positioning of the modified bases (LIN et al. 1995). It is believed that the enhanced binding properties of these analogs is due to extended stacking and increased hydrophobic interactions.

In general, as expected, modifications in the C-6 position of pyrimidines are highly duplex destabilizing (SANGHVI et al. 1993). Oligonucleotides containing 6-aza pyrimidines have been shown to reduce T_m by 1°C–2°C per modification, but to enhance the nuclease stability of oligonucleotides and to support RNase H-induced degradation of RNA targets (SANGHVI 1993).

2. Purine Modifications

Although numerous purine analogs have been synthesized, when incorporated into oligonucleotides, they usually have resulted in destabilization of duplexes. However, there are a few exceptions, where a purine modification had a stabilizing effect. A brief summary of some of these analogs is discussed below.

Generally, N1 modifications of purine moiety has resulted in destabilization of the duplex (MANOHARAN 1993). Similarly, C2 modifications have usually resulted in destabilization. However, 2-6-diaminopurine has been reported to enhance hybridization by approximately 1°C per modification when paired with T (SPROAT et al. 1991). Of the 3-position substituted bases reported to date, only the 3-deaza adenosine analog has been shown to have no negative effective on hybridization.

Modifications at the C-6 and C-7 positions have likewise resulted in only a few interesting bases from the point of view of hybridization. Inosine has been shown to have little effect on duplex stability, but because it can pair and stack with all four normal DNA bases, it behaves as a universal base and creates an ambiguous position in an oligonucleotide (MARTIN et al. 1985). Incorporation of 7-deaza inosine into oligonucleotides was destabilizing, and this was considered to be due to its relatively hydrophobic nature (SANTALUCIA et al. 1991). 7-Deaza guanine was similarly destabilizing, but when 8-aza-7-deaza guanine was incorporated into oligonucleotides, it enhanced hybridizations (SEELA et al. 1989). Thus, on occasion, introduction of more than one modification in a nucleobase may compensate for destabilizing effects of some modifications. Interestingly, 7-iodo 7-deazaguanine residue was recently incorporated into oligonucleotides and shown to enhance the binding affinity dramatically (ΔT_m 10.0°C/modification compared to 7-deazaguanine; SEELA et al. 1995). The increase in T_m value was attributed to (a) the hydrophobic nature of the modification, (b) increased stacking interaction, and (c) favorable pKa of the base.

In contrast, some C8 substituted bases have yielded improved nuclease resistance when incorporated in oligonucleotides, but seem to be somewhat destabilizing (SANGHVI 1993).

3. Oligonucleotide Conjugates

Although conjugation of various functionalities to oligonucleotides has been reported to achieve a number of important objectives, the data supporting some of the claims are limited and generalizations are not possible based on the data presently available.

a) Nuclease Stability

Numerous 3' modifications have been reported to enhance the stability of oligonucleotides in serum (MANOHARAN 1993). Both neutral and charged substituents have been reported to stabilize oligonucleotides in serum and, as a general rule, the stability of a conjugated oligonucleotide tends to be greater as bulkier substituents are added. Inasmuch as the principle nuclease in serum is a 3' exonuclease, it is not surprising that 5' modifications have resulted in significantly less stabilization. Internal modifications of base, sugar, and backbone have also been reported to enhance nuclease stability at or near the modified nucleoside (MANOHARAN 1993). In a recent study, thiono triester (adamantyl, cholesteryl and others) modified oligonucleotides have shown improved nuclease stability, cellular association, and binding affinity (ZHANG et al. 1995).

The demonstration that modifications may induce nuclease stability sufficient to enhance activity in cells in tissue culture and in animals has proven to be much more complicated because of the presence of 5' exonucleases and endonucleases. In our laboratory, 3' modifications and internal point modifications have not provided sufficient nuclease stability to demonstrate pharmacological activity in cells (HOKE et al. 1991). In fact, even a 5 nucleotide long phosphodiester gap in the middle of a phosphorothioate oligonucleotide resulted in sufficient loss of nuclease resistance to cause complete loss of pharmacological activity (MONIA et al. 1992).

In mice, neither a 5'-cholesterol nor 5'-C18 amine conjugate altered the metabolic rate of a phosphorothioate oligodeoxynucleotide in liver, kidney, or plasma (S.T. CROOKE et al. 1996). Furthermore, blocking the 3' and 5' termini of a phosphodiester oligonucleotide did not markedly enhance the nuclease stability of the parent compound in mice (SANDS et al. 1995). However, 3' modification of a phosphorothioate oligonucleotide was reported to enhance its stability in mice relative to the parent phosphorothioate (TEMSAMANI et al. 1993). Moreover, a phosphorothioate oligonucleotide with a 3'-hairpin loop was reported to be more stable in rats than its parent (ZHANG et al. 1995). Thus, 3' modifications may enhance the stability of the relatively stable phosphorothioates sufficiently to be of value.

b) Enhanced Cellular Uptake

Although oligonucleotides have been shown to be taken up by a number of cell lines in tissue culture, with perhaps the most compelling data relating to phosphorothioate oligonucleotides, a clear objective has been to improve cellular uptake of oligonucleotides (R.M. CROOKE 1991; S.T. CROOKE et al. 1994). Inasmuch as the mechanisms of cellular uptake of oligonucleotides are still very poorly understood, the medicinal chemistry approaches have been largely empirical and based on many unproven assumptions.

Because phosphodiester and phosphorothioate oligonucleotides are water soluble, the conjugation of lipophilic substituents to enhance membrane permeability has been a subject of considerable interest. Unfortunately, studies in this area have not been systematic and, at present, there is precious little information about the changes in physicochemical properties of oligonucleotides actually effected by specific lipid conjugates. Phospholipids, cholesterol and cholesterol derivatives, cholic acid, and simple alkyl chains have been conjugated to oligonucleotides at various sites in the oligonucleotide. The effects of these modifications on cellular uptake have been assessed using fluorescent, or radiolabeled, oligonucleotides or by measuring pharmacological activities. From the perspective of medicinal chemistry, very few systematic studies have been performed. The activities of short alkyl chains, adamantine, daunomycin, fluorescein, cholesterol, and porphyrin conjugated oligonucleotides were compared in one study (BOUTORINE et al. 1991). A cholesterol modification was reported to be more effective at enhancing uptake than the other substituents. It also seems likely that the effects of various conjugates on cellular uptake may be affected by the cell type and target studied. For example, we have studied cholic acid conjugates of phosphorothioate deoxyoligonucleotides or phosphorothioate 2'-methoxy oligonucleotides and observed enhanced activity against HIV and no effect on the activity of ICAM-directed oligonucleotides.

Additionally, polycationic substitutions and various groups designed to bind to cellular carrier systems have been synthesized. Although many compounds have been synthesized, the data reported to date are insufficient to draw firm conclusions about the value of such approaches or structure activity relationships (MANOHARAN 1993).

c) RNA Cleaving Groups

Oligonucleotide conjugates were recently reported to act as artificial ribonucleases, albeit with low efficiencies (DE MESMAEKER et al. 1995). Conjugation of chemically reactive groups such as alkylating agents, photoinduced azides, porphyrin, and psoralene have been utilized extensively to effect a cross-linking of oligonucleotide and the target RNA. In principle, this treatment may lead to translation arrest. In addition, lanthanides and complexes thereof have been reported to cleave RNA via a hydrolytic pathway. Recently, a novel europium complex was covalently linked to an oligonucleotide and shown to cleave 88% of the complementary RNA at physiological pH (HALL et al. 1994).

d) In Vivo Effects

To date, relatively few studies have been reported in vivo. The properties of a 5'-cholesterol and 5'-C-18 amine conjugates of a 20-mer phosphorothioate oligodeoxynucleotide have been determined in mice. Both compounds increased the fraction of an I.V. bolus dose found in the liver. The cholesterol conjugate, in fact, resulted in more than 80% of the dose accumulating in the liver. Neither conjugate enhanced stability in plasma, liver, or kidney (S.T.

CROOKE et al. 1996). Interestingly, the only significant change in the toxicity profile was a slight increase in effects on serum transamineses and histopathological changes indicative of slight liver toxicity associated with the cholesterol conjugate (HENRY et al. 1997e). A 5'-cholesterol phosphorothioate conjugate was also recently reported to have a longer elimination half-life, to be more potent, and to induce greater liver toxicity in rats (DESJARDINS et al. 1995).

4. Sugar Modifications

The focus of second-generation oligonucleotide modifications has centered on the sugar moiety. In oligonucleotides, pentofuranose sugar ring occupies a central connecting manifold that also positions the nucleobases for effective stacking. Recently, a symposium series volume has been published on the carbohydrate modifications in antisense research that covers this topic in great detail (SANGHVI and COOK 1994). Therefore, the content of the following discussion is restricted to a summary of the main events in this area.

A growing number of oligonucleotides in which the pentofuranose ring is modified or replaced have been reported (BRESLAUER et al. 1986). Uniform modifications at the 2' position have been shown to enhance hybridization to RNA, and in some cases, to enhance nuclease resistance (BRESLAUER et al. 1986). Chimeric oligonucleotides containing 2'-deoxyoligonucleotide gaps with 2'-modified wings have been shown to be more potent than parent molecules (MONIA et al. 1993).

Other sugar modifications include α-oligonucleotides, carbocyclic oligonucleotides and hexapyranosyl oligonucleotides (BRESLAUER et al. 1986). Of these, α-oligonucleotides have been most extensively studied. They hybridize in parallel fashion to single-stranded DNA and RNA and are nuclease resistant. However, they have been reported to be oligonucleotides designed to inhibit Ha-*ras* expression. All these oligonucleotides support RNase H and, as can be seen, a direct correlation between affinity and potency exists.

A growing number of oligonucleotides in which the C-2' position of the sugar ring is modified have been reported (MANOHARAN 1993; DE MESMAEKER et al. 1995). These modifications include lipophilic alkyl groups, intercalators, amphipathic amino-alkyl tethers, positively charged polyamines, highly electronegative fluoro or fluoro alkyl moities, and sterically bulky methylthio derivatives. The beneficial effects of a C-2' substitution on the antisense oligonucleotide cellular uptake, nuclease resistance, and binding affinity have been well documented in the literature. In addition, excellent review articles have appeared in the last few years on the synthesis and properties of C-2'-modified oligonucleotides (DE MESMAEKER et al. 1995; LAMOND and SPROAT 1993; SPROAT and LAMOND 1993; PARMENTIER et al. 1994).

Other modifications of the sugar moiety have also been studied, including other sites as well as more substantial modifications. However, much less is known about the antisense effects of these modifications (S.T. CROOKE 1995b).

2'-Methoxy-substituted phosphorothioate oligonucleotides have recently been reported to be more stable in mice than their parent compounds and to display enhanced oral bioavailability (ZHANG et al. 1995; AGRAWAL et al. 1995). The analogs displayed tissue distribution similar to that of the parent phosphorothioate.

Similarly, we have compared the pharmacokinetics of 2'-propoxy modified phosphodiester and phosphorothioate deoxynucleotides (S.T. CROOKE et al. 1996). As expected, the 2'-propoxy modification increased lipophilicity and nuclease resistance. In fact, in mice the 2'-propoxy phosphorothioate was too stable in liver or kidney to measure an elimination half-life.

Interestingly, the 2'-propoxy phosphodiester was much less stable than the parent phosphorothioate in all organs except in the kidney, where the 2'-propoxy phosphodiester was remarkably stable. The 2'-propoxy phosphodiester did not bind to albumin significantly, while the affinity of the phosphorothioate for albumin was enhanced. The only difference in toxicity between the analogs was a slight increase in renal toxicity associated with the 2'-propoxy phosphodiester analog (HENRY et al. 1997).

Incorporation of the 2'-methoxyethyoxy group into oligonucleotides increased the T_m by 1.1°C/modification when hybridized to the complement RNA. In a similar manner, several other 2'-O-alkoxy modifications have been reported to enhance the affinity (MARTIN 1995). The increase in affinity with these modifications was attributed to (a) the favorable gauche effect of side chain and (b) additional solvation of the alkoxy substituent in water.

More substantial carbohydrate modifications have also been studied. Hexose-containing oligonucleotides were created and found to have very low affinity for RNA (PITSCH et al. 1995). Also, the 4' oxygen has been replaced with sulfur. Although a single substitution of a 4'-thio modified nucleoside resulted in destabilization of a duplex, incorporation of two 4'-thio modified nucleosides increased the affinity of the duplex (BELLON et al. 1994). Finally, bicyclic sugars have been synthesized with the hope that preorganization into more rigid structures would enhance hybridization. Several of these modifications have been reported to enhance hybridization (SANGHVI and COOK 1994).

5. Backbone Modifications

Substantial progress in creating new backbones for oligonucleotides that replace the phosphate or the sugar-phosphate unit has been made. The objectives of these programs are to improve hybridization by removing the negative charge, enhance stability, and potentially improve pharmacokinetics.

For a review of the backbone modifications reported to date, please see S.T. CROOKE (1995b) and SANGHVI and COOK (1994). Suffice it to say that numerous modifications have been made that replace phosphate, retain hybridization, alter charge, and enhance stability. Since these modifications are now being evaluated in vitro and in vivo, a preliminary assessment should be possible shortly.

Replacement of the entire sugar-phosphate unit has also been accomplished and the oligonucleotide analogs produced have displayed very interesting characteristics. Peptide nuclei acid (PNA) oligomers have been shown to bind to single-stranded DNA and RNA with extraordinary affinity and high sequence specificity. They have been shown to be able to invade some double-stranded nucleic acid structures. PNA oligomers can form triple-stranded structures with DNA or RNA.

PNA oligomers were shown to be able to act as antisense and transcriptional inhibitors when microinjected in cells (HANVEY et al. 1992). PNA oligonucleotides appear to be quite stable to nucleases and peptidases as well.

In summary, then, in the past 5 years, enormous advances in the medicinal chemistry of oligonucleotides has been reported. Modifications at nearly every position in oligonucleotides have been attempted and numerous potentially interesting analogs have been identified. Although it is far too early to determine which of the modifications may be most useful for particular purposes, it is clear that a wealth of new chemicals is available for systematic evaluation and that these studies should provide important insights into the structure–activity relationships of oligonucleotide analogs.

III. Conclusions

Although there are many more unanswered than answered questions about antisense, progress has continued to be gratifying. Clearly, as more is learned, we will be in the position to perform progressively more sophisticated studies and to understand more of the factors that determine whether an oligonucleotide actually works via an antisense mechanism. We should also have the opportunity to learn a great deal more about this class of drugs as additional studies are completed in humans.

Acknowledgement. The author wishes to thank Donna Musacchia for excellent typographic and administrative assistance.

References

Abe J, Zhou W, Taguchi J, Takuwa N, Miki K, Okazaki H, Kurokawa K, Kumada M, Takuwa Y (1994) Suppression of neointimal smooth muscle cell accumulation in vivo by antisense CDC2 and CDK2 oligonucleotides in rat carotid artery. Biochem Biophys Res Commun 198:16–24

Adams JU, Chen XH, deRiel JK, Adler MW, Liu-Chen LY (1994) In vivo treatment with antisense oligodeoxynucleotide to kappa-opioid receptors inhibited kappa-agonist-induced analgesia in rats. Regul Pept 54:1–2

Agrawal S, Goodchild J, Civeira MP, Thornton AH, Sarin PS, Zamecnik PC (1988) Oligodeoxynucleoside phosphoramidates and phosphorothioates as inhibitors of human immunodeficiency virus. Proc Natl Acad Sci U S A 85:7079–7083

Agrawal S, Temsamani J, Tang JY (1991) Pharmacokinetics, biodistribution, and stability of oligodeoxynucleotide phosphorothioates in mice. Proc Natl Acad Sci U S A 88:7595–7599

Agrawal S, Zhang X, Lu Z, Zhao H, Tamburin JM, Yan J, Cai H, Diasio RB, Habus I, Jiang Z, Iyer RP, Yu D, Zhang R (1995) Absorption, tissue distribution and in vivo stability in rats of a hybrid antisense oligonucleotide following oral administration. Biochem Pharmacol 50(4):571–576

Akabayashi A, Wahlestedt C, Alexander JT, Leibowitz SF (1994) Specific inhibition of endogenous neuropeptide Y synthesis in arcuate nucleus by antisense oligonucleotides suppresses feeding behavior and insulin secretion. Mol Brain Res 21:55–61

Akino K, Ohtsuru A, Yano H, Ozeki S, Namba H, Nakashima M, Ito M, Matsumoto T, Yamashita S (1996) Antisense inhibition of parathyroid hormone-related peptide gene expression reduces malignant pituitary tumor progression and metastases in the rat. Cancer Res 56(1):77–86

Amaratunga A, Morin PJ, Kosik KS, Fine RE (1993) Inhibition of kinesin synthesis and rapid anterograde axonal transport in vivo by an antisense oligonucleotide. J Biol Chem 268(23):17427–17430

Ambuhl P, Gyurko R, Phillips MI (1995) A decrease in angiotensin receptor binding in rat brain nuclei by antisense oligonucleotides to the angiotensin AT1 receptor. Regul Pept 59(2):171–182

Azad RF, Driver VB, Tanaka K, Crooke RM, Anderson KP (1993) Antiviral activity of a phosphorothioate oligonucleotide complementary to RNA of the human cytomegalovirus major immediate-early region. Antimicrob Agents Chemother 37(9):1945–1954

Baker BF (1993) "Decapitation" of a 5'-capped oligoribonucleotide by o-penanthroline:Cu(II). J Am Chem Soc 115:3378–3379

Baker BF, Miraglia L, Hagedorn CH (1992) Modulation of eucaryotic initiation factor-4E binding to 5'-capped oligoribonucleotides by modified anti-sense oligonucleotides. J Biol Chem 267:11495–11499

Barton CM, Lemoine NR (1995) Antisense oligonucleotides directed against p53 have antiproliferative effects unrelated to effects on p53 expression. Br J Cancer 71:429–437

Bayever E, Iversen PL, Bishop MR, Sharp JG, Tewary HK, Arneson MA, Pirruccello SJ, Ruddon RW, Kessinger A, Zon G (1993) Systemic administration of a phosphorothioate oligonucleotide with a sequence complementary to p53 for acute myelogenous leukemia and myelodysplastic syndrome: initial results of a phase I trial. Antisense Res Dev 3(4):383–390

Bellon L, Leydier C, Barascut JL (1994) 4-Thio RNA: a novel class of sugar-modified B-RNA. In: Sanghvi YS, Cook PD (eds) Carbohydrate modifications in antisense research. American Chemical Society, Washington DC, pp 68–79

Bennett CF, Crooke ST (1996) Oligonucleotide-based inhibitors of cytokine expression and function. In: Henderson B, Bodmer MW (eds) Therapeutic modulation of cytokines. CRC Press, Boca Raton, pp 171–193

Bennett CF, Chiang MY, Chan H, Shoemaker JEE, Mirabelli CK (1992) Cationic lipids enhance cellular uptake and activity of phosphorothioate antisense oligonucleotides. Mol Pharmacol 41:1023–1033

Bennett CF, Chiang MY, Chan H, Grimm S (1993) Use of cationic lipids to enhance the biological activity of antisense oligonucleotides. J Liposome Res 3:85–102

Bennett CF, Kornbrust D, Henry S, Stecker K, Howard R, Cooper S, Dutson S, Hall W, Jacoby HI (1997) An ICAM-1 antisense oligonucleotide prevents and reverses dextran sulfate sodium-induced colitis in mice. J Pharmacol Exp Ther 280(2):988–1000

Boutorine A, Huet C, Saison T (1991) Cell penetration studies of oligonucleotide derivatized with cholesterol and porphyrins. Conference proceedings on nucleic acid therapeutics, pp 1–60.

Breslauer KJ, Frank R, Blocker H, Marky LA (1986) Predicting DNA duplex stability from base sequence. Proc Natl Acad Sci U S A 83:3746–3750

Burch RM, Mahan LC (1991) Oligonucleotides antisense to the interleukin 1 receptor mRNA block the effects of interleukin 1 in cultured murine and human fibroblasts and in mice. J Clin Invest 88:1190–1196

Burgess TL, Fisher EF, Ross SL, Bready JV, Qian YX, Bayewitch LA, Cohen AM, Herrera CJ, Hu SSF, Kramer TB, Lott FD, Martin FH, Pierce GF, Simonet L, Farrell CL (1995) The antiproliferative activity of c-myb and c-myc antisense oligonucleotides in smooth muscle cells is caused by a nonantisense mechanism. Proc Natl Acad Sci U S A 92:4051–4055

Campbell JM, Bacon TA, Wickstrom E (1990) Oligodeoxynucleoside phosphorothioate stability in subcellular extracts, culture media, sera and cerebrospinal fluid. J Biochem Biophys Methods 20:259–267

Castro-Alamancos MA, Torres-Aleman I (1994) Learning of the conditioned eye-blink response is impaired by an antisense insulin-like growth factor I oligonucleotide. Proc Natl Acad Sci U S A 91:10203–10207

Cazenave C, Stein CA, Loreau N, Thuong NT, Neckers LM, Subasinghe C, Helene C, Cohen JS, Toulme JJ (1989) Comparative inhibition of rabbit globin mRNA translation by modified antisense oligodeoxynucleotides. Nucleic Acids Res 17:4255–4273

Cha XY, Xu H, Ni Q, Partilla JS, Rice KC, Matecka D, Calderon SN, Porreca F, Lai J, Rothman RB (1995) Opioid peptide receptor studies. 4. Antisense oligodeoxynucleotide to the delta opioid receptor delineates opioid receptor subtypes. Regul Pept 59(2):247–253

Cheng Y, Gao W, Han F (1991) Phosphorothioate oligonucleotides as potential antiviral compounds against human immunodeficiency virus and herpes viruses. Nucleosides Nucleotides 10:155–166

Chiang MY, Chan H, Zounes MA, Freier SM, Lima WF, Bennett CF (1991) Antisense oligonucleotides inhibit intercellular adhesion molecule 1 expression by two distinct mechanisms. J Biol Chem 266:18162–18171

Chiasson BJ, Hooper ML, Murphy PR, Robertson HA (1992) Antisense oligonucleotide eliminates in vivo expression of c-fos in mammalian brain. Eur J Pharmacol 227:451–453

Chrisey LA, Walz SE, Pazirandeh M, Campbell JR (1993) Internalization of oligodeoxyribonucleotides by Vibrio parahaemolyticus. Antisense Res Dev 3:367–381

Cohen JS (1993) Phosphorothioate oligodeoxynucleotides. In: Crooke ST, Lebleu B (eds) Antisense research and applications. CRC Press, Boca Raton, pp 205–222

Colby CJ (1971) The induction of interferon by natural and synthetic polynucleotides. Prog Nucleic Acid Res Mol Biol 11:1–32

Cook PD (1993) Medicinal chemistry strategies for antisense research. In: Crooke ST, Lebleu B (eds) Antisense research and applications. CRC Press, Boca Raton, pp 149–187

Cornish KG, Iversen P, Smith L, Arneson M, Bayever E (1993) Cardiovascular effects of a phosphorothioate oligonucleotide to p53 in the conscious rhesus monkey. Pharmacol Commun 3:239–247

Cossum PA, Sasmor H, Dellinger D, Truong L, Cummins L, Owens SR, Markham PM, Shea JP, Crooke S (1993) Disposition of the ^{14}C-labeled phosphorothioate oligonucleotide ISIS 2105 after intravenous administration to rats. J Pharmacol Exp Ther 267:1181–1190

Cossum PA, Truong L, Owens SR, Markham PM, Shea JP, Crooke ST (1994) Pharmacokinetics of a ^{14}C-labeled phosphorothioate oligonucleotide, ISIS 2105, after intradermal administration to rats. J Pharmacol Exp Ther 269:89–94

Cowsert LM (1993) Antiviral activities of antisense oligonucleotides. In: Crooke ST, Lebleu B (eds) Antisense research and applications. CRC Press, Boca Raton, pp 521–533

Cowsert LM, Fox MC, Zon G, Mirabelli CK (1993) In vitro evaluation of phosphorothioate oligonucleotides targeted to the E2 mRNA of papillomavirus: potential treatment of genital warts. Antimicrob Agents Chemother 37:171–177

Crooke RM (1991) In vitro toxicology and pharmacokinetics of antisense oligonucleotides. Anticancer Drug Des 6:609–646

Crooke RM (1993a) Cellular uptake, distribution and metabolism of phosphorothioate, phosphodiester, and methylphosphonate oligonucleotides. In: Crooke ST, Lebleu B (eds) Antisense research and applications. CRC Press, Boca Raton, pp 427–449
Crooke RM (1993b) In vitro and in vivo toxicology of first generation analogs. In: Crooke ST, Lebleu B (eds) Antisense research and applications. CRC Press, Boca Raton, pp 471–492
Crooke RM, Graham MJ, Cooke ME, Crooke ST (1995) In vitro pharmacokinetics of phosphorothioate antisense oligonucleotides. J Pharmacol Exp Ther 275(1):462–473
Crooke RM, Crooke ST, Graham MJ, Cooke ME (1996) Effect of antisense oligonucleotides on cytokine release from human keratinocytes in an in vitro model of skin. Toxicol Appl Pharmacol 140:85–93
Crooke ST (1992) Therapeutic applications of oligonucleotides. Annu Rev Pharmacol Toxicol 32:329–376
Crooke ST (1993) Progress toward oligonucleotide therapeutics: pharmacodynamic properties. FASEB J 7:533–539
Crooke ST (1995a) Oligonucleotide therapeutics. In: Wolff ME (ed) Burger's medicinal chemistry and drug discovery, vol 1. Wiley, New York, pp 863–900
Crooke ST (1995b) Therapeutic applications of oligonucleotides. Landes, Austin
Crooke ST, Lebleu B (1993) Antisense research and applications. CRC Press, Boca Raton
Crooke ST, Grillone LR, Tendolkar A, Garrett A, Fratkin MJ, Leeds J, Barr WH (1994) A pharmacokinetic evaluation of ^{14}C-labeled afovirsen sodium in patients with genital warts. Clin Pharm Ther 56:641–646
Crooke ST, Lemonidis KM, Nielson L, Griffey R, Monia BP (1995) Kinetic characteristics of E. coli RNase H1: cleavage of various antisense oligonucleotides–RNA duplexes. Biochem J 312(2):599–608
Crooke ST, Graham MJ, Zuckerman JE, Brooks D, Conklin BS, Cummins LL, Greig MJ, Guinosso CJ, Kornbrust D, Manoharan M, Sasmor HM, Schleich T, Tivel KL, Griffey RH (1996) Pharmacokinetic properties of several novel oligonucleotide analogs in mice. J Pharmacol Exp Ther 277(2):923–937
Crouch RJ, Dirksen ML (1985) Ribonucleases H. In: Linn SM, Roberts RJ (eds) Nucleases. Cold Spring Harbor Laboratory Press, Cold Spring Harbor, NY, p 211
Crum C, Johnson JD, Nelson A, Roth D (1988) Complementary oligodeoxynucleotide mediated inhibition of tobacco mosaic virus RNA translation in vitro. Nucleic Acids Res 16(10):4569–4581
Dean NM, McKay R (1994) Inhibition of protein kinase C-alpha expression in mice after systemic administration of phosphorothioate antisense oligodeoxynucleotides. Proc Natl Acad Sci U S A 91:11762–11766
Dean NM, McKay R, Miraglia L, Howard R, Cooper S, Giddings J, Nicklin P, Meister L, Zeil R, Geiger T, Muller M, Fabbro D (1996) Inhibition of growth of human tumor cell lines in nude mice by an antisense oligonucleotide inhibitor of PKC-alpha expression. Cancer Res 56(15):3499–3507
De Clercq E, Eckstein F, Merigan TC (1969) Interferon induction increased through chemical modification of synthetic polyribonucleotide. Science 165:1137–1140
De Mesmaeker A, Haener R, Martin P, Moser HE (1995) Antisense oligonucleotides. Acc Chem Res 28(9):366–374
Desjardins J, Mata J, Brown T, Graham D, Zon G, Iversen P (1995) Cholesteryl-conjugated phosphorothioate oligodeoxynucleotides modulate CYP2B1 expression in vivo. J Drug Targeting 2:477–485
Dominski Z, Kole R (1993) Restoration of correct splicing in thalassemic pre-mRNA by antisense oligonucleotides. Proc Natl Acad Sci U S A 90:8673–8677
Donis-Keller H (1979) Site specific enzymatic cleavage of RNA. Nucleic Acids Res 7:179–192

Ecker DJ (1993) Strategies for invasion of RNA secondary structure. In: Crooke ST, Lebleu R (eds) Antisense research and applications. CRC Press, Boca Raton, pp 387–400

Ecker DJ, Vickers TA, Bruice TW, Freier SM, Jenison RD, Manoharan M, Zounes M (1992) Pseudo–half-knot formation with RNA. Science 257:958–961

Eder PS, Walder JA (1991) Ribonuclease H from K562 human erythroleukemia cells. J Biol Chem 266:6472–6479

Flanagan LM, McCarthy MM, Brooks PJ, Pfaff DW, McEwen BS (1993) Arginine vasopressin levels after daily infusions of antisense oligonucleotides into the supraoptic nucleus. Ann N Y Acad Sci 689:520–521

Freier SM (1993) Hybridization considerations affecting antisense drugs. In: Crooke ST, Lebleu B (eds) Antisense research and applications. CRC Press, Boca Raton, pp 67–82

Furdon PJ, Dominski Z, Kole R (1989) RNase H cleavage of RNA hybridized to oligonucleotides containing methylphosphonate, phosphorothioate and phosphodiester bonds. Nucleic Acids Res 17:9193–9204

Gagnor C, Bertrand JR, Thenet S, Lemaitre M, Morvan F, Rayner B, Malvy C, Lebleu B, Imbach JL, Paoletti C (1987) Alpha-DNA. VI: comparative study of alpha- and beta-anomeric oligodeoxyribonucleotides in hybridization to mRNA and in cell free translation inhibition. Nucleic Acids Res 15(24):10419–10436

Gagnor C, Rayner B, Leonetti JP, Imbach JL, Lebleu B (1989) Alpha-DNA. IX: parallel annealing of alpha-anomeric oligodeoxyribonucleotides to natural mRNA is required for interference in RNase H mediated hydrolysis and reverse transcription. Nucleic Acids Res 17(13):5107–5114

Galbraith WM, Hobson WC, Giclas PC, Schechter PJ, Agrawal S (1994) Complement activation and hemodynamic changes following intravenous administration of phosphorothioate oligonucleotides in the monkey. Antisense Res Dev 4(3):201–206

Gao WY, Han FS, Storm C, Egan W, Cheng YC (1992) Phosphorothioate oligonucleotides are inhibitors of human DNA polymerases and RNase H: implications for antisense technology. Mol Pharmacol 41:223–229

Giles RV, Tidd DM (1992) Increased specificity for antisense oligodeoxynucleotide targeting of RNA cleavage by RNase H using chimeric methylphosphonodiester/phosphodiester structures. Nucleic Acids Res 20:763–770

Giles RV, Spiller DG, Tidd DM (1995) Detection of ribonuclease H-generated mRNA fragments in human leukemia cells following reversible membrane permeabilization in the presence of antisense oligodeoxynucleotides. Antisense Res Dev 5:23–31

Gillardon F, Beck H, Uhlmann E, Herdegen T, Sandkohler J, Peyman A, Zimmermann M (1994) Inhibition of c-fos protein expression in rat spinal cord by antisense oligodeoxynucleotide superfusion. Eur J Neurosci 6:880–884

Glover JM, Leeds JM, Mant TGK, Kisner DL, Zuckerman J, Levin AA, Shanahan WR (in press) Phase I safety and pharmacokinetic profile of an ICAM-1 antisense oligodeoxynucleotide (ISIS 2302). J Pharmacol Exp Ther

Graham MJ, Freier SM, Crooke RM, Ecker DJ, Maslova RN, Lesnik EA (1993) Tritium labeling of antisense oligonucleotides by exchange with tritiated water. Nucleic Acids Res 21:3737–3743

Gutierrez AJ, Terhorst TJ, Matteucci MD, Froehler BC (1994) 5-heteroaryl-2′-deoxyuridine analogs. Synthesis and incorporation into high-affinity oligonucleotides. J Am Chem Soc 116:5540–5544

Gyurko R, Wielbo D, Phillips MI (1993) Antisense inhibition of AT1 receptor mRNA and angiotensinogen mRNA in the brain of spontaneously hypertensive rats reduces hypertension of neurogenic origin. Regul Pept 49:167–174

Haeuptle MT, Frank R, Dobberstein B (1986) Translation arrest by oligodeoxynucleotides complementary to mRNA coding sequences yields polypeptides of predetermined length. Nucleic Acids Res 14(3):1427–1448

Hall J, Husken D, Pieles U, Moser HE, Haner R (1994) Efficient sequence-specific cleavage of RNA using novel europium complexes conjugated to oligonucleotides. Chem Biol 1(3):185–190

Hanvey JC, Peffer NC, Bisi JE, Thomson SA, Cadilla R, Josey JA, Ricca DJ, Hassman CF, Bonham MA, Au KG, Carter SG, Bruckenstein DA, Boyd AL, Noble SA, Babiss LE (1992) Antisense and antigene properties of peptide nucleic acids. Science 258:1481–1485

Hartmann G, Krug A, Waller-Fontaine K, Endres S (1996) Oligodeoxynucleotides enhance lipopolysaccharide-stimulated synthesis of tumor necrosis factor: dependence on phosphorothioate modification and reversal by heparin. Mol Med 2(4):429–438

Hawley P, Gibson I (1996) Interaction of oligodeoxynucleotides with mammilian cells. Antisense Nucleic Drug Dev 6:185–195

Heilig M, Engel JA, Soderpalm B (1993) C-fos antisense in the nucleus accumbens blocks the locomotor stimulant action of cocaine. Eur J Pharmacol 236:339–340

Helene C, Toulme JJ (1989) Control of gene expression by oligonucleotides covalently linked to intercalating agents and nucleic acid-cleaving reagents. In: Cohen JS (ed) Oligonucleotides: antisense inhibitors of gene expression. CRC Press, Boca Raton, pp 137–172

Henry S, Giclas PC, Leeds J, Pangburn M, Auletta C, Levin AA, Kornbrust DJ (1997a) Activation of the alternative pathway of complement by a phosphorothioate oligonucleotide: potential mechansim of action. J Pharmacol Exp Ther 281:810–816

Henry S, Novotny W, Leeds J (1997b) Inhibition of clotting parameters by a phosphorothioate oligonucleotide (in press)

Henry SP, Grillone LR, Orr JL, Brunner RH, Kornbrust DJ (1997c) Comparison of the toxicity profiles of ISIS 1082 and ISIS 2105, phosphorothioate oligonucleotides, following subacute intradermal administration in Sprague-Dawley rats. Toxicology 116(1–3):77–88

Henry SP, Taylor J, Midgley L, Levin AA, Kornbrust DJ (1997d) Evaluation of the toxicity profile of ISIS 2302, a phosphorothioate oligonucleotide in a 4-week study in CD-1 mice. Antisense Nucleic Acid Drug Dev (in press)

Henry SP, Zuckerman JE, Rojko J, Hall WC, Harman RJ, Kitchen D, Crooke ST (1997e) Toxicologic properties of several novel oligonucleotide analogs in mice. Anticancer Drug Des 12(1):1–14

Hertl M, Neckers LM, Katz SI (1995) Inhibition of interferon-gamma-induced intercellular adhesion molecule-1 expression on human keratinocytes by phosphorothioate antisense oligodeoxynucleotides is the consequence of antisense-specific and antisense-non-specific effects. J Invest Dermatol 104:813–818

Higgins KA, Perez JR, Coleman TA, Dorshkind K, McComas WA, Sarmiento UM, Rosen CA, Narayanan R (1993) Antisense inhibition of the p65 subunit of NF-kappaB blocks tumorigenicity and causes tumor regression. Proc Natl Acad Sci U S A 90:9901–9905

Hijiya N, Zhang J, Ratajczak MZ, Kant JA, DeRiel K, Herlyn M, Zon G, Gewirtz AM (1994) Biologic and therapeutic significance of MYB expression in human melanoma. Proc Natl Acad Sci U S A 91(10):4499–4503

Hodges D, Crooke ST (1995) Inhibition of splicing of wild-type and mutated luciferase-adenovirus pre-mRNA by antisense oligonucleotides. Mol Pharmacol 48:905–918

Hoke GD, Draper K, Freier SM, Gonzalez C, Driver VB, Zounes MC, Ecker DJ (1991) Effects of phosphorothioate capping on antisense oligonucleotide stability, hybridization and antiviral efficacy versus herpes simplex virus infection. Nucleic Acids Res 19:5743–5748

Hooper ML, Chiasson BJ, Robertson HA (1994) Infusion into the brain of an antisense oligonucleotide to the immediate-early gene c-fos suppresses production of fos and produces a behavioral effect. Neuroscience 63:917–924

Hughes JA, Avrutskaya AV, Brouwer KLR, Wickstrom E, Juliano RL (1995) Radio-labeling of methylphosphonate and phosphorothioate oligonucleotides and evaluation of their transport in everted rat jejunum sacs. Pharm Res 12:817–824

Hutcherson SL, Palestine AG, Cantrill HL, Lieberman RM, Holland GN, Anderson KP (1995) Antisense oligonucleotide safety and efficacy for CMV retinitis in AIDS patients. Conference Proceedings of 35th Interscience Conference on Antimicrobial Agents and Chemotherapy (ICAAC), 17–20 September 1995, San Francisco, p 204

Inoue H, Imura A, Ohtsuka E (1985) Synthesis and hybridization of dodecadeoxyribonucleotides containing a fluorescent pyridopyrimidine deoxynucleoside. Nucleic Acids Res 13(19):7119–7128

Iversen P (1991) In vivo studies with phosphorothioate oligonucleotides: pharmacokinetics prologue. Anticancer Drug Des 6(6):531–538

Joos RW, Hall WH (1969) Determination of binding constants of serum albumin for penicillin. J Pharmacol Exp Ther 166:113

Katz SM, Browne B, Pham T, Wang ME, Bennett CF, Stepkowski SM, Kahan BD (1995) Efficacy of ICAM-1 antisense oligonucleotide in pancreatic islet transplanation. Transplant Proc 27(6):3214

Kawasaki AM, Casper MD, Freier SM, Lesnik EA, Zounes MC, Cummins LL, Gonzalez C, Cook PD (1993) Uniformly modified 2′-deoxy-2′-fluoro phosphorothioate oligonucleotides as nuclease-resistant antisense compounds with high affinity and specificity for RNA targets. J Med Chem 36:831–841

Kindy MS (1994) NMDA receptor inhibition using antisense oligonucleotides prevents delayed neuronal death in gerbil hippocampus following cerebral ischemia. Neurosci Res Commun 14:175–183

Kitajima I, Shinohara T, Bilakovics J, Brown DA, Xiao X, Nerenberg M (1992) Ablation of transplanted HTLV-1 tax-transformed tumors in mice by antisense inhibition of NF-KB. Science 258:1792–1795

Konradi C, Cole RL, Heckers S, Hyman SE (1994) Amphetamine regulates gene expression in rat striatum via transcription factor CREB. J Neurosci 14:5623–5634

Krieg AM, Yi AK, Matson S, Waldschmidt TJ, Bishop GA, Teasdale R, Koretzky GA, Klinman DM (1995) CpG motifs in bacterial DNA trigger direct B-cell activation. Nature 374:546–549

Kulka M, Smith CC, Aurelian L, Fishelevich R, Meade K, Miller P, Ts'o POP (1989) Site specificity of the inhibitory effects of oligo(nucleoside methylphosphonate)s complementary to the acceptor splice junction of herpes simplex virus type 1 immediate early mRNA 4. Proc Natl Acad Sci U S A 86:6868–6872

Kumasaka T, Quinlan WM, Doyle NA, Condon TP, Sligh J, Takei F, Beaudet AL, Bennett CF, Doerschuk CM (1996) The role of the intercellular adhesion molecule-1 (ICAM-1) in endotoxin-induced pneumonia evaluated using ICAM-1 antisense oligonucleotides, anti-ICAM-1 monoclonal antibodies, and ICAM-1 mutant mice. J Clin Invest 97(10):2362–2369

Kuramoto E, Yano O, Kimura Y, Baba M, Makino T, Yamamoto S, Yamamoto T, Kataoka T, Tokunaga T (1992) Oligonucleotide sequences required for natural killer cell activation. Jpn J Cancer Res 83(11):1128–1131

Lai J, Bilsky EJ, Rothman RB, Porreca F (1994) Treatment with antisense oligodeoxynucleotide to the opioid-δ receptor selectively inhibits δ2-agonist antinociception. Neuroreport 5:1049–1052

Lamond AI, Sproat BS (1993) Antisense oligonucleotides made of 2′-O-alkylRNA: their properties and applications in RNA biochemistry. FEBS Lett 325:123–127

LeMaitre M, Bayard B, Lebleu B (1987) Specific antiviral activity of a poly(L-lysine)-conjugated oligodeoxyribonucleotide sequence complementary to vesicular stomatitis virus N protein mRNA initiation site. Proc Natl Acad Sci U S A 84:648–652

Liebsch G, Landgraf R, Gerstberger R, Probst JC, Wotjak CT, Engelmann M, Holsboer F, Montkowski A (1995) Chronic infusion of a CRH1 receptor antisense oligodeoxynucleotide into the central nucleus of the amygdala anxiety-related behavior in socially defeated rats. Regul Pept 59(2):229–239

Lima WF, Crooke ST (1997) Binding affinity and specificity of Escherichia coli RNase H1: impact on the kinetics of catalysis of antisense oligonucleotide-RNA hybrids. Biochemistry 36(2):390–398

Lima WF, Monia BP, Ecker DJ, Freier SM (1992) Implication of RNA structure on antisense oligonucleotide hybridization kinetics. Biochemistry 31:12055–12061

Lima WF, Venkatraman M, Crooke ST (in press) The influence of antisense oligonucleotide-induced RNA structure on E. coli RNase H1 activity. J Biol Chem

Lin KY, Jones RJ, Matteucci M (1995) Tricyclic-2'-deoxycytidine analogs: synthesis and incorporation into oligodeoxynucleotides which have enhanced binding to complementary RNA. J Am Chem Soc 117:3873–3874

Lin PKT, Brown DM (1989) Synthesis and duplex stability of oligonucleotides containing cytosine-thymine analogues. Nucleic Acids Res 17:10373–10383

Loke SL, Stein CA, Zhang XH, Mori K, Nakanishi M, Subasinghe C, Cohen JS, Neckers LM (1989) Characterization of oligonucleotide transport into living cells. Proc Natl Acad Sci U S A 86:3474–3478

Maher LJ III, Wold B, Dervan PB (1989) Inhibition of DNA binding proteins by oligonucleotide-directed triple helix formation. Science 245:725–730

Maier JAM, Voulalas P, Roeder D, Maciag T (1990) Extension of the life-span of human endothelial cells by an interleukin-1α antisense oligomer. Science 249:1570–1574

Majumdar C, Stein CA, Cohen JS, Broder S, Wilson SH (1989) Stepwise mechanism of HIV reverse transcriptase: primer function of phosphorothioate oligodeoxynucleotide. Biochemistry 28:1340–1346

Mani SK, Blaustein JD, Allen JMC, Law SW, O'Malley BW, Clark JH (1994) Inhibition of rat sexual behavior by antisense oligonucleotides to the progesterone receptor. Endocrinology 135:1409–1414

Manoharan M (1993) Designer antisense oligonucleotides: conjugation chemistry and functionality placement. In: Crooke ST, Lebleu B (eds) Antisense research and applications. CRC Press, Boca Raton, pp 303–349

Martin FH, Castro MM, Aboul-ela F, Tinoco IJ (1985) Base pairing involving deoxyinosine: implications for probe. Nucleic Acids Res 13:8927–8938

Martin P (1995) Ein neuer Zugang zu 2'-O-Alkylribonucleosiden und Eigenschaften deren Oligonucleotide. Helv Chim Acta 78:486–489

McCarthy MM, Masters DB, Rimvall K, Schwartz-Giblin S, Pfaff DW (1994) Intracerebral administration of antisense oligodeoxynucleotides to GAD65 and GAD67 mRNAs modulate reproductive behavior in the female rat. Brain Res 636:209–220

McCarthy MM, Nielsen DA, Goldman D (1995) Antisense oligonucleotide inhibition of tryptophan hydroxylase activity in mouse brain. Regul Pept 59(2):163–170

McManaway ME, Neckers LM, Loke SL, Al-Nasser AA, Redner RL, Shiramizu BT, Goldschmidts WL, Huber BE, Bhatia K, Magrath IT (1990) Tumour-specific inhibition of lymphoma growth by an antisense oligodeoxynucleotide. Lancet 335:808–811

Messina JP, Gilkeson GS, Pisetsky DS (1991) Stimulation of in vitro murine lymphocyte proliferation by bacterial DNA. J Immunol 147 (6):1759–1764

Miller PS (1989) Non-ionic antisense oligonucleotides. In: Cohen JS (ed) Oligodeoxynucleotides: antisense inhibitors of gene expression. CRC Press, Boca Raton, p 79

Minshull J, Hunt T (1986) The use of single-stranded DNA and RNase H to promote quantitative "hybrid arrest of translation" of mRNA/DNA hybrids in reticulocyte lysate cell-free translations. Nucleic Acids Res 14:6433–6451

Mirabelli CK, Crooke ST (1993) Antisense oligonucleotides in the context of modern molecular drug discovery and development. In: Crooke ST, Lebleu B (eds) Antisense research and applications. CRC Press, Boca Raton, pp 7–35

Mirabelli CK, Bennett CF, Anderson K, Crooke ST (1991) In vitro and in vivo pharmacologic activities of antisense oligonucleotides. Anticancer Drug Des 6:647–661

Miyao T, Takakura Y, Akiyama T, Yoneda F, Sezaki H, Hashida M (1995) Stability and pharmacokinetic characteristics of oligonucleotides modified at terminal linkages in mice. Antisense Res Dev 5(2):115–121

Mizoguchi H, Narita M, Nagase H, Tseng LF (1995) Antisense oligodeoxynucleotide to a delta-opioid receptor blocks the antinociception induced by cold water swimming. Regul Pept 59(2):255–259

Monia B, Johnston JF, Sasmor H, Cummins LL (1996) Nuclease resistance and antisense activity of modified oligonucleotides targeted to Ha-ras. J Biol Chem 24(14):14533–14540

Monia BP, Johnston JF, Ecker DJ, Zounes M, Lima WF, Freier SM (1992) Selective inhibition of mutant Ha-ras mRNA expression by antisense oligonucleotides. J Biol Chem 267:19954–19962

Monia BP, Lesnik EA, Gonzalez C, Lima WF, McGee D, Guinosso CJ, Kawasaki AM, Cook PD, Freier SM (1993) Evaluation of 2'-modified oligonucleotides containing deoxy gaps as antisense inhibitors of gene expression. J Biol Chem 268:14514–14522

Monia BP, Johnston JF, Geiger T, Muller M, Fabbro D (1995) Antitumor activity of a phosphorothioate oligodeoxynucleotide targeted against C-raf kinase. Nature Med 2(6):668–675

Morishita R, Gibbons GH, Ellison KE, Nakajima M, Zhang L, Kaneda Y, Ogihara T, Dzau VJ (1993) Single intraluminal delivery of antisense CDC 2 kinase and proliferating-cell nuclear antigen oligonucleotides results in chronic inhibition of neointimal hyperplasia. Proc Natl Acad Sci U S A 90:8474–8478

Morishita R, Gibbons GH, Kaneda Y, Ogihara T, Dzau VJ (1994) Pharmacokinetics of antisense oligodeoxyribonucleotides (cyclin B1 and CDC 2 kinase) in the vessel wall in vivo: enhanced therapeutic utility for restenosis by HVJ-liposome delivery. Gene 149:13–19

Morris M, Li P, Barrett C, Callahan MF (1995) Oxytocin antisense reduces salt intake in the baroreceptor-denervated rat. Regul Pept 59(2):261–266

Morvan F, Rayner B, Imbach JL (1991) Alpha-oligonucleotides: a unique class of modified chimeric nucleic acids. Anticancer Drug Des 6(6):521–529

Nagel KM, Holstad SG, Isenberg KE (1993) Oligonucleotide pharmacotherapy: an antigene strategy. Pharmacotherapy 13(3):177–188

Neckers LM (1993) Cellular internalization of oligodeoxynucleotides. In: Crooke ST, Lebleu B (eds) Antisense research and applications. CRC Press, Boca Raton, pp 451–460

Nesterova M, Cho-Chung YS (1995) A single-injection protein kinase A-directed antisense treatment to inhibit tumor growth. Nature Med 1:528–533

Neumann I, Porter DWF, Landgraf R, Pittman QJ (1994) Rapid effect on suckling of an oxytocin antisense oligonucleotide administered into rat supraoptic nucleus. Am J Physiol Regul Integr Comp Physiol 267:R852–R858

Nikiforov TT, Connolly BA (1991) The synthesis of oligodeoxynucleotides containing 4-thiothymidine residues. Tetrahedron Lett 32(31):3851–3854

Nyce JW, Metzger WJ (1997) DNA antisense therapy for asthma in an animal model. Nature 385(6618):721–725

Offensperger WB, Offensperger S, Walter E, Teubner K, Igloi G, Blum HE, Gerok W (1993) In vivo inhibition of duck hepatitis B virus replication and gene expression by phosphorothioate modified antisense oligodeoxynucleotides. EMBO J 12:1257–1262

Ogo H, Hirai Y, Miki S, Nishio H, Akiyama M, Nakata Y (1994) Modulation of substance P/neurokinin-1 receptor in human astrocytoma cells by antisense oligodeoxynucleotides. Gen Pharmacol 25:1131–1135

Osen-Sand A, Catsicas M, Staple JK, Jones KA, Ayala G, Knowles J, Grenningloh G, Catsicas S (1993) Inhibition of axonal growth by SNAP-25 antisense oligonucleotides in vitro and in vivo. Nature 364:445–448

Parmentier G, Schmitt G, Dolle F, Luu B (1994) A convergent synthesis of 2′-O-methyl uridine. Tetrahedron 50(18):5361–5368

Perlaky L, Saijo Y, Busch RK, Bennett CF, Mirabelli CK, Crooke ST, Busch H (1993) Growth inhibition of human tumor cell lines by antisense oligonucleotides designed to inhibit p120 expression. Anticancer Drug Des 8:3–14

Phillips MI, Wielbo D, Gyurko R (1994) Antisense inhibition of hypertension: a new strategy for renin-angiotensin candidate genes. Kidney Int 46:1554–1556

Pisetsky DS, Reich CF (1994) Stimulation of murine lymphocyte proliferation by a phosphorothioate oligonucleotide with antisense activity for herpes simplex virus. Life Sci 54:101–107

Pitsch S, Krishnamurthy R, Bolli M, Wendeborn S, Holzner A, Minton M, Lesueur C, Schloenvogt I, Jaun B et al (1995) Pyranosyl-RNA ("p-RNA"): base-pairing selectivity and potential to replicate. Helv Chim Acta 78(7):1621–1635

Pollio G, Xue P, Zanisi M, Nicolin A, Maggi A (1993) Antisense oligonucleotide blocks progesterone-induced lordosis behavior in ovariectomized rats. Mol Brain Res 19:135–139

Qin ZH, Zhou LW, Zhang SP, Wang Y, Weiss B (1995) D_2 dopamine receptor antisense oligodeoxynucleotide inhibits the synthesis of a functional pool of D_2 dopamine receptors. Mol Pharmacol 48(4):730–737

Quartin RS, Brakel CL, Wetmur JG (1989) Number and distribution of methylphosphonate linkages in oligodeoxynucleotides affect exo- and endonuclease sensitivity and ability to form RNase H substrates. Nucleic Acids Res 17:7253–7262

Quattrone A, Papucci L, Schiavone N, Mini E, Capaccioli S (1994) Intracellular enhancement of intact antisense oligonucleotide steady-state levels by cationic lipids. Anticancer Drug Des 9:549–553

Rappaport J, Hanss B, Kopp JB, Copeland TD, Bruggeman LA, Coffman TM, Klotman PE (1995) Transport of phosphorothioate oligonucleotides in kidney: implications for molecular therapy. Kidney Int 47:1462–1469

Ratajczak MZ, Kant JA, Luger SM, Huiya N, Zhang J, Zon G, Gewirtz AM (1992) In vivo treatment of human leukemia in a scid mouse model with c-myb antisense oligodeoxynucleotides. Proc Natl Acad Sci U S A 89:11823–11827

Rosolen A, Whitesell L, Ikegaki N, Kennett RH, Neckers LM (1990) Antisense inhibition of single copy N-myc expression results in decreased cell growth without reduction of c-myc protein in a neuroepithelioma cell line. Cancer Res 50(19):6316–6322

Sakai RR, Ma LY, He PF, Fluharty SJ (1995) Intracerebroventricular administration of angiotensin type 1 (AT1) receptor antisense oligonucleotides attenuate thirst in the rat. Regul Pept 59(2):183–192

Sands H, Gorey-Feret LJ, Cocuzza AJ, Hobbs FW, Chidester D, Trainor GL (1994) Biodistribution and metabolism of internally ^{3}H-labeled oligonucleotides. I. Comparison of a phosphodiester and a phosphorothioate. Mol Pharmacol 45:932–943

Sands H, Gorey-Feret LJ, Ho SP, Bao Y, Cocuzza AJ, Chidester D, Hobbs FW (1995) Biodistribution and metabolism of internally ^{3}H-labeled oligonucleotides. II. 3′,5– blocked oligonucleotides. Mol Pharmacol 47:636–646

Sanghvi YS (1993) Heterocyclic base modifications in nucleic acids and their applications in antisense oligonucleotides. In: Crooke ST, Lebleu B (eds) Antisense research and applications. CRC Press, Boca Raton, pp 273–288

Sanghvi YS, Cook PD (1994) Carbohydrate modifications in antisense research. American Chemical Society, Washington DC (ACS Symposium Series no 580)

Sanghvi YS, Hoke GD, Freier SM, Zounes MC, Gonzalez C, Cummins L, Sasmor H, Cook PD (1993) Antisense oligodeoxynucleotides: synthesis, biophysical and biological evaluation of oligodeoxynucleotides containing modified pyrimidines. Nucleic Acids Res 21:3197–3203

SantaLucia J Jr, Kierzek R, Turner DH (1991) Functional group substitutions as probes of hydrogen bonding between GA mismatches in RNA internal loops. J Am Chem Soc 113:4313–4322

Saxena SK, Ackerman EJ (1990) Microinjected oligonucleotides complementary to the α-sarcin loop of 28 S RNA abolish protein synthesis in Xenopus oocytes. J Biol Chem 265:3263–3269

Sburlati AR, Manrow RE, Berger SL (1991) Prothymosin alpha antisense oligomers inhibit myeloma cell division. Proc Natl Acad Sci U S A 88:253–257

Schwab G, Chavany C, Duroux I, Goubin G, Lebeau J, Helene C, Saison-Behmoaras T (1994) Antisense oligonucleotides adsorbed to polyalkylcyanoacrylate nanoparticles specifically inhibit mutated Ha-ras-mediated cell proliferation and tumorigenicity in nude mice. Proc Natl Acad Sci U S A 91:10460–10464

Seela F, Kaiser K, Bindig U (1989) 2'-Deoxy-beta-D-ribofuranosides of N6-methylated 7-deazaadenine and 8-aza-7-deazaadenine: solid-phase synthesis of oligodeoxyribonucleotides and properties of self-complementary duplexes. Helv Chim Acta 72(5):868–881

Seela F, Ramzaeva N, Chen Y (1995) Oligonucleotide duplex stability controlled by the 7-substituents of 7-deazaguanine bases. Bioorg Med Chem Lett 5(24):3049–3052

Simons M, Edelman ER, DeKeyser JL, Langer R, Rosenberg RD (1992) Antisense c-myb oligonucleotides inhibit arterial smooth muscle cell accumulation in vivo. Nature 359:67–70

Simons M, Edelman ER, Rosenberg RD (1994) Antisense proliferating cell nuclear antigen oligonucleotides inhibit intimal hyperplasia in a rat carotid artery injury model. J Clin Invest 93:2351–2356

Skorski T, Nieborowska-Skorska M, Nicolaides NC, Szczylik C, Iversen P, Iozzo RV, Zon G, Calabretta B (1994) Suppression of Philadelphia leukemia cell growth in mice by BCR-ABL antisense oligodeoxynucleotide. Proc Natl Acad Sci U S A 91:4504–4508

Skutella T, Probst JC, Jirikowski GF, Holsboer F, Spanagel R (1994) Ventral tegmental area (VTA) injections of tyrosine hydroxylase phosphorothioate antisense oligonucleotide suppress operant behavior in rats. Neurosci Lett 167:55–58

Smith CC, Aurelian L, Reddy MP, Miller PS, Ts'o POP (1986) Antiviral effect of an oligo(nucleoside methylphosphonate) complementary to the splice junction of herpes simplex virus type 1 immediate early pre-mRNAs 4 and 5. Proc Natl Acad Sci U S A 83:2787–2791

Sproat BS, Lamond AI (1993) 2'-O-alkyloligoribonucleotides. In: Crooke ST, Lebleu B (eds) Antisense research and applications. CRC Press, Boca Raton, pp 351–362

Sproat BS, Lamond AI, Beijer B, Neuner P, Ryder U (1989) Highly efficient chemical synthesis of 2'-O-methyloligoribonucleotides and tetrabiotinylated derivatives; novel probes that are resistant to degradation by RNA or DNA specific nucleases. Nucleic Acids Res 17:3373–3386

Sproat BS, Iribarren AM, Garcia RG, Beijer B (1991) New synthetic routes to synthons suitable for 2'-0-allyloligoribonucleotide assembly. Nucleic Acids Res 19(4):733–738

Srinivasan SK, Tewary HK, Iversen PL (1995) Characterization of binding sites, extent of binding, and drug interactions of oligonucleotides with albumin. Antisense Res Dev 5(2):131–139

Stein CA, Cheng YC (1993) Antisense oligonucleotides as therapeutic agents – is the bullet really magical? Science 261:1004–1012

Stein CA, Neckers M, Nair BC, Mumbauer S, Hoke G, Pal R (1991) Phosphorothioate oligodeoxycytidine interferes with binding of HIV-1 gp120 to CD4. J Acquir Immune Defic Syndr 4:686–693

Stepkowski SM, Tu Y, Condon TP, Bennett CF (1994) Blocking of heart allograft rejection by intercellular adhesion molecule-1 antisense oligonucleotides alone or in combination with other immunosuppressive modalities. J Immunol 153:5336–5346

Stepkowski SM, Tu Y, Condon TP, Bennett CF (1995) Induction of transplantation tolerance by treatment with ICAM-1 antisense oligonucleotides and anti-LFA-1 monoclonal antibodies. Transplant Proc 27:113

Suzuki S, Pilowsky P, Minson J, Arnolda L, Llewellyn-Smith IJ, Chalmers J (1994) c-fos antisense in rostral ventral medulla reduces arterial blood pressure. Am J Physiol 266(4,2):R1418–R1422

Takakura Y, Mahato RI, Yoshida M, Kanamaru T, Hashida M (1996) Uptake characteristics of oligonucleotides in the isolated rat liver perfusion system. Antisense Nucleic Acid Drug Del 6:177–183

Tao LF, Marx KA, Wongwit W, Jiang Z, Agrawal S, Coleman RM (1995) Uptake, intracellular distribution, and stability of oligodeoxynucleotide phosphorothioate by Schistosoma mansoni. Antisense Res Dev 5(2):123–129

Temsamani J, Tang J, Padmapriya A, Kubert M, Agrawal S (1993) Pharmacokinetics, biodistribution, and stability of capped oligodeoxynucleotide phosphorothioates in mice. Antisense Res Dev 3:277–284

Thuong NT, Asseline U, Monteney-Garestier T (1989) Oligodeoxynucleotides covalently linked to intercalating and reactive substances: synthesis, characterization and physicochemical studies. In: Cohen JS (ed) Oligodeoxynucleotides: antisense inhibitors of gene expression. CRC Press, Boca Raton, p 25

Tischmeyer W, Grimm R, Schicknick H, Brysch W, Schlingensiepen KH (1994) Sequence-specific impairment of learning by c-jun antisense oligonucleotides. Neuroreport 5:1501–1504

Vasanthakumar G, Ahmed NK (1989) Modulation of drug resistance in a daunorubicin resistant subline with oligonucleoside methylphosphonates [published erratum appears in Cancer Commun 1990; 2 (8):295]. Cancer Commun 1(4):225–232

Vickers T, Baker BF, Cook PD, Zounes M, Buckheit RW Jr, Germany J, Ecker DJ (1991) Inhibition of HIV-LTR gene expression by oligonucleotides targeted to the TAR element. Nucleic Acids Res 19:3359–3368

Vlassov VV (1989) Inhibition of tick-borne viral encephalitis expression using covalently linked oligonucleotide analogs. Conference proceedings ••

Wagner RW, Matteucci MD, Lewis JG, Gutierrez AJ, Moulds C, Froehler BC (1993) Antisense gene inhibition by oligonucleotides containing C-5 propyne pyrimidines. Science 260:1510–1513

Wahlestedt C, Pich EM, Koob GF, Yee F, Heilig M (1993) Modulation of anxiety and neuropeptide Y-Y1 receptors by antisense oligodeoxynucleotides. Science 259:528–531

Walder RY, Walder JA (1988) Role of RNase H in hybrid-arrested translation by antisense oligonucleotides. Proc Natl Acad Sci U S A 85:5011–5015

Walker K, Elela SA, Nazar RN (1990) Inhibition of protein synthesis by anti-5.8 S rRNA oligodeoxyribonucleotides. J Biol Chem 265:2428–2430

Wang S, Lee RJ, Cauchon G, Gorenstein DG, Low PS (1995) Delivery of antisense oligodeoxyribonucleotides against the human epidermal growth factor receptor into cultured KB cells with liposomes conjugated to folate via polyethylene glycol. Proc Natl Acad Sci U S A 92:3318–3322

Weiss B, Zhou LW, Zhang SP, Qin ZH (1993) Antisense oligodeoxynucleotide inhibits D_2 dopamine receptor- mediated behavior and D_2 messenger RNA. Neuroscience 55:607–612

Westermann P, Gross B, Hoinkis G (1989) Inhibition of expression of SV40 virus large T-antigen by antisense oligodeoxyribonucleotides. Biomed Biochim Acta 48(1): 85–93

Whitesell L, Rosolen A, Neckers LM (1991) In vivo modulation of N-myc expression by continous perfusion with an antisense oligonucleotide. Antisense Res Dev 1:343–350

Wickstrom E (1986) Oligodeoxynucleotide stability in subcellular extracts and culture media. J Biochem Biophys Methods 13:97–102

Woodburn VL, Hunter JC, Durieux C, Poat JA, Hughes J (1994) The effect of C-FOS antisense in the formalin-paw test. Regul Pept 54(1,2):327–328

Wu H, MacLeod AR, Lima WF, Crooke ST (submitted) Identification and partial purification of human double-stranded RNase activity: a novel terminating mechanism for oligonucleotide antisense drugs

Wyatt JR, Vickers TA, Roberson JL, Buckheit RW Jr, Klimkait T, DeBaets E, Davis PW, Rayner B, Imbach JL, Ecker DJ (1994) Combinatorially selected guanosine-quartet structure is a potent inhibitor of human immunodeficiency virus envelope-mediated cell fusion. Proc Natl Acad Sci U S A 91:1356–1360

Yacyshyn B, Woloschuk B, Yacyshyn MB, Martini D, Tami J, Bennett F, Kisner D, Shanahan W (1997) Efficacy and safety of ISIS 2302 (ICAM-1 antisense oligonucleotide) treatment of steroid-dependent Crohn's disease. Annual meetings of the American Gastroenterological Association and the American Association for the Study of Liver Diseases, Washington DC

Yazaki T, Ahmad S, Chahlavi A, Zylber-Katz E, Dean NM, Rabkin SD, Martuza RL, Glazer RI (1996) Treatment of glioblastoma U-87 by systemic administration of an antisense protein kinase C – a phosphorothioate oligodeoxynucleotide. Mol Pharmacol 50(2):236–242

Zamecnik PC, Stephenson ML (1978) Inhibition of Rous sarcoma virus replication and cell transformation by a specific oligodeoxynucleotide. Proc Natl Acad Sci U S A 75:289–294

Zamecnik PC, Goodchild J, Taguchi Y, Sarin PS (1986) Inhibition of replication and expression of human T-cell lymphotropic virus type III in cultured cells by exogenous synthetic oligonucleotides complementary to viral RNA. Proc Natl Acad Sci U S A 83:4143–4146

Zerial A, Thuong NT, Helene C (1987) Selective inhibition of the cytopathic effect of type A influenza viruses by oligodeoxynucleotides covalently linked to an intercalating agent. Nucleic Acids Res 15(23):9909–9919

Zhang M, Creese I (1993) Antisense oligodeoxynucleotide reduces brain dopamine D_2 receptors: behavioral correlates. Neurosci Lett 161:223–226

Zhang R, Lu Z, Zhang X, Zhao H, Diasio RB, Liu T, Jiang Z, Agrawal S (1995a) In vivo stability and disposition of a self-stabilized oligodeoxynucleotide phosphorothioate in rats. Clin Chem 41(6,1):836–843

Zhang R, Yan J, Shahinian H, Amin G, Lu Z, Liu T, Saag MS, Jiang Z, Temsamani J, Martin RR, Schechter PJ, Agrawal S, Diasio RB (1995b) Pharmacokinetics of an anti-human immunodeficiency virus antisense oligodeoxynucleotide phosphorothioate (GEM 91) in HIV-infected subjects. Clin Pharmacol Ther 58:44–53

Zhang SP, Zhou LW, Weiss B (1994) Oligodeoxynucleotide antisense to the D_1 dopamine receptor mRNA inhibits D_1 dopamine receptor-mediated behaviors in normal mice and in mice lesioned with 6-hydroxydopamine. J Pharmacol Exp Ther 271:1462–1470

Zheng H, Sahai BM, Kilgannon P, Fotedar A, Green DR (1989) Specific inhibition of cell-surface T-cell receptor expression by antisense oligodeoxynucleotides and its effect on the production of an antigen-specific regulatory T-cell factor. Proc Natl Acad Sci U S A 86:3758–3762

Zhou LW, Zhang SP, Qin ZH, Weiss B (1994) In vivo administration of an oligodeoxynucleotide antisense to the D_2 dopamine receptor messenger RNA inhibits D_2 dopamine receptor-mediated behavior and the expression of D_2 dopamine receptors in mouse striatum. J Pharmacol Exp Ther 268:1015–1023

CHAPTER 2
Antisense Medicinal Chemistry

P.D. Cook

A. Introduction

Making drugs out of oligonucleotides is a relatively new drug-discovery endeavor. Intense efforts in this area were only initiated about 8 years ago by several new biotechnology companies. Various approaches within this area of research attempt to use oligonucleotides as novel drugs, i.e., the use of antisense strands, ribozymes, triplexes and double-stranded nucleic-acid decoys. In the antisense approach, RNA serves as a novel receptor, which will bind an oligonucleotide ligand in a Watson–Crick, sequence-specific manner. Phosphorothioate (PS) oligonucleotides, developed as first generation antisense agents, have been shown to operate by this novel drug mechanism and several PSs have progressed to late-stage clinical development. Much has been learned about the pharmacology, toxicology and pharmacokinetic properties of the first generation PS antisense oligonucleotides and, along with this, certain limitations as antisense agents have become apparent. While it appears certain that PS drugs will soon become available, in order to continually improve the oligonucleotide drug class and to overcome certain limitations, structural changes to PS are required. To date, a diverse range of modifications, at all possible modification sites of an oligonucleotide (Fig. 1), have been reported. This chemical data base should prove to be very valuable in continuing medicinal chemistry efforts toward enhancing the properties of PSs. A number of general reviews, primarily focusing on antisense oligonucleotide medicinal chemistry, have periodically been published as the area has grown (Zon 1988; Stein and Cohen 1988; Goodchild 1990; Uhlmann and Peyman 1990; Cook 1991, 1993; Milligan et al. 1993; Moser 1993; Varma 1993; Sanghvi and Cook 1994; Kiely 1994; De Mesmaeker et al. 1995; Agrawal 1996a; Matteucci 1996; Matteucci and Wagner 1996; Miller 1996; Sanghvi 1997).

B. Scope of the Review

Of the published reviews concerned with medicinal chemistry of oligonucleotides, two general approaches have been taken. One involves a listing of the possible oligonucleotide modifications, along with a discussion of how each

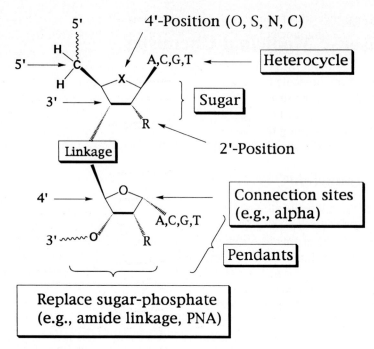

Fig. 1. Oligonucleotide modifications

modification might affect a desired antisense oligonucleotide property (COOK 1991). The other approach involves a listing of potential limitations (or problems) of oligonucleotides as antisense drugs and a discussion of the available modifications that might relate to the problem (COOK 1993). In this review, I will briefly focus on the merits and limitations of PS oligonucleotides as antisense drug candidates. I will then discuss key oligonucleotide modification research that has taken place to enhance PS drug properties. Although a large number of modifications have been reported, many of the resulting oligonucleotides have not been examined in meaningful assays and, for the most part, these will not be discussed in this review.

C. Phosphorothioates as Antisense Drug Candidates

About 19 PS oligonucleotides with antiviral, anticancer and antiinflammatory indications have been entered into human clinical trials (DEAN et al. 1996). The length of these PSs range from 15 to 25 bases. Most of the PSs are administered intravenously, with 3 mg/kg as the highest dose examined thus far. Formiversen, ISIS 2922, a 20-mer PS oligonucleotide, targets an immediate early mRNA in region 2 of the cytomegalovirus and inhibits the expression of two regulatory proteins (ANDERSON et al. 1996). It is administered locally by intravitreal injections for cytomegalovirus-induced retinitis in AIDS patients. It was the first oligonucleotide to exhibit a positive response after local administration. In pivotal phase-III clinical trials, Formiversen is the most advanced PS drug candidate. ISIS 2302, a 20-mer PS oligonucleotide targeting intracellular adhesion molecule one (ICAM-1) (BENNETT et al. 1996), is in five phase-II antiinflammatory drug trials. ISIS 2302 has recently demonstrated very exciting activity against Crohn's disease (YACYSHYN et al. 1997). This drug has the distinction of being the first oligonucleotide to exhibit systemic activity in humans. Other PSs in clinical trials will be described in other chapters in this volume.

A great deal of information has been learned about the pharmacology, pharmacokinetics and toxicology of PSs from the pioneering efforts to place 19 PSs into clinical trials. These major research and development areas, along with clinical development of PSs, will be addressed in several other chapters in this volume. The chemical development of PSs has also advanced to the point that PSs are expected to be cost effective drugs. Major advances have been accomplished in process research, large-scale synthesis, formulations and analytical controls and methods. Thus, the future of PSs as antisense drugs appears very bright, considering the chemical data base generated from introducing 19 PSs into clinical trials, the lack of serious toxicity and the evidence of clinical activity of locally (Formiversen, ISIS 2922) and systemically administered drugs (ISIS-2302). However, there are certain limitations to PS antisense oligonucleotides which may prevent them from being classified as broadly enabling drugs. The pharmacokinetic properties of PS oligonucleotides are a major concern; in particular, the lack of oral bioavailability and penetration of the blood–brain barrier are of concern. Furthermore, continually improving a new drug class, such as oligonucleotides, is an important drug development endeavor. To overcome current limitations (particularly pharmacokinetic issues) and to continually enhance oligonucleotide drug properties, I believe that modifications of the basic structure of PSs are required and that this is best accomplished by traditional structure–activity relationship (SAR) studies. A discussion of PS limitations is provided in Sect. D and the data base of available modifications is discussed in Sect. I.

D. Phosphorothioate Limitations

I. Pharmacodynamic Limitations

For every thiophosphate linkage in a PS oligonucleotide, a destabilization of the heteroduplex with complementary RNA of approximately $-0.7°C$ occurs relative to parent DNA. This amounts to approximately $-14°C$ destabilization in a 21-mer PS or approximately three to four orders of magnitude increase in K_d compared with wild type DNA hybridized to RNA (FREIER et al. 1992; FREIER 1993; LESNIK et al. 1993). Thus, PSs are not likely to invade dsRNA structures and are most often targeted to certain single-stranded sites. This low binding affinity per monomer unit of a PS limits the sites that can be effectively targeted by PSs. Even when PSs are targeted to single-stranded RNA and supposedly bind sequence-specifically, they are not likely to elicit biological activity simply because of this binding. The consensus mode of action of antisense PSs is that on sequence-specific binding to target RNA an endogenous, ubiquitous endonucleolytic enzyme, RNase H, binds the duplex and cleaves the RNA strand of the heteroduplex. Therefore, antisense PSs appear to require a RNase-H mechanism for activity and, thus far, this mode of action requires sulphur in the backbone. This sulphur is in the form of thiophosphates which provide resistance to degradation by nucleases. So, pharmacodynamically, PSs as antisense agents are limited in terms of the sites they can target in RNA, require sulphur for nuclease resistance and, thus far, require a RNase-H mechanism for biological activity. At higher concentrations, PSs bind to RNase H and prevent it serving as an effector of biological activity. This has not been demonstrated in animal studies.

II. Pharmacokinetic Limitations

Significant biological activity of orally administered PSs has not been reported. This may be the most serious shortcoming of oligonucleotides as therapeutic agents. Recent studies of PSs stabilized by $2'$-O-methyl (Me) modifications at the $3'$- and $5'$-flanks exhibit encouraging results (AGRAWAL et al. 1995, 1997b). Also, penetration of the blood–brain barrier has not been reported for PS. The pharmacokinetics of PSs, which are relatively large, negatively charged molecules, are dependent on the dose administered. This is because of the propensity of PSs to bind to serum proteins (GREIG et al. 1995). Saturation of protein binding sites may occur and cause variation in PS absorption and distribution. Modifications of PSs to alter the pharmacokinetic properties remain an important medicinal chemistry problem. A clear understanding of how PSs are taken into cells, absorbed by various tissues and transversed to various compartments, is also under investigation. Finally, PSs, once thought to be extremely stable to nucleolytic degradation, are now known to have half lives in the range less than 1 h to greater than 24 h in various animal models, depending on the tissue or organ examined. Other oligonucleotides with greater stability than PS will be essential for oral administration of oligomers (AGRAWAL 1996b) and will have important implications for improving parenteral administration.

III. Toxicological Limitations

Dose-limiting toxicities of PS in primates are clotting abnormalities and transient hypertension, which are likely related to an inhibition of the clotting cascade and activation of the complement pathway, respectively. Toxicities in rodents are quite different; immune stimulation is most prevalent (see Chap. 5 for a detailed description). The polyanionic nature of the oligonucleotides in particular the PSs, is often implicated as the cause of these and other toxicity parameters (GAO et al. 1992; WALLACE et al. 1996; HENRY et al. 1997; AGRAWAL et al. 1997). The current knowledge of oligonucleotide–protein interactions is sparse and, as more is learned, rationally modifying oligonucleotides will allow modulating protein–oligonucleotide interactions. Research directed to controlling the charge density, lipophilicity and sulphur content (replacement of the thiophosphate linkage) of oligomers is another important medicinal chemistry problem.

E. Types of Oligonucleotide Modifications

Figure 1 represents a dimer of an oligonucleotide and depicts all of the available subunits that may be modified. These subunits are composed of heterocycles, carbohydrates, linkages (backbones) and several types of connection sites, including conjugation sites and complete removal of the sugar-phosphate backbone. As an example of the extent of oligonucleotide modifications performed to enhance drug properties, consider the guanine–cytosine dimer of Fig. 1. This dimer has 26 positions that may be modified without directly interfering with Watson–Crick base-pair hydrogen bonding. All of these positions have been modified by oligonucleotide chemists in the past 8 years. Furthermore, all of the 26 available positions present in an adenine–thymine dimer have also been examined. The antisense concept has well-defined structural requirements for the oligonucleotide ligand that binds to a reasonably well-known receptor (RNA). Although this knowledge is highly valuable, it also provides limits to the scope of potential chemical modifications. A very significant structural modification data base, as depicted by Fig. 1, is available to search for enhanced drug properties of PSs.

F. Drug Properties of Antisense Oligonucleotides that May Be Altered by Chemical Modifications

Modifications of oligonucleotides may be expected to address essentially every facet of antisense drug properties. Biophysical and biochemical properties that may be affected by modifications include binding affinity, base-pair specificity, nuclease resistance, support of endonucleolytic cleavage of the RNA of a heteroduplex, chemical stability, lipophilicity and solubility. Drug properties that will be affected by altering biophysical and biochemical parameters would include the general areas of pharmacokinetics, pharmacodynamics and toxicology.

Most modification efforts are designed to enhance the binding affinity of an oligonucleotide to target RNA and/or enhance nuclease resistance. The binding affinity of an oligomer is a physical chemical property, determined by measuring a modified oligomer's sequence-specific interaction with its length-matched RNA complement. The hybridization or melting process is performed under a rather standard set of conditions designed to mimic an intracellular environment (FREIER et al. 1992). As this is a calculation taken under artificial conditions, it may not accurately represent the binding of an oligomer to native RNA inside the cell. Similarly, nuclease resistance of an oligonucleotide is determined in various assays, such as heat-inactivated fetal calf serum, cellular extracts or with purified exonucleases or endonucleases. These results are also likely to differ from the stability of an oligomer in an in vivo situation. However, these assays do allow structure–property/activity relationship studies to proceed and, thus, provide reasonable methods to compare various oligonucleotide modifications.

G. Why Is It Important to Continue to Design and Synthesize Antisense Oligonucleotides with Increased Affinity for Target RNA and with Greater Resistance to Nucleolytic Degradation?

In considering ligand–receptor theory for pharmacological activity, increasing the affinity of an oligonucleotide for its RNA target should increase potency. Oligonucleotide affinity, as measured by melting curves, is increased as the length of the oligonucleotide–RNA heteroduplex increases. This has been verified experimentally in that 15- to 25-mers are typically used in antisense experiments rather than shorter oligonucleotides which may have melting transitions (T_m) close to or below the physiological temperature of 37°C and, therefore, may only form low levels of the required heteroduplex. Few examples of short oligomers (12-mers or less) have exhibited interesting biological activity and these were modified to have high affinity per nucleotide unit (WAGNER et al. 1996). Several recent studies correlate biophysical properties of a series of 2'-O-modified oligonucleotides with increased in vitro and in *vivo* activity (KAWASAKI et al. 1993; MORVAN et al. 1993; ALTMANN et al. 1996a,b, 1997; MONIA et al. 1993, 1996a; CROOKE et al. 1996a).

A number of in *vivo* pharmacokinetic studies in several animal species indicate PS oligonucleotides are not as stable as initially thought (ZHANG et al. 1995, 1996; see Chap. 3). Although stability of PSs may be sufficient for many drug applications, greater stability of PS oligonucleotides would be helpful in expanding dosage regiments (longer duration of action relates to less frequent dosing) in efforts to develop oral bioavailability, and less degradation of modified oligomers will minimize metabolite toxicity. In summary, the preponderance of antisense biological data suggests that oligonucleotides with higher binding affinities and greater stabilities towards nucleases are important design features.

H. What Are the Standards for Binding Affinity, Nuclease Resistance, and Other Properties Recently Achieved by Structure–Property/Activity Relationship Studies of Oligonucleotides?

Due to intense oligonucleotide research in the past 8 years, a remarkable enhancement of several of the desired antisense drug properties has been achieved. The results achieved with modified oligonucleotides in binding affinity, base-pair specificity, nuclease resistance and support of RNase H cleavage of the targeted RNA are impressive. I believe that it is important in this review to provide an idea of the level of binding affinity (as represented by T_m), nuclease resistance (as represented by $t_{1/2}$) and support of a RNase-H mechanism that a new modification should possess to be of interest. A number of "winners" have been identified and a discussion of these modified oligonucleotides is provided in Sect. I.

Certain modifications provide an increase in T_m of greater than 1.5°C/modification relative to a PS oligonucleotide (~1.0°C/mod relative to phosphodiester [PO]) and nuclease resistance ($t_{1/2}$) of greater than 24 h with SVPD (snake venom phosphorodiesterase) (about the level of PS). In view of these values, it would seem that a novel modification should exhibit a $T_m > 1$°C/mod compared with its PS parent. In evaluation of the binding properties, the modified oligonucleotides should be hybridized with a RNA complement as this is the receptor required for the antisense approach. A clear correlation between T_ms derived from hybridizations to a DNA complement or RNA complement has not been established. Also, correlations of T_ms of oligonucleotides having just one modification (point or pendent modification) or several modifications distributed throughout the sequence or a contiguous placement of the modification in the sequence have not been established (COOK 1991; LESNIK et al. 1993; MATTEUCCI and VON KROSIGK 1996; BUHR et al. 1996). Since the application of a modification will likely require its uniform placement in the sequence or at least several in a row for a gapmer strategy (COOK 1993), measurements should be taken with these types of modified oligonucleotides. The modification must not compromise base-pair specificity. Thus far, data from a number of papers suggests that base-pair specificity actually increases as T_m values are increased. However, a specificity level comparable to PS-base mismatches would appear to be a useful standard.

Nuclease resistance of a novel modification should be at least at the level of uniformly modified PSs. As $t_{1/2}$ values, unlike T_m measurements, are performed under a variety of conditions, use of PS controls (standards) is necessary. In addition, several concentrations of the enzyme should be employed to minimize complications of inhibiting the nuclease (CUMMINS et al. 1995). Values of $t_{1/2}$ of ~24h are often reported for PSs in SVPD assays (see Chap. 3). When the 3' end of an oligonucleotide is modified to be sufficiently resistant to 3'-exonucleases, then endonucleolytic cleavage becomes evident. Thus, modifications should also protect against endonucleases. In a gapmer strategy, this is accomplished by a PS linkage which will also support a RNase H cleavage mechanism. Nuclease resistance of a modified oligomer, if not also provided by the modification, may in many cases be provided by employing a PS backbone.

In considering the relative importance of nuclease resistance of an antisense oligomer and its level of affinity to its RNA target, recent biological results suggest that stability of an oligonucleotide may be a more important property to enhance than binding affinity (CROOKE et al. 1996a). Modifications that provide high-binding oligonucleotides, but have low nuclease resistance do not provide significant biological activity. However, oligonucleotides such as PSs, having relatively low binding affinities but with considerable stability to nucleases, have demonstrated broad and significant biological activity. Although some modifications provide high binding affinities and high nuclease resistance, they may not exhibit useful antisense activities, because they do not support an RNase-H mechanism. A modification which will support an RNase H mode of action and also possess high T_m and $t_{1/2}$ values has not been reported. Thus, an ideal oligonucleotide modification would provide an oligomer that hybridizes to target RNA with high binding affinity and specificity, that is stable to nucleolytic degradation and would allow a RNase H cleavage of the RNA. This has lead to the concept that to optimize the antisense activity of an oligomer, a combination of oligonucleotide modifications is required (COOK 1991, 1993).

An additional standard that one should be aware of is the impact which oligonucleotide modifications may have on the cost of future antisense drugs. Again, as a standard, the cost of PS oligonucleotides should be considered. The cost of PSs has been dramatically reduced in the past 8 years due to improvements in the process, cost reduction of key reagents and advantages of larger-scale synthesis. Currently, the cost of goods for a 20-mer PS is less than $300 per gram and is projected to be less than $50 per gram as larger quantities are required. The 2'-O-(methoxyethyl)-modified oligonucleotides are derived from ribonucleosides from RNA and will eventually be substantially less expensive than modified oligonucleotides derived from deoxyribofuranosyl nucleosides. PS antisense drugs are expected to be cost competitive, when considering parenteral treatment, three times per week with a 1 mg/kg dose. Modified oligonucleotides must also be cost effective and should be less expensive to synthesize, have a dosing advantage (less often), have a greater therapeutic index or have other important advantages to offset an increase in cost of synthesis.

Another important consideration in the modification of oligonucleotides is to what extent proprietary protection of a modification can be obtained. Important patent positions for many types of modifications have been established, during the past 8 years, as making drugs out of oligonucleotides became interesting (CROOKE et al. 1996b; SHEFFERY and GORDON 1996).

I. Summary of Key Oligonucleotide Modifications

This review will not cover all of the modifications reported to enhance antisense drug properties, as several reviews are available. However, a brief summary of recently published and interesting modifications that have been directed at the sub-units is provided in Fig. 1. The general format will be to discuss what was expected of the modification (i.e., the purpose), how well this was accomplished (i.e., the results), liabilities of the modification (i.e., cost, toxicities, patent positions, etc.) and the underlying principles being exploited to obtain the desired oligonucleotide properties (i.e., the rationale). Modifications that do not approach the level of standards described above, typically, will not be discussed.

I. Heterocycles (Nucleobases)

A wide variety of heterocycle-modified oligonucleotides have been synthesized for the potential enhancement of antisense properties. Most of the modifications have been discussed in several general reviews or specific medicinal chemistry reviews (COOK 1991, 1993; SANGHVI 1993, 1997; MANOHARAN 1993; MEYER 1994). More recent modifications with binding affinities comparable to standards described in Sect. H or modifications having other interesting properties are discussed.

1. 5′-Position Pyrimidine- and Tricyclic Cytidine-Modified Oligonucleotides

The primary purpose for modifying the heterocycles of oligonucleotides is to enhance oligonucleotide binding affinity to target RNA. This has been accomplished in several cases as significant increases in T_m have been reported. 5-Propynyl-deoxyuridine (Structure 1, Fig. 2)- and 5-propynyl-deoxycytidine (Struct. 2)-modified oligonucleotides provide T_m increases of ~1.6°C/mod (FROEHLER et al. 1992, 1993; WAGNER et al. 1993). An even greater enhancement of binding affinity (2.2°C/mod) was observed when oligonucleotides were modified with 5-(Me-thiazolyl)-2-deoxyuridine (Struct. 3) (GUTIERREZ and FROEHLER 1996). Presumably, the increased binding affinity of Me-thiazoles is due to an increased base stacking of the planar heterocycle with adjacent base pairs (GUTIERREZ et al. 1994; GUTIERREZ and FROEHLER 1996).

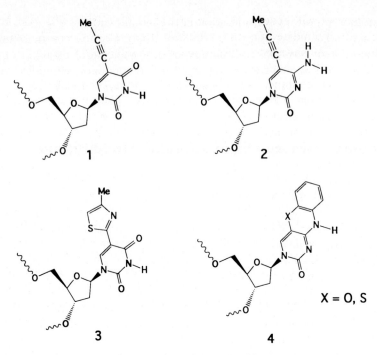

Fig. 2. Pyrimidine heterocycle-modified oligonucleotides

In the tricyclic 2′-deoxycytidine analogs, the greatest enhancement of T_m was observed when three "phenoxazine" cytidine mimics (Struct. 4) were contiguous (5.0°C/mod) (MATTEUCCI and VON KROSIGK 1996; LIN et al. 1995) (Fig. 2).

The thiazole ring is coplanar with the uracil ring system (pi–electron delocalization between the aromatic rings) leading to increased base stacking with adjacent base pairs. Furthermore, the thiazole ring has a dipole moment; the nitrogen atom is negative and the sulphur atom has a positive charge. Maximizing the pi–pi overlap between adjacent aromatic faces and dipole-induced London interactions increases the binding affinity (GUTIERREZ and FROEHLER 1996). The overlap of the tricyclic aromatic rings was supported by molecular modeling studies (LIN et al. 1995). The extended pi–pi interactions, provided by the propynyl moiety, as well as their symmetrical, hydrophobic nature is postulated to be responsible for their enhanced binding affinity (FROEHLER et al. 1992). As a contiguous placement of heterocycle modifications in an oligonucleotide may be required to obtain the high level of binding affinity, this series of modifications, as well as other heterocylic modifications designed to enhance binding affinity, may be limited to pyrimidine- or purine-rich sense sequences. Certainly, the most desirable application of antisense oligonucleotides will require the use of mixed base sequences which will require both pyrimidine and purine modifications. In this regard, propynyl-modified purines have been prepared and examined within a propynyl-modified pyrimidine sequence (BUHR et al. 1996).

As nuclease resistance is not significantly enhanced with the pyrimidine class of modifications, a PS backbone has been utilized for oligomer stability in antisense studies in cell culture (WAGNER et al. 1993, 1996; FENSTER et al. 1994). Cell culture antisense activity of microinjected 5-propynyl-modified oligonucleotides has been reported (WAGNER et al. 1993, 1996). In considering a mode of action of the 5-propynyl-modified oligonucleotides, pyrimidine-rich oligonucleotides of various lengths were targeted to a SV40 large T antigen-reporter gene. Activity was dependent on a contiguous series of 5-propynyl-modified deoxyuridine and deoxycytidines, which provided enhanced binding and a PS backbone for nuclease resistance and support of a RNase H activation as a mode of action. Even though the 5-propynyl-modified pyrimidines provided a significant increase in binding affinity, it was not sufficient for activity, independent of RNase H cleavage. This is evident by the lack of activity of the most active propynyl/PS modified sequences when uniformly modified by 2′-O-allyls. The allyl-modified oligomers significantly enhance the binding affinity over the parent deoxyribofuranosyl oligomers (76.5° to 89.5°C, ~0.7°C/mod). However, since uniform 2′ modifications do not support RNase H, biological activity was not observed. A gapmer strategy, which takes advantage of the enhanced binding of both the heterocyclic propynyls and the 2′-O-methoxyethyl sugar modifications, nuclease resistance by PSs and supports a RNase H mechanism, is described in Sect. II.

2. N^2-Purine-Modified Oligonucleotides

Modifications on the N^2-nitrogen of guanine, the C^2-carbon of adenine and the 3 position (3-deaza) of purines are unique in that these are the only heterocycle modifications which will allow a pendent group to reside in the minor groove of an A-form heteroduplex. The 2-keto group of pyrimidines is another possibility, but it is not amenable to reasonable chemical manipulation required to place moieties in the minor groove. Modifications of these positions are considered important sites to link and position potential cleavers for the RNA of the heteroduplex (CASALE and MCLAUGHLIN 1990; KIDO et al. 1992; WANG and BERGSTROM 1993a,b; RAMASAMY et al. 1993, 1994; BERGSTROM and GERRY 1994; HEEB and BENNER 1994; BAKER et al. 1996; MANOHARAN et al. 1996). It is clear from these studies that relatively large groups attached to the N^2 or C^2 positions of purines may be accommodated in the minor groove without significantly compromised binding. Enhancing binding affinity and nuclease resistance by employing several N^2-purine modifications in antisense sequences was of interest (COOK 1991; RAMASAMY et al. 1994).

N^2-Adenine-modified oligonucleotides have long been known to enhance binding, because the additional hydrogen bond contributed to its base-pairing with thymine or uridine (for recent results see GRYAZNOV and SCHULTZ 1994). N^2-Purine modifications may provide thermal stability in addition to that provided by the third base-pair hydrogen bond. The binding affinity of three contiguous N^2-imidazolylpropylguanines (Struct. 5) or N^2-imidazolylpropyl-2-aminoadenines (Struct. 6) provides about +1.2°C/mod (RAMASAMY et al. 1994) and a single N^2-aminopropyguanine (Struct. 7) increases an ICAM-1 sequence by +3°C (MANOHARAN et al. 1996) (Fig. 3). The attachment of a spermine to the C^2 position of hypoxanthine (Struct. 8) (SCHMID and BEHR 1995) and N^2-imidazolylpropylguanine (RAMASAMY et al. 1994) provided modified oligonucleotides with some of the strongest binders to DNA ever reported (+6.0° and 7.5°C/mod, respectively). 3′-End-capping 15-mers with three N^2-imidazolylpropylguanines or N^2-imidazolylpropyl-2-aminoadenines provided half-lives of 9 and 16 h in fetal calf serum, respectively, compared with greater than 1 h for the parent sequence (RAMASAMY et al. 1994). In an attempt to prepare RNA cleavers, HEEB and BENNER (1994) attached a 3-imidazolepropionic acid to the N^2 group of deoxyguanosine (Struct. 9) and

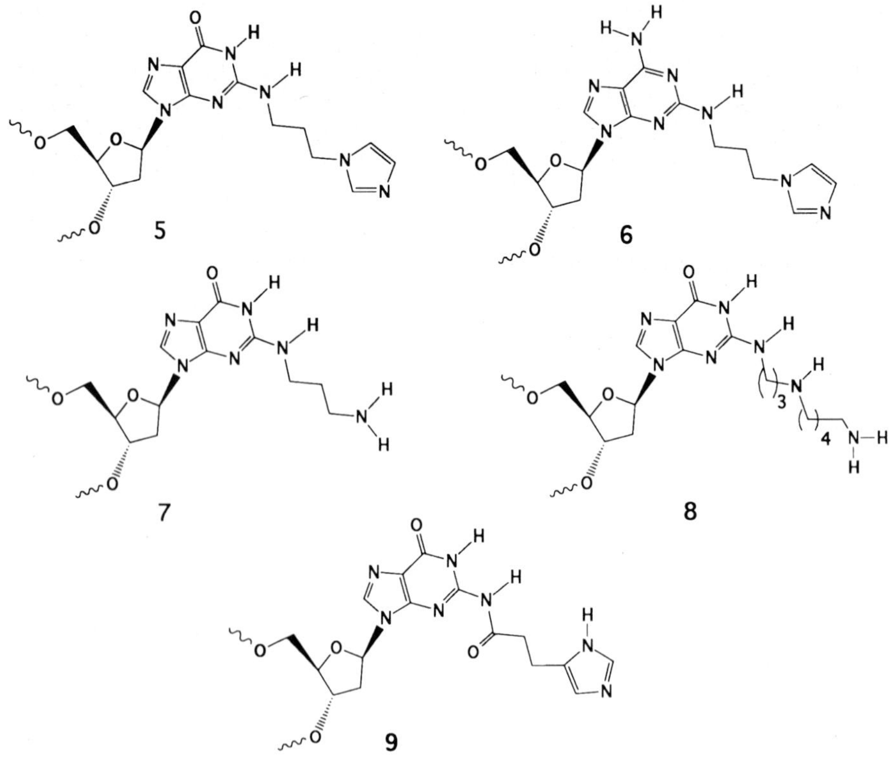

Fig. 3. N^2-Purine-heterocycle-modified oligonucleotides

incorporated it into a 15-mer oligonucleotide, dT_7HdT_8. No significant cleavage was observed on hybridization to $riboA_{12-18}$.

3. 7-Deaza-Modified Oligonucleotides

Various 7-deaza-7-substituted-guanines (Struct. 10) and -adenines (Struct. 11), and 7-substituted-7-deaza-8-aza-adenines (Struct. 12) and -guanines have been incorporated into antisense DNAs (SEELA et al. 1995, 1997; SEELA and THOMAS 1995) (Fig. 4). Six contiguous 7-deaza-adenine-7-substituted (-iodo, -bromo, -aminopropynyl or -hexynyl) modifications in 12-mers strongly bind to complementary DNA (4.5°, 3.7°, 2.8° and 2.8°C/mod, respectively). Similar high-binding affinities were observed with the same moieties placed in the pyrazolopyrimidine analog of adenine. The 7-deaza and 7-deaza-7-propynyl modifications of adenines and guanines were recently examined in 5-propynyl-pyrimidine-modified PSs . The T_ms, in this case taken with relevant RNA complements, were decreased by 0.25°–0.6°C/mod with the 7-deaza purine modifications, but were increased by ~0.85°C/mod with the corresponding propynyl-purines. The uniform propynyl-modified 15-mer PS (all guanines, uridines and cytosines are modified) exhibited an IC_{50} of 0.15 μM for the antisense inhibition of a TAg gene, targeted downstream of the start codon. The 18-mer PS with propynyl-modified adenines, uridines and cytidines and two unmodified guanines demonstrated an IC_{50} of >5 μM. The authors suggest that the large difference in activities may be due to the lack of contiguous propynyl-modified bases in the 18-mer, caused by having two unmodified guanines placed within the sequence. This may affect RNase H recruitment and cleavage of the 18-mer relative to the contiguous modified 15-mer. 7-Deaza-guanine, 7-deaza-7-Me-guanine and 7-iodo-7-deaza-guanine-modified DNAs provided little resistance to the 3′-exonuclease

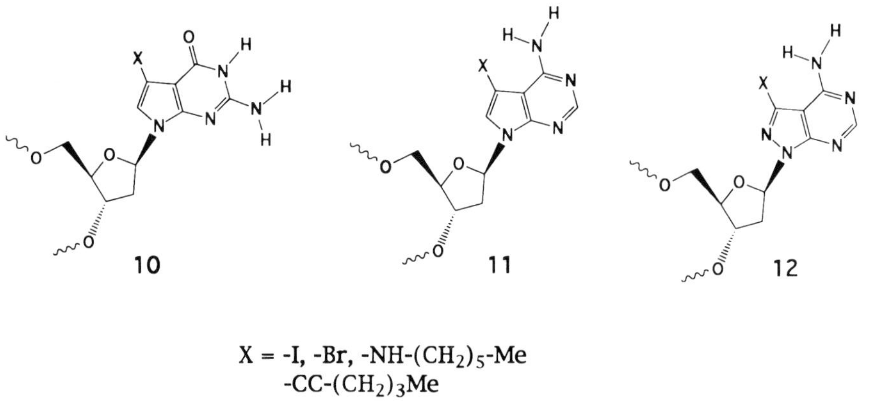

X = -I, -Br, -NH-(CH$_2$)$_5$-Me
-CC-(CH$_2$)$_3$Me

Fig. 4. 7-Deaza-purine-heterocycle-modified oliognucleotides

SV-phosphodiesterase ($t_{1/2} < 20$ min), whereas the same oligomers were quite stable with respect to the 5′-exonuclease CS-phosphodiesterase ($t_{1/2} \sim 20$ h) (SEELA et al. 1995).

4. Additional Advantages of Heterocycle-Modified Oligonucleotides

A desirable feature of heterocyclic modifications, along with providing enhanced oligonucleotide binding affinity to target RNA, is that the modified oligomers would also support a RNase H mode of action (COOK 1991; SANGHVI et al. 1993; MANOHARAN et al. 1996). This is potentially a very important feature since available backbone (other than PS) and carbohydrate modifications do not support RNase H mechanisms. Several heterocycle modifications have been reported to possess these interesting properties. 5-Propynyl-deoxyuridine- and 5-propynyl-deoxycytidine-modified oligonucleotides were shown to support *E. coli* RNase H degradation in a biochemical assay at the same level as the parent PS (CROOKE et al. 1995). WAGNER et al. (1996) has also demonstrated that propynyl-pyrimidine-modified oliognucleotides support cleavage of complementary RNA by HeLa nuclear extracts. Placing a N^6-imidazoylpropyladenine or a N^2-imidazoylpropyl-2-aminoadenine-modified base in the middle of a sequence affects cleavage depending on whether the pendent group will reside in the major or minor groove of the heteroduplex. *E. coli* cleavage of the RNA is 2.5-fold less with the minor groove placement (N^2-imidazoylpropyl) and is 1.2-fold greater with major groove placement (N^6-imidazoylpropyl), both compared with the parent PSs (CROOKE et al. 1995). In that *E. coli* RNase H has been shown to bind in the minor groove of a RNA-DNA heteroduplex (NAKAMURA et al. 1991), it may be expected that modifications emanating from the minor groove would interfere with RNase H binding.

A seemingly innocuous heterocyclic modification is a change in the ring system, such as an aza/deaza modification not involved in Watson-Crick base pairing. In one case, after replacing the eight thymine bases (which were dispersed throughout the sequence) by eight 6-azathymines (or 5-bromouracils), a 22-mer PO was hybridized to a complementary RNA target and was cleaved by *E. coli* RNase H and HeLa cell extracts (SANGHVI et al. 1993). Information concerning the susceptibility of RNase H cleavage of other pyrimidine-modified oligonucleotides and 7-deaza-7-substituted purine-modified oligonucleotides has not been reported. A heterocycle modification that would allow the use of PO oligonucleotides is of considerable interest, because they provide significant nuclease resistance, have increased binding affinities and would support an RNase H mechanism.

Other potential advantages of heterocyclic modifications are that pharmacokinetic properties are likely to be altered, although at this time, not in a predictable manner. Stabilization towards depurinations such as adenine may also be realized by certain modifications. Extensive conjugation of various moieties to the heterocycle sub-unit may affect a broad range of pharmacodynamic and pharmacokinetic properties (MANOHARAN 1993; MANOHARAN et al. 1996).

An interesting application of heterocyclic modifications, which relate to the fundamental function of the nucleobases, that is, specificity due to Watson-Crick base-pair hydrogen bonding, has been devised. Heterocyclic modifications that would prevent base pairing are an interesting twist, since for antisense applications, we design for enhanced specificity and affinity of Watson-Crick base pairing. KUTYAVIN et al. (1996) and WOO et al. (1996) constructed nucleobase analogs that would not base-pair in palindromic sequences, but would base-pair with their complementary bases in dsRNA or dsDNA. Thus, oligonucleotides bearing 2-amino A (Struct. 13) and 2-thiothymine (Struct. 14) (KUTYAVIN et al. 1996) will not base-pair with each other, but can hybridize to cytosine and adenine, respectively, in complementary strands of RNA and DNA (Fig. 5). In this manner, disruption of secondary structure (e.g., stems) may be enhanced by the use of two complementary oligonucleotides, having the non-hybridizing bases. In the same manner, deoxyinosine (Struct. 15) and 3-(2'-deoxy-B-D-ribofuranosyl)pyrrolo-[2, 3-d]pyrimidine-2-(3H)-one (Struct. 16) do not base-pair, but when placed in complementary strands, can effectively disrupt secondary structure (WOO et al. 1996). Another potential use of this type of heterocycle modification is to prevent internal secondary structure from the self-binding of partially complementary antisense sequences possessing high affinity modifications (CUMMINS et al. 1995).

5. Toxicity Liability of Heterocyclic Modifications

Several important issues must be considered with heterocyclic modifications of potential antisense oligonucleotides. The Watson-Crick base-pairing properties cannot be disrupted by modifying the heterocycle (COOK 1991). In this case, one should readily be able to assess whether a modified heterocycle will possess the required Watson-Crick binding, by examining its hydrogen-bond-donating and accepting features. Relatively simple studies of base mismatches in the complementary RNA strand will provide specificity information concerning heterocycle modifications (RAMASAMY et al. 1993; LIN et al. 1995; MATTEUCCI and VON KROSIGK 1996). Reduction of specificity of base-pair recognition with heterocycle modifications has not been reported. A more important issue with modifications of the heterocycles or monomers in general (modified nucleotides), which are used to prepare modified oligonucleotides, is an increased toxicity liability. Pharmacokinetic studies indicate that PSs are degraded, primarily by 3'-exonucleases (see Chap. 3) with various half-lives and, as noted above, heterocyclic modifications have not provided effective nuclease protection. Thus, the intracellular transport of modified oligonucleotides and their subsequent catabolic metabolism by exonucleases will provide

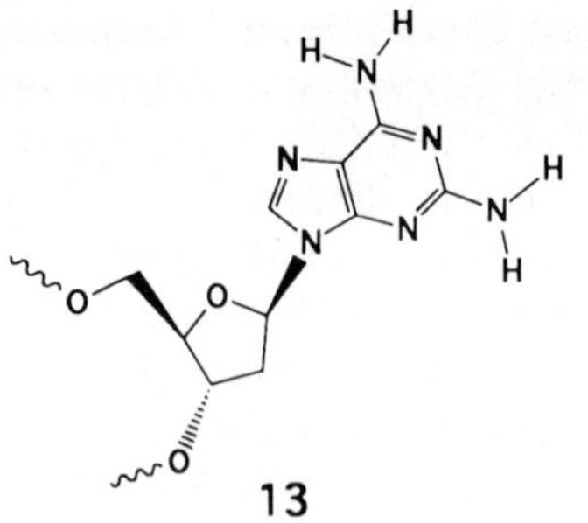

Fig. 5. Non-base-pairing heterocycles

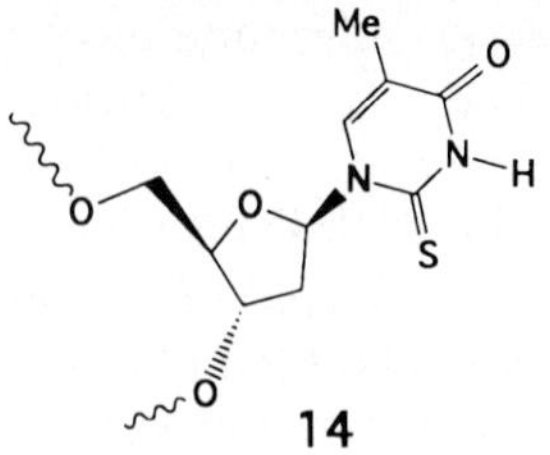

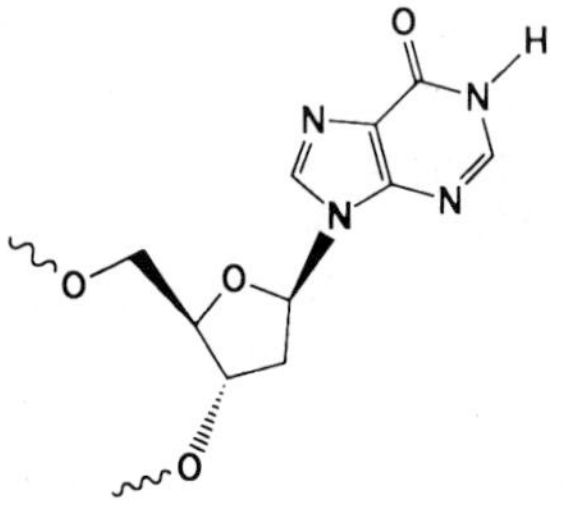

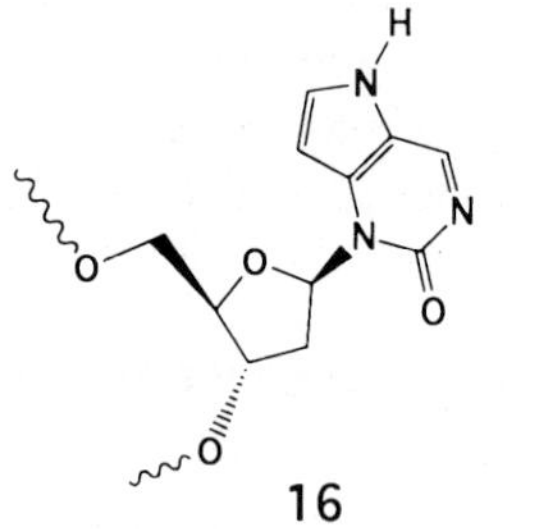

intracellular 5′-monothiophosphates and 5′-monophosphates of modified nucleosides. These mono-nucleotides may be toxic species as such and, if anabolized to di- and tri-phosphates, have the potential to exhibit an even broader range of toxicities. In fact, using an oligonucleotide composed of biologically active nucleosides as a pro-drug for transport into cells has been reported (Alderfer et al. 1985; Gmeiner et al. 1995). The fact that 5-position-modified nucleosides are cytotoxic suggests that the nucleoside is transported into cells and, on phosphorylation, to mono-, di- and tri-phosphates that generate toxic species. In an analogous situation, purine nucleosides and other moieties incorporated into an oligomer via phosphodiester linkages presents a potential toxicity liability. The toxicity of N^2-modified purine nucleosides, as described above, have not been reported.

II. Carbohydrate-Modified Oligonucleotides

The carbohydrate moiety of oligonucleotides has received considerable attention as an oligonucleotide sub-unit to modify to enhanced antisense drug properties. Several reviews specifically addressing medicinal chemistry of carbohydrate-modified oligonucleotides are available (Cook 1991, 1993; Sanghvi and Cook 1994; De Mesmaeker et al. 1995; Herdewijn 1996; Altmann et al.

1996a). The major modifications examined are the 4' and 2' positions, bicyclo connections of the sugar, acyclonucleoside connections of the sugar, and heterocycle and ribofuranose replacements with hexoses and morpholines.

1. Basis for Design

The primary purpose for modifying the carbohydrate sub-unit (Fig. 1), as in the heterocycle sub-unit modifications, has been to enhance the binding affinity to target RNA and, secondarily, to enhance nuclease resistance. The principle driving the carbohydrate modifications is to *pre-organize* the antisense oligomer into an *A*-form geometry by converting the ribofuranosyl moieties into a 3'-endo conformation and conforming to torsion angles mimicking dsRNA (reviewed by HERDEWIJN 1996; see Fig. 6). Successful modifications would provide a modified oligomer-RNA duplex, resembling the structure of a RNA-RNA duplex which is generally more stable than DNA-RNA or DNA-DNA duplexes (LESNIK and FREIER 1995).

2. 4'-Modified Oligonucleotides

Modifications in the 4'-sugar position of oligonucleotides (Fig. 1 and Struct. 17) include replacing the oxygen with nitrogen (aza, ALTMANN et al. 1994c), methylene (carbacyclic, PERBOST et al. 1989), substituted carbon (ALTMANN et al. 1995, 1996b, 1997), or sulphur (4'-thia-RNA, BELLON et al. 1994; LEYDIER et al. 1995; 4'-thia-DNA, JONES et al. 1996) (Fig. 7). Relative to certain other modifications, these 4' modifications have not significantly enhanced binding affinities or nuclease resistance, and do not support RNase H degradation. Considering these unfavorable properties and the relative difficulty in the synthesis of the 4' modification, it is unlikely that 4'-modified oligonucleotides will have important value as antisense agents.

3. Sugar Conformationally Restricted Oligonucleotides

Conformationally restricting oligonucleotides to enhance their antisense properties has received recent attention (reviewed by GRIFFEY et al. 1994;

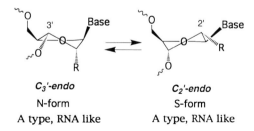

Fig. 6. Conformational equilibrium of furanose sugar in nucleic acids

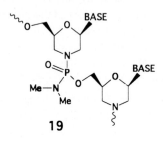

17

X = O, S, NH, NR, CH$_2$
4'-Sugar-modified oligonucleotides

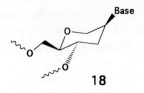

18

Sugar-conformationally
restricted oligonucleotides

19

Morpholino diamidate
modified-oligonucleotides

20

2'-Fluoro-modified
oligonucleotides

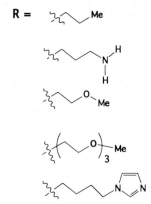

21

2'-Modified oligonucleotides

Fig. 7. Sugar-modified oligonucleotides

HERDEWIJN 1996; EGLI 1996). This somewhat rational approach to select sugar modifications is based on the knowledge of the structural properties of an *A*-form dsRNA. In particular, attention has been given preparing modified nucleosides possessing a high 3'-endo sugar conformation and mimicking key torsion angles of the sugar-phosphate backbone of dsRNA. Modified nucleosides incorporated into oligonucleotides are of three types: cyclonucleosides (connection between heterocycle and sugar), bicyclonucleosides (bridge within the sugar moiety, BOLLI et al. 1994; ALTMANN et al. 1994a,b; BÉVIERRE et al. 1994) and six-membered sugars (pyranoses, HERDEWIJN et al. 1994; HERDEWIJN 1996). Of the many sugar conformationally restricted oligonucleotides examined, only 1,5-anhydro-2,3-dideoxy-D-arabinohexitol modified DNA (HNA, Struct. 18) has demonstrated enhanced binding to RNA (1.0° to 5.0°C/mod; HERDEWIJN 1996; DE BOUVERE et al. 1997). Whether HNA oligonucleotides are resistant to nuclease degradation and/or support an RNase H mode of action has not been reported. Biological activity has not been reported with oligonucleotides bearing any of the conformationally constrained sugar modifications.

4. Morpholino Diamidate-Modified Oligomers

Morpholino oligomers are another example of the replacement of the deoxyribofuranose sugar. In this modification, the morpholino nucleosides can be connected by several types of linkers, but in particular, a dimethylphosphodiamidate linker (Struct. 19) has shown the greatest binding affinity (SUMMERTON and WELLER 1997). The binding affinity of a mixed base 22-mer, modified as a morpholino diamidate, a 2′-deoxy/PO and a 2′-O-Me/PO, provided a ΔT_m of +0.7°, +0.4° and 1.3°C/mod, relative to the PS parent. Thus, in this study, the morpholino oligomer demonstrates a binding affinity at about the level of DNA, but considerably less than the 2′-O-Me oligomer. Morpholino diamidates are stable to a wide range of nucleases and proteases, but do not activate RNase H (HUDZIAK et al. 1996). In cell culture experiments, morpholino diamidates directed to 3′-UTR sequences of HBV CORE/luciferase mRNA and rabbit α-globin RNA exhibited IC_{50}s of approximately 100nM when internalized into HeLa cells via a scrape-loading procedure (PARTRIDGE et al. 1996; SUMMERTON and WEILER 1997). Activity was not found when sequences further down stream of the leader sequence were targeted. This suggests that the sites to target in RNA may be restricted, since morpholino-RNA heteroduplexes do not support a RNase-H mechanism (see discussion in Sect. I.III.6). The basis for the design of morpholinos was a desire to prepare as cost effective antisense agents as possible. In this regard, morpholino monomers are chemically derived from naturally occurring ribonucleosides, which are considerably less expensive than 2′-deoxynucleosides (SUMMERTON 1992).

5. 2′-Modified Oligonucleotides

a) Structure–Property/Activity Relationship Studies

Of the available positions on the β-D-ribofuranosyl moiety to modify, the 2′ position (Fig. 1) has proven to be the most valuable, in that a number of modifications have been identified which markedly enhance several desirable antisense properties. Placement of a hydroxyl or a substituted hydroxyl (alkoxy) in the 2' position of a nucleoside has the general effect of shifting the equilibrium between the 2'-endo and 3'-endo to a more 3'-endo conformation. Heteroduplexes of such modified oligonucleotides with RNA resemble an *A*-form geometry. These type-modified oligonucleotides have been referred to as RNA mimics (Cook 1991). In addition, certain 2'-*O*-modified oligonucleotides possess a high level of nuclease resistance. The combination of T_m and $t_{1/2}$ values of these modifications have set high standards for future modification research. The identification of 2'-modified oligomers exhibiting enhanced antisense drug properties has resulted from rather extensive structure–property/activity relationship studies, performed in the past eight years (Guinosso et al. 1991; Cook 1991, 1993; Freier et al. 1992; Kawasaki et al. 1993; Morvan et al. 1993; Monia et al. 1993, 1996a; Lesnik et al. 1993; De Mesmaeker et al. 1995; Fraser et al. 1993; Hughes et al. 1993, 1995; Griffey et al. 1994, 1996; Cummins et al. 1995; Manoharan et al. 1995a–c; Martin 1995; Ross et al. 1994; Crooke et al. 1995, 1996a,b; McKay et al. 1996; Altmann et al. 1996a–c, 1997; Baker et al. 1997).

α) Binding Properties of 2'-*O*-Modified Oligonucleotides

When comparing a series of 2'-*O*-alkyls, it was found that an inverse relationship exists, relating the size of the alkyl chain to the binding affinity and the nuclease resistance; the smaller the alkyl group, the greater the T_m; whereas nuclease resistance was increased as the alkyl group size increased (Lesnik et al. 1993; Cummins et al. 1995). The 2'-fluoro-modified oligonucleotides (Struct. 20) exhibited the greatest increase in binding affinity ($\Delta°C/mod, +2.3°$ PS, $+2.7°$ PO). This correlated with the highest 3'-endo sugar conformation reported (Griffey et al. 1994). A study of the direct attachment of a carbon to the 2' position, that is, 2'-carbon–carbon-modified oligonucleotides, indicated that this modification is very destabilizing (Schmit et al. 1994). In another series of modifications (general Structure 21, Fig. 7), the 2'-*O*-alkyl chain was substituted with nitrogen and oxygen heteroatoms. 2'-*O*-Aminopropyls (AP), were shown to provide binding affinities about the level of POs ($\Delta T\ 0.0°C$ PO, $+0.7°$ PS). The most interesting modifications in this series were the oxygen modified 2'-*O*-alkyls, particularly the methylated ethylene glycols. The 2'-*O*-(Methoxyethyl) provided an increase of $+2.3°C/mod.$, PO (Altmann et al. 1996a,b).

The 2'-*O*-(methoxyethyl), besides providing a 3'-endo conformation, has an additional factor in that the oxygen atom of the ethylene glycol provides a gauche effect with the 2'-oxygen, which allows favorable positioning of the group within the minor groove of the heteroduplex. This has been thought to be the reason for the methoxyethyl-modified oligonucleotides having stronger binding affinities than closely related analogs such as methyl, propyl and methoxypropyl (ALTMANN 1996a,b). Even much longer ethylene glycols, such as 2'-(methoxytriethoxy), demonstrate high binding affinities (1.8°C/mod).

β) Stability of 2'-O-Modified Oligonucleotides

As more is learned about antisense oligonucleotides, it appears that the stability of the oligomer is emerging as a very crucial property. Focusing modification efforts in this area, while maintaining a reasonable level of binding affinity, may be more important than continuing to design and modify for higher binding affinity of PSs. It is well established that the sulphur in the backbone of PS DNA, as a thiophosphate linkage, primarily contributes nuclease resistance to the oligonucleotide. Whether this level of stability is sufficient for all drug applications of oligonucleotides has rarely been questioned. However, as more data is accumulated, it appears that the level of nuclease resistance of PSs is not as high as originally thought and that greater levels are desired. Furthermore, most of the limitations of PSs, as discussed in Sect. D., may be correlated to the thiophosphate linkage. Thus, the desire to entirely remove the sulphur or modulate the level of sulphur in antisense oligonucleotides, while increasing the level of nuclease resistance, is an interesting medicinal chemistry problem. 2'-O-modified oligonucleotides that provide greater nuclease resistant than the PS parent, yet provide useful binding affinities, may be a solution to this problem. Since PSs are destabilizing (ΔT_m/sulphur is approx. 0.7°C) and *2'-O modifications* significantly stabilize (compared with PSs), replacement of PS linkages with PO linkages and adding certain *2'-O modifications* should provide favorable binding affinities. The important issue is whether *2'-O modifications* can provide nuclease resistance such that phosphorodiester can be employed in place of PS linkages.

Enhanced resistance to degradation by nucleases is exhibited by certain *2'-O modifications*. For the simple alkyl groups, resistance is correlated to the length of the alkyl chain (CUMMINS et al. 1995). Thus, the greater the bulk emanating from the 2' position, the greater the $t_{1/2}$ nuclease assays. The relative stabilities of several 2'-O-alkyl-modified RNAs have the following order compared with the parent PS: *O*-Me < *O*-Pr < *O*-Pentyl ~PS. 2'-*O*-Me-modified PO oligonucleotides are generally about 10-fold less resistant than parent PSs (Table 1, CUMMINS et al. 1995). Nuclease resistance was not increased over that of the parent DNA with uniformly modified 2'-fluoro oligonucleotides. Certain substituted 2'-*O*-alkyl-modified PO oligonucleotides provide much greater nuclease resistance. The 2'-*O*-methyoxyethyl modification, along with its high binding affinity, exhibits nuclease resistance at approximately the same level as a PS linkage (Table 1). Further modifications of the alkyl chain has led to 2'-*O*-alkyl moieties bearing an amine or an imidazole in the chain which is protonated under physiological conditions. These modified oligomers exhibit greatly increased nuclease resistance compared with PSs (CUMMINS et al. 1995; GRIFFEY et al. 1996). 2'-O-(Imidazolybutyl) and 2'-O-(aminopropyl)-modified oligonucleotides have shown a $t_{1/2}$ 4-fold and 14-fold greater than the PS parent, respectively, with SVPD (Table 1).

A comparison of several 2' modifications in a uniform or capping motif of a PO or PS 9-mer is depicted in Table 1. Several points should be noted.

Only the 2'-O-methoxyethyl, 2'-O-(imidazolylbutyl) and 2'-O-(aminopropyl)-modified oligomers have sufficient nuclease resistance to allow replacement of the PS linkage with the natural PO linkage. Obviously, the same modifications in a PS backbone are extremely stable. The 2'-O-Me modifications will not support use of the natural PO backbone. The 2'-O modifications bearing a positive charge (aminopropoxy and imidazolylbutoxy) exhibit remarkable stability towards SVPD.

b) Modes of Action of 2'-O-Modified Oligonucleotides

α) Chimera 2'-O-Modified Oligonucleotides – RNase H-Dependent Mode of Action (Gapmer Technology)

Having developed high-binding, nuclease resistant 2'-O-modified oligonucleotides, it was rather disappointing that these oligomers were inactive or less active than their first generation parent PSs. It is now well known that uniformly 2'-O-modified-oligonucleotides do not support a RNase H mechanism. The 2'-O-modified oligonucleotide–RNA heteroduplex has been shown to present a structural conformation that is recognized by the enzyme, although cleavage is not supported (LIMA and CROOKE 1997; CROOKE et al. 1995). The lack of activity of 2'-O-modified oligonucleotides has led to the development of a chimeric strategy (gapmer technology) (reviewed by COOK 1991, 1993; MONIA et al. 1993).

Table 1. Half-life in minutes of modified oligonucleotides exposed to snake venom phosphorodiesterase (SVPD) @ 5×10^{-3} units/ml

Backbone → Sugar Modification →	Diester		Thioate	
	Capped	Full	Capped	Full
X = T, 2'-Deoxy		4		240
X = U, 2'-O-Methyl	10	15	450	>1440
X = U, 2'-O-Propyl	7	80	>1440	>>1440
X = U, 2'-O-Pentyl	7		720	
X = U, 2'-Fluoro	3			
X = U, 2'-O-(Methoxyethyl)	110			
X = U, 2'-O-(Propylamino)	1440			
X = U, 2'-O-(Imidazolylbutyl)	420			
Capped	TTT TTT TTT TTT TTX XXX T-3'			
Fully Modified	XXX XXX XXX XXX XXX XXX T-3'			

This approach focuses on the design of high-binding, nuclease resistant antisense oligonucleotides that are *gapped* with a contiguous sequence of 2'-deoxy PSs (Fig. 8). On hybridization to target RNA, a heteroduplex is presented that supports a RNase H-mediated cleavage of the RNA strand. The stretch of the modified oligonucleotide–RNA heteroduplex that is recognized by RNase H may be placed anywhere within the modified oligonucleotide (GRIFFEY et al. 1996; YU et al. 1996). The modifications in the flanking regions of the gap must not only provide nuclease resistance to exo- and endonucleases, but also must not compromise binding affinity and base-pair specificity.

Modifications of the phosphorus in natural PO linkages to provide Me-phosphonates, phosphorothioates and phosphoramidates destabilize heteroduplexes: $-0.7°$ to $-1.5°C$ for each modification (HOKE et al. 1991; DAGLE et al. 1991; AGRAWAL et al. 1990; FREIER et al. 1992). The decreased binding affinity of these modified oligonucleotides could be expected to reduce antisense effectiveness. In the case of chimeric 2'-methoxy or 2'-fluoro-modified oligo-

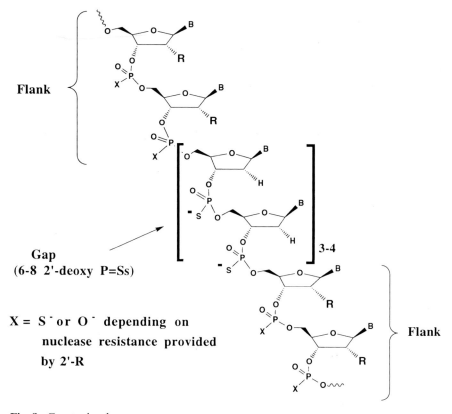

Flank

Gap
(6-8 2'-deoxy P=Ss)

X = S⁻ or O⁻ depending on nuclease resistance provided by 2'-R

Flank

Fig. 8. Gap technology

nucleotides, an enhancement in the binding affinity of about 2.0° to 2.3°C (compared with PS) for each modification is obtained (GUINOSSO et al. 1991; KAWASAKI et al. 1993; FREIER et al. 1992). However, it is now clear that 2'-O-Me and 2'-F-modified DNA are not sufficiently nuclease resistant to have an antisense value as PO backbones (Table 1; MILLER et al. 1991; SPROAT et al. 1989, 1993; MORVAN et al. 1993). The potential problem in this area can be circumvented by the use of 2'-O-Me or 2'-F-modified PSs in the flanking regions (GUINOSSO et al. 1991; KAWASAKI et al. 1993).

More recent research has focused on 2'-O modifications, such as methoxyethoxy (ALTMANN et al. 1996a,b) and aminopropoxy (GRIFFEY et al. 1996), which not only provide relatively high binding affinities but also a level of nuclease resistance that allows the replacement of the thiophosphate with the natural PO linkage. 2'-O modifications with a favorable combination of T_m and $t_{1/2}$ can be employed in a chimera strategy (gap technology, Fig. 8) which will allow a significant portion of the PS linkages to be replaced with PO linkages (Fig. 9). Just how many sulphurs can be replaced depends on the length of the oligomer and the gap size or RNase H cleavage site. Typically, a 21-mer with a 7-nucleotide gap has 65% of the PS linkages replaced with PO linkages. As noted in the limitations of PSs (Sect. D), removal of the sulphur

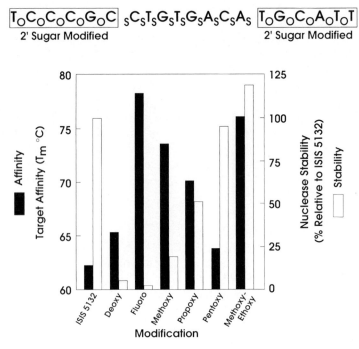

Fig. 9. Biophysical properties of 2'-modified (PS/PO) chimeric analogs of ISIS-5132

content in a PS oligonucleotide should have important implications in the pharmacokinetic and pharmacodynamic properties as well as the toxicity profile of oligonucleotides.

A very important aspect to the gapmer technology is that the gap or RNase H cleavage site must be protected from endonucleolytic cleavage. Phosphodiester linkages and even an alternating PS/PO motif in the gap does not provide a useful level of nuclease resistance for biological activity (HOKE et al. 1991; SANDS et al. 1995). A recent report of lack of activity of a *gapped* 3'-amidate is likely due to endonuclease degradation (HEIDENREICH et al. 1997). Uniform PSs are the only useful modification to allow a reasonable combination of binding affinity and nuclease resistance while also supporting an RNase H mechanism. Hence, as noted in Sect. D, most antisense activities require an RNase H mechanism which in turn requires sulphur in the form of thiophosphate somewhere in the chimera for nuclease resistance. Knowing that backbone (linkage) and sugar modifications do not support RNase H, it has been of interest to design heterocycle modifications which would have the desired properties described above (COOK 1991, 1993; RAMASAMY et al. 1994; SANGHVI et al. 1991; see Sect. I). Modifications of the purines may be directed into the major or minor grooves of the heteroduplex and, as noted in Sect. I will have different effects on RNase H. In the desire to remove sulphur from PSs to gain greater nuclease resistance and minimize toxicity parameters, the gap technology will be of great value.

An informative depiction of the concept of a combination of modifications and the balance or optimization between binding affinity and nuclease resistance using a gapmer strategy is presented in Fig. 9. The sequence examined is ISIS-5132: a 20-mer and a uniform 2'-deoxy/PS in phase II antitumor clinical trials. The present biophysical study examined the binding affinity (T_m) and the nuclease resistance ($t_{1/2}$) of a modified gapmer, that is, the flanks were POs rather than the usual PSs. The binding affinity of ISIS-5132 (full 2'-deoxy/PS) is about 62°C, the lowest of the oligomers in the study, but its $t_{1/2}$ is very high because of the PS linkages. In comparison, the ISIS-5132 sequence, as a full 2'-deoxy/PO, has higher binding affinity than the 5132/PS, as expected because of the removal of the destabilizing sulphur in the backbone, but this modified oligomer is rapidly degraded by SVPD ($t_{1/2}$ ~5 min). The 2'-fluoro modification in the flanks provides the highest binding affinity (2.3°C /mod) but the lowest nuclease resistance. As can be seen from Table 2, the best combination of binding affinity and nuclease resistance is exhibited by the 2'-*O*-(methoxyethyl) modification, where the nuclease resistance is the highest of the oligomers studied and the binding affinity is almost as high as the 2'-fluoro oligomer. It is interesting to note that the 2'-*O*-Me modification is not very stable as a modified gapmer and is unlikely to allow removal of the sulphur from antisense oligonucleotides. Hence, the 2'-*O*-Me modification requires a PS backbone. The 2'-*O*-propoxy also has a favorable combination of properties. In vitro and vivo antisense studies of the oligomers in this study correlated well with the observed biophysical properties (see discussion in Sect. I.III.5.c). It is also important to note that, in this modified gapmer, 53% of the sulphurs (10 of the 19 PSs) were converted to the natural PO linkages.

β) 2′-*O*-Modified Oligonucleotides with an RNase H-independent Mode of Action

Another mode of action for the high-binding, nuclease-resistant 2′-*O*-modified oligonucleotides is direct, sequence-specific binding at a site on the targeted RNA sequence. RNase H degradation of the targeted RNA is not involved in this case as uniformly 2′-*O*-modified oligonucleotide–RNA heteroduplexes, although binding to RNase H they do not allow cleavage by the enzyme (LIMA et al. 1997). The observed antisense activities result from the effects of binding to the target RNA and occupying an essential site required for ribosomal binding, altering essential secondary structure or perhaps activating the recruitment of other RNases (MACLEOD and CROOKE 1997). Since these modified oligonucleotides operate by simply occupying a site, sequence-specifically, both the selection of the target site and the affinity to the target site are critical. Hence, maximizing the stability of heteroduplexes that do not activate RNase H is an important design feature for a direct-binding mode of action. The available phosphorus atom modifications, such as Me-phosphonates, phosphoramidates and phosphotriesters, as noted above, significantly reduce, rather than enhance, binding affinities and have not exhibited useful antisense activity. Failure to obtain interesting activity with these low binding oligomers is likely due to the ability of the initiation factors to disrupt the heteroduplex structures, thus, displacing the antisense oligonucleotide (LIEBHABER et al. 1992; NASHIKURA 1992). Me-phosphonates (MILLER et al. 1993) and *alpha*-oligonucleotides (CHAIX et al. 1993) targeted to the 5′-cap region or in the vicinity of the AUG start codon have demonstrated antisense activities likely due to a direct-binding effect. These activities are obtained from very high concentrations relative to the RNase H sensitive PS (CAZENAVA and HELENE 1991). A morpholino *bis*-amidate, described above, has also exhibited biological activity when targeting sites in the capping region. A few positive antisense activities have resulted in targeting backbone-modified blocking oligonucleotides to an mRNA coding region (LIEBHABER et al. 1992; BOIZIAU et al. 1991).

The first report of a RNase H-independent antisense oligonucleotide, having greater activity than its parent P = Ss (which supports RNase H), describes targeting the AUG site of a core protein coding sequence of a human hepatocyte cell line with a uniformly modified 2′-*O*-(methoxyethoxy) PO 20-mer (HANECAK et al. 1996). Hepatitis C virus core protein levels were reduced as efficiently as the corresponding 2′-deoxy/PS. The binding affinity of the uniform 2′-O-methoxyethoxy PO was greater than 90°C or about 31°C greater than the corresponding 2′-deoxy/PS. Another recent example of an

RNase H-independent mode of action is the targeting of the 5'-cap region of human ICAM-1 transcript in HUVEC cells with a series of uniformly 2'-O-modified 20-mer oligonucleotides (BAKER et al. 1997). These RNase H-independent oligomers did not affect splicing or transport of the ICAM-1 mRNA, but instead selectively inhibited formation of the 80S translation initiation complex. The 2'-O-(methoxyethyl)/P = O oligomer demonstrated the greatest activity with a IC_{50} of 2.1 nM (T_m 87.1°C) and its P = S analog had an IC_{50} of 6.5 nM (T_m 79.2°C). Correlation of activity with binding affinity was not always followed as the 2'-F/P = S (T_m 87.9°C) was less active that the 2'-O-(methoxyethyl) P = S (T_m 79.2°C) by 4-fold. The 2'-O-methoxy/P = S analog had an IC_{50} of greater than 50 nM (T_m 76.5°C). The RNase H competent 2'-deoxy/P = S parent oligonucleotide (T_m 76.4°C) exhibited an IC_{50} of 41 nM.

c) Selected, Interesting Antisense Results with 2'-O-Modified Oligonucleotides in Gapmer Motifs

Several studies in vitro and in vivo demonstrate that methoxyethoxy (MOE) and aminopropoxy (AP)-modified oligonucleotides, when utilized in a gap strategy, possess substantially greater biological activity than their PS parents. It is important to note that greater than 50% of the sulphur has been removed from these modified oligomers. A 20-mer with nine 2'-O-AP/POs at the 3' end and ten 2'-deoxy/PSs at the 5' end (5'-TsCsCsCsGsCsCsTsGsTo–GoAoCoAoUoGoCoAoUs U) was electroporated into A549 cells. The AP-modified oligonucleotide inhibited C-raf mRNA (IC_{50} 250 nM at 24h) and C-raf protein production (IC_{50} 150 nM at 48h) compared with the PS parent (IC_{50} 2.5 mM and 300 nM, respectively) (GRIFFEY et al. 1996). The sulphur content (PS linkages) of the AP modification was reduced by 55%. As noted in Table 1, the $t_{1/2}$ (n to n-1) of the 3'-T of a 16-mer oligonucleotide against SVPD, when modified with four AP POs at the 3' end (n –1, –2, –3, –4) was greater than 24h, whereas the control PS parent was 6h. A 12-mer PO with eleven 2'-O-(aminopropoxy), 2'-O-Me- or 2'-O-propyl (3'-C is deoxy, 5'-GoAoGoCoUoCoCoAoGoGoC) exhibited an increase of +1.88°, +2.00°, and +2.13°C/mod, respectively, relative to PO (GRIFFEY et al. 1996).

The C-raf sequence ISIS-5132 was modified with 2'-O-(methoxyethyl)/PO sequences in the 3' and 5' ends with an internal gap of eight PS/2'-deoxys, a modified gapmer strategy (5'-ToCoCoCoGoCo–CsTsGsTsGsAsCsAs–ToGoCoAoTsT) and compared with the uniform 2'-deoxy/PS parent. The comparison of the binding affinities demonstrates a ΔT of 14°C or 1.2°C/mod and nuclease resistant studies with SVPD indicate the parent PS/2'-deoxy and the modified gapmer had about the same stability. The gapmer demonstrated 4-fold greater activity in reducing C-raf mRNA T24 cells (human bladder–carcinoma cell line) (ALTMANN et al. 1996b). About 50% of the PS linkages have been removed with this antisense sequence.

Another C-*raf* sequence, (5'-*TxTxCxCx*TsCsGsCsCsCsGsCs*TxCxCx*-*TxCxTxCxC*) was modified with 2'-*O*-Me, propyl, methoxyethyl, methoxydiethoxyethyl, or methoxyethyl (α-Me), in the flanking positions as denoted by italicized letters. This series of modified oligonucleotides were assayed for their ability to inhibit C-*raf* kinase mRNA expression in T24 cells (ALTMANN et al. 1997). The sequence was a standard gapmer (where x = PS, uniform PS) or a modified gapmer (where x = PPO, e.g., PO/PS/PO). The 2'-deoxy/PS parent oligomer exhibited a T_m of 68.6°C with an IC_{50} of 350 nM. The most active modified oligomers were the 2'-*O*-methoxyethyls in a gapmer or modified gapmer (modified gapmer has a T_m of 84.9°C) with an IC_{50} of 40 nM in both cases. Another interesting modification in this series is the high affinity heterocycle modifications, 5-propynyl-uridine and 5-propynyl-cytidine, combined with the 2'-*O*-methoxyethyl modification. These oligomers represented the highest binding affinity of the gapped sequences (standard gapmer, T_m 85.9°C; modified gapmer T_m 92.4° C) and exhibited IC_{50}s of 100 and 124 nM, respectively; these high-affinity, triply modified oligonucleotides were 2 to 3-fold less active than the corresponding doubly modified 2'-*O*-methoxyethyl/PS. Additional modifications of the interesting 2'-*O*-methoxyethyl series is placing a (*R*)-Me group on the β position in the ethyl group and extending the polyethylene glycol chain to a methoxytriethoxy. The (*R*)-Me modifications (T_m 78.1° and 82.6°C for standard and modified gapmer, respectively) exhibited an IC_{50} of 50 and 70 nM, respectively. The oligomers modified with the much larger methoxytriethoxy moiety also had enhanced binding affinities compared with the PS parent (T_m 78.2° and 84.3°C) and exhibited IC_{50}s of 150 and 350 nM, respectively. Generally, in the comparison of the standard gapmer (uniform PS) with the modified gapmer (PO flanks), the more nuclease-resistant standard oligomer exhibited the greater activity; hence, *the activities may correlate more closely with nuclease resistance rather than binding affinity*. Note that the modified gapmers have replaced nine PS linkages with the naturally occurring PO linkages (~50% removal of sulphur).

d) *Pharmacokinetic Properties of 2'-Modified Oligonucleotides*

As the binding affinities and nuclease-resistance properties of 2'-*O*-modified oligonucleotides are vastly improved compared with the first generation PSs and important modes of action have been delineated, attention is beginning to turn to enhancing the pharmacokinetic properties of the oligonucleotides. Since every repeating unit of an oligonucleotide has a 2' position, then one or more of these may be modified with groups that, along with enhancing T_m and $t_{1/2}$, will favorably alter the pharmacokinetics of the modified oligonucleotides. Pharmacokinetics of PSs in various animal species have generally demonstrated a very similar pattern of rapid, biphasic clearance from plasma into high-affinity tissues such as the kidney, liver, and spleen (CROOKE et al. 1996a).

The first report of *vivo* pharmacokinetics of 2'-*O*-modified oligonucleotides describes modifications of a mouse ICAM-1, 20-mer sequence uniformly modified with 2'-*O*-propyls in a PS and PO backbone, and a modified gapmer (PS/2'-deoxy gap, PO/2'-*O*-propyl flanks; CROOKE et al. 1996a). As a more detailed description of this large study is provided in Chap. 20 of this volume, only a few comments concerning the stability and distribution of the 2'-*O*-modified oligomers will be made. As would be expected, doubly modifying the sequence with a PS backbone and 2'-*O*-propyls at the 3' end significantly increases the stability towards 3'-exoncleases in the tissues examined (Table 1). A full length 2'-*O*-propyl/PS oligo was present in the kidney and liver at 24h whereas only 50% of the parent PS was present at 1h in the same tissues. The 3' end doubly modified oligomer was more than 75% intact at 2h in plasma compared with the parent PS which was less than 40% intact at this time. The modified gapmer and the uniformly modified 2'-*O*-propyl/PO, both having the 2'-*O*-propyl as a single modification at the 3' end, were degraded in plasma at about the rate of the parent PS, as was expected. However, these oligomers differ in their stability in liver and kidney. The uniformly modified oligomer, having no sulphur and nineteen 2'-*O*-propyls, compared with the modified gapmer, with 37% PSs (7 PS in the gap) and 58% propyls (11 in the flanks), demonstrated significantly greater stability in liver and kidney than the gapmer and the parent PS. A significant difference in tissue distribution of the three modified oligomers compared with parent PS was also observed. The 3'-end single-modified oligomers (propyls/PO) at 24h had approximately 20% and 5% of the dose in kidney and liver, respectively. This ratio was reversed with the doubly modified oligomer, 10% and 45% in the kidney and liver, respectively. The parent PS was present in approximately 15% in the kidney and liver at 24h. A significant percentage of the administered dose of the parent PS was present in the skin and skeletal muscle (~40%) at 24h. In comparison, the 3'-end singly modified oligomers had ~10% at 24h, whereas essentially none of the doubly modified oligomers accumulated in these tissues.

The pharmacokinetic behavior of the C-*raf* sequence ISIS 5132, described above as the parent PS and the modified gapmer (8 PS in the gap, 11 2'-*O*-methoxyethyls/P = O flanks), was examined in tumored nude mice (ALTMANN et al. 1996a). The gapmer was cleared much more rapidly as indicated by higher concentrations in the kidney and urine than the parent PS. Other organs demonstrated similar concentrations, especially at the early time points. The antitumor activity of the gapmer was substantially increased compared with the parent PS.

A series of 5'-fluorescein-labeled poly A PSs, 10-mers bearing various 2' modifications (nonyloxy, pentoxy, propoxy, methoxy, fluoro and hydrogen [deoxy]) with different lipophilicity, were examined for their ability to transverse synthetic liposomal membranes (HUGHES et al. 1993, 1995). The 2-*O*-propyl and 2'-fluoro analogs exhibited significantly shorter efflux half-lives across the membranes than the other modified oligomers. The greater lipophilic nonyl- and pentyl-modified oligomers, when encapsulated with [^{14}C] sucrose, demonstrated significant shorter efflux half-lives compared with the other analogs. The results of this study suggest that it is possible to modify the membrane permeation characteristics of oligonucleotides by means of 2' modifications.

The rank-order value of a series of modifications examined in tissue culture is important in selecting which oligonucleotides to move forward to in vivo studies. However, comparing the stability and distribution of isosequential oligomers, having one 2' modification used in various motifs (uniform or gapmer) and varying the backbone linkage (PO or PS), demonstrates that singly and doubly modified oligonucleotides will present very different pharmacokinetics within the modification and in comparison to the parent PS.

Thus, the pharmacokinetic properties are sure to change significantly depending on the modification and their in vivo activity may not correlate with in vitro data. Much is still to be learned when considering the scope of modifications that may be placed in the 2' position.

e) Toxicity Implications of 2'-Modified Oligonucleotides

One of the potential limitations of PSs may relate to toxicity issues. Considerable evidence has implicated the PS linkage as central to various toxicology parameters (see LEVIN et al. 1997; Chap. 5). On one hand, sulphur as a PS significantly enhances nuclease resistance of PS antisense oligonucleotides and is essential for oligonucleotide antisense biological activity by a RNase-H mechanism. On the other hand, the PS linkage has been suggested, in very early antisense research, to bind more tightly to proteins (GAO et al. 1992; GREIG et al. 1995) and, therefore, likely to cause toxic situations (see Chap. 5). However, to remove sulphur from antisense oligonucleotides to minimize potential toxicity problems would require their replacement with a nuclease-resistant modification. Only recently have antisense oligonucleotide modifications been discovered that will allow the removal of the PS linkage. The 2'-O modifications, such as aminopropyl, imidazolylbutyl and methoxyethyl, provide a level of nuclease resistance that will allow the removal of sulphur, i.e., by the replacement of the PS linkage with PO linkages. Recent reports have described lowering important toxicology parameters by 2'-O-modified oligonucleotides. A C-raf sequence modified with 2'-O-methoxyethyl)/PO sequence in the 3' and 5' ends and an internal gap of eight PS/2'-deoxys (5'-ToCoCoCoGoCo–CsTsGsTsGsAsCsAs–ToGoCoAoTsT) was examined in an assay to determine clotting time against concentration (APTT assay). This sequence with ten PS linkages replaced with POs, because of the sufficient nuclease resistance provided by the 2'-O-methoxyethoxys, exhibited a significantly reduced clotting time compared with the parent PS with 19PS linkages. The concentrations of the parent PS and the modified gapmer, required to double the clotting time, were 12.1 and more than 53 mM, respectively (ALTMANN et al. 1996a).

2'-O-(Methoxyethyl) derivatives of adenosine, guanosine, 5-Me-uridine and 5-Me-cytidine, examined as the nucleosides, did not exhibit any cellular toxicity (P. NICKLIN, personal communication). The lack of biological activity of the alkylated nucleoside may be expected, as the 2'-O-Me derivatives of the anticancer drugs, ara-C and ara-A, were not biologically active and did not exhibit toxicity (DE CLERCQ et al. 1975). This data suggests that sugar modifications, such as 2'-alkoxys, may have less toxicity liability with regards to metabolites than modified heterocycles, e.g., 5-propynylpyrimidines. However, it is uncertain how the cellular toxicity or lack of toxicity of modified nucleoside monomers relates to toxicity of the intracellular 5'-mononucleoside phosphoates which will result from 3'-exonuclease metabolism.

f) Summary of 2'-Sugar-Modified Oligonucleotides

Relatively simple chemistry derived from the much less expensive ribose rather than deoxyribose has been developed to prepare a variety of modifications in the 2' position. As certain of these modifications provides a superior combination of binding affinity and nuclease resistance and favorably alters pharmacokinetics, it is unlikely that the more "rational" approaches which attempt to constrain the ribofuranosyl ring to obtain a fixed 3'-endo conformation by rather difficult chemistry will provide an advantage over 2'-O modifications. Constraint modifications have not yielded molecules which enhance the properties desired for antisense drugs. An additional advantage of 2'-O modifications is that they present a means to modulate various properties by modifying the 2'-O positions. For example, in an oligonucleotide of 21 nucleotides, twenty 2'-O positions are available for modification which may affect a wide variety of properties such as T_m, $t_{1/2}$, lipophilicity, absorption and distribution, and toxicology.

In summary, of the sugar modifications, the 2'-O modifications are relatively easy to synthesize and will be less expensive than their deoxy counterparts. They offer high binding affinities due to a greater 3'-endo conformation and certain modifications provide superior nuclease resistance compared with PSs. Unlike other sub-units to be modified, a 2' modification motif has many sites available to be modified that will alter various antisense properties. Progress of antisense oligonucleotides is much more related to innovations in the 2'-O-modified sugars and their motifs than any other sub-unit depicted in Fig. 1. At this stage in the process of making drugs out of oligonucleotides, the 2' modification area has clearly demonstrated superior properties relative to modifications of other sub-units.

III. Backbone- or Linkage-Modified Oligonucleotides

Modifications of the backbone (Fig. 1) to enhance oligonucleotide drug properties and overcome limitations of PS have received considerably more attention than the heterocycle, carbohydrate, conjugations and connection site sub-units. As this is the unit that connects the monomers and also presents the weakest link due to biological or chemical degradation, it has been of great interest to medicinal chemists to modify. Some of the pioneering work was the preparation of "Teflon" DNA and "MATAGEN" oligomers (Ts'o et al. 1987). A number of reviews addressing backbone or linkage changes of oligonucleotides to enhance antisense properties have been published (Cook 1991, 1993; Sanghvi and Cook 1993, 1994; Varma 1993; De Mesmaeker et al. 1995; Herdewijn 1996; Matteucci 1996).

The primary motivation for pursuing a change in the linking phosphate is to minimize or completely prevent nuclease degradation. This medicinal chemistry effort has provided the first generation PS, Me-phosphonate and phosphoramidate (N-P) oligonucleotides. A relatively simple chemical conversion of a non-bonding PO oxygen into a sulphur, Me or substituted amine provided backbone (linkage)-modified oligonucleotides with greatly enhanced resistance to nuclease degradation when compared with unmodified DNA. This effort has been successful in providing many oligonucleotides for biological evaluations without the complications of degraded materials. *As now evident from the literature regarding these first generation modifications, only PSs possess the additional pharmacological and pharmacokinetic properties required to be drug candidates.* However, as discussed above, PSs do have limitations, which has spurred considerable ongoing oligonucleotide research with regard to modifying the backbone or linkage unit.

The linkage, as described in this review (Fig. 1), is the phosphoryl group and the 5'-methylene group of the deoxyribofuranosyl moiety. Thus, the unit being modified is a four-atom *linkage*, connecting adjacent sugar moieties (Teng and Cook 1994). Also, there is an obvious overlap between what is considered carbohydrate modifications and linkage modifications as most of the linkage changes require chemistry of the 3'- and 5'-carbons of adjacent sugars. I chose to describe the morpholino modifications in the carbohydrate section. A brief summary of the most advanced backbone modifications which could conceivably provide drug candidates is provided.

1. MMI- and Amide-3-Modified Oligonucleoside/tides

Two of the more interesting backbone modifications have a methylene group replacing the C3'-oxygen atom and the phosphodiester group replaced by a methylhydroxylamine [MMI, methylene(Me)imino, 3'-CH_2-N(Me)-OCH_2-5'; Sanghvi and Cook 1993] or a carboxamido group (Amide-3, 3'-$CH_2CONHCH_2$-5', 23; Lebreton et al. 1994) (Fig. 10). In other words, a hydroxylamine or an amide group connects the 3'- and 5'-methylenes, thus,

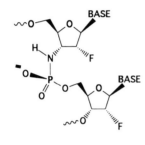

22 MMI bis-methoxy

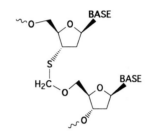

23 Amide-3 bis-methoxy

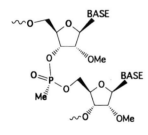

24 3'-Amidate bis-fluoro

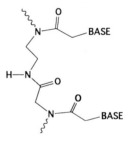

25 Thioformacetal

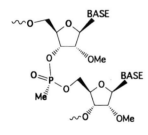

26 Chiral methylphosphonate bis-methoxy

27 PNA

Fig. 10. Backbone-modified oligonucleotides

maintaining a four-atom linkage between the deoxyribofuranosyl sugars, as with unmodified DNA. The C3' carbon-carbon bond generates a 3'-endo sugar pucker conformation relative to the normal sugar–PO linkage (35% to 68% Northern conformation in MMI; PEOC'H et al. 1997). This has the effect of pre-organizing the modified strand into an A-form conformation (geometry)

Table 2. Correlation of MMI binding affinities with sugar conformation

Backbone	DNA-(PO)	MMI		
2'-Modification X	H	H	F	OMe
% N upper unit	35	68	96	95
% N lower unit	28	31	96	76
$\Delta T_m °C$ per dimer (vs. P = S)	−0.9	+1.2	4.1	4.6

*Estimated from the $J_{1'2'}$ coupling constant.

(HEINEMANN et al. 1991) which significantly enhances binding affinities (+0.1°C/mod compared with DNA, +1.1°C compared with a PS dimer).

These modifications have been mainly studied as dimers being linked by POs or thiophosphates. A remarkable gain in binding affinity is obtained when MMI and Amide-3 are further modified by 2'-O-groups, which also increase the 3'-endo conformation, as noted above. The percentage of 3'-endo conformation of the two 2'-O-methoxy ribofuranosyl groups in a MMI bis-2'-O-methoxy dimer are 95 in the top and 76 in the bottom (compare with 35% and 28% for the parent PO dimer; PEOC'H et al. 1997). In this case, a MMI bis-2'-O-methoxy dimer provides the highest binding affinity reported thus far (+3.74°C/dimer compared with DNA or +4.6°C/dimer compared to a PS dimer) (Table 2). Thus, a PKC, a 20-mer having 10 MMI bis-2'-methoxy dimers, has a T_m of greater than 95°C or about 30°C higher than the parent PS. The situation is similar with Amide-3 bis-2'-methoxy. In summary, MMI and Amide-3, as their 2'-bis-methoxy modifications, provide extremely high binding affinities, have enhanced specificity compared with PSs, and are nuclease resistant (the modified dimer protects the adjacent PO linkage). In vivo pharmacology and pharmacokinetic studies of the these exciting modifications are in progress and will be reported in 1997.

2. 3'-Amidate-Modified Oligonucleotides

Another C3'-modified oligonucleotide, a C3'-amino (3'-NHP(O)$_2$OCH$_2$-5', 24), has exhibited interesting antisense properties (GRYAZNOV 1997). This is referred to as N3'-5' phosphoramidate or 3'-amidate. The binding affinities of 3'-amidates to RNA are increased up to 2.7°C/mod as the 2'-deoxy modification. Again, as found with the MMI and Amide-3 modifications, 2' modifications significantly enhance 3'-amidate binding affinity (3'-amidate-bis-2'-fluoro, +3.7°C/mod). This modification, still maintaining a phosphorus link, is more susceptible to SVPD degradation than certain 2'-O modifications (methoxyethyl and aminopropyl) and MMI and Amide-3. A 3'-amidate sequence was reported to possess a $t_{1/2}$ of 5.4h with SVPD, whereas the $t_{1/2}$ of a 2'-O-(methoxyethyl) modified sequence was greater than 110h. Furthermore, the chemical instability of 3'-amidate *bis*-2'-fluoro oligomers in acid conditions may hamper broad use of this modification. 3'-Amidate-modified oligonucleotides have several interesting features compared with PSs. These include the opportunity to remove sulphur from the oligomers; the resulting amidate linkage is not chiral, the binding affinity is greatly increased and the 3'-amidates have increased lipophilicity yet decreased protein binding (GRYAZNOV et al. 1996).

3. 3'-Thioformacetal- and Formacetal-Modified Oligomers

In this backbone modification, the 3'-thioformacetal has replaced the C3'-*O* with a sulphur atom, which connects a methyleneoxy to the 5'-Me group to afford a four-atom linker (3'-SCH$_2$OCH$_2$-5' 25; MATTEUCCI et al. 1991; JONES et al. 1993). In a similar manner, the formacetal modification has a methyleneoxy connecting the C3'-*O* with the 5'-Me group (3'-OCH$_2$OCH$_2$-5'; MATTEUCCI 1990). Thioformacetal-modified oligomers bind to complementary RNA with about the same stability as unmodified DNA; formacetal-modified sequences are less stable than the thioformacetals. 5-propynyl-modified pyrimidine oligomers bearing a formacetal or thioformacetal have been reported (LIN et al. 1994). Very little information concerning these modified oligomers is available. 2' modifications of the formacetals modification have not been reported.

4. 2'-Modified *Rp*-Me-Phosphonates

It is well established that Me-phosphonate-modified DNA hybridizes weakly to complementary sequences (KEAN et al. 1994; REYNOLDS et al. 1996). Use of the more stable Me-phosphonate *Rp* stereoisomer provides increased binding compared with racemic Me-phosphonates. However, their binding affinities are less than PSs. Subsequently, use of the 2'-*O*-Me modification with racemic Me-phosphonates has further enhanced the binding affinity compared with the *Rp* Me-phosphonate, but the stability is still less than PSs (MILLER et al. 1991; KEAN et al. 1995). The use of Me-phosphonate *Rp* 2'-*O*-Me (Struct. 26, Fig. 10) dimers connected by POs provides the highest T_ms for Me-phosphonates (REYNOLDS et al. 1996). The binding affinity (T_m) of Me-phosphonates with the chirally pure *Rp* stereoisomer and having 2'-*O*-Me groups is somewhat less than 2'-*O*-modified oligonucleotides.

5. Peptide Nucleic Acid

Peptide nucleic acids (PNA, Struct. 27) are DNA mimics with a pseudopeptide backbone composed of achiral and uncharged *N*-(2-aminoethyl)glycine units (NIELSEN et al. 1991). PNA has been shown to hybridize sequence-specifically with high affinity to complementary RNA and DNA (HYRUP and NIELSEN 1996).

6. General Observations About Backbone Research

The backbone modifications, briefly described above, present a number of interesting and overlapping antisense properties. Removal of all of the negatively charged phosphoryl groups or portions (dimer strategy) from oligonucleotides has been a primary goal of backbone modifications since making drugs out of oligonucleotides became of interest. There are several reasons for this: It has long been thought that oligonucleotides with a reduced charge may enhance pharmacokinetic properties, particularly membrane permeation (e.g., Me-phosphonates). Another long standing concern was that the chirality of the thiophosphoryl group had an important effect on binding affinity. Thus, "remove the sulphur" research has been actively pursued through backbone modifications. Rp or Sp stereochemistry of the thiophosphoryl group has been found not to significantly affect T_ms (KANEHARA et al. 1996), and how charge affects permeation of antisense oligonucleotides is still being investigated. A more important and recent reason to remove charge (and thus sulphur) from PSs is to modulate protein binding. Because of the propensity of PSs to bind to proteins (GREIG et al. 1995), their pharmacokinetics are dose-dependent which complicates drug development. Moreover, many of the undesired toxicological effects of PSs have been associated with the thiophosphate linkage (see Chap. 5). All of the backbone modifications described in this review remove sulphur and all modifications remove charge, except for the 3'-amidates. The MMI, Amide-3, PNA and the formacetal modifications have replaced the phosphoryl group with a non-ester linkage that will not serve as a substrate for esterases. High nuclease stability of the chiral 2'-O-Me phosphonates has been reported (REYNOLDS et al. 1996). The 3'-amidates and their 2'-flouro modification appear to be substantially less nuclease resistant than a PS, although the conditions of the SVPD assay were not reported (GRYAZNOV 1997). It is interesting that all of the sugar-based modifications, except for the formacetals, possess 2' modifications. This double modification motif is essential to achieving the extremely high binding affinities reported for these modifications. It is also important to note that none of these backbone modifications supports a RNase H mechanism. Thus, to exhibit antisense activities, these backbone-modified oligonucleotides must be prepared as gapmers which allows a RNase H mechanism or a mechanism that operates by direct binding. As noted above, antisense activities have rarely been found with oligonucleotides that do not activate RNase H (see BAKER et al. 1997). However, several of the above backbone-modified antisense oligomers have exhibited interesting activity. 3'-Amidates targeted to the c-*myb*, c-*myc*, and *bcr-abl* mRNAs have demonstrated sequence- and dose-dependent inhibition in the range 0.5–5.0 mM (GRYAZNOV et al. 1996). PNA has demonstrated very potent, sequence- and dose-dependent inhibition of translation in cell-free extracts and when microinjected into cells (HYRUP and NIELSEN 1996; KNUDSEN and NIELSEN 1996). The pharmacokinetics in mice of MMI *bis*-2'-*O*-methoxy antisense oligomers composed of dimers having either a PO or thio–PO linkage have been examined (SANGHVI et al. 1997). Considerable in vitro pharmacology of MMI *bis*-2'-*O*-methoxy oligomers in a gapmer motif has been accomplished (SANGHVI 1997). Biological data from the other backbone modifications has not been reported.

As further research takes place, the backbone/2' modification and PNA motifs are likely to offer the greatest opportunity for RNase H-independent activity because of their high binding affinities. In addition, the high binding per unit monomer of this type of oligomers may eventually allow significantly shorter oligomers to be developed. Shorter oligomers would be expected to have an important impact on drug properties and cost of antisense oligonucleotides.

Researchers in this area should note that the level of antisense properties (standards) of this selected series of backbone modifications is extremely high. Even with oligomers possessing these very attractive antisense properties, extensive biological evaluation, preferably in animals, has yet to be reported. The difficulty of synthesis of new oligomers in appropriate quantities and purity is primarily responsible for the slow progress. Several other important hurdles that are particular to backbone modification research have also been encountered. Removing charge from an oligomer has complicated rather than enhanced the absorption and distribution process, compared with PSs. The uncertainty of the mode of action of backbone-modified oligomers, for example, a RNase H-dependent or independent activity or by some unidentified action has also complicated antisense research.

IV. Oligonucleotide Pendants (Conjugates)

I consider the sugar, heterocycles and backbone (linkages) sub-unit modifications, as depicted in Fig. 1 and discussed in Sects. I–III, as core modifications of oligonucleotides. To further enhance the antisense properties of core-modified oligonucleotides, a variety of molecules (pendants) have been attached (conjugated) in a point modification motif (i.e., typically only one pendant in an antisense oligonucleotide, (COOK 1991). Although modification of this sub-unit has not received as much attention as modifications of the other sub-units depicted in Fig. 1, pendants have the potential to address problems remaining after optimizing core modifications. Pendent modifications have primarily been directed to enhance oligonucleotide uptake. Other potential applications of pendants include increased solubility, lipophilicity, nuclease resistance, binding affinity, means to attach synthetic cleavers and crosslinking and alkylating groups. Several reviews have discussed oligonucleotide pendants (GOODCHILD 1990; COOK 1991, 1993; MANOHARAN 1993; DE MESMAEKER et al. 1995).

Due to the importance of cellular uptake of antisense oligonucleotides, a variety of compounds have been conjugated to oligonucleotides and examined for enhanced permeation. Within the sub-unit of pendants, the cholesterol-modified oligonucleotides have received the most attention (Fig. 11). Unfortunately, this very active area of conjugate research has almost exclusively been

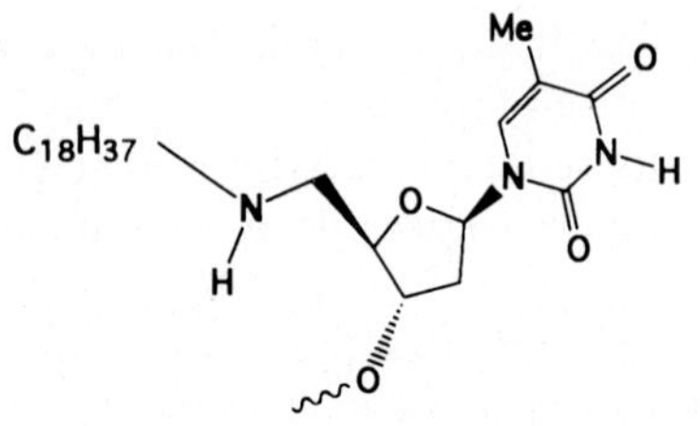

28 5'-C$_{18}$-Amino

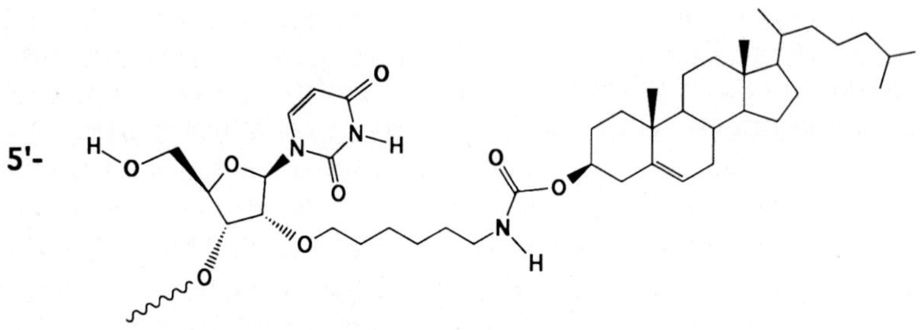

29 2'-Hexylaminocarbonyloxy-cholesteryl

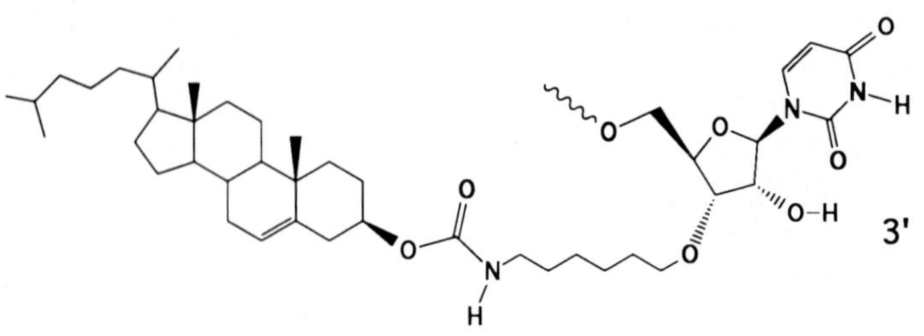

30 3'-Hexylaminocarbonyloxy-cholesteryl

Fig. 11. Conjugate-modified oligonucleotides

conducted in cell cultures. Since the primary purpose of conjugated oligonucleotides is directed toward enhanced absorption and distribution, in vivo studies should be employed to obtain the most meaningful information. Only very few studies of conjugated oligonucleotides in vivo have been reported (DE SMIDT et al. 1991; KRIEG et al. 1993; DESJARDINS et al. 1995; CROOKE et al. 1996; BIJSTERBOSCH et al. 1997; MANOHARAN et al. 1997) and these have all been cholesterol-modified oligonucleotides with the exception of a 5′-deoxy-5′-octadecylamine 20-mer PS. HDL and LDL lipoproteins in rats are high binding sites for cholesterol-modified oligonucleotides (DE SMIDT et al. 1991) and, thus, provide a means to target specific organs such as the liver which has specific receptors for lipoproteins. These in vivo studies will be discussed briefly.

An extensive pharmacokinetic study comparing five analogs of a 20-mer PS ICAM-1 mouse sequence with its parent PS was conducted in mice (CROOKE et al. 1996). Two of the analogs were 5′-lipophilic PS conjugates [5′-deoxy-5′-octadecylamine (Struct. 28, Fig. 11) and 2′-O-hexylaminocarbonyloxy-cholesterol uridine (Struct. 29) at the 5′ end] which were more lipophilic than the parent PS by 3- and 7-fold, respectively, but exhibited binding affinities similar to the parent PS. These modified oligonucleotides exhibited initial volumes of distribution less than parent PS and were cleared from the plasma slower than parent PS, as would be expected due to increased protein binding. The amount of the cholesterol-modified oligonucleotide was substantially increased in the liver compared with parent PS (60% of the administered dose compared with 15% of the parent PS, at 3 h) and the C_{18}-amino analog. Results of pharmacological evaluation of the ICAM-1 cholesterol conjugate in mice also correlated with the increased concentration of the antisense conjugate in the liver (MANOHARAN et al. 1997). Mice treated intravenously with a dose of 10 mg/kg demonstrated reduced induced ICAM-1 mRNA levels in the liver by ~50% whereas the parent PS had no effect up to a dose of 100 mg/kg. An additional advantage of cholesterol-modified oligonucleotides is that they may circumvent the need for cationic lipid formulations typically required for cell culture experiments (ALAHARI et al. 1996).

The ICAM-1 mouse sequence described above, was derivatized at the 3′-O of the 3′-terminal nucleotide with a cholesterol group (Struct. 30) (BIJSTERBOSCH et al. 1997). This attachment site has been shown to provide greater nuclease resistance than a 2′-O connection at the 5′-nucleotide (MANOHARAN et al. 1997). Greater than 75% of the dose was found associated with lipoproteins compared with less than 2% of the parent PS. The concentration of the conjugate PS in the endothelial, Kupffer and parenchymal cells was 53%, 18%, and 29%, respectively. A similar ratio of concentrations in the liver was found with the parent PS. As in other pharmacokinetic studies, the cholesterol-modified oligonucleotide was cleared more slowly from the plasma than parent PS. It was concluded in this study that uptake by scavenger receptors on endothelial cells provides a major metabolic pathway for

ICAM1 PSs. As noted, the cholesterol-modified oligonucleotides used in this study were attached at the 3'-O of the 3'-nucleotide (Struct. 30) by a modified amidite approach rather than standard–post conjugation. This connection, unlike typical 3'- or 5'- post-conjugations via PO linkages, is an alkyl carbamate linker which is not a substrate for nucleases and, thus, provides enhanced stability.

DESJARDINS et al. (1995) also report enhanced activity of a 5'-cholesterol-modified antisense oligonucleotide in rats compared with its unmodified parent PS. In these in vivo experiments, a 5'-cholesterol-modified oligonucleotide PS was evaluated for modulation of the expression of a cytochrome P450 gene (CYP2B1). Hexobarbital (HB) sleep times provide a measure of CYP2B1 enzyme activity in vivo. The cholesterol-modified oligonucleotide, at a dose of 0.1 mg/day for 2 days, increased HB sleep time 10-fold, compared with the parent PS.

These in vivo studies with cholesterol-modified PSs report altered pharmacokinetic properties compared with parent PSs and generally a 10-fold increase in antisense activity. The absorption of the conjugates are thought to be mediated by association with lipoprotein receptors rather than simply an increase in lipophilicity derived from the cholesterol molecule. Conjugation of cholic acid – a steroid of similar lipophilicity as cholesterol but which does not bind to lipoproteins – to oligonucleotides, in a similar manner as described with cholesterol, did not enhance conjugate absorption in vitro (MANOHARAN et al. 1997). Although the preponderance of pendent modifications has centered on cholesterol-assisted absorption and distribution, and indeed with encouraging results, a wide variety of molecules are available which, upon conjugation to antisense oligonucleotides, may enhance absorption and distribution by protein-mediated processes. Exploration of this sub-unit modification has received little attention, but offers great promise for enhancing oligonucleotide antisense properties, and in ways not likely to be achieved *via* modification of the other sub-units.

J. Conclusions

PS drugs will soon become available and in order to continually improve the oligonucleotide drug class and to overcome certain limitations, structural changes to PSs are required. In the past 8 years, a diverse range of modifications, at all possible modification sites of an oligonucleotide (Fig. 1) have been reported. This application of traditional medicinal chemistry (structure–activity-relationship studies) to drug discovery in antisense oligonucleotides has answered many important questions. For example, as a result of this rather intense effort, we are now aware of modifications that stabilize oligonucleotides towards nucleolytic degradation, modifications that greatly enhance binding affinities while maintaining base-pair specificity and modifications that support endonucleolytic cleavage by RNases. Although these are physiochemical and biophysical properties, a large volume of cellular and animal studies support the notion that enhancing these properties correlates with enhanced antisense biological activity. Unfortunately, a single modification providing high binding, nuclease-resistant antisense oligonucleotides which will support an RNase H mechanism is not available. A modification of this nature is of current interest. We are also aware that, thus far, changing the structure of PS oligonucleotides provides an opportunity to alter their pharmacokinetic profile. In structural changes which remove sulphur (as thiophosphate) and/or change lipophilicity (e.g., by 2'-O modifications), more favorable toxicity profiles have also resulted. Although we are aware of these important antisense properties (and there may be many more to learn about) and we have learned how to control them, we are unaware of the optimum values at which to aim our modifications. In addition, antisense oligonucleotides that are orally available and/or penetrate the blood–brain barrier present the most important deficiency of antisense oligonucleotides. Recent reports of certain modified oligonucleotides have provided encouraging results that these pharmacokinetic deficiencies will soon be solved by appropriate chemical modifications.

One should be aware of the level of accomplishments achieved in oligonucleotide medicinal chemistry research in the past 8 years. In this review, I have discussed these (binding affinities, nuclease resistance, support of RNase H and cost of synthesis) and suggest that they be considered (as standards) before initiating or continuing certain oligonucleotide modification research. In addition, understanding the proprietary patent positions that have been established is an important research consideration. I believe that, at this stage of antisense oligonucleotide medicinal chemistry, it is highly unlikely that a single modification will be discovered that will significantly impact all of the important antisense properties described above. The types of modified oligonucleotides currently being pursed (going beyond PSs) possess a combination of modifications, and this trend will certainly continue as pendants will be conjugated to oligonucleotides with optimized core sub-units to obtain a *completely* optimized antisense oligomer.

I view the current *winners* or the first modifications most likely to be incorporated into antisense oligonucleotides that will undergo clinical trials, as the RNA mimics, 2'-O-(methoxyethyl) and 2'-O-(aminopropyl) and the backbone modification, MMI *bis*-methoxys. These will likely be utilized in a gap strategy (gap technology). However, efforts to prepare uniform modifications, such as RNA mimics (2'-O modifications), MMI and PNA are of considerable interest, in that reliance on RNase H for a mode of action would not be required. In addition to these modifications which act, either by direct binding (RNase H independent) or RNase H, I believe the sub-unit, pendent modifications, e.g., cholesterol-conjugates, will become increasingly more important for optimizing multi-modified oligonucleotides.

Acknowledgements. The author thanks Drs. Stanley Crooke, Muthiah Manoharan, Richard Griffey, and Bruce Ross for helpful scientific discussions and Julie Walker for expert technical assistance.

References

Agrawal S (1996a) Methods in molecular medicine. In: Agrawal S (ed) Antisense therapeutics. Humana, Totowa, NJ

Agrawal S (1996b) Antisense oligonucleotides: towards clinical trials. Tibtech 14(10):376–38

Agrawal S, Mayrand SH, Zamecnik PC, Pederson T (1990) Site-specific excision from RNA by RNase H and mixed-phosphate-backbone oligodeoxynucleotides. Proc Natl Acad Sci USA 87:1401

Agrawal S, Zhang X, Lu Z, Zhao H, Tamburin JM, Yan J, Cai H, Diasio RB, Habus I, Jiang Z, Iyer RP, Yu D, Zhang R (1995) Absorption, tissue distribution and vivo stability in rats of a hybrid antisense oligonucleotide following oral administration. Biochem Pharmacol 50(4):571–576

Agrawal S, Jiang Z, Zhao Q, Shaw D, Sun D, Saxinger C (1997a) Mixed-backbone oligonucleotides containing phosphorothioate and Me-phosphonate linkages as second generation antisense oligonucleotide. Nucleosides Nucleotides (in press)

Agrawal S, Zhang X, Lu Z, Zhao H, Tan W, Jian Z, Yu D, Iyer RP, Zhang R (1997b) Comparative pharmacokinetics and metabolism of an oligonucleotide phosphorothioate and its end protected analogues in rats and mice following oral administration. Nucleosides Nucleotides Biochem Pharmacol (in press)

Alahari SK, Dean NM, Fisher MH, Delong R, Manoharan M, Tivel KL, Juliano RL (1996) Inhibition of expression of the multidrug resistance-associated P-glycoprotein by phosphorothioate and 5' cholesterol-conjugated phosphorothioate antisense oligonucleotides. Am Soc Pharmacol Exp Ther 50:808–819

Alderfer JL, Loomis RE, Soni SD, Sharma M, Bernacki R, Hughes R Jr (1985) Halogenated nucleic acids: biochemical and biological properties of fluorinated polynucleotides. Polym Mater Med 32:125–138

Altmann KH, Kesselring R, Francotte E, Rihs G (1994a) 4',6'-Methano carbocyclic thymidine: a conformationally constrained building block for oligonucleotides. Tetrahredron Lett 35(15):2331–2334

Altmann KH, Imwinkelried R, Kesselring R, Rihs G (1994b) 1',6'-Methano carbocyclic thymidine: synthesis, x-ray crystal structure, and effect on nucleic acid duplex stability. Terahedron Lett 35(41):7625–7628

Altmann KH, Freier SM, Pieles U, Winkler T (1994c) Synthesis of an azathymidine and its incorporation into oligonucleotides. Angew Chem Int Ed Engl 33(15–16):1654–1657

Altmann KH, Bévierre MO, De Mesmaeker A, Moser HE (1995) The evaluation of 2'- and 6'-substituted carbocyclic nucleosides as building blocks for antisense oligodeoxyribonucleotides. Bioorg Med Chem Lett 5:431–436

Altmann KH, Dean NM, Fabbro D, Freier SM, Geiger T, Häner R, Hüsken D, Martin P, Monia BP, Müller M, Natt F, Nicklin P, Phillips J, Pieles U, Sasmor H, Moser HE (1996a) Second generation of antisense oligonucleotides: from nuclease resistance to biological efficacy in animals. Chima 50:168–176

Altmann KH, Kesselring R, Pieles U (1996b) 6'-Carbon-substituted carbocyclic analogs of 2'-deoxyribonucleosides – synthesis and effect on DNA/RNA duplex stability. Tetrahedron 52(39):12699–12722

Altmann KH, Fabbro D, Dean NM, Geiger T, Monia BP, Muller M, Nicklin P (1996c) Second-generation antisense oligonucleotides: structure-activity relationships and the design of improved signal-transduction inhibitors. Biochem Soc Trans 24:630–637

Altmann KH, Martin P, Dean NM, Monia BP (1997) Second generation antisense oligonucleotides - inhibiton of pkc-a and c-raf kinase expression by chimeric oligonucleotides incorporating 6'-substituted carbocyclic nucleosides and 2'-O-ethylene glycol substituted ribonucleosides. Nucleosides Nucleotides (in press)

Anderson KP, Fox MC, Brown-Driver V, Martin MJ (1996) Inhibition of human cytomegalovirus immediate-early gene expression by an antisense oligonucleotide complementary to immediate-early RNA. Antimicrob Agents Chemother 40(9):-2004–2011

Baker BF, Ramasamy K, Kiely J (1996) Decapitation of a 5' capped RNA by an antisense copper complex conjugate. Bioorg Med Chem Lett 6(14):1647–1652

Baker BF, Lot S, Condon TP, Cheng-Flourney S, Lesnik E, Sasmor H, Bennett CF (1997) 2'-O-(2-Methyoxy)ethyl-modified anti-intercellular adhesion molecule 1 (ICAM-1) oligonucleotides selectively increase the ICAM-1 mRNA level and inhibit formation of the ICAM-1 translation initiation complex in human umbilical vein endothelial cells. J Biol Chem 272:11994–20000

Bellon L, Leydier C, Barascut JL, Maury G, Imbach JL (1994) 4'-thio-RNA: a novel class of sugar-modified b-RNA. In: Sanghvi YS, Cook PD (eds) Carbohydrate modifications in antisense research. American Chemical Society, Washington DC, Chap. 5 (ACS symposium series 580)

Bennett CF, Dean N, Ecker DJ, Monia BP (1996) Pharmacology of antisense therapeutic agents. In: Agrawal S (ed) Methods in molecular medicine: antisense therapeutics. Humana Press, Totowa, NJ, Chap. 2, pp 13–46

Bergstrom DE, Gerry NP (1994) Precision sequence-specific cleavage of a nucleic acid by a minor-groove-directed metal-binding ligand linked through N-2 of deoxyguanosine. J Am Chem Soc 116:12067–12068

Bévierre MO, DeMesmaeker A, Wolf RM, Freier SM (1994) Synthesis of 2'-O-methyl-6, 3'-ethanouridine and its introduction into antisense oligonucleotides. Bioorg Med Chem Lett 4:237–240

Bijsterbosch MK, Manoharan M, Tivel KL, Rump ET, Biessen EAL, De Vrueh RLA, Cook PD, van Berkel TJC (1997) Predominant uptake of phosphorothioate antisense oligonucleotides by scavenger receptors on endothelial cells. Nucleosides Nucleotides (in press)

Boiziau C, Kurfurst R, Cazenave C, Roig V, Thuong NT, Toulme JJ (1991b) Inhibition of translation initiation by antisense oligonucleotides via an RNase-H independent mechanism. Nucleic Acids Res 19:1113–1119

Bolli M, Lubini P, Tarköy M, Leumann C (1994) a-Bicyclo-DNA: synthesis, characterization, and pairing properties of a-DNA-analogues with restricted conformational flexibility in the sugar-phosphate backbone, medicinal chemistry strategies for antisense research. In: Crooke ST, Lebleu B (eds) Antisense research and applications, Chap. 7. CRC Press, Boca Raton, FL

Buhr CA, Wagner RW, Grant D, Froehler BC (1996) Oligodeoxynucleotides containing C-7 propyne analogs of 7-deaza-2'-deoxyguanosine and 7-deaza-2'-deoxyadenosine. Nucleic Acids Res 24(15):2974–2980

Casale R, McLaughlin LW (1990) Synthesis and properties of an oligodeoxynucleotide containing a polycyclic aromatic hydrocarbon site specifically bound to the N^2 amino group of a 2'-deoxyguanosine residue. J Am Chem Soc 112:5264–5271

Cazenave C, Helene C (1991) Antisense oligonucleotides. In: Mol JNM, van der Drol AR (eds) Antisense nucleic acids and proteins: fundamentals and applications. Dekker, New York

Chaix C, Toulmé JJ, Morvan F, Rayner B, Imbach JL (1993) a-Oligonucleotides. In: Crooke ST, Lebleu B (eds) Antisense research and applications. CRC Press, Boca Raton, FL, Chap. 12

Cook PD (1991) Medicinal chemistry of antisense oligonucleotides – future opportunities. Anti Cancer Drug Design 6(6):585–607
Cook PD (1993) Medicinal chemistry strategies for antisense research. In: Crooke ST, Lebleu B (eds) Antisense research and applications. CRC Press, Boca Raton, FL, pp 149–187
Crooke ST, Lemonidis KM, Neilson L, Griffey R, Lesnik EA, Monia BP (1995) Kinetic characteristics of Escherichia coli RNase H1: cleavage of various antisense oligo nucleotide-RNA duplexes. J Biochem 312:599–608
Crooke ST, Graham MJ, Zuckerman JE, Brooks D, Conklin BS, Cummins LL, Greig MJ, Guinosso CJ, Kornbrust D, Manoharan M, Sasmor HM, Schleich T, Tivel KL, Griffey RH (1996a) Pharmacokinetic properties of several novel oligonucleotide analogs in mice. J Pharmacol Exp Ther 277:923–937
Crooke ST, Bernstein LS, Boswell H (1996b) Progress in the development and patenting of antisense drug discovery technology. Exp Opin Ther Patents 6(9):855–870
Cummins LL, Owens SR, Risen LM, Lesnik EA, Freier SM, McGee D, Guinosso CJ, Cook PD (1995) Characterization of fully 2′-modified oligoribonucleotide hetero- and homoduplex hybridization and nuclease sensitivity. Nucleic Acids Res 23(11):2019–2024
Dagle JM, Andracki ME, DeVine RJ, Walder JA (1991) Physical properties of oligonucleotides containing phosphoramidate-modified internucleoside linkages. Nucleic Acids Res 19:1805
Dean NM, McKay R, Miraglia L, Geiger T, Muller M, Fabbro D, Bennett CF (1996) Antisense oligonucleotides as inhibitors of signal transduction: development from research tools to therapeutic agents. Biochem Soc Trans 24(3):623–629
De Bouvere B, Kerremans L, Hendrix C, De Winter H, Schepers G, Van Aerschot A, Herdewijn P (1997) Hexitol nucleic acids (HNA): synthesis and properties. Nucleosides Nucleotides (in press)
De Clercq E, Darzynkiewicz E, Shugar D (1975) Antiviral activity of O'-alkylated derivatives of cytosine arabinoside. Biochem Pharmacol 24:523–527
De Mesmaeker A, Häner R, Martin P, Moser HE (1995) Antisense oligonucleotides. In: Accounts of chemical research: antisense oligonucleotides. Acc Chem Res 28:366–374
Desjardins J, Mata J, Brown T, Graham D, Zon G, Iversen P (1995) Cholesteryl-conjugated phosphorothioate oligodeoxynucleotides modulate CYP2B1 expression vivo. J Drug Targeting 2:477–485
de Smidt P, Doan T, de Falco S, van Berkel T (1991) Association of antisense oligonucleotides with lipoproteins prolongs the plasma half-life and modifies the tissue distribution. Nucleic Acid Res 19:4695–4700
Egli M (1996) Structural aspects of nucleic acid analogs and antisense oligonucleotides. Angew Chem Int Ed Engl 35:1894–1909
Fenster SD, Wagner RW, Froehler BC, Chin DJ (1994) Inhibition of human immunodeficiency virus type-1 env expression by C-5 propyne oligonucleotides specific for rev-response element stem-loop V. Biochemistry 33:8391–8398
Fraser A, Wheeler P, Cook PD, Sanghvi YS (1993) Synthesis and conformational properties of 2′-deoxy-2′-methylthio -pyrimidine and -purine nucleosides: potential antisense applications. J Heter Chem 30(5):1277–1287
Freier S (1993) Hybridization: considerations affecting antisense drugs. In: Crooke ST, Lebleu B (eds) Antisense research and applications. CRC Press, Boca Raton, FL, pp 67–82
Freier SM, Lima WF, Sanghvi YS, Vickers T, Zounes M, Cook PD, Ecker DJ (1992) Thermodynamics of antisense oligonucleotide hybridization. In: Erickson RP, Izant JG (eds) Gene regulation: biology of antisense RNA and DNA (series: Molecular and cellular biology). Raven, New York, pp 95–107
Froehler BC, Wadwani S, Terhorst TJ, Gerrard SR (1992) Oligodeoxynucleotides containing C-5 propyne analogs of 2′-deoxyuridine and 2′-deoxycytidine. Tetrahedron Lett 33(37):5307–5310

Froehler BC, Jones RJ, Cao X, Terhorst TJ (1993) Oligonucleotides derived from 5-(1-propynyl)-2′-O-allyl-uridine and 5-(1-propynyl)-2′-O-Cytidine: synthesis and RNA duplex formation. Tetrahedron Lett 34(16):1003–1006

Gao WY, Han FS, Storm C, Egan W, Cheng YC (1992) Phosphorothioate oligonucleotides are inhibitors of human DNA polymerases and RNase H: implications for antisense technology. Mol Pharmacol 41:223–229

Gmeiner WH, Sahasrabudhe P, Pon RT, Sonntag J, Srinivasan S, Iversen PL (1995) Preparation of oligomeric 2′-deoxy-5-fluorouridylate of defined length and backbone composition: a novel pro-drug form of the potent anti-cancer drug 2′-deoxy-5-fluorouridylate (1995). Nucleosides Nucleotides 14(1,2):243–253

Goodchild J (1990) Conjugates of oligonucleotides and modified oligonucleotides: a review of their synthesis and properties. Bioconjugate Chem 1:166–187

Greig MJ, Gaus H, Cummins LL, Sasmor H, Griffey RH (1995) Measurement of macromolecular binding using electrospray mass spectrometry. Determination of dissociation constants for oligonucleotide - serum albumin complexes. J Am Chem Soc 117:10765–10766

Griffey RH, Lesnik E, Freier S, Sanghvi YS, Teng K, Kawasaki A, Guinosso C, Wheeler P, Mohan V, Cook PD (1994) New twists on nucleic acids: structural properties of modified nucleosides incorporated into oligonucleotides. In: Sanghvi YS, Cook PD (eds) Carbohydrate modifications in antisense research. American Chemical Society, Washington DC, pp 212–224 (ACS symposium series no 580)

Griffey RH, Monia BP, Cummins LL, Freier S, Greig MJ, Guinosso CJ, Lesnik E, Manalili SM, Mohan V, Owens S, Ross BR, Sasmor H, Wancewicz E, Weiler K, Wheeler PD, Cook PD (1996) 2′-O-aminopropyl ribonucleotides: a zwitterionic modification that enhances the exonuclease resistance and biological activity of antisense oligonucleotides. J Med Chem 39:5100–5109

Gryaznov S, Schultz RG (1994) Stabilization of DNA:DNA and DNA:RNA duplexes by substitution of 2′-deoxyadenosine with 2′-deoxy-2-aminoadenosine. Tetrahedron Lett 35(16):2489–2492

Gryaznov S, Skorski T, Cucco C, Nieborowska-Skorska M, Chiu CY, Lloyd D, Chen JK, Koziolkiewicz M, Calabretta B (1996) Oligonucleotide N3′–>P5′ phosphoramidates as antisense agents. Nucleic Acids Res 24(8):1508–1514

Gryaznov SM (1997) Synthesis and properties of the oligonucleotide N3′–>P5′ phosphoramidates. Nucleosides Nucleotides (in press)

Guinosso CJ, Hoke GD, Freier SM, Martin JF, Ecker DJ, Mirabelli CK, Crooke ST, Cook PD (1991) Synthesis and biophysical and biological evaluation of 2′ modified antisense oligonucleotides. Nucleoside Nucleotides 10:259

Gutierrez AJ, Froehler BC (1996) RNA duplex formation by oligodeoxynucleotides containing C-5 alkyne and C-5 thiazole substituted deoxyuridine analogs. Tetrahedron Lett 37(23):3959–3962

Gutierrez AJ, Terhorst TJ, Matteucci MD, Froehler BC (1994) 5-Heteroaryl-2′-deoxyuridine analogs. Synthesis and incorporation into high-affinity oligonucleotides. J Am Chem Soc 116:5540–5544

Hanecak R, Brown-Driver V, Fox MC, Azad RF, Furusako S, Nozaki C, Ford C, Sasmor H, Anderson KP (1996) Antisense oligonucleotide inhibition of hepatitis C virus gene expression in transformed hepatocytes. J Virol 70(8):5203–5212

Heeb NV, Benner SA (1994) Guanosine derivatives bearing an N^2-3-imidazolepropionic acid. Tetrahedron Lett 35(19):3045–3048

Heidenreich O, Gryaznov S, Nerenberg M (1997) RNase-H independent antisense activity of oligonucleotide N3′–>P5′ phosphoramidates. Nucleic Acids Res 25(4):776–780

Heinemann U, Rudolph L-N, Alings C, Morr M, Heikens W, Frank R, Blocker H (1991) Effect of a single 3′-methylene phosphonate linkage on the conformation of an A-DNA octamer double helix. Nucleic Acids Res 19(3):427–433

Henry SP, Zuckerman JE, Rojko J, Hall WC, Harman RJ, Kitchen D, Crooke ST (1997) Toxicological properties of several novel oligonucleotide analogs in mice. Anti Cancer Drug Design 12:1–14

Herdewijn P (1996) Targeting RNA with conformationally restricted oligonucleotides. Liebigs Ann 1337–1348

Herdewijn P, De Winter H, Doboszewski B, Verheggen I, Augustyns K, Hendrix C, Saison-Behmoaras T, De Ranter C, Van Aerschot A (1994) Hexopyranosyl-like Oligonucleotides. In: Sanghvi YS, Cook PD (eds) Carbohydrate modifications in antisense research. American Chemical Society, Washington DC, Chap. 6 (ACS symposium series no 580)

Hoke GD, Draper K, Freier SM, Gonzalez C, Driver VB, Zounes MC, Ecker DJ (1991) Effects of phosphorothioate capping on antisense oligonucleotide stability, hybridization and antiviral efficacy versus herpes simplex virus infection. Nucleic Acids Res 19:5743

Hudziak RM, Barofsky E, Barofsky DF, Weller DL, Huang S-B, Weller DD (1996) Resistance of morpholino phosphorodiamidate oligomers to enzymatic degradation. Antisense Nucleic Acid Drug Dev 6:267–272

Hughes JA, Bennett CF, Cook PD, Guinosso CJ, Mirabelli CK, Juliano RL (1993) Lipid membrane permeability of 2'-modified derivatives of phosphorothioate oligonucleotides. J Pharm Sci 83(4):597–600

Hughes J, Avroutskaya A, Sasmor HM, Guinosso CJ, Cook PD, Juliano RL (1995) Oligonucleotide transport across membranes and into cells: effects of chemical modifications. In: Akhtar S (ed) Delivery strategies for antisense oligonucleotide therapeutics. CRC Press, Boca Raton

Hyrup B, Nielsen PE (1996) Peptide Nucleic Acids (PNA): synthesis, properties and potential applications. Bioorg Med Chem 4(1):5–23

Jones GD, Lesnik EA, Owens SR, Risen LM, Walker RT (1996) Investigation of some properties of oligodeoxynucleotides containing 4'-thio-2'-deoxynucleotides: duplex hybridization and nuclease sensitivity. Nucleic Acids Res 24(21):4117–4122

Jones RL, Lin KY, Milligan JF, Wadwani S, Matteucci MD (1993) Synthesis and binding properties of pyrimidine oligonucleoside analogs containing neutral phosphodiester replacements: the formacetal and 3'-thioformacetal internucleoside linkages. J Org Chem 58:2983–2991

Kanehara H, Wada T, Mizuguchi M, Makino K (1996) Influence of a thiophosphate linkage on the duplex stability – does Sp configuration always lead to higher stability than Rp? Nucleosides Nucleotides 15(6):1169–1178

Kawasaki AM, Casper MD, Freier SM, Lesnik EA, Zounes MC, Cummins LL, Gonzalez C, Cook PD (1993) Uniformly modified 2'-deoxy-2'-fluoro phosphorothioate oligonucleotides as nuclease-resistant antisense compounds with high affinity and specificity for RNA targets. J Med Chem 36:831–841

Kean JM, Cushman CD, Kang H, Leonard TE, Miller PS (1994) Interactions of oligonucleotide analogs containing methylphosphonate internucleotide linkages and 2'-O-methylribonucleosides. Nucleic Acids Res 22(21):4497–4503

Kean JM, Kip SA, Miller PS, Kulka M, Aurelian L (1995) Inhibition of Herpes Simplex Virus replication by antisense oligo-2'-O-methylribonucleoside methylphosphonates. Biochemistry 34(45):14617–14620

Kido K, Inoue H, Ohtsuka E (1992) Sequence-dependent cleavage of DNA by alkylation with antisense oligodeoxyribonucleotides containing a 2-(N-iodoacetylaminoethyl)thio-adenine. Nucleic Acids Res 20(6):1339–1344

Kiely JS (1994) Recent advances in antisense technology. Ann Rep Med Chem 29:297

Knudsen H, Nielsen PE (1996) Antisense properties of duplex- and triplex-forming PNAs. Nucleic Acids Res 24(3):494–500

Krieg A, Tonkinson J, Matson S, Zhao Q, Saxon M, Zhang L, Bhanja U, Yakubov L, Stein C (1993) Modification of antisense phosphodiester oligodeoxynucleotides by a 5' cholesteryl moiety increases cellular association and improves efficacy. Proc Natl Acad Sci USA 90:1048–1052

Kutyavin IV, Rhinehart RL, Lukhtanov EA, Gorn VV, Meyer RB Jr, Gamper HB Jr (1996) Oliognucleotides containing 2-aminoadenine and 2-thiothymine act as selectively binding complementary agents. Biochemistry 35(34):11170–11176

Lebreton J, Waldner A, Lesueur C, De Mesmaeker A (1994) Antisense oligonucleotides with alternating phosphodiester "Amide-3" linkages. Synlett 2:137–140

Lesnik EA, Freier SM (1995) Relative thermodynamic stability of DNA, RNA, and DNA:RNA hybrid duplexes: relationship with base composition and structure. Biochemistry 34(34):10807–10815

Lesnik EA, Guinosso CJ, Kawasaki AM, Sasmor H, Zounes M, Cummins LL, Ecker DJ, Cook PD, Freier SM (1993) Oligodeoxynucleotides containing 2′-O-modified adenosine: synthesis and effects on stability of DNA:RNA duplexes. Biochemistry 32:7832–7838

Levin AA, Henry SP, Bennett CF, Cole DL, Hardee GE, Srivatsa GS (1997) Preclinical development of oligonucleotide therapeutics. In: Crooke ST (ed) Handbook of experimental pharmacology (in press)

Leydier C, Bellon L, Barascut JL, Morvan F, Rayner B, Imbach JL (1995) 4′-Thio-RNA: synthesis of mixed base 4′-thio-oligoribonucleotides, nuclease resistance, and base-pairing properties with complementary single and double strand. Antisense Res Dev 5(3):167–174

Liebhaber SA, Russel JE, Cash FE, Eshlemann SS (1992) Inhibition of mRNA translation by antisense sequences. In: Erickson RP, Izant JG (eds) Gene regulation: biology of antisense RNA and DNA. Raven, New York

Lima WF, Crooke ST (1997) Binding affinity and specificity of Escherichia coli RNase H1: impact on the kinetics of catalysis of antisense oligonucleotide-RNA hybrids. Biochemistry (in press)

Lima WF, Mohan V, Crooke ST (1997) The influence of antisense oligonucleotide-induced RNA structure on E coli RNase H1 activity. J Biol Chem (in press)

Lin KY, Pudlo JS, Jones RJ, Bischofberger N, Matteucci MD, Froehler BC (1994) Oligodeoxynucleotides containing 5(1-propynyl)-2′-deoxyuridine formacetal and thioformacetal dimer synthons. Bioorg Med Chem Lett 4(8):1061–1064

Lin KY, Jones RJ, Matteucci M (1995) Tricyclic 2′-deoxycytidine analogs: syntheses and incorporation into oligodeoxynucleotides which have enhanced binding to complementary RNA. J Am Chem Soc 117:3873–3874

MacLeod AR, Crooke ST (1997) Cleavage of single strand RNA adjacent to RNA-DNA duplex regions by Escherichia coli RNase H1. Science (submitted)

Manoharan M (1993) Designer antisense oligonucleotides: conjugation chemistry and functionality placement. In: Crooke ST, Lebleu B (eds) Antisense research and applications. CRC Press, Boca Raton, FL, pp 303–349

Manoharan M, Tivel KL, Andrade LK, Cook PD (1995a) 2′-O- and 3′-O pyrimidine aminotether-containing oligonucleotides: synthesis and conjugation chemistry. Tetrahedron Lett 36(21):3647–3650

Manoharan M, Tivel KL, Cook PD (1995b) Lipidic nucleic acids. Tetrahedron Lett 36(21):3651–3654

Manoharan M, Tivel KL, Andrade LK, Mohan V, Condon TP, Bennett CF, Cook PD (1995c) Oligonucleotides conjugates: alteration of the pharmacokinetic properties of antisense agents. Nucleosides Nucleotides 14(3–5):969–973

Manoharan M, Ramasamy KS, Mohan V, Cook PD (1996) Oligonucleotides bearing cationic groups: N^2-(3-aminopropyl)deoxyguanosine. Synthesis, enhanced binding properties and conjugation chemistry. Tetrahedron Lett 37(43):7675–7678

Manoharan M, Tivel KL, Condon TP, Mohan V, Graham MJ, Bennett CF, Crooke ST, Cook PD (1997) Synthesis, molecular modeling, structure and vitro and vivo function of ICAM-1 antisense oligonucleotides conjugated to cholesterol and other lipidic molecules. J Med Chem (submitted)

Martin P (1995) A new access to 2′-O-alkylated ribonucleosides and properties of 2′-O-alkylated oligoribonucleotides, Helv Chim Acta 78:486–504

Matteucci M (1990) Deoxyoligonucleotide analogs based on formacetal linkages. Tetrahedron Lett 31(17):2385–2388
Matteucci M (1996) Structural modifications toward improved antisense oligonucleotides. Perspect Drug Discovery Design 4:1–16
Matteucci M, Lin KY, Butcher S, Moulds C (1991) Deoxyoligonucleotides bearing neutral analogues of phosphodiester linkages recognize duplex DNA via triple-helix formation. J Am Chem Soc 113:7767–7768
Matteucci MD, von Krosigk U (1996) Hybridization properties of oligonucleotides bearing a tricyclic 2'-deoxycytidine analog based on a carbazole ring system. Tetrahedron Lett 37(29):5057–5060
Matteucci MD, Wagner RW (1996) In pursuit of antisense. Nature 384 [6604, Suppl]:20–22
McKay RA, Cummins LL, Graham MJ, Lesnik EA, Owens SR, Winniman M, Dean NM (1996) Enhanced activity of an antisense oligonucleotide targeting murine protein kinase C-α by the incorporation of 2'-O-propyl modifications. Nucleic Acids Res 24(3):411–417
Meyer RB Jr (1994) Incorporation of modified bases into oligonucleotides. In: Agrawal S (ed) Protocols for oligonucleotide conjugates. Humana Press, Totowa, NJ, Chap. 2 (Methods in molecular biology, vol 26)
Miller PS (1996) Antisense/antigene oligonucleotides. In: Hecht SM (ed) Bioorganic chemistry: nucleic acids. Oxford University Press, New York, pp 347–374
Miller PS, Bhan P, Cushman CD, Kean JM, Levis JT (1991) Antisense oligonucleoside methylphosphonates and their derivatives. Nucleosides Nucleotides 10(1–3):37–46
Milligan JF, Matteucci MD, Martin JC (1993) Current concepts in antisense drug design. J Med Chem 36(14):1923–1937
Monia BP, Lesnik EA, Gonzalez C, Lima WF, McGee D, Guinosso, CJ, Kawasaki AM, Cook PD, Freier SM (1993) Evaluation of 2'-modified oligonucleotides containing 2'-deoxy gaps as antisense inhibitors of gene expression. J Biol Chem 268 (19):14514–14522
Monia BP, Johnston JF, Sasmor H, Cummins LL (1996a) Nuclease resistance and antisense activity of modified oligonucleotides targeted to Ha-ras. J Biol Chem 271(24):14533–14540
Monia BP, Sasmor H, Johnston JF, Freier SM, Lesnik EA, Muller M, Geiger T, Altmann KH, Moser H, Fabbro D (1996b) Sequence-specific antitumor activity of a phosphorothioate oligodeoxyribonucleotide targeted to human C-raf kinase supports an antisense mechanism of action vivo. Proc Natl Acad Sci USA 93(26):15481–15484
Morvan F, Porumb H, Degols G, Lefebvre I, Pompon A, Sproat S, Rayner B, Malvy C, Lebleu B, Imbach JL (1993) Comparative evaluation of seven oligonucleotide analogues as potential antisense agents. J Med Chem 36:280–287
Moser HE (1993) Strategies and chemical approaches towards oligonucleotide therapeutics. In: Testa B, Fuhrer W, Kyburz E, Giger R (eds) Perspectives in medicinal chemistry. Helvetica Chimica Acta, Basel, pp 275–297
Nakamura H, Oda Y, Iwai S, Inoue H, Ohtsuka E, Kanaya S, Kimura S, Katsuda C, Katayanagi K, Morikawa K, Miyashiro H, Ikehara M (1991) How does RNase H recognize a DNA-RNA hybrid? Proc Natl Acad Sci USA 88:11535
Nashikura K (1992) A cellular activity that modifies and alters the structure of double-stranded RNA. In: Erickson RP, Izant JG (eds) Gene regulation: biology of antisense RNA and DNA. Raven, New York
Nielsen PE, Engholm M, Berg RH, Buchardt O (1991) Sequence-selective recognition of DNA by strand displacement with a thymine-substituted polyamide. Science 254:1497–1500
Partridge M, Vincent A, Matthews P, Puma J, Stein D, Summerton J (1996) A simple method for delivering morpholino antisense oligos into the cytoplasm of cells. Antisense Nucleic Acid Drug Dev 67:169–175

Peoc'h D, Swayze EE, Bhat B, Dimock S, Griffey R, Sanghvi YS (1997) Synthesis and Evaluation of 2'-modified MMI linked dimers in antisense constructs. Nucleosides Nucleotides (in press)

Perbost M, Lucas M, Chavis C, Pompon A, Baumgartner H, Rayner B, Griengl H, Imbach JL (1989) Sugar modified oligonucleotides I. Carbo-oligodeoxynucleotides as potential antisense agents. Biochem Biophys Res Commun 165(2):742–747

Ramasamy KS, Bakir F, Baker B, Cook PD (1993) Synthesis of 2'-deoxyguanosine containing a N2-polyamine. J Heter Chem 30:1373–1377

Ramasamy KS, Zounes, M, Gonzalez C, Freier SM, Lesnik EA, Cummins LL, Griffey RH, Monia BP, Cook PD (1994) Remarkable enhancement of binding affinity of Heterocycle-modified DNA to DNA and RNA. Synthesis, characterization and biophysical evaluation of N2-imidazolylpropylguanine and N2-imidazolylpropyl-2-aminoadenine modified oligonucleotides. Tetrahedron Lett 35:215–218

Reynolds MA, Hogrefe RI, Jaeger JA, Schwartz DA, Riley TA, Marvin WB, Daily WJ, Vaghefi MM, Beck TA, Knowles SK, Klem RE, Arnold L Jr (1996) Synthesis and thermodynamics of oligonucleotides containing chirally pure Rp methylphosphonate linkages. Nucleic Acids Res 24(22):4584–4591

Ross BS, Springer RH, Vasquez G, Andrews RS, Cook PD, Acevedo OL (1994) General preparative synthesis of 2'-O-methylpyrimidine ribonucleosides. J Hetercycl Chem 34(4):765–769

Sands H, Gorey-Feret LJ, Ho SP, Bao Y, Cocuzza AJ, Chidester D, Hobbs FW (1995) Biodistribution and metabolism of internally ^{3}H-labeled oligonucleotides. II. 3', 5'-blocked oligonucleotides. Am Soc Pharm Exp Ther 47:636–646

Sanghvi Y (1993) Heterocyclic base modifications in nucleic acids and their applications in antisense oligonucleotides. In: Crooke ST, Lebleu B (eds) Antisense research and applications. CRC Press, Boca Raton, FL, pp 273–288

Sanghvi YS, Cook PD (1993) Towards second-generation synthetic backbones for antisense oligonucleosides. In: Chu CK, Baker D (eds) Nucleosides and nucleotides as antitumor and antiviral agents. Plenum, New York, pp 311–324

Sanghvi Y, Cook PD (1994) Carbohydrates: synthetic methods and applications in antisense therapeutics: an overview. In: Sangvhi YS, Cook PD (eds) Carbohydrate modifications in antisense research. American Chemical Society, Washington DC, pp 1–22 (ACS symposium series no 580)

Sanghvi YS, Hoke GD, Zounes MC, Freier SM, Martin JF, Chan H, Acevedo OL, Ecker DJ, Mirabelli CK, Crooke ST, Cook PD (1991) Synthesis and biological evaluation of antisense oligonucleotides containing modified pyrimidines. Nucleosides Nucleotides 10:345

Sanghvi YS, Hoke GD, Freier SM, Zounes MC, Gonzalez C, Cummins L, Sasmor H, Cook PD (1993) Antisense oligodeoxynucleotides: synthesis, biophysical and biological evaluation of oligodeoxynucleotides containing modified pyrimidines. Nucleic Acids Res 21:3197–3203

Sanghvi YS (1997) DNA with altered backbone in antisense applications. In: Barton DHR, Nakanishi K (eds) DNA and aspects of molecular biology. Pergamon, New York (Comprehensive natural products chemistry, vol 7)

Sanghvi YS, Swayze EE, Peoc'h D, Bhat B, Dimock S (1997) Concept, discovery and development of MMI linkages: story of a novel linkages for antisense constructs. Nucleosides Nucleotides (in press)

Schmid N, Behr JP (1995) Recognition of DNA sequences by strand replacement with polyamino-oligonucleotides. Tetrahedron Lett 36(9):1447–1450

Schmit C, Bévierre MO, De Mesmaeker A, Altmann KH (1994) The effects of 2'- and 3'-alkyl substituents on oligonucleotide hybridization and stability. Bioorg Med Chem Lett 4(16):1969–1974

Seela F, Thomas H (1995) Duplex stabilization of DNA: oligonucleotides containing 7-substituted 7-deazaadenines. Helv Chim Acta 78:94

Seela F, Ramzaeva N, Chen Y (1995) Oligonucleotide duplex stability controlled by the 7-substituents of 7-deazaguanine bases. Biorg Med Chem Lett 5(24):3049–3052

Seela F, Ramzaeva N, Zulauf M (1997) Duplex stability of oligonucleotides containing 7-substituted 7-deaza- and 8-aza-7-deazapurine nucleosides. Nucleosides Nucleotides (in press)

Sheffery M, Gordon CL (1996) Leadership positions in antisense patents. Company report. Mehta and Isaly Equity Research, NY

Sproat BS, Lamond AI, Beijer B, Neuner P, Ryder U (1989) Highly efficient chemical synthesis of 2′-O-methyloligoribonucleotides and terabiotinylated derivatives: novel probes that are resistant to degradation by RNA or DNA specific nucleases. Nucleic Acids Res 17:3373

Sproat BS, Lamond A I, Beijer B, Neuner P, Ryder U (1993) 2′-O-Alkyloligoribonucleotides. In: Crooke ST, Lebleu B (eds) Antisense research and applications. CRC Press, Boca Raton, FL, Chap. 18

Stein CA, Cohen JS (1988) Oligonucleotides as inhibitors of gene expression: a review. Cancer Res 48:2659–2668

Summerton JL (1992) Cost-effective antisense structures. Biotechnology International. Century, London, pp 73–77

Summerton JL, Weller D (1997) Antisense properties of morpholino oligomers. Nucleosides Nucleotides (in press)

Teng K, Cook PD (1994) Nucleic acid mimics. Synthesis of ethylene glycol- and propoxy-linked thymidyl-tetrahydrofuranylthymine dimers via a Vorbrüggen-type glycosylation reaction. J Organ Chem 59:278–280

Ts'o POP, Miller PS, Aurelian L, Murakami A, Agris C, Blake KR, Lin AB, Lee BL, Smith CC (1987) An approach to chemotherapy based on base sequence information and nucleic acid chemistry. Matagen (masking tape for gene expression) USA. Ann NY Acad Sci 507:220–241

Uhlmann E, Peyman, A (1990) Antisense oligonucleotides: a new therapeutic principle. Chem Rev 90(4):543–584

Varma RS (1993) Synthesis of oligonucleotide analogues with modified backbones. Synlett 9:621–637

Wagner RW, Matteucci MD, Lewis JG, Gutierrez AJ, Moulds C, Froehler BC (1993) Antisense gene inhibition by oligonucleotides containing C-5 propyne pyrimidines. Science 260:1510–1513

Wagner RW, Matteucci MD, Grant D, Huang T, Froehler BC (1996) Potent and selective inhibition of gene expression by an antisense heptanucleotide. Nature Biotechnol 14:840–844

Wallace TL, Bazemore SA, Kornbrust DJ, Cossum PA (1996) I. Single-dose hemodynamic toxicity and pharmacokinetics a partial phosphorothioate anti-HIV oligonucleotide (AR177) after intravenous infusion to cynomolgus monkeys. J Pharmacol Exp Ther 278:1306–1312

Wang G, Bergstrom DE (1993a) Synthesis of oligonucleotides containing N^2-(5-carboxypentyl)-2′-deoxyguanosine and 5-[2(4′-methyl-2, 2′-dipyrid-4-yl-carboxamido)ethylthio]-2′-deoxyuridine. Tetrahedron Lett 34(42):6721–6724

Wang G, Bergstrom DE (1993b) Synthesis of oligonucleotides containing N^2-[2-(imidazol-4-ylacetamido)ethyl]-2′-deoxyguanosine. Tetrahedron Lett 34(42):6725–6728

Woo J, Meyer RB Jr, Gamper HB Jr (1996) G/C-modified oligodeoxynucleotides with selective complementarity: synthesis and hybridization properties. Nucleic Acids Res 24(13):2470–2475

Yacyshvn B, Woloschuk B, Yacyshyn MB, Martini D, Tami J, Bennett F, Kisner D, Shanahan W (1997) Efficacy and safety of ISIS 2302 (ICAM-1 antisense oligonucleotide) treatment of steroid-dependent Crohn's Disease. N Engl J Med (submitted)

Yu D, Iyer RP, Shaw DR, Lisziewicz J, Li Y, Jiang Z, Roskey A, Agrawal S (1996) Hybrid oligonucleotides: synthesis, biophysical properties, stability studies, and biological activity. Bioorg Med Chem 4(10):1685–1692

Zhang R, Lu Z, Zhao H, Zhang X, Diasio RB, Habus I, Jiang Z, Iyer RP, Yu D, Agrawal S (1995) In vivo stability, disposition and metabolism of a "hybrid" oligonucleotide phosphorothioate in rats. Biochem Pharmacol 50(4):545–556

Zhang R, Iyer RP, Yu D, Tan W, Zhang X, Lu Z, Zhao H, Agrwal S (1996) Pharmacokinetics and tissue disposition of a chimeric oligodeoxynucleoside phosphorothioate in rats after intravenous administration. J Pharmacol Exp Ther 278(2):971–979

Zon G (1988) Oligonucleotide analogues as potential chemotherapeutic agents. Pharm Res 5(9):539–549

CHAPTER 3
In Vitro Cellular Uptake, Distribution, and Metabolism of Oligonucleotides

R.M. CROOKE

A. Introduction

Eighteen years ago, Zamecnik and Stephenson used synthetic antisense oligonucleotides to inhibit Rous sarcoma virus replication and RNA translation in a cellular system (ZAMECNIK and STEPHENSON 1978). Since that time, enormous progress has been made towards the development of antisense oligonucleotides as therapeutic agents against a wide variety of host and viral disease targets (CROOKE and LEBLEU 1993; ST CROOKE 1995a; CROOKE and BENNETT 1996). Like other pharmacological agents, these novel compounds have both specific and nonspecific effects (STEIN and CHENG 1993; CROOKE and BENNETT 1996; ST CROOKE 1996). However, constantly emerging data from properly performed and appropriately controlled in vitro and in vivo experiments, as well as preliminary clinical data, suggest that antisense therapeutics do work via an antisense mechanism, i.e., by altering intermediary RNA metabolism and ultimately decreasing production of disease-associated gene products (CROOKE and BENNETT 1996; CROOKE 1996).

Even though these compounds represent a novel paradigm in drug discovery and development, they are nonetheless, as mentioned above, pharmacological agents and, as such, certain criteria must be met for these agents to be effective in experimental cell-based systems and the clinical setting (RM CROOKE 1991, 1993a). First, these compounds must be sufficiently stable in vivo and in vitro in extra- and intracellular environments in order to traverse negatively charged cellular membranes, reach their specific intracellular nucleic acid targets, and achieve effective therapeutic concentrations. Second, antisense oligonucleotides must selectively hybridize to their specific RNA target sequences with significant affinity in order to be efficacious. The non-specific effects produced by antisense oligonucleotides, which are usually seen well above in vitro therapeutic concentrations, result from oligomer interactions with intracellular organelle, protein, nucleic acid, and lipid binding sites (RM CROOKE 1993b; STEIN and CHENG 1993; WAGNER 1994).

In this chapter, I would like to discuss the in vitro pharmacokinetics of the first-generation antisense oligonucleotides, i.e., phosphodiesters, methylphosphonates, and phosphorothioates, by breaking down the uptake process into specific steps. Specifically, I will describe the physicochemical properties of the three oligomer types, their stability in the extracellular milieu, and the uptake, subcellular distribution, efflux, and intracellular stability of these compounds in a variety of cells. I will also include a short section briefly describing some methods to enhance in vitro cellular uptake and, finally, will summarize what is known about oligonucleotide in vitro uptake from the body of literature and our laboratories.

B. Stability in the Extracellular Milieu

The physical and chemical characteristics of antisense oligonucleotides, as well as any other drug, will influence cellular uptake. Some of the characteristics that affect pharmacokinetics include molecular size and shape, lipid solubility, and degree of ionization (BENET and SHEINER 1985; BRADLEY et al. 1992). Oligonucleotides are fairly flexible, rodlike molecules ranging in size from 15 to 30 nucleotides possessing molecular weights of 4500–10 000 Da (RM CROOKE 1991, 1993a). Phosphodiesters, which resemble natural nucleic acids, and phosphorothioate oligomers, which are formed by a sulfur for oxygen atom substitution at the phosphorous, are negatively charged and hydrophilic (GOODCHILD 1990; ZON 1989; COHEN 1993) and, in the biological milieu, these anionic charges are probably masked by counterions and other types of molecules, such as polyamines (TABOR and TABOR 1984). Methylphosphonates, which are synthesized by attaching a methyl group to the phosphorous of the internucleotide bond, are uncharged and relatively lipophilic (ST CROOKE 1995b; MILLER et al. 1993). It should be noted that the pharmacokinetic parameters and cellular interactions of chemically modified and conjugated oligonucleotides will vary as a function of the introduced modifications (COOK 1993; NECKERS 1993).

I. Phosphodiester Oligonucleotides

During the past 10 years, much has been learned about the stability of unmodified phosphodiester antisense oligonucleotides in various experimental systems (WICKSTROM 1986; EDER et al. 1991; COOK 1993; ST CROOKE 1992; RM 1993a). These compounds are extremely sensitive to nucleases, the same enzymes that degrade DNA and RNA in vitro and in vivo, and it is this instability that makes their use as therapeutic agents problematic. Calf serum, which is used by many groups to supplement tissue culture media, contains predominantly 3′ exonucleases (RM CROOKE 1993a; HOKE et al. 1991; TIDD 1990), although there has been one report by TIDD and WARENIUS (1989) suggesting that fetal calf serum (FCS) also possesses some endonucleolytic activity. In general, the stability of phosphodiesters varies as a function of the length and sequence of the compounds and also depends on the type of sera evaluated (human vs. calf vs. rabbit), with half-lives ($t_{1/2}$) ranging from 15 to 60 min (WICKSTROM 1986; HOKE et al. 1991). Even though heating serum to 55°–65°C for 30–60 min is thought to inactivate many nucleases, thereby making the serum suitable for use in experimental cell systems, this is not always the case due to lot-to-lot variations in sera and enzyme activities.

The stability of phosphodiesters in serum has been shown to be enhanced by various modifications to the oligomers. For example, SHAW et al. (1991), using 5′ or 3′ end-capped oligonucleotides with phosphoroamidate linkages, found that the 3′-modified phosphodiester remained stable in serum for up to 7 days, while the stability of 5′-phosphoramidate remained unchanged. HAWLEY and GIBSON (1992) also reported enhanced serum stability of phosphodiester–phosphorothioate oligonucleotides that were modified with 2,3′-phosphorothioate internucleoside linkages. Another strategy for improving the stability of compounds involved using α-anomeric phosphodiesters, which are generally more stable than β- or natural nucleosides (CHAIX et al. 1993). GOTTIKH et al. (1994) showed that this was the case by demonstrating that $\alpha\beta$ chimeric oligomers were more resistant to nucleolytic hydrolysis in calf serum than unmodified compounds.

The structure of oligonucleotides also influences resistance to degradation. BISHOP et al. (1996) established that guanine-rich phosphodiesters that form quartet structures have a much longer $t_{1/2}$ (up to 4 days) than nonstructured control oligomers. Although the stability of hairpin or double-stranded dumbbell phosphodiesters was not tested in serum (CLUSEL et al. 1993; AGUILAR et al. 1996), one would predict that these structures would also confer additional nuclease resistance.

Physically protecting phosphodiesters with cationic lipids or liposomes has also been reported to increase stability of these compounds against serum nucleases. THIERRY and DRITSCHILO (1992) showed that phosphodiesters encapsulated in minimum volume entrapment (MVP) liposomes were stable in tissue culture medium containing 10% heat-inactivated FCS for more than 5 days, whereas the unprotected compounds were approximately 87% degraded within 1 h. CAPACCIOLI et al. (1993) also reported partial protection against serum nucleases after phosphodiesters were incubated with the cationic lipids DOTMA (N-[1-(2,3-dioleyloxy)-propyl]-N,N,N-trimethylammonium chloride) and DOTAP (N-[N-[1-(2,3,-dioleoyloxy)-propyl]-N,N,N-trimethylammonium methyl sulfate).

II. Methylphosphonate Oligonucleotides

Methylphosphonate oligonucleotides have been reported to be extremely stable against the nucleases found in FCS and in tissue culture media containing serum (MILLER et al. 1993; ST CROOKE 1995b; AKHTAR et al. 1992). MILLER et al. (1993) reported that intact methylphosphonates had been recovered after a 3- to 4-day incubation period with 10% FCS. TIDD and WARENIUS (1989) also demonstrated that the serum stability of phosphodiester oligomers (15- to 20-mers) with two or three methylphosphonate linkages at the 3′ terminus can be significantly enhanced.

III. Phosphorothioate Oligonucleotides

Phosphorothioate oligonucleotides have been thought to be the most promising of the first-generation antisense compounds for development as therapeutic agents, in part due to their stability against nucleases in a variety of experimental systems using media, sera, urine, and cerebrospinal fluid (CAMPBELL et al. 1990; RM CROOKE 1993a; RM CROOKE et al. 1995; COHEN 1993; HOKE et al. 1991). The $t_{1/2}$ of many phosphorothioates in biological fluids has been reported to be longer than 24h (ST CROOKE 1992; RM CROOKE et al. 1995). Several factors contribute to the enhanced stability of phosphorothioates against enzymatic degradation in vitro. A well-documented physicochemical characteristic of these compounds that influences the in vitro stability and, ultimately, their cellular uptake is their ability to nonspecifically adsorb and interact with various proteins. This "stickiness" of phosphorothioates is a consequence of the negative charge delocalization at the internucleotide thioate bond (RM CROOKE et al. 1995; GAO et al. 1993; WOOLF et al. 1990). As a result, these compounds are competitive inhibitors of nucleases, the enzymes responsible for their degradation and metabolism, as well as other purified and intracellular enzymes (RM CROOKE et al. 1995; GAO et al. 1992; BROWN et al. 1994).

Phosphorothioates have also been shown to bind to various anionic binding proteins found in serum and plasma, including bovine, rat, and human serum albumin and α_2-macroglobulin (LEEDS et al. 1994; KUMAR et al. 1995; GESELOWITZ and NECKERS 1992). These interactions in in vitro systems and in the in vivo setting will obviously have important consequences on the pharmacokinetics and ultimately the pharmacodynamics of these new therapeutic agents (KUMAR et al. 1995; ST CROOKE 1992).

Finally, the stereochemistry of these compounds has also been shown to influence nuclease resistance. Phosphorothioates, unlike phosphodiesters, are chiral and commonly synthesized as a mixture of R_p and S_p isomers (ECKSTEIN 1985; COHEN 1993; ST CROOKE 1992). Several groups have shown that the R_p diastereomer is digested more rapidly by nucleases than the S_p diastereomer and that various purified nucleases (S1 and P1 nucleases, snake venom phosphodiesterases) display differing sensitivities to the two configurations (ECKSTEIN 1985; COHEN 1993; STEIN et al. 1988a).

C. In Vitro Uptake, Intracellular Distribution, and Efflux

I. Phosphodiester Oligonucleotides

1. Uptake

Even though phosphodiester antisense oligonucleotides are extremely sensitive to nucleolytic degradation in tissue culture medium containing serum and within cells, many groups over the past 10 years have examined the pharmacokinetics and efficacy of these compounds against numerous targets. Generally, the uptake of phosphodiesters has been shown to be dependent on time, temperature, cell line, concentration, and energy. Several groups, as exemplified by WU-PONG et al. (1992, 1994), ZAMECNIK et al. (1994), and NAKAI et al. (1996), have examined uptake in serum-free media because of the lability of the compounds. The actual amount of phosphodiesters internalized varied from group to group and has been reported to range from 0.1% to 11% (HAWLEY and GIBSON 1992; RM CROOKE 1993a; ZAMECNIK et al. 1994).

Cell association of phosphodiesters has been reported to be length dependent in HL60 cells (STEIN et al. 1988b; LOKE et al. 1989). STEIN et al. (1988b), using 5'-acridine-linked phosphodiester homopolymers, demonstrated that shorter-length conjugates $d(T)_7$ were taken up into cells more rapidly than longer-length oligomers $d(T)_{15}$ or $d(T)_{20}$. LOKE et al. (1989), using 3'-acridine-labeled deoxythymidine homopolymers ranging from three to 20 nucleotides in length, corroborated Stein's data. It should be noted that, in both of these studies, the integrity or stability of the acridine-conjugated phosphodiesters was never demonstrated, so it may be possible that some of the observed uptake could represent degradates or acridine itself. More recently, NAKAI et al. (1996) examined the length dependence of phosphodiester uptake using ^{32}P 5' end-labeled 8-, 15-, 20-, and 30-mer heterosequences and also noted that the rate of uptake of smaller-length oligomers was faster than that seen for the large compounds. This group did, however, demonstrate intracellular degradation of their compounds over their experimental time course.

The uptake of phosphodiester and phosphorothioate congeners has also been compared. STEIN et al. (1988b), in HL60 cells, and NAKAI et al. (1996), in HCT-15 cells, a human colorectal adenocarcinoma cell line, suggested that phosphorothioates associated with cells more slowly than their phosphodiester analogues. In our laboratories (Fig. 1), we demonstrated that the uptake of an internally ^{32}P-labeled 20-mer phosphodiester (ISIS 1047) in Dulbecco's minimum essential medium (DMEM) containing 10% heat-inactivated FCS was similar to its identically labeled phosphorothioate (ISIS 1080) congener in HeLa cells. Analysis of the compounds by gel electrophoresis over the course of the experiment showed that ISIS 1047 was rapidly degraded within the cells, while ISIS 1080 remained relatively stable (RM CROOKE 1991). The reported differences in uptake between our group and others probably represent different assay conditions, including cells, media, lots of serum, sequences of oligomers, and methods of labeling the compounds.

The mechanism of phosphodiester uptake has not been conclusively established. Several laboratories reported that uptake occurred via a receptor-mediated endocytotic process (STEIN et al. 1988b, 1993b; LOKE et al. 1989; YAKUBOV et al. 1989; NAKAI et al. 1996). LOKE et al. (1989), using acridine-linked homothymidine oligomers in HL60 cells, demonstrated that uptake was dependent on time, temperature, and length and was saturable and specific. This cell association was shown to be mediated by an 80-kDa cell surface protein that was affinity purified with oligo-dT cellulose beads. NAKAI et al.

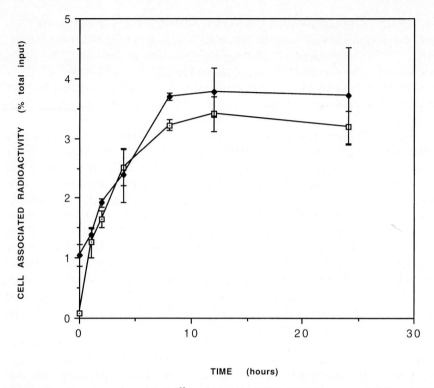

Fig. 1. Kinetics of uptake of $5\,\mu M$ ^{32}P-end-labeled ISIS 1047 (*open symbols*) and ISIS 1080 (*closed symbols*) in HeLa cells. Values at each time point represent means and standard deviations from three replicates

(1996), using many known inhibitors of uptake and metabolism, also suggested that uptake of a 20-mer phosphodiester in HCT-15 cells was mediated by endocytosis with plasma membrane binding sites. GOODARZI et al. (1991) additionally identified, under acidic pH conditions, a 34-kDa membrane protein in the membranes of human T lymphoblasts (HUT78 cells) and HepG2 cells that specifically bound ^{32}P 5' end-labeled 12- to 55-mer phosphodiesters.

YAKUBOV et al. (1989), using dT homopolymers derivatized to a reactive alkylating group or radiolabeled with ^{32}P, identified two cell surface proteins, 79 and 90 kDa, in L929 and Krebs ascite cells that were thought to be involved in an endocytotic process. In L929 cells, the number of oligonucleotide receptors was calculated to be approximately 1.2×10^5 per cell. However, the same group also suggested that internalization was very complex, consisting of both receptor-mediated endocytosis and pinocytosis. More recent work by STEIN et al. (1993b) also suggests that the internalization of fluoresceinated phosphodiester homooligomers (dT) in HL60 cells occurred in a manner consistent with pinocytosis, or fluid-phase endocytosis, and that this internalization was dependent upon protein kinase C activity.

HAWLEY and GIBSON (1992), using 5' ^{32}P-labeled 3'-phosphorothioate end-capped phosphodiester 16-mers, detected several proteins that bound oligomers. At 4°C, oligomers bound primarily to a 46-kDa protein. This binding decreased in the presence of excess unlabeled oligomer and heparin, but not ATP. At 37°C, the binding pattern was altered, with oligonucleotides associating with cytoplasmic and membrane proteins, the most prominent being 30 kDa.

Specific phosphodiester membrane binding proteins have also been identified. GESELOWITZ and NECKERS (1992, 1995), using a photoreactive radiolabeled crosslinker, have identified a 75-kDa protein in HL60 cells that binds to phosphodiesters, and presumably other anionic compounds, to be bovine serum albumin (BSA). They suggested that BSA, which can block overall cell association, was the main cell surface protein that binds oligonucleotides.

ZAMECNIK et al. (1994) studied the uptake of phosphodiesters labeled with ^{3}H thymidine in mouse 3T3, CEF, and HeLa cells. They determined that cellular internalization was linear and rapid and could be inhibited by dinitrophenol. Based on electron micrographic evidence, they suggested that uptake was mediated by potocytosis, a process where small molecules and other compounds are transported across cellular membranes in membrane invaginations called caveolae (ANDERSON 1993).

Finally, WU-PONG et al. (1994) studied the uptake of 21-mer phosphodiesters in Rauscher Red 5-1.5 erythroleukemia cells, a cell line that displays minimal pinocytotic activity. Experiments showed that internalization was partially dependent upon cellular energy and a trypsin-sensitive component. Because several inhibitors of endocytosis (chloroquine, monensin, phenylarsine oxide) were ineffective in inhibiting uptake and because the process was nonspecifically inhibited by a number of charged molecules, including salmon sperm DNA, ATP, a variety of random oligomers, and dextran sulfate, it was suggested that endocytosis was not the primary mechanism of internalization. These investigators suggested that uptake in these cells was very complex and that the mechanism of phosphodiester cellular internalization may depend on the cell line.

2. Intracellular Distribution

Once internalized, phosphodiester antisense oligonucleotides are distributed within various compartments in numerous cell types. Many of the earlier studies performed by LOKE et al. (1989) and YAKUBOV et al. (1989) using radiolabeled or fluorescently labeled compounds demonstrated faint distribution within most cellular compartments, including the nucleus, but with a tendency for the majority of labeled compound to accumulate in discrete, cytoplasmic vesicles. NAKAI et al. (1996), in HCT-15 cells and using confocal microscopy, also determined that fluorescein isothiocyanate (FITC)- and tetramethylrhodamine-5-isothiocyanate (TRITC) -labeled phosphodiesters were distributed primarily in intracellular vesicles. This localization of oligonucleotides in vesicles or endosomes suggests an endocytotic mechanism of cellular uptake.

Other laboratories have demonstrated alternative intracellular distribution patterns. For example, Wu-Pong et al. (1994) found diffuse cytoplasmic and nuclear fluorescence of internally labeled FITC and biotin-streptavidin-FITC oligomers (0.1 μM) in Rauscher Red 5-1.5 cells after a 45-min incubation period, suggesting that the phosphodiesters were not sequestered in vesicles. Thierry and Dritschilo (1992) observed weak cytoplasmic fluorescence in MOLT-3 cells after a 24-h incubation period with $2\mu M$ of their FITC-labeled phosphodiester. They also noted that their compound was extensively degraded over the experimental time course.

Noonberg and associates (1993) used confocal microscopy to compare the intracellular distribution of FITC-labeled phosphodiesters in primary, human keratinocytes and various established cell lines, including HeLa, MCF-7, and HaCat cells, an immortalized keratinocyte cell line. They determined that primary human keratinocytes can internalize oligonucleotides without accumulation in vesicles or endosomes and that distribution was primarily nuclear. In contrast, FITC phosphodiesters were concentrated in discrete, punctate vesicles in the immortalized cell lines. The data were confirmed using fluorescein and Bodipy, two different fluorophores in live and fixed cell populations.

The data presented by Zamecnik and colleagues (1994) are consistent with those of Noonberg et al. (1993). Using electron micrography autoradiography, they determined in 3T3, CEF, and Hela cells that ^{3}H thymidine-labeled phosphodiesters distributed throughout the cytosol and then into the nucleus of cells within 5 min of incubation in serum-free media. Within 60 min, silver grains were associated with euchromatin, not nucleoli. These data are suggestive of a potocytotic, and not an endocytotic mechanism of oligomer uptake.

3. Efflux

One would expect that cell association of antisense oligonucleotides at equilibirium, like other drugs, would consist of uptake and efflux components. This has been shown to be the case for phosphodiester oligomers (Stein et al. 1993b; Tonkinson and Stein 1994; Aguilar et al. 1996). Stein and colleagues studied in great detail the compartmentalization, trafficking, and efflux of 5′-fluorescein-labeled phosphodiesters and phosphorothioates in HL60 cells using flow cytometry. Both classes of oligonucleotides, which they previously showed to be localized in intracellular vesicles, were effluxed from two different compartments. There appeared to be differences in rates depending upon the backbone, i.e., phosphodiesters effluxed more rapidly from cells than their phosphorothioate congeners, with a $t_{1/2}$ of approximately 5 min. The authors also noted that the internalized phosphodiester was extensively degraded.

AGUILAR et al. (1996) studied efflux of single-stranded and structured phosphodiesters in Jurkat cells, a T-lymphoblastic cell line. Regardless of structure, all the phosphodiesters were shown to be effluxed rapidly from these cells, with a $t_{1/2}$ of 10–15 min. These data are consistent with those obtained by TONKINSON and STEIN (1994). AGUILAR et al. (1996) also followed the kinetics of exocytosis by examining, by confocal microscopy, the subcellular efflux distribution of FITC-labeled compounds. The FITC phosphodiesters distributed within endocytic vesicles, but also displayed both perinuclear and nuclear fluorescence. During the 2-h efflux process, which was independent of oligomer stability, they determined that the oligomer was removed from all three compartments over time.

II. Methylphosphonate Oligonucleotides

1. Uptake

As described in previous sections, methylphosphonate oligonucleotides were designed to improve in vitro and, ultimately, in vivo pharmacokinetics (MILLER et al. 1993; ST CROOKE 1995a; AKHTAR et al. 1992). The majority of the methylphosphonate oligomers hybridize to target DNA and RNA as effectively as their charged oligodeoxynucleotide congeners (MILLER et al. 1993). Even though these compounds are nonionic, they are soluble in aqueous solutions, including tissue culture media (MILLER et al. 1993; ST CROOKE 1995b; AKHTAR et al. 1992).

Like the other first-generation antisense compounds, methylphosphonates of varying lengths have been shown to be taken up into a variety of cells in culture. SPILLER and TIDD (1992), MILLER et al. (1981), MARCUS-SEKURA et al. (1987), VASANTHAKUMAR and AHMED (1989), SHOJI et al. (1991), and LEVIS et al. (1995), using radiolabeled and fluorescently labeled compounds, determined that the cellular uptake process was linear over time. LEVIS et al. (1995) suggested that uptake was dependent upon the cell type being studied. For example, it was determined that a ^{3}H T_8 methylphosphonate was taken up into HL60 cells, but not isolated human erythrocytes, over a 48-h experimental time course.

Since methylphosphonates are nonionic and lipophilic, it was thought for many years that the mechanism of uptake of these compounds was passive diffusion. However, several lines of evidence from Juliano's laboratories suggest that the uptake mechanism is much more complex (AKHTAR et al. 1991, 1992; SHOJI et al. 1991). Liposomes or phospholipid membranes were used as model systems to examine the mechanism by which ^{32}P- or fluorescently labeled phosphodiesters, phosphorothioates, methylphosphonates, and methylphosphonate–phosphodiester chimeras were effluxed across cellular membranes. All of the compounds were transported across the membranes very slowly, with an efflux $t_{1/2}$ generally longer than 4 days. Since biological activity of the methylphosphonates had been usually observed over shorter time periods, it was suggested that simple diffusion was probably not the mechanism of uptake of these compounds.

Shoji et al. (1991) examined the mechanism of uptake in greater detail using ^{32}P- or fluorescently labeled 15-mer methylphosphonates in CH^{R}C5 cells. Several lines of evidence suggested that the mechanism involved fluid-phase or adsorptive endocytosis and not receptor-mediated endocytosis. First, the uptake of their compounds was sensitive to temperature incubation conditions, with much more 15-mer methyphosphonate associating with cells at 37°C than at 4°C. Second, labeled compounds were not easily competed off cells with excess unlabeled material, suggesting that specific cell surface receptors were not involved. Finally, acidification of the CH^{R}C5 cytoplasm, a procedure which blocks receptor-mediated endocytosis, had no effect on uptake.

Levis et al. (1995) studied the uptake of ^{3}H T$_8$ and T$_{16}$ methylphosphonates and a methylphosphonate with a single phosphodiester linkage labeled with ^{32}P and an ethylenediamine conjugate on the 5′ end of the molecule into HL60, K562, mouse L929 fibroblast cells, and human erythrocytes over a 48-h period. They reported that methylphosphonates were not passively diffused across cellular membranes and that uptake was independent of cell surface receptors, results that are consistent with those presented by Shoji et al. (1991). Although their data suggest that the majority of oligomer appeared to enter cells in the fluid phase, fluid-phase pinocytosis or endocytosis was not the sole mechanism of uptake, since sucrose, which was taken up via fluid-phase pinocytosis, was internalized much faster than their labeled methylphosphonates. They suggest that the slower rate of methylphosphonate uptake may result from oligonucleotides nonspecifically adsorbing to cellular membranes.

2. Intracellular Distribution

Once methylphosphonates are internalized, Shoji et al. (1991), using digitized video imaging, demonstrated that fluorescently (TRITC)-labeled compounds in CH^{R}C5 cells showed a punctate cytoplasmic distribution pattern, with some perinuclear staining, suggesting localization in endosomal or lysosomal vesicles. The TRITC-labeled methylphosphonates also colocalized with FITC-labeled dextran, an endosomal marker. The subcellular distribution of fluorescently labeled methylphosphonate–phosphodiester chimeras was studied in MOLT-4 cells by Giles et al. (1993). These studies indicated that the addition of central phosphodiester regions in a methylphosphonate molecule can alter distribution from endosomes to the cytoplasmic and nuclear compartments.

3. Efflux

Levis et al. (1995), once again using the ^{3}H T$_8$ and T$_{16}$ methylphosphonates, found that these compounds, like other drugs, were effluxed from tissue culture cells. Although sucrose and methylphosphonate uptake differed in their system, the efflux rates of the two types of compounds were similar, suggesting that methylphosphonates exited cells via fluid-phase exocytosis (LEVIS et al. 1995).

III. Phosphorothioate Oligonucleotides

1. Uptake

Phosphorothioates, like phosphodiesters and methylphosphonates, are taken up into a variety of cell lines. The uptake of these compounds has been shown to be dependent on time, temperature, cell line, concentration, and energy. The actual amount of phosphorothioate internalized varies, as was seen with the phosphodiesters, but can range from 1% to 20%, obviously depending upon the sequence of the oligomer, the cell types being examined, and experimental conditions (RM CROOKE 1993a; RM CROOKE et al. 1995; JAROSZEWSKI et al. 1990). For example, in our laboratories using radiolabeled ^{3}H or ^{35}S 17- to 21-mers, we have detected between 0.5% and 20% of total radiolabeled phosphorothioate associating with a variety of cell lines, including HeLa, NHDF, C127, and I38 cells, over a 24-h period (RM CROOKE 1995). However, JAROSZEWSKI et al. (1990), using ^{35}S-labeled dC_{14} or dC_{28} in MCF-7 WT and adriamycin-resistant MCF human breast cancer cells, found that only 1.6%–2.2% of the radiolabeled oligomers associated with cells over a 4-day period.

The amount of time it takes for uptake to plateau varies in different experimental systems (MARTI et al. 1992; HO et al. 1991; RM CROOKE et al. 1995; GAO et al. 1990; TEMSAMANI et al. 1994; SMETSERS et al. 1994). MARTI et al. (1992), using a fluorescently labeled 28-mer, demonstrated that the compound reached steady state in H9 and a variety of hematopoietic cells between 0.5 and 2 h, while HO et al. (1991), using ^{32}P 5' end-labeled compounds in HL60 cells, showed that a plateau was achieved after approximately 1 h of incubation. Several groups have determined that the uptake plateau occurred at longer incubation times, i.e., between 12 and 24 h (RM CROOKE et al. 1995; GAO et al. 1990; TEMSAMANI et al. 1994; SMETSERS et al. 1994). For example, SMETSERS et al. (1994), using fluorescein-labeled oligomers in a Philadelphia chromosome-positive cell line (BV173 cells), found a linear and rapid increase in cell accumulation of the phosphorothioate and a plateau between 18 and 50 h of incubation.

Phosphorothioate antisense oligonucleotides have been shown to be concentrated within cells, i.e., the intracellular concentration is higher than the input medium (TEMSAMANI et al. 1994; GAO et al. 1990; IVERSEN 1991; RM CROOKE 1993a). TEMSAMANI et al. (1994) found the intracellular concentration of a 20-mer ^{35}S-labeled compound to be $25\,\mu M$, which was 100 times higher than the input, or extracellular concentration of oligomer ($250\,nM$) in the medium. Our laboratories also demonstrated that the phosphorothioates we

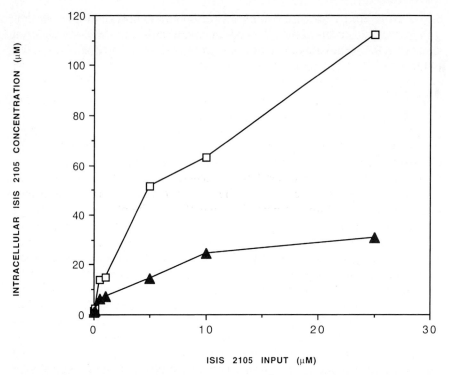

Fig. 2. Concentration-dependent cell association of ^{3}H ISIS 2105. C127 (*open symbols*) and I38 cells (*closed symbols*) were incubated for 24 h at 4°C and 37°C with increasing concentrations of radiolabeled ISIS 2105. Values at each point represent calculated intracellular concentrations from six replicates and were determined by subtracting data obtained at 4°C from those obtained at 37°C

studied are concentrated within cells. The data in Fig. 2 indicate that ^{3}H ISIS 2105, a 20-mer phosphorothioate that inhibits the production of a gene product essential to the growth of human papillomavirus, is concentrated within C127 and I38 cells after a 24-h incubation period with varying amounts of the compound. For example, the intracellular concentration of the oligomer after incubation with 10 μM of the compound in C127 cells was approximately 70 μM, i.e., sevenfold higher than the input concentration, while in I38 cells, the intracellular concentration was only approximately 30 μM, i.e., three times higher than the input amount. This study also underscores the cell line differences in various pharmacokinetic parameters.

Not all groups report concentration of internalized oligonucleotides. MARTI et al. (1992), using fluoresceinated phosphorothioates in a variety of cell lines, demonstrated that, after 1 h of incubation in medium containing 25 μM of compound, the mean intracellular concentration of oligomer was calculated to be only approximately 0.5 μM.

As described in Sect. C.I.1, phosphodiesters and phosphorothioates have been compared in terms of cell association by several groups (STEIN et al. 1988; NAKAI et al. 1996; RM CROOKE 1991; HO et al. 1991). STEIN et al. (1988) and NAKAI et al. (1996) demonstrated that phosphodiesters were taken up into cells much more rapidly than phosphorothioates, while data from our laboratories suggest that there were no differences in uptake between congeners (RM CROOKE 1991). HO et al. (1991) and ZHAO et al. (1993) present a different scenario, with phosphorothioates being taken up much more rapidly and to a greater extent than phosphodiesters. For example, HO et al. (1991) demonstrated that their ^{32}P-radiolabeled phosphorothioate rapidly associated with HL60 cells, with plateau levels being reached within 1 h of incubation. In contrast, the same sequence phosphodiester was taken up very slowly over a 6-h period, with total uptake being less that 10% of the phosphorothioate.

Phosphorothioate sequence-specific differences in cellular uptake have been noted. HUGHES et al. (1994), using 5′ fluorescently tagged 10-mer phosphorothioate homopolymers (A, T C, G) in CH^{R}C5 cells, showed that the poly(G) oligomer associated to the greatest degree with cells and bound with the highest affinity to liposomes, which were used as model membranes. Using four heterosequences (ISIS 1082, ISIS 1570, ISIS 2105, and ISIS 2922) that were tritiated at the C8H position of purines within each molecule (GRAHAM et al. 1993), our laboratories demonstrated that uptake varied as a function of sequence (RM CROOKE et al. 1995). As shown in Fig. 3, association of phosphorothioates of similar length was significantly different in one cell line (NHDF cells), and these differences could not be solely explained by variations in base composition. In our system, ISIS 1570, the oligomer with the greatest number of G's, had the poorest uptake. Experimental conditions, including cell line, media, and differences between hetero- and homooligomer sequences, probably account for the discrepancy between the data obtained by HUGHES et al. (1994) and our own data.

Phosphorothioate length may also affect pharmacokinetics, although few studies comment upon this factor. Our laboratory compared the uptake of two compounds of different length using an ^{35}S-labeled 15-mer (ISIS 1788) and 30-mer (ISIS 1790). As shown in Fig. 4, a greater amount of the longer-length compound associated with HL60 cells than its shorter congener. HO et al. (1991) indirectly addressed this length effect and suggested, based upon inhibitory activity of their phosphorothioate oligomers against the human transferrin receptor, that the 30-mers were more potent than the 17-mers in their studies.

We also determined that factors besides cell line and oligonucleotide sequence can affect phosphorothioate uptake (RM CROOKE et al. 1995). The uptake of ^{3}H ISIS 2922 was compared in NHDF cells using two different media (Fig. 5). The data suggest that threefold more radiolabel associated with cells incubated in DMEM containing 10% FCS than NHDF cells which were maintained in fibroblast growth medium (FGM), a defined medium containing 0.2% FCS.

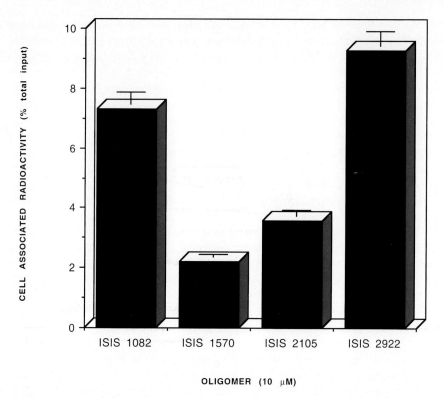

Fig. 3. Sequence-specific differences in cellular uptake of $10 \mu M$ tritiated phosphorothioates in NHDF cells after a 24-h incubation period. Values at each point represent the means and standard error of the means from six to nine replicates and from two to three experiments

2. Intracellular Distribution

Once internalized, phosphorothioates are distributed within various cellular compartments depending, once again, on cell type and experimental conditions. For example, GAO et al. (1990), using ^{35}S dC_{28} in post-herpes simplex virus (HSV)-2-infected cells, determined that 30% of the intracellular oligomer associated with the nuclear and 70% with the cytoplasmic fractions of cells. While the cytoplasmic-associated phosphorothioate remained in the same compartment from 6 to 24h, the amount of nuclear-associated oligomer increased over time. IVERSEN (1991), using ^{35}S and fluorescently labeled phosphorothioates, also determined that the oligonucleotides localized to the nuclear, mitochondrial, and cytosolic compartments of cells, with the majority of the labeled oligomers associating with the membrane/cytosolic fraction, which contained lysosomes.

Our laboratories also examined the intracellular distribution of the four phosphorothioate heterosequences described earlier in NHDF cells, a primary

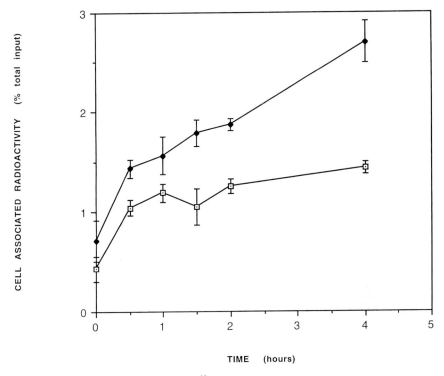

Fig. 4. Length-dependent uptake of ^{35}S radiolabeled oligonucleotide in HL60 cells. Values represent means and standard deviations from three replicates. *Open symbols*, ISIS 1788; *closed symbols*, ISIS 1790

human dermal fibroblast cell line after a 24-h exposure (RM CROOKE et al. 1995). Using cell fractionation and differential centrifugation to separate cells into plasma membrane, nuclear, mitochondrial, microsomal, and cytoplasmic fractions, the overall pattern of distribution in NHDF cells was similar for all of the oligonucleotides, with the majority of the radiolabel associating with the cytoplasmic fraction of cells (Fig. 6). TEMSAMANI et al. (1994), ZHAO et al. (1993), and JAROSZEWSKI et al. (1990) also reported that the majority of their phosphorothioate oligomers resided in the cytoplasmic fraction.

NESTLE et al. (1994), using FITC-labeled ISIS 2302, a 20-mer phosphorothioate, determined that cultured human keratinocytes localized oligomer primarily in the nucleus and nucleoli. ROBINSON et al. (1995) confirmed cellular nuclear localization in established non-small-cell lung cancer cell lines. JANSEN et al. (1995) noted nuclear accumulation of an FITC phosphorothioate in SK-2 cells, but also showed some cytoplasmic/endosomal accumulation. Other groups (BELTINGER et al. 1995; SMETSERS et al. 1994) reported majority localization of phosphorothioates in vesicular structures that are consistent with endosomes.

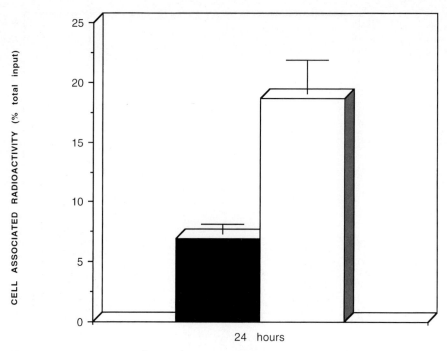

Fig. 5. Cell association of ^{3}H ISIS 2922 in NHDF cells after a 24-h incubation period. Values at each point represent the means and standard error of the means from six to nine replicates and from two to three experiments. *Black bar*, NHDF cells incubated with fibroblast growth medium (FGM); *white bar*, NHDF cells incubated with 10% DMEM

As with the phosphodiesters, the exact mechanism of uptake is still not defined, although many groups believe that internalization occurs via an endocytotic process. BELTINGER et al. (1995) suggested that cell association results from fluid-phase endocytosis when high concentrations of phosphorothioates are incubated with cells (>1 μM), while at lower concentrations (<1 μM) receptor-mediated endocytosis was the predominant mechanism. Cross-linking studies identified five major cell surface phosphorothioate-binding proteins ranging in size from 20 to 143 kDa. Glycosylation of the proteins is also important for binding of oligomers. In K562 cells, the number of binding sites was calculated to be approximately 200,000 sites per cell. The number of sites varied in Jurkat and EL-4 cells, although the receptor affinity stayed the same. STEIN et al. (1991) have also suggested that phosphorothioates and some derivatives bind to the CD4 receptor of T cells, and other cellular surface proteins, such as MAC-1 (STEIN et al. 1993a).

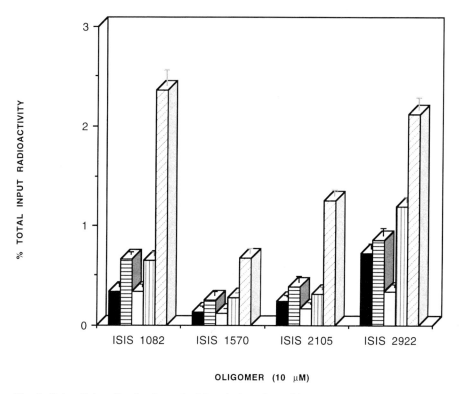

Fig. 6. Subcellular distribution of tritiated phosphorothioates at 37°C in NHDF cells after a 24-h incubation period. *Black bar*, nuclear fraction; *horizontal shading*, membrane fraction; *white bar*, mitochondrial fraction; *vertical shading*, microsomal fraction; *oblique shading*, cytosolic fraction

3. Efflux

Time- and temperature-dependent phosphorothioate efflux from various cell lines has been shown by several groups (RM CROOKE 1993a; MARTI et al. 1992; GAO et al. 1993; TONKINSON and STEIN 1994; TEMSAMANI et al. 1994). TEMSAMANI et al. (1994) and GAO et al. (1993) suggested that the efflux of phosphorothioate oligomers was biphasic, with a rapid efflux in the first phase ($t_{1/2}$, 10–30 min), followed by a much slower phase. Data from our laboratories also confirm these observations. Figure 7 shows the efflux of ^{35}S ISIS 1082 from HeLa cells after a 4-h pulse. Our data demonstrate that efflux of the radiolabeled 21-mer was dependent on both time and temperature, with most of the oligomer being effluxed from cells within 30 min.

MARTI et al. (1992) and TONKINSON and STEIN (1994) compared the efflux of phosphorothioates and phosphodiesters. As described above, TONKINSON and STEIN (1994) demonstrated that phosphodiesters effluxed more rapidly

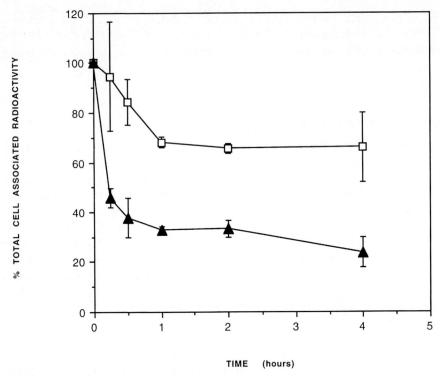

Fig. 7. Efflux of ^{35}S ISIS 1082 from HeLa cells. Cells were incubated with $5\,\mu M$ ISIS 1082 for 4 h. After this incubation period, the media was removed and fresh media without radiolabeled compound was added to cells. Values represent means and standard deviations from three replicates. *Open symbols*, 4°C; *closed symbols*, 37°C

from HL60 cells than their phosphorothioate congeners. MARTI et al. (1992) corroborated these data by showing that the efflux of phosphorothioates was five times slower in H9 cells than the phosphodiester congener. Much more phosphorothioate was also retained within cells than the phosphodiesters (50% vs. 10%, respectively).

D. Intracellular Stability

Nucleases reside in numerous intracellular compartments. These enzymes have various functions in metabolically active cells, including the biosynthesis of nucleic acids, replication, repair, and recombination (KORNBERG 1980). Even though specific nucleases interact only with DNA (deoxyribonucleases) or RNA (ribonucleases), in many cases, depending upon the experimental conditions, some enzymes will use both types of nucleic acids as substrates

(LLOYD and LINN 1993). Exonucleases are enzymes that require a terminus (either 3′ or 5′) for hydrolysis of the phosphodiester bond and that degrade their substrates processively, while endonucleases do not require a terminus (KORNBERG 1980).

I. Phosphodiester Oligonucleotides

As mentioned above in Sect. B, unmodified phosphodiesters are extremely sensitive to nucleolytic degradation in the extracellular medium and within cells (ST CROOKE 1992; COOK 1993). Very early in vitro pharmacokinetic studies did not report on the intracellular integrity of oligonucleotides (LOKE et al. 1989; YAKUBOV et al. 1989). However, most groups now commonly characterize the intracellular fates of their oligomers. Several laboratories (NAKAI et al. 1996; LEROY et al. 1996; CAPACCIOLI et al. 1993; THIERRY and DRITSCHILO 1992; BISHOP et al. 1996), using standard phenol/chloroform extraction and polyacrylamide gel analytical techniques and/or high-performance liquid chromatography (HPLC), readily demonstrate the instability of unmodified phosphodiesters within a variety of cells types, with a $t_{1/2}$ usually less than 1 h.

The typical phosphodiester intracellular degradation pattern results in a ladder on polyacrylamide gels, indicative of processive, exonuclease activity (NAKAI et al. 1996; HO et al. 1991; RM CROOKE et al. 1995). However, there have been reports suggestive of endonucleolytic activity (NAKAI et al. 1996; HO et al. 1991; HOKE et al. 1991; RM CROOKE et al. 1995). For example, experiments by HO et al. (1991) examining the degradation of their 17-mer phosphodiester in HL60 cells showed two distinct bands instead of the usual exonucleolytic processive laddering.

The intracellular stability of phosphodiesters can be altered by modifying the compounds structurally (BISHOP et al. 1996) or chemically (COOK 1993; GOTTIKH et al. 1994; THIERRY and DRITSCHILO 1992) or by encapsulation in various liposomal preparations (CAPACCIOLI et al. 1993; THIERRY and DRITSCHILO 1992). BISHOP et al. (1996), using unmodified structured G-quartet phosphodiesters, demonstrated that their 17-mer phosphodiester had a $t_{1/2}$ in HeLa cells of less than 1 h, whereas the structured compound remained approximately 80% intact after a 24-h incubation period. GOTTIKH et al. (1994) determined that $\alpha\beta$ chimeric phosphodiester oligomers displayed greater nuclease resistance in DUNNI mouse fibroblast cell lysates than the parent compounds and also suggested predominantly endonucleolytic activity. THIERRY and DRITSCHILO (1992) determined that phosphodiesters end-capped with phosphorothioates displayed increased stability in MOLT-3 cells. Finally, CAPACCIOLI et al. (1993) and THIERRY and DRITSCHILO (1992) showed that liposomal encapsulation of phosphodiesters of varying lengths can enhance nuclease resistance in CCRF-CEM/VLB, leukemic T lymphoblasts, and leukemic MOLT-3 cells, respectively.

II. Methylphosphonate Oligonucleotides

Even though methylphosphonate oligomers are thought to be very resistant to nucleases, very few groups have actually analyzed stability against intracellular nucleolytic degradation. SHOJI et al. (1991) incubated CH^RC5 cells with 5' ^{32}P-labeled 15-mer methylphosphonates for 3 h and determined, after analysis by denaturing gel electrophoresis, that the compounds were completely undegraded over the experimental time course. LEVIS et al. (1995) determined that ^3H T_8 and T_{16} methylphosphonates were degraded in tissue culture medium containing 10% FCS over a 2-h period. However, the methylphosphonate with a single phosphodiester linkage labeled with ^{32}P and an enthylenediamine conjugate on the 5' end of the molecule was stable both in media and within mouse L929 cells for 24 h.

III. Phosphorothioate Oligonucleotides

During the early years of antisense research, it was believed that oligonucleotide stability in plasma and tissue culture media containing various types of sera would predict stability of oligomers within cells and organs (COOK 1993). Since phosphorothioates displayed much greater nuclease resistance in these systems than their phosphodiester congeners, it was suggested that these compounds would be extremely stable against intracellular nucleolytic degradation. More recent in vitro and in vivo experiments have demonstrated that, while phosphorothioates are relatively more stable than their phosphodiester equivalents, they are still metabolized by cellular nucleases. In general, the $t_{1/2}$ for phosphorothioate stability within various cells and tissues has been reported to be between 12 and 24 h (RM CROOKE 1993a; RM CROOKE et al. 1995; CROOKE and BENNETT 1996). As described above, typical $t_{1/2}$ values for phosphodiester oligonucleotides have been reported to be less than 1 h.

The intracellular metabolism of phosphorothioate oligomers has been examined by analytical polyacrylamide gel electrophoresis in some tissue culture systems after varying amounts of incubation. For example, TEMSAMANI et al. (1994), using 20% polyacrylamide gels, reported that, after a 16-h incubation period in various cell lines, including human 293, HeLa S_3, and NIH 3T3 cells, their 20-mer phosphorothioate was stable in the medium but approximately 30% degraded within the cytoplasmic, but not nuclear, fraction of cells. GAO et al. (1993), using 15% polyacrylamide gels, reported that their ^{35}S 28-mer homopolymer C was greater than 90% intact in HepG2 and hepatoma 2215 cells over a 72-h period.

The pattern of degradation observed in many systems suggests primarily exonuclease activity (ST CROOKE 1995a; CROOKE and BENNETT 1996). However, there may be some contribution of endonucleases to the degradation of these compounds. HOKE et al. (1991), when examining the stability of various oligonucleotides, including natural and capped phosphodiesters and phosphorothioates in HeLa cell extracts, determined that the predominant enzymatic activity was endonucleolytic, as evidenced by the lack of nonprocessive laddering of their extracted oligomers. COUTURE and CHOW (1992) have actually purified and characterized a mammalian DNase with has both exo- and endonucleolytic activity. This 65-kDa monomeric enzyme was isolated from untransformed African green monkey kidney cells (CV-1 cells) and cells that were transformed with SV–40T antigen (COS-1 cells).

Recent work in our laboratories has confirmed and extended the observations described above (RM CROOKE et al. 1995, 1997). We analyzed the stability of several phosphorothioates in NHDF and bEND cells using multiple analytical techniques, including anion-exchange HPLC and liquid scintillation counting (Fig. 8), direct autoradiography (Fig. 9), 5′ end-labeling using γ^{32}P-ATP and T4 polynucleotide kinase (Fig. 10), and capillary gel electrophoresis (Fig. 11) after oligonucleotides were phenol/chloroform extracted from cultured cells and purified using strong anion-exchange and reverse-phase columns (RM CROOKE et al. 1995, 1997; LEEDS et al. 1996; GRAHAM et al. 1997). Our data demonstrated that phosphorothioates were metabolized in a time- and concentration-dependent manner and more rapidly than previously suggested by earlier in vitro pharmacokinetic experiments performed in our and other laboratories (HOKE et al. 1991; ST CROOKE 1992; RM CROOKE 1993a; TEMSAMANI et al. 1994; GAO et al. 1993). The data are, however, more consistent with in vivo studies examining the pharmacokinetics of several phosphorothioates in various species, including mice, rats, monkeys, and humans, which suggest that phosphorothioates are eliminated by metabolic processes involving exo- and endonucleolytic degradation (LEEDS and GEARY 1997; CROOKE and BENNETT 1996; SANDS et al. 1994; COSSUM et al. 1994a,b; ST CROOKE et al. 1994).

Several factors may help explain the discrepancy between the observations described above and previous experimental results. First, as has been noted in all aspects of in vitro cellular uptake, pharmacokinetic parameters, including uptake, intracellular distribution, and efflux, are significantly influenced by the cell type being examined. Our data suggest that this is also true for in vitro metabolism. As shown in Fig. 12, the nuclease activity in HeLa and NHDF cell homogenates against ISIS 1049, a 21-mer phosphodiester, varied over a 30-min incubation period at 37°C. The amount of time required to degrade the compound by 50% for HeLa and NHDF cells was 23 min and more than 30 min, respectively, under these experimental conditions.

Second, our metabolism data suggest that phosphorothioates and their metabolites actually inhibit nucleases (RM CROOKE et al. 1995). Figure 13 shows the effect of increasing concentrations of ISIS 2105 in NHDF cells after a 24-h incubation period. The data demonstrate that the compound was more rapidly degraded in these cells as the input concentration decreased. At $0.1\,\mu M$, no full-length ISIS 2105 could be detected, while at $10\,\mu M$, approximately 10% of the material extracted from cells was intact. HPLC profiles showing the time course of degradation of ISIS 2105 (Fig. 8) also indicated that the rate of metabolism was reduced from 24 to 72 h, suggesting loss of

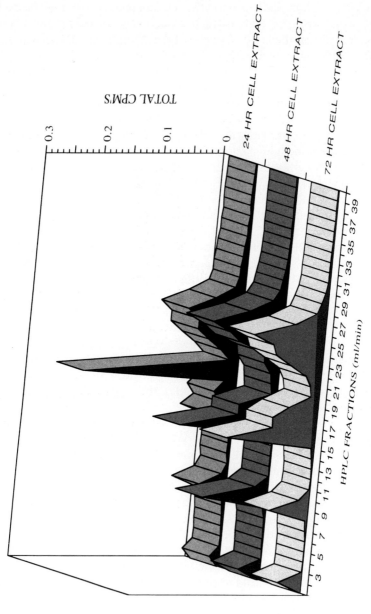

Fig. 8. Stability of ³H ISIS 2105 extracted from NHDF cells after a 24-, 48-, and 72-h incubation period. The profile at each time point represents the percentage of the total measured cpm derived from anion-exchange high-performance liquid chromatography (HPLC)-separated fractions analyzed by liquid scintillation counting

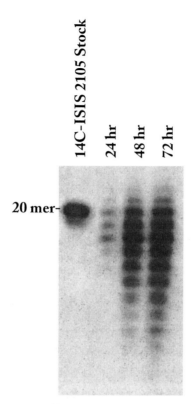

Fig. 9. Direct autoradiography of ^{14}C ISIS 2105 present in NHDF cells after 24, 48, and 72 h. The radiolabeled oligomer was extracted by phenol/chloroform extraction and ethanol precipitation before 20% polyacrylamide gel electrophoretic analysis. 10X stock represents unincubated oligomer whose integrity was determined to be more than 85% of the full length, as assessed by densitometry

enzyme activity. Our data are consistent with reports indicating that phosphorothioates inhibited a variety of enzymes (RM CROOKE 1993a; GAO et al. 1993; OLSEN et al. 1990) and two reports suggesting that higher concentrations of phosphorothioate oligomers specifically inhibited nuclease activity (SHAW et al. 1991; WOOLF et al. 1990). Earlier data suggesting much greater phosphorothioate stability might have resulted from studies being performed with inhibitory concentrations of oligomer.

Third, we established that degradation of phosphorothioates may depend upon oligonucleotide sequence. For example, as shown in Fig. 14, ISIS 2922 and ISIS 1082 are more stable in NHDF cells after a 24-h period than ISIS 2105. This increased stability may be directly related to the variability seen in cell association in Fig. 3, where higher amounts of radiolabeled ISIS 2922 and ISIS 1082 were taken up by cells. More ISIS 2105 might be degraded, since less of the compound is available to inhibit intracellular nu-

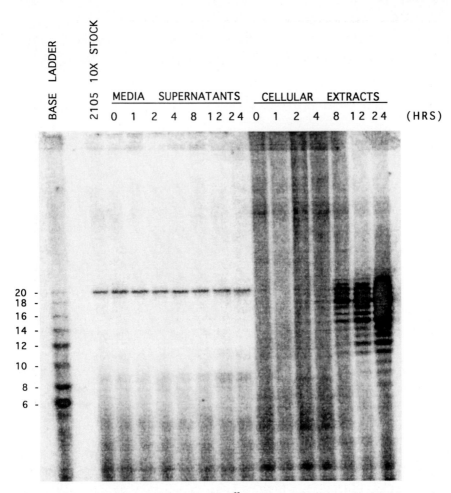

Fig. 10. 5' End-labeling of ISIS 2105 using $\gamma^{32}P$ ATP and T4 polynucleotide kinase after incubation in media or NHDF cells over a 24-h period. The shortmer standards were synthesized to represent successive two-base truncations of the ISIS 2105 sequence from the 3' end. Approximately 1 pmol 10X stock and media supernatants from each time point was 5' end-labeled. Oligonucleotide present in NHDF cells was phenol/chloroform extracted and ethanol precipitated prior to end-labeling

cleases. Alternatively, cellular nucleases might prefer certain stretches of nucleotides or oligonucleotides of varying lengths (RM CROOKE et al. 1995; STEIN et al. 1988).

The fourth and final factor involves the improvement in analytical techniques for detection of metabolism after exposure of cells and tissues to oligonucleotides (RM CROOKE et al. 1995; LEEDS et al. 1996). The availability of the four techniques described above (anion-exchange HPLC with liquid

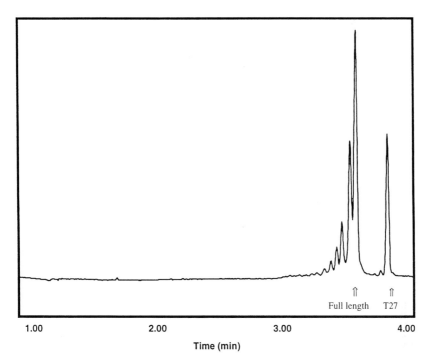

Fig. 11. Capillary gel electropherogram of ISIS 1082 (1 μM) extracted from bEND cells after a 24-h treatment. T27 represents the internal phosphorothioate standard added to the extracted material

scintillation counting, direct autoradiography, 5′ end-labeling using γ^{32}P-ATP and T4 polynucleotide kinase, and capillary gel electrophoresis) provides for much more sensitive detection of oligomer metabolism than was possible in earlier studies and allows the use multiple analytical systems to confirm experimental data. Quantitative and qualitative results from the four analytical techniques are remarkably consistent and suggest that, in various cell types, the actual amount of full-length phosphorothioate represents only 10%–20% of total extracted oligonucleotide after a 24-h incubation period (RM CROOKE et al. 1995).

In our laboratories, we currently prefer to analyze oligomers using capillary gel electrophoresis (LEEDS et al. 1996; RM CROOKE et al. 1997; GRAHAM et al. 1997). This technique is extremely sensitive, with single nucleotide resolution, and permits quantitation of phosphorothioate oligonucleotides within specific cells and subcellular fractions from tissue culture cell lines and various organs in vivo. This analytical method is also extremely useful since the metabolism of first- and multiple-generation oligonucleotides with varying chemistries, both in vitro and in vivo, can be examined without the use of radiolabeled compounds.

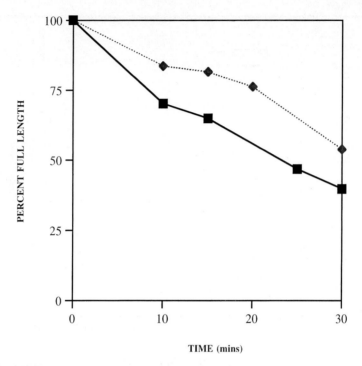

Fig. 12. Stability of ISIS 1049, a 21-mer phosphodiester, in 100 μg HeLa and NHDF cell homogenates over a 30-min period at 37°C. The oligonucleotide was phenol/chloroform extracted from cellular homogenates and analyzed by capillary gel electrophoresis. *Solid line*, HeLa cell homogenate; *dotted line*, NHDF cell homogenate

E. Methods of Uptake Enhancement

Although it has been shown that antisense oligonucleotides are taken up by a variety of cells, a great deal of effort has been involved in enhancing cellular uptake or altering intracellular distribution. Various physical methods to increase intracellular levels of oligomer include membrane permeabilization with streptolysin O (SPILLER and TIDD 1995), scrape loading (PARTRIDGE et al. 1996), electroporation (BERGAN et al. 1993), calcium phosphate precipitation (TOLOU 1993), microinjection (FISHER et al. 1993; CHIN et al. 1990; LEONETTI et al. 1991), association with polycations (ST CROOKE 1995b), adsorption of oligomers onto polyisohexylcyanoacrylate nanoparticles (CHAVANY et al. 1994), and starburst polyamidoamine dendrimers (KUKOWSKA-LATALLO et al. 1996).

Liposomes and various liposomal preparations have also been used to enhance the cellular internalization of oligonucleotides as well as DNA, RNA, and proteins (BENNETT 1995; RM CROOKE 1991; WANG et al. 1995). Several types of cationic lipids, including lipofectin, lipofectamine, lipofectace, transfectam, and GS2888 or cytofectin, have been used by our and other

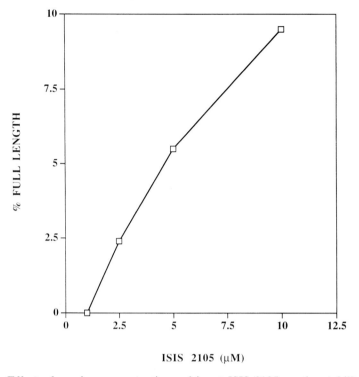

Fig. 13. Effect of varying concentrations of input ISIS 2105 on the stability of the compound in NHDF cellular extracts after a 24-h incubation period. The percentage of full-length oligonucleotide relative to the starting material at each dose is plotted. Values at each point represent a single sample obtained from two 175-cm^2 flasks

laboratories (BENNETT et al. 1992; PICKERING et al. 1996; ALBRECHT et al. 1996; LEWIS et al. 1996). BENNETT et al. (1992) demonstrated that cationic liposomes composed of a DOTMA/DOPE (dioleoylphophatidylethanolamine) mixture enhanced the activity of a phosphorothioate oligomer targeted to intercellular adhesion molecule (ICAM)-1 in human umbilical vein endothelial cells by a factor of approximately 1000, while the amount of ^{35}S-radiolabeled oligonucleotide associating with cells increased 6- to 18-fold. In the presence of cationic lipids, the intracellular distribution of fluorescently tagged compound was altered in that more oligomer distributed to nucleus. While cationic liposomes are useful in some systems, the efficiency and efficacy of transfection can vary as a function of cell type, media, and oligonucleotide chemistry (BENNETT et al. 1992; PICKERING et al. 1996; ALBRECHT et al. 1996; LEWIS et al. 1996; ST CROOKE 1995b).

Antisense oligonucleotides have also been modified with various chemical moieties to enhance uptake, such as acridine (ASSELINE et al. 1984; ZERIAL et al. 1987; BIRG et al. 1990), hydrophobic ligands, e.g., cholesterol and other

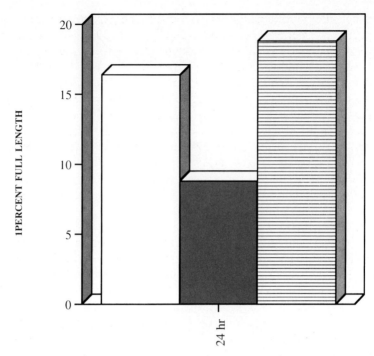

Fig. 14. Stability of ISIS oligonucleotides in NHDF cells after a 24-h incubation period. The percentage of full-length oligonucleotide relative to the starting material at each dose is plotted. Values at each point represent a single sample obtained from two 175-cm^2 flasks. *White bar*, ISIS 1082; *black bar*, ISIS 2105; *shaded bar*, ISIS 2922

hydrophobic groups (CHOW et al. 1994; BISHOP et al. 1995; NECKERS 1993; AKHTAR and JULIANO 1992; ST CROOKE 1995; BUDKER et al. 1992), alkyl-derivatizing groups (RM CROOKE 1991; COOK 1993), poly(L-lysine) and other peptides (DIBAISE et al. 1994; RM CROOKE 1991; NECKERS 1993), cyclic polysaccharides, e.g., cyclodextrins (ZHAO et al. 1995), and various moieties that target specific cell membrane receptors, e.g., mannose (LIANG et al. 1996) or epidermal growth factor (AKHTAR and JULIANO 1992). As described above, chemically modifying oligonucleotides may not only alter uptake, but also other pharmacokinetic parameters such as stability, metabolism, and intracellular distribution.

F. Conclusions

It is obvious from this review that many groups have studied the pharmacokinetics of a variety of antisense oligonucleotides in numerous cell lines and that comparative analysis of this uptake data is problematic because experimental conditions have varied widely. However, some generalizations can be made concerning the pharmacokinetics of either fluorescently or radiolabeled phosphodiester, methylphosphonate, and phosphorothioate oligomers. In vitro cellular uptake was shown to be dependent on time, temperature, and energy and to be dynamic, consisting of two distinct phases: (1) cell association and internalization and (2) efflux or exocytosis. The amount internalized generally varied between 0.5% and 20% of the input material. Experimental conditions, including cell type, media, the amount and type of serum used to supplement media (if any), the presence of certain ions such as calcium and magnesium, and pH have also been shown to influence cellular uptake. The specifics of uptake obviously varied depending upon the physicochemical characteristics of each oligomer, including chemistry, sequence, length, whether the compounds were radiolabeled or fluorescently labeled, and the location of the label within the molecules.

As mentioned above, there are caveats associated with the use of labeled oligonucleotides, and proper controls must be performed in order to unequivocally state that the compound whose uptake and distribution is being followed is intact over the experimental time course and that various reporter groups have not influenced pharmacokinetic parameters. As described above, many groups 5' end-label oligonucleotides with ^{32}P. We (RM CROOKE 1991, 1993a) and others (SAISON-BEHMOARAS et al. 1991; HAWLEY and GIBSON 1992; WU-PONG et al. 1994; NAKAI et al. 1996) have suggested that 5' ^{32}P end-labeled compounds are susceptible to 5'-phosphatase activity. Our date and that obtained by others indicate that the cell-associated radioactivity in many studies performed with this type of labeled compound represents uptake of free phosphate, not intact full-length oligonucleotide. NAKAI et al. (1996) compared the stability and uptake of the same sequence oligomer that was ^{32}P end-labeled at both termini (3' and 5'). They demonstrated that the 3'-labeled compound was more stable and that uptake was threefold higher than for the 5' congener.

Several protocols have been developed to circumvent the lability of the radiolabel, i.e., internal labeling with ^{32}P or ^{3}H (SANDS et al. 1994) or, for phosphorothioates, uniform labeling with ^{35}S within the sulfur of the molecule (STEIN et al. 1990), at the 2' position of pyrimidines with ^{14}C (SASMOR et al. 1995), or with ^{3}H at the C_8H position of purines (GRAHAM et al. 1993).

Concerns about the influence of fluorochromes, including acridine, fluorescein, and rhodamine, on uptake and intracellular distribution have been expressed. Many groups, including those in our laboratories (BUTLER et al. 1997), have recently found that FITC and TRITC (rhodamine) groups do not appear to alter activity or intracellular localization. In contrast, acridine has been shown to influence various pharmacokinetic parameters, including stability and intracellular distribution (ASSELINE et al. 1984; RM CROOKE 1993a; SAISON-BEHMOARAS et al. 1991).

A number of other factors inherent in the use of cultured cells should be considered when evaluating in vitro pharmacokinetic data. For example, it has been demonstrated that dead cells accumulate large amounts of oligomers

(SMETSERS et al. 1994; ZHAO et al. 1993). FARRELL et al. (1995) have also shown that uptake varied between primary cultured rabbit smooth muscle cells, rabbit arterial cells, and arterial cells that have been injured by balloon catheterization. The age (AKHTAR et al. 1996), state of differentiation (ZHAO et al. 1994), stage in cell cycle (ZAMECNIK et al. 1994), and state of activation (ZHAO et al. 1996) influenced uptake. Additionally, it has been shown that, within specific cultures, uptake was heterogeneous (SMETSERS et al. 1994; NOONBERG et al. 1994).

Cellular uptake, internalization, and distribution, as well as activity, also vary depending on whether the cells are primary or immortalized (NOONBERG et al. 1993; NESTLE et al. 1994; KRIEG et al. 1991; FARRELL et al. 1995; ZHAO et al. 1994). NOONBERG et al. (1993) and NESTLE et al. (1994) demonstrated that human primary keratinocytes, unlike immortalized cells, including a keratinocyte line, HeLa cells, and MCF-7 cells, take up oligonucleotide without internalization into cytoplasmic endosomes. Krieg's group (KRIEG et al. 1991; ZHAO et al. 1994, 1996) have also observed that uptake into primary hematopoietic cells varied and is heterogeneous. For example, pre-pro-B and early B cells take up very low levels of oligonucleotides, whereas late pro-B and pre-B cells have increased cellular accumulation. B cells take up more oligomer than T cells, whereas myeloid/macrophage cells accumulate more than B or T cells. Finally, neutrophils take up very little compound compared to all other cell types. Intracellular localization of FITC-labeled oligomers will also vary depending upon the hematopoietic cell type. These investigators also compared uptake between freshly isolated B cells and those cultured for 48 h and determined that uptake by B cells increased as a function of time in tissue culture.

The actual mechanism of uptake appears to vary according to oligonucleotide class and types of cells being examined. As shown by the large amount of conflicting data, the exact mechanism is complex and probably involves several processes, including endocytosis, pinocytosis, and potocytosis. A number of studies suggest that receptor-mediated endocytosis is likely for phosphodiesters and phosphorothioates, whereas methylphosphonates are internalized by a fluid-phase or absorptive endocytosis. Even though some putative "receptors" have been isolated, the molecular weights have varied depending upon the oligomer class, type of cell, and experimental conditions. Additionally, the true function of most of these proteins is unclear.

Based upon the in vitro data described above, one might predict that some antisense oligonucleotides would be unable either to cross cellular membranes efficiently or to distribute within cells to sites where they might exert their pharmacological effects in vivo. Additionally, one might also believe that oligonucleotide sequence would be an important parameter determining in vivo distribution, since significant sequence-specific differences in uptake in cells in tissue culture were noted (RM CROOKE et al. 1995). However, in recent years, numerous phosphorothioate oligonucleotides have been administered to multiple species, including mice, rats, monkeys, and humans (COSSUM et al.

1994a,b; ST Crooke et al. 1994; Sands et al. 1994; Leeds and Geary 1997; Leeds et al. 1996). The data from these studies suggest that phosphorothioates, regardless of sequence, displayed similar pharmacokinetic parameters, i.e., were well absorbed from various sites of administrations, widely distributed to multiple organs, and eliminated by metabolism. Additionally, sequence-specific in vivo pharmacological activity has been reported by multiple groups; it is thus obvious that these compounds, once inside cells, are stable enough and present in sufficient quantities in target sites to be efficacious (Dean and McKay 1994; Stepkowski et al. 1994; Crooke and Bennett 1996).

The aggregate in vitro data and studies from our laboratories (RM Crooke et al. 1995) suggest that predicting in vivo pharmacokinetic behavior based on studies in only one or even a few cell types may be tenuous due to multiple experimental artifacts, including cell type, media, media additions, method of oligomer labeling, and type of intracellular oligomer localization. Future in vitro studies should utilize existing in vivo data to design experiments which begin to address the many remaining questions about in vitro uptake. These include the following:

1. What are the actual mechanisms of cellular uptake?
2. If receptor-mediated endocytosis is a principal mechanism of uptake, where and what are the receptors? Are there ubiquitous receptors for oligomers, and are they related to the DNA receptors described by RM Bennett (1993)?
3. What are the steps involved in releasing oligonucleotides from endocytic vesicles or lysosomes to their nuclear or cytoplasmic target sites?
4. Why is there in vivo uptake and efficacy when many cells in culture do not take up oligonucleotides or display activity without the use of various transfecting agents?
5. Why is there nuclear distribution and oligomer activity in some cultured cells, e.g., primary keratinocytes, but not in others, e.g., immortalized keratinocytes?

Nonetheless, while it is clear that overextrapolation of data from the in vitro to the in vivo setting is problematic, pharmacokinetic studies in tissue culture are necessary and provide valuable information, ranging from delineating the behavior of oligonucleotides in cells where in vitro efficacy is examined to providing useful models for evaluating the utility and efficacy of the exciting next generations of antisense oligonucleotides.

References

Aguilar L, Hemar A, Dautry-Varsat A, Blumenfeld M (1996) Hairpin, dumbbell, and single-stranded phosphodiester oligonucleotides exhibit identical uptake in T lymphocyte cell lines. Antisense Nucleic Acid Drug Dev 6:157–163

Akhtar S, Juliano RL (1992) Cellular uptake and intracellular fate of antisense oligonucleotides. Trends Cell Biol 2:139–144

Akhtar S, Basu S, Wickstrom E, Juliano RL (1991) Interactions of antisense DNA oligonucleotide analogs with phospholipid membranes (liposomes). Nucleic Acids Res 19:555–5559

Akhtar S, Shoji Y, Juliano RL (1992) Pharmaceutical aspects of the biological stability and membrane transport characteristics of antisense oligonucleotides. In: Erickson RP, Izant JG (eds) Gene regulation: biology of antisense RNA and DNA. Raven, New York, pp 133–145

Akhtar S, Beck GF, Hawley P, Irwin WJ, Gibson I (1996) The influence of polarized epithelial (Caco-2) cell differentiation on the cellular binding of phosphodiester and phosphorothioate oligonucleotides. Antisense Res Dev 6:19–206

Albrecht T, Schwab R, Peschel C, Engels HJ, Fischer T, Huber C, Aulitzky WE (1996) Cationic lipid mediated transfer of c-abl and bcr antisense oligonucleotides to immature normal myeloid cells: uptake, biological effects and modulation of gene expression. Ann Hematol 72:73–79

Anderson RGW (1993) Caveolae: where incoming and outgoing messengers meet. Proc Natl Acad Sci USA 90:10909–10913

Asseline U, Toulme F, Thoung NT, Delarue M, Montenay-Garestier T, Helen C (1984) Oligodeoxynucleotides covalently linked to intercalating dyes as base sequence-specific ligands. Influence of dye attachment. EMBO J 3:795–800

Beltinger C, Saragovi HU, Smith RM, LeSauteur L, Shah N, DeDionisio L, Christensen L, Raible A, Jarett L, Gewirtz AM (1995) Binding, uptake and intracellular trafficking of phosphorothioate-modified oligodeoxynucleotides. J Clin Invest 95:1814–23

Benet LA, Sheiner LB (1985) Pharmacokinetics: the dynamics of drug absorption, distribution and elimination. In: Gilman AG, Goodman LS, Rall TW, Murad F (eds) The pharmacological basis of therapeutics, 7th edn. Macmillan, New York, pp 3–13

Bennett RM (1993) As nature intended? The uptake of DNA and oligonucleotides by eukaryotic cells. Antisense Res Dev 3:235–241

Bennett CF (1995) Intracellular delivery of oligonucleotides with cationic liposomes. In: Akhtar S (ed) Delivery strategies for antisense oligonucleotide therapeutics. CRC Press, Boca Raton, pp 223–232

Bennett CF, Chiang M-Y, Chan H, Shoemaker JE, Mirabelli CK (1992) Cationic lipids enhance cellular uptake and activity of phosphorothioate antisense oligonucleotides. Mol Pharmacol 41:1023–1033

Bergan R, Connell Y, Fahmy B, Neckers L (1993) Electroporation enhances c-myc antisense oligodeoxynucleotide efficacy. Nucleic Acids Res 21:3567–3573

Birg F, Praseuth D, Zerial A, Thuong NT, Asseline U, LeDoan T, Helene C (1990) Inhibition of simian virus 40 DNA replication in CV-1 cells by an oligodeoxynucleotide covalently linked to an intercalating agent. Nucleic Acids Res 18:2901–2908

Bishop WP, Lin J, Stein CA, Krieg AM (1995) Interruption of a transforming growth factor α autocrine loop in Caco-2 cells by antisense oligodeoxynucleotides. Gastroenterology 109:1882–1889

Bishop JS, Guy-Caffey JK, Ojwang JO, Smith SR, Hogan ME, Cossum PA, Rando RF, Chaudhary N (1996) Intramolecular G-quartet motifs confer nuclease resistance to a potent anti-HIV oligonucleotide. J Biol Chem 271:5698–5703

Bradley MO, Chrisey LA, Hawkins JW (1992) Antisense therapeutics. In: Erickson RP, Izant JG (eds) Gene regulation: biology of antisense RNA and DNA. Raven, New York, p 285

Brown DA, Kang S-H, Gryzanov SM, De Dionisio L, Heidenreich L, Sullivan S, Xu X, Nerenberg MI (1994) Effect of phosphorothioate modification of oligodeoxynucleotides on specific protein binding. J Biol Chem 269:26801–26805

Budker VG, Knorre DG, Vlassov VV (1992) Cell membranes as barriers for antisense constructions. Antisense Res Dev 2:177–184

Butler M, Stecker K, Bennett CF (1997) Distribution of phosphorothioate oligodeoxynucleotides in normal rodent tissues. Lab Invest 77:379–388

Campbell JM, Bacon TA, Wickstrom E (1990) Oligodeoxynucleotide phosphorothioate stability in subcellular extracts, culture media, sera and cerebrospinal fluid. J Biochem Biophys Methods 20:259–267

Capaccioli S, Di Pasquale G, Mini E , Mazzei T Quattrone A (1993) Cationic lipids improve antisense oligonucleotide uptake and prevent degradation in cultured cells and human serum. Biochem Biophys Res Commun 197:818–825

Chaix C, Toulme J-J, Morvan F, Rayner B, Imbach J-L (1993) α-Oligonucleotides: an entry to a challenging class of antisense molecules. In: Crooke ST, Lebleu B (eds) Antisense research and applications. CRC Press, Boca Raton, pp 223–234

Chavany C, Saison-Behmoaras T, LeDoan T, Puisieux F, Couvreur P, Helene C (1994) Adsorption of oligonucleotides onto polyisohexylcyanoacrylate nanoparticles protects them against nucleases and increases their cellular uptake. Pharm Res 11:1370–1378

Chin DJ, Green GA, Zon G, Szoka FC Jr, Straubinger RM (1990) Rapid nuclear accumulation of injected oligodeoxyribonucleotides. New Biol 2:1091–1100

Chow TY-K, Juby C, Brousseau R (1994) Specific targeting of antisense oligonucleotides to neutrophils. Antisense Res Dev 4:81–86

Clusel C, Ugarte E, Enjolras NO, Vasseur M, Blumenfeld M (1993) Ex vivo regulation of specific gene expression by nanomolar concentrations of double-stranded dumbbell oligonucleotides. Nucleic Acids Res 21:3405–3411

Cohen JS (1993) Phosphorothioate oligodeoxynucleotides. In: Crooke ST, Lebleu B (eds) Antisense research and applications. CRC Press, Boca Raton, pp 206–221

Cook PD (1993) Medicinal chemistry strategies for antisense research. In: Crooke ST, Lebleu B (eds) Antisense research and applications. CRC Press, Boca Raton, pp 149–187

Cossum PA, Sasmor H, Dellinger D, Truong L, Cummins L, Owens S, Markham PM, Shea JP, Crooke ST (1994a) Disposition of ^{14}C-labeled phosphorothioate oligonucleotide ISIS 2105 after intravenous administration to rats. J Pharmacol Exp Ther 267:1181–1190

Cossum PA, Truong L, Owens S, Markham PM, Shea JP, Crooke ST (1994b) Pharmacokinetics of ^{14}C-labeled phosphorothioate oligonucleotide ISIS 2105 after intradermal administration to rats. J Pharmacol Exp Ther 269:89–94

Couture C, Chow T-Y (1992) Purification and characterization of a mammalian endoexonuclease. Nucleic Acids Res 20:4355–4361

Crooke RM (1991) In vitro toxicology and pharmacokinetics of antisense oligonucleotides. Anti Cancer Drug Design 6:609–646

Crooke RM (1993a) Cellular uptake, distribution and metabolism of phosphorothioate, phosphodiester and methylphosphonate oligonucleotides. In: Crooke ST, Lebleu B (eds) Antisense research and applications. CRC Press, Boca Raton, pp 428–229

Crooke RM (1993b) In vitro and in vivo toxicology of first generation analogs. In: Crooke ST, Lebleu B (eds) Antisense research and applications. CRC Press, Boca Raton, pp 472–492

Crooke RM, Graham MJ, Cooke ME, Crooke ST (1995) In vitro pharmacokinetics of phosphorothioate antisense oligonucleotides. J Pharmacol Exp Ther 275:462–473

Crooke RM, Graham MJ, Martin MM, Griffey R, Cummins L (1997) Characterization of in vitro antisense oligonucleotide metabolism in rat liver homogenates. J Pharmacol Exp Ther (submitted)

Crooke ST (1992) Therapeutic applications of oligonucleotides. Annu Rev Pharmcol Toxicol 32:329–276

Crooke ST (1995a) Therapeutic applications of oligonucleotides. Landes, Austin

Crooke ST (1995b) Oligonucleotide therapeutics. In: Wolff ME (ed) Burger's medicinal chemistry and drug discovery, 5th edn. 1. Principles and practice. Wiley, New York

Crooke ST (1996) Proof of mechanism of antisense drugs. Antisense Nucleic Acid Drug Dev 6:145–147
Crooke ST, Bennett CF (1996) Progress in antisense oligonucleotide therapeutics. Annu Rev Pharmacol Toxicol 36:107–129
Crooke ST, Lebleu B (eds) (1993) Antisense research and applications. CRC Press, Boca Raton
Crooke ST, Grillone LR, Tendolkar A, Garrett A, Fratkin M, Leeds JM, Barr WH (1994) A pharmacokinetic evaluation of ^{14}C-labeled afovirsen sodium in genital wart patients. Pharmacol Ther 56:641–646
Dean NM, McKay R (1994) Inhibition of protein kinase C-α expression in mice after systemic administration of phosphorothioate antisense oligodeoxynucleotides. Proc Natl Acad Sci USA 91:11762–11766
DiBaise JK, Ebadi M, Iversen PL (1994) Patterns of cellular uptake and effects on cell survival using antimetallothionein oligodeoxyribonucleotide conjugates in vitro. Biol Sign 3:140–149
Eder PS, DeVine EJ, Dagle JM, Walder JA (1991) Substrate specificity and kinetics of degradation of antisense oligonucleotides by a 3' exonuclease in plasma. Antisense Res Dev 1:141–151
Eckstein F (1985) Nucleoside phosphorothioates. Annu Rev Biochem 54:367–402
Farrell CL, Bready JV, Kaufman SA, Qian Y-X, Burgess TL (1995) The uptake and distribution of phosphorothioate oligonucleotides into vascular smooth muscle cells in vitro and in rabbit arteries. Antisense Res Dev 5:175–183
Fisher TL, Terhorst T, Cao X, Wagner RW (1993) Intracellular disposition and metabolism of fluorescently labeled unmodified and modified oligonucleotides microinjected into mammalian cells. Nucleic Acids Res 21:3857–3865
Gao W-Y, Jaroszewski JW, Cohen JS, Cheng Y-C (1990) Inhibition of herpes simplex virus type 2 growth by 28-mer phosphorothioate oligodeoxycytidine. J Biol Chem 265:21072–20178
Gao W-Y, Han FS, Storm C, Egan W, Cheng Y-C (1992) Phosphorothioate oligonucleotides are inhibitors of human DNA polymerases and RNase H: implications for antisense technology. Mol Pharmacol 41:223–229
Gao W-Y, Storm C, Egan W, Cheng Y-C (1993) Cellular pharmacology of phosphorothioate homooligodeoxynucleotides in human cells. Mol Pharmacol 43:45–50
Geselowitz DA, Neckers LM (1992) Analysis of oligonucleotide binding, internalization and intracellular trafficking utilizing a novel radiolabeled crosslinker. Antisense Res Dev 2:17–25
Geselowitz DA, Neckers LM (1995) Bovine serum albumin is a major oligonucleotide-binding protein found on the surface of cultured cells. Antisense Res Dev 5:213–217
Giles RV, Spiller DG, Tidd DM (1993) Chimeric oligodeoxynucleotide analogues: enhanced cell uptake of structures which direct ribonuclease H with high specificity. Anti Cancer Drug Des 8:33–51
Goodarzi G, Watabe M, Watabe K (1991) Binding of oligonucleotides to cell membranes at acidic pH. Biochem Biophys Res Commun 181:1343–1351
Goodchild J (1990) Conjugates of oligonucleotides and modified oligonucleotides: a review of their synthesis and properties. Bioconj Chem 1:165–187
Gottikh M, Bertrand, J-R, Baud-Demattei M-V, Lescot E, Giorgi-Renault S, Shabarova Z, Malvy C (1994) $\alpha\beta$ Chimeric antisense oligonucleotides: synthesis and nuclease resistance in biological media. Antisense Res Dev 4:251–258
Graham MJ, Freier SM, Crooke RM, Ecker DJ, Maslova RN, Lesnik EA (1993) Tritium labeling of antisense oligonucleotides by exchange with tritiated water. Nucleic Acids Res 21:3737–3743
Graham MJ, Lemonidis KM, Monteith DM, Cooper S, Crooke ST, Crooke RM (1997) In vivo distribution and metabolism of a phosphorothioate oligonucleotide within the rat liver after intravenous administration. J Pharmacol Exp Ther (submitted)

Hawley P, Gibson I (1992) The detection of oligodeoxynucleotide molecules following uptake into mammalian cells. Antisense Res Dev 2:119–127

Ho PTC, Ishiguro K, Wickstrom E, Sartorelli AC (1991) Non-sequence-specific inhibition of transferrin receptor express in HL-60 leukemia cells by phosphorothioate oligodeoxynucleotides. Antisense Res Dev 1:329–342

Hoke GD, Draper K, Freier SM, Gonzalez C., Driver VB, Zounes MC, Ecker DJ (1991) Effects of phosphorothioate capping on antisense oligonucleotide stability, hybridization and antiviral efficacy versus herpes simplex virus infection. Nucleic Acids Res 19:5743–5748

Hughes JA, Avrutskaya AV, Juliano RL (1994) Influence of base composition on membrane binding and cellular uptake of 10-mer phosphorothioate oligonucleotides in Chinese hamster ovary (CHRC5) cells. Antisense Res Dev 4:211–215

Iversen P (1991) In vivo studies with phosphorothioate oligonucleotides: pharmacokinetic prologue. Anti Cancer Drug Des 6:531–538

Jansen B, Wadl H, Inoue SA, Trulzsch B, Selzer E, Duchene M, Eichler H-G, Wolff K, Pehamberger H (1995) Phosphorothioate oligonucleotides reduce melanoma growth in a SCID-hu mouse model by a nonantisense mechanism. Antisense Res Dev 5:271–277

Jaroszewski JW, Kaplan O, Syi J-L, Sehested M, Faustino PJ, Cohen JS (1990) Concerning antisense inhibition of the multiple drug resistance gene. Cancer Commun 2:287–294

Kornberg A (1980) DNA replication. Freeman, San Francisco

Krieg AM, Gmelig-Meyling F, Gourley MF, Kisch WJ, Chrisey LA, Steinberg AD (1991) Uptake of oligodeoxyribonucleotides by lymphoid cells is heterogeneous and inducible. Antisense Res Dev 1:161–171

Kukowska-Latallo JF, Bielinska AU, Johnson J, Spindler R, Tomalia DA, Baker JR Jr (1996) Efficient transfer of genetic material into mammalian cells using Starburst polyamidoamine dendrimers. Proc Natl Acad Sci USA 93:4897–4902

Kumar S, Srinivasan H, Tewary HK, Iversen PL (1995) Characterization of binding sites, extent of binding, and drug interactions of oligonucleotides with albumin. Antisense Res Dev 5:131–139

Leeds JM, Geary RS (1997) Pharmacokinetic properties of phosphorothioate oligonucleotides in humans. In: Crooke ST (ed) Antisense research and application. Springer, Berlin Heidelberg New York

Leeds JM, Truong LA, Cossum P, Prowse C, Crooke ST, Kornbrust D (1994) Interaction of phosphorothioate oligonucleotides with plasma proteins. Pharmacol Res 11:S-352

Leeds JM, Graham MJ, Truong L, Cummins LL (1996) Quantitation of phosphorothioate oligonucleotides in human plasma. Anal Biochem 235:36–43

Leonetti J-P, Mechti N, Degols C, Gagnor C, Lebleu B (1991) Intracellular distribution of microinjected antisense oligonucleotides. Proc Natl Acad Sci USA 88:2702–2706

LeRoy C, Leduque P, Dubois PM, Saez JM, Langlois D (1996) Repression of transforming growth factor $\beta 1$ protein by antisense oligonucleotide induced increase of adrenal cell differentiated functions. J Biol Chem 271:11027–11033

Levis JT, Butler WO, Tseng BY, Ts'o POP (1995) Cellular uptake of oligodeoxyribonucleoside methylphosphonates. Antisense Res Dev 5:251–259

Lewis JG, Lin K-Y, Kothavale A, Flanagan WM, Matteucci MD, DePrince RB, Mook RA Jr, Hendren RW, Wagner RW (1996) A serum-resistant cytofectin for cellular delivery of antisense oligodeoxynucleotides and plasmic DNA. Proc Natl Acad Sci USA 93:3176–3181

Liang W, Shi Z, Deshpande D, Malanga CJ, Rojanasakul Y (1996) Oligonucleotide targeting to alveolar macrophages by mannose receptor-mediated endocytosis. Biochem Biophys Acta 1279:227–234

Lloyd RS, Linn S (1993) Nucleases involved in DNA repair. In: Linn SM, Lloyd RS, Roberts RJ (eds) Nucleases, 2nd edn, monograph 25. Cold Spring Harbor Laboratory Press, Cold Spring Harbor, pp 263–316

Loke SL, Stein CA, Zhang XH, Mori K, Nakanishi M, Subasinghe S, Cohen JS, Neckers LM (1989) Characterization of oligonucleotide transport into living cells. Proc Natl Acad Sci USA 86:3474–3478

Marcus-Sekura CJ, Woerner AM, Shinozuka K, Zon G, Quinnan GV, Jr (1987) Comparative inhibition of chloramphenical acetyltransferase gene expression by antisense oligonucleotide analogues having alkyl phosphotriester, methylphosphonate and phosphorothioate linkages. Nucleic Acids Res 15:5749–5963

Marti G, Egan W, Noguchi P, Zon G, Matsukura M, Broder S (1992) Oligodeoxyribonucleotide phosphorothioate fluxes and localization in hematopoietic cells. Antisense Res Dev 2:27–39

Miller PS, McParland KB, Jayaraman K, Ts'o POP (1981) Biochemical and biological effects of nonionic nucleic acid methylphosphonates. Biochemistry 20:1874–1880

Miller PS, Ts'o POP, Hogrefe RI, Reynolds MA, Arnold LA Jr (1993) Anticode oligonucleoside methylphosphonates and their psoralen derivatives. In: Crooke ST, Lebleu B (eds) Antisense research and applications. CRC Press, Boca Raton, pp 190–203

Nakai D, Sewita T, Iwasa T, Aiwasa S, Shoji Y, Mizushima Y, Sugiyama Y (1996) Cellular uptake mechanism for oligonucleotides: involvement of endocytosis in the uptake of phosphodiester oligonucleotides by a human colorectal adenocarcinoma cell line, HCT-15. J Pharmacol Exp Ther 278:1362–1372

Neckers LM (1993) Cellular internalization of oligodeoxynucleotides. In: Crooke ST, Lebleu B (eds) Antisense research and applications. CRC Press, Boca Raton, pp 452–460

Nestle FO, Mitra RS, Bennett CF, Chan H, Nickoloff BJ (1994) Cationic lipid is not required for uptake and selective inhibitory activity of ICAM-1 phosphorothioate antisense oligonucleotides in keratinocytes. J Invest Dermatol 103:569–575

Noonberg SB, Garovoy MR, Hunt CA (1993) Characteristics of oligonucleotide uptake in human keratinocyte cultures. J Invest Dermatol 101:727–731

Olsen DB, Kotzorek G, Eckstein F (1990) Investigation of the inhibitory role of phosphorothioate internucleotide linkages on the catalytic activity of the restriction endonuclease EcoRV. Biochemistry 29:9546–9551

Partridge M, Vincent A, Matthews P, Puma J, Stein D, Summerton JS (1996) A simple method for delivering morpholino antisense oligos into the cytoplasm of cells. Antisense Nucleic Acid Drug Dev 6:169–175

Pickering JG, Isner JM, Ford CM, Weir L, Lazarovits A, Rocnik EF, Chow LH (1996) Processing of chimeric antisense antisense oligonucleotides by human vascular smooth muscle cells and human atherosclerotic plaque. Implications for antisense therapy of restonosis after angioplasty. Circulation 93:772–780

Robinson LA, Smith, LJ, Fontaine MP, Kay HD, Mountjoy CP, Pirruccello SJ (1995) c-myc antisense oligodeoxynucleotides inhibit proliferation of non-small cell lung cancer. Ann Thorac Surg 60:1583–1591

Saison-Behmoaras T, Tocque B, Rey I, Chassignol M, Thoung NT, Helene C (1991) Short modified antisense oligonucleotides directed against Ha-ras point mutation induce selective cleavage of the mRNA and inhibit T24 cell proliferation. EMBO J 10:1111–1118

Sands, H, Feret G, Cocuzza AJ, Hobbs FW, Chidester D, Trainor GL (1994) Biodistribution and metabolism of internally ^{3}H-labeled oligonucleotides. I. Comparison of a phosphodiester and a phosphorothioate. Mol Pharmacol 45:932–943

Sasmor HM, Dellinger DJ, Zenk P, Lee LP (1995) A practical method for the synthesis and purification of ^{14}C-labeled oligonucleotides. J Lab Cpds Rad Pharm 36:15–31

Shaw J-P, Kent K, Bird J, Fishback, J, Froehler B (1991) Modified deoxyoligonucleotides stable to exonuclease degradation in serum. Nucleic Acids Res 19:747–750

Shoji Y, Akhtar A, Periasamy A, Herman B, Juliano RL (1991) Mechanism of cellular uptake of modified oligodeoxynucleotides containing methylphosphonate linkages. Nucleic Acids Res 19:5543–5550

Smetsers TFCM, Skorski T, van de Locht LTF, Wessels HMC, Pennings AHM, deWitte T, Calabretta B, Mensink EJBM (1994) Antisense BCR-ABL oligonucleotides induce apoptosis in the Philadelphia chromosome-positive cell line BV173. Leukemia 8:129–140

Spiller DG, Tidd DM (1992) The uptake kinetics of chimeric oligodeoxynucleotide analogues in human leukaemia MOLT-4 cells. Anti Cancer Drug Des 7:115–129

Spiller DG, Tidd DM (1995) Nuclear delivery of antisense oligodeoxynucleotides though reversible permeabilization of human leukemia cells with streptolysin O. Antisense Res Dev 5:13–21

Stein CA, Cheng Y-C (1993) Antisense oligonucleotides as therapeutic agents – is the bullet really magical? Science 261:1004–1012

Stein CA, Subasinghe C, Shinozuka K, Cohen JS (1988a) Physicochemical properties of phosphorothioate oligodeoxynucleotides. Nucleic Acids Res 16:3206–3221

Stein CA, Mori K, Loke SL, Subasinghe C, Shinozuka K, Cohen JS, Neckers LM (1988b) Phosphorothioate and normal oligonucleotides with 5'-linked acridine: characterization and preliminary kinetics of cellular uptake. Gene 72:333–341

Stein CA, Iversen PL, Subasinghe C, Cohen JS, Stec WJ, Zon G (1990) Preparation of ^{35}S-labeled polyphosphorothioate oligodeoxyribonucleotides by use of hydrogen phosphonate chemistry. Anal Biochem 188:11–16

Stein CA, Neckers LM, Nair BC, Mumbauer S, Hoke G, Pal R (1991) Phosphorothioate oligodeoxycytidine interferes with binding of HIV-1 gp20 to CD4. J Acquir Immune Defic Syndr 4:686–693

Stein CA, Cleary AM, Yakubov L, Lederman S (1993a) Phosphorothioate oligodeoxynucleotides bind to the third variable loop domain (v3) of human immunodeficiency virus type 1 gp120. Antisense Res Dev 3:19–31

Stein CA, Tonkinson JL, Zhang L-M, Yakubov L, Gervasoni J, Taub R, Rotenberg SA (1993b) Dynamics of the internalization of phosphodiester oligodeoxynucleotides in HL60 cells. Biochemistry 32:4855–4861

Stepkowski SM, Tu Y, Condon TP, Bennett CF (1994) Blocking of heart allograft rejection by intercellular adhesion molecule-1 antisense oligonucleotides alone or in combination with other immunosuppressive modalities. J Immunol 153:5336–5346

Tabor CW, Tabor H (1984) Polyamines. Annu Rev Biochem 53:749–790

Temsamani J, Kubert M, Tang J, Padmapriya A, Agrawal S (1994) Cellular uptake of oligodeoxynucleotide phosphorothioates and their analogs. Antisense Res Dev 4:35–42

Thierry AR, Dritschilo A (1992) Intracellular availability of unmodified, phosphorothioated and liposomally encapsulated oligodeoxynucleotides for antisense activity. Nucleic Acids Res 20:5691–5698

Tidd DM (1990) A potential role for antisense oligonucleotide analogues in the development of oncogene targeted cancer chemotherapy. Anticancer Res 10:1169–1182

Tidd DM, Warenius HM (1989) Partial protection of oncogene, anti-sense oligodeoxynucleotides against serum nuclease degradation using terminal methylphosphonate groups. Br J Cancer 60:343–350

Tolou H (1993) Administration of oligonucleotides to cultured cells by calcium phosphate precipitation method. Anal Biochem 215:156–158

Tonkinson JL, Stein CA (1994) Patterns of intracellular compartmentalization, trafficking and acidification of 5' fluorescein labeled phosphodiester and phosphorothioate oligodeoxynucleotides in HL60 cells. Nucleic Acids Res 22:4268–4275

Vasanthakumar G, Ahmed NA (1989) Modulation of drug resistance in a daunorubicin resistant subline with oligonucleoside methylphosphonates. Cancer Commun 1:225–232

Wagner RW (1994) Gene inhibition using antisense oligodeoxynucleotides. Nature 372:333–335

Wang S, Lee RJ, Cauchon G, Gorenstein DG, Low PS (1995) Delivery of antisense oligodeoxyribonucleotides against the human epidermal growth factor receptor into cultured KB cells with liposomes conjugated to the folate via polyethylene glycol. Proc Natl Acad Sci USA 92:3318–3322

Wickstrom E (1986) Oligodeoxynucleotide stability in subcellular extracts and culture media. J Biochem Biophys Methods 13:97–102

Woolf TM, Jennings CGB, Rebagliati M, Melton DA (1990) The stability, toxicity and effectiveness of unmodified and phosphorothioate antisense oligodeoxynucleotides in Xenopus oocytes and embryos. Nucleic Acids Res 18:1763–1769

Wu-Pong S, Weiss TL, Hunt CA (1992) Antisense c-myc oligodeoxyribonucleotide cellular uptake. Pharmacol Res 9:1010–1017

Wu-Pong S, Weiss TL, Hunt AC (1994) Antisense c-myc oligonucleotide cellular uptake and activity. Antisense Res Dev 4:155–163

Yakubov LA, Deeva EA, Zarytova VF, Ivanova EM, Ryte AS, Yurchenko LV, Vlassov VV (1989) Mechanism of oligonucleotide uptake by cells: Involvement of specific receptors? Proc Natl Acad Sci USA 86:6454–6458

Zamecnik PC, Stephenson ML (1978) Inhibition of Rous sarcoma virus replication and cell transformation by a specific oligodeoxynucleotide. Proc Natl Acad Sci USA 75:280–288

Zamecnik P, Aghajanian J, Zamecnik M, Goodchild J, Witman G (1994) Electron micrographic studies of transport of oligodeoxynucleotides across eukaryotic cell membranes. Proc Natl Acad Sci USA 91:3156–3160

Zerial A, Thuong NT, Helene C (1987) Selective inhibition of the cytopathic effect of type A influenza viruses by oligodeoxynucleotides covalently linked to an intercalating agent. Nucleic Acids Res 15:9909–9919

Zhao Q, Matson S, Herrera CJ, Fisher E, Yu H, Krieg AM (1993) Comparison of cellular binding and uptake of antisense phosphodiester, phosphorothioate, and mixed phosphorothioate and methylphosphonate oligonucleotides. Antisense Res Dev 3:53–66

Zhao Q, Waldeschmidt T, Fisher R, Herrera CJ, Krieg AM (1994) Stage-specific oligonucleotide uptake in murine bone marrow B-cell precursors. Blood 84:3660–3666

Zhao Q, Temsamani J, Agrawal S (1995) Use of cyclodextrin and its derivatives as carriers for oligonucleotide delivery. Antisense Res Dev 5:185–192

Zhao Q, Song X, Waldschmidt T, Fisher E, Krief AM (1996) Oligonucleotide uptake in human hematopoietic cells is increased in leukemia and is related to cellular activation. Blood 88:1788–1795

Zon G (1989) Oligonucleotide analogues as potential chemotherapeutic agents. Pharmacol Res 5:539–549

CHAPTER 4

Pharmacokinetic Properties of Phosphorothioates in Animals – Absorption, Distribution, Metabolism and Elimination

P.L. NICKLIN, S.J. CRAIG, and J.A. PHILLIPS

A. Introduction

Antisense oligonucleotides can hybridise with target mRNA and cause protein-specific translation arrest (ZAMENCNIK and STEVENSON 1978; HELENE and TOULME 1990). This "antisense concept" is a simple, generic and rational approach to drug discovery which has stimulated considerable interest in the potential of oligonucleotides as therapeutic agents (CROOKE 1995). Indeed, there are now many examples where antisense oligonucleotides have demonstrated efficacy against disease-relevant targets, notably for tumour (HIGGINS et al. 1993; MONIA et al. 1996; DEAN et al. 1996) and viral (AGRAWAL 1992; WAGNER and FLANAGAN 1997) indications. The rapid emergence of this class of compounds has placed new demands on the preclinical profiling of therapeutic candidates, particularly with respect to their in vivo behaviour.

For all drug molecules, the relationship between drug administration and pharmacological effect is characterised by two phases: pharmacokinetics and pharmacodynamics. The pharmacokinetic phase encompasses the processes involved in achieving and maintaining pharmacological availability at the target site (absorption, distribution, metabolism and elimination), while the pharmacodynamic phase describes the nature of drug–receptor interaction. In the context of antisense oligonucleotides (see Fig. 1), the factors influencing pharmacokinetics include nuclease degradation, protein binding, plasma clearance, tissue distribution, cellular uptake and subcellular localisation. Target mRNA acts as the "receptor", and the factors affecting pharmacodynamics include target copy number, sequence accessibility (secondary and tertiary mRNA structure), affinity, selectivity and the mechanism of translation arrest (RNase H activation, steric blockade, disruption of mRNA structure). CGP 64128A (ISIS 3521; Fig. 1A) and CGP 69846A (ISIS 5132; Fig. 1B) are 20-mer phosphorothioate oligodeoxynucleotides which target the 3′-untranslated region of human protein kinase C-α and human c-raf-1 kinase, respectively. They cause a potent (IC$_{50}$, approximately 50–100 nM) and specific reduction of target mRNA in vitro and have anti-tumour effects in human tumour xenograft nude mouse models (see Chap. 13). Moreover, there is compelling evidence that their actions are effected through an antisense mechanism (MONIA et al. 1996; DEAN et al. 1996). Evidently, CGP 64128A and CGP

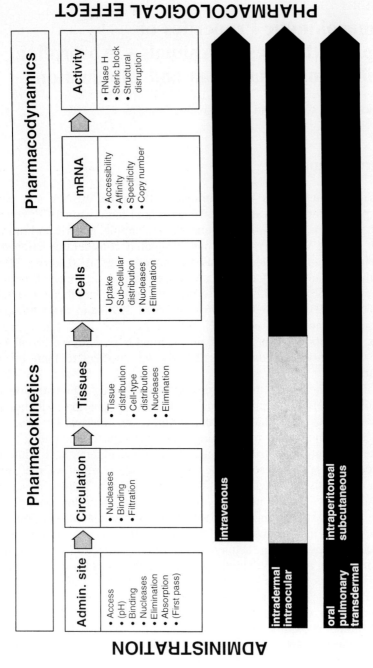

Fig. 1. Pharmacokinetic and pharmacodynamic barriers to the pharmacological efficacy of oligonucleotides

69846A are able to overcome a range of pharmacokinetic and pharmacodynamic hurdles which are presented in vivo; they resist nuclease degradation, have appropriate blood and tissue kinetics, localise to target tissue, penetrate cells and are available in sufficient quantity and duration to specifically hybridise with target mRNA and cause protein-specific translation arrest. Presently, the absorption, distribution, metabolism and elimination of phosphorothioate oligodeoxynucleotides are described with particular reference to these two compounds.

I. Phosphorothioate Modification

Naturally occurring deoxyribonucleic acid contains phosphodiester linkages between adjacent nucleotides and is susceptible to nuclease attack. Phosphodiester antisense oligonucleotides are therefore rapidly degraded in biological environments (SANDS et al. 1994; AGRAWAL 1995a), making them unsuitable as therapeutic candidates. The phosphorothioate modification, in which one of the non-bridging oxygen atoms is replaced by sulphur (see Fig. 2), confers increased biological stability and retains the ability to activate RNase H. These beneficial characteristics must be balanced against the introduction of an inter-nucleotide stereocentre (resulting in a diastereoisomeric mixture of 2^n compounds, where n is the number of phosphorothioate linkages), a lower duplex melting temperature and increased protein-binding

A GsTsTsCsTsCsGsCsTsGsGsTsGsAsGsTsTsTsCsA

B TsCsCsCsGsCsCsTsGsTsGsAsCsAsTsGsCsAsTsT

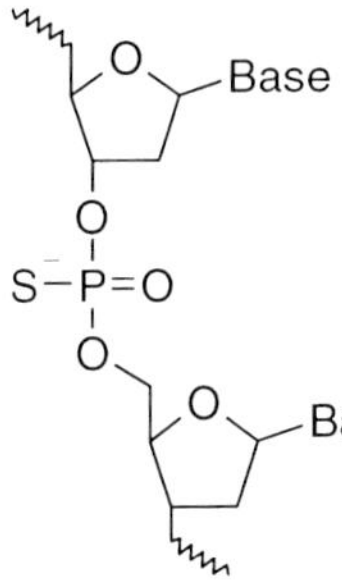

Fig. 2. A CGP 64128A (ISIS 3521), protein kinase C-α. **B** CGP 69846A (ISIS 5132), c-*raf*-1 kinase. *s*, phosphorothioate

characteristics, which may be responsible for the acute haemodynamic and cardiovascular side-effects reported in monkeys (GALBRAITH et al. 1994; SHAW et al. 1997) and for some non-specific effects (STEIN 1996).

II. Oligonucleotide Detection and Analysis

Early pharmacokinetic studies preceded direct, specific and sensitive methods for the analysis of oligonucleotides and their metabolites in biological tissues. Synthetic and post-synthetic radiolabelling strategies were therefore developed to provide radioactive tracers for in vivo experiments. Synthetic approaches include the incorporation of ^{35}S into the backbone of phosphorothioates by the oxidation of H-phosphonate linkages using ^{35}S$_8$ (AGRAWAL et al. 1991) or the introduction of ^{14}C-labelled phosphoramidite synthons during oligonucleotide synthesis (COSSUM et al. 1993). High-quality radiolabelled oligonucleotides can be produced by either method. The ^{35}S approach has the advantage that the metabolic status of the oligonucleotide in tissue extracts can be qualitatively assessed by polyacrylamide gel electrophoresis (PAGE) autoradiography. This method of radiolabelling, however, is restricted to oligonucleotides containing phosphorothioate linkages and may be susceptible to metabolic exchange of ^{35}S in vivo (see Sect. C). Post-synthetic radiolabelling strategies such as T4 polynucleotide kinase-mediated transfer of a ^{32}P phosphate group from [γ32P]ATP on to the 5' terminus of the oligonucleotide (GOODARZI et al. 1992; INAGAKI et al. 1992), enzymatic methylation (SANDS et al. 1994) and ^{3}H-labelling by tritium exchange at the C$_8$ position of purine bases (GRAHAM et al. 1993) have also been employed. The T4 polynucleotide kinase has little utility in in vivo pharmacokinetic studies. The parent molecule is modified by an additional phosphate group, which is readily removed by ubiquitous phosphatases in vivo and is therefore prone to giving unreliable results. In contrast, enzymatic radiolabelling at internal positions of oligonucleotides, such as *Hha*I methylation, can be very useful, but the general application of this approach is restricted by the substrate and sequence specificity of these reactions. The tritium-exchange reaction can be used to radiolabel any oligonucleotide containing purine bases and produces high-purity tritiated compounds with good specific activities which are readily quantitated in biological samples. Tritium does not back-exchange from the C$_8$ position with an appreciable rate at temperatures below 90°C, and the radiolabel is considered to be physically stable under in vivo experimental conditions. Moreover, the low energy of tritium decay makes it ideal for localising oligonucleotides in tissues using autoradiography.

Oligonucleotides radiolabelled by the above methods have contrasting qualities, but share the disadvantage of relying on indirect quantitation methods. A growing body of evidence suggests that the apparent biological handling (especially elimination) of oligonucleotides based on radioactivity measurements is influenced by the nature of the radiolabel. For example, phosphorothioate oligonucleotides labelled with ^{35}S at the 5' end consistently exhibit greater urinary excretion (AGRAWAL et al. 1991; IVERSEN et al. 1994) than those having internally placed ^{3}H or ^{14}C radioisotopes (COSSUM et al. 1993; SANDS et al. 1994; PHILLIPS et al. 1997). Radioactivity from oligonucleotides labelled with ^{14}C at the C_2 position of the thymine ring is largely cleared via expired air as $^{14}CO_2$ (COSSUM et al. 1993). These observations do not preclude the use of tracer isotopes for oligonucleotide pharmacokinetic studies, but advocate the use of supportive analytical methods whenever possible. Many analytical methods, e.g. PAGE (AGRAWAL et al. 1991), high-performance liquid chromatography (HPLC; COSSUM et al. 1993; SANDS et al. 1994), hybridisation-based assays (DE SERRES et al. 1996), capillary gel electrophoresis (LEEDS et al. 1996; CROOKE et al. 1996; PHILLIPS et al. 1997) and electrospray mass spectrometry (GAUS et al. 1997; PHILLIPS et al. 1997), have been developed to support oligonucleotide pharmacokinetic studies. Of these, capillary gel electrophoresis has the requisite resolving capability and sensitivity for reliable quantitation of parent compound and its metabolites in plasma and tissue samples. In our laboratories, capillary gel electrophoresis is routinely used to support pharmacokinetic studies with tritium-labelled oligonucleotides. Electrospray mass spectrometry is also used when more specific information about the in vivo biotransformation of oligonucleotides is required.

B. Distribution

I. Blood Kinetics

Phosphorothioate oligodeoxynucleotides are rapidly cleared from blood (AGRAWAL et al. 1991; IVERSEN 1991; GOODARZI et al. 1992; COSSUM et al. 1993; SANDS et al. 1994; ZHANG et al. 1995a; RIFAI et al. 1996; PHILLIPS et al. 1997). In vivo studies using CGP 64128A show biphasic blood kinetics (Fig. 3A) comprising a rapid distribution phase (0–240 min; $t_{1/2a}$, 19.6 min) followed by a prolonged elimination phase (240 min onwards; $t_{1/2b}$, 693 min), during which circulating tritium levels account for less than 1% of the administered dose. Total blood clearance (Cl) was 1.57 ml/min per kg with a volume of distribution (V_d) of 391 ml (or 1564 ml/kg, approximately two-fold greater than the total body water volume of the rat). Although CGP 64128A was metabolised during the initial phase (see Sect. C), its distribution to tissues was the primary mechanism for clearance from the vascular compartment. While biexponential blood kinetics are most frequently observed for phosphorothioate oligodeoxynucleotides (COSSUM et al. 1993), multi-component kinetics have been reported from more detailed studies using a large number of time points with narrow sampling intervals.

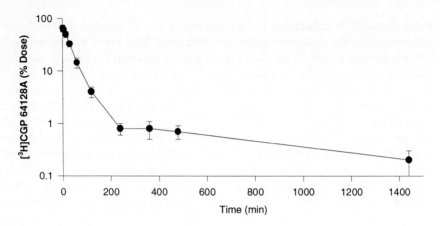
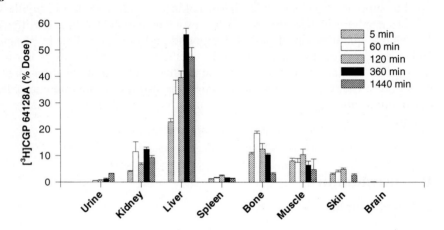
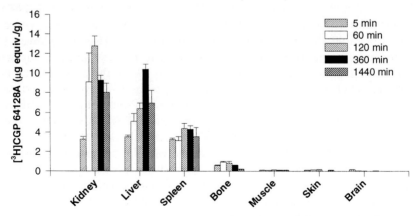

Fig. 3. A Blood kinetics and **B,C** time-dependent tissue distribution of [^{3}H]CGP 64128A (0.6 mg/kg) following intravenous administration to rats (mean ± SEM, n = 4)

II. Tissue Distribution

Phosphorothioate oligodeoxynucleotides have a distinctive tissue distribution pattern, as exemplified by CGP 64128A (Fig. 3B,C, Table 1). Comparison of its concentration in tissues with that of blood (at 120 min when 95% of the dose is cleared from the blood) show several tissues to have tissue:blood ratios greater than unity. These tissues – kidney, liver, spleen, pancreas, adrenal gland, salivary gland, mesenteric lymph nodes, duodenum and ileum – are able to accumulate CGP 64128A, suggesting that they bind to tissue matrix and/or that active uptake into cells occurs within these organs. The principle accumulating organs are the kidney and liver, which concentrate CGP 64128A 37-fold and 17.3-fold, respectively. The highest concentration is achieved in the kidney, whereas the liver accounts for the greatest proportion of the dose by virtue of its greater tissue mass. Interestingly, CGP 64128A is not evenly distributed within organs; in the kidney it is localised to the renal cortex (see also Cossum et al. 1993; Sands et al. 1994), and in bone it is present in the marrow. It is also widely distributed to, but not accumulated by, most other tissues except the brain and fat. The post-distribution clearance of phosphorothioate oligodeoxynucleotides from tissues occurs very slowly (Fig. 3B,C).

Comparison of these observations with those for other phosphorothioate oligodeoxynucleotides is complicated by numerous experimental variables (such as oligonucleotide length and sequence, radiolabel, detection methods, dose and animal species); however, they reflect the consensus of scientific literature (Agrawal et al. 1991; Iversen 1991; Goodarzi et al. 1992; Cossum et al. 1993; Sands et al. 1994; Zhang et al. 1995a; Rifai et al. 1996; Phillips et al. 1997).

III. Cellular Uptake In Vivo

Unequivocal evidence for cellular uptake of [^{3}H]CGP 64128A in vivo was provided by light microscopy autoradiography of tissues 360 min after an intravenous dose of 20 mg/kg (130 μCi) (Williamson et al. 1995). In the kidney, [^{3}H]CGP 64128A-derived radiolabel was concentrated in the renal cortex, with only low levels being associated with the renal medulla. High levels were present in the cells of the proximal tubules, but not distal tubules, where it had a perinuclear localisation (Fig. 4A). Only very low levels were associated with glomeruli at 360 min. Analogous studies with [^{3}H]CGP 69846A (ISIS 5132) in mice showed high concentrations of radioactivity in the Bowman's capsule at 2 min (R.A. Christian, unpublished observations). This is consistent with the observations by other workers (Rifai et al. 1996; Oberbauer et al. 1996; Carome et al. 1997) and provides direct evidence for renal filtration of phosphorothioate oligodeoxynucleotides. In the liver, [^{3}H]CGP 64128A-derived radiolabel was concentrated within specific cells lining the hepatic sinusoids (i.e. endothelial cells or Kupffer cells). Much lower levels were

Table 1. Tissue distribution of [^{3}H]CGP 64128A (0.6 mg/kg) at 120 min after intravenous administration to rats

Tissue	[^{3}H]CGP 64128A content		Mean tissue to blood concentration ratio	Tissue accumulation[a]
	Percent dose	μg equiv/g		
Kidney	13.9 ± 1.2	13.7 ± 1.0	37.0	High
Liver	39.6 ± 2.4	6.4 ± 0.6	17.3	High
Spleen	2.5 ± 0.3	4.4 ± 0.5	11.9	High
Pancreas	0.4 ± 0.06	1.16 ± 0.18	3.1	High
Adrenal gland	0.04 ± 0.01	1.07 ± 0.1	2.9	High
Bone	12.6 ± 2.0	0.83 ± 0.18	2.2	High
Salivary gland	0.23 ± 0.01	0.78 ± 0.13	2.1	High
Mesenteric lymph nodes	0.05 ± 0.01	0.74 ± 0.11	2.0	High
Duodenum	0.4 ± 0.07	0.7 ± 0.2	1.9	High
Ileum	3.4 ± 0.8	0.6 ± 0.2	1.6	High
Blood	4.8 ± 0.8	0.37 ± 0.08	1.0	n.a.
Lung	0.2 ± 0.03	0.23 ± 0.04	0.6	Low
Colon	0.3 ± 0.06	0.2 ± 0.03	0.5	Low
Skin	5.0 ± 0.5	0.17 ± 0.02	0.5	Low
Heart	0.09 ± 0.01	0.16 ± 0.02	0.4	Low
Muscle	10.5 ± 2.1	0.13 ± 0.02	0.4	Low
Thymus	0.03 ± 0.01	0.13 ± 0.01	0.4	Low
Gonads	0.2 ± 0.02	0.13 ± 0.02	0.4	Low
Eye	0.01 ± 0.00	0.09 ± 0.02	0.2	Low
Fat	1.0 ± 0.1	0.05 ± 0.1	0.1	Negligible
Brain	0.03 ± 0.00	0.03 ± 0.00	0.1	Negligible
Urine	0.9 ± 0.1	n.d.	n.d.	n.d.

n.a., not applicable; n.d., not determined.
[a] High, >1.0; low, <1.0; negligible, 0.1.

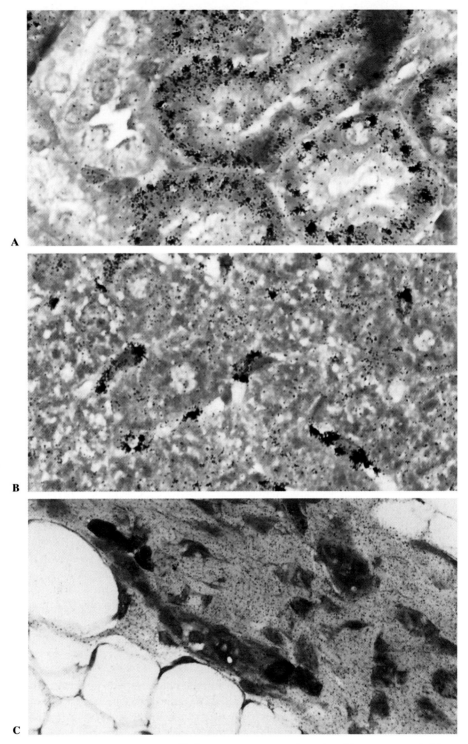

Fig. 4A–C. Tissue uptake of [^{3}H]CGP 64128A in vivo. **A** Kidney cortex. **B** Liver. **C** Skin

associated with the cytoplasm of parenchymal cells (Fig. 4B). This agrees with the results of INAGAKI et al. (1992), who reported preferential distribution of a phosphorothioate oligodeoxynucleotide to the non-parenchymal cells of liver. In skin, an abundant and diffuse localisation in the dermal region was observed; radioactivity was associated with the extracellular matrix rather than a specific cell type (Fig. 4C).

IV. Dose Dependence

The blood kinetics and tissue distribution of phosphorothioate oligodeoxynucleotides are dose dependent (PHILLIPS et al. 1996; RIFAI et al. 1996). This is shown for [^{3}H]CGP 64128A in Fig. 5. As the dose of CGP 64128A increased, it was disproportionately distributed between organs. The percentage of the dose associated with high-accumulation tissues decreased, leading to an extended blood circulation time and an increased percentage of the dose detected in low-accumulation tissues. For instance, as the intravenous dose increased from 0.06 to 0.6 to 6.0 mg/kg, the percentage renal and hepatic accumulation decreased from $16.9 \pm 2.6\%$ to $13.9 \pm 1.2\%$ to $6.1 \pm 0.3\%$ and from $59.1 \pm 3.7\%$ to $39.6 \pm 2.4\%$ to $25.0 \pm 0.7\%$, respectively. In concert, the plasma half-life increased from 20 min to 28 min to 50 min and the percentage of dose associated with skeletal muscle increased from $7.2 \pm 5.8\%$ to $10.5 \pm 2.1\%$ to $17.6 \pm 4.1\%$. Renal and hepatic uptake therefore appears to be the primary clearance mechanism for CGP 64128A from the blood, and this shows saturation at higher doses. Cross-inhibition studies between [^{3}H]CGP 69846A (ISIS 5132) and a range of polyanions provide further evidence for saturable uptake mechanisms by several high-accumulation tissues in vivo. Moreover, the cross-inhibition profile suggests that the "scavenger receptor" has an important role in its uptake by the liver. Hepatic uptake was significantly reduced by polyguanylic acid, polyinosinic acid and fucoidan, which are known polyanionic substrates for the scavenger receptor, but not by the closely related polyanions polycytidylic acid or chondroitin, which are not substrates (A. STEWARD and P.L. NICKLIN, unpublished observations). A similar role for the scavenger receptor has been shown in the kidney (SAWAI et al. 1996). The emerging picture is that the non-linear pharmacokinetics of phosphorothioate oligodeoxynucleotides in rodents results from a saturation of receptor-mediated cellular uptake processes, which are involved in their clearance from blood, as the dose increases.

V. Sequence Dependence

The phosphorothioate backbone is the dominant characteristic governing the pharmacokinetic handling of phosphorothioate oligodeoxynucleotides; however, tissue distribution is also affected by their base sequence. It is difficult to appreciate this sequence dependence when reviewing the current scientific literature, since it is hidden among other experimental variables such as the

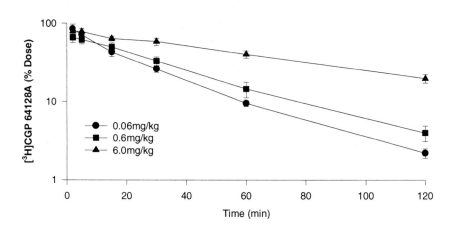
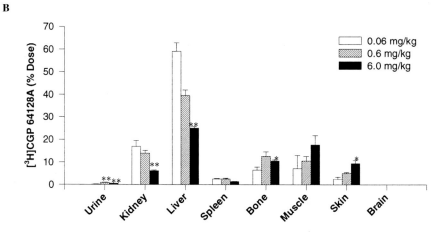
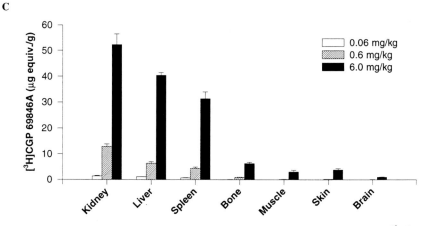

Fig. 5. A Dose-dependent blood kinetics and **B,C** tissue distribution of [^{3}H]CGP 64128A (0.06, 0.6 and 6.0 mg/kg) following intravenous administration to rats (mean ± SEM, $n = 3$–4)

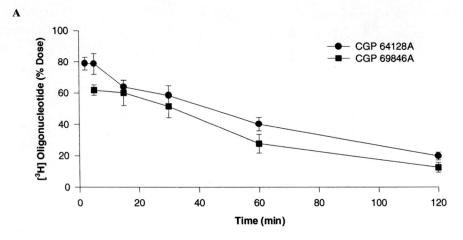

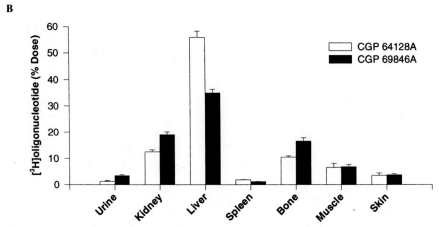

Fig. 6. Sequence-dependent blood kinetics and tissue distribution of phosphorothioate oligodeoxynucleotides in rats (mean ± SEM, $n = 3$)

detection method, dose and animal species. In our laboratories, we compared two 20-mer phosphorothioate oligodeoxynucleotides of different sequence (CGP 64128A and CGP 69846A, see Fig. 1) in rats at the same dose, over the same time and using the same detection methods. We observed similar blood kinetics, but different tissue distributions (Fig. 6). [^{3}H]CGP 64128A had a higher hepatic and splenic accumulation, whereas [^{3}H]CGP 69846A had a greater urinary excretion and distribution to kidney, bone and fat. The differential renal and hepatic accumulation for these two compounds was confirmed by capillary gel electrophoretic quantitation (data not shown). At this stage, there is no rational explanation for the apparent sequence dependence, although it may be dependent upon the relative affinity and spectrum of oligonucleotide–protein interactions.

VI. Multiple Dosing

Repeat dosing does not influence the pharmacokinetics or tissue distribution of phosphorothioate oligodeoxynucleotides. IVERSEN et al. (1994) and AGRAWAL et al. (1995a) administered ^{35}S-labelled phosphorothioate oligodeoxynucleotides by 28-day continuous subcutaneous infusion and repeated intravenous administration daily for 8 days, respectively, to rats and reported no plasma accumulation or saturation of tissue uptake. The stability of the ^{35}S-labelled oligonucleotide tracers over these time scales in vivo is questionable and makes meaningful interpretation of these data difficult. In more detailed studies, the blood kinetics and tissue distribution of CGP 64128A were unaffected by a multiple dose regimen. Rats were dosed by daily tail vein injection for 14 days with saline alone (groups 1 and 3) or with CGP 64128A at a dose of 0.6mg/kg (group 2) or 6.0mg/kg (group 4) mg/kg in 100μl saline. On day 15, a dose of 0.6mg [^{3}H]CGP 64128A per kg was administered to rats in groups 1 and 2, and 6.0mg [^{3}H]CGP 64128A per kg was administered to rats in groups 3 and 4. The saline groups 1 and 3 therefore represented single dosing, whereas groups 2 and 4 represent multiple dosing at their respective dose levels. The blood kinetics of [^{3}H]CGP 64128A were comparable in rats treated with single or multiple doses for 0.6mg/kg (group 1 vs. 2) and 6.0 mg/kg (group 3 vs. 4) (Fig. 7A). The tissue distribution of [^{3}H]CGP 64128A was statistically equivalent for rats dosed once (group 1) or daily for 14 days (group 2) with 0.6mg CGP 64128A per kg. Similarly, the single or multiple dose regimens at the 6.0mg/kg level (groups 2 and 4, respectively) had statistically equivalent tissue distributions (Fig. 7B,C). The dose-dependent blood kinetics and tissue distribution for [^{3}H]CGP 64128A were also evident in rats after dosing for 14 days and were similar to those observed after a single dose. In summary, the blood kinetics and tissue distribution of CGP 64128A were unaffected by daily administration of doses up to 6.0mg/kg over 14 days. The recent development of capillary gel electrophoresis for the quantitation of oligonucleotides and their metabolites in tissues (LEEDS et al. 1996; CROOKE et al. 1996; PHILLIPS et al. 1997) will allow the multiple-dose pharmacokinetic properties and tissue accumulation of phosphorothioate oligonucleotides to be studied in more detail.

C. Metabolism

Metabolic chain shortening of phosphorothioate oligodeoxynucleotides in vivo has been widely observed (AGRAWAL et al. 1991, 1995a; ZHANG et al. 1995a; CROOKE et al. 1996) and attributed to 3'-exonuclease-mediated cleavage through indirect experimental evidence. TEMSAMANI et al. (1993) showed that 3' capping, but not 5' capping, of a phosphorothioate oligonucleotide increased its in vivo stability compared to the uncapped compound. Furthermore, phosphorothioate oligodeoxynucleotides self-stabilised with a 3' hairpin loop were more stable in vitro (TANG et al. 1993) and in vivo (TANG et al. 1993;

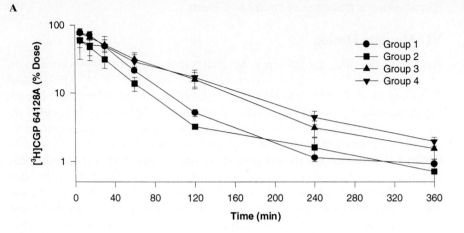
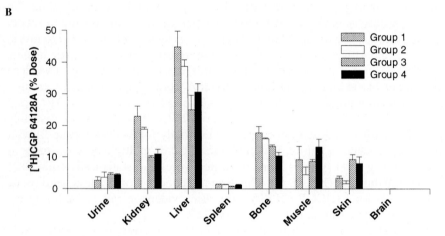
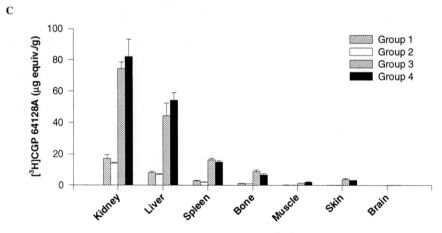

Fig. 7. A Blood kinetics and **B,C** tissue distribution of [³H]CGP 64128A (*groups 1 and 2*, 0.6 mg/kg; *groups 3 and 4*, 6.0 mg/kg) following intravenous administration to rats dosed for the previous 14 days with saline (controls, *groups 1 and 3*) or CGP 64128A (0.6 mg/kg or 6.0 mg/kg, *groups 2 and 4*, respectively) (mean ± SEM, $n = 3$)

ZHANG et al. 1995b) than their linear counterparts. At present, the metabolism of phosphorothioate oligodeoxynucleotides in vivo has been illustrated for CGP 64128A and CGP 69846A. Oligonucleotides were extracted from plasma and tissues (using methods described previously; LEEDS et al. 1996; CROOKE et al. 1996; PHILLIPS et al. 1997), and their metabolic fate was assessed using capillary gel electrophoresis and electrospray mass spectrometry.

I. Plasma

Intact CGP 64128A was detected in plasma extracts using capillary gel electrophoresis at 2.5, 10, 30, 60 and 120 min after a 6.0 mg/kg intravenous dose. In addition to the parent compound (N), a series of chain-shortened metabolites (N_{-n} nucleotides) were observed to migrate at a faster rate than CGP 64128A. The metabolite series expanded with time; however, the parent compound was the predominant circulating species throughout. For reasons that are unclear, metabolism occurred rapidly at first, but was followed by an apparent decrease in the rate of degradation (i.e. a 55.5 ± 0.8% loss of CGP 64128A within 10 min, but only a 60.7 ± 3.3% loss by 120 min). Possible explanations include substrate inhibition of nucleases in vivo (which has previously been observed in vitro; CROOKE 1995) and/or the random distribution of R_p and S_p phosphorothioate diastereoisomers, with their significantly different nuclease resistance (SPITZER and ECKSTEIN 1988), within the oligonucleotide. Electrospray mass spectrometry revealed these chain-shortened metabolites to be parent compound having successive nucleotide deletions mainly from its 3' end, with possible 5' end nucleotide deletions detected at later time points (Fig. 8). In plasma, therefore, metabolism principally occurred by processive degradation by 3'-exonuclease activity. In addition, mass spectrometry resolved a series of more subtle metabolites which had multiples of 16 Da mass units lower than that of the parent compound or its principle metabolites. For instance, the mass spectrum for CGP 64128A (calculated M_r, 6431.6 Da) contained two peaks; the principle peak of fully the thioated compound (measured M_r, 6430.4 Da) and a minor second peak with a mass 16 Da lower (measured M_r, 6415.9 Da), representing the monophosphodiester full-length material present as a synthetic by-product. The relative abundance of this minor peak increased in extracts of plasma collected between 2.5 min and 60 min after dosing. In addition, further peaks with masses 32 Da and 48 Da lower than the parent compound were also observed at later time points (Fig. 8E). It can be speculated that these metabolites may arise through an exchange of sulphur with oxygen (i.e. metabolic oxidation) at the phosphorothioate linkage.

II. Tissues

CGP 64128A and a series of chain-shortened metabolites were detected in kidney and liver extracts at 10, 60, 120 and 1440 min after a 6.0 mg/kg intravenous dose (Fig. 9). At the earliest time point, metabolites may have originated

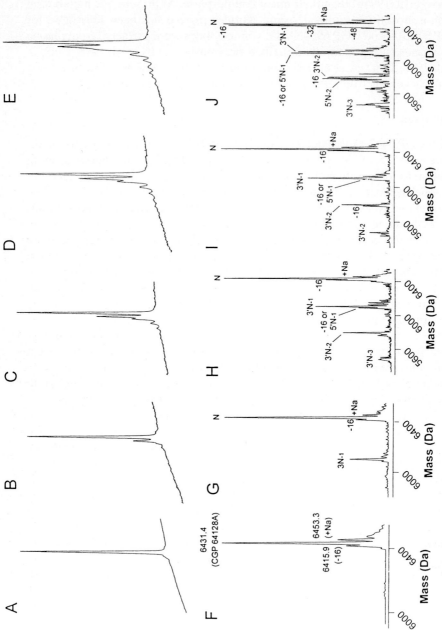

Fig. 8. A–E Capillary gel electrophoresis (Beckman P/ACE 5010 with Beckman replaceable gel system) and **F–J** electrospray mass spectrometry (Micromass Platform II, with the addition of 2% tripropylamine to extracts) of CGP 64128A-derived oligonucleotides in rat plasma extracts. **A,F** 0 min. **B,G** 2.5 min. **C,H** 10 min. **D,I** 30 min. **E,J** 60 min

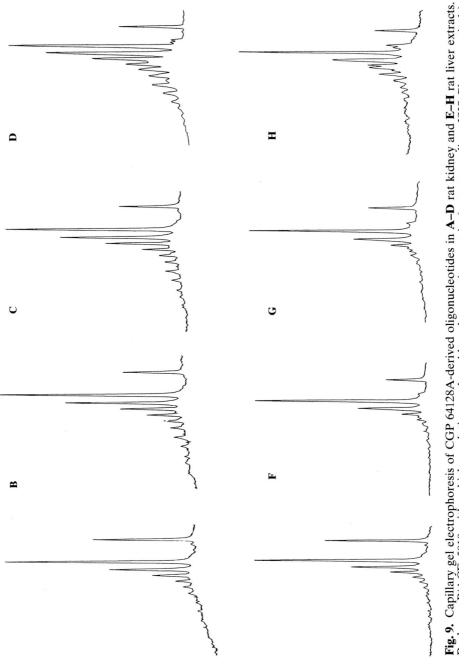

Fig. 9. Capillary gel electrophoresis of CGP 64128A-derived oligonucleotides in **A–D** rat kidney and **E–H** rat liver extracts. Beckman P/ACE 5010 with a high-resolution replaceable gel separation matrix (patent pending, ISIS Pharmaceuticals). **A,E** 10 min. **B,F** 120 min. **C,G** 360 min. **D,H** 1440 min

from the tissue itself or may have been present as circulating metabolites within residual blood of tissue samples. At later time points, when circulating levels were no longer detectable, oligonucleotides must have originated from the tissues. Comparison of the oligonucleotide profile in these two tissues showed the kidney to contain a more complex array of metabolites (as seen previously with other phosphorothioate oligodeoxynucleotides; Cossum et al. 1993). This could either result from the kidney extracting higher-order (shorter) metabolites more efficiently or could represent a greater rate of biotransformation in this tissue. We have seen no evidence for its renal metabolism in analogous studies with CGP 69846A (Phillips et al. 1997); the former explanation therefore appears to be the most plausible.

Hepatic metabolism of CGP 69846A was also examined by electrospray mass spectrometry (Fig. 10, Table 2). By comparison with calculated molecular weights, metabolites having masses consistent with 3′ cleavage alone, 5′ cleavage alone and both 3′ and 5′ cleavage were observed. These data suggest that, in addition to the assumed 3′ degradation, 5′-exonucleases are capable of degrading this phosphorothioate oligonucleotide in rat liver. A series of more subtle metabolites which had a mass 16 Da lower than that of the parent compound or its principle metabolites were also observed. Once again, it can be speculated that these occur by an exchange of sulphur with

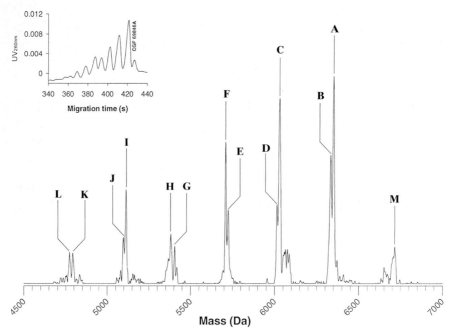

Fig. 10. Electrospray mass spectrometry of CGP 69846A-derived oligonucleotides in rat liver extracts

Table 2. Assignment of mass peaks corresponding to CGP 69846A metabolites in rat liver extracts

Peak	M$_r$ measured	Possibilities	3'-loss	5'-loss	Other	M$_r$ calculated
A	6343.7 ± 1.3	(i)				6344.6
B	6329.0 ± 2.4	(i)			−16 Da	6328.6
C	6024.4 ± 0.8	(i)	T			6023.6
		(ii)		T		6023.6
D	6007.8 ± 0.4	(i)	T		−16 Da	6007.6
		(ii)		T	−16 Da	6007.6
E	5719.4 ± 0.8	(i)		TC		5718.5
F	5702.8 ± 1.5	(i)	TT			5703.5
		(ii)		TC	−16 Da	5702.5
		(iii)	T	T		5702.5
G	5409.9 ± 0.4	(i)		TCC		5413.5
H	5375.8 ± 3.5	(i)	ATT			5374.5
I	5109.3 ± 1.5	(i)		TCCC		5108.5
J	5092.8 ± 0.4	(i)	T	TCC		5092.5
		(ii)		TCCC	−16 Da	5092.5
K	4786.3 ± 2.0	(i)	TT	TCC		4787.4
		(ii)	T	TCCC		4787.4
L	4771.3 ± 1.7	(i)	TT	TCC	−16 Da	4771.4
			T	TCCC	−16 Da	4771.4
M	6725.4 ± 0.4	(i)			+382 Da	n.a.

oxygen (i.e. metabolic oxidation) at the phosphorothioate linkage. The principle elimination of [^{14}C]ISIS 2105 as $^{14}CO_2$ in expired air (51% of administered radioactivity over 10 days) suggests that phosphorothioate oligonucleotide-derived bases are ultimately substrates for metabolism (Cossum et al. 1993).

In addition to chain-shortened and lower molecular weight metabolites, peaks which migrated more slowly than CGP 64128A (denoted N$_{+x}$; Fig. 9) were also detected in kidney and liver extracts. The N$_{+x}$ metabolites increased with time and were more abundant in the liver than kidney. Similar metabolites of phosphorothioate oligodeoxynucleotides have been reported previously (Agrawal et al. 1991), and others have suggested that they correspond to the parent compound with ribonucleotide adducts (Gaus et al. 1997).

D. Elimination

A simple relationship between blood and tissue levels does not exist for phosphorothioate oligodeoxynucleotides. They are rapidly distributed from the vascular to tissue compartments, from which they are slowly cleared. Significant tissue levels persist well beyond the time when circulating oligonucleotides can be detected. Cossum et al. (1993) showed that a [^{14}C]-labelled

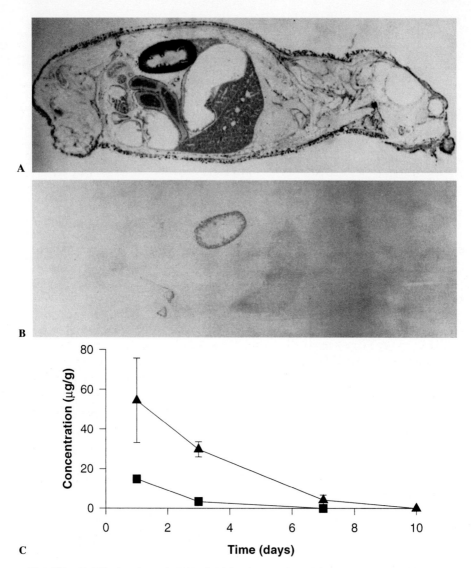

Fig. 11A–C. Elimination of CGP 69846A from mice. Whole-body autoradiography: [^{35}S]CGP 69846A (10 mg/kg) was intravenously administered to mice, which were killed at **A** 60 min and **B** 7 days and then processed for whole-body autoradiography. **C** Capillary gel electrophoresis. CGP 69846A (6 mg/kg) was intravenously administered to mice, which were killed at defined post-dose intervals. CGP 69846A-derived oligonucleotides in kidney (*triangles*) and liver (*squares*) extracts were quantified

20-mer phosphorothioate oligodeoxynucleotide (ISIS 2105 at 3.6 mg/kg) was slowly cleared from tissues, with low levels still remaining 10 days after a single intravenous dose. Clearance of a [^{35}S]-labelled 25-mer phosphorothioate oligodeoxynucleotide (GEM 91 at 30 mg/kg) appeared to occur more slowly, particularly from the kidney (ZHANG et al. 1995a). Similarly, post-distribution tissue levels of [^{3}H]CGP 64128A remained relatively constant between 6 and 24 h, with urinary and faecal excretion accounting for less than 6% of the administered dose over 24 h. Clearance studies with radioactivity as a readout must be interpreted cautiously, since catabolism of the radioactive tracer and subsequent anabolism of the radioisotope could generate misleading results. The clearance of CGP 69846A from mice that received a single intravenous dose was assessed using whole-body autoradiography (Fig. 11A,B) and capillary gel electrophoretic analysis of selected tissues (Fig. 11C). The combination of these methods enabled the clearance to be monitored simultaneously from a wide range of tissues and specific information to be obtained for the clearance of CGP 69846A and its metabolites from kidney and liver. Whole-body tissue-associated radioactivity was much reduced after 7 days; low levels were still present in the kidney cortex, and trace levels were associated with the liver. The clearance of unlabelled CGP 69846A from kidney and liver, as shown by capillary gel electrophoresis, reflected that observed for whole-body autoradiography elimination and also showed a progressive metabolism of CGP 69846A with time. These data support the notion that metabolism is an important clearance mechanism for phosphorothioate oligodeoxynucleotides. The rate of tissue clearance for CGP 69846A was similar to that for ISIS 2105 and greater than that for GEM 91. The reasons for this difference are not clear, but it may be related to dose dependence or the length of the parent compound and hence its major metabolites.

E. Absorption

Pharmacokinetic hurdles relating to the administration and absorption of antisense oligonucleotides are among the most important from a clinical perspective (see Fig. 2) and yet have received very little attention to date. Preclinical studies in monkeys have shown that phosphorothioate oligodeoxynucleotides can elicit acute haemodynamic (anticoagulation, complement activation) and cardiovascular (hypotension) side-effects once plasma levels exceed a "threshold" concentration (GALBRAITH et al. 1994; SHAW et al. 1997). If these effects could be circumvented by alternative delivery strategies which (a) maintain peak plasma levels below threshold concentrations and (b) are convenient for patients and health-care professionals, the therapeutic scope for this class of compounds would be greatly enhanced. The obstacles and opportunities associated with various parenteral, local and non-parenteral delivery routes are considered below.

I. Parenteral Administration

Bolus intravenous administration of antisense oligonucleotides for the treatment of systemic conditions is the most straightforward delivery mode from a pharmacokinetic perspective. Unfortunately, phosphorothioates cannot be administered in this way, since clinically relevant doses would result in peak plasma levels exceeding the putative threshold concentration for acute side-effects. Slow intravenous infusion, an approach adopted for the majority of clinical trials with phosphorothioate oligodeoxynucleotides, results in much lower but sustained peak levels and potentially circumvents these effects. It also incurs additional inconvenience to patients and health-care professionals.

Subcutaneous administration of phosphorothioate oligodeoxynucleotides (AGRAWAL et al. 1995a; PHILLIPS et al. 1997) is an alternative approach to reducing the rate of dose input. This is illustrated in Fig. 12 for a 0.6mg/kg subcutaneous dose of [^{3}H]CGP 64128A to rats. When compared to the same intravenous dose, a systemic bioavailability of 39.6% was achieved, peak blood concentrations were reduced five-fold and the tissue distribution pattern was identical. In addition, bioavailability increases with higher doses (see Table 3); the systemic appearance of metabolites is only marginally increased following subcutaneous administration (PHILLIPS et al. 1997), and this route has been shown to be efficacious for antisense oligonucleotides in the human tumour xenograft model (M. MÜLLER, unpublished observations). The subcutaneous route may therefore offer a viable alternative to intravenous infusion, and its clinical application warrants further consideration.

Table 3. Bioavailability of phosphorothioate oligodeoxynucleotides from parenteral and nonparenteral routes in rats

Administration route	Compound	Dose (mg/kg)	Bioavailability (%)
Parenteral			
Subcutaneous	CGP 64128A	0.6	39.6
	CGP 69846A	0.6	41.0
		6.0	42.9
		60.0	101.0
Intraperitoneal	CGP 64128A	0.6	27.8
Nonparenteral			
Oral gavage	CGP 64128A	6.0	0.3
Intrajejunal	CGP 64128A	6.0	1.9
Intraileal	CGP 64128A	6.0	0.2
Rectal	CGP 69846A	6.0	1.6
Intratracheal (pulmonary)	CGP 64128A	0.06	2.9
		0.6	14.0
		6.0	39.4

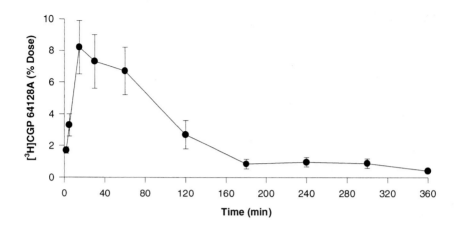
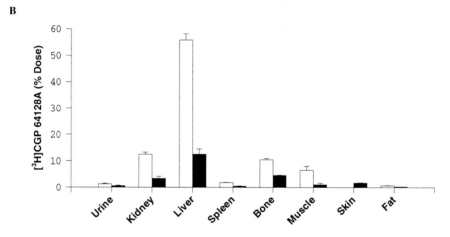
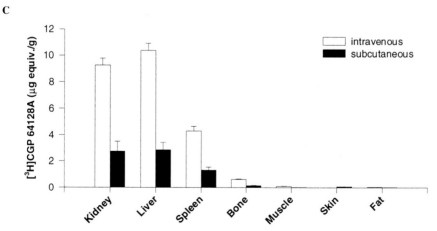

Fig. 12. A Blood kinetics following subcutaneous administration and **B,C** tissue distribution of [^{3}H]CGP 64128A (0.6 mg/kg) following intravenous and subcutaneous administration of [^{3}H]CGP64128A (0.6 to rats (mean ± SEM, $n = 4$)

II. Local Administration

Local administration has great potential for dose retention and therefore reduction of systemic side-effects by minimising distribution to non-target tissues. At least three oligonucleotides have entered clinical trials with local administration regimens. ISIS 2105 (Afovirsen, targeted against human papillomavirus-induced genital warts) is administered by intra-dermal injection directly below excised lesions. The failure to achieve statistically significant clinical activity may have resulted from poor retention at the site of intra-dermal administration, as observed in rats (Cossum et al. 1994). ISIS 2922 (fomivirsen, targeted against cytomegalovirus retinitis) is administered intra-ocularly. This site combines excellent retention with a low nuclease environment, allowing local therapeutic levels to be maintained by infrequent dosing and resulting in minimal exposure for non-target tissues. This compound is currently undergoing phase III clinical evaluation (Kisner 1996). GEM 132 (UGGGGCTTACCTTGCGAACA, a uniform phosphorothioate backbone with additional 2-O-methyl modification on underscored bases), the first 2'-modified antisense oligonucleotide to undergo clinical evaluation, is also being tested in patients with cytomegalovirus retinitis. Autoradiography studies with [^{35}S]GEM 132 showed radioactivity to be retained in the retina for longer than 1 month after a single intra-ocular injection (Schechter et al. 1996). In summary, local administration is a highly attractive approach, but is only applicable to a limited number of diseases.

III. Non-parenteral Administration

Non-parenteral administration, particularly oral dosing, is considered to be the "holy grail", since it combines clinical convenience with patient acceptability. The requirement for efficient transepithelial or transepidermal transport of oligonucleotides presents a formidable challenge.

The oral bioavailability of CGP 64128A in fasted rats was negligible. Less than 1% of [^{3}H]CGP 64128A was absorbed following administration to the gastrointestinal tract by gavage or direct introduction into isolated duodenal or ileal segments. Even this poor bioavailability probably represents an overestimation of the actual oral bioavailability, since partial degradation of the [^{3}H]CGP 64128A in the gastrointestinal tract (data not shown) would generate smaller and potentially more absorbable radioactive fragments. The small amount of absorbed radioactivity was distributed to the liver and kidney and probably represented partially intact oligonucleotides, since radioactive mononucleotides are accumulated in the spleen (Sands et al. 1994). There are many possible explanations for the poor systemic availability after oral administration, including the following: (a) precipitation in the acidic gastrointestinal environment, (b) binding to luminal contents, (c) metabolic instability in the lumen or at the absorptive epithelium, (d) binding to the mucosal surface, (e) low permeability of the gastrointestinal epithelium or (f) first-pass hepatic clearance. The physico-chemical properties of phosphorothioate oligodeoxynucleotides, i.e. high molecular weight (M_r, approximately 5–8kDa), multiple negative charges and hydrophilia ($\log D_{(octanol/water)}$, approximately −3.5), predict that that the gastrointestinal epithelium should present a significant physical barrier to their absorption. This is supported by the very low apparent permeability coefficient (P_{app}) for the apical-to-basolateral (mucosal-to-serosal) transport of [^{3}H]CGP 64128A across the Caco-2 monolayer model of the human gastrointestinal epithelium (PINTO et al. 1983; NICKLIN et al. 1992, 1995; BECK et al. 1996). Its P_{app} of 2.0×10^{-8} cm/s (R. FOX and P.L. NICKLIN, unpublished observations) predicts an oral absorption of less than 1% according to the correlation described by ARTURSSON and KARLSSON (1991). This is not surprising in light of its physicochemical properties. In summary, these data suggest that the oral route (including rectal administration, see Table 3) is not feasible for phosphorothioate oligodeoxynucleotides. Interestingly, significant oral bioavailability of a metabolically stabilised oligonucleotide has been reported (AGRAWAL et al. 1995b), and this needs to be examined further.

Pulmonary administration was the most promising non-parenteral delivery route, resulting in significant, dose-dependent systemic bioavailability of CGP 64128A (see Table 3; PHILLIPS and NICKLIN 1996). Relatively high pulmonary bioavailabilities have been reported previously for peptides (MACKAY et al. 1994), and the present work shows that this is equally true for antisense oligonucleotides. After intra-tracheal dosing of [^{3}H]CGP 64128A, blood levels increased slowly to a peak 20-fold lower than after intravenous administration and were maintained over extended time periods, indicating absorption-limiting pharmacokinetics. The systemic availability was confirmed by the detection of intact CGP 64128A and a typical array of metabolites in kidney and liver. Possible reasons for the dose-dependent bioavailability include the following: (a) a dose-related toxicity which increased the permeability of the pulmonary epithelium and (b) saturation of absorption-limiting tissue-binding at higher doses. Experiments with permeability markers and histological examination of the pulmonary tissues showed that the highest dose of CGP 64128A had no effect on the functional or morphological integrity of tracheal or lung epithelia, supporting the latter hypothesis. The dose-dependent characteristic can be viewed in two ways: low doses of oligonucleotide are retained at the administration site, while systemic availability can be achieved at higher doses. The potential of the pulmonary route for the local or systemic delivery of oligonucleotides should therefore be considered in more detail.

F. Conclusion

The pharmacokinetics of phosphorothioate oligodeoxynucleotides are well characterised. After intravenous administration, they are rapidly cleared from the blood compartment. There is a distinctive distribution amongst high (e.g. kidney, liver and spleen), low (e.g. skeletal muscle and skin) and negligible (e.g. brain) tissues. The pharmacokinetics are dose- and sequence-dependent but are unaffected by multiple dosing. Phosphorothioate oligodeoxynucleotides are primarily metabolised by 3'-exonucleases, however, 5'-exonuclease-mediated degradation also occurs. It appears that the modulation of pharmacokinetic parameters by alternative delivery routes may offer attractive opportunities for their use in the clinical setting.

References

Agrawal S (1992) Antisense oligonucleotides as antiviral agents. Trends Biotechnol Sci 10:152–157

Agrawal S, Temsamani J, Tang JY (1991) Pharmacokinetics, biodistribution and stability of oligodeoxynucleotide phosphorothioates in mice. Proc Natl Acad Sci USA 88:7595–7599

Agrawal S, Temsamani J, Galbraith W, Tang J (1995a) Pharmacokinetics of antisense oligonucleotides. Clin Pharmacokinet 28:7–16

Agrawal S, Zhang X, Lu Z, Hui Z, Tamburin JM, Yan J, Cai H, Diasio RB, Habus I, Jiang Z, Iyer RP, Yu D, Zhang R (1995b) Absorption, tissue distribution and in vivo stability in rats of a hybrid antisense oligonucleotide following oral administration. Biochem Pharmacol 50:571–576

Artursson P, Karlsson J (1991) Correlation between oral absorption in humans and apparent drug permeability coefficients in human intestinal epithelial (Caco-2) cell culture. Biochem Biophys Res Commun 175:880–885

Beck GF, Irwin WJ, Nicklin PL, Akhtar S (1996) Interactions of phosphodiester and phosphorothioate oligonucleotides with intestinal epithelial Caco-2 cells. Pharmacol Res 13:1028–1037

Carome MA, Kang YH, Bohen EM, Nicholson DE, Carr FE, Kiandoli LC, Brummel SE, Yuan CM (1997) Distribution of the cellular uptake of phosphorothioate oligodeoxynucleotides in the rat kidney in vivo. Nephron 75:82–87

Cossum PA, Sasmor H, Dellinger D, Truong L, Cummins LL, Owens SR, Markham PM, Shea JP, Crooke ST (1993) Disposition of the ^{14}C-labeled phosphorothioate oligonucleotide ISIS 2105 after intravenous administration to rats. J Pharmacol Exp Ther 267:1181–1190

Cossum PA, Sasmor H, Dellinger D, Truong L, Cummins L, Owens SR, Markham PM, Shea JP, Crooke ST (1994) Disposition of the ^{14}C-labeled phosphorothioate oligonucleotide ISIS 2105 after intradermal administration to rats. J Pharmacol Exp Ther 269:89–94

Crooke ST (1995) Therapeutic applications of oligonucleotides. Landes, Austin, Texas

Crooke ST, Graham MJ, Zuckerman JE, Brooks D, Conklin BS, Cummins LL, Greig MJ, Kornburst D, Manoharan M, Sasmor H, Schleich T, Tivel KL, Griffey R (1996) Pharmacokinetic properties of several oligonucleotide analogs in mice. J Pharmacol Exp Ther 277:923–937

Dean NM, McKay R, Miraglia L, Howard R, Cooper S, Giddings J, Nicklin PL, Miester L, Ziel R, Geiger T, Müller M, Fabbro D (1996) Inhibition of human tumor cell lines in nude mice by an antisense inhibitor of PKC-α expression. Cancer Res 56:3499–3507

De Serres M, McNulty MJ, Christensen L, Zon G, Findlay JWA (1996) Development of a novel scintillation proximity competitive hybridisation assay for the determination of phosphorothioate antisense oligonucleotide plasma concentrations in a toxicokinetic study. Anal Biochem 233:228–233

Galbraith WM, Hobson WC, Giclas PC, Schechter PJ, Agrawal S (1994) Complement activation and hemodynamic changes following intravenous administration of phosphorothioate oligonucleotides in the monkey. Antisense Res Dev 4:201–206

Gaus HJ, Owens SR, Winniman M, Cooper S, Cummins LL (1997) On-line HPLC electrospray mass spectrometry of phosphorothioate oligonucleotide metabolites. Anal Chem 69(3):313–319

Goodarzi G, Watabe M, Watabe K (1992) Organ distribution and stability of phosphorothioated oligodeoxyribonucleotides in mice. Biopharm Drug Dispos 13:221–227

Graham MJ, Freier SM, Crooke RM, Ecker DJ, Maslova RN, Lesnik EA (1993) Tritium labeling of antisense oligonucleotides by exchange with tritiated water. Nucleic Acids Res 21:3737–3743

Helene C, Toulme JJ (1990) Specific regulation of gene expression by antisense, sense and antigene nucleic acids. Biochim Biophys Acta 1049:99–125

Higgins KA, Perez JR, Coleman TA, Dorshkind K, McComas WA, Sarmiento UM, Rosen CA, Narayanan R (1993) Antisense inhibition of the p65 subunit of NF- B blocks tumorigenicity and causes tumor regression. Proc Natl Acad Sci USA 90:9901–9905

Inagaki M, Togawa K, Carr BL, Ghosh K, Cohen JS (1992) Antisense oligonucleotides: inhibition of liver cell proliferation and in vivo disposition. Transplant Proc 24:2971–2972

Iversen P (1991) In vivo studies with phosphorothioate oligonucleotides: pharmacokinetics prologue. Anti Cancer Drug Des 6:531–538

Iversen P, Mata J, Tracewell WG, Zon G (1994) Pharmacokinetics of an antisense phosphorothioate oligodeoxynucleotide against rev from human immunodeficiency virus type 1 in the adult male rat following single injections and continuous infusions. Antisense Res Dev 4:43–52

Kisner D (1996) Development of phosphorothioate oligonucleotides. Proceedings of the XII international roundtable: nucleosides, nucleotides and their biological applications, La Jolla, CA

Leeds JM, Graham MJ, Truong L, Cummins LL (1996) Quantification of phosphorothioate oligonucleotides in human plasma. Anal Biochem 235:36–43

Mackay M, Phillips JA, Steward A, Hastewell JG (1994) Pulmonary absorption of therapeutic peptides and proteins. In: Byron PR, Dalby RN, Farr SJ (eds) Respiratory drug delivery IV. Buffalo Grove, IL, Interpharm Press, pp 31–37

Monia BP, Johnston JF, Geiger T, Müller M, Fabbro D (1996) Antitumor activity of a phosphorothioate antisense oligodeoxynucleotide targeted against c-raf kinase. Nature Med 2:668–675

Nicklin PL, Irwin WJ, Hassan IF, Williamson I, Mackay M (1992) Permeable support type influences the transport of compounds across Caco-2 cells. Int J Pharmacol 83:197–209

Nicklin PL, Irwin WJ, Hassan IF, Mackay M, Dixon HBF (1995) The transport of acidic amino acids and their analogues across monolayers of human intestinal absorptive (Caco-2) cells in vitro. Biochim Biophys Acta 1269:176–186

Oberbauer R, Schreiner GF, Meyer TW (1995) Renal uptake of an 18-mer phosphorothioate oligonucleotide. Kidney Int 48:1226–1232

Oberbauer R, Murer H, Schreiner GF, Meyer TW (1996) Antisense and the kidney. Kidney Blood Press Res 19:221–224

Phillips JA, Nicklin PL (1996) The pulmonary route: potential for the non-parenteral delivery of oligonucleotides? In: Proceedings of the international congress: therapeutic oligonucleotides, Rome, Italy, 10–12 June 1996

Phillips JA, Craig SJ, Nicklin PL (1996) Dose-dependent disposition of a 20-mer phosphorothioate oligonucleotide (CGP 69846A) after intravenous administration. Proceedings of the international congress: therapeutic oligonucleotides, Rome, Italy

Phillips JA, Craig SJ, Bayley D, Christian RA, Geary R, Nicklin PL (1997) Pharmacokinetics, metabolism and elimination of a 20-mer phosphorothioate oligodeoxynucleotide (CGP 69846A) after intravenous and subcutaneous administration. Biochem Pharmacol 54:657–668, 1997

Pinto M, Robine-Leon S, Appay MD, Kedinger M, Triadou N, Dussalux E, Lacroix B, Assman S, Haffen P, Fogh J, Zweibaum A (1983) Enterocyte-like differentiation and polarisation of the human colon carcinoma cell line Caco-2 in culture. Biol Cell 47:323–330

Rifai A, Brysch W, Fadden K, Clark J, Schlingensiepen K-H (1996) Clearance kinetics, biodistribution and organ saturability of phosphorothioate oligodeoxynucleotides in mice. Am J Pathol 149:717–725

Sands H, Gorey-Feret LJ, Cocuzza AJ, Hobbs FW, Chidester D, Trainor GL (1994) Biodistribution and metabolism of internally ^{3}H-labeled oligonucleotides. 1. Comparison of a phosphodiester and a phosphorothioate. Mol Pharmacol 45:932–943

Sawai K, Mahato RI, Oka Y, Takakura Y, Hashida M (1996) Disposition of oligonucleotides in isolated perfused rat kidney: involvement of scavenger receptors in their renal uptake. J Pharmacol Exp Ther 279:284–290

Schechter PJ, Martin RR, Grindel JM (1996) Clinical results of drug candidates. Proceedings of the XII international roundtable: nucleosides, nucleotides and their biological applications, La Jolla, CA

Shaw DR, Rustagi PK, Kandimalla ER, Manning AN, Jiang Z, Agrawal S (1997) Effect of synthetic oligonucleotides on human complement and coagulation. Biochem Pharmacol 53:1123–1132

Spitzer S, Eckstein F (1988) Inhibition of deoxyribonucleases by phosphorothioate groups in oligodeoxyribonucleotides. Nucleic Acids Res 16:11691–11704

Stein CA (1996) Phosphorothioate antisense oligodeoxynucleotides: questions of specificity. Trends Biotechnol 14:147–149

Tang JY, Temsamani J, Agrawal S (1993) Self-stabilized antisense oligodeoxynucleotide phosphorothioates: properties and anti-HIV activity. Nucleic Acids Res 21:2729–2735

Temsamani J, Tang JY, Padmapriya A, Kubert M, Agrawal S (1993) Pharmacokinetics, biodistribution and stability of capped oligodeoxynucleotide phosphorothioates in mice. Antisense Res Dev 3:277–284

Wagner RW, Flanagan M (1997) Antisense technology and prospects for therapy of viral infections and cancer. Mol Med Today 3:31–38

Williamson I, Phillips JA, Nicklin PL (1995) Pharmacokinetics, organ distribution and cellular uptake of phosphorothioate oligonucleotides in vivo. Proceedings of therapeutic oligonucleotides from cell to man, Selliac, France

Zamecnik PC, Stevenson ML (1978) Inhibition of Rous sarcoma virus replication and cell transformation by a specific oligodeoxynucleotide. Proc Natl Acad Sci USA 75:280–284

Zhang R, Diasio RB, Lu Z, Liu T, Jiang Z, Galbraith WM, Agrawal S (1995a) Pharmacokinetics and tissue distribution in rats of an oligodeoxynucleotide phosphorothioate (GEM-91) developed as a therapeutic agent for human immunodeficiency virus type-1. Biochem Pharmacol 49:929–939

Zhang R, Lu Z, Zhang X, Zhao H, Diasio RB, Liu T, Jiang Z, Agrawal S (1995b) In vivo stability and disposition of a self-stabilized oligodeoxynucleotide phosphorothioate in rats. Clin Chem 41:836–843

CHAPTER 5
Toxicity of Oligodeoxynucleotide Therapeutic Agents

A.A. LEVIN, D.K. MONTEITH, J.M. LEEDS, P.L. NICKLIN, R.S. GEARY,
M. BUTLER, M.V. TEMPLIN, and S.P. HENRY

A. Introduction

One of the most exciting aspects of antisense therapeutic agents is their remarkable pharmacologic specificity. Because antisense activity depends on Watson and Crick base-pairing rules, sequence specificity ensures that antisense agents have high selectivity for the intended mRNA target. The selective inhibition of the expression of disease-related genes is the pharmacologic equivalent of laser surgery, well aimed and presumably with minimal inadvertent effects. This specificity suggests that these agents will have favorable therapeutic indices.

Antisense oligodeoxynucleotides as a class promise to be among the first group of antiviral and anticancer agents that successfully separate pharmacologic activity from toxicologic activity. For example, in traditional cancer chemotherapy, drugs produce their pharmacologic effects by killing rapidly dividing cells. Unfortunately, this is also the mechanism of toxicity. Similarly, most antiviral therapy agents inhibit DNA polymerases, which is the mechanism of both pharmacologic and toxicologic activities. Because of the overlapping of mechanisms of pharmacologic and toxicologic activities, it is difficult to have significant separation between therapeutic doses and toxic doses with these classes of drugs. The therapeutic index of antisense drugs is partially determined by the selected target. The pharmacologic activity of antisense therapeutic agents is the result of inhibition of translation of specific mRNAs. A specific mRNA may be targeted because it is inappropriately expressed in a disease state. Successful antisense therapy results in a temporary "knockdown" of the target gene. Inhibiting the expression of some key genes may be one of the potential mechanisms of toxicity with antisense therapeutic agents. In this case, the adverse effect would be associated with the intended antisense mechanism, and the therapeutic index of the antisense agent might be more akin to traditional drugs. For antisense therapeutic agents that produce toxicity through gene expression knockout, toxicity will be avoided through careful control of dose, in much the same way agents with low therapeutic indices are employed today. For toxicity studies in animals, we have administered oligodeoxynucleotides that were directed to either human mRNA targets or species-specific antisense molecules for the same targets. In our experience to date, there have been no instances in which we or others observed toxicities that were considered to be related to an inhibition of the target gene (HENRY et al. 1976, 1997a,e,g; SARMIENTO et al. 1994), despite the fact that inhibition of the target gene expression was demonstrated in a number of animal models (BENNETT et al. 1997; MONIA et al. 1992, 1995, 1996; STEPKOWSKI et al. 1994). The absence of toxicity through an antisense mechanism of action is probably related to the prudent selection of mRNA targets. For obvious reasons, mRNA targets that would dramatically affect cell viability have been avoided. (Note that developmental biologists using the antisense technology as part of efforts to dissect out the role of gene expression in morphogenesis have produced what might referred to as sequence-related toxicities; AUGUSTINE et al. 1995; SADLER et al. 1995; see also Sect. C.IV).

Other potential toxic effects that are caused by antisense-mediated mechanisms are those which occur through cross-hybridization of the antisense agent and some mRNA that is not the intended target mRNA. In theory, this type of event should be exceedingly rare. The specificity of the antisense agent is designed into the sequence and is a function of the number of bases in the antisense compound. With an oligodeoxynucleotide 15–20 bases in length, there is little probability that hybridization with an unintended target could occur. In practice, this probability is diminished even further, given the fact that hybridization affinity decreases significantly if there are mismatches and that only a fraction of perfectly matched antisense oligodeoxynucleotides have significant pharmacologic activity. There is a growing body of data at Isis Pharmaceuticals (Carlsbad, USA) on the relative activities of 20-mer oligodeoxynucleotide sequences that span an entire gene from 3′-untranslated regions to 5′-untranslated regions. Depending on the gene target, many if not most oligodeoxynucleotides in this type of "gene walk" are relatively weak inactivators of gene expression (MIRAGLIA et al. 1996; BENNETT et al. 1996, 1997; MONIA et al. 1995, 1996; CHIANG et al. 1991; STEPKOWSKI et al. 1994; BENNETT 1994; DEAN et al. 1994). This apparent fastidiousness for specific target sites may be related to accessibility of the antisense construct to targeted mRNA sequence. Only sites on mRNA that are accessible to the hybridizing antisense molecule will be "hot spots" for antisense inhibition. Depending on the mRNA and its secondary structure, there may be a limited number of ideal (or active) target sites for antisense inhibition. This accessibility factor further reduces the probability that there would be unintended hybridization. In addition, other studies have demonstrated that there is a significant loss of pharmacologic activity when the antisense compound has one or more mismatched bases, suggesting that potent pharmacologic activity requires near-perfect sequence matches (MONIA et al. 1996). Taken together, these data suggest that the probability of toxicity occurring as a result of unintended cross-hybridization is vanishingly small.

While toxicity is unlikely to be the result of hybridization to unintended targets, there are other potential mechanisms of toxicity unrelated to the antisense mechanism. Antisense therapeutic agents, like all other xenobiotics, have their own inherent toxicities, which stem from the physical or chemical characteristics of the compounds. In contrast to antisense-mediated toxicities, these toxicities would generally be less dependent on sequence and more dependent on the chemical class. The class-related toxicities can be thought of as effects that are mediated through mechanisms other than hybridization, e.g., oligodeoxynucleotide–protein interactions. The first generation of antisense agents, the phosphorothioate oligodeoxynucleotides, are water-soluble, polyanionic molecules with a length of 20–25 nucleotides. These compounds are known to bind to proteins with affinities that span the micromolar to the millimolar ranges (Brown et al. 1994; Srinivasan et al. 1995). These protein-binding affinities suggest that there can be significant interactions with proteins at plasma levels attained in toxicity studies. Some specific examples of toxicities induced by protein binding will be discussed below. Note that if the physical and chemical properties of phosphorothioate oligodeoxynucleotides are the predominant driving forces in these protein interactions, then it would be expected that sequence or length would not be particularly strong influences on toxicity, because small changes in sequence and length have little effect on the physical chemical properties of an antisense agent. In fact, the results from preclinical toxicity studies support this hypothesis.

The phosphorothioate oligodeoxynucleotides have now been examined in a full range of acute, chronic, and reproductive studies in rodents, lagomorphs, and primates. It is clear from these studies and clinical trials that phosphorothioate oligodeoxynucleotides can be administered safely to animals and humans at pharmacologically relevant doses for extended periods of time. At high doses, there is a distinctive pattern of toxicity that is common to all phosphorothioate oligodeoxynucleotides (Henry et al. 1996, 1997a,e,g; Sarmiento et al. 1994). The remarkable similarity in toxicity with different oligodeoxynucleotides suggests that, for this class of antisense agents, toxicity is independent of sequence and is the result of non-antisense-mediated mechanisms. These toxicities arise from unintended interactions of the antisense compounds with plasma or cellular components independent of base-pairing rules. The most probable mechanism of toxicity is the binding of oligodeoxynucleotides to proteins. These non-antisense-mediated pathways are thought to be responsible for most, if not all, of the toxicities associated with the administration of these compounds to laboratory animals. This conclusion is borne out in studies in which little or no differences in toxicity is observed between pharmacologic active and inactive sequences (Henry et al. 1997a,e,g; Sarmiento et al. 1994).

Although there are similarities in the patterns of toxicologic response that are independent of sequence, there are occasional differences in the relative potency between sequences. This spectrum of potency is thought to be due to specific sequence motifs that might enhance the protein binding and ultimately toxicity. To date, however, only quantitative differences exist, and qualitatively the toxicities are similar (Henry et al. 1997a,e,g; McIntyre et al. 1993;

SARMIENTO et al. 1994). Thus there is an apparent conundrum regarding toxicities that are generally independent of sequence hybridization, but that may be enhanced by specific sequences. Sequence-related modulation of toxicities is best exemplified by the immunostimulatory effects of oligodeoxynucleotides.

This chapter focuses on the non-antisense-mediated toxicities of the phosphorothioate oligodeoxynucleotides and will present an initial review of the toxicities of the next generation of antisense agents, modified phosphorothioate oligodeoxynucleotides. A comparison of the toxicologic profiles of the different chemistries suggests that some of the class-related toxicities of phosphorothioate oligodeoxynucleotides can be ameliorated by chemical modifications. In contrast to the similarities in toxicity of phosphorothioate oligonucleotides from sequence to sequence, the toxicities varied greatly from species to species. There are characteristic responses for rodents and a different set of responses in primates. This chapter will also review the species differences in responses and the relationship between preclinical data and current clinical data.

The goal of this chapter is to review the existing data on the preclinical toxicity in order to demonstrate that the safety profile of these compounds supports their continued use as therapeutic agents. Additional data on novel oligonucleotide chemistries and their toxicities are presented to provide insight into the improved safety profiles of the next generation of antisense therapeutic agents. (Note that a listing of all of the sequences mentioned in this chapter is included in Table 1).

Table 1. Sequences and structures of oligonucleotides

Name	Sequence/linkages
ISIS 3521	GsTsTsCsTsCsGsCsTsGsGsTsGsAsGsTsTsTsCsA
ISIS 2105	TsTsGsCsTsTsCsCsAsTsCsTsTsCsCsTsCsGsTsC
ISIS 2922	GsCsGsTsTsTsGsCsTsCsTsCsTsTsCsTsTsCsTsTsGsCsG
ISIS 2302	GsCsCsCSCsAsAsGsCsTsGsGsCsAsTsCsCsGsTsCsA
ISIS 1082	GsCsCsGsAsGsGsTsCsCsAsTsGsTsCsGsTsAsCsGsC
ISIS 3082	TsGsCsAsTsCsCsCsCsCsAsGsGsCsCsAsCsCsAsT
CGP 69846A diester	ToCoCoCoGoCoCoToGoToGoAoCoAoToGoCoAoToT
ISIS 5132	TsCsCsCsGsCsCsTsGsTsGsAsCsAsTsGsCsAsTsT.
CGP 71849A	tsc′sc′sc′sgsc′sCsTsGsTsGsAsCsAstsgsc′sastst
CGP69845A	toc′oc′oc′ogoc′sCsTsGsTsGsAsCsAstogoc′oaotot
ISIS 12449	AsCsCsGsAsTsAsAsCsGsTsTsGsCsCsGsGsTsGsAsCsG
rel a	GsAsCsGsGsGsAsAsAsCsAsGsAsTsCsGsTsCsCsAsTsGsGsT
rel a SENSE	AsCsCsAsTsGsGsAsCsGsAsTsCsTsGsTsTsTsCsCsCsTsC
GEM 91	CsTCsTsCsGsCsACsCsCsAsTsCsTsCsTsCsTsCsCsTsTsCsCsT
GEM 91-C	<u>CTCTC</u>sGsCsAsCsCsAsTsCsTsCsTsCsTsCs<u>CTTCT</u>
GEM 91-H	**CsUsCsUsCsGsCsAsCsCsCsAsTsCsTsCsTsCsTsCsCsUsUsCsU**
En-1	TsTsAsGsCsTsTsCsCsTsGsGsTsGsCsGsGsTsGsGsA
AR177	GsToGoGoToGoGoGoGoToGoGoGoGoToGoGoGs

s, phosphorothioate linkage; o, phosphodiester linkage; c′, 5-methyl C; lower case bold, 2′-methoxyethoxy; upper case bold, 2′-methoxy; underlined, methylphosphonate linkage.

B. Pharmacokinetics and Metabolism

The toxicity and pharmacokinetics of phosphorothioate oligodeoxynucleotides were characterized in mice, rats, and monkeys exposed from 2 to 26 weeks. Some of these data are reviewed elsewhere in this volume (see Chaps. 4, 6, this volume), but a brief summary of the pertinent information is presented to aid in the interpretation of toxicity data.

Existing data on the pharmacokinetics and toxicity is limited to parenteral routes of exposure, including intravenous, intradermal, subcutaneous, intraperitoneal, and intravitreal injections (see Chap. 4, this volume; AGRAWAL et al. 1991; ALTMANN et al. 1996; COSSUM et al. 1993, 1994; IVERSEN 1991; LEEDS et al. 1996a, 1997a; SAIJO et al. 1994; SANDS et al. 1994; SRINIVASAN and IVERSEN 1995), because the oral bioavailability of the present generation of oligodeoxynucleotide therapeutic agents is limited (AGRAWAL et al. 1995b). Phosphorothioate oligodeoxynucleotides are absorbed from parenteral sites of administration rapidly, and there is significant systemic distribution of intact phosphorothioate oligodeoxynucleotides from all these routes of administration (see Chap. 4, this volume; ALTMANN et al. 1996; AGRAWAL et al. 1991; COSSUM et al. 1993, 1994; IVERSEN 1991; LEEDS et al. 1996a, 1997a; SANDS et al. 1994; SRINIVASAN and IVERSEN 1995; SAIJO et al. 1994), except for intravitreal injection, where systemic exposure is negligible (see Sect. C.V; LEEDS et al. 1997b). While there are some species differences, in general, phosphorothioate oligodeoxynucleotides in plasma are cleared with a half-life of 1 h or less (depending on dose). Both distribution of oligodeoxynucleotide to tissues and metabolism are factors in the clearance from plasma, but it appears that distribution is the predominant factor.

Phosphorothioate oligodeoxynucleotides can be detected in nearly all tissues and organs within minutes of intravenous, subcutaneous, and intradermal doses. The exceptions to the broad tissue distribution include brain and testes, which appear to have significant barriers for the transport of these types of molecules and are not target sites for direct toxicity.

Accumulation in other organs may predispose them to toxicity with repeated exposure from high doses. Following subcutaneous, intravenous, and intradermal administration, the liver and kidney are the major sites of accumulation of phosphorothioate oligodeoxynucleotides, with the spleen, bone marrow, and lymph nodes having somewhat lower levels of oligodeoxynucleotide. Other tissues and organs have some levels of oligonucleotide, but these are lower than the above. This pattern of distribution is similar in mice, rats, and monkeys (Fig. 1). Plasma kinetics are not linear with dose, and this nonlinearity may be related to organ distribution. Plasma concentrations increase proportionally to dose, but the increases are greater than would be predicted on the basis of dose alone (see Chap. 4, this volume; DE SERRES et al. 1996; RIFAI et al. 1996; SRINIVASAN and IVERSEN 1995). Part or all of the nonlinearity in plasma kinetics may be related to the saturation of tissue uptake. Liver and kidney concentrations of oligodeoxynucleotide appear to

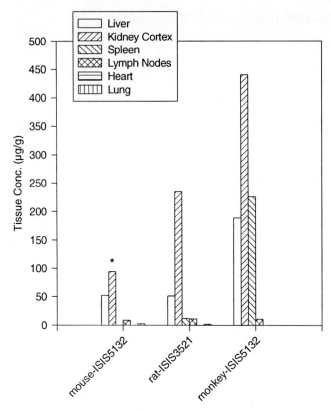

Fig. 1. Comparative tissue distribution of phosphorothioate oligodeoxynucleotides for mouse, rat, and monkey treated with 4, 3, and 3 mg/kg, respectively, every other day for 28 days. Whole kidney was measured in mice (*asterisk*). The mean values for three to six animals per group is plotted. (From GEARY et al. 1997b)

saturate as dose is increased (Fig. 2). In this experiment, there was marked saturation of liver uptake over the dose range. Renal concentrations of oligonucleotides also appear to saturate, increasing only fourfold over a tenfold dose range, but the degree of saturation is not as apparent as that in the liver. Preliminary data suggest that the degree of saturation observed at these dose levels would be altered if the doses were administered over longer periods of time. In fact, prolonging the infusion times allows for increased accumulation by liver and kidney, suggesting that the nonlinearity is both a function of dose rate and of dose (R.S. GEARY and A.A. LEVIN, unpublished observation). These saturation phenomena have significant implications for understanding the relationships between dose and toxicity. If tissue saturation controls plasma pharmacokinetics, then it is possible that changes in the organ distribution could ultimately affect plasma concentrations and alter the predictable patterns now established between dose and peak plasma levels.

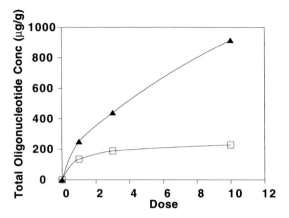

Fig. 2. Total oligodeoxynucleotide concentration in liver (*white squares*) and renal cortex (*black triangles*) of monkeys treated with ISIS 5132. Mean oligodeoxynucleotide concentrations determined by capillary gel electrophoresis from the tissues obtained from monkeys treated for 4 weeks by 2-hour intravenous infusion at 1, 3, and 10 mg/kg every other day. The values represent the means of six animals per group

The metabolism of oligodeoxynucleotides and rate of clearance from tissues determine tissue accumulations and ultimately may influence toxicity. Phosphorothioate oligodeoxynucleotides are cleared from tissues and organs by nuclease-mediated metabolism with half-lives that vary from 20 to 120 h depending on the tissue or organ (Cossum et al. 1993, 1994; Geary et al. 1997). Following repeated administration, the extensive partitioning into tissues, particularly liver and kidney, combined with the relatively slow metabolic clearance rates results in significant accumulation of oligodeoxynucleotide in these tissues. Using organ half-life data from single-dose studies, we have obtained estimates of the degree of accumulation. The extent of accumulation at steady state varies with half-life; steady state tissue levels are shown in Table 2. As shown, there is accumulation in tissues even after every-other-day dosing. These values indicate that repetitive dosing on a daily basis is not necessary to maintain tissue concentrations of phosphorothioate oligodeoxynucleotides, and pharmacologically active concentrations might be maintained with less frequent administration. Thus many toxicity studies with phosphorothioate oligodeoxynucleotides have employed every-other-day dosing regimens. On the basis of accumulation and exposure, liver and kidney would be the most likely target organs for toxicity. In fact, histopathologic and functional changes have been observed in these organs receiving high doses at the end of subchronic and chronic dosing (see below).

To understand the toxicologic profile of these compounds, it is necessary to understand the fate of oligodeoxynucleotides in tissues. Techniques such as fluorescent labeling, immunohistochemistry, and autoradiography have been employed to localize oligodeoxynucleotides in tissues (Oberbauer et al. 1995;

Table 2. Tissue clearance half-lives and estimates of multiple-dose accumulation in different tissues following intravenous administration of ISIS 3521 to the monkey

Tissue	Half-life (h) (ISIS 3521)	Accumulation (multiple of single dose)
Kidney cortex	120	4.1
Kidney medulla	74	2.8
Liver	66	2.5
Spleen	91	3.3
Bone marrow	32	1.5
Lymph nodes	22	1.3

Tissue half-lives are expressed as the time (in h) to clear parent compound (ISIS 3521) from various tissues. The multiple-dose accumulation factor is a variable dependent on the half-life that represents the n-fold increase in tissue concentrations at steady state versus the tissue concentrations after a single dose. The steady state tissue concentration is a function of the half-life for each tissue such that, as clearance slows, the steady state concentration increases. Therefore, the accumulation in tissue is highest for those tissues that exhibit the longest clearance half-lives.

PLENAT et al. 1994; RAPAPORT et al. 1996; RIFAI et al. 1996; SANDS et al. 1994; BUTLER et al. 1997). In these studies, the specific cell types that take up oligodeoxynucleotides after in vivo administration have been characterized. In the livers of mice treated with phosphorothioate oligodeoxynucleotide, Kupffer cells are enlarged and contain basophilic granules that are thought to contain oligodeoxynucleotide (SARMIENTO et al. 1994). We have also observed oligodeoxynucleotide in hepatocytes, albeit at lower levels (BUTLER et al. 1997).

The most significant accumulation of oligodeoxynucleotides is within cells of the proximal tubules in the kidney. In proximal tubule cells, it is possible, with routine histology, to visualize granular inclusions that contain oligodeoxynucleotide (SARMIENTO et al. 1994; BUTLER et al. 1997). With autoradiography, it is possible to identify the granules containing oligodeoxynucleotide-derived material and to differentiate exposures between cortex and medulla, with the cortex having significantly higher concentrations (OBERBAUER et al. 1995; PLENAT et al. 1994; RAPPAPORT et al. 1995; RIFAI et al. 1996; SANDS et al. 1994). Micropuncture studies have demonstrated that the proximal tubule accumulates phosphorothioate oligodeoxynucleotides as a result of tubular reabsorption (OBERBAUER et al. 1995; RIFAI et al. 1996). The transport mechanisms into cells and the sites of oligodeoxynucleotide binding are under investigation, but may be the result of binding to the scavenger receptor (SAWAI et al. 1995; WU-PONG et al. 1994). The critical information obtained from these studies is that oligonucleotides are taken up by cells in animals treated with phosphorothioate oligodeoxynucleotides. In contrast to observations made in vitro, where cell uptake is dependent on the presence of cationic lipid (BENNETT et al. 1992), cell uptake of oligonucleotide in vivo does not require the presence of cationic lipids or other absorption enhancers (OBERBAUER et al. 1995; PLENAT et al. 1994; RAPAPORT et al. 1996; RIFAI et al. 1996; SANDS et al. 1994; BUTLER et al. 1997). Thus the toxicities resulting from the administration of phosphorothioate oligodeoxynucleotides may be the result of both extracellular or intracellular actions of these compounds.

The successive removal of bases from the 3' end of the oligodeoxynucleotide is the major pathway for metabolic degradation (see Chap. 4, this volume; CUMMINS et al. 1996; TEMSAMANI et al. 1993) in plasma, and both 5'- and 3'-nuclease excision may occur in tissues. Hours after dosing rodents with oligodeoxynucleotide, Nicklin et al. demonstrated the presence of metabolites shortened from either the 3' or 5' ends using electrospray mass spectral analysis (see Chap. 4, this volume). Similar oligonucleotide metabolites were detected in tissues (liver and kidney) following subcutaneous administration of phosphorothioate oligodeoxynucleotides (CUMMINS et al. 1997).

The resultant products of nuclease degradation are nucleotides either with a phosphorothioate or phosphodiester group. From metabolism studies with ^{14}C thymine labeled at the C_2 position, it appears that the ultimate metabolic fate of the liberated nucleotides is via a metabolic pathway similar to that for endogenous nucleotides. Thus, CO_2 is the final product of the C-2-labeled thymidine, and presumably other nucleotides from phosphorothioate oligodeoxynucleotides are similarly metabolized. These same patterns of single-base excision and chain-shortened metabolites have been reported for all species from mouse to humans, suggesting that the metabolic pathways are similar.

We have now collected data on the plasma pharmacokinetics from a number of different oligodeoxynucleotides administered to a variety of species, including humans. There are similarities in the plasma kinetics and metabolic patterns between species. The maximal plasma concentrations achieved after 2-h intravenous infusions of 1 mg/kg to cynomolgus monkeys or humans are nearly equivalent (Table 3). The log clearance values and the plasma half-lives are also similar from species to species. Plotting clearance against log body weight yields an almost straight line as body size increases from mouse to human. The greatest deviation from this relationship is that of the mouse. The data from a number of different phosphorothioate oligodeoxynucleotides indicate that the mouse clears oligodeoxynucleotide from plasma more rapidly than rats or primates (Fig. 3). However, it is apparent that species-to-species comparisons of dose generally can be scaled on the basis of body weight, not surface area.

It follows from Fig. 3 that exposure of humans at 1 mg/kg is nearly equivalent to exposure of monkeys (Table 3) or rats at 1 mg/kg. Because the kinetics of distribution and the plasma levels are scalable from species to species on the

Table 3. Maximum concentration in plasma (C_{max}) and area under the plasma concentration–time curve (AUC) for both the cynomolgus monkey and humans given equivalent doses of 1 mg/kg, infused i.v. over a 2-h period (average ± standard deviation, $n = 3$–6)

		C_{max} (µg/ml)	AUC (µg min/ml)
ISIS 2302	Monkey	4.59 ± 0.16	580 ± 11.4
	Human	3.96 ± 1.02	506 ± 56.4
ISIS 5132	Monkey	3.27 ± 0.89	369 ± 85
	Human	3.78 ± 0.61	612[a]
ISIS 3521	Monkey	4.30 ± 0.75	486 ± 63
	Human	4.78 ± 1.33	624[a]

[a] $n = 2$.

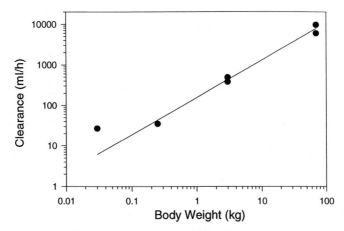

Fig. 3. Clearance (ml/h) versus body weight for ISIS 5132 in different species. Log clearance (dose/area under curve, AUC) is plotted versus body weight (experimental data for mice, rats, and monkeys and estimated to be 70 kg for humans). Clearance values for the parent drug were obtained over the range of 0.5–1.0 mg/kg. The line represents a least square allometric fit of the data. The number of subjects ranged from three to six

basis of body weight, this allows for good interspecies extrapolations. For the purpose of understanding the relationships between dose, plasma concentrations, tissue accumulations, and toxicity, it is important for us to be able to relate plasma and tissue kinetics in experimental models to the plasma kinetics in humans. The similarities in plasma pharmacokinetics between the animal models and humans suggest that the tissue kinetics may be similar, putting us in a position to predict target organ concentrations of oligodeoxynucleotides in humans. Toxicity studies in animals have provided information on the relationship between organ concentrations and morphologic changes. Predicting human tissue levels from animal data allows us to relate human exposures to known thresholds for toxicity in animals. Ultimately, this information should allow us to use pharmacokinetic models obtained in animals to design dose regimens for humans that avoid thresholds for toxicity. Much of the recent work at Isis Pharmaceuticals has been aimed at characterizing these thresholds for toxicity (see below). In our experience, this interspecies predictability facilitates the development of these compounds as therapeutic agents.

C. Toxicity of Phosphorothioate Oligodeoxynucleotides

The systemic toxicities of phosphorothioate oligodeoxynucleotides have been studied extensively in rodents, rabbits, and monkeys for a number of different sequences (CORNISH et al. 1993; CUMMINS et al. 1996; GALBRAITH et al. 1994; HENRY et al. 1997a,e,g; MCINTYRE et al. 1993; SARMIENTO et al. 1994; SRINIVASAN and IVERSEN 1995; TEMSAMANI et al. 1993). Because the toxicity of phosphorothioate oligodeoxynucleotides appears to be more closely related to the class of the molecule rather than the specific sequence, it is possible to generalize about the toxicologic profiles of these compounds. With different patterns of toxicity observed between rodents and primates, understanding the mechanisms behind these differences is crucial information for understanding which species best predicts the potential human effects. The remainder of the chapter will be divided into sections each discussing a specific aspect of the toxic effects of antisense therapeutic agents and our understanding of the mechanism of toxicity and structure activity relationships.

I. Genetic Toxicity

A number of phosphorothioate oligodeoxynucleotides have been examined in one or more of the following battery of genotoxicity assays: Ames test, in vitro chromosomal aberrations, in vitro mammalian cell mutation (HGPRT locus and mouse lymphoma), in vitro unscheduled DNA synthesis tests, and in vivo mouse micronucleus. In all of these assays the results were negative, and there was no evidence of mutagenicity or clastogenicity of these compounds.

II. Acute Toxicities

In rodents, the acute toxicity of phosphorothioate oligodeoxynucleotides has been characterized as part of an effort to determine the maximum tolerated doses for in vivo genotoxicity assays. The doses of three phosphorothioate oligodeoxynucleotides required to produce 50% lethality (LD_{50}) were estimated to be approximately 750mg/kg (Table 4). These data were obtained from groups of mice (ten of each sex) treated by bolus intravenous injections

Table 4. Estimated dose required to produce 50% mortality in mice injected intravenously

Sequence	LD_{50} (mg/kg)
ISIS 2105	890
ISIS 2922	720
ISIS 2302	>1000

Groups of 20 mice (ten of each sex) were injected intravenously with phosphorothioate oligodeoxynucleotides in phosphate buffer saline via the tail vein. The mice were observed for 7 days for lethality. The LD_{50} was estimated using probit analysis.

of oligodeoxynucleotides at doses of 600, 800, and 1000 mg/kg in phosphate-buffered saline. There is only a small degree of variability in the estimated LD_{50} values from sequence to sequence, indicating that the LD_{50} in mice is independent of sequence. With relatively high LD_{50} values, these compounds appeared to have a low potential to induce acute toxicity in mice. The exact cause of the lethal effects has not been determined, but the mice show no acute changes pathognomonic of some specific mechanism of toxicity. However, studies in primates have demonstrated a potential for acute toxicity markedly different than in rodents.

In primates, the acute dose-limiting toxicities are a transient inhibition of the clotting cascade and the activation of the complement cascade (GALBRAITH et al. 1994; HENRY et al. 1997b; WALLACE et al. 1996b). Both of these toxicities are thought be related to the polyanionic nature of the molecules and the binding of these compounds to specific protein factors in plasma.

1. Complement Activation

Of the two acute toxicities, the activation of the complement cascade has the potential to produce the most profound effects. In some monkeys, treatment with high doses over short infusion times resulted in marked hematologic effects and marked hemodynamic changes that are thought to be secondary to complement activation. Hematologic changes are characterized by transient reductions in neutrophil counts, presumably due to margination, followed by neutrophilia with abundant immature, nonsegmented neutrophils (bands; Fig. 4). The increase in immature cells is probably the result of neutrophils being recruited by the release of chemotactic factors. In a small fraction of monkeys, complement activation was accompanied by marked reductions in heart rate, blood pressure, and subsequently cardiac output. In some animals, these hemodynamic changes can be lethal (CORNISH et al. 1993; GALBRAITH et al. 1994; HENRY et al. 1997b).

There is an association between cardiovascular collapse and complement activation. All treated monkeys demonstrating some degree of cardiovascular

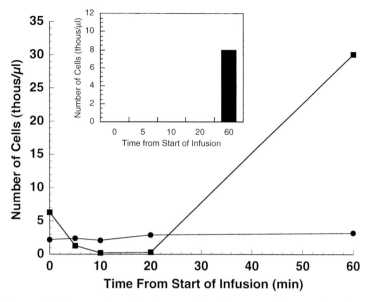

Fig. 4. Neutrophil (*squares*) and leukocyte (*circles*) counts (thousands/ml) with time (in min) after the start of a 10-min infusion of ISIS 2302 to a cynomolgus monkey (representative data). The *inset* shows the number of nonsegmented neutrophils (*bands*) at various time points after the start of dosing. (From HENRY et al. 1997b)

collapse or hemodynamic changes had markedly elevated levels of complement split products. However, the converse is not true, in that only a fraction of the animals with activated complement had cardiovascular functional changes. This observation suggests that there may be sensitive subpopulations or predisposing factors within individual animals that make them susceptible to the physiologic sequelae of complement activation. Because of these hemodynamic changes, primate studies to monitor for these effects have become part of the normal evaluation of these compounds (BLACK et al. 1993, 1994).

Complement activation was characterized by reduction of serum hemolytic potential (CH_{50} analysis) and concomitant increases in the liberation of complement split products Bb, C3a, and C5a (Fig. 5). Notably absent from this pattern of split product liberation is an increase in the complement split product C4a, which is demonstrative of classical pathway activation. The absence of C4a and the presence of Bb are consistent with the activation of complement occurring through the alternative pathway.

The exact mechanism for complement activation is not known, but there are some potential explanations that need to be further explored. Under normal conditions, the alternative pathway of complement is constituitively active, albeit at low levels. The pathway consists of six glycoproteins, C3, and factors B, D, H, and I. During activation, C3 is converted to C3b, which binds

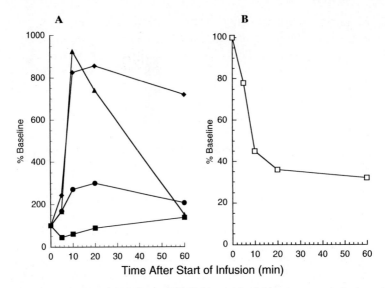

Fig. 5A,B. Complement activation produced by a 10-min intravenous infusion of 20 mg ISIS 2302 per kg in cynomolgus monkeys. **A** Plasma concentrations of the complement split products C5a (*triangles*), C3a (*circles*), C4a (*squares*), and Bb (*diamonds*) are represented at a percentage of baseline concentrations for each animal ($n = 3$). **B** Total hemolytic activity in serum, i.e., CH_{50}, presented as a percentage of baseline activity. (From HENRY et al. 1997b)

to the nearest surface. If the surface is not an activator of the alternative pathway, then C3b is rapidly deactivated by circulating deactivators. If the C3b binds an activating surface and remains active long enough to interact with factor B to form the C3bB complex, then activation occurs. Under most circumstances, the low level of activity is retained in control by the circulating negative regulatory proteins, factor H and factor I. Results reported recently suggest that interactions of phosphorothioate oligodeoxynucleotides with factor H may play a role in the activation. Treatment of cynomolgus monkeys with ISIS 2302 results in a reduction in factor H (HENRY et al. 1997b). Reductions in factor H produced by another polyanion, dextran sulfate, have been reported to activate the alternative pathway in vitro (BITTER-SUERMANN et al. 1981). Such a reduction in regulatory factors would lead to uncontrolled activation of the alternative complement cascade similar to that described in monkeys treated with high doses of phosphorothioate oligodeoxynucleotides. How circulating phosphorothioate oligodeoxynucleotides cause the removal, masking, or sequestration of factor H in vivo has not been established. However, phosphorothioate oligodeoxynucleotides can bind to factor H (HENRY et al. 1997b), and in some way this may result in its removal, masking, or sequestration.

Although effects on factor H have been observed, it is conceivable that there could be other mechanisms through which circulating phosphorothioate oligodeoxynucleotides could activate the alternative pathway. In one scenario, phosphorothioate oligodeoxynucleotides could directly contribute to the activation of C3 to promote the cascade. Alternatively, interactions of phosphorothioate oligodeoxynucleotides could reduce the effectiveness of the inhibitory factors either through an interaction with the C3bB complex or with the inactivator factors H and I. This interaction could reduce their binding and inactivation of the C3bB complex, resulting in formation of $C3b_nBb$ and ultimately C5a and the full complement cascade. While the mechanism of activation needs to be further elucidated, it is clear that decreasing the affinity of oligodeoxynucleotide–protein (or glycoprotein) interactions might reduce the ability of oligodeoxynucleotides to activate complement.

To characterize phosphorothioate oligodeoxynucleotide-induced complement activation, we selected the complement split product Bb as a marker of activation. While the levels of C5a may be more closely related to the ultimate biologic effects, Bb, with its longer circulating half-life, was selected because elevation of this split product would be less transient and more readily apparent (see Fig. 5). Thus Bb levels were considered a more reliable indicator of complement activation for exploratory studies. The relationship between dose, plasma oligodeoxynucleotide concentration, and activation was studied with a series of intravenous infusions of various doses over varying intervals. These data demonstrate that the complement cascade is activated only at high plasma concentrations of phosphorothioate oligodeoxynucleotides and that by controlling the dose and rate of infusion, it is possible to administer phosphorothioate oligodeoxynucleotides in such a way as to avoid complement activation (Fig. 6). The toxicity associated with complement activation is probably secondary to the physiologic responses to the formation and release of biologically active complement split products C3a and C5a.

Although complement activation at high doses is consistent and predictable between animals, there is currently little appreciation for the variability in the severity of the associated hemodynamic changes. While Bb levels were selected as the most reliable index of complement activation, it is C5a that is the most biologically active split product. Preliminary data obtained relating response to complement split product levels indicate that C5a levels are elevated more significantly in some of the more affected animals (S.P. HENRY and A.A. LEVIN, unpublished observations).

There may be other mechanisms of toxicity, in addition to complement activation, that lead to changes in hemodynamic parameters. However, no other plausible mechanisms of toxicity have been put forward. The present understanding is that complement activation is generally necessary, but not sufficient to produce cardiovascular effects in primates. The goal of toxicity studies is to characterize the toxicity of compounds and to establish a framework upon which clinical safety studies can be built. In this regard, it is useful to examine the relationship between plasma concentrations and the activation of complement, because clearly in clinical studies it is critical to minimize the potential for complement activation.

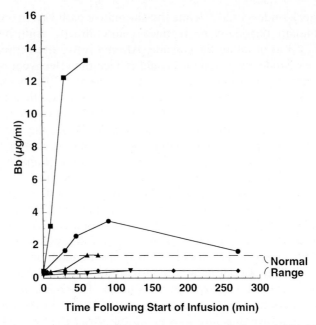

Fig. 6. Complement activation following intravenous infusions of ISIS 2302 at various doses and rates. *Squares*, 10 mg/kg over 10 min (C_{max}, 160 µg/ml); *circles*, 10 mg/kg over 30 min (C_{max}, 98 µg/ml); *upward triangles*, 10 mg/kg over 60 min (C_{max}, 70 µg/ml); *diamonds*, 6 mg/kg over 120 min (C_{max}, 49 µg/ml); *downward triangles*, 3.3 mg/kg over 60 min (C_{max}, 35 µg/ml). Complement activation is represented as the concentration of the split product Bb (µg/ml). The values represent mean values of at least three animals. The *dashed line* represents the upper limit of the normal range of variability. (From HENRY et al. 1997b)

The relationships between dose, infusion rates, and plasma concentration of both oligodeoxynucleotide and Bb were characterized in a series of studies in cynomolgus monkeys (HENRY et al. 1997b). Monkeys were treated with various doses and infusion regimens with one of three phosphorothioate oligodeoxynucleotides, and Bb concentrations were assayed at various times during infusion. When Bb concentrations were plotted against the concurrent plasma concentrations of oligodeoxynucleotide, it was apparent that complement was only activated at concentrations of phosphorothioate oligodeoxynucleotides that exceed a threshold value of 40–50 µg/ml. Bb levels remained unchanged from control values at plasma concentrations below the threshold. Remarkably, this threshold concentration is similar for three 20-mer phosphorothioate oligodeoxynucleotides (Fig. 7) and for an 8-mer phosphorothioate oligodeoxynucleotide that forms a tetrad complex (ISIS 5320, data not shown). When concentrations exceeded the threshold values, there was a rapid and marked rise in Bb, indicative of complement activation. If these data from primates are predictive of the responses of humans, it is clear that clinical dose regimens should be designed to avoid plasma

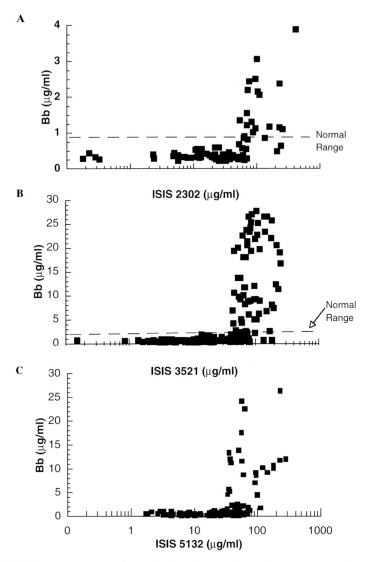

Fig. 7A–C. Plasma concentrations of oligodeoxynucleotides versus complement split product Bb in cynomolgus monkeys treated with various regimens of **A** ISIS 2302, **B** ISIS 3521, or **C** ISIS 5132. Only plasma samples collected during intravenous infusions were plotted. The *dashed line* represents the upper limit of the normal range. (From HENRY et al. 1997d)

oligodeoxynucleotide concentrations that exceed 40–50 µg/ml. To this end, the similarities in plasma pharmacokinetics between monkeys and humans have allowed the design of dose regimens with the desired plasma profiles.

One approach for staying below the plasma thresholds for complement activation is to reduce the rate of dosing by substituting prolonged infusions for bolus injections. In clinical trials with ISIS compounds, phosphorothioate oligodeoxynucleotides are administered either as 2-h infusions or as constant 24-h infusions. At a rate of infusion of 2 mg/kg over 2 h, the C_{max} was 8–15 μg/ml, i.e., well below the threshold for complement activation. The concentrations of intact ISIS 2302 in clinical trials at doses ranging from 0.1 to 2.0 mg/kg are reported elsewhere in this volume (see Chap. 18; this volume). In all cases, there is a substantial margin of safety between concentrations achieved in patients and volunteers and the known thresholds for complement activation in monkeys. In fact, phosphorothioate oligodeoxynucleotides have been administered by intravenous infusion to more than 120 patients as part of clinical trials without any significant indication of activation of the alternative complement cascade. Through the careful characterization of the concentration–response relationship for complement activation and the understanding of the plasma kinetics in animals and humans, it is possible to maintain plasma concentrations under the threshold by prolonging infusions and minimizing peak plasma concentrations.

There may be additional factors in determining the response of monkeys to treatment with phosphorothioate oligodeoxynucleotides. For example, the stress associated with the 2-h constant infusion procedure in monkeys may exacerbate the biologic effects associated with complement activation. This hypothesis is supported by two observations. The first is that cynomolgus monkeys could easily tolerate intravenous administration of bolus injections of 50 mg/kg on alternate days for a month with no indication of cardiovascular collapse, and only a few incidences of transient lethargy were noted after the first dose (HENRY et al. 1997a). This dose would be expected to produce plasma concentrations that transiently exceed 500 μg/ml; however, no gross cardiovascular sequelae were observed. The second observation is that cardiovascular collapse has been observed in a control monkey after a 2-h constant infusion of saline in the (verified) absence of any drug exposure (M.V. TEMPLIN, S.P. HENRY, and A.A. LEVIN, unpublished observations). The relationship between the area under the plasma concentration curve (AUC) and complement activation needs to be further investigated to determine whether the prolonged elevations in plasma levels of phosphorothioate oligodeoxynucleotides associated with infusions are more prone to activate complement than bolus injections; if so, then there may be more convenient ways to administer these drugs.

a) Effects of Oligodeoxynucleotide Chemistries

Chemical modifications to phosphorothioate oligodeoxynucleotides may reduce the potential to activate complement. In one study, cynomolgus monkeys were administered intravenous infusion over a 10-min period with a 5, 20, or 50 mg/kg dose of a 17-mer phosphodiester oligodeoxynucleotide, Ar177, that had phosphorothioate caps on the 3′ and 5′ termini (WALLACE et al. 1996a,b). This oligodeoxynucleotide is known to have a complex secondary structure. In this experiment, although there was a dose-related increase in plasma concentrations of Bb, the magnitudes of the increases were small, with less than a doubling of the values (WALLACE et al. 1996b). A full phosphorothioate oligodeoxynucleotide of the same length would be expected to produce increases in Bb on the order of tenfold or more over baseline values (see Figs. 6, 7). Whether this diminished potential to activate the complement cascade is related to the reduction in phosphorothioate linkages or whether it is due to the configuration of the oligodeoxynucleotide was not established by these experiments. Some insight into this question was obtained in a second series of experiments performed with oligodeoxynucleotides that contained methoxyethyl at the 2′ position. Cynomolgus monkeys were treated by 10-min intravenous infusion with single doses of 1, 5, or 20 mg/kg of a 20-mer oligodeoxynucleotide that was either fully thiated (CGP 71849A) or had phosphodiester wings and a thioate central region (nine linkages, CGP 69845A). The wings of both compounds had six 2′-modified residues. A third unmodified phosphorothioate oligodeoxynucleotide (ISIS 1082) with a different sequence was included as a positive control. The unmodified compound, ISIS 1082, produced marked increases in Bb and severe cardiovascular effects at a dose of 5 mg/kg (30- to 60-fold over baseline). The modified compounds produced no significant changes in Bb at a dose of 5 mg/kg. At the high dose, the compound with the mixed phosphodiester–phosphorothioate backbone (fewer thioates) was somewhat less potent in activating complement than the fully thiated derivative, suggesting that reduction in the number of phosphorothioate linkages reduced the potential for complement activation. However, the more important difference was that both 2′-methoxyexthoxy compounds were markedly less potent in activating complement than an unmodified oligodeoxynucleotide (D.K. MONTEITH, P.L. NICKLIN, and A.A. LEVIN, unpublished observations).

It is also possible to reduce the complement activation liability of phosphorothioate oligodeoxynucleotides by modifying the dosage form. Encapsulating the phosphorothioate oligodeoxynucleotides in liposomes allowed monkeys to be dosed with up to 60 mg/kg in a 30-min intravenous infusion with only minimal evidence of complement activation. In contrast, the same oligodeoxynucleotide injected without the liposomal formulation was a potent activator of the complement cascade (M.V. TEMPLIN and A.A. LEVIN, unpublished observations). These data suggest that the acute safety profile of phosphorothioate oligodeoxynucleotides can be enhanced through formulation technologies and that prolonged intravenous infusions currently employed in clinical studies may be replaced in the future by bolus injections of oligodeoxynucleotides in delivery vehicles.

2. Hemostasis

In addition to the complement, the other effect occurring concurrently with administration of phosphorothioate oligodeoxynucleotides is a transient prolongation of clotting times. A number of different phosphorothioate oligodeoxynucleotides alter the clotting cascade, as indicated by a concentration-dependent prolongation of activated partial thromboplastin times (APTT; GRIFFIN et al. 1993; HENRY et al. 1994, 1997a; NICKLIN et al. 1997; WALLACE et al. 1996b). In all studies in animals, APTT was prolonged more than prothrombin times (PT), indicating that the effect was more pronounced in the intrinsic pathway than in the extrinsic pathway. Typically, at the end of a 2-h constant intravenous infusion of 10 mg/kg in cynomolgus monkeys, there is an increase of roughly 50% in APTT and an increase of less than 20% in PT. The prolongation of APTT is highly transient and directly proportional to plasma concentrations of oligodeoxynucleotide and therefore parallels the plasma drug concentration curves with various dose regimens (Fig. 8; HENRY et al. 1994, 1997f; NICKLIN et al., 1997; WALLACE et al. 1996b). As drug is cleared from plasma, the inhibition diminishes such that there is complete reversal within hours of dosing. With repeated administration, there is no evidence of residual inhibition since prolongation of APTT is similar on days 1 and 28 of dosing. Similar observations have been made in clinical studies. The action of phosphorothioate oligodeoxynucleotides on

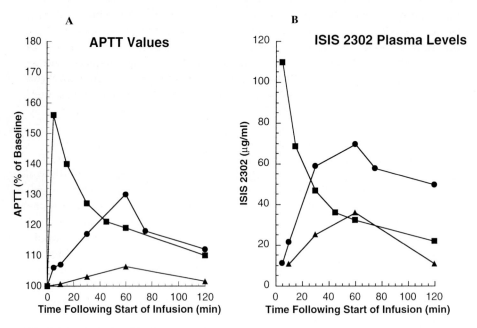

Fig. 8A,B. Relationship between plasma concentrations and activated partial thromboplastin time (*APTT*) in cynomolgus monkeys following various regimens. *Squares*, 10mg/kg over 2min; *circles*, 10mg/kg over 60min; *triangles*, 3.3mg/kg over 60min. **A** APTT values expressed as a percentage of baseline plotted versus the time (min) after the start of infusion. **B** Plasma concentrations of ISIS 2302 (μg/ml) plotted versus time after the start of infusion. (From HENRY et al. 1997f)

clotting is thought to be the result of the interaction of oligodeoxynucleotides with proteins and in this regard similar to the activation of complement.

It is well known that polyanions are inhibitors of clotting, and phosphorothioate oligodeoxynucleotides may act through similar mechanisms. In fact, a phosphodiester oligodeoxynucleotide that binds with thrombin has been proposed for use as an anticlotting agent (GRIFFIN et al. 1993). If phosphorothioate oligodeoxynucleotides inhibit the clotting cascade as a result of their polyanionic properties, then binding and inhibition of thrombin would be a likely mechanism of action (reviewed in ROSENBERGER 1989). However, the greater sensitivity of the intrinsic pathway to inhibition by phosphorothioate oligodeoxynucleotides suggests that there are other clotting factors specific to this pathway that may be inhibited as well.

Prolongation of APTT has been observed in all species examined to date, including human, monkey, and rat (HENRY et al. 1997d,f; NICKLIN et al., 1997). In vitro analyses have made it possible to test the effects of altering sequence, length, and chemistry (ALTMANN et al. 1996). Inhibition is generally independent of sequence, although G-rich oligodeoxynucleotides may be slightly more potent than other oligodeoxynucleotides (NICKLIN et al. 1997). There appears to be little or no effect of sequence on the potency of the inhibitory effects. Length or number of phosphorothioate linkages has also been implicated as a potential modulating factor in the prolongation of APTT. As length increased, the potency of the inhibitory effect also increases, and thus the concentration required to double APTT decreases (Fig. 9). In this series of oligonucleotides presented, for each additional base added there is approximately a 2.8-μM

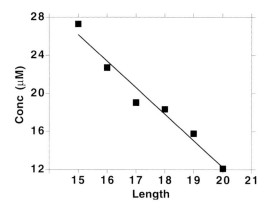

Fig. 9. Effect of oligodeoxynucleotide length on clotting times. Phosphorothioate oligodeoxynucleotides ranging in length from 15 to 20 bases in length were incubated in human plasma from two pools (of three subjects each), and the activated partial thromboplastin time (APTT) was determined in duplicate. Data plotted as length versus concentration (μM) required to double APTT

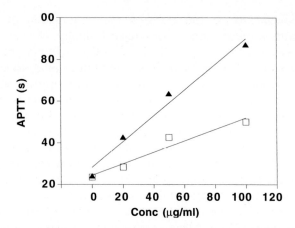

Fig. 10. Oligodeoxynucleotide-induced prolongation of activated partial thromboplastin time (*APTT*) in monkey (*white squares*) and human plasma (*black triangles*) in vitro. ISIS 2302 was added to citrated monkey and human plasma at the indicated concentrations and APTT was determined. Each point represents the mean of triplicate assays from pooled plasma. The line is a least squares regression fit to the data points for each species

decrease in the concentration of oligodeoxynucleotide required to produce a doubling of APTT.

Some subtle oligodeoxynucleotide sequence effects may be overlooked in vivo because increases in APTT seem to be independent of sequence, but it appears that in vitro there is some sequence-related variability in the extent of inhibitory effects (P.L. NICKLIN, unpublished observations). Human plasma is slightly more sensitive to the inhibitory effects of phosphorothioate oligodeoxynucleotides in vitro than monkey plasma. When citrated human or monkey plasma was treated with ISIS 2302 in vitro, there was a concentration-dependent increase in APTT (Fig. 10). There was an approximately twofold difference in the sensitivity of human compared to monkey plasma (0.61 vs. 0.27 s per µg/ml). While the relationship between concentration may not be strictly linear, these slope values provide a means to estimate the increases in APTT over baseline with increasing concentrations of oligodeoxynucleotide.

In clinical trials with ISIS 2302, normal volunteers and patients were dosed with 2 mg/kg infused over 2 h. This regimen produced total oligonucleotide concentrations of 10–15 µg/ml and an increase in APTT of approximately 50% (GLOVER et al., 1997), which correlates well with in vitro human and animal data (Fig. 10). The transient and reversible nature of APTT prolongation, combined with the relatively small magnitude of the change, makes these effects clinically insignificant for the current treatment regimens.

a) Effects of Oligodeoxynucleotide Chemistries

Structural modifications to phosphorothioate oligodeoxynucleotides may reduce the potency for APTT prolongation. When various structural modifications of the GEM 91 sequence were examined for their potential for prolonging APTT, the potency of the inhibition was reduced by either replacing eight of the phosphorothioate linkages with methyl phosphonates (GEM 91-C) or by adding four 2'-methoxy groups to each end of the molecule (GEM 91-H). In this series, replacing the phosphorothioate linkages with methylphosphonates was a more effective strategy for reducing the potency of inhibition (AGRAWAL et al. 1995a). These findings were extended (NICKLIN et al., 1997) in studies in which a series of backbone and sugar modifications were made on the ISIS 5132 sequence. Modifications that increased the inhibitory effects on APTT included 2'-fluoro substituents and substitution of 5-methyl cytosine (5-methyl C) for cytosine. As in the studies with GEM 91 (AGRAWAL et al. 1995a), the potency of inhibition could be reduced by decreasing the number of phosphorothioate linkages and, to a lesser extent, by the addition of 2'-alkoxy groups (methoxyethoxy or propoxy) on the sugars (NICKLIN et al., 1997).

The number of thioate linkages may influence the potency of the inhibitory effects on APTT (reported here as the concentration required to produce a doubling of APTT). In human plasma, the concentrations required to double APTT were 31.8, 12.1, 16.6, and more than 53.0 μM for the unmodified phosphodiester, the unmodified phosphorothioate, the fully thiated 2'-methoxyethoxy (CGP 71849A), and the phosphodiester–phosphorothioate 2'-methoxyethoxy (CGP 69845A), respectively. Addition of the 2'-methoxyethoxy groups (and additions of 5-methyl C residues in the modified ends) also decreased the potency of the inhibitory effects for the fully thiated compounds (12.1 vs. 16.6 μM, unmodified vs. modified). A more significant change in the potency was obtained by reducing the number of phosphorothioate linkages from 19 to nine (16.6 vs. more than 53.3 μM, unmodified vs. modified). These in vitro results correlated well with in vivo results from a study in cynomolgus monkeys in which CGP 71849A and CGP 69845A were compared. The fully thiated compound was more potent than the compound with a mixed phosphodiester–phosphorothioate backbone (D.K. MONTEITH, P.L. NICKLIN, and A.A. LEVIN, unpublished observations). After a 10-min infusion with 5 mg/kg, the APTT values were 41 and 33 s for the fully thiated CGP 71849A and the partially thiated CGP69845A, respectively. Under the same experimental conditions, an unmodified phosphorothioate oligodeoxynucleotide produced an APTT of 72 s. These data suggest that reduction in the numbers of phosphorothioate linkages reduced the inhibitory effects on APTT.

Although the safety profile of phosphorothioate has proven satisfactory, the acute safety profile of the next generation of phosphorothioate oligodeoxynucleotides may be improved by reductions in phosphorothioate linkages and by modification of the 2' position with an alkoxy group.

III. Toxicologic Effects Associated with Chronic Exposure

The toxicologic consequences of repeated administration of phosphorothioate oligodeoxynucleotides continue to be investigated. Pharmacokinetics studies have determined that there is no plasma accumulation of phosphorothioate oligodeoxynucleotides with repeated daily or every-other-day dosing, but, as discussed above, there are specific tissues that appear to accumulate oligodeoxynucleotide with these regimens. Recent studies have identified a number of specific target organs and tissues that are responsive to high-dose administration of phosphorothioate oligodeoxynucleotides and a number of other organs that are unaffected by subchronic or chronic treatment with phosphorothioate oligodeoxynucleotides. These data are summarized in Table 5.

1. Immune Stimulation

One of the characteristic toxicities observed with repeated exposure of rodents to phosphorothioate oligodeoxynucleotides is a profile of effects that can be described as immune stimulation. These are characterized by splenomegaly, lymphoid hyperplasia, and diffuse multiorgan mixed mononuclear cell infiltrates (Fig. 11). The severity of these changes was dose dependent and most notable at doses equal to or exceeding 10 mg/kg. The mixed mononuclear cell infiltrates consisted of monocytes, lymphocytes, and fibroblasts and were particularly notable in liver, kidney, heart, lung, thymus, pancreas, and periadrenal tissues (HENRY et al. 1997e,g; MONTEITH et al. 1997; SARMIENTO et al. 1994). Maintenance of normal cellular architecture and the relatively innocuous infiltrate seen in the parenchyma or in perivascular regions are characteristic of this lesion (Fig. 11). The cellular infiltrate was not associated with fibrotic changes, as would be expected in a classical inflammatory response, and differed also in that it was partially or fully reversible upon the cessation of treatment (HENRY et al. 1997e,g). In general, there was no tissue damage associated with the infiltrate, and its presence did not worsen with prolonged exposure for up to 6 months (M.V. TEMPLIN, unpublished results). Marginal increases in cellular infiltrates in some organs (e.g., liver) have been observed at doses as low as 0.8 mg/kg administered every other day for 28 days (HENRY et al. 1997g), although the degree of infiltration at this low dose was difficult to distinguish from the background incidence of hepatic inflammation usually present in CD-1 mice.

Significant increases in spleen weights have been observed in mice and rats treated with a number of different phosphorothioate oligodeoxynucleotides. However, there appear to be quantitative differences in the potency of different sequences for inducing this effect. Compared to controls, spleen weights have been reported to more than double at high doses (≥25 mg/kg; BRANDA et al. 1993; HENRY et al. 1997e,g; MCINTYRE et al. 1993; MONTEITH et al. 1997; SARMIENTO et al. 1994). Because it is a reproducible and easy marker for assessing the degree of immune stimulation, we have used spleen

Table 5. Organ/system summary of phosphorothioate oligodeoxynucleotide toxicities

Organ/system	Species	Results
Bone/connective tissue	Mouse, rat, monkey	No significant findings
CNS/nervous	Mouse, rat, monkey	No significant findings
Cardiovascular	Mouse, rat	No significant findings except for mononuclear cell infiltrates
	Monkey	Cardiovascular collapse related to complement activation; associated with plasma levels that exceed 40–50 µg/ml; no histologic evidence of direct cardiotoxicity
Endocrine	Mouse, rat, monkey	No treatment-related lesions except mononuclear cell infiltrates in rodents
Gastrointestinal	Mouse, rat	Presence of mononuclear cell infiltrates
	Monkey	No significant effects
Hematopoiesis/blood	Mouse	Reductions in platelets after repeated administration of 50 mg/kg; extramedullary hematopoiesis in spleen and liver; increases in WBC
	Monkey	Occasional transient acute reductions in platelet numbers during infusion; transient increases in APTT
Hepatic	Mouse	Single cell necrosis, hepatocytomegaly with no evidence of hepatic dysfunction at doses greater than 50 mg/kg and accompanied by increases in transaminases; mononuclear cell infiltrates
	Monkey	Slight increases in transaminases, little or no evidence of direct hepatoxicity
Immune	Mouse, rat, monkey	Dose-related increases in B cells in lymphoid organs; in mice and rats, increased spleen and lymph node size and diffuse multiorgan infiltrates of mononuclear cells
Integument	Monkey, mouse, rat, rabbit	Local inflammatory responses to intradermal or SC injections
Kidney	Mouse	No significant alterations at clinically relevant doses; minimal to mild reductions in proximal tubular brush border height and appearance of enlarged active nuclei; at doses of 100 mg/kg, tubular degeneration; mononuclear cell infiltrates present
	Rat	No significant alterations at clinically relevant doses; tubular degenerative changes at doses of 40 mg/kg or greater; functional changes only in the presence of significant histopathologic changes; mononuclear cell infiltrates

Table 5. *Continued*

Organ/system	Species	Results
	Monkey	No significant alteration at clinically relevant doses; minimal to mild changes in the proximal tubular epithelium at 10mg/kg and more prominent lesions at 40 and 80mg/kg, with alteration in functional markers observed at higher doses; dose-related tubular degenerative changes at higher doses (approx. 80mg/kg)
Reproductive	Mouse, rat, rabbit, monkey	No treatment-related lesions observed; presence of mononuclear cell infiltrates in uterus and ovaries in rodents; no changes in testes
Pulmonary	Mouse, rat	Presence of mononuclear cell infiltrates
	Monkey	No treatment-related lesions

weight as a biomarker in studies designed to examine the "structure–activity relationships" with different sequences, backbones, and chemistries (HENRY et al. 1997c; MONTEITH et al. 1997). The use of spleen weight as a marker is based upon the correlation of spleen weight with the potential for oligodeoxynucleotides to stimulate [^{3}H]thymidine incorporation into splenocytes or peripheral blood mononuclear cells in vitro (BRANDA et al. 1993; MONTEITH et al. 1997). When the spleens of treated mice are examined microscopically, the relationship between increased spleen weight and B cell proliferation is further strengthened. The spleens of treated mice show evidence of B cell hyperplasia, histiocytosis in the connective tissue capsule, and stromal hyperplasia. In addition, the oligodeoxynucleotide-treated mice often have marked extramedullary hematopoiesis, which could also contribute to splenomegaly. Other lymphoid organs were also enlarged by treatment. For example, lymph nodes have areas of B cell proliferation and histiocytosis, but they were not as readily weighed and thus were not ideal markers to quantitate stimulation. Despite the increase in the proliferation of B cells and the attendant increases in spleen weight, proliferation of B cells was not accompanied by apparent increases in T cell areas.

The phenomena of B cell stimulation may be better understood by examining the responses of cells in vitro. Incubation of B cells with oligodeoxynucleotides in vitro stimulated the secretion of interleukin (IL)-6, IL-12, and interferon (INF)-γ (KLINMAN et al. 1996; KRIEG and STEIN 1995a; YI et al. 1996), which may stimulate B cell proliferation. It is also possible that oligodeoxynucleotide treatment can stimulate cytokine production by other cell types which could modulate B cell activity and have a profound influence on the nature of the responses to this class of compounds. For example, primary cultures of human keratinocytes secrete IL-1α in response to treat-

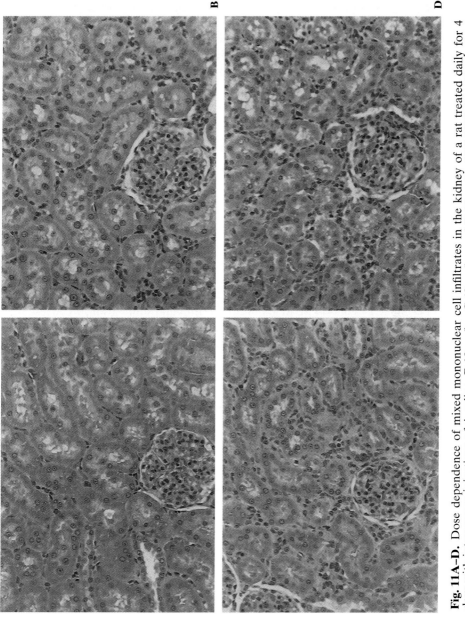

Fig. 11A–D. Dose dependence of mixed mononuclear cell infiltrates in the kidney of a rat treated daily for 4 days with intravenous injections of **A** saline, **B** 10mg/kg, **C** 40mg/kg, and **D** 80mg/kg ISIS 2105. Paraffin sections, H&E, ×200

ment with a phosphorothioate oligodeoxynucleotide. The release of cytokines by non-lymphoid-derived cells following exposure to oligodeoxynucleotides is a possible explanation for the cellular infiltrate observed in many organs and tissues in rodent toxicity studies (CROOKE et al. 1996).

Following systemic administration of phosphorothioate oligodeoxynucleotides, B cells are not the only infiltrating cell type found in greater abundance in liver, kidney, and other tissues. A large percentage of the infiltrating cells in tissues consists of monocytes, and it follows that the circulating cell populations reflect this increase as well. Mice treated with phosphorothioate oligodeoxynucleotides have increased total circulating leukocytes, especially neutrophil and monocyte counts (HENRY et al. 1997e; SARMIENTO et al. 1994).

Although immune stimulation is thought to be a class effect of phosphorothioate oligodeoxynucleotides and not dependent on hybridization, sequence is an important factor in determining immunostimulatory potential (BRANDA et al. 1993; KRIEG et al. 1995; YAMAMOTO et al. 1994). Immunostimulatory motifs have been described in the literature, e.g., palindromic sequences and CG (i.e., cytosine–guanosine) motifs (KRIEG et al. 1995; YAMAMOTO et al. 1994).

Using spleen weight as an index of in vivo immune stimulation, we investigated the sequence specificity of immunostimulatory effects (MONTEITH et al. 1997) . Fifteen different 20- or 21-mer phosphorothioate oligodeoxynucleotides were administered to mice at doses up to 100 mg/kg for 2–4 weeks. Within the series, there were marked differences in the immunostimulatory effects. One oligodeoxynucleotide, ISIS, 12449 contained a six-nucleotide palindromic element (AACGTT) identified as being a highly immunostimulatory sequence motif, as well as three other CG combinations also thought to be favorable to stimulation (BRANDA et al. 1993; KRIEG et al. 1995; YAMAMOTO et al. 1994). In in vivo studies, this oligodeoxynucleotide was more potent than a series of other oligodeoxynucleotides that contained from zero to three CG combinations. Notable differences in potency of immune stimulation were also observed in studies in which mice that were dosed with a 24-mer phosphorothioate oligodeoxynucleotide designed to hybridize to *Rel a*, the p65 subunit of NF-κB, or the sense construct to that sequence (*Rel a* SENSE). Mice injected daily intraperitoneally for 3 days with 25 mg/kg of the sense sequence had increases in spleen weights greater than twofold over control values. In contrast, the mice injected with the antisense sequence under identical conditions had a 1.3-fold increase in spleen weights. Although potency differed, histologic examination of the spleens demonstrated that, qualitatively, the effects in the spleens were similar (MCINTYRE et al. 1993).

There is no clearly defined rule for establishing the potency of different sequences. Part of the confusion stems from the in vivo observation that all of the sequences of unmodified phosphorothioate oligodeoxynucleotides display some potential for inducing splenomegaly, regardless of whether or not there is a CG in the sequence, suggesting that these motifs are not the sole determinants of immune stimulation (MONTEITH et al. 1997). What can be concluded from these studies is that the degree of B cell stimulation follows a complex set of rules that define the interactions of oligodeoxynucleotide with lymphocytes. Clearly, CG and palindromic elements containing CG are important in the potent stimulation of both B cell proliferation and cytokine release. High-affinity receptors have been proposed that might control these interactions (BENNETT et al. 1985), but it is still not clear whether the receptor for this effect is on the cell surface, as indicated using human B cells (LIANG et al. 1996a) or related to internalization of the oligodeoxynucleotide in rodent cells. Identification and structural mapping of high-affinity binding sites on the lymphocytes will obviously aid in the understanding of this interaction.

Among the most remarkable features of oligodeoxynucleotide-induced immune stimulation are the species differences. Rodents are highly susceptible to this generalized immune stimulation, whereas primates appear to be relatively insensitive to the effect at equivalent doses. Even 6 months of treatment of cynomolgus monkeys with 10mg/kg every other day with a 20-mer oligodeoxynucleotide produces only a relatively mild increase in B cell numbers in spleen and lymph nodes and does not produce changes in organ weights. The mixed monocellular infiltrates in liver and other organs that are so characteristic of the response in rodents are absent even after long-term exposure in monkeys (M.V. TEMPLIN, and A.A. LEVIN, unpublished observation). We have examined the pharmacokinetics of phosphorothioate oligodeoxynucleotides in rodents and primates and were unable to identify any differences in plasma kinetics or tissue distribution that would explain the species differences in sensitivity to immune stimulation (ISIS PHARMACEUTICALS, unpublished observations). It is known that rodents are more susceptible to the stimulatory effects of lipopolysaccharides, and much of the immune stimulation produced by oligodeoxynucleotides shares characteristics withlipopolysaccharide stimulation. Therefore, we conclude that rodents are markedly more sensitive to phosphorothioate oligodeoxynucleotide-induced immune stimulation than primates. Assuming results obtained in monkeys can be used to predict stimulation in humans, then the immunostimulatory effects may not be a prominent adverse effect in humans. In clinical trials with ISIS 2302, there are no indications of immune stimulation with intravenous doses of 2mg/kg administered for 1 month.

The mechanism of immune stimulation might be related to the polyanionic nature of the oligodeoxynucleotides. Polyanions including bacterial DNA are known to stimulate B cell proliferation (DIAMANTSTEIN et al. 1971a,b; DIAMANTSTEIN and BLISTEIN-WILLINGER 1978; TALMADGE et al. 1985). The polyanionic phosphorothioate oligodeoxynucleotides appear to behave in the same way.

The stimulation of B cell proliferation is associated with an increase in the secretion of immunoglobulins in both in vivo and in vitro studies (BRANDA et al. 1993, 1996; LIANG et al. 1996b). Culturing human B cells or mouse splenocytes with phosphorothioate oligodeoxynucleotides results in an increase in the secretion of IgG, IgM, and IgA. Thus stimulation of immunoglobulin production is polyclonal (BRANDA et al. 1993, 1996; LIANG et al. 1996b), and there is no indication that the immunoglobulins secreted are specifically directed toward the oligodeoxynucleotide, but it is possible that treatment with oligodeoxynucleotides could have an adjuvant-like effect. In vivo, the coadministration of phosphorothioate oligodeoxynucleotides with a tetanus toxin vaccine results in a more pronounced IgG and IgM response compared to the response in mice that receive tetanus toxin alone (BRANDA et al. 1996). From our experiences, trying to produce monoclonal antibodies to oligonucleotide, it is known that these compounds are weak antigens and need to be conjugated with a hapten for successful antibody production. The stimulation of Ig production could be enhanced in vitro by the addition of IL-2 or INF-γ (LIANG et al. 1996b; YI et al. 1996), but INF-γ did not modulate the proliferation of B cells, suggesting that Ig production is not directly linked to proliferation (YI et al. 1996).

Despite these in vitro observations, we have observed little or no increase in total globulin concentrations following 1 month of treatment of monkeys and mice with the phosphorothioate oligonucleotides ISIS 2105, ISIS 2302, ISIS 3521, and ISIS 5132. Because phosphorothioate oligodeoxynucleotides are clearly immunostimulatory in some species, we investigated whether treatment with phosphorothioate oligodeoxynucleotides induced the expression of anti-oligodeoxynucleotide antibodies. To examine this, plasma from monkeys and rats treated by intradermal injections with ISIS 2105 every other day for 4 weeks was analyzed for the presence of antibodies to ISIS 2105. An enzyme-linked immunosorbent assay (ELISA) was setup whereby a 96-well plate was coated with ISIS 2105 and serial dilutions of plasma from the animals in the study were added. The presence of antibodies was detected by an anti-monkey or anti-rat antibody linked with alkaline phosphatase using a p-nitrophenylphosphate substrate. A mouse monoclonal antibody directed against ISIS 2105 conjugated to keyhole lympet hemocyanin (KLH) was used as a positive control to ensure that the ISIS 2105 adhered to the plates was immunoreactive. There was an increase in the mean optical density (approximately 50%) in rat plasma diluted 1:10; however, this increase correlated with the expected slight increases in immunoglobulin, and there was no significant increase in optical density in wells with rat plasma diluted 1:100. Taken together, these data indicate that there was a very slight increase in antibodies, but no indication of a significant increase in anti-ISIS 2105 antibodies. In cynomolgus monkeys, no antibodies to ISIS 2105 were detected after intradermal administration of up to 10mg/kg every other day for 4 weeks (J.M. LEEDS and L. TRUONG, unpublished data).

Because no anti-oligodeoxynucleotide antibodies were detected in animal studies, it was of interest to examine plasma from human subjects. Volunteers were administered four doses of ISIS 2302 at 0, 0.2, 0.5, 1.0, or 2.0mg/kg by 2-h constant infusion every other day for 8 days. Plasma samples were collected before and 22 or 29 days after the initiation of dosing. An ELISA assay similar to that described above was employed using goat anti-human IgM or IgG antibodies conjugated with alkaline phosphatase to visualize the response. The results showed that there were no detectable IgG or IgM antibodies specific to ISIS 2302 (J.M. LEEDS, L. TRUONG, and A.A. LEVIN, unpublished data). Thus we conclude that treatment of rats, monkeys, or humans with phosphorothioate oligodeoxynucleotides does not result in antioligodeoxynucleotide antibodies, but we are continuing to monitor for this response preclinically and clinically.

It is evident that there are both species and sequence differences involved in immune stimulation and that specific sequences should, if possible, be excluded from oligodeoxynucleotides. In long-term toxicity studies in rodents, the constant cell proliferation associated with immune stimulation may have promoter-like effects and may thus complicate the interpretation of rodent carcinogenicity studies. At this time, there are no reports of toxicity studies longer than 6 months, and the long-term sequelae of immune stimulation in rodents are at present merely speculation. More importantly, immune stimulation following systemic administration of phosphorothioate oligodeoxynucleotides does not appear to be clinically relevant.

a) Effects of Oligodeoxynucleotide Chemistries

Structural modifications to oligodeoxynucleotides have been shown to reduce the potency of immune stimulation. The simplest modification with remarkable activity for reducing the immunostimulatory effects of oligodeoxynucleotides is the replacement of cytosine with 5-methyl cytosine. The methylation of a single cytosine residue in a CpG motif reduced [^{3}H]uridine incorporation and IgM secretion by mouse splenocytes. Methylation of a cytosine not in a CpG motif did not reduce the immunostimulatory potential (KRIEG et al. 1995). In another series of experiments, the three indices of immune stimulation (i.e., induction of proliferation and IgG and IgM secretion) by mouse splenocytes were compared for three sequences synthesized with or without methylation of CpG motifs (BRANDA et al. 1996). Methylation markedly reduced the proliferative effects for two of the three sequences. While all of the oligomers stimulated IgG and IgM secretion, the methylation of cytosine reduced the potency for IgM secretion. For IgG secretion, methylated sequences were more stimulatory for some sequences and less stimulatory for the other(s). In our experience with mice, when sequences with 5-methyl cytosine are compared with the same sequence without methylation, the methylated sequence has a lower potency for inducing immune stimulation, as determined by spleen weights (HENRY et al. 1997c).

Substitution of methyl phosphonate linkages for phosphorothioate linkages on each of the 3' and 5' termini have also been reported to reduce the proliferative effects and the secretion of IgG and IgM compared to the full phosphorothioate oligodeoxynucleotide analogue of this sequence, suggesting that this modification can also be used to ameliorate immune stimulation

(ZHAO et al. 1995). Determining the role of the length or number of phosphorothioate linkages in immune stimulation has been problematic because of sequence- related effects; thus it is not clear whether reduction in the number of phosphorothioate linkages alone can explain the reduction in potency with methyl phosphonate substitutions.

The addition of four 2'-methoxy substituents (GEM 91-H) and substitution of uridine for thymidine (and addition of four methyphosphonate linkages) in the termini of GEM 91 sequence (GEM 91-C) also reduced immunostimulatory potential, but the relative contribution of the uridine substitution and the 2'-methoxy substitution could not be dissected in this experiment (ZHAO et al. 1995). How 2'-alkoxy modifications alter the immunostimulatory potential of oligodeoxynucleotides needs to be further investigated.

2. Hepatic Effects of Oligodeoxynucleotide Treatment

In toxicity studies with phosphorothioate oligodeoxynucleotides, a variety of hepatic changes have been observed. The immune-mediated cellular infiltrates in rodent livers were discussed above. With high-dose administration of oligodeoxynucleotides in all species examined, there was a hypertrophic change in Kupffer cells accompanied by inclusions of basophilic material that was observed with hematoxylin and eosin staining. Immunohistochemical analysis of liver using monoclonal antibodies directed toward oligodeoxynucleotides conjugated with KLH demonstrates that these inclusions were composed of oligodeoxynucleotide (BUTLER et al. 1997). Furthermore, it was demonstrated that the presence of these inclusions was related to dose. The pattern of staining changes over time after a single dose, with decreasing amounts of oligodeoxynucleotide being found in the hepatocytes and increasing amounts in Kupffer cells (Fig. 12).

At doses of phosphorothioate oligodeoxynucleotide administered to rodents in the range of 100–150mg/kg, there are distinct hepatocellular toxicities. Histologic examination revealed multifocal hepatocellular degeneration or single cell necrosis. This hepatic change was associated with increases in serum transaminases (alanine aminotransferase, ALT; aspartate aminotransferase, AST), and decreased levels of albumin and cholesterol consistent with some degree of mild hepatic dysfunction (HENRY et al. 1997e,g; SARMIENTO et al. 1994). These changes were related to dose; for example, slight increases in ALT and AST were observed in mice treated by intradermal injection with ISIS 2105 for 4 weeks at 21.7mg/kg per day, but no changes were found at lower doses (HENRY et al. 1997g).

Hepatocellular changes were not a prominent feature of toxicity in primates. In cynomolgus monkeys, 50mg ISIS 2302 per kg administered every other day for 4 weeks by intravenous injection produced no morphologic indication of liver toxicity, although there was a slight (1.5-fold) increase in AST in this group (HENRY et al. 1997a). Following subcutaneous doses of ISIS

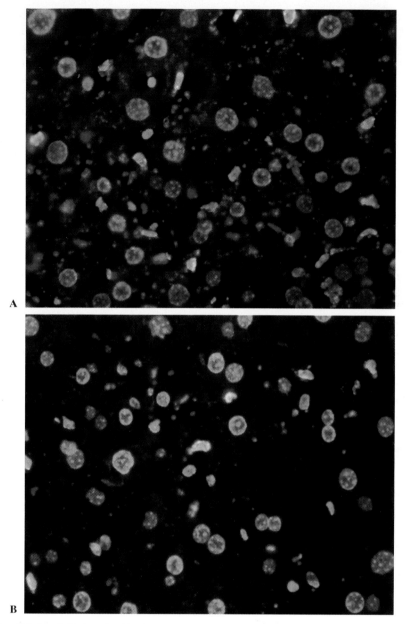

Fig. 12A,B. Cellular distribution of ISIS 2105 in the livers of mice **A** 2h and **B** 24h after an intravenous dose of 20mg ISIS 2105 per kg. The localization of a rhodamine-labeled oligodeoxynucleotide is show by fluorescence microscopy. From 2 to 24h, the signal dissipates from a punctate pattern in hepatocytes, while signal is retained in Kupffer cells. ×500

3521 and ISIS 5132 of up to 80 mg/kg every other day for four doses, there was Kupffer cell hypertrophy and periportal cell vacuolation, but no indication of necrosis and only a very slight increase in ALT. After 4 weeks of alternate-day dosing with 10 mg/kg via 2-h intravenous infusion of either ISIS 3521 or ISIS 5132, there were no alterations in AST or ALT, suggesting that at clinically relevant doses of these compounds, there was no evidence for hepatic pathology or transaminemia. In clinical trials with ISIS 2302, ISIS 3521, and ISIS 5132 at doses of 2 mg/kg administered by 2-h infusion on alternate days for 3–4 weeks, there was no indication of hepatic dysfunction, nor was there any evidence of transaminemia. In contrast, reports on GEM 91 indicate that transaminemias were observed in some acquired immunodeficiency syndrome (AIDS) patients treated with repeated administration of the phosphorothioate oligodeoxynucleotide, although these serum enzyme changes were transient and reversed with continued treatment (R.R. MARTIN, verbal communication at the Antisense Therapeutics 1997 Meeting, San Diego, CA, February 1997).

In summary, with high concentrations of oligodeoxynucleotide accumulating in the liver, it was thought that this organ would be a potential target organ for toxicity. In the mouse, at doses in the range of 50–150 mg/kg, there is an indication of hepatotoxicity, but it is related to dose and is associated with concentrations of total oligodeoxynucleotide that exceed $350\,\mu g/g$. At clinically relevant doses, there is no evidence for direct hepatocellular injury in monkeys. The concentrations of oligodeoxynucleotide in the liver at the end of 4 weeks of dosing (with 3 mg/kg every other day) in monkeys can be in the range of 200–$400\,\mu g/g$ tissue, but there is no hepatotoxicity with these compounds at clinically relevant doses. In clinical studies in humans at similar doses, no evidence for hepatotoxicity has been observed either. The effects of chemical modifications of phosphorothioates oligonucleotides on hepatotoxicity are currently being investigated.

3. Renal Effects of Oligodeoxynucleotide Treatment

Tissue distribution studies have shown that, like the liver, the kidney is also a major site of deposition of phosphorothioate oligodeoxynucleotide. Both Kupffer cells and proximal tubule epithelial cells contained oligodeoxynucleotide, as demonstrated by autoradiographic studies and immunohistochemistry as discussed above (OBERBAUER et al. 1995; PLENAT et al. 1994; SANDS et al. 1994) and by the use of special histologic stains (HENRY et al. 1997a). The appearance of basophilic inclusions is dose dependent (Fig. 13), and at higher doses the distribution of oligodeoxynucleotide appears to become more widespread, with concentrations in the renal medulla becoming more obvious.

Significant renal toxicity can be induced by extremely high doses. For example, doses of 100 or 150 mg phosphorothioate oligodeoxynucleotide per kg directed against *Rel-a* administered three times a week for 2 weeks produced renal proximal tubular degeneration in mice (SARMIENTO et al. 1994).

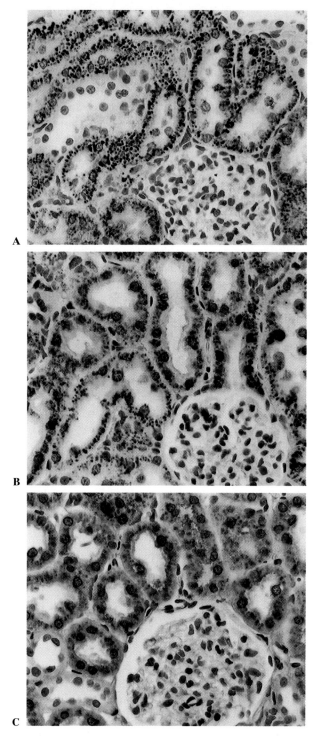

Fig. 13A–C. Dose-related differences in the distribution of ISIS 2105 in kidney following subcutaneous injections of **A** 5, **B** 40, or **C** 80 mg/kg. Monkeys were treated every other day with four doses of 5 mg/kg and were sacrificed 4 h after the last dose. Immunohistochemical localization of the oligonucleotide with a monoclonal antibody to ISIS 2105 detected with an antimouse antibody conjugated with peroxidase

The extent of tubular degeneration in this study was not sufficient to induce elevations in blood urea nitrogen or creatine; however, higher doses in rats and monkeys (up to 80 mg/kg subcutaneously) have induced both histologic and serum chemistry changes (D.K. MONTEITH and A.A. LEVIN, unpublished observations). A series of studies in rats and monkeys was performed to assess the relationship between histopathologic changes and the alterations in serum or urinary markers of renal toxicity (HORNER et al. 1997; MANZA et al. 1997). In both species, there was no histopathologic change after four doses of 10 mg/kg or below, but after four doses of 40–80 mg/kg subcutaneously there is mild to moderate renal tubular degeneration. Even in the presence of histopathologic changes, there is no increase in serum urea nitrogen or creatinine in either species. The absence of serum chemistry changes is not unexpected in light of the relatively mild degree and extent of the tubular involvement. In monkeys, however, there are increases in urinary protein concentrations, but only at doses associated with histopathologic changes. There are no changes in other urinary markers of renal function, including β_2-micro globulin, N-acetylglucosaminidase (NAG), and retinol-binding protein. In the rat, at doses associated with histopathologic changes, there are increases in urinary protein and NAG levels. These data suggest there is a clear dose–response relationship in the renal effects of phosphorothioate oligodeoxynucleotides. At clinically relevant doses, there was no indication of renal dysfunction. Even at doses associated with some morphologic renal changes, there is only a marginal effect on renal function, as indicated by markers in serum and urine. In 4-week or 6-month toxicity studies with phosphorothioate oligodeoxynucleotides, we observed a much more subtle type of morphologic change in the kidney. At a dose of 10 mg/kg on alternate days, there was a decrease in the height of the brush border and enlarged nuclei in some proximal tubule cells. These changes have been characterized as minimal to mild tubular atrophic and regenerative changes. At a dose of 3 mg/kg and below, these changes were only infrequently observed (if at all).

In light of the renal changes observed in toxicity studies, we are in the process of characterizing the relationship between morphologic renal changes and exposure to oligodeoxynucleotide. To assess exposure we have examined the concentrations of oligodeoxynucleotides in the renal cortex of samples obtained from subchronic toxicity studies. Renal concentrations increase with increasing doses (see Sect. B, Fig. 2). The concentration in the renal cortex associated with minimal to mild (and not clinically relevant) renal tubular atrophy or regenerative changes is approximately 1000 μg/g tissue. The cortex concentrations of total oligodeoxynucleotide that are associated with moderate degenerative changes after subcutaneous doses of 40–80 mg/kg are greater than 2000 μg/g. At a clinically relevant dose of 3 mg/kg every other day, the steady state concentration of total oligodeoxynucleotide in the kidney is in the range of 400–500 μg/g, demonstrating a significant margin of safety between the clinical doses and those doses associated with even the most minimal morphologic renal changes. Application of clearance and steady state pharmacokinetic models suggests that continued administration of oligodeoxynucleotide at this dose should never achieve the renal concentrations associated with dysfunction or even minimal morphologic changes (GEARY et al. 1997).

The effects of novel chemistries on the potential to induce renal toxicity have not been characterized at the time of writing.

4. Hematopoiesis

Morphologic changes in the bone marrow of mice were observed after 2 weeks of treatment (three doses/week) with 100–150 mg *Rel a* antisense phosphorothioate oligodeoxynucleotide per kg. There was a reduction in the number of megakaryocytes that was accompanied by a reduction of approximately 50% in circulating platelet counts (SARMIENTO et al. 1994). Reductions in platelets have been observed in rats treated with 21.7 mg ISIS 2105 per kg every other day (HENRY et al. 1997g), but were not observed in primates administered 10 mg/kg. Similarly, a reduction in platelets was observed in mice, but not in monkeys treated for 4 weeks with ISIS 2302 at doses of 100 and 50 mg/kg every other day, respectively. There were indications of reduced numbers of megakaryocytes in the bone marrow (SARMIENTO et al. 1994). These data suggest that the mouse may be more sensitive to these subchronic effects on platelets than nonhuman primates.

In acute studies in primates, we have occasionally observed transient reductions in platelets that occur acutely during 2-h infusions of doses of 10 mg/kg. These reductions reverse after completion of the infusion and have not been associated with any measurable change in platelet number 24 or 48 h after subchronic or chronic treatment regimens. Thrombocytopenia has been observe in AIDS patients treated with GEM 91 (reported in PLENAT 1996). In rodents at high doses, a slight anemia has been observed and an increase in monocytes has been previously discussed. Other meaningful changes in hematologic parameters have not been described.

IV. Reproductive Effects of Antisense Oligodeoxynucleotides

Little has been reported in the literature regarding the reproductive effects of phosphorothioate oligodeoxynucleotides. Antisense compounds have shown activity in in vitro models of embryonic development and have been used to study the effects of altering gene transcription on morphogenesis.

SADLER and colleagues used antisense oligodeoxynucleotides as a tools to knock out the expression of specific gene product, Engrailed-1 (*En-1*), in cultured mouse embryos (AUGUSTINE et al. 1995; SADLER et al. 1995). These experiments demonstrated that sequence-dependent alterations in developing mice embryos could be produced by a single intra-yolk sac injection of 15–25 μM oligodeoxynucleotide. At these concentrations, no malformations were observed in embryos treated with control oligodeoxynucleotide. The malformations produced by this direct injection technique were reminiscent of the malformations observed in transgenic mice lacking *En-1*. Yolk sac injections in mouse embryos of an antisense phosphorothioate oligodeoxynucleotide (final concentration, 25 μM) directed to retinol-binding protein produced apparent reductions in retinol, while a rearranged control oligodeoxynucleotide did not (BAVIK et al. 1996). These data suggest that phosphorothioate oligodeoxynucleotides can have activity in embryos when they are exposed directly at high concentrations and at the critical times of gestation. The effects on developing embryos were sequence specific.

When antisense oligonucleotides are administered to pregnant animals the results may differ. A phosphorothioate oligodeoxynucleotide directed toward the developmentally significant gene *Sry* or control oligodeoxynucleotides were administered to pregnant mice at various stages of gestation at approximately 18 mg/kg, no abnormalities were observed (GAUDETTE et al. 1993). Recent results with ISIS 2302 in rabbits and mice and with a mouse-specific anti-intercellular adhesion molecule (ICAM) oligodeoxynucleotide, ISIS 3082, in mice indicate that there were no malformations in either species when given at doses of 12 mg/kg per day during the critical periods of embryogenesis. The data also demonstrate that there were no effects on reproductive performance or fertility in male and female mice at these doses. (Note that ISIS 3082 is pharmacologically active in mice and was used to test whether alterations in ICAM-1 expression might have developmental consequences.) The only alteration observed was a reduction in fetal weight in rabbits that was associated with a reduction in maternal weight and evidence of slight maternal toxicity. Taken together, these data suggest that, at the doses administered, there was no evidence of class-related alterations in development or fertility. Whether or not an antisense oligodeoxynucleotide has developmental effects will need to be assessed for each compound and will depend on the target gene, the time of administration, and the pharmacokinetics of oligodeoxynucleotides in the test species.

V. Ocular Toxicity of Antisense Therapeutic Agents

In light of the potential usefulness of antisense agents in a number of disease states, including viral diseases, we and others have investigated the utility of an antisense approach in treating viral diseases in the eye. Human cytomegalovirus (HCMV) is a ubiquitous virus that is generally nonpathogenic except in extremely immunocompromised patients. While HCMV is a systemic disease, the ocular manifestations of the disease are most dramatic and can rapidly cause blindness in these patients. ISIS 2922 is an antisense molecule that is complementary to mRNA transcripts of the major immediate-early region (IE2) of HCMV (ANDERSON et al. 1996). This region of the HCMV genome encodes several proteins responsible for regulation of virus gene expression, which are essential for production of infectious HCMV. The synthesis of these proteins is inhibited by binding of ISIS 2922 to the mRNA, subsequently preventing immediate-early protein production and thus inhibiting virus replication. This antisense oligodeoxynucleotide is being developed for the treatment of HCMV retinitis in AIDS patients. ISIS 2922 is being used in phase III clinical trials on the basis of phase I/II trials in which the compound demonstrated promising antiviral activity (LIEBERMAN et al., submitted; PALESTINE et al. 1994, 1995).

ISIS 2922 is administered by intravitreal injection at weekly intervals for the induction phase of treatment and every other week for the maintenance phase. Preclinical studies in rabbits and monkeys have been used to evaluate ocular pharmacokinetics and tolerability for this compound. In order to compare dose levels across species, it is necessary to express dosage on the basis of initial vitreal concentrations of ISIS 2922, assuming vitreal volumes of 1.1 and 1.5 ml for rabbits and monkeys, respectively. The clinically used doses of 150 or 330 µg per eye yields concentrations of approximately 5 and 10 µM, respectively.

1. Ocular Pharmacokinetics

Following intravitreal administration to rabbits and monkeys, ISIS 2922 is distributed from vitreous to retina, the site of activity, and is retained in the retina with a long residence time. To test the hypothesis that vitreal concentrations could be estimated on the basis of dose (in micrograms) and vitreal volumes, we sacrificed rabbits shortly after dosing. Four hours after a single injection of 66 µg ISIS 2922 in rabbits, the concentration of intact oligodeoxynucleotide in the vitreous was 3.9 µM, consistent with the theoretical initial concentration of 4.6 µM. In rabbits, ISIS 2922 was cleared from the vitreous by 10 days with a half-life ($t_{1/2}$) of 60 h (2.5 days).

Clearance from the vitreous is driven primarily by slow distribution to surrounding tissues, including retina, where peak concentrations as high as 5.0 µM occurred 3–4 days after injection of 66 µg (4.6 µM) in rabbits (LEEDS et al. 1997b). Retina was cleared of ISIS 2922 with an estimated $t_{1/2}$ of approximately 96 h (4 days). Ten days after dosing, the concentration of intact ISIS 2922 in the retina diminished to approximately 1.8 µM, still well above the range of therapeutically active concentrations (EC_{50}, 0.1–0.5 µM).

The ocular kinetics of ISIS 2922 were characterized in cynomolgus monkeys at several dose levels designed to achieve vitreal concentrations of approximately 1, 5, and 10 µM (11, 57, and 115 µg, respectively). Vitreal concentrations of ISIS 2922 (measured 2 days following treatment) were proportional to dose (0.1–1.2 µM), and were eliminated with a $t_{1/2}$ of approximately 22 h, independent of dose (LEEDS et al. 1996b).

In monkeys, retinal concentrations (determined 2 days after dosing) increased with dose, although there was only a small difference between 5 and 10 µM doses (0.8 and 1.0 µM, respectively), suggesting saturation of tissue uptake. Elimination of ISIS 2922 from retina was slow and dose-dependent. As the dose increased from 5 to 10 µM, the $t_{1/2}$ increased from approximately 44 to 74 h.

Multiple-dose studies on ocular pharmacokinetics in monkeys indicated that both 5 and 10 μM doses, administered either weekly or every other week, maintained retinal concentrations at or above the estimated therapeutic range. Following three doses administered every other week, there was retinal accumulation of ISIS 2922 with every-other-week administration of 10 μM (144% of retinal concentration following a single dose), but accumulation was not observed with every-other-week doses of 5 μM, consistent with faster retinal elimination at this dose. However, weekly administration of 5 μM ISIS 2922 did result in slight retinal accumulation (170% of retinal concentration following a single dose). Thus retinal accumulation of ISIS 2922 is dependent on both dose and schedule. The pharmacokinetic data in both the rabbit and the monkey suggest that this compound can be administered on an infrequent basis (every other week) and still maintain sufficient concentrations for antiviral activity.

Due to the local route of administration (intravitreal injection) only microgram total amounts of drug need to be administered. Of the small amount of drug administered, much is retained in the eye or is metabolized locally by the retina. Small amounts of oligodeoxynucleotide that escape from the eye can be further metabolized in plasma and tissues. Together, these factors result in little or no systemic exposure to ISIS 2922 following clinically relevant dose regimens. In cynomolgus monkeys, levels of oligodeoxynucleotide in the plasma, liver, kidney, ovaries, and uterus following intravitreal injection at a clinically relevant dose of 115 μg ISIS 2922 (10 μM vitreal concentration) every other week for 3 months (eight doses in total) were below the levels of detection (10 nM) for both plasma and tissue. These detection limits are at least 1000-fold lower than tissue concentrations known to be associated with toxicities for related phosphorothioate oligodeoxynucleotides following intravenous administration. The absence of detectable exposure of ISIS 2922 in the systemic circulation is not surprising, given that the total equivalent systemic dose following intravitreal injection of 115 μg ISIS 2922 in both eyes of a 4-kg monkey is only 0.06 mg/kg. Consistent with the lack of systemic exposure, there have been no systemic toxicities observed following repeated intravitreal administration of ISIS 2922.

2. Ocular Toxicity Profile

ISIS 2922 has been injected intravitreally in rabbits at doses designed to achieve vitreal concentrations of 1–160 μM. Monkeys have received multiple intravitreal injections, either weekly or every other week, for up to 12 weeks at doses targeted to achieve initial vitreal concentrations of 0.4–10 μM. The primary ophthalmic observation in both species was ocular inflammation, i.e., cyclitis. In rabbits, the severity of ocular inflammation was dose-dependent and was observed at all doses examined, with 100% incidence at concentrations greater than or equal to 4 μM. Inflammation was largely reversible at concentrations up to $10\,\mu M$, peaking around day 8 and resolving by day 15. Ocular inflammation was observed in monkeys, but the incidence was lower (generally 15%–40% of treated eyes), was observed equally at all dose levels, and was also occasionally observed in control eyes, suggesting some attendant cause such as the injection technique or procedure. In the monkey, inflammation was associated with a number of physiologic responses, including vasculitis, posterior synechia, neovascularization of iris, hypopyon and closure of the drainage angle resulting in increased ocular pressure, and changes in the pigmentation pattern of retinal pigment epithelium (RPE) cells.

Subconjunctival injection of corticosteroids in rabbits markedly decreased, but did not eliminate inflammation associated with intravitreal ISIS 2922. However, inflammation could be completely prevented with coadministration of corticosteroids in monkeys. In the absence of inflammation, there were no effects on the retina or the ciliary body, as determined by either electroretinography or microscopic examination. These data suggest that, at clinically relevant concentrations of ISIS 2922, there was no direct retinal toxicity and that all the observed ocular effects were secondary to inflammation. Ocular inflammation or cyclitis has been observed in some patients in clinical trials, but is generally reversible and well managed with standard clinical practice, i.e., administration of topical corticosteroids, enabling patients to remain on the study.

Recent advances in medicinal chemistry have provided a second-generation HCMV antisense oligodeoxynucleotide that promises to improve the overall therapeutic profile relative to ISIS 2922. A novel viral inhibitor has the same base sequence and phosphorothioate linkages as ISIS 2922; however, it has the additional modification of 2′-methoxyethoxy substituents at the 3′ and 5′ terminal residues. Early preclinical evaluation of this compound indicates that it has comparable antiviral activity to ISIS 2922. The real advantage of this novel compound appears to be in the greater resistance to nuclease degradation, which will result in slow elimination from the retina and may allow for longer treatment intervals, i.e., dosing once a month or less. In addition to improved ocular pharmacokinetics, this compound also has an improved ocular tolerability profile with a much reduced incidence of ocular inflammation. The combination of these properties may result in an alternative to ISIS 2922 that has fewer side effects and a more convenient treatment regimen.

These data suggest that phosphorothioate oligodeoxynucleotides can be safely used as local therapy for HCMV retinitis and that they have a manageable profile as far as side effects are concerned. The lower potential for immune stimulation of 2′-alkoxy-modified phosphorothioate oligodeoxynucleotides and the greater resistance to nuclease-mediated degradation suggest that, while ISIS 2922 is effective and has an attractive treatment interval (every other week), the second generation of anti-HCMV antisense molecule may have an even more attractive safety profile and may allow an even longer interval between dosing.

References

Agrawal S, Temsamani J, Tang JY (1991) Pharmacokinetics, biodistribution, and stability of oligodeoxynucleotide phosphorothioates in mice. Proc Natl Acad Sci USA 88:7595–7599

Agrawal S, Rustagi P, Shaw DR (1995a) Novel enzymatic and immunological responses to oligonucleotides. In: Reed DJ (ed) Proceedings of the international congress of toxicology VII. Elsevier, Amsterdam, pp 431–434

Agrawal S, Zhang X, Lu Z, Zhao H, Tamburin JM, Yan J, Cai H, Diasio RG, Habus I, Jlang Z, Iyer RP, Yu D, Zhang R (1995b) Absorption, tissue distribution and in vivo stability in rats of a hybrid antisense oligonucleotide following oral administration. Biochem Pharmacol 50:571–576

Altmann K-H, Dean NM, Fabbro D, Freier SM, Geiger T, Haner R, Husken D, Martin P, Monia BP, Muller M, Natt F, Nicklin P, Phillips J, Pieles U, Sasmor H, Moser HE (1996) Second generation of antisense oligonucleotides: from nuclease resistance to biological efficant in animals. Chimia 50:168–176

Anderson KP, Fox MC, Brown-Driver V, Martin MJ, Azad RF (1996) Inhibition of human cytomegalovirus immediate-early gene expression by an antisense oligonucleotide complementary to immediate-early RNA. Antimicrob Agents Chemother 40:2004–2011

Augustine KA, Liu ET, Sadler TW (1995) Antisense inhibition of engrailed genes in mouse embryos reveals roles for these genes in craniofacial and neural tube development. Teratology 51:300–310

Bavik C, Ward SJ, Chambon P (1996) Developmental abnormalities in cultured mouse embryos deprived of retinoic acid by inhibition of yolk-sac retinol binding protein synthesis. Proc Natl Acad Sci USA 93:3110–3114

Bennett CF (1994) Inhibition of cell adhesion molecule expression with antisense oligonucleotides: activity in-vitro and in-vivo. J Immunol 152:3530–3540

Bennett CF, Chiang M-Y, Chan H, Shoemaker JEE, Mirabelli CK (1992) Cationic lipids enhance cellular uptake and activity of phosphorothioate antisense oligonucleotides. Mol Pharmacol 41:1023–1033

Bennett CF, Dean N, Ecker DJ, Monia BP (1996) Pharmacology of antisense therapeutic agents: cancer and inflammation. In: Agrawal S (ed) Methods in molecular medicine: antisense therapeutics. Humana, Totowa, New York, pp 13–46

Bennett CF, Kornbrust D, Henry S, Stecker K, Howard R, Cooper S, Dutson S, Hall W, Jacoby HI (1997) An ICAM-1 antisense oligonucleotide prevents and reverses dextran sulfate sodium-induced colitis in mice. J Pharmacol Exp Ther 280:988–1000

Bennett RM, Gabor GT, Merritt MM (1985) DNA binding to human leukocytes. J Clin Invest 76:2182–2190

Bitter-Suermann D, Burger R, Hadding U (1981) Activation of the alternative pathway of complement: efficient fluid-phase amplification by blockade of the regulatory complement protein b1 h through sulfated polyanions. Eur J Immunol 11:291–295

Black LE, DeGeorge JJ, Cavagnaro JA, Jordan A, Ahn C-H (1993) Regulatory considerations for evaluating the pharmacology and toxicology of antisense drugs. Antisense Res Dev 3:399–404

Black LE, Farrelly JG, Cavagnaro JA, Ahn C-H, DeGeorge JJ, Taylor AS, DeFelice AF, Jordan A (1994) Regulatory considerations of oligonucleotide drugs: updated recommendations for pharmacology and toxicology studies. Antisense Res Dev 4:299–301

Branda RF, Moore AL, Mathews L, McCormack JJ, Zon G (1993) Immune stimulation by an antisense oligomer complementary to the rev gene of HIV-1. Biochem Pharmacol 45:2037–2043

Branda RF, Moore AL, Lafayette AR, Mathews L, Hong R, Zon G, Brown T, McCormack JJ (1996) Amplification of antibody production by phosphorothioate oligodeoxynucleotides. J Lab Clin Med 128:329–338

Brown DA, Kang S-H, Gryaznov SM, DeDionisio L, Heidenreich O, Sullivan S, Xu X, Nerenberg MI (1994) Effect of phosphorothioate modification of oligodeoxynucleotides on specific protein binding. J Biol Chem 269:26801–26805

Butler M, Stecker K, Bennett CF (1997) Histologic localization of phosphorothioate deoxyoligonucleotides in normal rat tissues. Lab Invest 77(4):379–388

Chiang M-Y, Chan H, Zounes MA, Freier SM, Lima WF, Bennett CF (1991) Antisense oligonucleotides inhibit intercellular adhesion molecule 1 expression by two distinct mechanisms. J Biol Chem 266:18162–18171

Cornish KG, Iversen P, Smith L, Arneson M, Bayever E (1993) Cardiovascular effects of a phosphorothioate oligonucleotide with sequence antisense to p53 in the conscious rhesus monkey. Pharmacol Commun 3:239–247

Cossum PA, Sasmor H, Dellinger D, Troung L, Cummins L, Owens SR, Markham PM, Shea JP, Crooke S (1993) Disposition of the 14C-labeled phosphorothioate oligonucleotide ISIS 2105 after intravenous administration to rats. J Pharmacol Exp Ther 267:1181–1190

Cossum PA, Troung L, Owens SR, Markham PM, Shea JP, Crooke ST (1994) Pharmacokinetics of a 14C-labeled phosphorothioate oligonucleotide, ISIS 2105, after intradermal administration to rats. J Pharmacol Exp Ther 269:89–94

Crooke RM, Crooke ST, Graham MJ, Cooke ME (1996) Effect of antisense oligonucleotides on cytokine release from human keratinocytes in an in vitro model of skin. Toxicol Appl Pharmacol 140:85–93

Cummins LL, Leeds JM, Greig M, Griffey RJ, Graham MJ, Crooke R, Gaus HJ (1997) Capillary gel electrophoresis and mass spectrometry: powerful tools for the analysis of antisense oligonucleotides and their metabolites. In: Cook PD, Manoharan M (eds) Nucleosides and nucleotides. Marcel Decker, New York (in press)

Dean NM, McKay R, Condon TP, Bennett CF (1994) Inhibition of protein kinase C-α expression in human A549 cells by antisense oligonucleotides inhibits induction of intercellular adhesion molecule 1 (ICAM-1) mRNA by phorbol esters. J Biol Chem 269:16416–16424

de Serres M, McNulty MJ, Christensen L, Zon G, Findlay JWA (1996) Development of a novel scintillation proximity competitive hybridization assay for the determination of phosphorothioate antisense oligonucleotide plasma concentrations in a toxicokinetic study. Anal Biochem 233:228–233

Diamantstein T, Blistein-Willinger E (1978) Specific binding of poly (I)–poly (C) to the membrane of murine B lymphocyte subsets. Eur J Immunol 8:896–899

Diamantstein T, Wagner B, Beyse I, Odenwald MV, Schultz G (1971a) Stimulation of humoral antibody formation by polyanions. I. The effect of polyacrylic acid on the primary immune response in mice immunized with sheep red blood cells. Eur J Immunol 1:335–340

Diamantstein T, Wagner B, Beyse I, Odenwald MV, Schultz G (1971b) Stimulation of humoral antibody formation by polyanions. II. The influence of sulfate esters of polymers on the immune response in mice. Eur J Immunol 1:340–343

Galbraith WM, Hobson WC, Giclas PC, Schechter PJ, Agrawal S (1994) Complement activation and hemodynamic changes following intravenous administration of phosphorothioate oligonucleotides in the monkey. Antisense Res Dev 4:201–206

Gaudette MF, Hampikian G, Metelev V, Agrawal S, Crain W (1993) Effect on embryos of injection of phosphorothioate-modified oligonucleotides into pregnant mice. Antisense Res Dev 3:391–397

Geary R, Leeds J, Henry SP, Monteith DK, Levin AA (1997) Antisense oligonucleotide inhibitors for the treatment of cancer. 1. Pharmacokinetic properties of phosphorothioate oligodeoxynucleotides. Anti Cancer Drug Des 12:383–393

Glover JM, Leeds JM, Mant TGK, Amin D, Kisner DL, Zuckerman J, Levin AA, Shanahan WR (1997) Phase 1 safety and pharmacokinetic profile of an ICAM-1 antisense oligodeoxynucleotide (ISIS 2302). J Pharmacol Exp Ther 282:1173–1180

Griffin LC, Tidmarsh GF, Bock LC, Toole JJ, Leung LLK (1993) In vivo anticoagulant properties of a novel nucleotide-based thrombin inhibitor and demonstration of regional anticoagulation in extracorporeal circuits. Blood 81:3271–3276

Henry SP, Larkin R, Novotny WF, Kornbrust DJ (1994) Effects of ISIS 2302, a phosphorothioate oligonucleotide, on in vitro and in vivo coagulation parameters. Pharmaceut Res 11:S-353

Henry SP, Leeds J, Giclas PC, Gillett NA, Pribble JP, Kornbrust DJ, Levin AA (1996) The toxicity of ISIS 3521, a phosphorothioate oligonucleotide, following intravenous (iv) and subcutaneous (sc) administration in cynomolgus monkeys. Toxicologist 30:112

Henry SP, Bolte H, Auletta C, Kornbrust DJ (1997a) Evaluation of the toxicity of ISIS 2302, a phosphorothioate oligonucleotide, in a 4-week study in cynomolgus monkeys. Toxicology 120:145–155

Henry SP, Giclas PC, Leeds J, Pangborn M, Auletta C, Levin AA, Kornbrust DJ (1997b) Activation of the alternative pathway of complement by a phosphorothioate oligonucleotide: potential mechanism of action. J Pharmacol Exp Ther 281:810–816

Henry SP, Monteith DK, Bennett CF, Levin AA (1997c) Toxicologic and pharmacokinetic parameters of chemically modified antisense oligonucleotide inhibitors of PKC-alpha and C-raf kinase. Anti Cancer Drug Des 12:409–420

Henry SP, Monteith DK, Levin AA (1997d) Antisense oligonucleotide inhibitors for treatment of cancer. 2. Toxicologic properties of phosphorothioate oligodeoxynucleotides. Anti Cancer Drug Des 12:395–408

Henry SP, Taylor J, Midgley L, Levin AA, Kornbrust DL (1997e) Evaluation of the toxicity of ISIS 2302, a phosphorothioate oligonucleotide, in a 4-week study in CD-1 mice. Antisense Nucleic Acid Drug Dev 7:473–481

Henry SP, Novotny W, Leeds J, Auletta C, Kornbrust DJ (1997f) Inhibition of coagulation by a phosphorothioate oligonucleotide. Antisense Nucleic Acid Drug Dev 7:503–510

Henry SP, Orr JL, Brunner RH, Kornbrust DJ (1997g) Comparison of the toxicity profiles of ISIS 1082 and ISIS 2105, phosphorothioate oligonucleotides, following subacute intradermal administration in Sprague-Dawley rats. Toxicology 116:77–88

Horner MJ, Monteith DK, Gillett NA, Butler M, Henry SP, Bennett CF, Levin AA (1997) Evaluation of the renal effects of phosphorothioate oligonucleotides in monkeys. Fundam Appl Toxicol 36:147

Iversen P (1991) In vivo studies with phosphorothioate oligonucleotides: pharmacokinetics prologue. Anti Cancer Drug Des 6:531–538

Klinman DM, AE-Kyung Y, Beaucage SL, Conover J, Krieg AM (1996) CpG motifs present in bacterial DNA rapidly induce lymphocytes to secrete interleukin 6, interleukin 12, and interferon g. Proc Natl Acad Sci USA 93:2879–2883

Krieg AM, Stein CA (1995) Phosphorothioate oligodeoxynucleotides: antisense or anti-protein? Antisense Res Dev 5:241

Krieg AM, Yi A-K, Matson S, Waldschmidt TJ, Bishop GA, Teasdale R, Koretzky GA, Klinman DM (1995) CpG motifs in bacterial DNA trigger direct B-cell activation. Nature 374:546–549

Leeds JM, Kiorpes A, Geary RS, Henry SP, Goldensoph C, Levin AA (1996a) Single dose subcutaneous bioavailability of ISIS 2302, a phosphorothioate oligonucleotide. In: Research P (ed) Cynomolgus monkeys, vol 13. Plenum, Seattle, p S395

Leeds JM, Williams K, Scherrill S, Levin AA, Bistner S, Henry SP (1996b) Potential for retinal accumulation of a phosphorothioate oligonucleotide (ISIS 2922) after intravitreal injection in cynomolgus monkeys. Fundam Appl Toxicol 30:40

Leeds JM, Geary RS, Henry SP, Glover J, Shanahan W, Fitchett J, Burckin T, Truong L, Levin AA (1997a) Pharmacokinetic properties of phosphorothioate oligonucleotides. In: Cook PD, Manoharan M (eds) Nucleosides and nucleotides. Marcel Decker, New York 16(7–9):1689–1693

Leeds JM, Henry SP, Truong L, Zutsi A, Levin AA, Douglas KJ (1997b) Pharmacokinetics of a potential human cytomegalovirus therapeutic, A phosphorothioate oligonucleotide after intravitreal injection in the rabbit. Drug Metab Dispos 25(8):921–926

Liang H, Nishioka Y, Reich CF, Pisetsky DS, Lipsky PE (1996a) Activation of human B cells by phosphorothioate oligodeoxynucleotides. J Clin Invest 98:1119–1129

Liang WW, Shi X, Deshpande D, Malanga CJ, Rojanasakul Y (1996b) Oligonucleotide targeting to alveolar macrophages by mannose. Biochim Biophys Acta 1279:227–234

Lieberman R, Boyer D, Terry B, Antoszyk A, Park S, Cantrill H, Ai E, Holland G, Palestine A. (submitted) Antisense oligonucleotide dose response for treatment of cytomegalovirus retinitis. Am J Ophthal

Manza LL, Butler MM, Bennett CF, Levin AA, de Peyster AA, Monteith DK (1997) Renal toxicity of phosphorothioate oligonucleotides in rats. Fundam Appl Toxicol 36:147

McIntyre KW, Lombard-Gillooly K, Perez JR, Kunsch C, Sarmiento UM, Larigan JD, Landreth KT, Narayanan R (1993) A sense phosphorothioate oligonucleotide directed to the initiation codon of transcription factor NF-κb p65 causes sequence-specific immune stimulation. Antisense Res Dev 3:309–322

Miraglia L, Geiger T, Bennett CF, Dean N (1996) Inhibition of interleukin-1 type receptor expression in human cell-lines by an antisense phosphorothioate oligonucleotide. Int J Immunopharmacol 18:227–240

Monia BP, Johnston JF, Ecker DJ, Zounes M, Lima WF, Freier SM (1992) Selective inhibition of mutant Ha-ras mRNA expression by antisense oligonucleotides. J Biol Chem 267:19954–19962

Monia BP, Johnston JF, Geiger T, Muller M, Fabbro D (1995) Antitumor activity of a phosphorothioate oligodeoxynucleotide targeted against C-raf kinase. Nature Med 2:668–675

Monia BP, Sasmor H, Johnston JF, Freier SM, Lesnik EA, Muller M, Geiger T, Altmann K-H, Moser H, Fabbro D (1996) Sequence-specific antitumor activity of a phosphorothioate oligodeoxyribonucleotide targeted to human c-raf kinase supports an antisense mechanism of action in vivo. Proc Natl Acad Sci USA 93:15481–15484

Monteith DK, Henry SP, Howard RB, Flournoy S, Levin AA, Bennett CF, Crooke ST (1997) Immune stimulation – a class effect of phosphorothioate oligonucleotides in rodents. Anti Cancer Drug Des 12:421432

Nicklin PL, Ambler J, Mitchelson A, Bayley D, Phillips JA, Craig SJ, Monia BP (1997) Preclinical profiling of modified oligonucleotides: anticoagulant and pharmacokinetic properties. In: Cook PD, Manoharan M (eds) Nucleosides and nucleotides. Marcel Decker, New York 16(7–9):1145–1153

Oberbauer R, Schreiner GF, Meyer TW (1995) Renal uptake of an 18-mer phosphorothioate oligonucleotide. Kidney Int 48:1226–1232

Palestine AG, Cantrill HL, Luckie AP, Ai E (1994) Intravitreal treatment of CMV retinitis with an antisense oligonucleotide, ISIS 2922, 10th international conference on AIDS and the international conference on STD, Yokohama, Japan, August 1994, vol 2, p 404

Palestine AG, Cantrill H, Ai E, Liberman R (1995) Treatment of cytomegalovirus retinitis with ISIS 2922. Invest Ophthalmol Vis Sci 36:S181

Plenat F (1996) Animal models of antisense oligonucleotides: lessons for use in humans. Mol Med Today 2:250–257

Plenat F, Klein-Monhoven N, Marie B, Vignaud J-M, Duprez A (1994) Cell and tissue distribution of synthetic oligonucleotides in healthy and tumor-bearing nude mice. Am J Pathol 147:124–135

Rappaport J, Hanss B, Kopp JB, Copelend TD, Bruggeman LA, Coffman TM, Klotman PE (1995) Transport of phosphorothioate oligonucleotide in kidney: implications for molecular therapy. Kidney Int 47:1462–1469

Rapaport E, Levina A, Metelev V, Zamecnik PC (1996) Antimycobacterial activities of antisense oligodeoxynucleotide phosphorothioates in drug-resistant strains. Proc Natl Acad Sci USA 93:709–713

Rifai A, Brysch W, Fadden K, Clark J, Schlingensiepen KH (1996) Clearance kinetics, biodistribution, and organ saturability of phosphorothioate oligodeoxynucleotides in mice. Am J Pathol 149:717–725

Rosenberger RD (1989) Biochemistry of heparin antithrombin interactions, and the physiologic role of this natural anticoagulant mechanism. Am J Med 87:2S–3S

Sadler TW, Liu ET, Augustine KA (1995) Antisense targeting of Engrailed-1 causes abnormal axis formation in mouse embryos. Teratology 51:292–299

Saijo Y, Perlaky L, Wang H, Busch H (1994) Pharmacokinetics, tissue distribution, and stability of antisense oligodeoxynucleotide phosphorothioate ISIS 3466 in mice. Oncol Res 6:243–249

Sands H, Gorey-Feret LJ, Cocuzza AJ, Hobbs FW, Chidester D, Trainor GL (1994) Biodistribution and metabolism of internally 3H-labeled oligonucleotides. I. Comparison of a phosphodiester and phosphorothioate. Mol Pharmacol 45:932–943

Sarmiento UM, Perez JR, Becker JM, Narayanan R (1994) In vivo toxicological effects of rel A antisense phosphorothioates in CD-1 mice. Antisense Res Dev 4:99–107

Sawai K, Miyao T, Takakura Y, Hashida M (1995) Renal disposition characteristics of oligonucleotides modified at terminal linkages in the perfused rat kidney. Antisense Res Dev 5:279–287

Srinivasan SK, Iversen P (1995a) Review of in vivo pharmacokinetics and toxicology of phosphorothioate oligonucleotides. J Clin Lab Anal 9:129–137

Srinivasan SK, Tewary HK, Iversen PL (1995b) Characterization of binding sites, extent of binding and drug interactions of oligonucleotides with albumin. Antisense Res Dev 5:131–139

Stepkowski SM, Tu Y, Condon TP, Bennett CF (1994a) Blocking of heart allograft rejection by intercellular adhesion molecule-1 antisense oligonucleotides alone or in combination with other immunosuppressive modalities. J Immunol 153:5336–5346

Talmadge JE, Adams J, Phillips H, Collins M, Lenz B, Schneider M, Schlick E, Ruffmann R, Wiltrout RH, Chirigos MA (1985) Immunomodulatory effects in mice of polyinosinic-polycytidylic acid complexed with poly-L-lysine and carboxymenthycellulose1. Cancer Res 45:1058–1065

Temsamani J, Tang J-Y, Padmapriya A, Kubert M, Agrawal S (1993) Pharmacokinetics, biodistribution, and stability of capped oligodeoxynucleotide phosphorothioates in mice. Antisense Res Dev 3:277–284

Wallace TL, Bazemore SA, Kornbrust DJ, Cossum PA (1996a) Repeat-dose toxicity and pharmacokinetics of a partial phosphorothioate anti-HIV oligonucleotide (AR177) after bolus intravenous administration to cynomolgus monkeys 2. J Pharmacol Exp Ther 278:1313–1317

Wallace TL, Bazemore SA, Kornbrust DJ, Cossum PA (1996b) Single-dose hemodynamic toxicity and pharmacokinetics of a partial phosphorothioate anti-HIV oligonucleotide (AR177) after intravenous infusion to cynomolgus monkeys 1. J Pharmacol Exp Ther 278:1306–1312

Wu-Pong S, Weiss TL, Hunt AC (1994) Antisense c-myc oligonucleotide cellular uptake and activity. Antisense Res Dev 4:155–163

Yamamoto T, Yamamoto S, Kataoka T, Tokunaga T (1994) Ability of oligonucleotides with certain palindromes to induce interferon production and augment natural killer cell activity is associated with their base length. Antisense Res Dev 4:119–122

Yi A-K, Chace JH, Cowdery JS, Krieg AM (1996) IFN-gamma promotes IL-6 and IgM secretion in response to CpG motifs in bacterial DNA and oligonucleotides. Immunology 156:558–564

Zhao Q, Temsamani J, Iadarola PL, Jiang Z, Agrawal S (1995) Effect of different chemically modified oligodeoxynucleotides on immune stimulation. Biochem Pharmacol 51:173–182

CHAPTER 6
Pharmacokinetic Properties of Phosphorothioate Oligonucleotides in Humans

J.M. LEEDS and R.S. GEARY

A. Introduction

The use of antisense molecules to inhibit the expression of disease-causing proteins represents a new paradigm in disease treatment. The molecular target for the antisense molecules, mRNA, is chemically and biologically well defined. This well-defined target is the same for all different possible therapeutic applications, whether the application is antiviral, anticancer, or anti-inflammatory. The specificity with which antisense molecules are capable of inhibiting gene expression, exemplified by the isozyme selectivity demonstrated by DEAN and MCKAY (1994), has allowed entirely novel molecular targets to be explored for potential therapeutic applications. Thus antisense molecules have presented an opportunity to exploit a single target for multiple therapeutic indications as well as to explore the utility of inhibiting targets which traditional therapeutic agents could not specifically inhibit.

Antisense oligonucleotides are now emerging from research into clinical development for a wide range of therapeutic indications (Table 1). The first antisense molecules to enter clinical trials have all been phosphorothioate oligodeoxynucleotides, in which a single nonbridging oxygen of the internucleotide linkage has been replaced by a sulfur. This simple modification imparts increased nuclease resistance to the DNA analogues, increasing the in vitro and in vivo half-life (STEIN et al. 1988; WICKSTROM 1986; CAMPBELL et al. 1990; HOKE et al. 1991; SANDS et al. 1994).

The justification for movement of phosphorothioate oligonucleotides into clinical trials has come from promising antisense activities seen in animal pharmacology studies (CROOKE and BENNETT 1996). In one recently published study, an oligonucleotide complementary to human c-*raf* kinase mRNA inhibited tumor growth in human tumor xenograft mouse models and c-*raf* kinase RNA levels in implanted tumors, while a mismatch–control oligonucleotide had no effect (MONIA et al. 1996). The oligonucleotide inhibited tumor growth in mice at a dose of as low as 0.06mg/kg. This antisense oligonucleotide, ISIS 5132/CGP 69846A, is currently being evaluated in phase I trials in patients with solid cancers (Table 1). Antisense activities seen in animal pharmacology studies appear to be sequence dependent, with a small number of exceptions (AGRAWAL 1996).

Table 1. Antisense phosphorothioate oligodeoxynucleotides evaluated in human pharmacokinetic studies

Compound	Molecular target	Therapeutic target	Doses studied (mg/kg)	Administration route	Analytical methodology
ISIS 2105[a]	HPV-6/11 (E2)	Antiviral	~0.06	Intradermal	^{14}C/HPLC
GEM 91[b]	HIV (*gag*/pol)	Antiviral	0.10	2-h IV infusion	^{35}S/PAGE
OL(1)p53[c]	p53	Cancer	1.2–6.0	Continuous IV infusion	Post-extraction ^{32}P-labeling
ISIS 2302[d]	ICAM-1	Inflammation	0.06–2.0	2-h IV infusion	CGE
G3139[e]	BCL-2	Cancer	2.0	Continuous subcutaneous infusion	HPLC
ISIS 3521	PKC-α	Cancer	0.15–0.6	2-h IV infusion	CGE
ISIS 5132	c-RAF kinase	Cancer	0.5–1.0	2-h IV infusion	CGE

HPV, human papillomavirus; HIV, human immunodeficiency virus; ICAM, intercellular adhesion molecule; PKC, protein kinase C; HPLC, high-performance liquid chromatography; PAGE, polyacrylanride gel electrophoresis; CGE, capillary gel electrophoresis.
[a] CROOKE et al. (1994).
[b] ZHANG et al. (1995).
[c] BISHOP et al. (1996).
[d] GLOVER et al. (submitted).
[e] LEBEDEV et al. (1997).

The probability of successfully inhibiting a given molecular target in vivo will be dependent on the therapeutic agent reaching the necessary site of action, and that will be determined by the pharmacokinetic properties of the compound. Early clinical trials of phosphorothioate oligonucleotides are evaluating both the pharmacokinetics and the safety profile of the compounds. The scope of this review is to examine the results of both published and unpublished studies of the pharmacokinetics of phosphorothioate oligodeoxynucleotides in humans. In this examination, an effort will be made, where possible, to compare results for different sequences in order to obtain a picture of the pharmacokinetics of this class of compounds.

B. Human Pharmacokinetics

Published pharmacokinetic studies of phosphorothioate oligonucleotides from clinical trials indicate that the approaches being used both to administer and to follow the compounds are divergent (Table 1). ISIS 2105, designed to inhibit the growth of human papilloma virus HPV-6 and HPV-11 was studied after intradermal administration of the ^{14}C-radiolabeled compound in humans (CROOKE et al. 1994). The pharmacokinetics of GEM 91, designed to inhibit expression of the *gag*/pol gene of human immunodeficiency virus (HIV)-1, were evaluated after systemic administration of a 0.1 mg/kg dose of ^{35}S-radiolabeled oligonucleotide (ZHANG et al. 1995). BISHOP et al. (1996) evaluated the pharmacokinetics of OL(1)p53, a phosphorothioate oligonucleotide, in humans administered as a 10-day continuous intravenous infusion of doses ranging from 1.2 to 6.0 mg/kg per day. That study did not use radiolabeled oligonucleotide; rather, extracted oligonucleotide was end-labeled enzymatically using [^{32}P]ATP (BISHOP et al. 1996). ISIS 2302, an antisense oligonucleotide designed to inhibit the expression of intercellular adhesion molecule (ICAM)-1, has been evaluated in multiple human clinical trials after administration via a 2-h intravenous infusion. Analysis of the plasma ISIS 2302 concentrations used gel-filled capillary electrophoresis (CGE) (LEEDS et al. 1996) rather than relying on radiolabel for analysis. Two additional phosphorothioate oligonucleotides currently in phase I trials, ISIS 3521/CGP 64128A and ISIS 5132/CGP 69846A, are being evaluated for anticancer activity. ISIS 3521/CGP 64128A inhibits the expression of protein kinase C (PKC)-α (DEAN and McKAY 1994), while ISIS 5132/CGP 69846A inhibits the expression of c-*raf* kinase (MONIA et al. 1996). Both of those compounds are being evaluated after administration of either 2-h intravenous infusions given three times weekly or continuous intravenous therapy. The plasma oligonucleotide levels in these studies are also being quantitated using CGE following solid-phase extraction of the oligonucleotide (LEEDS et al. 1996). The pharmacokinetics of a continuous subcutaneous infusion of an oligonucleotide, G3139, designed to inhibit the expression of the apoptosis-inhibiting protein BCL-2, have been studied. Oligonucleotide levels were quantitated using anion-exchange high-performance liquid chromatography (HPLC) (LEBEDEV et al. 1997). When the data from these studies are compared, a consistent picture of phosphorothioate oligonucleotide pharmacokinetics emerges.

I. Absorption

An early published report of the pharmacokinetics of phosphorothioate oligonucleotides in humans followed radiolabel after the intradermal administration of [2-^{14}C-thymidine]-ISIS 2105 to humans (CROOKE et al. 1994). It was hoped that the compound, designed as an antisense inhibitor of HPV-6 and HPV-11, the cause of genital warts (COWSERT et al. 1993), would remain in the wart where it had been injected and would inhibit viral replication. However, radioactivity in plasma peaked 1 h after the intradermal injection, indicating that the compound was readily absorbed from the intended site of pharmacologic action. Based on the plasma radioactivity level at that time point, it was estimated that 65% of the administered dose had been absorbed. By 24 h, the radioactivity remaining in the warts had decreased to less than 10% of the radioactivity seen

1 h after the injection. The total systemic bioavailability was estimated to be at least 95%. The rapidity and extent of absorption of the locally injected oligonucleotide suggested that oligonucleotides could be administered systemically by routes other than the intravenous one.

During the continuous subcutaneous infusion of 2 mg G3139 per kg per day, the plasma concentrations were 2.7 ± 0.5 and 1.6 ± 0.3 µg/ml on days 2 and 7 of the continuous infusion, respectively (LEBEDEV et al. 1997). By comparison, after the continuous intravenous infusion of OL(1)p53 at a lower dose (1.2 mg/kg per day), the peak plasma concentration was 3.03 ± 0.15 µg/ml (BISHOP et al. 1996). A comparison of the peak plasma concentrations suggests that a substantial portion of the subcutaneously administered dose is not being absorbed into the systemic circulation. From the limited data available, it is not possible to estimate the fraction of the dose absorbed after subcutaneous administration.

II. Plasma Pharmacokinetics

1. Peak Plasma Concentrations

Many of the ongoing clinical studies have administered the phosphorothioate oligonucleotide using a 2-h intravenous infusion (Table 1). The mean peak plasma concentrations seen in clinical studies have ranged from 0.285 ± 0.040 µg/ml at the end of a 2-h intravenous infusion of 0.1 mg GEM 91 per kg (ZHANG et al. 1995) to 14.54 ± 4.85 µg/ml for a 2-h infusion of 2.0 mg ISIS 2302 per kg (Isis Study No. 2302-CS7). Because so many of the oligonucleotides are being administered using the same regimen, comparisons across different oligonucleotide sequences is possible. Measured peak plasma concentrations of phosphorothioate oligonucleotides appeared to be directly related to dose. The peak plasma concentrations of the compounds being administered by 2-h intravenous infusions were plotted versus the dose administered (Fig. 1). The peak plasma concentrations increased slightly more than would be predicted from a strict linear response. The curve with the best fit had an r^2 value of 0.963. The graph contains peak plasma concentrations from four different oligonucleotide sequences, three of which are composed of 20 nucleotides, while the fourth has 25 nucleotide residues. The doses administered ranged from 0.06 to 2.0 mg/kg. The strong correlation between the dose administered and the peak plasma concentration of these different sequences suggests that the plasma pharmacokinetics in humans are independent of sequence.

In addition to dose, the peak plasma concentrations appeared to be related to the duration of the infusion. Plasma concentrations seen during continuous infusions would be predicted to be lower than those seen after a 2-h infusion for a given dose. When a total daily dose of 1.2 mg OL(1)p53 per kg was administered by a continuous intravenous infusion, the peak plasma concentration was determined to be 3.03 µg/ml, 25% lower than the 4.01 ± 0.74 µg/

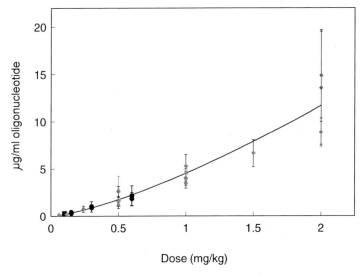

Fig. 1. Peak plasma concentrations measured at the end of 2-h intravenous infusions of four different phosphorothioate oligonucleotides. The values shown are the mean plasma concentrations for cohorts of three patients, except for GEM 91, which represents the mean peak plasma concentration from a cohort of six patients. r^2 of the nonlinear fit was 0.963. *Squares*, GEM 91; *circles*, ISIS 3521; *triangles*, ISIS 5132; *diamonds*, ISIS 2302

ml peak plasma concentration observed for a 2-h intravenous infusion of 1 mg ISIS 2302 per kg in a phase I study of healthy volunteers (GLOVER et al., in press). As the dose was increased sixfold, the peak plasma concentrations of OL(1)p53 which were observed during continuous intravenous infusion increased by less than 50% (BISHOP et al. 1996). These results suggest that, in contrast to the greater than linear response in peak plasma concentrations seen after a 2-h intravenous infusion, peak plasma concentrations during a continuous infusion may increase less than would be predicted based on a linear dose response.

2. Calculated Parameters

With data currently emerging from clinical trials, pharmacokinetic parameters calculated for phosphorothioate oligonucleotides are beginning to be published. One parameter, the plasma half-life, has been reported in two published reports. The initial distribution half-life, $t_{1/2}\alpha$, reported from the 2-h intravenous infusion of GEM 91 was 0.18h, a relatively short $t_{1/2}\alpha$ (ZHANG et al. 1995). In contrast, the plasma $t_{1/2}\alpha$ reported after intradermal injection of ^{14}C-ISIS 2105 was 0.9h (CROOKE et al. 1994). Although the total administered doses were similar (0.10mg GEM 91 per kg vs. approximately 0.06mg ISIS 2105 per kg), the $t_{1/2}\alpha$ values were quite different. Both calculations were made following total radioactivity rather than intact oligomer, and the radiolabel used differed (^{35}S vs. ^{14}C). However, the most likely cause for the difference in the $t_{1/2}\alpha$ is the different route of administration. Prolonged absorption from the intradermally administered compound likely increased the apparent $t_{1/2}\alpha$. By following the ^{14}C radiolabel, $t_{1/2}\beta$ and $t_{1/2}\gamma$ values were estimated to be 4.9 and 147.1h, respectively. In the GEM 91 study, however, by following the ^{35}S radiolabel, the $t_{1/2}\beta$ was estimated to be 26.7h.

Recent clinical investigations have allowed the estimation of other pharmacokinetic parameters, in addition to $t_{1/2}$ values, for several phosphorothioate oligonucleotides over a range of doses (GEARY et al. 1997). In one recently completed study, patients with steroid-refractory Crohn's disease were administered 13 doses every other day of either 0.5, 1.0, or 2.0 mg/kg. The pharmacokinetic parameters were determined from plasma concentrations of intact oligonucleotide measured during and after administration of the first dose (Table 2). The estimated $t_{1/2}$ increased with increasing dose. (*Note:* Because CGE is less sensitive than radiolabel, only single $t_{1/2}$ values are obtained when CGE analysis is used to follow plasma levels of administered oligonucleotides.) Coincident with the increase in $t_{1/2}$ seen with increasing dose, the clearance values (Cl) decreased, as did the steady state volume of distribution (V_{ss}) in patients. Both the maximum plasma concentration (C_{max}) and the area under the plasma concentration–time curve (AUC) values increased greater than the fourfold dose range studied; C_{max} varied eightfold, while the AUC values varied more than sevenfold over the fourfold dose range. This dose dependency of the pharmacokinetics was consistent with phase I studies in normal volunteers (GLOVER et al., in press) and, taken together, indicate that the plasma pharmacokinetics of phosphorothioate oligonucleotides in humans are dose dependent.

These pharmacokinetic parameters calculated in humans are similar to those observed in monkeys. The recent completion of a number of pharmaco-

Table 2. Dose-dependency of human plasma pharmacokinetics

Pharmacokinetic parameter	Dose (mg/kg)		
	0.5	1.0	2.0
AUC (μgh/ml)	3.44 ± 1.68	8.44 ± 0.94	24.51 ± 5.62
C_{max} (μg/ml)	1.68 ± 0.89	3.96 ± 1.02	14.84 ± 4.85
Cl (ml/h per kg)	167.2 ± 67.53	119.5 ± 13.94	84.43 ± 18.69
V_{ss} (ml/kg)	153.0 ± 51.02	143.5 ± 15.26	110.7 ± 11.37
MRT (h)	0.99 ± 0.34	1.22 ± 0.26	1.34 ± 0.23
$t_{1/2}$ (h)	0.58 ± 0.10	0.79 ± 0.20	1.08 ± 0.32

Pharmacokinetic parameters derived from plasma concentration time curves following administration of ISIS 2302 of the doses shown. The oligonucleotide was administered as a 2-h intravenous infusion. The samples analyzed were collected on the first day of a 1-month study.
AUC, area under plasma concentration–time curve; C_{max}, maximum concentration in plasma; Cl, clearance; V_{ss}, steady state volume of distribution.

Table 3. Maximum concentration in plasma (C_{max}) and area under plasma concentration–time curve (AUC) for both cynomolgus monkey and humans given equivalent doses based on weight (1 mg/kg) infused i.v. over a 2-h period (average ± standard deviation, $n = 3$–6)

Compound	Species	C_{max} (µg/ml)	AUC (µg/min per ml)
ISIS 2302	Monkey	4.59 ± 0.16	580 ± 11.4
	Human	3.96 ± 1.02	506 ± 56.4
ISIS 5132/CGP 69846A	Monkey	3.27 ± 0.82	369 ± 82
	Human	3.78 ± 0.61	612[a]
ISIS 3521/CGP 64128A	Monkey	4.30 ± 0.75	486 ± 63
	Human	4.78 ± 1.33	624[a]

[a] $n = 2$.

kinetic studies in monkeys allowed the comparison of some pharmacokinetic parameters in monkeys with those being defined for the same oligonucleotide sequences in humans (Table 3; GEARY et al. 1997). The peak plasma concentrations and AUC values for ISIS 2302, ISIS 5132/CGP 69846A, and ISIS 3521/CGP 64128A which were seen at the end of a 2-h intravenous infusion are consistent between monkeys and human for all three oligonucleotides. Additionally, the C_{max} and AUC values are consistent across the three different oligonucleotide sequences. These results indicate that the pharmacokinetic parameters for phosphorothioate oligonucleotides are nearly equivalent in humans and monkeys. Additionally, the pharmacokinetic parameters are again shown to be independent of sequence.

To date the pharmacokinetics of phosphorothioate oligonucleotides in humans appear remarkably similar to those observed in monkeys, and seem to be generally consistent across all animal species studied. Plasma pharmacokinetics of phosphorothioate oligonucleotides in humans are dose-dependent, with C_{max} plasma concentrations and AUC values increasing more than would be predicted from a linear response to increasing dose. Concomitantly, half-lives increase with increasing dose, and clearance rates decrease. This nonlinear response has been previously observed in multiple animal species (GEARY et al. 1997) and is believed to be due to saturation of uptake into the liver, and at higher doses, into the kidney (GEARY et al. 1997; LEEDS et al. 1997). The dose-dependent pharmacokinetics are predictable, reproducible and independent of sequence. This has been observed in all species studied to date, including human.

III. Metabolism

Animal studies have shown that clearance of phosphorothioate oligonucleotides from plasma is the result of both metabolism and extravascular distribution (COSSUM et al. 1993; CROOKE et al. 1996). Presumably, plasma clearance in humans is a combination of the same two factors. Analysis of plasma extracts by polyacrylamide gel electrophoresis (PAGE) showed that GEM 91 was metabolized rapidly after administration. When the zero time sample was found to have substantial levels of chain-shortened oligomers, oligonucleotides smaller than the parent by one or more nucleotides, the intact molecule was run through the extraction procedure to ensure that the degradation did not occur during the extraction. These controls confirmed that degradation of the oligonucleotide had not occurred during the extraction process (ZHANG et al. 1995). Rather, the oligonucleotide was metabolized very quickly in vivo. Urine analyzed by HPLC contained no intact GEM 91. The metabolite or metabolites seen were not identified, but coeluted with much lower molecular weight species.

Most human pharmacokinetic studies have not examined tissue for the presence of intact oligonucleotides or their metabolites. In one study, genital warts were excised from patients and extracted and analyzed by HPLC after the intradermal injection of [2-^{14}C-thymidine]-ISIS 2105 (CROOKE et al. 1994). Several warts were taken for analysis of radioactivity at 1, 24, 48, 72, and 144h after the intradermal injection. A single wart was taken for HPLC analysis of the extracted radioactivity at 24, 48, 72, and 144h after the intradermal injection. Anion-exchange HPLC analysis of the radioactivity extracted from an excised wart indicated that only 27% of that radioactivity coeluted with intact ISIS 2105 on HPLC 24h after the intradermal injection. By 72h after the intradermal injection, the percentage of radioactivity coeluting with intact ISIS 2105 on HPLC had decreased to 21%. The other radioactive material that was present coeluted with low molecular weight metabolites and chain-shortened oligonucleotides. These radioactive profiles on HPLC were consistent with those seen in a rat study using ^{14}C-labeled ISIS 2105 (COSSUM et al. 1993). The results of the wart analysis indicated that the phosphorothioate oligonucleotides in tissues were metabolized slowly, but steadily, so that intact oligonucleotide disappeared over several days.

More recently, analyses of metabolites of phosphorothioate oligonucleotides have been done using CGE (LEEDS et al. 1996; CROOKE et al. 1996). CGE separates oligonucleotides based on length. Quantitation can be achieved with the use of the appropriate internal standard. In animal studies, phosphorothioate oligonucleotide levels from both tissues and plasma have been quantitated (GEARY et al. 1996, 1997; LEEDS et al. 1996). The pattern of metabolites in human plasma seen by CGE analysis is virtually identical to that seen in monkey and mouse plasma (Fig. 2). The most abundant oligonucleotide is almost always the parent compound at all time points. Chain-shortened oligonucleotides are present in progressively smaller quantities; thus, in plasma, oligonucleotides more than four nucleotides smaller than the parent are present at very low concentrations relative to the parent phosphorothioate oligonucleotides at all times.

CGE analysis of plasma extracts has been used to examine metabolic profiles over the course of infusion of oligonucleotides in humans. At 30min

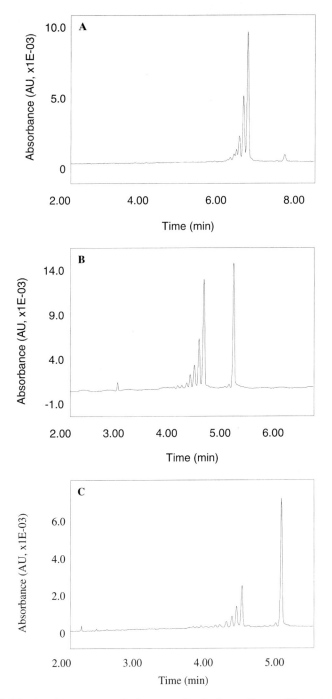

Fig. 2A–C. Electropherograms of plasma extracts from three different species for three different phosphorothioate oligonucleotides. **A** ISIS 3082 extracted from mouse plasma. **B** ISIS 5132/CGP 69846A extracted from monkey plasma. **C** ISIS 2302 extracted from human plasma. The single late migrating peak is the internal standard. The clusters of peaks are the parent oligonucleotide preceded by lower concentrations of chain-shortened metabolites (n-1, n-2, and n-3). AU, absorbance unit

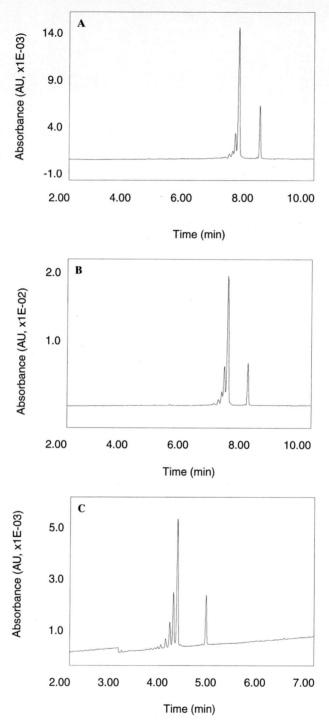

Fig. 3A–F. Electropherograms of extracts from human plasma from a subject receiving a 2-h infusion of 2.0 mg ISIS 2302 per kg during and after the end of the infusion. The internal standard concentration in all the extracts was 250 nM or 0.25 μM. The time points shown are **A** 30 min, **B** 2 h, **C** 150 min, **D** 3 h, **E** 4 h, and **F** 6 h after the beginning of the infusion. The single late migrating peak is the internal standard. The clusters of peaks are the parent oligonucleotide preceded by lower concentrations of chain-shortened metabolites (n-1, n-2, and n-3)

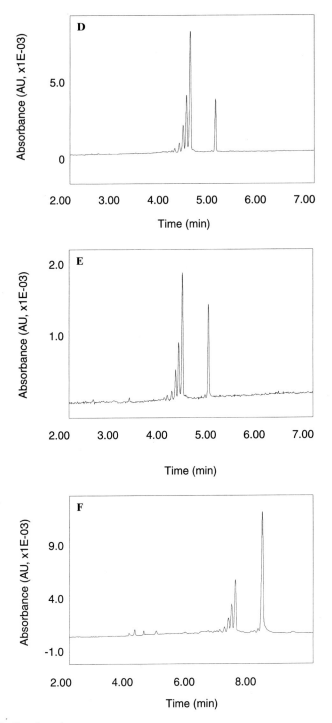

Fig. 3A–F. *Continued*

after the beginning of a 2-h intravenous infusion of 2.0 mg ISIS 2302 per kg, there were already abundant levels of chain-shortened metabolites (Fig. 3A), consistent with the early metabolites detected with the infusion of GEM 91 (ZHANG et al. 1995). At the end of infusion (Fig. 3B) where the plasma oligonucleotide concentration was highest, the full-length oligonucleotide represented 66.8% of total detected oligonucleotide (GLOVER et al., in press). After the end of the infusion, as no new intact oligonucleotide was infused, the metabolic pattern remained very similar, although the percentage of oligonucleotide representing full-length compound continued to drop slowly (Fig. 3C–E). Six hours after the beginning of the infusion (Fig. 3F), the last time point at which oligonucleotide was detected, 50% of the phosphorothioate oligonucleotide was still full-length compound.

Quantitation and identification of phosphorothioate oligonucleotides by CGE is based on ultraviolet (UV) detection at 260 nm and a relative migration time that corresponds to an oligonucleotide of a certain length. It does not identify which residues have been removed from the parent compound. Mass spectrometry has recently been used to identify metabolites from animal studies (GAUS et al., 1997). Analysis of extracts from the plasma, liver, and kidney of rats showed that the metabolites corresponded to chain-shortened oligonucleotides which had nucleotides removed from both the 5' and the 3' end of the molecule. Metabolites observed in plasma extracts were chain-shortened exclusively from the 3' end. Additional masses identified were consistent with depurinated molecules. Standards extracted and analyzed were not depurinated. Mass spectrometry analyses of oligonucleotide extracts from mice and monkeys showed a similar profile of metabolites. As the CGE profiles of the oligonucleotide extracts from rat plasma were consistent with those seen in mice, monkeys, and humans, human metabolites in plasma may also be derived from 3'-exonucleolytic cleavage of the parent oligonucleotide.

Metabolites in human urine have also been examined by CGE (GLOVER et al., in press). The pattern and relative abundance of chain-shortened metabolites is quite different from that seen in plasma, with a higher proportion of shorter oligonucleotides present in urine (Fig. 4). At the doses administered in humans, the smaller, chain-shortened oligonucleotides are excreted in the urine more readily than the intact oligonucleotides. It may be that the increased propensity of smaller oligonucleotides to be excreted in the urine is the reason why the predominant compound seen in plasma at all time points is the parent compound.

IV. Elimination

Reports of the levels of radioactivity derived from phosphorothioate oligonucleotides excreted in the urine vary widely. After intradermal administration of [2-^{14}C-thymidine]-ISIS 2105, more than 30% of the administered ^{14}C-labeled ISIS 2105 was eliminated as $^{14}CO_2$ via the lungs over a period of 140 h after the injection (CROOKE et al. 1994). The C_2 position of the thymine

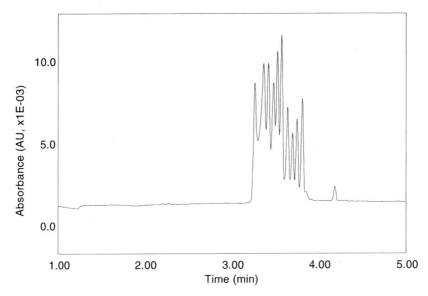

Fig. 4. Electropherogram of extract from urine of subject after a dose of 2.0 mg ISIS 2302 per kg. The profile of oligonucleotides is quite different from that seen in plasma. The predominant metabolites are oligonucleotides comigrating with standards of eight nucleotides in length through approximately 15 nucleotides in length. (The smaller oligonucleotides elute first). The single late migrating peak is the internal standard

base, where the oligonucleotide contained the ^{14}C label, is metabolized to CO_2. Only 10% of the total radioactivity administered was detected in urine over the 6-day study. In another study, 49.2% of the radioactivity from ^{35}S-labeled GEM 91 was recovered in the urine in the first 24 h after administration, and by the end of the 96 h, 70.4% of the total administered radioactivity was recovered in the urine (ZHANG et al. 1995). None of that radiolabel coeluted with intact GEM 91 on HPLC. Although the doses in those two studies were different (approximately 0.06 mg/kg for ISIS 2105 vs. 0.10 mg/kg for GEM 91), the most likely explanation for the differences in the amount of radiolabel excreted in the urine is the difference in the type and placement of the radiolabels. The ^{14}C label used in the study of ISIS 2105 was on the C2 of the thymine base, which is readily broken down to CO_2, while the ^{35}S label of the GEM 91 was part of the phosphate backbone of the oligonucleotide. The molecular identity of the ^{35}S excreted has not yet been determined, but was not identified as intact oligonucleotide.

Analysis of urine from the continuous intravenous infusion of OL(1)p53 indicated that between 17% and 59% of the total administered dose was excreted in the urine as intact oligonucleotide, based on analysis done after extraction and end-labeling (BISHOP et al. 1996). In additional, 36% of the radioactivity in urine from a patient who received ^{35}S-labeled OL(1)p53 coeluted with intact oligonucleotide, while 53% coeluted with highly degraded material. In a more recent study, urine was extracted and analyzed for ISIS 2302 and metabolites after a 2-h infusion of 2 mg/kg (Fig. 4). Urine was collected during the first 6 h of the study and analyzed. Very low concentrations of intact oligonucleotide were measured, while the concentration of chain-shortened metabolites was higher (GLOVER et al., in press). It was estimated that less than 0.5% of the total administered dose was excreted in urine over the first 6 h after dosing.

What is clear from the results of these studies is that disposition of radiolabel after administration of a macromolecule will depend on the location of the label and on how, as well as where, the oligonucleotide is metabolized throughout the body. In two of three studies which used radiolabeled oligonucleotide to determine excretion, urinary excretion appeared to be a minor pathway of elimination. In the case of the study on f OL(1)p53 oligonucleotide, a much greater proportion of the total administered dose appeared to have been excreted in the urine (BISHOP et al. 1996). It is not possible to say whether this higher proportion of urinary excretion is due to the different dosing regimen or whether it is related to the different analytical methodology used in that study. CGE analysis, which does not rely on radiolabel and is thus a very reliable determination of oligonucleotide concentrations, indicated that the proportion of administered oligonucleotide being excreted in the urine as intact drug or as chain-shortened oligonucleotide metabolites was quite small (GLOVER et al., in press).

Although tissues take up considerable amounts of phosphorothioate oligonucleotides, almost as large of a fraction of radiolabel – derived from the oligonucleotide – is rapidly eliminated from the body. The apparent route of oligonucleotide elimination depends on the method used to track the oligonucleotide. Whereas most of the ^{35}S radiolabel was excreted in the urine, albeit not intact (ZHANG et al. 1995), most of the ^{14}C radiolabel from ^{14}C-ISIS 2105 was excreted as ^{14}C-CO_2 in expired air (CROOKE et al. 1994). The difference in disposition of radiolabeled metabolites is illuminating in so far as it provides information about excretion of the phosphorothioate backbone (urine) and the metabolized base (CO_2 in expired air). Thus, urinary excretion of phosphorothioate oligonucleotides is a very minor pathway of elimination in humans. The differential disposition of different radiolabels in humans suggests that the major pathway for elimination of phosphorothioate oligonucleotides in humans is via metabolism.

C. Pharmodynamics

As data from clinical evaluation of phosphorothioate oligonucleotides emerges, both clinicians and researchers will want to understand any correlation between the disposition of the compound and its pharmacologic effect, this is, its pharmacodynamics. Antisense technology has allowed the development of inhibitors of molecular targets which the pharmaceutical industry previously had been unable to exploit (i.e., isotypic pharmacology) because of a lack of specificity. It will be important to distinguish between antisense targets that are not useful therapeutically because inhibiting that particular molecular target does not have the intended pharmacologic effect, and molecules for which phosphorothioate antisense molecules fail to reduce expression because they do not reach the necessary site for activity. Understanding which compounds belong to which group will help clinicians and researchers make rational choices about the next generation of chemistries to be developed, as well as good targets to pursue.

The pharmacokinetics of phosphorothioate oligonucleotides in humans are dose-dependent, but predictable, reproducible, and sequence independent. The growing database of knowledge about pharmacokinetics strongly suggests that, as has been observed in animal studies, the pharmacokinetics of these compounds is driven by the chemical class, rather than by the sequence. As the clinical trials progress, additional pharmacokinetic research will be published, adding to the nascent database currently available.

References

Agrawal S (1996) Antisense oligonucleotides: towards clinical trials. Trends Biotech 14:376–387

Bishop MR, Iversen PL, Bayever E, Sharp JG, Greiner TC, Copple BL, Ruddon R, Zon G, Spinolo J, Arneson M, Armitage JO, Kessinger A (1996) Phase I trial of an antisense oligonucleotide OL(1)p53 in hematologic malignancies. J Clin Oncol 14:1320–1326

Campbell JM, Bacon TA, Wickstrom E (1990) Oligodeoxynucleoside phosphorothioate stability in subcellular extracts, culture media, sera and cerebrospinal fluid. J Biochem Biophys Methods 20:259–267

Cossum PA, Sasmor H, Dellinger D, Troung L, Cummins L, Owens SR, Markham PM, Shea JP, Crooke S (1993) Disposition of the ^{14}C-labeled phosphorothioate oligonucleotide ISIS 2105 after intravenous administration to rats. J Pharmacol Exp Ther 267:1181–1190

Cowsert LM, Fox MC, Zon G, Mirabelli CK (1993) In vitro evaluation of phosphorothioate oligonucleotides targeted to the E2 mRNA of papillomavirus: potential treatment for genital warts. Antimicrob Agents Chemother 37:171–177

Crooke ST, Bennett CF (1996) Progress in antisense oligonucleotide therapeutics. Annu Rev Pharmacol Toxicol 36:107–129

Crooke ST, Grillone LR, Tendolkar A, Garrett A, Fratkin MJ, Leeds J, Barr WH (1994) A pharmacokinetic evaluation of ^{14}C-labeled afovirsen sodium in patients with genital warts. Clin Pharmacol Ther 56:641–646

Crooke ST, Graham MJ, Zuckerman JE, Brooks D, Conklin BS, Cummins LL, Greig MJ, Guinosso CJ, Kornbrust D, Manoharan M, Sasmor HM, Schleich T, Tivel KL, Griffey RH (1996) Pharmacokinetic properties of several novel oligonucleotide analogs in mice. J Pharmacol Exp Ther 277:923–937

Dean NM, McKay R (1994) Inhibition of protein kinase C-α expression in mice after systemic administration of phosphorothioate antisense oligodeoxynucleotides. Proc Natl Acad Sci USA 91:11762–11766

Gaus H, Owens SR, Winniman M, Cooper S, Cummins LL. (1997) On-line HPLC electrospray mass spectrometry of phosphorothioate oligonucleotide metabolites. Anal Chem 69(3):313–319

Geary RS, Leeds JM, Shanahan W, Glover J, Pribble J, Truong L, Fitchett J, Burckin T, Nicklin P, Philips J, Levin AA (1996) Sequence independent plasma and tissue pharmacokinetics for 3 antisense phosphorothioate oligonucleotides: mouse to man. Pharm Res 13:S395

Geary RS, Leeds JM, Henry SP, Monteith D, Truong L, Fitchett J, Burckin T, Levin AA (1997) Pharmacokinetic properties of phosphorothioate oligonucleotides. In: Cook PD, Manoharan M (eds) Nucleosides and nucleotides. Marcel Decker, New York (in press)

Glover JM, Leeds JM, Mant TGK, Amin D, Kisner DL, Zuckerman J, Geary RS, Levin AA, Shanahan WR (in press) Phase 1 safety and pharmacokinetic profile of an ICAM-1 antisense oligodeoxynucleotide (ISIS 2302). J Pharmacol Exp Ther

Hoke GD, Draper K, Freier SM, Gonzalez C, Driver VB, Zounes MC, Ecker DJ (1991) Effects of phosphorothioate capping on antisense oligonucleotide stability, hybridization and antiviral efficacy versus herpes simplex virus infection. Nucleic Acids Res 19:5743–5748

Lebedev AV, Raynaud E, M, Beck T, Jaeger JA, Brown BD, Cunningham D, Webb A, McCampbell E, Riley T, Judson I, Woodle MC (1997) Anti-BCL2 phosphorothioate G3139 clinical pharmacology: plasma levels during continuous subcutaneous infusion by HPLC assay. In: Cook PD, Manoharan M (eds) Nucleosides and nucleotides. Marcel Decker, New York (in press)

Leeds JM, Graham MJ, Truong , Cummins LL (1996) Quantitation of phosphorothioate oligonucleotides in human plasma. Anal Biochem 235:36–43

Leeds JM, Geary RS, Henry SP, Glover J, Shanahan W, Fitchett J, Burckin T, Truong L, Levin AA (1997) Pharmacokinetic properties of phosphorothioate oligonucleotides. In: Cook PD, Manoharan M (eds) Nucleosides and nucleotides. Marcel Decker, New York (in press)

Monia BP, Johnston JF, Geiger T, Muller M, Fabbro D (1996) Antitumor activity of a phosphorothioate antisense oligodeoxynucleotide targeted against C-raf kinase. Nature Med 2:668–675

Sands H, Gorey-Feret LJ, Cocuzza AJ, Hobbs FW, Chidester D, Trainor GL (1994) Biodistribution and metabolism of internally ^{3}H-labeled oligonucleotides. I. Comparison of a phosphodiester and phosphorothioate. Mol Pharmacol 45:932–943

Stein CA, Subasinghe C, Shinozuka K, Cohen JS (1988) Physicochemical properties of phosphorothioate oligodeoxynucleotides. Nucleic Acids Res 16:3209–3221

Wickstrom E (1986) Oligodeoxynucleotide stability in subcellular extracts and culture media. J Biochem Biophys Methods 13:97–102

Zhang R, Yan J, Shahinian H, Amin G, Lu Z, Liu T, Saag MS, Jiang Z, Temsamani J, Martin R, Schechter PJ, Agrawal S, Diasio RB (1995) Pharmacokinetics of an antihuman immunodeficiency virus antisense oligodeoxynucleotide phosphorothioate (GEM 91) in HIV-infected subjects. Clin Pharmacol Ther 58(1):45–53

CHAPTER 7
Safety and Tolerance of Phosphorothioates in Humans

P.J. Schechter and R.R. Martin

A. Introduction

Antisense oligonucleotides represent a promising, rational, and seductive approach to drug discovery and development. Several early attempts to develop therapeutic agents using this technology targeted clinical conditions which could be treated by local administration of the oligonucleotide, thereby avoiding their systemic use. However, in order for antisense oligonucleotides to reach their full potential as therapeutic entities across a wide variety of disease applications, systemic use is likely to be necessary; identifying and solving any problems associated with systemically administered oligonucleotides is therefore of major importance.

Phosphorothioates are the most widely applied oligonucleotides for antisense drug development. There have been multiple examples of phosphorothioate oligonucleotides being developed for a variety of gene targets, and at least ten of these molecules have been investigated in clinical trials. GEM 91 is a 25-mer phosphorothioate oligodeoxynucleotide targeted at the gag site of the human immunodeficiency virus (HIV) gene (Agrawal and Tang 1992; Lisziewicz et al. 1994). To date, GEM 91 has been administered parenterally to over 250 subjects, making it probably the most widely used and best studied systemically administered phosphorothioate in humans. It will be employed as the prototype phosphorothioate oligonucleotide in this chapter.

B. Cardiovascular Safety

Early reports of cardiovascular complications, including death, of intravenously administered phosphorothioate oligonucleotides in primates (Cornish et al. 1993; Black et al. 1994; Galbraith et al. 1994) effectively retarded clinical development of these compounds. Rapid intravenous injection of phosphorothioate oligonucleotides of 20- to 30-mer size produces mild, transient hypertension followed by more profound and prolonged hypotension in anesthetized monkeys (Galbraith et al. 1994). Heart rate shows reciprocal changes to blood pressure, suggesting that cardiovascular reflexes remain intact. Effects on blood pressure are dose dependent and, at least at the lower doses, reversible. No direct effects on cardiac function are observed with treatment. The dose–effect relationship can be shifted rightwards by slowing the infusion rate; for example, a 10-min infusion of 5 mg GEM 91 per kg in rhesus monkeys produces a similar degree of hypotension as 80 mg/kg over 120 min.

Due to these findings in primates, initial multiple-dose clinical studies of GEM 91 in the United States were required to include continuous monitoring by cardiac telemetry. In a blinded, placebo-controlled study of GEM 91 via continuous intravenous infusion for 8 to 14 days with escalating doses ranging from 0.1 to 4.4 mg/kg per day (six subjects treated with GEM 91 and three with placebo in each dose group), this intensive cardiac monitoring did not reveal any evidence of cardiotoxicity associated with GEM 91. In these subjects, vital signs were monitored frequently throughout the treatment course and for several weeks thereafter. Here again, there was no evidence of treatment-associated hemodynamic instability.

In a phase I trial of OL(1) p53, a 20-mer phosphorothioate targeted at p53 mRNA, in subjects with hematological malignancies, one patient died of congestive heart failure during treatment (Bishop et al. 1996). This individual had previously been treated with daunorubin and was described as having anthrocycline-induced cardiac failure at autopsy.

Thus the dose regimens of phosphorothioate oligonucleotides used clinically have not been associated with cardiovascular intolerance. Based on data in nonhuman primates, however, it is advisable to avoid rapid intravenous injections of these compounds.

C. Hematological Safety

I. Coagulation

Single-stranded DNA is known to bind to and inhibit human thrombin (Bock et al. 1992). In addition, diverse polyanionic compounds, e.g., dextran sulfate (Hjort and Stormorken 1957), curdlan sulfate (Gordon et al. 1994), various sulfated alkyl oligosaccharides (Nakashima et al. 1995), and heparin, have nonspecific, dose-dependent anticoagulant properties. In fact, serial measurement of activated partial thromboplastin time (aPTT) has been used to titrate the infusion rate of dextran sulfate in some studies (Flexner et al. 1991). On the other hand, a reversible prolongation of aPTT in monkeys was also observed with a 17-mer phosphodiester, partial phosphorothioate, non-antisense oligonucleotide having few anionic charges (Wallace et al. 1996), albeit only after very high plasma concentrations were reached.

In a dose-rising study of single doses of GEM 91 ranging from 0.1 to 2.5 mg/kg over 2h as intravenous infusions, a prolongation of aPTT was noted at doses of 1 mg/kg or higher (Sereni et al. 1994). This prolongation occurred at peak plasma oligonucleotide concentration and rapidly returned towards pretreatment values as the plasma concentrations declined. The maximum prolongation of aPTT was linearly correlated with the peak plasma GEM 91

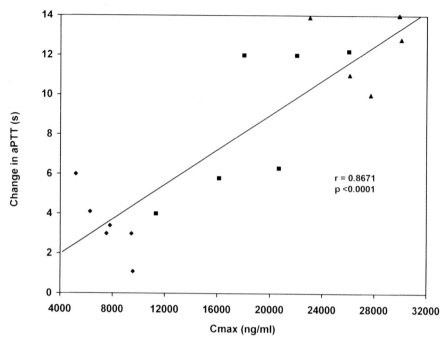

Fig. 1. Linear correlation relating maximum increases in activated partial thromboplastin time (*aPTT*) and maximum plasma GEM 91 concentrations following single 2-h intravenous infusions of GEM 91 at doses of 1.0 (*diamonds*), 2.0 (*squares*), or 2.5 mg/kg (*triangles*) to human immunodeficiency virus (HIV)-1-positive individuals

concentrations (Fig. 1). The maximum prolongation of aPTT did not exceed 45% over baseline control values in any patient. There were no changes in prothrombin time, platelet counts, fibrinogen, or factors V, VIII, IX, X, XI, or XII (TOULON et al. 1995). Not surprisingly, therefore, there was no evidence of bleeding.

Similarly, ISIS 2302, a 20-mer antisense phosphorothioate oligonucleotide targeted at intercellular adhesion molecule (ICAM)-1, produced transitory aPTT prolongation with intravenous infusion of doses ranging from 0.5 to 2.0 mg/kg over 2h (SHANAHAN et al. 1996).

II. Platelets

Thrombocytopenia is among the most prevalent dose-dependent effects observed in toxicological studies of phosphorothioate oligonucleotides administered chronically and systemically to rodents (SARMIENTO et al. 1994). This effect is less prominent in monkeys.

A decrease in platelets has been observed in patients following 1 to 2 weeks of treatment with intravenously administered GEM 91 at doses of 2 mg/kg per day or higher (Fig. 2). This decrease is promptly reversible upon

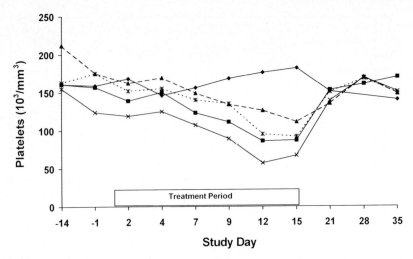

Fig. 2. Platelet counts in five human immunodeficiency virus (HIV)-1-positive patients treated with GEM 91 via 2-h intravenous infusion of 1 mg/kg every 8 h for 2 weeks. Each symbol represents a separate individual

discontinuation of treatment. To date, no patient has had a platelet count of 50000/mm^3 or less following GEM 91 administration. Although platelet abnormalities were noted in all five patients treated intravenously for 10 days with a 20-mer phosphorothioate oligonucleotide complementary to p53 mRNA, these changes were attributed to the underlying leukemia or myelodysplastic syndrome being treated (BAYEVER et al. 1993).

Thrombocytopenia associated with phosphorothioate oligonucleotides can be attributed, at least in part, to their polyanionic structure. Thus a reversible decrease in platelets is found following treatment with heparin (NELSON et al. 1978; KING and KELTON 1984; POWERS et al. 1984) and dextran sulfate (HJORT and STORMORKEN 1957). In fact, in a study of intravenous dextran sulfate in HIV-positive patients, all subjects who received treatment for 3 days or more developed profound, but reversible thrombocytopenia (FLEXNER et al. 1991). No evidence for a change in platelet survival was found.

III. Red Blood Cells

Decreases in parameters of peripheral red blood cell numbers are not uncommon with the usual intense monitoring in phase I studies and the consequent vigorous blood sampling over a short time period. Nevertheless, anemia has not been associated with GEM 91 therapy (ZHANG et al. 1995c). An increase in severity of anemia observed in several patients treated with OL(1) p53 was not related to the dose administered, but was attributed to their underlying hematological malignancies (BISHOP et al. 1996). Doses up to 2mg/kg every other day for 13 doses of ISIS 2302, a 20-mer antisense phosphorothioate oligonucleotide targeted at ICAM-1, had no effect on red blood cells counts (W. Shanahan, personal communication).

D. Complement Activation

Evidence of phosphorothioate oligonucleotide-associated activation of serum complement was first noted in studies in monkeys (GALBRAITH et al. 1994). Rapid intravenous infusion of GEM 91 produced activation of C5 complement without a clear dose–response relationship, i.e., at doses of 5 mg/kg or higher infused over 10 min, C5a split products of complement increased markedly, whereas at lower doses no increase was observed. The higher the dose, however, the earlier the increase was noted. This activation of complement was accompanied by transient decreases in neutrophil and total white blood cell counts and changes in blood pressure and heart rate (see above).

Evidence of activation of the alternate complement pathway in monkeys, with or without all of the above-described accompanying phenomena, has been reported following single intravenous doses of other oligonucleotides (WALLACE et al. 1996).

In clinical studies with GEM 91 and other phosphorothioate oligonucleotides, complement has been monitored routinely. No drug-related changes in blood complement were found at any dose up to 4.4 mg GEM 91 per kg per day administered intravenously by continuous infusion over 8 days or up to 0.5 mg GEM 132 (a 20-mer advanced-chemistry, hybrid phosphorothioate oligonucleotide targeted at human cytomegalovirus) per kg over 2 h, the highest doses of either oligonucleotide tested to date. On the other hand, single intravenous doses of ISIS 2302 in normal volunteers were reported to increase complement C3a, but not C5a in blood (SHANAHAN et al. 1996).

E. Specific Organ Toxicity

Kidneys and liver are preferential target organs for systemic phosphorothioate oligonucleotide toxicity in animals (SARMIENTO et al. 1994; FIELD and GOODCHILD 1995; SRINIVASAN and IVERSEN 1995). These organs correspond to the sites of highest drug concentration following parenteral administration of these compounds (ZHANG et al. 1995a,b; SRINIVASAN and IVERSEN 1995). Thus renal tubular degeneration and necrosis and hepatocellular degeneration with Kupffer cell hyperplasia are consistent findings with toxic doses of phosphorothioate oligonucleotides in rodent and primate toxicology studies.

As a consequence, careful monitoring of renal and hepatic function has been routine in clinical studies of systemic oligonucleotides. No evidence for renal intolerance, as indicated by changes in blood urea nitrogen levels, serum creatinine concentrations, creatinine clearance, urine specific gravity, or routine urinalysis, has been found with GEM 91 (SERENI et al. 1995). A single incident of transient nonoliguric renal failure was reported following intravenous OL(1)p53 infusion, but this occurred after the initiation of vancomycin coadministration (Bishop et al. 1996). Transitory increases in serum transaminase activity have been noted in some patients treated with intravenous doses of GEM 91 of 2 mg/kg per day or more for periods of 1 week or longer (Fig. 3). Serum glutamate pyruvate transaminase (SGPT; alanine aminotransferase) was more affected than serum glutamic-oxaloacetic transaminase (SGOT; aspartate aminotransferase). In general, the effect of GEM 91 was greater in individuals whose serum transamine activities were above the upper limit of normal at baseline. The enzyme elevations were transitory in all cases and often would begin to reverse towards baseline even before discontinuation of therapy. Values greater than four times the upper limit of the normal range were seldom seen, and no associated changes in serum bilirubin concentrations, alkaline phosphatase activity, or lactate dehydrogenase activity were noted.

Of 16 patients treated with intravenous infusion of OL(1)p53, transitory elevation of serum transaminases in two patients and mild transitory increase in serum total bilirubin in three patients were reported (Bishop et al. 1996).

The reversible increases in serum transaminase activity observed with phosphorothioate oligonucleotides can also be attributed, at least in part, to

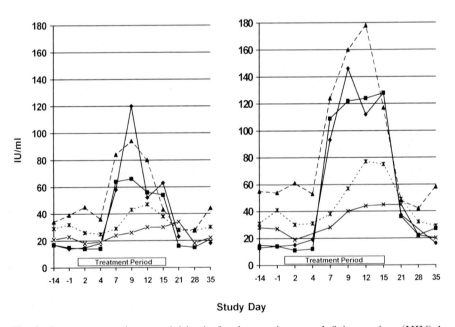

Fig. 3. Serum transaminase activities in five human immunodeficiency virus (HIV)-1-positive patients treated with GEM 91 via 2-h intravenous infusion of 1 mg/kg every 8 h for 2 weeks. *Left*, serum glutamic-oxaloacetic transaminase (SGOT). *Right*, serum glutamate pyruvate transaminase (SGPT). Each symbol represents a separate individual

their polyanionic structure. Other polyanions, such as dextran (FLEXNER et al. 1991) and heparin (MONREAL et al. 1989; OLSSON et al. 1978; DUKES et al. 1984), increase these enzymes transiently.

F. Conclusions

There are few publications describing clinical use of systemically administered phosphorothioate oligonucleotides and the tolerance of such treatment. The potentially adverse effects observed following systemic phosphorothioate oligonucleotide administration in humans include prolongation of aPPT, thrombocytopenia, and increased serum transaminase activities. These effects are readily reversible upon termination of therapy, are rarely associated with clinical consequence, and, to date, have seldom necessitated stopping oligonucleotide treatment. These abnormalities can be attributed, at least in part, to the polyanionic structure of phosphorothioate oligonucleotides, since other polyanions produce many of the same effects.

Cardiovascular effects of phosphorothioate oligonucleotides, initially feared due to toxicological findings in nonhuman primates, have not been reported in clinical studies. Likewise, anemia and renal toxicity, often observed in rodent and monkey toxicology studies, have not been associated with clinical use of these agents.

In summary, although clinical experience with systemic administration of phosphorothioate oligonucleotides is limited, it is clear that they can be used over a wide range of doses and for a duration of at least several weeks with good safety. This tolerance profile should allow clinical proof of the concept for systemic antisense therapeutics and potential wide use in a variety of clinical conditions.

References

Agrawal S, Tang JY (1992) GEM 91 – an antisense oligonucleotide phosphorothioate as a therapeutic agent for AIDS. Antisense Res Dev 2:261–266

Bayever E, Iversen PL, Bishop MR, Sharp JG, Tewary HK, Arneson MA, Pirruccello SJ, Ruddon RW, Kessinger A, Zon G, Armitage JO (1993) Systemic administration of a phosphorothioate oligonucleotide with a sequence complementary to p53 for acute myelogenous leukemia and myelodysplastic syndrome: initial results of a phase I trial. Antisense Res Dev 3:383–390

Bishop MR, Iversen PL, Bayever E, Sharp JG, Greiner TC, Copple BL, Ruddon R, Zon G, Spinolo J, Arneson M, Armitage JO, Kessinger A (1996) Phase I trial of an antisense oligonucleotide OL(1) p53 in hematologic malignancies. J Clin Oncol 14:320–1326

Black LE, Farrelly JG, Cavagnaro JA, Ahn C-H, DeGeorge JJ, Taylor AS, DeFelice AF, Jordan A (1994) Regulatory considerations for oligonucleotide drugs: update recommendations for pharmacology and toxicology studies. Antisense Res Dev 4:299–301

Bock LC, Griffin LC, Latham JA, Vermaas EH, Toole JJ (1992) Selection of single-stranded DNA molecules that bind and inhibit human thrombin. Nature 355:564–566

Cornish KG, Iversen P, Smith L, Arneson M, Bayever E (1993) Cardiovascular effects of a phosphorothioate oligonucleotide with sequence antisense to p53 in the conscious Rhesus monkey. Pharmacol Commun 3:239–247

Dukes GE, Sanders SW, Russo J, Swenson E, Burnakis TG, Saffle JR, Warden GD (1984) Transaminase elevations in patients receiving bovine or porcine heparin. Ann Intern Med 100:646–650

Field AK, Goodchild J (1995) Antisense oligonucleotides: rational drug design for genetic pharmacology. Exp Opin Invest Drugs 4:799–821

Flexner C, Barditch-Crovo PA, Kornhauser DM, Farazadegan H, Nerhood LJ, Chaisson RE, Bell KM, Lorentsen KJ, Hendrix CW, Petty BG, Lietman PS (1991) Pharmacokinetics, toxicity and activity of intravenous dextran sulfate in human immunodeficiency virus infection. Antimicrob Agents Chemother 35: 2544–2550

Galbraith WM, Hobson WC, Giclas PC, Schechter PJ, Agrawal S (1994) Complement activation and hemodynamic changes following intravenous administration of phosphorothioate oligonucleotides in the monkey. Antisense Res Dev 4:201–206

Gordon M, Guralnik M, Kaneko Y, Mimura T, Baker M, Lang W (1994) A phase I study of curdlan sulfate – an HIV inhibitor. Tolerance, pharmacokinetics and effects on coagulation and CD4 lymphocytes. J Med 25:163–180

Hjort P, Stormorken H (1957) A study of the vivo effects of a synthetic heparin-like anticoagulant: dextran sulphate. Scand J Clin Lab Invest 9:6–81

King DJ, Kelton JG (1984) Heparin-associated thrombocytopenia. Ann Intern Med 100:535–540

Lisziewicz J, Sun D, Wiechold FF, Thierry AR, Lusso P, Tang J, Gallo RC, Agrawal S (1994) Antisense oligodeoxynucleotide phosporothioate complementary to gag mRNA blocks replication of human immunodeficiency virus type 1 in human peripheral blood cells. Proc Natl Acad Sci U S A 91:7942–7946

Monreal M, Lafoz E, Salvador R, Roncales J, Navarro A (1989) Adverse effects of three different forms of heparin therapy: thrombocytopenia, increased transaminases and hyperkalemia. Eur J Clin Pharmacol 37:415–418

Nakashima H, Inazawa K, Iciyama K, Ito M, Ikushima N, Shoji T, Katsuraya K, Uryu T, Yamamoto N, Juodawikis AS, Schinazi RF (1995) Sulfated alkyl oligosaccharides inhibit human immunodeficiency virus in vitro and provide sustained drug levels in mammals. Antiviral Chem Chemother 6:271–280

Nelson JC, Lerner RG, Goldstein R, Cagin NA (1978) Heparin-induced thrombocytopenia. Arch Intern Med 138:548–552

Olsson R, Korsan-Bengstein BM, Korsan-Bengstein K, Lennartsson J, Waldestrom J (1978) Serum aminotransferases after low-dose heparin treatment. Acta Med Scand 204:229–230

Powers PJ, Kelton JG, Carter CJ (1984) Studies on the frequency of heparin-associated thrombocytopenia. Thromb Res 33:439–443

Sarmiento UM, Perez JR, Becker JM, Narayanan R (1994) In vivo toxicological effects of rel A antisense phosphorothioates in CD-1 mice. Antisense Res Dev 4:99–107

Sereni D, Katlama C, Gouyette A, Re M, Lascoux C, Tubiana R, Tournerie C (1994) Open-label safety and pharmacokinetic study of single intravenous or subcutaneous ascending doses of GEM 91 in untreated, adult, HIV positive, asymptomatic human volunteer patients. In: Proceedings of the 34th interscience conference on antimicrobial agents and chemotherapy, Orlando, 4–7 October 1994, p 217

Sereni D, Katlama C, Vilde JL, Bouvet E, Brun-Vezinet F, Gouyette A, Tournerie C (1995) An open-label, multiple-dose safety, pharmacokinetic and preliminary antiviral activity study of GEM 91 administered by intermittent intravenous infusion in HIV-positive adults. In: Proceedings of the 35th interscience conference on antimicrobial agents and chemotherapy, San Francisco, 17–20 September 1995, p 236

Shanahan W, Glover J, Mant T, Amin D, Leeds J, Zuckerman J, Levin A (1996) Phase I study of an ICAM-1 antisense oligonucleotide (ISIS 2302). Clin Pharmacol Ther 59:208

Srinivasan SK, Iversen P (1995) Review of vivo pharmacokinetics and toxicology of phosphorothioate oligonucleotides. J Clin Lab Anal 9:129–137

Toulon P, Ankri A, Tournerie C, Sereni D (1995) Effect of GEM 91, a phosphorothioate on the coagulation system. Results of ex vivo and vitro studies. Blood 86:889a

Wallace TL, Bazemore SA, Kornbrust DJ, Cossum PA (1996) Single-dose hemodynamic toxicity and pharmacokinetics of a partial phosphorothioate anti-HIV oligonucleotide (AR177) after intravenous infusion to cynomolgus monkeys. J Pharmacol Exp Ther 59:208

Zhang RW, Diasio RB, Lu ZH, Liu TP, Jiang ZW, Galbraith WM, Agrawal S (1995a) Pharmacokinetics and tissue distribution in rats of an oligodeoxynucleotide phosphorothioate (GEM 91) developed as a therapeutic agent for human immunodeficiency virus type 1. Biochem Pharmacol 49:929–939

Zhang RW, Lu ZH, Zhang XS, Zhao H, Diasio RB, Lui TP, Jiang ZW, Agrawal S (1995b) In vivo stability and disposition of a self-stabilized oligodeoxynucleotide phosphorothioate in rats. Clin Chem 41:836–843

Zhang R, Yan J, Shahinian H, Amin G, Lu Z, Liu T, Saag MS, Jiang Z, Temsamani J, Martin RR, Schechter PJ, Agrawal S, Diasio RB (1995c) Pharmacokinetics of an anti-human immunodeficiency virus antisense oligodeoxynucleotide phosphorothioate (GEM 91) in HIV-infected subjects. Clin Pharmacol Ther 58:44–53

CHAPTER 8
Immune Stimulation by Oligonucleotides

A.M. KRIEG

A. Sequence-Independent Immune Effects of the DNA Backbone

As has been pointed out previously (STEIN and CHENG 1993), DNA is a polyanion. Like other polyanions, DNA can have a broad range of biologic effects. Native or modified DNA backbones can cause immune activation which is concentration- and length-dependent and sequence-independent. However, the different backbones currently under development for antisense applications differ greatly in the degree to which they induce immune effects. As might be expected, neutral backbones have minimal, if any, immune effects (A.M. KRIEG, unpublished observation). Among the charged backbones that have been carefully investigated, the greatest magnitude of nonspecific effects are seen with phosphorothioate oligodeoxynucleotides (PS-ODNs) (STEIN and CHENG 1993).

Among its nonsequence-specific activities, PS-ODN can induce cytokine secretion. For example, Crooke and colleagues have found that at very high concentrations (>100 μM) PS oligonucleotides nonspecifically stimulate keratinocyte cytokine secretion (CROOKE et al. 1996). Perez and colleagues have recently reported that PS-ODNs induce DNA binding activity of the transcription factor Sp1 in a nonsequence-specific manner (PEREZ et al. 1994). Sp1 activity was increased in human and murine lymphoid cell lines and in primary spleen cells in vitro as well as in vivo in mice. Induction of Sp1 activity occurred at concentrations of 2 μM; far lower than those found to induce keratinocyte cytokine production. Although the mechanism of this effect is obscure, it was surprisingly rapid and was apparent within 30 min. The central role of Sp1 in the transcriptional regulation of multiple genes in leukocytes raises concerns that PS-ODNs may have a wide range of immune effects due to their activation of this or other nuclear binding proteins. Sp1 typically interacts with other transcription factors, suggesting that some immune activities of PS-ODNs might only become apparent in the context of other activation signals. In other words, PS-ODNs may be found to synergize or antagonize other lymphocyte signaling pathways.

Recent studies have started to confirm the ability of PS-ODNs to synergize with other immune signals. Even at very low concentrations (100 nM), PS oligonucleotides can synergize with lipopolysaccharide (LPS) to

List of Abbreviations

Ag	Antigen
BCG	Bacille Calmette-Guérin
ELISA	Enzyme-linked immunosorbent assay
EMSA	Electrophoretic mobility shift assays
FITC	Fluorescein isothiocyanate
GM-CSF	Granulocyte macrophage colony-stimulating factor
ICAM	Intercellular adhesion molecule
IFN	Interferon
Ig	Immunoglobulin
IL	Interleukin
LPS	Lipopolysaccharide
NF	Nuclear factor
NK	Natural killer
ODN	Oligodeoxynucleotide
PBMC	Peripheral blood mononuclear cells
PMA	Phorbol myristate acetate
PS	Phosphorothioate
PS-ODN	Phosphorothioate oligodeoxynucleotide
SCID	Severe combined immunodeficiency disease
TNF	Tumor necrosis factor

increase tumor necrosis factor (TNF)-α secretion by fresh human peripheral blood mononuclear cells (PBMC) (HARTMANN et al. 1996). This effect of PSs appeared to be mediated through interactions with the cell surface and was reversed by the polyanion heparin. This study suggests that administration of PS oligonucleotides to patients with circulating endotoxin (such as in the sepsis syndrome) may induce secretion of large amounts of TNF-α, which would be very hazardous.

Some of the immune effects of PS-ODN may be species-specific. Toxicologic analyses of PS-ODNs in mice and rats have generally shown splenomegaly, which can be extreme (reviewed in AGRAWAL 1996). While the degree of splenomegaly is partially sequence-specific, it can occur with any PS-ODN (in a length-dependent fashion). However, splenomegaly is minimal with any PS-ODN in monkeys (HENRY et al. 1997a). Recently, Crooke and colleagues observed that subcutaneous administration of PS-ODN to humans can cause local lymphadenopathy in a concentration- and time-dependent manner (S.T. CROOKE, personal communication). The mechanism and significance of this (if any) is still unclear.

B. Sequence-Specific Immune Effects of CpG Motifs in Oligodeoxynucleotides

I. Immune Activation by Bacterial DNA and the Identification of Stimulatory Palindromes

The history of the discovery of the CpG motif is interesting and will be reviewed here briefly. There is a well-established literature on the fact that certain polynucleotides cause immune stimulation. Perhaps the best example is poly(I,C) which is a potent inducer of interferon (IFN) production, as well as a macrophage activator and inducer of natural killer (NK) activity (TALMADGE et al. 1985; WILTROUT et al. 1985; KROWN 1986; EWEL et al. 1992). The immune activation of poly(I,C) is backbone-specific, since poly(dI,dC) is nonstimulatory. Deoxypolynucleotides can also have immune activating effects. For example, poly(dG,dC) also induces IFN production and activates NK cells (TOKUNAGA et al. 1988). B cell proliferation and immunoglobulin (Ig) secretion have also been reported in response to poly(dG),(dC) and poly(dG,dC) (MESSINA et al. 1993).

Attenuated mycobacteria stimulate IFN secretion and NK lytic activity, and have been used in tumor immunotherapy for many years with some success. TOKUNAGA et al. (1984) were the first to report that the mycobacterial DNA is sufficient to induce this antitumor effect. Tumor regression is accompanied by induction of IFN secretion and NK lytic activity (YAMAMOTO et al. 1988). In further studies, they showed that injection of DNA from virtually any bacteria can cause tumor regression, but that vertebrate DNA does not (YAMAMOTO et al. 1992a). The induction of IFN by bacterial DNA can be prevented by antibodies to interleukin (IL)-12 and TNF-α, indicating a role for these cytokines in mediating the effect (HALPERN et al. 1996). More recently it has been shown that in addition to its antitumor properties, DNA from bacteria (but not from vertebrates) can also induce marked B cell activation and Ig secretion (MESSINA et al. 1991). The mechanism(s) of these stimulatory effects of bacterial DNA had previously been unclear.

To identify the sequences in mycobacterial DNA which induced the production of IFN and NK lytic activity, Tokunaga and colleagues cloned and analyzed fragments of bacille Calmette-Guérin (BCG) genes. Through these studies they eventually identified a number of 45-mer single-stranded ODNs which retained immune activating properties (YAMAMOTO et al. 1992b). This immune activation was attributed to hexamer palindrome sequences present in the stimulatory ODNs (YAMAMOTO et al. 1992b). These palindromic sequences, such as AACGTT and GACGTC, were all noted to contain CpG dinucleotides and are somewhat more common in bacterial genomes than those of vertebrates (KURAMOTO et al. 1992). Although these investigators were aware that there are also other structural differences between bacterial and vertebrate DNAs, such as methylation of CpG dinucleotides (BIRD 1987;

see Sect. B.VII), they concluded that the differential immune activation by bacterial DNA as compared to vertebrate DNA resulted from the presence of CpG-containing palindromes rather than from methylation or any non-palindrome sequences (KURAMOTO et al. 1992). Perhaps because of their interest in native DNA and in the anticancer applications of activating NK cells, these investigators did not explore the immune effects of modified DNA backbones or report activation of other types of leukocytes.

II. Immune Activation by Antisense and Control Oligodeoxynucleotides

An independent and at first apparently unrelated type of DNA-induced immune activation was discovered by several groups of investigators working in the antisense field. The common theme of these later reports was the identification of ODN as the trigger of unexpected sequence-specific stimulation of B cell proliferation with or without induction of Ig secretion. Firstly, Tanaka and colleagues reported in 1992 that, to their surprise, an ODN that was synthesized as an antisense ODN to an Ig sequence actually increased the RNA levels of the target gene and caused a profound induction of B cell proliferation (TANAKA et al. 1992). This proliferative effect did not appear to be due to an antisense mechanism of action since a variety of missense ODNs had the same effect. Ig secretion was not induced; some missense ODNs lacked the stimulatory property, and the structural feature (if any) responsible for the stimulation remained unclear. In particular, there was no association with palindrome sequences.

In the year following Tanaka's observation, there were three more reports of unexpected induction of lymphocyte proliferation by ODNs. BRANDA et al. (1993) reported that an antisense ODN to the rev gene of HIV caused marked immune stimulation in mice. Only a 2-h exposure to the ODN was required to induce B cell proliferation. In contrast to the results of Tanaka and colleagues, the anti-rev ODN (which also had no palindrome) induced the production of immunoglobulin. Effects on non-B cells were not observed. Moreover, mice treated with the anti-rev ODN developed massive splenomegaly, but only if the ODN had a PS backbone – not if it was phosphodiester. The effects were relatively sequence-specific, since a PS antisense ODN to p53 and a homo-deoxycytidine 28 mer caused no B cell activation.

McINTYRE et al. (1993) described very similar immune activation with a sense ODN to nuclear factor (NF)-κB p65. The sense ODN to p65, but not the antisense, caused profound B cell proliferation and Ig secretion in vitro, and splenomegaly in vivo in mice. It is of note that these authors also reported that the sense ODN to NF-κB caused the rapid activation of NF-κB binding activity (McINTYRE et al. 1993), providing a possible clue to the ODN mechanism of action. The third report in 1993 on unexpected induction of B cell proliferation in vitro was by PISETSKY and REICH (1993), who were using a PS-ODN antisense for the herpes simplex virus. It is understandable that these antisense investigators did not make any association between their findings and the earlier observations of Tokunaga and colleagues since the types of immune stimulation were different (B cells vs NK cells), the DNA backbones were different, and none of the B cell-activating ODNs had palindromes. The B cell investigators were aware of each others' results, but did not identify any common motif between the different stimulatory ODNs that could tie the results together (McIntyre et al. 1993).

III. Discovery of Immune Activation by CpG Motifs

In 1987, I started experiments using antisense ODNs against a murine endogenous retroviral sequence (Krieg et al. 1989). Since the endogenous retroviral gene contained an immunosuppressive domain and was highly expressed in lymphocytes, I expected that the antisense might release the lymphocytes from constitutive repression, allowing their proliferation. My very first experiment with antisense ODNs appeared to be a complete success; both of the overlapping antisense oligos caused B cell proliferation and Ig secretion, but neither of the controls did (Krieg et al. 1989). This observation proved to be highly reproducible with these particular oligo sequences (Krieg et al. 1993; Mojcik et al. 1993), but as we later started to synthesize and study additional controls, we found a handful of "control" phosphodiester and PS-ODNs that also mediated B cell activation. Like the immune activating ODNs described above, these "control" ODNs induced splenic B cells to proliferate and to secrete immunoglobulin. It was clear that there was some dependence on the DNA backbone for the effect, since some ODN sequences caused stimulation when they had a phosphodiester backbone, but not if the backbone was modified to a PS. On the other hand, with a good stimulatory sequence, the strongest stimulation was seen with the PS backbone, presumably because of its greater stability.

The magnitude of the immune stimulation was remarkable and with some ODNs exceeded that induced by LPS. I, therefore, decided to pursue the possibility that one or more structural or sequence motifs in ODNs may trigger B cell activation. Comparing my stimulatory sequences and those described in the previous sections, I could discern no single sequence homology that could explain all of the effects. However, by chance, both the phosphodiester control ODN that was the strongest B cell mitogen and the best antisense ODN had the potential to form a stem-loop structure. I synthesized multiple variations of these ODNs to determine whether the stem loop was important for the effect. The first mutant ODN synthesized that could not form a stem loop lost mitogenic activity, but others in which I restored a different loop gained activity. At least half the time the stimulation appeared to correlate with the presence of a stem-loop structure, initially leading me to suspect that there were multiple different sequence elements that could mediate B cell activation, one of which was the stem loop. However, after making and testing several hundred ODNs, I finally realized that the apparent association with stem-loop structures had been spurious and due to the coincidence that I had often been introducing or modifying a subtly different type of sequence motif in these ODNs. On careful examination, it became apparent that all of our results could be explained by the presence of a single, rather simple motif based on a CpG dinucleotide in a particular sequence context which included the two or three flanking bases on both the 5′ and 3′ sides.

To confirm that this CpG motif present in the stimulatory ODN was responsible for the stimulation observed, we switched the CpG or other bases in the ODNs to either eliminate or increase the CpG dinucleotides present. Eliminating the CpG invariably abolished stimulation by phosphodiester ODNs (KRIEG et al. 1995). Changes in the ODN sequence that did not affect the CpG or the immediate flanking bases did not affect the level of stimulation. Increasing the number of CpG motifs generally increased the level of B cell stimulation. However, addition of a CpG to the 5′ end of an ODN or addition of a CpG in an unfavorable sequence context could actually reduce B cell proliferation induced by a stimulatory CpG ODN.

Our studies of over 300 ODNs ranging in length from five to 42 bases that contained CpG dinucleotides in different sequence contexts show that the bases flanking the CpG dinucleotide play a critical role in determining the B cell proliferation induced by an ODN. The formula for the consensus optimal B cell mitogenic CpG motif is: $R_1R_2CGY_1Y_2$, where R_1 is a purine (mild preference for G), R_2 is a purine or T (preference for A; T can be substituted with minimal loss of activity if the rest of the motif is intact), Y_1 is a pyrimidine (preference for T), and Y_2 is a pyrimidine (Table 1). The effects of the motif

Table 1. Effects of flanking bases on immune stimulation by CpG ODN

ODN	Sequence (5′-3′)	IL-6 (pg/ml)[a]	Proliferation[b]	IgM (ng/ml)[c]	NK LU[d]
512	TCCATGT<u>CG</u>GTCCTGATGCT	1300.8 ± 106.4	5.8 ± 0.3	3875 ± 141	0.61
1637	. . . . . . C	136.5 ± 27.4	1.7 ± 0.2	166 ± 37	0.00
1615	. . . . . . G	1201.8 ± 155.2	3.7 ± 0.3	2956 ± 411	0.04
1614	. . . . . . A	1533.7 ± 321.1	10.8 ± 0.6	6184 ± 284	2.75
1636	. . . . A	1181.6 ± 76.5	5.4 ± 0.4	3372 ± 268	2.42
1634	. . . . C	1049.6 ± 223.5	9.2 ± 0.9	5171 ± 1153	2.41
1619	. . . . T	1555.7 ± 304.9	12.5 ± 1.0	7632 ± 606	4.14
1618	. . . . . . A . . T	2109.4 ± 291.6	12.9 ± 0.7	ND	3.37
1639	. . . . . . AA . . T	1827.6 ± 83.2	11.5 ± 0.4	8068 ± 399	2.22
1707	. . . . . . A . . TC	ND	4.0 ± 0.2	2267 ± 562	0.52
1708	. . . . . CA . . TG	ND	1.5 ± 0.1	131 ± 80	2.14

Dots indicate identity; CpG dinucleotides are underlined; ND, not determined.
[a] Level of IL-6 ± Std. Dev. of triplicate samples in supernatants of the murine B cell line WEHI-231 cultured for 24 hr with the indicated ODN at 20 μM and measured by ELISA as described (YI et al. 1996c).
[b] B cell proliferation was measured using T cell depleted DBA/2 spleen cells by 3H uridine incorporation as described (YI et al. 1996c); numbers are stimulation index compared to untreated cells which had 2323 cpm.
[c] Measured by ELISA as described (YI et al. 1996c).
[d] Natural killer lytic activity measured in lytic units (LU) as described (Ballas et al. 1996).

are enhanced when it is preceded by a T. Mutations of ODNs that bring the CpG motif closer to this ideal improve stimulation, while mutations that disturb the motif reduce stimulation. Our studies clearly show that palindromes are not required and do not contribute to the mitogenic effect of an ODN containing a CpG motif.

It is important to note that repeating CGs alone are not stimulatory – the effect requires a motif similar to this formula. A factor that had made it difficult to identify the motif at first was that any single one of the positions may be switched for another base with only a modest decrease in B cell stimulation (except that a C on the 5' side of the CG or a G on the 3' side greatly reduces the effect). Significant stimulation can occur even when two bases differ from the formula (depending on the specific substitutions and the ODN backbone), indicating a remarkable degree of flexibility in the recognition of this sequence motif (KRIEG et al. 1995; YI et al. 1996c). The sequence specificity of the CpG effects have been confirmed by BOGGS et al. (1997).

In retrospect, all of the results of Tokunaga and colleagues described in Sect. B.I can be explained by the fact that the palindromes in their stimulatory ODN contain CpG motifs that fit the above formula. Likewise, the unexpected immune activating effects of various PS-ODNs described in Sect. B.II were due to the coincidental presence of CpG motifs, which differed from one another to a great enough extent that no obvious sequence homology could be detected on examination of the primary sequences. Our review of the antisense literature has revealed a number of cases, including an ODN that is currently in human clinical trials, in which effects that had been attributed to an antisense mechanism of action can in fact be attributed to the coincidental presence of CpG motifs which had not been accounted for in controls (KRIEG 1997). In this regard it is of interest that a recent analysis of the frequency of different nucleotide combinations in several hundred reported antisense ODNs found that CpG sequences were present at greater than the expected random frequency (SMETSERS et al. 1996). Possible interpretations of this overabundance of CpG dinucleotides among antisense sequences include the possibility that they may mark sites in the mRNA that make good targets for antisense ODNs for some structural reason, and the possibility that these dinucleotides have independent biologic effects such as those described above.

Not all results of the studies described above can be readily accommodated under a single CpG motif model. For example, although all of the ODNs described in Sect. B.II cause B cell proliferation, not all of them induce B cell Ig secretion. Specifically, TANAKA et al. (1992) reported that their antisense ODN did not induce B cell Ig secretion. Several other investigators who studied Tanaka's ODN found the same result, indicating that this is not just a minor difference between experimental systems (MOJCIK et al. 1993; McINTYRE et al. 1993). The reason for the loss of Ig induction in this ODN is obscure. While the significance of this is not yet clear, it raises the possibility that there may be several mechanisms through which CpG ODNs may work, and that not all CpG ODN motifs will have all of these effects. We hypothesize that there may be multiple CpG binding proteins which have slightly different binding specificities and which specifically and independently trigger B cell proliferation, B cell Ig synthesis, and/or monocyte production of cytokines leading to NK cell activation.

IV. Immune Effects of CpG Motifs

1. Mitogenic and Anti-apoptotic Effects of Oligodeoxynucleotides

Before considering the mitogenic effects of ODNs, an important technical point should be raised regarding the interpretation of [^{3}H]thymidine incorporation assays performed on cells treated with phosphodiester ODNs. In this case, thymidine released from degraded ODNs competes with the [^{3}H]thymidine, thereby giving spuriously low proliferation values (MATSON and KRIEG 1992). This effect is worst when the thymidine nucleotides are present at the 3' end of an ODN. Apart from including appropriate controls for this, investigators may wish to use alternative assays, nuclease resistant ODNs, or assays for RNA production, such as [^{3}H]uridine incorporation.

a) Virtually All B Cells Can Respond to the Mitogenic Effects of CpG Oligodeoxynucleotides

B cells can be divided into several subsets with different properties. For example, the low-affinity IgE receptor, CD23, appears to distinguish the B1 and B2 B cell lineages (WALDSCHMIDT et al. 1991). To determine whether CpG ODNs may have differential effects on these B cell subpopulations, splenic B cells were sorted by flow cytometry into the CD23$^-$ and CD23$^+$ subpopulations and then used in stimulation assays with ODNs containing or lacking the CpG motif. Both B cell subsets showed similar stimulatory responses. In other experiments B cells were fractionated over Percoll gradients to compare the response in dense resting B cells to that in lighter activated cells. B cell fractions from the 60%, 63%, and 70% Percoll layers all gave very similar stimulation indexes in response to CpG ODNs, providing evidence that the response to CpG ODNs is not limited to either the resting or activated B cell subsets (A.M. KRIEG et al., unpublished data).

Cell cycle analysis was used to obtain additional information regarding the proportion of B cells activated by CpG ODNs. Optimal concentrations of the B cell mitogen LPS induced about 80% of B cells to enter the cell cycle. Remarkably, CpG ODNs induced cycling in more than 95% of B cells, indicating that essentially all B cells become activated. However, at low concentrations B cell activation by CpG DNA synergizes with signals through antigen (Ag)-specific activation pathways (KRIEG et al. 1995). Thus, CpG DNA could promote the development of an Ag-specific response to an antigen given at the same time. Studies by BRANDA et al. (1996a) suggest the utility of CpG DNA as a vaccine adjuvant.

The above studies were performed with murine B cells, but human B cells also are induced to proliferate by PS CpG ODNs (BRANDA et al. 1996b; LIANG et al. 1996). Thus, the mitogenic effects of CpG ODNs do not appear to be limited to rodent species. However, as noted above, several investigators have noted that splenomegaly occurs after repeated injection of PS CpG ODNs in rodents but not in monkeys, indicating that there may be species differences in the immune effects of CpG motifs (HENRY et al. 1997a).

b) CpG DNA Blocks B Cell Apoptosis

Depending on their stage of maturation, cross-linking of the B cell Ag receptor may result in activation or apoptosis. Mature B lymphocytes proliferate and secrete Ig in response to Ag receptor cross-linking (DEFRANCO 1987). In contrast, Ag receptor cross-linking on immature B lymphocytes causes cell death, which is thought to be an important mechanism in the maintenance of immune tolerance (NOSSAL et al. 1979).

Because of its characteristic growth arrest and apoptosis in response to surface IgM cross-linking, the B cell line WEHI-231 has been a useful model system for studies of Ag receptor-mediated apoptosis (JAKWAY et al. 1986). WEHI-231 cells are rescued from growth arrest by certain mitogenic stimuli such as LPS and by the CD40 ligand (TSUBATA et al. 1993). Since CpG ODNs are strong B cell mitogens, we evaluated whether they rescue WEHI-231 cells from anti-IgM-induced cell cycle arrest and apoptosis. Indeed, CpG ODNs protected WEHI-231 cells from anti-IgM-mediated apoptosis, even if their addition was delayed for up to 3 h after the anti-IgM treatment (YI et al. 1996b). To determine the mechanism of this protection, we investigated whether CpG DNA affected the normal decrease in WEHI-231 expression of c-*myc* and *bcl-x_L* following IgM crosslinking. These studies revealed that CpG DNA reversed anti-IgM-induced down-regulation of c-*myc* expression in WEHI-231 and up-regulated *myn*, *bcl$_2$* and *bcl-x_L* mRNA expression (YI et al. 1996b). These results suggest that CpG DNA protection of WEHI-231 cells from anti-IgM-induced apoptosis may be mediated by specific and/or cooperative interactions of multiple genes. Moreover, to the extent that this model mimics the normal mechanisms of programmed cell death that are important in maintaining self tolerance, these data suggested the possibility that treatment with CpG DNA could lead to the induction of autoantibody production and autoimmunity (KRIEG 1995).

c) Induction of B Cell Cytokine and Ig Secretion by CpG DNA

B cells respond to CpG DNA by producing IL-6 which promotes B cell differentiation by co-stimulating CpG DNA-activated IgM production (YI et al. 1996c). IL-6 is a multifunctional regulator of immune responses, hematopoiesis, and inflammation which plays a critical role in the immune and acute phase responses to control bacterial and viral infections (KOPF et al. 1994). The production of IL-6 is rapidly induced in response to many types of bacterial and viral infection. IL-6 mRNA expression in the spleen, liver, thymus, and bone marrow is significantly increased within 30 min following IV injection of 200 μg of a CpG PS-ODN (YI et al. 1996c). IL-6 mRNA levels peak at 2 h and then decline. Likewise, the level of IL-6 protein in sera increases within 1 h, peaks at 2 h after stimulation, and returns to an undetectable base level within 12 h. The in vivo induction of IL-6 production by CpG motifs occurs in both B cells and macrophages, and is as rapid as that by LPS and other cytokines (YI et al. 1996c).

To investigate whether signals through the B cell surface Ag receptor could synergize with CpG DNA to activate B cell IL-6 production, B cells were simultaneously treated with anti-IgM and CpG DNA. We showed that like other B cell responses, such as cell proliferation and Ig secretion, CpG DNA also provides costimulatory signals for IL-6 secretion (YI et al. 1996c). To determine the biologic role of this IL-6 expression, we used neutralizing anti-mouse IL-6 antibodies. This confirmed that the IL-6 secretion induced by the CpG motif mediates Ig production but is not required for B cell proliferation (YI et al. 1996c). This effect was associated with an increase in the transcriptional activity of the IL-6 promoter.

Murine B cell IL-12 secretion is also increased within 4 h after in vivo treatment with CpG DNA (KLINMAN et al. 1996). The optimal sequence context of the CpG motif is identical for B cell proliferation, Ig secretion, and IL-6 and IL-12 cytokine secretion and is unrelated to the presence of palindromes (KLINMAN et al. 1996; YI et al. 1996c).

2. Induction of Monocyte Cytokine Secretion by CpG DNA

In addition to inducing cytokine secretion from B cells, CpG DNA also directly induces cytokine secretion from primary monocytes, macrophages, and monocytic cell lines (YI et al., to be published; CHACE et al., to be published). The monocytic cytokine production following CpG exposure includes TNF-α, IL-6, IL-12, IFN-α, and granulocyte-macrophage colony-stimulating factor (GM-CSF) (KLINMAN et al. 1996; SATO et al. 1996; YI et al., to be published; CHACE et al., to be published).

We have recently found that CpG oligonucleotides also induce cytokines from human cells in a sequence-specific manner. This cytokine expression includes TNF-α, IL-6, IL-12, IFN-γ, GM-CSF, and chemokines (KRIEG et al., unpublished data). As with mouse cells, human IFN-γ production appears to be from NK cells, while the other cytokines are produced by mononuclear cells (KRIEG et al., manuscript in preparation). It is noteworthy that monocyte cytokine induction by CpG DNA can cause synergistic toxicity with endotoxin, even when the CpG DNA has a phosphodiester backbone (COWDERY et al. 1996).

3. Induction of Natural Killer Interferon-γ Secretion and Lytic Activity by CpG DNA

Spleen cells are rapidly induced by CpG DNA to secrete IFN-γ (KLINMAN et al. 1996; COWDERY et al. 1996). In principle, the source of this IFN-γ could be either T cells or NK cells. It is well established that IL-12 and other monocyte-derived cytokines are potent stimuli for NK IFN-γ secretion. Indeed, NK cells increase IFN-γ production within 4 h after CpG DNA stimulation in mice in vivo (KLINMAN et al. 1996; COWDERY et al. 1996). The IL-12 production precedes and is at least partially responsible for the subsequent IFN-γ secretion induced by CpG DNA (KLINMAN et al. 1996; COWDERY et al. 1996). There appears to be little or no role for T cells in the acute IFN-γ production after CpG DNA stimulation (COWDERY et al. 1996). Studies with highly purified cell populations demonstrate that CpG DNA induces the direct activation of monocytes to secrete IL-12 (and other cytokines) which secondarily drive NK cells to secrete IFN-γ (CHACE et al., to be published). Remarkably, although CpG DNA has no direct effect on inducing NK cytokine production, it synergizes with low concentrations of IL-12 to induce far higher levels of IFN-γ secretion than would be induced by any concentration of IL-12 alone (CHACE et al., to be published).

In addition to cytokine secretion, NK lytic activity is increased in murine splenic and human PBMC within 24 h of exposure to CpG DNA (BALLAS et al. 1996). Neither B nor T cells are required as similar NK activation is seen in severe combined immunodeficiency disease (SCID) mice (BALLAS et al. 1996). This fits well with the model described above of a primary role for monocytic cells in enabling the NK response to CpG DNA.

IFN-γ is an important immune regulatory cytokine which regulates cell-mediated immunity, humoral immunity, and NK function. Since its production is induced by CpG DNA within 4 h in vitro and in vivo (KLINMAN et al. 1996; COWDERY et al. 1996), it could influence the CpG motif-activated B cell differentiation. IFN-γ does not induce B cell activation by itself. However, depending on the experimental system, other investigators have reported that it can act either as a costimulatory factor to promote B cell differentiation or can inhibit LPS-activated IgM secretion. We have found that IFN-γ is a costimulatory factor that augments CpG DNA-induced B cell IL-6 and IgM secretion (YI et al. 1996a). In our studies exogenous rIFN-γ doubles the level of IgM secretion induced by CpG DNA. By comparing the level of IgM induced by CpG ODNs in normal mice to that in mice with disrupted IFN-γ genes, we demonstrated that IFN-γ produced after CpG DNA stimulation also promotes B cell differentiation under normal conditions (YI et al. 1996a). Previous studies have shown that IFN-γ can enhance IL-6 expression, which is important in driving IgM secretion, suggesting to us the possibility that it might also costimulate CpG DNA-induced B cell IL-6 expression. Indeed, we showed that IFN-γ synergistically increases the IL-6 secretion induced by CpG DNA (YI et al. 1996a).

V. Role of the Phosphorothioate Backbone in the Immune Effects of CpG Oligodeoxynucleotides

Our studies show that the constraints of sequence context are more rigid for PS-ODNs than for phosphodiester (BALLAS et al. 1996). For example, we have noted that ODNs which match the consensus CpG element are extremely mitogenic for B cells when synthesized as PSs. However, an ODN, which has a pyrimidine at position R_2 and a purine at position Y_1 has no mitogenic effect as a PS, despite reasonably good stimulation as a phosphodiester. Similar backbone-dependent effects of the variations to the CpG motif have recently been described by BOGGS et al. (1997). As compared to phosphodiester (PO) ODNs, PS CpG ODNs are often far stronger B cell activators than NK cell activators. These data suggest that in different cell types there may be more than one protein which binds CpG ODNs, and that at least one of the CpG binding proteins which transduce the stimulatory signal has a higher affinity for the phosphodiester backbone.

Other investigators who have found immune stimulation by PS sequences with CpG motifs have reported that these sequences showed little or no stimulation with a phosphodiester backbone (BOGGS et al. 1997). Is this in conflict with our observation above that phosphodiester DNA can mediate the CpG B cell activation? I think not, since the studies that found no effect with phosphodiester DNA did not specify whether the fetal calf serum used was heat-inactivated or otherwise screened to ensure low nuclease content. We routinely use fetal calf serum heat-inactivated to 65°C for 30 min to reduce nuclease, and see little or no stimulation by PO CpG ODNs in in vitro cultures if this step is omitted.

On the other hand, some immune effects are probably backbone-specific. Only in vivo in mice do CpG ODNs with a PS backbone cause the massive splenomegaly (the spleen is often enlarged ten-fold compared to normal) and B cell proliferation described above. This splenomegaly is most marked with CpG ODNs, and is greatly reduced in ODN-bearing 5-methyl C substitutions (HENRY et al. 1997b). As will be reviewed below, phosphodiester DNA with CpG motifs can stimulate large amounts of cytokines in vivo if the DNA is in a form that can resist the 3' exonuclease activity of serum [such as in genomic *Escherichia coli* DNA; COWDERY et al. (1996)]. However, we have never observed splenomegaly with any dose of *E. coli* DNA, even after repeated administration (A. M. KRIEG, unpublished observation).

VI. Mechanism of Leukocyte Activation by CpG Oligodeoxynucleotides

In principle, leukocyte activation by CpG ODNs may be mediated through binding to a CpG-specific cell surface receptor. Indeed, evidence that CpG ODNs can stimulate human B cells through binding a surface receptor was recently published (LIANG et al. 1996). To determine whether cellular uptake of CpG ODNs is required for cell activation, these investigators cultured human B cells in tissue culture wells in which biotinylated CpG ODNs had been bound to avidin proteins on the plastic substrate. Following this, 48-h tritiated thymidine incorporation assays were then performed to determine whether the cells were still induced to proliferate, as they were in wells in which the ODNs were not biotinylated. Since similar cell activation was still observed, the investigators concluded that cellular uptake was not required.

We have performed studies using a similar methodology, but have come to the opposite conclusion (KRIEG et al. 1995). In our studies, we found that ODN was released from plates after 48 h of culture, perhaps due to protease digestion of the avidin used to attach the ODN to the plastic tissue culture wells. Therefore, we used a 4-h assay for cell stimulation since there was minimal release of ODN at this early time point. Under these conditions, we found that ODN bound to the tissue culture wells caused no detectable induction of proliferation of mouse B cells (KRIEG et al. 1995). Furthermore, in extensive studies using both human and murine B cells and monocytes, we found no detectable difference in binding of CpG and non-CpG ODNs to cell membranes (KRIEG et al., unpublished data). Thus, we concluded from our studies that there is no CpG-specific cell membrane receptor and that cellular uptake of CpG ODNs is required for cell activation. Our findings are in agreement with those of other investigators (D. HUME, personal communication).

Alternatively, since ODNs are taken up by cells into endosomes and subsequently some fraction of the intracellular ODN enters the nucleus, the stimulatory effects may be mediated through an interaction with cell proteins or other molecules in the endosomes or in the nucleus. To detect possible cell surface membrane proteins with binding specificity for CpG ODNs, we performed flow cytometry measuring the surface binding, uptake, and egress of fluorescein isothiocyanate (FITC) ODNs with or without a CpG motif. These studies showed no difference between the ODNs, suggesting that specificity does not lie at the level of cell binding or uptake. CpG ODNs appeared to require cell uptake for their activity, since ODNs covalently linked to a solid Teflon support were nonstimulatory, as were biotinylated ODNs linked to either avidin beads or avidin-coated petri dishes (A.M. KRIEG, unpublished data). CpG ODNs conjugated to either FITC or biotin retained full mitogenic properties, indicating no apparent steric hindrance. Finally, unlike antigens that trigger B cells through their surface Ig receptor, CpG ODNs did not induce any detectable Ca^{2+} flux, changes in protein tyrosine phosphorylation, or phosphatidylinositol-3-kinase activation. B cell stimulation by CpG ODNs generally encountered no competition from nonstimulatory ODNs, even at a five-fold molar excess. However, if the nonstimulatory ODN was complementary to the CpG ODN, then competition was observed, indicating that the stimulatory pathway requires single-stranded intracellular CpG motifs.

We hypothesize that an endosomal or nuclear protein may specifically bind single-stranded CpG DNA, and directly or indirectly transduce a stimulatory signal. To identify this hypothetical CpG-specific protein, we used electrophoretic mobility shift assays (EMSA) with murine B cell and monocyte nuclear and cytoplasmic extracts. The nucleus seemed an especially likely place to find a sequence-specific ssDNA binding protein, since it is the home of many transcription factors and other nucleic acid-binding proteins. Indeed, we detected such a CpG-specific factor in B and monocyte cell line nuclear extracts (TUETKEN et al., manuscript in preparation). This factor also appears to be present in cytoplasmic extracts, so the site of action of CpG DNA may be in the cytoplasm, organelles, or endosomes. It is not yet clear whether the B and macrophage CpG-specific factors are the same or not.

Whatever the pathway of activation, transcriptional activation by CpG ODN is detectable within 15 min, indicating that the signal is transduced quite rapidly. MCINTYRE et al. (1993) previously noted that some ODNs that caused unexpected immune stimulation induced NF-κB activity. The activation of NF-κB by DNA containing CpG motifs has recently also been demonstrated by others (STACEY et al. 1996). Our preliminary studies indicate that this NF-κB activation is required for the subsequent downstream effects of CpG DNA (YI et al., to be published). These data support the hypothesis that CpG DNA may exert its mitogenic and cytokine-inducing effects through interacting in some way with one or more intracellular proteins.

VII. Teleologic Significance of Immune Activation by CpG Motifs in DNA

The ability of the immune system to be activated by unmethylated CpG motifs is not a bizarre artifact, but rather appears to be an innate immune defense system that enables it to distinguish microbial DNA from self DNA. Vertebrate and bacterial DNA have marked differences in the frequency and methylation of CpG dinucleotides. While CpG dinucleotides are present at the expected frequency in bacterial DNA (one per 16 dinucleotides), they are only about one fourth as prevalent in vertebrate genomes (BIRD 1987). Furthermore, 50%–90% of the cytosines present in CpG dinucleotides are methylated at the 5' position in vertebrates, but other cytosines are very rarely methylated (BIRD 1987). Although many bacteria methylate certain bases in their DNA, they have no particular specificity for CpG dinucleotides. Finally, the most common base that precedes CpG in our genome is C, which greatly reduces the stimulatory effect (see Table 1), and provides an added safeguard against immune activation by self DNA.

The observation that bacterial DNA causes B cell activation was first reported by MESSINA et al. (1991), and followed earlier reports that microbial DNA activates NK cell function (TOKUNAGA et al. 1984). Messina and colleagues showed that the mitogenic effect of bacterial DNA is abolished by DNase digestion, but not by absorption with polymyxin B (which removes endotoxin contaminants), suggesting that a particular structure or other characteristic of bacterial DNA is responsible for its ability to trigger B cell activation. Our observations on the effects of CpG dinucleotides in ODNs suggested to us the possibility that the unmethylated CpG dinucleotides present in bacterial DNA may be responsible for the previously observed B cell activation. To test this hypothesis, we methylated bacterial DNA with CpG methylase, which abolished its mitogenicity (KRIEG et al. 1995). Thus, the specific mitogenic effects of bacterial DNA result from the presence of unmethylated CpG dinucleotides.

Of course, synthetic ODNs are normally produced with unmethylated cytosines, and so would appear to the immune system as microbial DNA. To determine the role of CpG methylation in ODN-mediated B cell proliferation, we tested the B cell mitogenicity of ODNs in which cytosines in CpG motifs or elsewhere were replaced by 5-methylcytosine. ODNs containing methylated CpG motifs caused no mitogenic effect (KRIEG et al. 1995). ODNs in which other cytosines were methylated retained their stimulatory properties.

Teleologically, therefore, it appears likely that lymphocyte activation by the CpG motif represents an immune defense mechanism which enables the host to recognize this simple yet characteristic pattern in microbial DNA. The presence of microbial DNA in host tissues would serve as a danger signal to activate innate immune defense mechanisms that will contain the infection while specific immune defenses are induced. Since the CpG pathway synergizes with B cell activation through the Ag receptor (KRIEG et al. 1995), B cells bearing Ag receptors specific for bacterial antigens would receive one activation signal through cell membrane Ig and a second signal from bacterial DNA, and would therefore tend to be preferentially activated. Thus, the interrelationships of this pathway with other pathways of B cell activation provide a physiologic mechanism that uses a polyclonal mitogen to induce protective antigen-specific immune responses.

Bacteria such as *E. coli*, which are well adapted to survive in the gut, have the expected frequency of CpG dinucleotides, indicating that they have been under no evolutionary pressure to avoid this immune defense. Indeed, since such bacteria do not normally invade tissue, CpG suppression would confer no apparent benefit, and may even be detrimental by harming their host. In this regard, it is of interest that most viruses and retroviruses, which of course must replicate within host cells, show dramatic CpG suppression (KARLIN et al. 1994a). Bacteriophage genomes have the expected frequency of CpG dinucleotides, suggesting that CpG suppression in viruses that infect vertebrates is not a consequence of genome size, but rather may reflect evolutionary pressure to avoid the CpG-induced host defense. Likewise, the genomes of pathogenic protists such as *Trypanosoma brucei* and *Plasmodium falciparum* also show significant and unexplained CpG suppression (KARLIN et al. 1994b). We hypothesize that this CpG suppression arose to avoid activating the CpG-mediated defense mechanisms. The molecular mechanisms of this will likely prove to be most interesting, since such suppression would probably not have occurred one motif at a time. Most likely there is some enzyme pathway that has been actively mutating mitogenic CpG motifs.

C. Sequence-Specific Immune Effects of Poly(G) Motifs in Oligodeoxynucleotides

Poly(G) sequences have a wide range of biologic and immunologic effects (Yaswen et al. 1993), including induction or facilitation of cytokine expression. The level of IFN induction by a CpG ODN can be enhanced by poly(G) sequences at the ends of the ODN (KIMURA et al. 1994). On the other hand, an ODN containing poly(G) sequences alone can block induction of IFN secretion by a CpG ODN (KIMURA et al. 1994). Poly(G) ODN can also block the production of IFN-γ induced by the mitogens concanavalin A, bacterial DNA, or the combination of phorbol myristate acetate (PMA) and the calcium ionophore A23187 (HALPERN and PISETSKY 1995). This inhibition, which could occur with ODNs as short as five residues, was only seen with the PS backbone.

In addition to blocking the inducible production of IFN-γ, poly(G)-rich ODNs can also block the downstream effects of IFN-γ. A poly(G) oligo has been shown to inhibit the binding of IFN-γ to its receptor (RAMANATHAN et al. 1994a; LEE et al. 1996), which prevents the normal enhancement of MHC class I and the intercellular adhesion molecule (ICAM)-1 in response to interferon-γ. Such ODNs do not inhibit induction of MHC-I expression by interferon-α or interferon-β (RAMANATHAN et al. 1994b). Interferon-γ-induced ICAM-1 expression has also been found to be inhibited by G-rich ODNs in keratinocytes by HERTL et al. (1995). In this system, phosphodiester ODNs were also ineffective. It is of note that the inhibition by the G-rich ODNs was reversible when a complementary sense ODN was added, suggesting that the poly(G) effect can only be exerted in a single-stranded form. We have recently reported that poly(G) regions can reduce the immune effect of a CpG motif in a PS oligo, suggesting the possibility of complex interactions between various types of sequence motifs present in ODNs (BALLAS et al. 1996).

Acknowledgements. The author thanks Vickie McCauley and Tilese Arrington for secretarial assistance. The author was supported in part by grants from the Department of Veterans Affairs, the RGK Foundation, and NIH R29-AR42556-01 and P01-CA66570.

References

Agrawal S (1996) Antisense oligonucleotides: towards clinical trials. TIB Tech 14:376–387
Ballas ZK, Rasmussen WL, Krieg AM (1996) Induction of natural killer activity in murine and human cells by CpG motifs in oligodeoxynucleotides and bacterial DNA. J Immunol 157:1840–1845
Bird AP (1987) CpG islands as gene markers in the vertebrate nucleus. Trends Genet 3:342–347
Boggs RT, McGraw K, Condon T, Flournoy S, Villiet P, Bennett CF, Monia BP (1997) Characterization and modulation of immune stimulation by modified oligonucleotides. Antisense Nucleic Acid Drug Dev 7:461–472

Branda RF, Moore AL, Mathews L, McCormack JJ, Zon G (1993) B-cell proliferation and differentiation in common variable immunodeficiency patients produced by an antisense oligomer to the rev gene of HIV-1. Biochem Pharmacol 45:2037–2043

Branda RF, Moore AL, Lafayette AR, Mathews L, Hong R, Zon G, Brown T, McCormack JJ (1996a) Amplification of antibody production by phosphorothioate oligodeoxynucleotides. J Lab Clin Med 128:329–338

Branda RF, Moore AL, Hong R, McCormack JJ, Zon G, Cunningham-Rundles C (1996b) B-cell proliferation and differentiation in common variable immunodeficiency patients produced by an antisense oligomer to the rev gene of HIV-1. Clin Immunol Immunopathol 79:115–121

Cowdery JS, Chace JH, Krieg AM (1996) Bacterial DNA induces in vivo interferon-γ production by NK cells and increases sensitivity to endotoxin. J Immunol 156:4570

Crooke RM, Crooke ST, Graham MJ, Cooke ME (1996) Effect of antisense oligonucleotides on cytokine release from human keratinocytes in an in vitro model of skin. Toxicol Appl Pharmacol 140:85–93

DeFranco AL (1987) Molecular aspects of B-lymphocyte activation. Annu Rev Cell Biol 3:143

Ewel CH, Urba SJ, Kopp WC, Smith JW II, Steis RG, Rossio JL, Longo DL, Jones MJ, Alvord WG, Pinsky CM, Beveridge JM, McNitt KL, Creekmore SP (1992) Polyinosinic-polycytidylic acid complexed with poly-L-lysine and carboxymethylcellulose in combination with interleukin 2 in patients with cancer: clinical and immunological effects. Cancer Res 52:3005–3010

Halpern HD, Pisetsky DS (1995) In vitro inhibition of murine IFN-γ production by phosphorothioate deoxyguanosine oligomers. Immunopharmacology 29:47–52

Halpern MD, Kurlander RJ, Pisetsky DS (1996) Bacterial DNA induces murine interferon-g production by stimulation of interleukin-12 and tumor necrosis factor-α. Cell Immunol 167:72–78

Hartmann G, Krug A, Waller-Fontaine K, Endres S (1996) Oligodeoxynucleotides enhance lipopolysaccharide-stimulated synthesis of tumor necrosis factor: dependence on phosphorothioate modification and reversal by heparin. Mol Med 2:429–438

Henry S, Monteith D, Levin AA (1997a) Antisense oligonucleotide inhibitors for the treatment of cancer. 2. Toxicologic properties of phosphorothioate oligodeoxynucleotides. Anticancer Drug Des 12:395–408

Henry SP, Monteith D, Bennett F, Levin AA (1997b) Toxicologic and pharmacokinetic properties of chemically modified antisense oligonucleotide inhibitors of PKC-α and C-raf kinase. Anticancer Drug Des 12:409–420

Hertl M, Neckers LM, Katz SI (1995) Inhibition of interferon-γ-induced intercellular adhesion molecule-1 expression on human keratinocytes by phosphorothioate antisense oligodeoxynucleotides is the consequence of antisense-specific and antisense-non-specific effects. J Invest Dermatol 104:813–818

Jakway JP, Unsinger WR, Gold MR, Mishell RI, DeFranco AL (1986) Growth regulation of the B lymphoma cell line WEHI-231 by anti-immunoglobulin, lipopolysaccharide, and other bacterial products. J Immunol 137:2225–2231

Karlin S, Doerfler W, Cardon LR (1994a) Why is CpG suppressed in the genomes of virtually all small eukaryotic viruses but not in those of large eukaryotic viruses? J Virol 68:2889–2897

Karlin S, Ladunga I, Blaisdell BE (1994b) Heterogeneity of genomes: measures and values. Proc Natl Acad Sci U S A 91:12837–12841

Kimura Y, Sonehara K, Kuramoto E, Makino T, Yamamoto S, Yamamoto T, Kataoka T, Tokunaga T (1994) Binding of oligoguanylate to scavenger receptors is required for oligonucleotides to augment NK cell activity and induce IFN. J Biochem 116:991–994

Klinman D, Yi A-K, Beaucage SL, Conover J, Krieg AM (1996) CpG motifs expressed by bacterial DNA rapidly induce lymphocytes to secrete IL-6, IL-12 and IFN. Proc Natl Acad Sci U S A 93:2879–2883

Kopf M, Baumann H, Freer G, Freudenberg M, Lamers M, Kishimoto T, Zinkernage R, Bluethmann H, Kohler G (1994) Impaired immune and acute-phase responses in interleukin-6-deficient mice. Nature 368:339–342

Krieg AM, Gause WC, Gourley MF, Steinberg AD (1989) A role for endogenous retroviral sequences in the regulation of lymphocyte activation. J Immunol 143:2448–2451

Krieg A, Tonkinson J, Matson S, Zhao Q, Saxon M, Zhang L-M, Bhanja U, Yakubov L, Stein CA (1993) Modification of antisense phosphodiester oligodeoxynucleotides by a 5' cholesteryl moiety increases cellular association and improves efficacy. Proc Natl Acad Sci U S A 90:1048–1052

Krieg AK, Yi A-K, Matson S, Waldschmidt TJ, Bishop GA, Teasdale R, Koretzky G, Klinman D (1995) CpG motifs in bacterial DNA trigger direct B-cell activation. Nature 374:546–549

Krieg AM (1995) CpG DNA: a pathogenic factor in systemic lupus erythematosus? J Clin Immunol 15:284–292

Krieg AM (1997) Leukocyte stimulation by oligonucleotides. In: Stein C, Krieg AM (eds) Applied antisense oligonucleotide technology. Wiley, New York (in press)

Krown SE (1986) Interferons and interferon inducers in cancer treatment. Semin Oncol 13:207–217

Kuramoto E, Yano O, Kimura Y, Baba M, Makino T, Yamamoto S, Yamamoto T, Kataoka T, Tokunaga T (1992) Oligonucleotide sequences required for natural killer cell activation. Jpn J Cancer Res 83:1128–1131

Lee PP, Ramanathan M, Hunt CA, Garovoy MR (1996) An oligonucleotide blocks interferon-γ signal transduction. Transplantation 62:1297–1301

Liang H, Nishioka Y, Reich CF, Pisetsky DS, Lipsky PE (1996) Activation of human B cells by phosphorothioate oligodeoxynucleotides. J Clin Invest 98:1119–1129

Matson S, Krieg AM (1992) Nonspecific suppression of ^{3}H-thymidine incorporation by "control" oligonucleotides. Antisense Res Dev 2:325–330

McIntyre KW, Lombard-Gillooly K, Perez JR, Kunsch C, Sarmiento UM, Larigan JD, Landreth KT, Narayanan R (1993) A sense phosphorothioate oligonucleotide directed to the initiation codon of transcription factor NF-κB p65 causes sequence-specific immune stimulation. Antisense Res Dev 3:309–322

Messina JP, Gilkeson GS, Pisetsky DS (1991) Stimulation of in vitro murine lymphocyte proliferation by bacterial DNA. J Immunol 147:1759–1764

Messina JP, Gilkeson GS, Pisetsky DS (1993) The influence of DNA structure on the in vitro synthetic polynucleotide antigens. Cell Immunol 147:148–157

Mojcik C, Gourley MF, Klinman DM, Krieg AM, Gmelig-Meyling F, Steinberg AD (1993) Administration of a phosphorothioate oligonucleotide antisense to murine endogenous retroviral MCF env causes immune effects in vivo in a sequence-specific manner. Clin Immunol Immunopathol 67:130–136

Nossal GJ, Pike BL, Battye FL (1979) Mechanisms of clonal abortion tolerogenesis. II. Clonal behavior of immature B cells following exposure to anti-mu chain antibody. Immunology 37:203–215

Perez JR, Li Y, Stein CA, Majumder S, Van Oorschot A, Narayanan R (1994) Sequence-independent induction of Sp1 transcription factor activity by phosphorothioate oligodeoxynucleotides. Proc Natl Acad Sci U S A 91:5957–5961

Pisetsky DS, Reich CF (1993) Stimulation of murine lymphocyte proliferation by a phosphorothioate oligonucleotide with antisense activity for herpes simplex virus. Life Sci 54:101–107

Ramanathan M, Lantz M, MacGregor RD, Garovoy MR, Hunt CA (1994a) Characterization of the oligodeoxynucleotide-mediated inhibition of interferon-γ-induced major histocompatibility complex class I and intercellular adhesion molecule-1. J Biol Chem 269:24564–24574

Ramanathan M, Lantz M, MacGregor RD, Huey B, Tam S, Ki Y, Garovoy MR, Hunt CA (1994b) Inhibition of interferon-γ-induced major histocompatibility complex class I expression by certain oligodeoxynucleotides. Transplantation 57:612–615

Sato Y, Roman M, Tighe H, Lee D, Corr M, Nguyen MD, Silverman GJ, Lotz M, Carson DA, Raz E (1996) Immunostimulatory DNA sequences necessary for effective intradermal gene immunization. Science 273:352–354.

Smetsers TFCM, Boezeman JBM, Mensink EJBM (1996) Bias in nucleotide composition of antisense oligonucleotides. Antisense Nucleic Acid Drug Dev 6:63–67

Stacey KJ, Sweet MJ, Hume DA (1996) Macrophages ingest and are activated by bacterial DNA. J Immunol 157:2116–2122

Stein CA, Cheng Y-C (1993) Antisense oligonucleotides as therapeutic agents – is the bullet really magical? Science 261:1004–1012

Talmadge JE, Adams J, Phillips H, Collins M, Lenz B, Schneider M, Schlick E, Ruffmann R, Wiltrout RH, Chirogos MA (1985) Immunomodulatory effects in mice of polyinosinic-polycytidylic acid complexed with poly-L-lysine and carboxymethylcellulose. Cancer Res 45:1058–1065

Tanaka T, Chu CC, Paul WE (1992) An antisense oligonucleotide complementary to a sequence in Ic2b increases c2b germline transcripts, stimulates B cell DNA synthesis, and inhibits immunoglobulin secretion. J Exp Med 175:597–607

Tokunaga T, Yamamoto H, Shimada S, Abe H, Fukuda T, Fujisawa Y, Furutani Y, Yano O, Kataoka T, Sudo T, Makiguchi N, Suganuma T (1984) Antitumor activity of deoxyribonucleic acid fraction from mycobacterium bovis GCG. I. Isolation, physicochemical characterization, and antitumor activity. JNCI 72:955–962

Tokunaga T, Yamamoto S, Namba K (1988) A synthetic single-stranded DNA, poly(dG,dC), induces interferon-α/β and -γ, augments natural killer activity, and suppresses tumor growth. Jpn J Cancer Res 79:682–686

Tsubata T, Wu J, Honjo T (1993) B-cell apoptosis induced by antigen receptor crosslinking is blocked by a T-cell signal through CD40. Nature 364:645–648

Waldschmidt TJ, Kroese FGM, Tygrett LT, Conrad DH, Lynch RG (1991) The expression of B cell surface receptors. III. The murine low-affinity IgE Fc receptor is not expressed on Ly 1 or "Ly 1-like" B cells. Int Immunol 3:305–315

Wiltrout RH, Salup RR, Twilley TA, Talmadge JE (1985) Immunomodulation of natural killer activity by polyribonucleotides. J Biol Respir Mod 4:512–517

Yamamoto S, Kuramoto E, Shimada S, Tokunaga T (1988) In vitro augmentation of natural killer cell activity and production of interferon-α/β and -γ with deoxyribonucleic acid fraction from mycobacterium bovis BCG. Jpn J Cancer Res 79:866–873

Yamamoto S, Yamamoto T, Shimada S, Kuramoto E, Yano O, Kataoka T, Tokunaga T (1992a) DNA from bacteria, but not from vertebrates, induces interferons, activates natural killer cells and inhibits tumor growth. Microbiol Immunol 36:983–997

Yamamoto S, Yamamoto T, Kataoka T, Kuramoto E, Yano O, Tokunaga T (1992b) Unique palindromic sequences in synthetic oligonucleotides are required to induce INF and augment INF-mediated natural killer activity. J Immunol 148:4072–4076

Yaswen P, Stampfer MR, Ghosh K, Cohen JS (1993) Effects of sequence of thioated oligonucleotides on cultured human mammary epithelial cells. Antisense Res Dev 3:67–77

Yi A-K, Chace JH, Cowdery JS, Krieg AM (1996a) IFN-γ promotes IL-6 and IgM secretion in response to CpG motifs in bacterial DNA and oligodeoxynucleotides. J Immunol 156:558–564

Yi A-K, Hornbeck P, Lafrenz DE, Krieg AM (1996b) CpG DNA rescue of murine B lymphoma cells from anti-IgM induced growth arrest and programmed cell death is associated with increased expression of c-myc and bcl-XL. J Immunol 157:4918–4925

Yi A-K, Klinman DM, Martin TL, Matson S, Krieg AM (1996c) Rapid immune activation by CpG motifs in bacterial DNA: systemic induction of IL-6 transcription through an antioxidant-sensitive pathway. J Immunol 157:5394–5402

Zhao Q, Song X, Waldschmidt T, Fisher E, Krieg AM (1996) Oligonucleotide uptake in human hematopoietic cells is increased in leukemia and is related to cellular activation. Blood 88:1788–1795

CHAPTER 9
Pharmacological Inhibition of Dopaminergic and Other Neurotransmitter Receptors Using Antisense Oligodeoxynucleotides

G. DAVIDKOVA and B. WEISS

A. Introduction

In recent years a great number of studies have been carried out that apply antisense oligonucleotide technology to investigate neurobiological systems (for recent reviews, see WEISS et al. 1996, 1997a,b; ZON 1995). The antisense approach has been particularly useful in characterizing the pharmacological properties and biological functions of receptors for neurotransmitters. This is due to the rapid discovery of the molecular structure of new subtypes of neurotransmitter receptors, the pharmacological and biological properties of which remain largely unknown. Antisense compounds, by hybridizing specifically to the nucleic acids encoding the different receptor subtypes, have provided a highly selective means to reduce the expression, and thereby the levels, of individual receptors, an effect that is not attainable with traditional pharmacological antagonists. In addition to providing a highly selective means to study various neurobiological events, antisense compounds have the potential to be used as therapeutic agents in the management of neuropsychiatric and neurodegenerative disorders. An antisense strategy to reduce the function of neuroreceptors might have a further distinct advantage over traditional pharmacological antagonists in that antisense agents, unlike the conventional pharmacological antagonists, might not induce the upregulation of the very receptors they are intended to inhibit (BURT et al. 1977; HYTTEL 1986; ROGUE et al. 1991).

This chapter will summarize the application of antisense oligodeoxynucleotides to receptors for neurotransmitters, both in vitro and in vivo, and will address the factors to consider in designing antisense oligodeoxynucleotides as tools to inhibit the expression of neurotransmitter receptors. It will not discuss the use of antisense oligodeoxynucleotides to inhibit the expression of other important biological receptors and nonreceptor proteins such as opioid receptors, estrogen receptors, neuropeptides, calmodulin, immediate-early genes and oncogenes, microtubule-associated proteins, synaptic proteins, and growth factors. This has been reviewed recently elsewhere (DAVIDKOVA et al. 1997). It will also not discuss at any length the use of antisense RNA expression vectors to generate antisense agents intracellularly, as this topic has also been reviewed recently (WEISS et al. 1997a). Nor will it consider the numerous potential other therapeutic applications of antisense strategies, such as their potential use in treating cancer, immunodeficiency syndromes, cardiovascular, infectious, and central nervous system diseases, topics which have been considered recently elsewhere (WEISS 1997; AGRAWAL 1996; CROOKE 1995; WICKSTOM 1991). It will, however, consider the various controls which should be employed in order to ascertain the specificity of the antisense approach. Finally, it will outline some of the obstacles that have to be overcome before antisense oligodeoxynucleotides can be put into practical use.

B. Advantages of Antisense Oligodeoxynucleotides over Traditional Pharmacological Antagonists

Antisense oligodeoxynucleotides directed to neurotransmitter receptors possess several distinct advantages over traditional pharmacological antagonists. The most important advantage is the exclusive selectivity with which antisense oligodeoxynucleotides can bind to and block the expression of only one subtype of neurotransmitter receptor [e.g., antisense oligodeoxynucleotides to the different subtypes of the dopamine receptor (WEISS et al. 1993, 1996; ZHOU et al. 1994; ZHANG et al. 1994; ZHANG and CREESE 1993; SILVIA et al. 1994; NISSBRANDT et al. 1995)], or a single subunit of a neurotransmitter receptor [e.g., antisense oligodeoxynucleotides to the different subunits of the γ-aminobutyric acid-A ($GABA_A$) receptor (ZHU et al. 1996; KARLE et al. 1995)]. Such high selectivity has been difficult to achieve with traditional pharmacological antagonists which usually interact with more than one subtype of

List of Abbreviations

AMP	Adenosine monophosphate
AS	Antisense
CNS	Central nervous system
CSF	Cerebrospinal fluid
FITC	Fluorescein isothiocyanate
FNM	Fluphenazine-N-mustard
GABA	γ-Aminobutyric acid
GFAP	Glial fibrillary astrocytic protein
5-HT	5-Hydroxytryptamine
IPSC	Inhibitory postsynaptic current
NMDA	N-methyl-D-aspartate
6-OHDA	6-Hydroxydopamine
PKA	Protein kinase A
PLC	Phospholipase C
TBPS	Tert-butyl-bicyclo-phosphorothionate
7TMR	Seven transmembrane spanning regions

neurotransmitter receptor or even with several different neurotransmitter receptors. For example, most typical antipsychotic drugs bind, although with different affinities, to all known subtypes of the D_2 dopamine receptor family. In addition, both the typical and atypical antipsychotic drugs such as clozapine [which exhibit preferential binding to the D_4 dopamine receptor (SEEMAN et al. 1997)] bind to α-adrenergic, muscarinic and serotonergic receptors (MELTZER 1994).

Another advantage of the use of antisense oligodeoxynucleotides, especially as potential therapeutic agents, is the possibility of selectively inhibiting the expression of neurotransmitter receptors only in those brain regions believed to be responsible for the particular pathological process to be treated. Indeed, blockade of a specific subtype of a specific neuroreceptor in discrete brain regions is the ultimate goal of neuropharmacologists. For example, most antipsychotic drugs are thought to produce their therapeutic effects by blocking dopamine receptors in the mesolimbic and mesocortical regions of the brain. However, their blockade of dopamine receptors in the nigrostriatal and tuberoinfundibular regions induces motor and endocrine side effects (CASEY 1991; CARLSSON and PIERCEY 1995; SEEMAN 1988; KANE 1995). Since there are different distributions of the various subtypes of neurotransmitters in various brain areas, the use of antisense oligodeoxynucleotides offers the possibility of inhibiting the dopamine receptors in those brain areas believed to be responsible for the disease process without inducing the side effects caused by blocking dopamine receptors in other brain areas. In addition, it might be possible to target antisense oligodeoxynucleotides to the dopamine receptors expressed only in limbic areas of the brain by encapsulating the oligodeoxynucleotides in immunoliposomes, coated on the outer surface with monoclonal antibodies specific to the limbic system-associated protein (LEVITT 1984).

Persistent blockade of neurotransmitter receptors with traditional pharmacological antagonists often induces a compensatory increase in their synthesis resulting in receptor upregulation, an effect that leads to tolerance to these drugs or, in the case of most neuroleptics, to debilitating motor effects such as tardive dyskinesia (CASEY 1991; SEEMAN 1988; KANE 1995). Antisense oligodeoxynucleotides apparently do not upregulate the receptors they are designed to inhibit and therefore should not produce these effects. This has been shown to be due to the entirely different mechanism of action of antisense oligodeoxynucleotides as these compounds not only inhibit the synthesis of neuroreceptors, they appear to have a preferential effect on the pool of functional neurotransmitter receptors (QIN et al. 1995).

Other advantages of the antisense oligodeoxynucleotides include: (a) the relative ease with which they can be designed on the basis of the structure of the nucleic acids encoding the neurotransmitter receptor, (b) the ease with which they can be synthesized at a low cost, and (c) the shorter overall time required to develop antisense oligodeoxynucleotides in comparison with traditional pharmacological antagonists.

C. Factors to Consider in the Development of Antisense Oligodeoxynucleotides Targeted to Neurotransmitter Receptors

I. Structure of Antisense Oligodeoxynucleotides: Optimal Recognition Site on the Targeted Transcript and Oligodeoxynucleotide Length and Chemical Structure

Several issues have to be taken into consideration in the development of antisense oligodeoxynucleotides to neurotransmitter receptors. Some are general issues in the development of antisense oligodeoxynucleotides and some depend on the intended use (in vitro or in vivo). Issues directly related to the antisense oligonucleotides are the structure (intended recognition site within the targeted receptor transcript, the optimal length, any chemical modifications of the oligodeoxynucleotides), the optimal oligodeoxynucleotide dose, and the selection of appropriate controls to ascertain the specificity of the antisense effects. The issues related to the structure of the oligodeoxynucleotides have been reviewed in detail elsewhere (UHLMANN and PEYMAN 1990; CROOKE 1992; LEONETTI et al. 1993; DAVIDKOVA et al. 1997) and will not be discussed further.

Based on the studies with antisense oligonucleotides to neurotransmitter receptors (see Tables 1 and 2), it is evident that most of the effective oligonucleotides have been targeted to a region spanning or near the translation start site (WEISS et al. 1997b; NECKERS and WHITESELL 1993). Although this has been a very effective target, it is difficult to predict in advance the regions which will be most susceptible to inhibition by oligodeoxynucleotides. To determine the most effective target site, the effects of several oligonucleotides complementary to different regions of the targeted transcript must be examined. Screening may also be done for oligodeoxynucleotides targeted to the same region, but possessing different chemical structures. For example, we have compared the in vivo effects of two D_2 antisense oligodeoxynucleotides targeted to the same site on the dopamine receptor mRNA (nucleotides –10 to +10), one being phosphorothioate-modified and the other, amidate-modified (WEISS et al. 1997b). Both oligodeoxynucleotides were equally effective at inhibiting quinpirole-induced rotational behavior in mice lesioned with 6-hydroxydopamine.

From Tables 1 and 2 it is evident that most of the useful oligodeoxynucleotides used in in vitro and in vivo studies have a length which varies between 15 and 20 bases and that most of the studies have been done with phosphorothioate-modified oligodeoxynucleotides. The latter are especially suitable for in vivo studies because of their greater resistance to metabolism by nucleases. It should be emphasized, however, that the dose of antisense oligodeoxynucleotides has to be carefully optimized as antisense oligodeoxy-

Table 1. In vitro effects of antisense oligodeoxynucleotides on the levels and functions of neurotransmitter receptors

Receptor type	Cell type	Oligo structure (type and length)	Target site on receptor mRNA	Effective oligo concentration (μM)	Delivery system	Treatment regimen	Effect on target protein	Effect on target mRNA	Effect on function	Reference
ACh m_1 muscarinic	Rat superior cervical ganglion cells	S 16-mer	Coding region (nucleotides 4–19)	100 μM	Dissolved in culture medium	4 days	Decrease	Decrease		Zang et al. 1994
ACh m_2 muscarinic	Primary rat cerebellar granule cells	N 15-mer	Coding region	5–25 μM	Dissolved in culture medium	1–6 days	Decrease		Prevention of the m_2-mediated inhibition of cyclic AMP formation	Holopainen and Wojcik 1993
ACh m_1 and m_3 muscarinic	Xenopus (oocytes)	N 36–46mer	Coding region	0.5–10 ng/oocyte	Dissolved in distilled water	Response assayed 2.5–5h after injection			Decrease in membrane responsiveness to ACh	Davidson et al. 1991
ACh m_1 and m_3 muscarinic	Xenopus (oocytes)	N 36–50mer	Coding region	9.6 ng/oocyte	Dissolved in distilled water	Response assayed 2.5–5h after injection			Decrease in responsiveness to Ach	Matus-Leibovitch et al. 1992
ACh nicotinic (α3, α4, α7 subunits)	Embryonic chicken sympathetic neurons	N 15-mer	Coding region	10 μM	Dissolved in culture medium	48h			Decrease in agonist-gated channel opening	Listerud et al.1991
ACh nicotinic (α3 subunit)	Embryonic chicken sympathetic neurons	N 15-mer	Translation start	10 μM	Dissolved in culture medium	24–48h			Decrease in agonist-gated channel opening	Yu et al. 1993
Adrenergic (α2)	PC 124D cells (express rat α2a receptor); NIH3T3 cells (express rat α2c receptor)	N 21-mer (α2a) 17-mer (α2c)	Translation start	5 μM	Dissolved in culture medium	3 days	Decrease			Mizobe et al. 1996
Dopamine D_{1a}	LTK fibroblast cells	S 21-mer	Translation start	5 μM	Dissolved in culture medium	2 days	Decrease		Blocks fenoldopam-induced increase in PLC-γ activity	Yu et al. 1996

Table 1. Continued

Receptor type	Cell type	Oligo structure (type and length)	Target site on receptor mRNA	Effective oligo concentration (μM)	Delivery system	Treatment regimen	Effect on target protein	Effect on target mRNA	Effect on function	Reference
Dopamine D_2	WERI-27 retinoblastoma cells	S 17-mer	Translation start (nucleotides 1–17)	$2\,\mu M$	Complexed with lipofectin in serum-free medium	Two 6-h treatments separated by an 18-h interval	Decrease			Silvia et al. 1994
Dopamine D_2	Primary rat pituitary cells	S 18-mer	Spanning the translation start (nucleotides −7 to +11)	$2\,\mu M$	Dissolved in tissue culture medium	7-day exposure with treatments on days 0, 2, 4, and 6	Decrease	Decrease	Disappearance of D_2 dopamine receptor-mediated signal transduction events	Valerio et al. 1994
$GABA_A$ (α subunits)	Organotypic slices of rat primary visual cortex	N 15-mer	Coding region	$20\,\mu M$	Dissolved in serum-free medium	Application with medium; refreshing on 8, 10, 12, and 14 days in culture			Inhibition of evoked GABAergic inhibitory postsynaptic currents	Brussard and Baker 1995
$GABA_A$ ($\alpha 6$ subunit; $\gamma 2$ subunit)	Primary rat cerebellar granule cells	S 18-mer	Spanning the translation start	$5\,\mu M$	Dissolved in culture medium	24–48 h	Decrease		Changes in $GABA_A$-mediated electrophysiological events	Zhu et al. 1996
$GABA_B$	Primary rat cerebellar granule cells	N 15-mer	Coding region	5–$25\,\mu M$	Dissolved in culture medium	1–6 days			Prevention of the $GABA_B$-mediated inhibition of cAMP formation	Holopainen and Wojcik 1993
NMDA-R1	Primary rat cerebral cortical neurons	N 18-mer	Coding region and spanning the translation start	$1\,\mu M$	Dissolved in serum-free medium	Application every 2 days to the culture medium	Decrease		Block calcium ion flux through the NMDA-gated channel; protect against NMDA-induced neurotoxicity	Wahlestedt et al. 1993

ACh, acetylcholine; cAMP, cyclic adenosine monophosphate; GABA, γ-aminobutyric acid; mRNA, messenger ribonucleic acid; N, natural (unmodified) oligodeoxynucleotides; NMDA-R, N-methyl-D-aspartate receptor; PLC, phospholipase C; S, phosphorothioate-modified oligodeoxynucleotides.

Table 2. In vivo effects of antisense oligodeoxynucleotides on the levels and functions of neurotransmitter receptors

Receptor type	Species	Administration site and delivery system	Oligo structure (type and length)	mRNA target site	Oligo concentration	Treatment regimen	Effect on target protein	Effect on target mRNA	Effect on function	Reference
ACh m$_1$ muscarinic	Rat	Subcutaneously implanted Alzet micro-osmotic pump (i.c.v.)	S 16-mer	Translation start	5 μg/m per hour	3-day infusion	Decrease	Decrease	Increases locomotion	ZANG et al. 1994
Adrenergic (α2)	Rat	Administration in locus coeruleus via cannula	N 21-mer (α2a) 17-mer (α2c)	Translation start	5 nmol/0.2 μl	Injection on days 1, 3, and 5			Decreases the duration of the hypnotic response to dexmedetomidine	MIZOBE et al. 1996
Dopamine D$_1$	Mouse	i.c.v.	S 20-mer	Spanning translation start	2.5 nmol/2 μl per injection	Multiple injections over 7 days	Decrease		Inhibits D$_1$ behavior (grooming in response to SKF 38393)	ZHANG et al. 1994
Dopamine D$_2$	Mouse	i.c.v.	S 20-mer	Spanning translation start	2.5 nmol/2 μl per injection	Multiple injections over 8 days	Decrease	Decrease	Inhibits D$_2$ behavior (rotations in response to quinpirole)	WEISS et al. 1993
Dopamine D$_2$	Mouse	i.c.v.	S 20-mer	Spanning translation start	2.5 nmol/2 μl per injection	Multiple injections over 8 days	Decrease	Decrease	Inhibits D$_2$ behavior (rotations in response to quinpirole)	ZHOU et al. 1994
Dopamine D$_2$	Mouse	i.c.v.	S 20-mer	Spanning translation start	2.5 nmol/2 μl per injection	Multiple injections over 2–8 days	Decrease		Potentiates catalepsy and inhibits stereotypic behavior in response to quinpirole; inhibits synthesis of D$_2$ receptors	QIN et al. 1995
Dopamine D$_2$	Mouse	i.c.v.	S 20-mer	Spanning translation start	2.5 nmol/2 μl per injection	Multiple injections over 2 days			Inhibits decrease in body temperature	WEISS et al. 1997b

Table 2. Continued

Receptor type	Species	Administration site and delivery system	Oligo structure (type and length)	mRNA target site	Oligo concentration	Treatment regimen	Effect on target protein	Effect on target mRNA	Effect on function	Reference
Dopamine D_2	Mouse	Corpus striatum	S 20-mer	Spanning translation start	2.5 nmol/2 μl per injection	Multiple injections over 2–4 days		Decrease	Affects D_2 behaviors (inhibits contralateral rotations in supersensitive mice; induces ipsilateral rotations in normal mice)	Zhou et al. 1996
Dopamine D_2	Mouse	i.c.v.	S 20-mer	Spanning translation start	2.5 nmol/2 μl per injection	Multiple injections over 2 days			Increases TH activity in striatum; increases ALAAD activity in striatum, olfactory tubercule and frontal cortex	Hadjiconstantinou et al. 1996
Dopamine D_2	Rat	i.c.v.	S 19-mer	Spanning translation start	10 μg/hr	Infusion via Alzet osmotic minipump			Affects D_2 behaviors (inhibits quinpirole-induced locomotion, inhibits spontaneous locomotion, induces catalepsy)	Zhang and Creese 1993
Dopamine D_2	Rat	Substantia nigra (via implanted cannula)	S 17-mer	Translation start (nucleotides 1–17)	0.04 nmol/dose	5 doses every 12 h	Decrease		Induces contralateral rotations in response to cocaine; enhances striatal release of dopamine in response to electrical stimulation	Silvia et al. 1994

Target	Species	Route	Type	Length	Location	Dose	Duration	Effect		Outcome	Reference
Dopamine D_3	Mouse	i.c.v.	S	19-mer	Coding region (nucleotides 3–22)	2.5 nmol/2 μl per injection	Multiple injections over 7 days			Increase in locomotion	Weiss et al. 1996
Dopamine D_3	Rat	i.c.v.	S	15-mer	Spanning translation start	10 μg/h	5-day infusion via subcutaneously implanted Alzet osmotic minipumps	Decrease		Increases DA synthesis	Nissbrandt et al. 1995
$GABA_A$ (τ2 subunit)	Rat	Hippocampus subcutaneously implanted Alzet micro-osmotic pump	S	18-mer	Spanning the translation start	1.7 μg/μl	3 or 5-day infusion	Decrease		Induces neuronal cell death	Karle et al. 1995
NMDA-R1	Rat	i.c.v.	N	18-mer	Coding region (nucleotides 4–21)	80–120 nmol	2- or 3-day infusion	Decrease	No change	Reduces cell death	Wahlestedt et al. 1993
NMDA-R1	Rat	i.c.v.	N	18-mer	Coding region (nucleotides 4–21)	60 nmol	2 days			Protects from brain injury	Sun and Faden 1995
Serotonin 5-HT6	Rat	i.c.v.	S	18-mer	Translation start site	24–48 μg	2 days		Decrease	Induces yawning, stretching, and chewing	Bourson et al. 1995

ACh, acetylcholine; ALAAD, aromatic L-amino acid decarboxylase; DA, dopamine; GABA, γ-aminobutyric acid; i.c.v., intracerebroventricular; N, natural (unmodified) oligodeoxynucleotides; S, phosphorothioate-modified oligodeoxynucleotides; TH, tyrosine hydroxylase.

nucleotides might give rise to toxic or nonspecific effects at high doses (GURA 1995; STEIN and CHENG 1993).

II. In Vitro and In Vivo Substrates for Studying Antisense Oligodeoxynucleotides

Other issues to consider in studies with antisense oligodeoxynucleotides which are targeted to neurotransmitter receptors are related to the cell type expressing the targeted neurotransmitter receptor. Ideally, in in vitro studies these cells should grow easily in culture, express sufficient levels of the targeted receptor, and have an easily assayable function mediated by the studied receptor. This, however, is usually not the case with primary cultures of neuronal cells. Generally, primary neuronal cultures can be obtained in relatively small amounts and are particularly sensitive to the lack of serum in the culture medium (which might be necessary if unmodified oligodeoxynucleotides are used) and potentially toxic substances (such as cationic lipids or high concentrations of oligodeoxynucleotides). This is why continuous cell lines stably transfected with different subtypes of neurotransmitter receptors have been developed (e.g., HEK 293 human embryonic kidney cells, CHO cells, Ltk⁻ fibroblast cells) which express the relevant receptor at high levels and can be propagated easily in culture for biochemical and functional assays (FILTZ et al. 1993, 1994; ZHANG et al. 1994; JOHANSSON and WESTLIND-DANIELSON 1994). Although such cell lines can be very useful in determining the specificity of antisense oligodeoxynucleotides, they can have potential drawbacks. Since these cell lines are usually not derived from neuronal cells, they may not be suitable for testing the biological functions mediated by the receptor. In addition, in the central nervous system, the biological functions mediated by a particular neurotransmitter depend on complex interactions between several neurotransmitter systems and, therefore, the results obtained from in vitro systems are not always applicable in vivo.

In order for an animal species to be an appropriate in vivo model for targeting neurotransmitter receptors using antisense oligodeoxynucleotides, the animal should express the neurotransmitter receptor of interest, the cDNA sequence of this receptor must be known, and there must be a suitable assay to determine any changes in the levels of the receptors (as the function that the newly-discovered receptors subserve may not always be known). If the purpose of the studies is to evaluate the potential of antisense oligodeoxynucleotides as therapeutic agents, it would be desirable that an animal model of the human disease of interest exists. Rodents have been used most often as in vivo models in studies with antisense oligodeoxynucleotides. They have proved to be very suitable models because many neurotransmitter receptors expressed in rodent brain have been cloned, and the pattern of their expression has been well characterized [e.g., the regional, developmental, and age-related expression of D_2 dopamine receptors in mice have been extensively characterized (WEISS et al. 1990, 1992)]. Moreover, some neurotransmitter receptors have been initially cloned from rodents [e.g. the rat D_2 dopamine receptor (Bunzow et al. 1988)] and have been subsequently used to identify the homologous receptor in humans. Among the receptors cloned thus far, the neurotransmitter receptors expressed in rodent brain share a very high degree of nucleotide and amino acid sequence homology with their human counterparts (Mack et al. 1991). They also have a similar anatomical distribution and often are linked to identical signal transduction pathways.

In our laboratory mice have been the preferred species for studying antisense oligodeoxynucleotides to the dopamine receptors because their smaller size requires the use of smaller amounts of oligodeoxynucleotides. This is an important practical issue to be considered as oligodeoxynucleotides are expensive. The function of dopaminergic receptors has also been well characterized in this species.

III. Delivery Systems

Another issue related to the use of antisense oligodeoxynucleotides is the means of delivering these compounds into cells. Cationic lipids such as lipofectin have been used widely to enhance the uptake of nucleic acids (including antisense oligodeoxynucleotides) in many cell lines and in in vivo preparations (Felgner et al. 1987, 1995; Cao et al. 1995; Ono et al. 1990; Lewis et al. 1996; Thierry et al. 1992). From Tables 1 and 2 it is evident, however, that only isolated studies have been performed using lipofectin (Silvia et al. 1994), and in most cases the oligodeoxynucleotides have been dissolved in the cell culture medium or have been directly administered in vivo into brain. This is probably due in part to the ease with which oligodeoxynucleotides penetrate into cells and brain tissue and cells (discussed in more detail in Sect. D) and because high concentrations of lipofectin may be toxic to cells (Muller et al. 1990); primary neurons in culture may be particularly sensitive to this lipid. An important issue that has to be considered if the oligodeoxynucleotides are intended to be used to inhibit receptors in the central nervous system (CNS) in vivo is the penetration across the blood–brain barrier (see below and Agrawal et al. 1995).

D. Uptake and Distribution of Antisense Oligodeoxynucleotides

The uptake and distribution of antisense oligodeoxynucleotides has been characterized using appropriately labeled oligodeoxynucleotides (radioactive, fluorescent, or biotin labels) in different cell types grown in vitro (Holopainen and Wojcik 1993; Loke et al. 1989), including primary neuronal cells (Yu et al. 1993) and in brain tissue in vivo (Yee et al. 1994; Szklarczyk and Kaczmarek 1995; Sommer et al. 1993; Whitesell et al. 1993; Leonetti et al. 1991). These studies have shown that oligodeoxynucleotides of 15–25 nucleotides in size rapidly penetrate into living cells in quantities sufficient to exert a biological effect, probably via receptor-mediated endocytosis (LEONETTI et al. 1991; YU et al. 1993; HOLOPAINEN and WOJCIK 1993). The uptake of antisense oligodeoxynucleotides is saturable and time- and temperature-dependent. For example, in embryonic chicken sympathetic neurons maintained in vitro, the amount of a 15-mer unmodified oligodeoxynucleotide antisense to the nicotinic acetylcholine receptor subunit $\alpha 3$ that is taken up into the cells continuously increases at the early time periods following the addition in the culture media (1–12 h) and gradually reaches a plateau at about 20 h (YU et al. 1993). The maximal amount of undegraded oligodeoxynucleotide is observed by 6 h. At 48 h of incubation with the oligonucleotide the total uptake of oligodeoxynucleotide is maximal and intact oligodeoxynucleotide can still be detected in the cells (YU et al. 1993). Another study showed that a 15-mer unmodified oligodeoxynucleotide antisense to the second membrane spanning region of muscarinic m_2 and $GABA_B$ receptors was readily taken up into primary cultures of cerebellar granule cells, the highest uptake being observed during the first 24 h after addition to the cell culture (HOLOPAINEN and WOJCIK 1993).

The uptake, distribution, and stability of antisense oligodeoxynucleotides in brain has been characterized in several laboratories (ZHANG et al. 1996; YEE et al. 1994; SZKLARCZYK and KACZMAREK 1995; SOMMER et al. 1993; WHITESELL et al. 1993; LEONETTI et al. 1991). These studies have concluded that after intracerebral administration, the oligodeoxynucleotides easily penetrate into various brain regions, the degree of penetration being dependent on the administration site. For example, in studies from our laboratory (ZHANG et al. 1996) we found that a 20-mer fluorescein isothiocyanate (FITC)-labeled phosphorothioate D_2 antisense oligodeoxynucleotide penetrates into many brain areas after intracerebroventricular injection, both in proximity to and well removed from the lateral ventricle. Thus, the oligodeoxynucleotide was found in the corpus striatum, septal nuclei, hippocampal pyramidal cells and dentate granular cells, thalamus, hypothalamus, central gray, superior and

Fig. 1A–D. Distribution of intracerebroventricularly administered fluorescein isothiocyanate-labeled D_2 antisense oligodeoxynucleotide (D_2 AS-FITC) in mouse brain. Mice were injected intracerebroventricularly with D_2 AS-FITC (2 nmol/2 µl) once daily for 2 days. At 24 h following the last injection, the mice were perfused intracardially with 4% paraformaldehyde, the brains were removed, and serial coronal cryostat sections (16 mm) were cut at different levels. The figure shows that there was differential uptake of the antisense in different brain regions, brain regions that contained cell bodies (e.g., ST, SP, AC, SC) evidencing higher fluorescent signal intensity than areas that consisted primarily of white matter (e.g., CC; corpus callosum). *ST*, striatum; *SP*, septum; *CC*, corpus callosum; *AC*, anterior commissure; *CA*, CA pyramidal cell layer of hippocampus; *TH*, thalamus; *HYP*, hypothalamus; *DG*, dentate gyrus; *SC*, superior colliculus; *IC*, inferior colliculus; *CG*, central gray matter; *SN*, substantia nigra; *CB*, cerebellum. (From ZHANG et al. 1996)

Pharmacological Inhibition of Dopaminergic 275

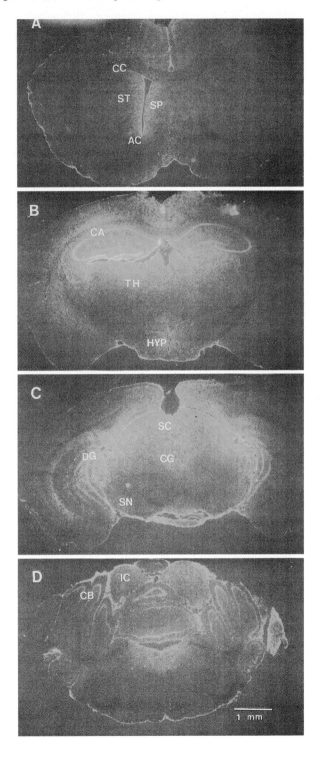

inferior colliculi, some areas of the cerebellum, the cerebral cortex, and the nucleus accumbens. Particularly high signals were found in those brain regions in direct or close proximity to the ventricles (Fig. 1). In another study (SOMMER et al. 1993), a 15-mer FITC-labeled oligodeoxynucleotide antisense to c-*fos* was injected into the corpus striatum of rats and was found to be distributed over large parts of the neostriatum. The corpus callosum seemed to be a barrier for further diffusion.

Another conclusion reached from the studies on the uptake of antisense oligodeoxynucleotides in brain is that the oligodeoxynucleotides are differentially taken up into the different cell types (ZHANG et al. 1996; YEE et al. 1994; SZKLARCZYK and KACZMAREK 1995). The oligodeoxynucleotides are found both in the cell bodies and the nucleus, and there appears to be a different subcellular distribution of the antisense oligodeoxynucleotides. For example, we consistently observed a greater amount of fluorescent signal in brain areas rich in cell bodies compared with areas of the brain consisting predominantly of white fiber tracts such as the corpus callosum and anterior commissure (Fig. 1; ZHANG et al. 1996). Furthermore, oligodeoxynucleotides are taken up into neuronal and nonneuronal cells such as ependymal cells (ZHANG et al. 1996), astrocytes, pericytes, and macrophages (YEE et al. 1994). As expected, the amount of D_2 antisense that is taken up into brain cells increases with increasing the number of injections. Thus, we found a markedly greater fluorescent signal in brains of mice receiving four injections of FITC-D_2 antisense oligodeoxynucleotide compared with that in animals receiving a single injection (ZHANG et al. 1996). Insofar as their stability is concerned, the phosphorothioated antisense oligodeoxynucleotides appear to remain unmetabolized for approximately 1 day, then become degraded (ZHANG et al. 1996; YU et al. 1993; HOLOPAINEN and WOJCIK 1993; SZKLARCZYK and KACZMAREK 1995).

Although the antisense oligodeoxynucleotides are rapidly taken up and well distributed in brain after intracerebral administration, a problem which remains to be overcome before they can be applied therapeutically is their relative inability to penetrate into brain from peripheral sites (AGRAWAL et al. 1995).

E. Mechanism of Action of Antisense Oligodeoxynucleotides

I. Mechanism of Inhibition of Gene Expression

It is known that antisense oligodeoxynucleotides may inhibit the expression of the targeted protein by one or a combination of several different mechanisms: they can interfere with transcription; they can inhibit several steps in the processing of the primary transcript (capping, methylation, polyadenylation, and splicing) or the nucleo-cytoplasmic transport of the primary transcript; they can directly inhibit translation by hybridization arrest; and they can activate the degradation of the RNA/DNA hybrids by RNase H (HELENE and TOULME 1990). Due to the different possible mechanisms of action, the reduction in the levels of a neurotransmitter receptor may not always be accompanied by a corresponding reduction in the levels of the targeted transcript (CHIANG et al. 1991; LANDGRAF et al. 1995; WAHLESTEDT et al. 1993). The absolute requirement, however, in confirming that the biological effect observed with an antisense oligodeoxynucleotide is due to a true antisense effect would be to determine that it reduces the levels of the targeted receptor, although a quantitative correspondence between the observed effects and the changes in the levels of receptor is dependent upon many factors. Namely, the turnover of the receptors may be slow and, therefore, it may take a long time for changes in the levels of the receptors to be detected; there may be a relatively small pool of functional receptors, the reduction of which causes pronounced changes in function; the behaviors used to measure the receptor function are frequently influenced by the interaction of several neurotransmitter systems localized in different brain regions.

II. Appropriate Controls to Determine the Specificity of the Antisense Effects

Regardless of the mechanism by which the antisense oligodeoxynucleotides block the expression of the targeted neurotransmitter receptor, it is necessary to determine that the observed changes are indeed the result of a specific antisense effect and are not due to nonspecific or toxic effects. This requires that a number of controls be included in the experiments in order to prove that the antisense oligodeoxynucleotides act by a true antisense mechanism: the inclusion of sense, mismatch, and random oligodeoxynucleotides in the studies; the determination whether the antisense oligodeoxynucleotide has any effect on the expression or function of closely related proteins (e.g., another subtype of neurotransmitter receptor from the same receptor family); the determination whether the antisense oligodeoxynucleotide has any effect on the expression of nonrelated proteins (e.g., changes in the expression of a housekeeping gene). These issues will be discussed in greater detail below and specific examples will be given in the section describing the applications of antisense oligodeoxynucleotides to study the dopamine receptors in the CNS.

III. Functional, Biochemical, and Molecular Consequences of Antisense Inhibition

The specific changes induced by the antisense oligodeoxynucleotides may occur at several different levels: functional (e.g., behaviors, signal transduction events, electrophysiological changes), biochemical (receptor levels), and molecular (mRNA levels). In all antisense experiments it is important to determine the effects of antisense treatment on these parameters and, whenever possible, to determine whether a positive correlation exists between the functional, biochemical, and molecular changes observed. For example, if an antisense oligodeoxynucleotide to the D_2 dopamine receptor subtype reduces the levels of D_2 dopamine receptors and the levels of D_2 receptor mRNA, and affects behaviors mediated by the D_2 receptor, but does not have effects on any of these parameters measured for the D_1 dopamine receptor, it is likely that the observed effects are the result of a true antisense mechanism. These controls are of particular importance if the antisense oligodeoxynucleotides are intended not only as tools to study the function of neurotransmitter receptors, but are intended as therapeutic agents to block the function of neurotransmitters in the CNS.

IV. Antisense Oligodeoxynucleotides Inhibit a Pool of Functional Neurotransmitter Receptors

Numerous studies have shown that there is a quantitative discordance between the reduction in the levels of receptors induced by the antisense oligodeoxynucleotides in vivo and the alterations in receptor-mediated behaviors, in that there were greater effects of the antisense oligodeoxynucleotides on the functions mediated by the receptors in comparison with the effect on the levels of the receptors. For example, repeated treatment with a D_2 antisense oligodeoxynucleotide caused a large reduction in rotational behavior in response to challenge injections with the D_2/D_3 agonist quinpirole and a statistically significant, but comparatively small reduction in the levels of the D_2 receptors in the dorsolateral portion of the striatum (WEISS et al. 1993; ZHOU et al. 1994). In another study, continuous infusion of a D_2 antisense oligodeoxynucleotide in rats caused a complete blockade of quinpirole-induced locomotor behavior, whereas the D_2 dopamine receptors were reduced by about 50% in the corpus striatum and by 70% in the nucleus accumbens (ZHANG and CREESE 1993). Similarly, intranigral administration of a D_2 antisense oligodeoxynucleotide in rats produced a several-fold increase in cocaine-induced rotational behavior, whereas it reduced the D_2 receptors by 57% in the substantia nigra (SILVIA et al. 1994). An antisense oligodeoxynucleotide to the 5-HT_6 receptor when administered intracerebroventricularly into rats produced a several-fold increase in yawning behavior, but only a 30% reduction in the levels of 5-HT_6 receptors (BOURSON et al. 1995).

To provide one possible explanation for these discrepancies, we hypothesized (QIN et al. 1995; WEISS et al. 1997b) that there are different pools of receptors: a large pool of nonfunctional receptors (perhaps located in the cytosol) and a relatively small pool of functional receptors that are membrane-associated and likely coupled to G-proteins (Fig. 2). Furthermore, it is possible that the functional receptors turn over more rapidly than the nonfunctional receptors. Therefore, it may be necessary to inhibit the synthesis of only the

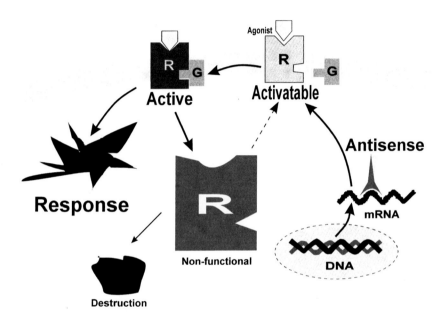

Fig. 2. Life cycle of a neurotransmitter receptor. The figure depicts the hypothesized different states of a neurotransmitter receptor and the site of action of antisense oligodeoxynucleotides. The newly-synthesized receptors translated from the mRNA (activatable receptors) are inserted into the cell membrane where they are functionally coupled to G-proteins and become the pool of functionally active receptors upon interaction with the agonist; this active form of the receptor then produces its characteristic response. These receptors then invaginate into the interior of the cell becoming part of a relatively large pool of nonfunctional receptors which are eventually destroyed. It is possible that a small portion of the nonfunctional receptors are reinserted into the cell membrane and become part of the pool of activatable receptors (*dotted line*). All forms of the receptor (activatable, active, and nonfunctional) are detected by the usual receptor ligand binding assays. We hypothesize that an antisense oligodeoxynucleotide, by blocking the synthesis of the activatable pool of receptors, greatly reduces the function of the receptors without producing a proportional change in the total levels of receptors, as measured by receptor ligand binding assays. *Heavy arrows* and *light arrows* suggest relatively rapid or slow rates, respectively. The relative sizes of the different pools of receptors are suggested by the sizes in the diagram. The scheme attempts to explain how a comparatively small change in the total number of receptors can produce marked changes in their function. (From WEISS et al. 1997b)

relatively small pool of functional receptors in order to profoundly inhibit the biological functions mediated by the receptors. Because this pool of functional receptors comprises only a small percentage of the total pool of receptors, the reductions in their number would be difficult to detect using the receptor radioligand-binding techniques (which do not discriminate between functional and nonfunctional receptors).

To obtain experimental evidence for this hypothesis, we first inhibited the total pool of D_2 dopamine receptors in mice by administering the irreversible D_2 antagonist fluphenazine-N-mustard (FNM) (WINKLER et al. 1987; QIN et al. 1994; CHEN et al. 1994). In mice this treatment induces catalepsy and inhibition of the stereotypy in response to challenge injections of quinpirole. The recovery of D_2 receptors and D_2-mediated behaviors (disappearance of the cataleptic response, and the appearance of stereotypic behavior after challenge injections with quinpirole) would be a result of the recovery of the pool of newly-synthesized receptors. Measuring the relative rates at which the levels and the function of the D_2 receptors recover would provide information on whether the newly-synthesized receptors constitute the functional pool. Following the injection of FNM we inhibited the synthesis of D_2 dopamine receptors by intraventricular injections of a D_2 antisense oligodeoxynucleotide and determined the rates of recovery of D_2-mediated behaviors and the rates of recovery of D_2 dopamine receptors. We found that there was a better correlation between the recovery of dopamine receptor-mediated behaviors and the levels of newly-synthesized receptors than there was between the dopamine-mediated behaviors and the total complement of receptors (QIN et al. 1995).

The study using FNM to inactivate the total pool of D_2 receptors and a D_2 antisense oligodeoxynucleotide to preferentially block the synthesis of D_2 receptors is an example of how antisense oligodeoxynucleotides have been valuable tools in solving fundamental problems in neurotransmitter receptor biology. Based on this data, as well as on data showing the role of G-proteins in neurotransmitter receptor function, we developed a model of the life-cycle of the D_2 dopamine receptor (WEISS et al. 1997b). According to this hypothesis, the newly-synthesized functional receptors are inserted into the plasma membrane where they are able to interact with the G-protein subunits. At some point, perhaps after being activated by a dopaminergic agonist, these receptors are internalized into the cytosol where they become a part of the relatively large pool of slowly-turning-over nonfunctional receptors (see Fig. 2). Based on this model one would predict that events that increase the synthesis of D_2 receptors, such as treatment with 6-hydroxydopamine (MANDEL et al. 1993; CREESE et al. 1977), would preferentially increase the functional pool of receptors, resulting in a large increase in function as opposed to a relatively small increase in the total pool of receptors. Conversely, inhibiting the synthesis of D_2 receptors by means of D_2 antisense oligodeoxynucleotide would cause a relatively large decrease in the number of functional receptors, and therefore marked changes in the biological function subserved by these receptors, but only a relatively small decrease in the total number of receptors.

F. In Vitro and In Vivo Effects of Antisense Oligodeoxynucleotides Targeted to the Transcripts Encoding Neurotransmitter Receptors

Studies of the effects of antisense oligodeoxynucleotides targeted to the transcripts encoding several neurotransmitter receptors are summarized below to illustrate the different neurobiological problems that have been addressed with antisense technology. These studies have also yielded relevant information related to the potential therapeutic application of antisense oligodeoxynucleotides as novel agents to reduce the function of neurotransmitter receptors in brain.

I. Acetylcholine Receptors

Molecular cloning studies have revealed that the muscarinic and nicotinic acetylcholine receptors each consist of several distinct receptor subtypes and subunits encoded by homologous genes. Antisense oligodeoxynucleotides have been employed to study both muscarinic and nicotinic acetylcholine receptors using in vitro and in vivo systems (Tables 1 and 2). In vitro, *Xenopus* oocytes have been used as a model system to investigate the patterns of expression and the physiological responses mediated by the m_1-like and m_3-like muscarinic receptors (HUMPHREY et al. 1993; MATUS-LEIBOVITCH et al. 1992). These giant cells provide a convenient model system to investigate the characteristics of intrinsic and acquired cell membrane receptors because they allow biochemical and electrophysiological manipulations at the single-cell level, and because the receptors that they express couple to an intrinsic signal transduction pathway. *Xenopus* oocytes have been shown to exhibit two types of muscarinic membrane electrical responses which have several qualitative and quantitative differences. Antisense oligodeoxynucleotides to m_1- or m_3-like receptor transcripts were used to determine whether the different types of muscarinic responses are due to the expression of different muscarinic receptor subtypes (DAVIDSON et al. 1991). These studies showed that the microinjection of antisense oligodeoxynucleotides into *Xenopus* oocytes caused specific inhibition of the synthesis of the corresponding receptor subtypes and significantly decreased muscarinic responses at 2.5–5 h after the microinjection (DAVIDSON et al. 1991). Taken together, the studies have shown (a) that the majority of oocyte donors express m_3-like muscarinic receptors and only a small fraction of the donors express m_1-like receptors (DAVIDSON et al. 1991) and (b) that the m_1 and m_3 receptor isoforms can be expressed at different densities at the animal and vegetal hemispheres of *Xenopus* oocytes where they mediate qualitatively different physiologic responses (MATUS-LEIBOVITCH et al. 1992).

The acetylcholine nicotinic receptors, as well as other neurotransmitter-gated ion channels [e.g., GABA receptors, *N*-methyl-D-aspartate (NMDA) receptors] are multimeric proteins composed of several subunits encoded by homologous genes (DENERIS et al. 1991; MEGURO et al. 1992; MONYER et al. 1992; BURT and KAMATCHI 1991; MACDONALD and OLSEN 1994; SMITH and OLSEN 1995). Antisense oligodeoxynucleotides have been useful in delineating the patterns of expression and the functional role of the individual subunits of the acetylcholine nicotinic receptor in neurons (LISTERUD et al. 1991; YU et al. 1993). These studies have also yielded valuable information on the mechanism of uptake and distribution of antisense oligodeoxynucleotides in neuronal cells (YU et al. 1993).

The relationship among different neurotransmitter receptors has also been studied using antisense oligodeoxynucleotides (HOLOPAINEN and WOJCIK 1993). A 15-mer unmodified antisense oligodeoxynucleotide that binds to a highly homologous amino acid sequence (LACADL) in the second membrane-spanning region of receptors containing seven transmembrane spanning regions (7TMR), has been added to cerebellar granule neuron cultures. This treatment decreased the total number of muscarinic receptor binding sites by about 40% and completely inhibited the expression of muscarinic m_2 receptors as determined by the inhibition of cyclic AMP formation. This treatment also caused a decrease in the levels of $GABA_B$ receptors and prevented the $GABA_B$-mediated inhibition of cyclic AMP formation. These effects appeared to be specific since another 15-mer antisense oligodeoxynucleotide sequence having four mismatches with the 7TMR-specific oligodeoxynucleotide did not have an effect on either m_2 muscarinic or $GABA_B$ receptor-mediated responses.

An antisense approach has also been used to study the physiological and pharmacological functions of m_1 muscarinic receptors in vivo (ZANG et al. 1994). Intraventricular infusion of a 16-mer phosphorothioate oligodeoxynucleotide to the m_1 receptor in rats for 3 days decreased the density of muscarinic receptors in m_1, but not m_2 receptor-rich brain regions. Locomotor activity of the treated rats, as measured by recording photocell interruptions in a photocell cage, was significantly higher than in the control saline-treated rats. These observations are consistent with reports showing hyper-locomotion after administration of the nonselective muscarinic receptor antagonist scopolamine (ZANG et al. 1994).

II. Adrenergic Receptors

Antisense oligodeoxynucleotide technology has not been used widely in studies on adrenoreceptors in the CNS. In a recently reported study (MIZOBE et al.

1996), antisense oligodeoxynucleotides specific for the α_{2A} and α_{2C} adrenoreceptor subtypes have been used in order to determine which receptor subtype mediates the sedative/hypnotic properties of anesthetic drugs. Initially, the antisense oligodeoxynucleotides were added to stably transfected cell lines (PC124D for rat α_{2A} and NIH3T3 for rat α_{2C} adrenoreceptors) to confirm their specificity of action. These antisense oligodeoxynucleotides were then administered into the locus coeruleus of chronically cannulated rats, and their effect on the hypnotic response to dexmedetomidine (an α_{2A} agonist) was determined. Only the α_{2A} antisense oligodeoxynucleotides significantly changed the hypnotic response, causing both an increase in the latency to, and a decrease in the duration of, the loss of the righting reflex following dexmedetomidine, thereby suggesting a role for the α_{2A} adrenoreceptor in the onset of hypnosis.

III. Dopamine Receptors

Dopamine is one of the major neurotransmitters in brain and regulates a number of behavioral, motor, and endocrine events in the CNS through its interaction with one of several subtypes of receptor proteins. In mediating these events, the dopaminergic system has been shown to interact with other neurotransmitter systems, including cholinergic (Dawson et al. 1990; Wang et al. 1993; Zhou et al. 1993), glutamatergic (Carlsson and Carlsson 1990), GABAergic (Ebadi and Hama 1988), and various peptidergic systems (Ebadi and Hama 1988; Garver et al. 1991). The dopamine receptors have been divided into two subfamilies, the D_1-like and the D_2-like subfamilies, on the basis of molecular biological criteria, pharmacological criteria, and second messenger coupling (Kebabian and Calne 1979; Gingrich and Caron 1993; Sibley et al. 1993; Sokoloff and Schwartz 1995). The D_1-like subfamily is comprised of the D_1 and D_5 receptor subtypes and the D_2-like subfamily is comprised of the D_2, D_3, and D_4 receptor subtypes (Gingrich and Caron 1993; Sibley et al. 1993; Sokoloff and Schwartz 1995). While the gene structure of the dopamine receptor subtypes has been well characterized, the functional properties of the individual receptor subtypes are not well understood. Moreover, although the D_1-like subfamily has been positively linked to the activation of adenylate cyclase, and the D_2-like subfamily has been negatively linked to this enzyme, the signal transduction pathways mediated by the individual receptor subtypes have not been characterized in detail.

The elucidation of the molecular biological structure of the dopamine receptor subtypes has made the investigation of the distribution of transcripts encoding these receptors possible. In turn, the information on the expression patterns of the dopamine receptor transcripts has allowed studies using dopamine receptor antisense oligodeoxynucleotides to investigate the function of the dopamine receptor subtypes. The D_1 and D_2 subtypes are most abundantly expressed in brain and are found both in nigrostriatal and mesolimbic areas. For example, high levels of D_1 mRNA have been found in the caudate putamen, nucleus accumbens, and olfactory tubercule (WEINER et al. 1991; MANSOUR et al. 1990). The D_2 mRNA has also been found to be highly concentrated in the caudate putamen (particularly in the dorsolateral and ventrolateral portions), nucleus accumbens, and olfactory tubercule, with lower levels in globus pallidus, substantia nigra, and ventral tegmental area (WEISS et al. 1990). In addition, it has been detected in the pituitary gland, in agreement with the neuroendocrine function of the dopaminergic system (O'MALLEY et al. 1990). Unlike the distribution of D_1 and D_2 mRNA, the D_3, D_4, and D_5 receptor mRNAs have a more restricted localization in brain, being found predominantly in limbic regions. Thus, D_3 mRNA is highly expressed in the olfactory tubercule, nucleus accumbens, and islands of Calleja, with relatively lower levels in the corpus striatum (SOKOLOFF et al. 1990). It has also been detected in other limbic areas such as the hippocampus, septum, and mammillary nuclei of the hypothalamus (SOKOLOFF et al. 1990; BOUTHENET et al. 1991). This distribution suggests the importance of D_3 receptors for cognitive, emotional, and endocrine functions. The D_4 receptor mRNA has been localized to mesocorticolimbic areas, i.e., frontal cortex, midbrain, hypothalamus, and medulla (VAN TOL et al. 1991). High levels of the D_5 receptor mRNA have been found in the hippocampus, hypothalamus, and mammillary and pretectal nuclei (SUNAHARA et al. 1990; TIBERI et al. 1991).

The mRNAs encoding the dopamine receptor subtypes have a differential expression during ontogeny (CHEN and WEISS 1991) and generally decrease with advanced age (WEISS et al. 1992). The dopamine receptor transcripts are also differentially regulated both physiologically (WEISS et al. 1992) and pharmacologically (CHEN et al. 1993; WEISS et al. 1990).

Elucidation of the functions of the different subtypes of dopamine receptor has important therapeutic implications, since disturbances of dopaminergic neurotransmission have been associated with a number of psychiatric and neurological disorders such as schizophrenia (SEEMAN 1987, 1993; REYNOLDS 1996), Parkinson's disease (CARLSSON and PIERCEY 1995), tardive dyskinesia (SEEMAN 1988; CASEY 1991; JESTE and CALIGIURI 1993), and Huntington's chorea (CARLSSON and PIERCEY 1995). In addition, increasing evidence supports the involvement of dopamine in the addictive properties of cocaine, amphetamine, alcohol, nicotine, and opiates via the activation of positive reinforcement pathways (NEVO and HAMON 1995; KUHAR et al. 1991; KOOB et al. 1992; MCMILLEN 1983).

The functional role that the individual dopamine receptors subserve has been examined using classical pharmacological receptor agonists and antagonists (STARR et al. 1995; ZHANG et al. 1994), and several dopamine antagonists have been used to treat disorders associated with dopaminergic hyperactivity (CREESE et al. 1976; NORDSTROM et al. 1993; KANE 1995). However, with the discovery of the multiple dopamine receptor subtypes, the selectivity of many of these agents has been questioned. For example, the antagonist SCH 23390, which was believed to be selective for the D_1 receptors, was found to have high affinity for the D_5 receptors as well (SUNAHARA et al. 1990). Quinpirole, which was used as a prototypical D_2 agonist, was shown to possess high affinity for the D_3 receptors (SOKOLOFF et al. 1990). Although newer pharmacological ligands, which bind with high selectivity to individual dopamine receptor subtypes have been developed (SEEMAN et al. 1997; MELTZER 1994), the most selective approach to block the functions of individual dopamine receptors is by means of antisense compounds.

Most of the studies using antisense oligodeoxynucleotides to the dopamine receptor have been performed in vivo since the main purpose of these studies has been to explore the biological functions of the various receptor subtypes. However, several studies have also been performed in vitro to examine the ability of antisense oligodeoxynucleotides to block the expression of dopamine receptors in cell lines expressing high levels of the studied receptor, or to investigate the signal transduction mechanisms mediated by the dopamine receptor. These studies, which will be summarized below, have not only yielded important general information on the design and application of antisense oligodeoxynucleotides targeted to the transcripts encoding neurotransmitter receptors, but have added new information concerning the specific behaviors mediated by distinct dopamine receptor subtypes, and have even suggested a basis for the potential development of molecular therapeutic agents for the treatment of neuropsychiatric disorders associated with dopaminergic hyperactivity.

1. In Vivo Models for Studying the Effects of Dopamine Receptor Antisense Oligodeoxynucleotides

Several years ago studies were initiated in our laboratory on the behavioral and molecular correlates of dopaminergic responses using antisense oligodeoxynucleotides to the D_1, D_2, and D_3 dopamine receptor subtypes. Two models were most frequently employed in our studies: mice with unilateral 6-hydroxydopamine (6-OHDA) lesions of the corpus striatum (WEISS et al. 1993; ZHOU et al. 1994, 1996), a treatment that renders the animals dopaminergically supersensitive, and normal mice, i.e., nonlesioned (ZHANG et al. 1994; ZHOU et al. 1996). In the first model, unilateral injections of 6-OHDA into the striatum of mice destroy the dopaminergic nerve terminals and create a denervation supersensitivity on the lesioned side. These animals exhibit contralateral rotational behavior in response to acute challenge injections of D_1-like (e.g., SKF 38393) and D_2-like (e.g., quinpirole) dopamine receptor agonists as well as to muscarinic cholinergic (oxotremorine) agonists (ZHOU et al. 1994). Accordingly, using this model, it is possible to examine the specificities of the antisense oligodeoxynucleotides by determining their ability to inhibit behaviors induced by a variety of neurotransmitter receptor agonists. In the second model, treating normal mice with D_1-like or D_2-like dopamine agonists elicits characteristic behaviors which can subsequently be blocked by the administration of antisense oligodeoxynucleotides. For example, treatment of normal mice with the D_1 receptor agonist SKF 38393 induces grooming behavior which can be blocked by the administration of a D_1 antisense oligodeoxynucleotide (ZHANG et al. 1994). Normal mice have also been used to determine the effects of antisense oligodeoxynucleotides on changes in autonomic functions such as body temperature, which are believed to be mediated by the dopamine receptor, described below.

2. Design and Synthesis of Antisense Oligodeoxynucleotides Targeted to the Dopamine Receptor Intended for In Vivo Use

All dopamine receptor antisense oligodeoxynucleotides developed in our laboratory, including the D_1, D_2, or D_3 receptor antisense oligodeoxynucleotides, were 20-mers, phosphorothioate-modified, and targeted to a region bridging the translation start. Thus, when comparing the specificity of the effects of the different dopamine receptor antisense agents, differences in the effects of the antisense due to different target sites on the respective transcripts are minimized. The controls for all dopamine antisense oligodeoxynucleotides (i.e., D_1, D_2, and D_3) were phosphorothioate-modified random oligodeoxynucleotides, which had the same bases as the corresponding antisense, but placed in a random sequence. All oligonucleotides were compared with the known cDNA sequences in the Gene Bank database using the GenePro program (Bainbridge Island, WA), in order to assure there was no significant sequence homology with other known sequences. The oligodeoxynucleotides were synthesized using standard phosphoroamidate procedures and were purified by reverse-phase high-performance liquid chromatography. The oligodeoxynucleotides were dissolved in sterile artificial cerebrospinal fluid (CSF). For the nucleotide sequences of the different dopamine antisense and random oligodeoxynucleotides refer to WEISS et al. (1993, 1996), ZHOU et al. (1994), and ZHANG et al. (1994).

The following sections will detail some of the results obtained from in vivo studies using antisense oligonucleotides to the dopamine receptor. They also provide a brief summary of some in vitro studies using cultured cell lines, all of which have been depicted in Tables 1 and 2.

3. D_1 Dopamine Receptor Antisense

a) In Vivo Studies

The D_1 antisense used in our laboratory was targeted to nucleotides −8 to +12 of the D_1 receptor cDNA sequence. This antisense was administered intracerebroventricularly (2.5nmol/2μl) into 6-OHDA-lesioned mice twice daily for 2 days. The control mice received similar injections of vehicle (2μl of artificial CSF) or a D_1 random oligodeoxynucleotide (2.5nmol/2μl). Contralateral rotational behavior in response to acute challenge injections of SKF 38393, quinpirole, and oxotremorine was measured after the last injection of the antisense oligodeoxynucleotide or the random control oligodeoxynucleotide. Intracerebroventricular treatment with the D_1 antisense (and not the controls) caused a blockade of D_1-mediated rotational behavior, but it did not block the rotational behavior induced by challenge injections of a D_2 dopamine agonist or a muscarinic cholinergic agonist (Fig. 3B; ZHANG et al. 1994). Thus, these experiments demonstrated that the D_1 antisense effect was specific for the D_1 receptor.

Multiple injections of the D_1 antisense were also administered intracerebroventricularly (at the same dose indicated above) in normosensitive mice (ZHANG et al. 1994). After the last injection, the mice received acute challenge injections of D_1 and D_2 dopaminergic agonists. These studies showed that D_1 antisense treatment blocked the grooming behavior induced by SKF 38393, but did not block the stereotyped behavior induced by quinpirole. The reduction in grooming was related to the amount and length of time the D_1 antisense was given, with significant reductions in grooming behavior observed within 2 days of repeated injections of the D_1 antisense (ZHANG et al. 1994). There was complete recovery from the inhibition of SKF-induced grooming behavior after cessation of the treatment with the D_1 antisense. This showed that the effect of the D_1 antisense was reversible and that continuous treatment did not induce any adaptive changes in the D_1 dopamine receptor. The observed behavioral effects of the D_1 antisense in vivo were correlated with the effects on the levels of D_1 receptors (WEISS et al. 1996). Mice were continuously infused with D_1 antisense or vehicle (artificial CSF) into the lateral cerebral ventricles using Alzet osmotic minipumps for 5 days. The levels of D_1 receptors were determined by autoradiography using [^{3}H] SCH 23390 as a ligand. It was found that the D_1 antisense caused a reduction in the levels of D_1 receptors in the corpus striatum, nucleus accumbens, and olfactory tubercule (WEISS et al. 1996).

It should be noted that after the application of D_1 antisense in vivo (as well as after the injection of any of the other oligodeoxynucleotide compounds tested in our laboratory), some nonspecific behavioral effects were observed initially. These effects included a type of barrel rolling and increased locomotor activity. However, those effects were short-lasting (less than 4 h after the first injection; ZHANG et al. 1994) and appeared to diminish after repeated injections. To avoid these initial nonspecific behaviors, in all studies we have waited at least 10–12 h before determining dopamine receptor-mediated behaviors, a time at which no unusual behaviors were apparent. Further proof for the specificity of action of the dopamine receptor antisense compounds used was obtained from behavioral (lack of effects on behaviors mediated by other closely related dopamine receptor subtypes) and biochemical (reduction in the levels of only the targeted receptor and receptor mRNA) studies.

From our studies using the D_1 antisense in vivo, the following conclusions, which were confirmed with the studies using antisense oligodeoxynucleotides

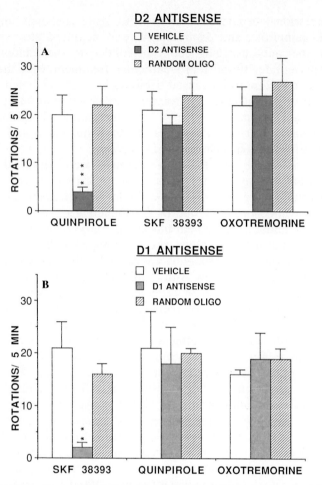

Fig. 3A,B. Effect of intraventricular administration of D_2 antisense and D_1 antisense oligodeoxynucleotides on rotational behavior induced by various neuroreceptor agonists in 6-hydroxydopamine (6-OHDA)-lesioned mice. Mice with unilateral striatal 6-OHDA lesions were administered intraventricular injections of vehicle, a random oligodeoxynucleotide (2.5 nmol) and either a D_2 antisense oligodeoxynucleotide (2.5 nmol) (**A**), or a D_1 antisense oligodeoxynucleotide (2.5 nmol) (**B**) twice daily for 2 days. Contralateral rotational behavior induced by challenge injections of the D_2 dopamine agonist quinpirole, the D_1 dopamine agonist SKF 38393, or the muscarinic cholinergic agonist oxotremorine was measured 10 h after the last injection of vehicle or oligomers. Each column represents the mean value from four to seven mice. *Vertical brackets* indicate the S.E., ** = $p < 0.01$, *** = $p < 0.001$, compared with vehicle for random oligodeoxynucleotide-treated mice. The data show that treatment with the D_2 and D_1 antisense oligodeoxynucleotides caused selective blockade of D_2 and D_1 agonist-mediated behaviors, respectively, in dopaminergically-supersensitive mice. (From WEISS et al. 1997b)

to other dopamine receptor subtypes, could be made: Antisense oligodeoxynucleotides targeted to the transcripts encoding the dopamine receptor subtypes can be used to specifically block the functions mediated by individual subtypes of this receptor; the effects of the antisense agents are completely reversible upon cessation of the antisense treatment; the effects of the antisense are dose- and time-dependent. These studies also suggested that antisense oligodeoxynucleotides may be used to uncover whether dopamine agonists that have multiple behavioral actions produce their effects on the same or different types of the dopamine receptor. Examining the effect of selective dopamine antisense oligodeoxynucleotides on various behaviors produced by a dopamine agonist should provide the answer to the question of which subtype of dopamine receptor is responsible for each behavior.

The fact that the antisense effect lasted for only a relatively short period of time and that multiple injections of the antisense were necessary in order to achieve changes in behavior and receptors, suggested that multiple applications of dopamine antisense would be necessary in vivo for therapeutic purposes. This could be a major obstacle to the therapeutic use of oligodeoxynucleotides in neuropsychiatric disorders, since these compounds do not readily penetrate into the CNS from peripheral sites (AGRAWAL et al. 1995).

b) In Vitro Studies

The effects of antisense oligodeoxynucleotides to the D_1 receptor have been studied in vitro in Ltk$^-$ fibroblast cells stably transfected with the D_{1a} receptor (D_{1a} has been used in this paper to denote the rat analog of the human D_1 receptor; YU et al. 1996) using a sense oligodeoxynucleotide as a control. Stimulation of these cells with the selective D_1 agonist fenoldopam (SKF 82526) caused an increase in phospholipase C (PLC) activity, whereas this treatment did not effect the PLC activity in nontransfected Ltk$^-$ cells. Fenoldopam also increased the PLC-τ1 protein isoform in a time-dependent manner. Since the currently available D_1 antagonists (e.g., SKF 83742) do not have receptor subtype specificity, the involvement of the D_{1a} dopamine receptor subtype in these changes was determined using a D_{1a} antisense oligodeoxynucleotide. D_1 antisense or sense oligonucleotides were incubated with Ltk$^-$ cells transfected with the D_1a receptor, for 2 days. The cells were then treated with fenoldopam. The increase in fenoldopam-induced PLC-τ protein was blocked by the D_{1a} antisense. The observed effect was specific because it was not observed with the sense control oligonucleotides; the sense oligonucleotides rather seemed to increase the levels of PLC-τ1 protein. Moreover, the reduced response to fenoldopam correlated with a reduced expression of the D_{1a} receptors in the transfected Ltk$^-$ cells as determined by immunocytochemistry using anti-D_{1a} antibodies (YU et al. 1996). Subsequent, more detailed studies of these effects provided evidence that the D_{1a}-mediated stimulation of PLC occurs as a result of protein kinase A (PKA) activation, and that PKA then stimulates PLC-τ in the cytosol and membrane via activation of PKC (Yu et al. 1996).

4. D_2 Dopamine Receptor Antisense

a) In Vivo Studies

The D_2 antisense used in our studies was targeted to nucleotides -10 to $+10$ of the D_1 receptor cDNA sequence. Initially, the antisense was administered intracerebroventricularly into 6-OHDA-lesioned mice according to the same regimen as that described for the D_1 antisense (see Fig. 3). The control mice received similar injections of vehicle or a D_2 random oligodeoxynucleotide. Intracerebroventricular treatment with the D_2 antisense (but not with the controls) caused a blockade of D_2-mediated rotational behavior revealed by challenge injections with quinpirole. The reduction in quinpirole-induced rotational behavior was related to the amount and length of time the antisense was given and was reversible within 2 days of cessation of the antisense treatment (Zhou et al. 1994). The D_2 antisense effect was specific because it did not block the rotational behavior induced by challenge injections of a D_1 or muscarinic cholinergic agonists (Zhou et al. 1994; Fig. 3A). As noted earlier, similar injections of a D_1 antisense blocked rotational behavior induced by the D_1 agonist but not that induced by a D_2 agonist or by a muscarinic cholinergic agonist (Fig. 3B).

Intracerebroventricular treatment of normal mice with a D_2 antisense oligodeoxynucleotide, administered according to the same regimen as that described for the D_1 antisense, blocked the stereotypic behavior in response to challenge injections of quinpirole. This effect was specific because D_2 antisense treatment did not have an effect on the grooming behavior in response to challenge injections of SKF 38393 (Weiss et al. 1993; Zhou et al. 1994). To examine further the specificity of the antisense treatment, mice were administered the D_1 antisense oligodeoxynucleotide and similar behavioral measurements were made. In this case, treatment with the D_1 antisense blocked D_1 agonist-induced grooming behavior but not D_2 agonist-induced stereotypy (Zhang et al. 1994). Taken together, the specific inhibition of D_2-mediated behaviors by the D_2 antisense and the specific inhibition of D_1-mediated behaviors by the D_1 antisense in two different model systems (supersensitive and normo-sensitive mice) strongly suggest that the effects of the D_2 and D_1 antisense oligodeoxynucleotides are specific for their respective target transcripts.

To determine further the specificity of the D_2 antisense effects, the changes in the levels of D_2 dopamine receptors and the D_2 dopamine receptor mRNA were determined (Zhou et al. 1994). Repeated treatment with the D_2 antisense significantly reduced the levels of D_2 dopamine receptors and D_2

dopamine receptor mRNA, but not the levels of D_1 receptors or D_1 receptor mRNA, in the dorsolateral areas of the lesioned striatum. It should be emphasized that the degree of reduction in the levels of D_2 receptors in a portion of the striatum was relatively small in comparison with the large reduction in D_2 dopamine-mediated behavior (ZHOU et al. 1994). As already discussed in our consideration of the mechanism of action of antisense oligodeoxynucleotides

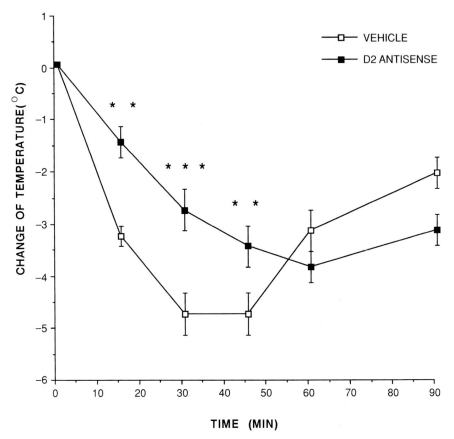

Fig. 4. Effect of intraventricular administration of a D_2 antisense oligodeoxynucleotide on the reduction of body temperature induced by quinpirole in mice. Mice were administered intracerebroventricular injections of vehicle or D_2 antisense (2.5 nmol) three times at 12-h intervals. Quinpirole (5 μmol/kg, s.c.) was administered 10 h after the last injection of D_2 antisense, and rectal temperatures were measured at varying times thereafter. Each point represents the average change in body temperature taken from nine to 11 mice. *Vertical brackets* indicate the S.E., ** = $p < 0.01$, *** = $p < 0.001$ compared with values from vehicle-treated mice. The data show that treatment with the D_2 antisense oligodeoxynucleotide significantly reduced the decrease in body temperature induced by the D_2 dopamine receptor agonist quinpirole. (From WEISS et al. 1997b)

(see Sect. E), we believe this apparent discrepancy is explained by our hypothesis that the antisense agents inhibit the synthesis of a relatively small pool of functionally-active D_2 receptors (QIN et al. 1995; WEISS et al. 1997b).

The D_2 antisense was administered intracerebroventricularly into normal mice in order to study the participation of D_2 receptors in another dopamine-mediated event, the regulation of body temperature (WEISS et al. 1997b). The rationale for this study was that D_2 agonists decrease body temperature (FAUNT and CROOKER 1987). Figure 4 shows that injections of D_2 antisense into the lateral ventricle of mice blocked the decrease in body temperature induced by quinpirole, supporting the view that D_2 receptors are involved in the regulation of body temperature (WEISS et al. 1997b).

The effects of D_2 antisense compounds have also been studied following their administration into discrete brain regions (ZHOU et al. 1996; SILVIA et al. 1994). Local administration of antisense compounds in brain would be useful to inhibit neurotransmitter receptors expressed in discrete brain areas, as intracerebroventricularly administered antisense oligodeoxynucleotides penetrate rapidly into many brain regions (ZHANG et al. 1996). Antisense compounds have also been administered locally into rat hippocampus in order to study the $\tau 2$ subunit of the $GABA_A$ receptors (KARLE et al. 1995). In our studies, a D_2 antisense was repeatedly administered into one of the corpus striatum of 6-OHDA-lesioned mice or normal mice (ZHOU et al. 1996). In the 6-OHDA-lesioned mice, intrastriatally administered D_2 antisense into the lesioned striatum specifically and reversibly blocked only D_2-mediated contralateral rotational behavior (induced by parenteral challenge injections of quinpirole).

Following repeated intrastriatal injections of D_2 antisense into normal mice, parenteral administration of quinpirole caused rotational behavior ipsilateral to the side in which the D_2 antisense was injected. This behavior was specific since it was not observed when the mice were similarly challenged with SKF 38393 or oxotremorine. Repeated administration of the D_2 antisense also caused a significant reduction in the levels of D_2, but not D_1, dopamine receptors in striatum, as determined by receptor autoradiography (ZHOU et al. 1996). These studies demonstrate that it is possible to use dopamine antisense compounds to uncover behaviors mediated by other subtypes of the dopamine receptor.

Studies on the biochemical effects produced when the synthesis of D_2 receptors is inhibited by repeated administration of D_2 antisense oligodeoxynucleotides revealed that intracerebroventricular administration of the D_2 antisense in mice increased the activity of tyrosine hydroxylase in the corpus striatum, but not in the olfactory tubercule or in the frontal cortex, whereas the activity of aromatic L-amino acid decarboxylase increased in all three brain regions (HADJICONSTANTINOU et al. 1996). The treatment with the D_2 antisense also elevated the mRNA levels for the two enzymes in the midbrain (HADJICONSTANTINOU et al. 1996). The effects of D_2 antisense treatment on the enzyme activities or their respective mRNAs were not observed after similar injections of a random oligodeoxynucleotide or vehicle. Taken together, these observations suggest that treatment with the D_2 antisense, by interfering with the synthesis and function of the D_2 receptors, upregulates the expression of the synthetic enzymes for dopamine. The reason for the differential effects on enzyme activity in various brain regions is not presently clear. One possible explanation is that tyrosine hydroxylase and aromatic L-amino acid decarboxylase are regulated differentially in these particular brain regions.

Other in vivo studies using D_2 antisense oligodeoxynucleotides (ZHANG and CREESE 1993; SILVIA et al. 1994) support further the possibility of specifically blocking the expression of D_2 receptors in order to characterize D_2-mediated events. Thus, intracerebroventricular infusion of a D_2 antisense into normo-sensitive rats inhibited quinpirole-induced locomotor activation without altering the grooming behavior induced by SKF 38393 (ZHANG and CREESE 1993). Furthermore, D_2 antisense treatment caused the appearance of behaviors characteristic of the inhibition of D_2 receptors, i.e., catalepsy and reduced spontaneous locomotor activity. These D_2-specific behaviors were accompanied by a corresponding reduction of the levels of D_2 receptors in striatum. That no significant changes in the levels of D_1, muscarinic cholinergic, or 5-HT_2 serotonergic receptors were noted attests to the specificity of action of the D_2 antisense (ZHANG and CREESE 1993).

A D_2 antisense approach has also been taken in order to examine the functions of nigrostriatal D_2 autoreceptors (SILVIA et al. 1994). After unilateral D_2 antisense treatment into the substantia nigra of rats, cocaine administration resulted in marked rotational behavior contralateral to the treated side. This finding was interpreted to be consistent with the loss of autoreceptor-mediated feedback inhibition and subsequently greater dopamine release on the treated side. Thus, it was shown that D_2 antisense treatment led to a reduction in nigral D_2 autoreceptor binding sites (but no changes in striatal D_2 receptor binding sites) and to an enhanced release on dopamine in striatum in response to electrical stimulation. Interestingly, there was no spontaneous rotational behavior in the D_2 antisense-treated mice, suggesting that the autoreceptor effects are masked by compensatory mechanisms during normal behavior.

b) In Vitro Studies

Studies have also been performed using D_2 antisense in vitro (Table 1). In the study of SILVIA et al. (1994) described above, the WERI-27 retinoblastoma cell line, which normally expresses D_2 receptors, was used initially to determine the efficacy and specificity of D_2 antisense oligodeoxynucleotides intended for in vivo use. Treatment of WERI-27 cells with a 17-mer D_2 antisense reduced the levels of D_2 receptors by 57% after 3 days, an effect which was not observed after parallel treatment with a 17-mer sense or a G-C content mismatched oligodeoxynucleotide.

A D_2 antisense oligodeoxynucleotide was also added to primary cultures of rat pituitary cells in order to study the role of D_2 receptors in the pituitary gland (VALERIO et al. 1994). This study provided further support for the hypothesis that the D_2 receptors in the pituitary are negatively coupled to adenylate cyclase and that they are functionally linked to the inhibition of prolactin synthesis. They also showed that prolactin release might be controlled, in part, by a dopamine receptor that is distinct from the D_2 dopamine receptor subtype.

5. D_3 Dopamine Receptor Antisense

The usefulness of an antisense oligodeoxynucleotide strategy for the study of the D_3 dopamine receptor subtype has been demonstrated in vivo. This receptor subtype is of particular interest since there is relatively little data concerning its function and because it is localized largely to limbic regions of the brain. Since the bulk of the available evidence suggests that D_3 receptors mediate locomotor activity, we examined the effect on locomotion by treating mice with a D_3 antisense oligodeoxynucleotide (WEISS et al. 1996). For this purpose quinpirole (a D_2/D_3 dopamine agonist) was infused continuously into mice via subcutaneously implanted Alzet minipumps, a treatment that produces significant locomotor behavior within 2 days. This behavior is not blocked by acute injections of the D_2 dopamine receptor antagonist sulpiride and continues for the 6 days of infusion (WEISS et al. 1996). To determine whether this locomotor behavior might have resulted from an alteration of the D_3 receptors in the nucleus accumbens, mice were administered intracerebroventricular injections of D_1, D_2, or D_3 antisense, vehicle, or random oligodeoxynucleotides for 7 days (WEISS et al. 1996). After 2 days of oligodeoxynucleotide treatment, the mice were infused continuously with quinpirole or vehicle via Alzet minipumps. Figure 5 shows that treating mice with the D_3 antisense, but not with the D_1 or D_2 antisense, significantly increased the locomotor behavior seen at 2, 3, and 4 days after continuously infusing quinpirole. These results suggest that the locomotor behavior induced by infusing quinpirole might be due to the down-regulation of D_3 dopamine receptors and that D_3 antisense may potentiate this decreased function of the D_3 dopamine receptors by reducing their synthesis.

Intraventricular infusion of a D_3 antisense into rats significantly reduced the binding of the D_2/D_3 ligand [^{3}H]-spiperone in the limbic forebrain, where the D_3 receptors are abundant, but not in the caudate putamen, where the D_3 receptors are sparse (NISSBRANDT et al. 1995). The D_3 antisense treatment also caused an increase in dopamine synthesis in the nucleus accumbens, but did not have an effect on dopamine synthesis in the caudate putamen. These results were interpreted as an indication of the possible role of D_3 receptors in the synthesis of dopamine in certain brain areas.

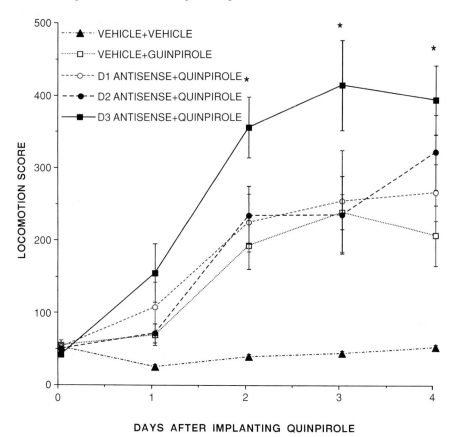

Fig. 5. Effect of intracerebroventricular administration of a D_3 antisense oligodeoxynucleotide on locomotor behavior induced by continuously infusing quinpirole in mice. Mice were administered intracerebroventricular injections of D_1, D_2, or D_3 antisense oligodeoxynucleotides (2.5 nmol), vehicle, or random oligomer twice daily for 7 days. After five such injections, mice were implanted subcutaneously with Alzet osmotic minipumps designed to deliver quinpirole at a rate of 2.5 µmol/kg per hour. Control mice were implanted with minipumps containing the vehicle. Locomotor behavior was assessed at various times after implanting quinpirole. Each point represents the mean locomotor score from five mice. *Vertical brackets* indicate the S.E., * = $p < 0.05$, compared with values obtained from mice treated with vehicle or quinpirole alone at the same time points. The data show that treatment with the D_3 antisense oligodeoxynucleotide significantly increased the locomotor behavior induced by continuously infusing quinpirole in mice. By contrast, treatment with the D_1 or D_2 antisense oligodeoxynucleotides had no significant effect on locomotor behaviors. (From WEISS et al. 1996)

IV. GABA Receptors

GABA is the principal inhibitory neurotransmitter in the CNS. The $GABA_A$ receptor belongs to the superfamily of ligand-gated ion channels and is believed to be a pentamer which is assembled from several homologous subunits.

Molecular cloning studies have revealed five major classes of receptor subunits, which may contain several isoforms (α1–6, β1–3, τ1–3, and ρ1–2). It has been shown that marked regional differences exist in the expression of the different $GABA_A$ receptor subunits in brain and that the functional and pharmacological properties of the $GABA_A$ receptor depend on the underlying subunit composition (BURT and KAMATCHI 1991; MACDONALD and OLSEN 1994; OLSEN et al. 1992; OLSEN and TOBIN 1990).

That GABA receptors play a wide and varied role in physiological and pharmacological events is attested to by the evidence that important interactions exist between the GABAergic and the dopaminergic neurotransmitter systems (EBADI and HAMA 1988), and that the $GABA_A$ receptor has been shown to be modulated by several classes of therapeutic drugs, such as benzodiazepines, barbiturates, steroids, and anesthetics (BURT and KAMATCHI 1991; MACDONALD and OLSEN 1994; LEWOHL et al. 1996). Due to the extensive heterogeneity in the structure and distribution of the $GABA_A$ receptor, antisense oligodeoxynucleotides have provided a valuable tool for analyzing the biological roles of the $GABA_A$ receptor subtypes. Antisense oligonucleotides may also provide the basis for the development of therapeutic strategies based on isoform-selective antagonists.

An antisense approach has been used to examine the spatial distribution and function of the $GABA_A$ receptor α1 and α2 subunits in cultured rat visual cortex slices (BRUSSAARD and BAKER 1995). Neurons from visual cortex slices, explanted at postnatal day 6 and maintained in a serum-free medium, evinced current clamp characteristics typical for stellate cells. Between 7 and 21 days in culture, both GABA- and glutamate-mediated postsynaptic currents were observed. Long-term culturing of these slices in the presence of a 15-mer antisense oligodeoxynucleotide directed against the transcripts of all α subunit genes of the $GABA_A$ receptor caused a dose-dependent reduction of the evoked GABA synaptic currents. This reduction was specific, because it was not observed after application of a random oligonucleotide. Moreover, there were no effects on the evoked glutamatergic excitatory postsynaptic currents. A 15-mer antisense directed to the α1 subunit reduced the amplitude of the evoked GABAergic inhibitory postsynaptic currents (IPSCs) in some cells by 50%–75%, whereas in other cells from the same slices, there was little or no effect. In contrast, an antisense oligodeoxynucleotide to the α2 subunit gave a consistent 80% reduction of the evoked GABAergic IPSCs. A 15-mer three-base mismatch oligodeoxynucleotide against α2 had no effect. From these studies it was concluded that the α2 subunit functions in postsynaptic $GABA_A$ receptors located on or close to the cell bodies of stellate cells. Although it was more difficult to make a conclusion about the role of the α1 subunit, this subunit seems spatially differentiated.

An antisense strategy has been employed in vivo in rats in order to study the τ2 subunit of the $GABA_A$ receptor (KARLE and NIELSEN 1995; KARLE et al. 1995). The τ2 receptor subunit has been shown to be important for benzodiazepine binding. Thus, only when coexpressing the τ2 subunit with α and β

subunits do the receptors exhibit high affinity binding to benzodiazepines. Intracerebroventricular bolus infusion of a τ2 antisense oligodeoxynucleotide induced a small but statistically significant reduction (9%–15%) of benzodiazepine binding in the cortex and in the striatum of rats (KARLE and NIELSEN 1995). In a related study, τ2 antisense was infused continuously by means of Alzet minipump into a defined brain region, the hippocampus (KARLE et al. 1995), as this region is regarded as important for the pathophysiology of disease states related to GABAergic neurotransmission, such as epilepsy and anxiety. Infusion of an 18-mer phosphorothioate τ2 antisense oligodeoxynucleotide into rat hippocampus induced a reduction of benzodiazepine radioligand binding by 43% compared to the sense or untreated controls, although in one experiment a mismatch oligodeoxynucleotide also induced a small but significant reduction of benzodiazepine binding compared to the untreated controls (KARLE et al. 1995). The reduction in benzodiazepine binding was paralleled by a decrease in the number of $GABA_A$ receptors as assessed by the binding of the radioligands [^{35}S]tert-butyl-bicyclophosphorothionate (TBPS)(51% reduction) and [^{3}H]muscimol (37%). Therefore, it was concluded that the τ2 antisense oligodeoxynucleotide induces a specific reduction of the τ2 subunit, probably as a result of the inhibited expression of the τ2 polypeptide. The small reduction in radioligand binding in the mismatch-treated controls may be a nonspecific effect of the mismatch oligodeoxynucleotide (KARLE et al. 1995). Behaviorally, the antisense-treated animals appeared more excited and hostile than the control animals, and structurally, the antisense-treated hippocampi exhibited an altered macroscopic appearance (swollen and pale), reduced protein content, neuronal loss and glial cell proliferation, findings which were not observed in the animals treated with the control sense or mismatch oligodeoxynucleotides. These results suggested that a reduced synthesis of $GABA_A$ receptors may lead to neuronal cell death.

The role of the α6 and τ2 subunits in the pharmacological properties of the $GABA_A$ receptor has been studied in primary cultures of rat cerebellar granule cells using antisense oligodeoxynucleotides (ZHU et al. 1996). Addition of antisense oligodeoxynucleotides to the α6 and τ2 subunits inhibited the expression of the corresponding subunit proteins as determined by Western blot analysis. These effects were specific as reduction in protein levels was not observed in the control mismatch-treated cultures. Moreover, the antisense oligodeoxynucleotides did not have an effect on the protein levels of a nontarget α2 receptor subunit. Inhibition of GABA-gated currents by furosemide, a selective inhibitor of $GABA_A$ receptors containing α6 subunits, was attenuated after the α6 antisense treatment. The effects of furosemide were examined in parallel in transfected cells expressing various combinations of the α1 and α6 subunits, yielding results which showed that the relative abundance of these subunit mRNAs determines the extent of furosemide-induced inhibition of GABA-gated currents. Compared with the control nontreated or mismatch oligodeoxynucleotide-treated cultures, treatment with the α6 antisense oligodeoxynucleotide caused a decrease in the membrane currents generated in response to activation with GABA. This treatment also enhanced flunitrazepam-induced potentiation of GABA activated currents. In contrast, the $\tau 2$ antisense increased the receptor sensitivity to GABA and decreased the response to flunitrazepam. These studies suggested that the $\alpha 6$ and $\tau 2$ subunits are crucial determinants of the pharmacological properties of the $GABA_A$ receptors in cerebellar granule cells.

V. NMDA Receptors

Glutamate is the most prevalent excitatory neurotransmitter in brain. It exerts its effects by interacting with several different receptors, which have been subdivided into two general categories: ionotropic, or ion channel-forming receptors, and metabotropic receptors, which are linked to PLC activation and adenylate cyclase inhibition. The NMDA receptors belong to the ionotropic receptors. Several NMDA receptor subunits have been identified: the NMDA-R1 receptor family contains subunits A-G and the NMDA-R2 receptor family contains the subunits A-D. The NMDA receptors appear to be heteromeric complexes formed by different combinations of the R1 and R2 subunits (Mansour et al. 1990; Monyer et al. 1992). The NMDA receptor is involved in processes related to memory formation and in neuropathological states that result from physical and physiological insults to the brain. It has been proposed that neuronal toxicity may depend on the subunit composition of the NMDA receptor.

To determine the involvement of the NMDA receptor in neuronal cell death following cerebral vascular ischemia, NMDA-R1 receptors have been inhibited with antisense oligodeoxynucleotides (Wahlestedt et al. 1993). Initially, the effectiveness of an NMDA-R1 antisense in protecting neurons against NMDA-induced neurotoxicity was determined in vitro on neuron-enriched cultures from rat embryos. On the basis of the in vitro experiments, the NMDA-R1 antisense was examined in vivo in rats for its ability to reduce the size of focal ischemic infarctions induced by occlusion of the middle cerebral artery (Wahlestedt et al. 1993). Spontaneously hypertensive rats were treated intracerebroventricularly with NMDA-R1 antisense, or sense oligodeoxynucleotides, or vehicle for 2 days. The middle cerebral artery was occluded 2 days after the first injection. The antisense-treated animals did not display any behavioral abnormalities upon gross examination. However, treatment with the NMDA-R1 antisense significantly reduced the volume of the ischemic infarction. This was correlated with a significant reduction in the levels of cortical NMDA receptors, measured by the binding of the high-affinity competitive NMDA receptor antagonist [^{3}H]CGS 19755. In contrast, the steady-state concentration of cortical NMDA-R1 mRNA was unchanged by the antisense treatment. These data were interpreted to suggest that the antisense primarily suppresses the translation of the NMDA-R1 protein and does not act through a RNase H mechanism. These experiments have provided evidence for the participation of NMDA-R1 receptors in the neurotoxicity elicited by focal cerebral ischemia and have established the feasibility of an antisense oligodeoxynucleotide strategy for the treatment of focal ischemic infarctions.

The neuroprotective role of NMDA-R1 antisense was determined in another in vivo model of brain injury in rats (SUN and FADEN 1995). NMDA-R1 antisense or sense oligodeoxynucleotides were administered intracerebroventricularly into rats prior to traumatic brain injury. Treatment with antisense oligodeoxynucleotides did not affect the physiological variables or the motor function prior to the trauma. However, this treatment significantly decreased the mortality and improved the behavioral recovery at 2 weeks after the trauma as compared to animals treated with the corresponding sense oligonucleotides. Moreover, astrocyte activation, as reflected by glial fibrillary astrocytic protein (GFAP) immunocytochemistry was significantly reduced in the antisense treated animals. Although the specificity of the antisense effect in this study was determined by comparison with sense oligodeoxynucleotide or saline vehicle-treated controls, further studies on the corresponding changes in the levels of NMDA-R1 receptors or mRNA would provide more details on the mechanism by which the antisense agent exerts its effects (e.g., does it act though an RNase H mechanism).

VI. Serotonin Receptor

The effects of the neurotransmitter serotonin, 5-hydroxytryptamine (5-HT), are mediated through a number of receptor subtypes (HUMPHREY et al. 1993; LOWENBERG et al. 1993; MATTHES et al. 1993; MONSMA et al. 1993). Among these, the more recently identified 5-HT receptors, such as the $5-HT_{5A}$, $5-HT_{5B}$, $5-HT_6$, and $5-HT_7$ receptors, have not been well characterized. For example, relatively little is known about the expression of the $5-HT_6$ receptors in brain. The $5-HT_6$ receptor has been shown to have little homology with the other 5-HT receptor subtypes. Interestingly, it has been shown to have high affinity for certain pharmacological compounds such as the atypical antipsychotic agent clozapine. However, as these compounds also interact with other neurotransmitter receptors (SEEMAN et al. 1997), it has been difficult to determine the in vivo role of the $5-HT_6$ receptor.

Antisense oligodeoxynucleotides to the $5-HT_6$ receptor have been employed to study the physiological functions of this receptor in vivo (BOURSON et al. 1995). Repeated intracerebroventricular administration of a $5-HT_6$ antisense oligodeoxynucleotide in rats induced a specific behavioral syndrome of yawning, stretching, and chewing and caused a 30% reduction in the number of [^{3}H]-lysergic acid diethylamide binding sites, a ligand often used to quantify the levels of 5-HT receptors. These effects were not observed after treatment with a random oligodeoxynucleotide. Neither the antisense nor the random oligodeoxynucleotides had any effect on other parameters measured, such as locomotor activity, body weight, food intake, body temperature, and nociception. The specific behavioral syndrome did not appear to be caused by modulation of dopaminergic neurotransmission since there were no changes in the tissue levels of either dopamine or its metabolites, and the dopamine receptor antagonist, haloperidol, did not reduce the number of yawns or stretches. However, an increase in cholinergic transmission may have been involved since the behavioral syndrome was dose-dependently antagonized by the cholinergic antagonist atropine. The results from this study suggested that the 5-HT$_6$ receptors are functionally expressed in brain and may be involved in the control of cholinergic neurotransmission.

G. Future Directions: Comparison Between the Properties of Antisense Oligodeoxynucleotides and Antisense RNA Produced by Mammalian Expression Vectors

Currently, the major drawbacks associated with the use of antisense oligodeoxynucleotides to inhibit neuroreceptors are their poor penetration into the CNS from peripheral sites and their relatively short duration of action, resulting in the need for repeated administration of these compounds in order to achieve biological effects. This can be a serious impediment to the development of these compounds as therapeutic agents for diseases of the CNS. In addition, the relative ease with which the oligodeoxynucleotides penetrate into several brain areas after intracerebral injection may present a problem when it is desirable to target a discrete brain region. Some of these issues may be obviated by the use of expression vectors which continuously synthesize antisense RNA inside the cells. A major advantage of such an approach would be the far longer duration of action achieved after a single administration of the antisense RNA expression vector. For example, in an attempt to achieve long-lasting inhibition of the D$_2$ dopamine receptor, we have administered intrastriatally into mice a eukaryotic expression vector producing an antisense RNA to the D$_2$ dopamine receptor transcript (DAVIDKOVA et al. 1997; WEISS et al. 1997c). A single injection of this D$_2$ antisense vector (complexed with a cationic lipid) produced persistent inhibition of the D$_2$ receptors which lasted for about 1 month. Another advantage of producing antisense RNA by an expression vector would be the possibility to express the antisense RNA only in certain cell types by including in the vector a tissue-specific or inducible promoter. For example, by placing a limbic system-associated protein promoter (ZACCO et al. 1990) upstream to the dopamine receptor antisense sequence, it may be possible to express the dopamine receptor antisense RNA only in neurons localized in the limbic system. This would be of special interest in view of the proposed role of mesolimbic dopamine receptors in certain neuropsychiatric disorders (e.g., schizophrenia, drug abuse).

H. Conclusions

Antisense oligodeoxynucleotides have many characteristics which render them attractive as tools to study the function of neurotransmitter receptors or as therapeutic agents to modify disorders involving neurotransmitter receptors. Among these characteristics are the rational basis with which the antisense compounds can be designed and the relative ease with which they can be synthesized and with which they can be delivered to cells in culture or brain tissue in vivo. Oligodeoxynucleotides of approximately 20-mer in size are rapidly taken up into neuronal cells in culture and into brain tissue in vivo when administered intracerebroventricularly. In brain the degree of uptake is highest near the administration site and its uptake is time- and cell type-dependent. Multiple administration results in a higher degree of uptake. In vivo phosphorothioate oligodeoxynucleotides remain stable for at least 1 day.

A significant number of studies suggest that antisense oligodeoxynucleotides can be extremely valuable tools to selectively block the expression of neurotransmitter receptors, provided that appropriate controls are carried out to exclude nonspecific effects. Thus, the specificity should be ascertained with control oligodeoxynucleotides of similar length, but differing structure, and by measuring corresponding changes in the targeted protein, targeted mRNA, and the biological functions mediated by the targeted neurotransmitter receptor. In addition, lack of any changes in these parameters measured for a closely related nontarget neurotransmitter receptor would strongly support the specificity of the antisense effects. A characteristic property of antisense oligodeoxynucleotides to neurotransmitters, which may be a consequence of their mechanism of action (i.e., the inhibition of only the pool of newly-synthesized functional receptors), is the relatively small reduction in the levels of the targeted receptor in comparison with the reduction in biological functions mediated by the receptor. This should be taken into consideration in the interpretation of data from in vivo studies examining the effects of antisense oligodeoxynucleotides targeted to the transcripts encoding neurotransmitter receptors.

Future efforts in the use of antisense strategies to reduce the synthesis of neurotransmitter receptors should be directed to improving the delivery of these compounds to selected brain areas and to developing methods for achieving prolonged effects after a single administration. Such advances may be achieved through the use of antisense RNA expression vectors housing cDNA sequences that encode antisense RNA targeted to specific neurotransmitter receptor transcripts.

Acknowledgements. This work was supported, in part, by a grant awarded by the National Institute of Mental Health, MH42148 and by funds from START Technology.

References

Agrawal S (ed) (1996) Antisense therapeutics. Humana, Totowa, NJ
Agrawal S, Temsamani J, Galbraith W, Tang J (1995) Pharmacokinetics of antisense oligonucleotides. Clin Pharmacokinet 28:7–16
Bourson A, Borroni E, Austin HR, Monsma FJ Jr, Sleight AJ (1995) Determination of the role of the 5-HT$_6$ receptor in the rat brain: a study using antisense oligodeoxynucleotides. J Pharmacol Exp Ther 274:173–180
Bouthenet M-L, Souil E, Martres M-P, Sokoloff P, Giros B, Schwartz J-C (1991) Localization of dopamine D$_3$ receptor mRNA in the rat brain using in situ hybridization histochemistry: comparison with dopamine D$_2$ receptor mRNA. Brain Res 564:203–219
Brussaard AB, Baker RE (1995) Antisense oligodeoxynucleotide-induced block of individual GABA A receptor α subunits in cultured visual cortex slices reduces amplitude of inhibitory postsynaptic currents. Neurosci Lett 191:111–115
Bunzow JR, Van Tol HHM, Grandy DK, Albert P, Salon J, Macdonald C, Machida CA, Neve KA, Civelli O (1988) Cloning and expression of a rat D$_2$ dopamine receptor cDNA. Nature 336:783–787
Burt DR, Kamatchi GL (1991) GABA (A) receptor subtypes: from pharmacology to molecular biology. FASEB J 5:2916–2923
Burt DR, Creese I, Snyder SH (1977) Antischizophrenic drugs: chronic treatment elevates dopamine receptor binding in brain. Science 196:326–328
Cao L, Zheng Z-C, Zhao Y-C, Jiang Z-H, Liu Z-G, Chen S-D, Zhou C-F, Liu X-Y (1995) Gene therapy of Parkinson disease model rat by direct injection of plasmid DNA-lipofectin complex. Human Gene Ther 6:1497–1501
Carlsson A, Piercey M (1995) Dopamine receptor subtypes in neurological and psychiatric diseases. Clin Neuropharmacol 18 [Suppl 1] (entire issue)
Carlsson M, Carlsson A (1990) Interactions between glutamatergic and monoaminergic systems within the basal ganglia-implications for schizophrenia and Parkinson's disease. Trends Neurosci 13:272–276
Casey DE (1991) Neuroleptic drug-induced extrapyramidal syndromes and tardive dyskinesia. Schizophr Res 4:109–120
Chen JF, Weiss B (1991) Ontogenetic expression of D$_2$ dopamine receptor mRNA in rat corpus striatum. Dev Brain Res 63:95–104
Chen JF, Aloyo VJ, Weiss B (1993) Continuous treatment with the D$_2$ dopamine receptor agonist quinpirole decreases D$_2$ dopamine receptors, D$_2$ dopamine receptor messenger RNA and proenkephalin messenger RNA, and increases mu opioid receptors in mouse striatum. Neuroscience 54:669–680
Chen JF, Aloyo VJ, Qin Z-H, Weiss B (1994) Irreversible blockade of D$_2$ dopamine receptors by fluphenazine-N-mustard increases D$_2$ dopamine receptor mRNA and proenkephalin mRNA and decreases D$_1$ dopamine receptor mRNA and mu and delta opioid receptors in rat striatum. Neurochem Int 25:355–366
Chiang M-Y, Chan H, Zounes MA, Freier S, Lima W, Bennett CF (1991) Antisense oligonucleotides inhibit intercellular adhesion molecule 1 expression by two distinct mechanisms. J Biol Chem 266:18162–18171
Creese I, Burt DR, Snyder SH (1976) Dopamine receptor binding predicts clinical and pharmacological potencies of antischizophrenic drugs. Science 192:481–483
Creese I, Burt DR, Snyder SH (1977) Dopamine receptor binding enhancement accompanies lesion-induced behavioral supersensitivity. Science 197:596–598
Crooke ST (1992) Therapeutic applications of oligonucleotides. Annu Rev Pharmacol Toxicol 32:329–376

Crooke ST (ed) (1995) Therapeutic applications of oligonucleotides. Landes, Austin, TX
Davidkova G, Zhang S-P, Zhou L-W, Nichols RA, Weiss B (1997) Use of antisense RNA expression vectors in neurobiology. In: Weiss B (ed) Antisense oligodeoxynucleotides and antisense RNA: novel pharmacological and therapeutic agents. CRC Press, Boca Raton, FL, pp 213–241
Davidson A, Mengod G, Matus-Leibovitch N, Oron Y (1991) Native Xenopus oocytes express two types of muscarinic receptors. FEBS Lett 284:252–256
Dawson VL, Dawson TM, Wamsley JK (1990) Muscarinic and dopaminergic receptor subtypes on striatal cholinergic interneurons. Brain Res Bull 25:903–912
Deneris ES, Connolly J, Rogers SW, Duvoisin R (1991) Pharmacological and functional diversity of neuronal nicotinic acetylcholine receptors. Trends Pharmacol Sci 29:34–38
Ebadi M, Hama Y (1988) Dopamine, GABA, cholecystokinin and opioids in neuroleptic-induced tardive dyskinesia. Neurosci Biobehav Rev 12:179–187
Faunt JE, Crooker AD (1987) The effects of selective dopamine receptor agonists and antagonists on body temperature in rats. Eur J Pharmacol 133:243–247
Felgner P, Tsai Y, Sukhu L, Wheeler C, Manthorpe M, Marshall J, Cheng S (1995) Improved cationic lipid formulations for in vivo gene therapy. Ann N Y Acad Sci U S A 772:126–138
Felgner PL, Gadek TR, Holm M, Roman R, Chan HW, Wenz M, Northrop JP, Ringold GM, Danielson M (1987) Lipofection: a novel highly efficient lipid mediated DNA transfection procedure. Proc Natl Acad Sci U S A 84:7413–7417
Filtz T, Guan W, Artymyshyn RP, Pacheco M, Ford C, Molinoff P (1994) Mechanisms of up-regulation of D_{2L} dopamine receptors by agonists and antagonists in transfected HEK-293 cells. J Pharmacol Exp Ther 271:1574–1582
Filtz TM, Artymyshyn RP, Guan W, Molinoff PB (1993) Paradoxical regulation of dopamine receptors in transfected 293 cells. Mol Pharmacol 44:371–379
Garver DL, Bissette G, Yao JK, Nemeroff CB (1991) CSF neurotensin concentrations in psychosis: relationship to symptoms and drug response. Am J Psychiatr 148:485–488
Gingrich J, Caron M (1993) Recent advances in the molecular biology of dopamine receptors. Annu Rev Neurosci 16:299–231
Gura T (1995) Antisense has growing pains. Science 270:575–577
Hadjiconstantinou M, Neff NH, Zhou L-W, Weiss B (1996) D_2 dopamine receptor antisense increases the activity and mRNA of tyrosine hydroxylase and aromatic L-amino acid decarboxylase in mouse brain. Neurosci Lett 217:105–108
Helene C, Toulme J (1990) Specific regulation of gene expression by antisense, sense and antigene nucleic acids. Biochim Biophys Acta 1049:99–125
Holopainen I, Wojcik WJ (1993) A specific antisense oligodeoxynucleotide to mRNAs encoding receptors with seven transmembrane spanning regions decreases muscarinic m_2 and gamma-aminobutyric acid$_B$ receptors in rat cerebellar granule cells. J Pharmacol Exp Ther 264:423–430
Humphrey PPA, Hartig P, Hoyer D (1993) A proposed new nomenclature for 5-HT receptors. Trends Pharmacol Sci 14:233–236
Hyttel J (1986) Effect of prolonged treatment with neuroleptics on dopamine D_1 and D_2 receptor density in corpus striatum of mice. Acta Pharmacol Toxicol 59:387–391
Jeste DV, Caligiuri MP (1993) Tardive dyskinesia. Schizophr Bull 19:303–315
Johansson MH, Westlind-Danielson A (1994) Forskolin-induced up-regulation and functional supersensitivity of dopamine D_2 long receptors expressed by Ltk$^-$ cells. Eur J Pharmacol 269:149–155
Kane JM (1995) Current problems with the pharmacotherapy of schizophrenia. Clin Neuropharmacol 18:S154–S161
Karle J, Nielsen M (1995) Modest reduction of benzodiazepine binding in rat brain in vivo induced by antisense oligonucleotide to GABA$_A$ receptor gamma2 subunit subtype. Eur J Pharmacol Mol Pharmacol 291:439–441

Karle J, Witt MR, Nielsen M (1995) Antisense oligonucleotide to $GABA_A$ receptor gamma2 subunit induces loss of neurones in rat hippocampus. Neurosci Lett 202:97–100

Kebabian JW, Calne DB (1979) Multiple receptors for dopamine. Nature 277:93–96

Koob GF, Maldonado R, Stinus L (1992) Neural substrates of opiate withdrawal. Trends Neurosci 15:186–191

Kuhar MJ, Ritz MC, Boja JW (1991) The dopamine hypothesis of the reinforcing properties of cocaine. Trends Neurosci 14:299–302

Landgraf R, Gerstberger R, Montkowski A, Probst JC, Wotjak C, Holsboer F, Engelmann M (1995) V1 vasopressin receptor antisense oligodeoxynucleotide into septum reduces vasopressin binding, social discrimination abilities, and anxiety-related behavior in rats. J Neurosci 15:4250–4258

Leonetti J, Degols G, Clarenc J, Mechti N, Lebleu B (1993) Cell delivery and mechanisms of action of antisense oligodeoxynucleotides. Prog Nucleic Acids Res Mol Biol 44:143–154

Leonetti JP, Mechti N, Degols G, Gagnor C, Lebleu B (1991) Intracellular distribution of microinjected antisense oligonucleotides. Proc Natl Acad Sci U S A 88:2702–2706

Levitt P (1984) A monoclonal antibody to limbic system neurons. Science 223:229–301

Lewis JG, Lin K-Y, Kothavale A, Flanagan WM, Matteuci MD, DePrince R, Mook R Jr, Hendren W, Wagner RW (1996) A serum-resistant cytofectin for cellular delivery of antisense oligonucleotides and plasmid DNA. Proc Natl Acad Sci U S A 93:3176–3181

Lewohl J, Crane DI, Dodd PR (1996) Alcohol, alcoholic brain damage, and GABA A receptor isoform gene expression. Neurochem Int 29:677–684

Listerud M, Brussaard A, Devay P, Colman D, Role L (1991) Functional contribution of neuronal AChR subunits revealed by antisense oligodeoxynucleotides. Science 254:1518–1521

Loke SL, Stein CA, Zhang XH, Mori K, Nakanishi M, Subasinghe C, Cohen JS (1989) Characterization of oligonucleotide transport into living cells. Proc Natl Acad Sci U S A 86:3474–3478

Lowenberg TW, Baron BM, De Lecea L, Miller JD, Prosser RA, Rea MA, Foye PE, Racke M, Slone A, Siegel BW, Danielson PE, Sutcliffe JG, Erlander MG (1993) A novel adenyl cyclase-activating serotonin receptor ($5-HT_7$) implicated in the regulation of mammalian circadian rhythms. Neuron 11:449–458

Macdonald RL, Olsen RW (1994) GABA (A) receptor channels. Annu Rev Neurosci 17:569–602

Mack KJ, Todd R, O'Malley K (1991) The mouse dopamine D_{2a} receptor gene: sequence homology with the rat and human genes and expression of alternative transcripts. J Neurochem 57:795–801

Mandel RJ, Hartgraves SL, Severson JA, Woodward JJ, Wilcox RE, Randall PK (1993) A quantitative estimate of the role of striatal D-2 receptor proliferation in dopaminergic behavioral supersensitivity: the contribution of mesolimbic dopamine to the magnitude of 6-OHDA lesion-induced agonist sensitivity in the rat. Behav Brain Res 59:53–64

Mansour A, Meador-Woodruff JH, Bunzow JR, Civelli O, Akil H, Watson SJ (1990) Localization of dopamine D_2 receptor mRNA and D_1 and D_2 receptor binding in the rat brain and pituitary: an in situ hybridization-receptor autoradiographic analysis. J Neurosci 10:2587–2600

Matthes H, Boschert U, Amlaiky N, Grailhe R, Plassat J-L, Muscatelli F, Mattei M-G, Hen R (1993) Mouse 5-hydroxytryptamine 5B receptors define a new family of serotonin receptors: cloning, functional expression and chromosomal localization. Mol Pharmacol 43:313–319

Matus-Leibovitch N, Mengod G, Oron Y (1992) Kinetics of the functional loss of different muscarinic receptor isoforms in Xenopus oocytes. Biochem J 285:753–758

McMillen BA (1983) CNS stimulants: two distinct mechanisms of action for amphetamine-like drugs. Trends Pharmacol Sci 4:429–432

Meguro H, Mori H, Araki H, Kushiya E, Kutsuwada T, Yamazaki M (1992) Functional characterization of a heteromeric NMDA receptor channel expressed from cloned cDNAs. Nature 357:70–74

Meltzer HY (1994) An overview of the mechanism of action of clozapine. J Clin Psychiatr Suppl B55:47–52

Mizobe T, Maghsoudi K, Sitwala K, Tianzhi G, Ou J, Maze M (1996) Antisense technology reveals the alpha 2A adrenoreceptor to be the subtype mediating the hypnotic response to the highly selective agonist, dexmedetomidine, in the locus coeruleus of the rat. J Clin Invest 98:1076–1080

Monsma FJ, Shen Y, Ward RP, Hamblin MW, Sibley DR (1993) Cloning and expression of a novel serotonin receptor with high affinity for tricyclic psychotropic drugs. Mol Pharmacol 43:320–327

Monyer H, Sprengel R, Schoepfer R, Herb A, Higuchi M, Lomeli H, Burnashev N, Sakman B, Seeburg P (1992) Heterotrimeric NMDA receptors: molecular and functional distribution of subtypes. Science 256:1217–1221

Muller S, Sullivan P, Clegg D, Feinstein S (1990) Efficient transfection and expression of heterologous genes in PC12 cells. DNA Cell Biol 9:221–229

Neckers L, Whitesell L (1993) Antisense technology: biological utility and practical considerations. Am J Physiol 265:L1–L12

Nevo I, Hamon M (1995) Neurotransmitter and neuromodulatory mechanisms involved in alcohol abuse and alcoholism. Neurochem Int 26:305–336

Nissbrandt H, Ekman A, Eriksson E, Heilig M (1995) Dopamine D_3 receptor antisense influences dopamine synthesis in rat brain. Neuroreport 6:573–576

Nordstrom A-L, Farde L, Wiesel F-A, Forslund K, Pauli S, Halldin C, Uppfeldt G (1993) Central D_2-dopamine receptor occupancy in relation to antipsychotic drug effects: a double-blind PET study of schizophrenic patients. Biol Psychiat 33:227–235

Olsen RW, Tobin AJ (1990) Molecular biology of GABA (A) receptors. FASEB J 4:1469–1480

Olsen RW, Bureau MH, Endo S, Smith GB, Brecha N, Sternini C, Tobin AJ (1992) GABA (A) receptor subtypes identified by molecular biology, protein chemistry and binding. Mol Neuropharmacol 2:129–133

O'Malley KL, Mack KJ, Gandelman KY, Todd RD (1990) Organization and expression of the rat D_{2A} receptor gene: identification of alternative transcripts and a variant donor splice site. Biochemistry 29:1367–1371

Ono T, Fujino Y, Tsuchiya T, Tsuda M (1990) Plasmid DNAs directly injected into mouse brain with lipofectin can be incorporated and expressed by brain cells. Neurosci Lett 117:259–263

Qin Z-H, Zhou L-W, Weiss B (1994) D_2 dopamine receptor messenger RNA is altered to a greater extent by blockade of glutamate receptors than by blockade of dopamine receptors. Neuroscience 60:97–114

Qin Z-H, Zhou L-W, Zhang S-P, Wang Y, Weiss B (1995) D_2 dopamine receptor antisense oligodeoxynucleotide inhibits the synthesis of a functional pool of D_2 dopamine receptors. Mol Pharmacol 48:730–737

Reynolds GP (1996) Dopamine receptors and schizophrenia. Biochem Soc Transact 24:202–205

Rogue P, Hanauer A, Zwiller J, Malviya AN, Vincendon G (1991) Up-regulation of dopamine D_2 receptor mRNA in rat striatum by chronic neuroleptic treatment. Eur J Pharmacol 207:165–168

Seeman P (1987) Dopamine receptors and the dopamine hypothesis of schizophrenia. Synapse 1:133–152

Seeman P (1988) Tardive dyskinesia, dopamine receptors, and neuroleptic damage to cell membranes. J Clin Psychopharmacol 8:3S–9S

Seeman P (1993) Schizophrenia as a brain disease. Arch Neurol 50:1093–1095

Seeman P, Corbett R, Van Tol HHM (1997) Atypical neuroleptics have low affinity for dopamine D_2 receptors or are selective for D_4 receptors. Neuropsychopharmacol 16:93–115

Sibley D, Monsma F, Shen Y (1993) Molecular neurobiology of D_1 and D_2 dopamine receptors. In: Waddington JL (ed) D_1:D_2 dopamine receptor interactions. Academic, San Diego, pp 1–17

Silvia CP, King GR, Lee TH, Xue Z-Y, Caron MG, Ellinwood EH (1994) Intranigral administration of D_2 dopamine receptor antisense oligodeoxynucleotides establishes a role for nigrostriatal D_2 autoreceptors in the motor actions of cocaine. Mol Pharmacol 46:51–57

Smith GB, Olsen RW (1995) Functional domains of GABA (A) receptors. Trends Pharmacol Sci 16:162–167

Sokoloff P, Schwartz J-C (1995) Novel dopamine receptors half a decade later. Trends Pharmacol Sci 16:270–275

Sokoloff P, Giros B, Martres MP, Bouthenet ML, Schwartz JC (1990) Molecular cloning and characterization of a novel dopamine receptor (D_3) as a target for neuroleptics. Nature 347:146–151

Sommer W, Bjelke B, Ganten D, Fuxe K (1993) Antisense oligonucleotide to c-fos induces ipsilateral rotational behaviour to d-amphetamine. Neuroreport 5:277–280

Starr S, Kozell L, Neve K (1995) Drug-induced up-regulation of dopamine D_2 receptors on cultured cells. J Neurochem 65:569–577

Stein CA, Cheng Y-C (1993) Antisense oligonucleotides as therapeutic agents-Is the bullet really magical. Science 261:1004–1012

Sun F-Y, Faden AI (1995) Pretreatment with antisense oligodeoxynucleotides directed against the NMDA-R1 receptor enhances survival and behavioral recovery following traumatic brain injury in rats. Brain Res 693:163–168

Sunahara RK, Niznik HB, Weiner DM, Storman TM, Brann MR, Kennerdy JL, Gelernter JE, Rozmahel R, Yang Y, Israel Y, Seeman P, O'Dowd BF (1990) Human dopamine D_1 receptor encoded by an intronless gene on chromosome 5. Nature 347:80–83

Szklarczyk A, Kaczmarek L (1995) Antisense oligodeoxyribonucleotides: stability and distribution after intracerebral injection into rat brain. J Neurosci Methods 60:181–187

Thierry AR, Rahman A, Dritschilo A (1992) Liposomal delivery as a new approach to transport antisense oligodeoxynucleotides. In: Erickson RP, Izant JG (eds) Gene regulation: biology of antisense RNA and DNA. Raven, New York, pp 147–159

Tiberi M, Jarvie KR, Silvia C, Falardeau P, Gingrich JA, Godinot N, Bertrand L, Yang-Feng TL, Fremeau RT Jr, Caron MG (1991) Cloning, molecular characterization, and chromosomal assignment of a gene encoding a second D_1 dopamine receptor subtype: differential expression pattern in rat brain compared with the D_{1A} receptor. Proc Natl Acad Sci U S A 88:7491–7495

Uhlmann E, Peyman A (1990) Antisense oligodeoxynucleotides: a new therapeutic principle. Chem Rev 544:579

Valerio A, Alberici A, Tinti C, Spano P, Memo M (1994) Antisense strategy unravels a dopamine receptor distinct from the D_2 subtype, uncoupled with adenylyl cyclase, inhibiting prolactin release from rat pituitary cells. J Neurochem 62:1260–1266

Van Tol HHM, Bunzow JR, Guan HC, Sunahara RK, Seeman P, Niznik HB, Civelli O (1991) Cloning of a human dopamine D_4 receptor gene with high affinity for the antipsychotic clozapine. Nature 350:610–614

Wahlestedt C, Golanov E, Yamamoto S, Yee F, Ericson H, Yoo H, Inturrisi CE, Reis DJ (1993) Antisense oligodeoxynucleotides to NMDA-R1 receptor channel protect cortical neurons from excitotoxicity and reduce focal ischaemic infarctions. Nature 363:260–263

Wang H-Y, Zhou L-W, Friedman E, Weiss B (1993) Differential regulation of release of acetylcholine in the striatum in mice following continuous exposure to selective D_1 and D_2 dopaminergic agonists. Neuropharmacology 32:85–91

Weiner DM, Levey AI, Sunahara RK, Niznik HB, O'Dowd BF, Seeman P, Brann MR (1991) D_1 and D_2 dopamine receptor mRNA in rat brain. Proc Natl Acad Sci U S A 88:1859–1863

Weiss B (ed) (1997) Antisense oligodeoxynucleotides and antisense RNA: novel pharmacological and therapeutic agents. CRC Press, Boca Raton, FL

Weiss B, Zhou L-W, Chen JF, Szele F, Bai G (1990) Distribution and modulation of the D_2 dopamine receptor mRNA in mouse brain: molecular and behavioral correlates. Adv Biosci 77:9–25

Weiss B, Chen JF, Zhang S, Zhou L-W (1992) Developmental and age-related changes in the D_2 dopamine receptor mRNA subtypes in rat brain. Neurochem Int 20 [Suppl]:49S–58S

Weiss B, Zhou L-W, Zhang S-P, Qin Z-H (1993) Antisense oligodeoxynucleotide inhibits D_2 dopamine receptor-mediated behavior and D_2 messenger RNA. Neuroscience 55:607–612

Weiss B, Zhou L-W, Zhang S-P (1996) Dopamine antisense oligodeoxynucleotides as potential novel tools for studying drug abuse. In: Raffa RB, Porreca F (eds) Antisense strategies for the study of receptor mechanisms. Landes, Georgetown, TX, pp 71–89

Weiss B, Davidkova G, Zhang S-P (1997a) Antisense strategies in neurobiology. Neurochem Int 31:321–348

Weiss B, Zhang S-P, Zhou L-W (1997b) Antisense strategies in dopamine receptor pharmacology. Life Sci 60:433–455

Weiss B, Davidkova G, Zhou L-W, Zhang S-P, Morabito M (1997c) Expression of D_2 dopamine receptor antisense RNA in brain inhibits D_2-mediated behaviors. Neurochem Int 31:571–580

Whitesell L, Geselowitz D, Chavany C, Fahmy B, Walbridge S, Alger JR, Neckers LM (1993) Stability, clearance, and disposition of intraventricularly administered oligodeoxynucleotides: implications for therapeutic application within the central nervous system. Proc Natl Acad Sci U S A 90:4665–4669

Wickstom E (ed) (1991) Prospects for antisense nucleic acid therapy of cancer and AIDS. Wiley-Liss, New York

Winkler JD, Thermos K, Weiss B (1987) Differential effects of fluphenazine-N-mustard on calmodulin activity and on D_1 and D_2 dopaminergic responses. Psychopharmacology 92:285–291

Yee F, Ericson H, Reis DJ, Wahlestedt C (1994) Cellular uptake of intracerebroventricularly administered biotin- or digoxigenin-labeled antisense oligodeoxynucleotides in the rat. Cell Mol Neurobiol 14:475–486

Yu C, Brussaard AB, Yang X, Listerud M, Role LW (1993) Uptake of antisense oligonucleotides and functional block of acetylcholine receptor subunit gene expression in primary embryonic neurons. Dev Genet 14:296–304

Yu P-Y, Eisner G, Yamaguchi I, Mouradian M, Felder RA, Jose PA (1996) Dopamine D_{1A} receptor regulation of phospholipase C isoform. J Biol Chem 271:19503–19508

Zacco A, Cooper V, Chantler PD, Fisher-Hyland S, Horton H-L, Levitt P (1990) Isolation, biochemical characterization and ultrastructural analysis of the limbic system-associated membrane protein (LAMP), a protein expressed by neurons comprising functional neural circuits. J Neurosci 10:73–90

Zang Z, Florijn W, Creese I (1994) Reduction in muscarinic receptors by antisense oligodeoxynucleotide. Biochem Pharmacol 48:225–228

Zhang LJ, Lachowicz JE, Sibley DR (1994) The D_{2S} and D_{2L} dopamine receptor isoforms are differentially regulated in Chinese hamster ovary cells. Mol Pharmacol 45:878–889

Zhang M, Creese I (1993) Antisense oligodeoxynucleotide reduces brain dopamine D_2 receptors: behavioral correlates. Neurosci Lett 161:223–226

Zhang S-P, Zhou L-W, Weiss B (1994) Oligodeoxynucleotide antisense to the D_1 dopamine receptor mRNA inhibits D_1 dopamine receptor-mediated behaviors in normal mice and in mice lesioned with 6-hydroxydopamine. J Pharmacol Exp Ther 271:1462–1470

Zhang S-P, Zhou L-W, Morabito M, Lin RCS, Weiss B (1996) Uptake and distribution of fluorescein-labeled D_2 dopamine receptor antisense oligodeoxynucleotide in mouse brain. J Mol Neurosci 7:13–28

Zhou L-W, Zhang S-P, Connell TA, Weiss B (1993) AF64A lesions of mouse striatum result in ipsilateral rotations to D_2 dopamine agonists but contralateral rotations to muscarinic cholinergic agonists. J Pharmacol Exp Ther 264:824–830

Zhou L-W, Zhang S-P, Qin Z-H, Weiss B (1994) In vivo administration of an oligodeoxynucleotide antisense to the D_2 dopamine receptor mRNA inhibits D_2 dopamine receptor-mediated behavior and the expression of D_2 dopamine receptors in mouse striatum. J Pharmacol Exp Ther 268:1015–1023

Zhou L-W, Zhang S-P, Weiss B (1996) Intrastriatal administration of an oligodeoxynucleotide antisense to the D_2 dopamine receptor mRNA inhibits D_2 dopamine receptor-mediated behavior and D_2 dopamine receptors in normal mice and in mice lesioned with 6-hydroxydopamine. Neurochem Int 29:583–595

Zhu WJ, Wang JF, Vicini S, Grayson DR (1996) Alpha 6 and gamma 2 subunit antisense oligodeoxynucleotides alter gamma-aminobutyric acid receptor pharmacology in cerebellar granule neurons. Mol Pharmacol 50:23–33

Zon G (1995) Brief overview of control of genetic expression by antisense oligonucleotides and in vivo applications. Mol Neurobiol 10:219–229

CHAPTER 10
Pharmacological Effects of Antisense Oligonucleotide Inhibition of Immediate-Early Response Genes in the CNS

B.J. CHIASSON, M.O. HEBB, and H.A. ROBERTSON

A. Introduction

Stimuli that activate the cells of the central nervous system (CNS) can have permanent or semi-permanent effects on the functioning of the brain. In many cases the stimuli responsible for this change in brain function also activate transcription factors (TF), some of which are of the immediate-early gene (IEG) family. Stimuli of both physiological and pathophysiological significance have been shown to activate the prototypical IEG, c-*fos*. Consequently, studies attempting to examine the role of IEGs in the CNS abound. In this chapter we describe studies which have associated IEGs with brain function and demonstrate the emerging role that antisense technology has played in this field and other areas of CNS pharmacology.

B. Some Correlative Studies Suggesting a Role of Immediate-Early Genes in Brain Function

Addictive psychostimulant drugs, such as amphetamine and cocaine, are pharmacological stimuli which lead to long-term changes in brain function, as well as overt behavioural modification. These drugs produce activation of IEGs (e.g. c-*fos*, *jun*-B, *egr*-1) in what are believed to be the neural substrates of addiction (BERETTA et al. 1993; GRAYBIEL et al. 1990; MORATELLA et al. 1992, 1993; for a review, see NESTLER et al. 1993), suggesting a potential link between IEGs and the process of addiction. In animal models of neuroplasticity, such as kindling, long-term potentiation (LTP) and long-term depression (LTD), activation of IEG TFs is observed following delivery of the inducing stimuli (ABRAHAM et al. 1994; CHIASSON et al. 1995; COLE et al. 1989; DRAGUNOW and ROBERTSON 1987; DRAGUNOW et al. 1988; SINOMATO et al. 1991; TESKEY et al. 1991; WISDEN et al. 1990; for reviews see DRAGUNOW et al. 1989 and ROBERTSON 1992a). In all of these experimental models of plasticity the correlations between IEG activation and the extent of plasticity are relatively strong. Likewise, non-invasive stimuli, such as a simple light pulse given to animals in a darkened environment, will also activate IEGs. Gene expression under these darkened conditions is restricted to the suprachiasmatic nucleus (SCN), an area of the brain believed to be largely responsible for circadian rhythms.

List of Abbreviations

AD	Afterdischarge
cAMP	Cyclic AMP
CCK	Cholecystokinin
CNS	Central nervous system
CREB	cAMP Response element binding protein
CSF	Cerebrospinal fluid
CTA	Conditioned taste aversion
Enk	Preproenkephalin
FBJ-MSV	Finkel-Biskis-Jinkins murine osteosarcoma virus
Fos-LI	Fos-like immunoreactivity
GABA	γ-Aminobutyric acid
GAD	Glutamic acid decarboxylase
IEG	Immediate-early gene
LTD	Long-term depression
LTP	Long-term potentiation
NGF	Nerve growth factor
NMDA	N-methyl-D-aspartate
NPY	neuropeptide Y
PKA	Protein kinase A
PKC	Protein kinase C
SCN	Suprachiasmatic nucleus
SNAP-25	Synaptosomal-associated protein 25
TF	Transcription factor
TH	Tyrosine hydroxylase
TRH	Thyrotropin-releasing hormone

Interestingly, when exposed to a light pulse during their subjective night, these animals demonstrated a shift in their circadian rhythms (Rusak et al. 1990). These are only a few examples which illustrate the association between stimuli that lead to changes in brain function and the activation of IEGs in the CNS. In these examples an important finding has been that the activation of IEGs (or the extent thereof) is stimulus-linked and that the cells demonstrating induction of IEGs are generally accepted to play a role in the process being studied. Thus, IEG activation demonstrates both a temporal and spatial correlation with brain-altering stimuli.

Observations such as these have led investigators to speculate that inducible TFs, such as c-*fos*, initiate the molecular events leading to medium- and long-term changes (hours to years) in brain function. However, until recently the evidence for this had been circumstantial (Dragunow et al. 1989; Robertson 1992a; Sheng and Greenberg 1990). Studies performed in vitro substantiate the notion that the Fos/Jun dimers regulate the expression of various neuropeptides and trophic molecules, such as nerve growth factor (NGF), and thus provide evidence that IEGs may be involved in neuroplasticity (HENGERER et al. 1990; SONNENBERG et al. 1989). However, the consequences of either acutely or chronically altering the expression of IEG TFs in vivo are far less obvious, but strong predictions can be made based on our current knowledge of the molecular cascades involving IEGs. The following text provides a brief background on the molecular basis of stimulus-transcription coupling, the mechanism by which IEG TF may effect the functioning of the CNS.

C. Immediate-Early Genes and Stimulus-Transcription Coupling

I. A Model for Change

The study of stimulus-transcription coupling within the nervous system is relatively recent and represents a major breakthrough in our way of perceiving the workings of the brain. Largely based on earlier work performed by investigators in several fields of medical research, a possible mechanism was proposed to account for long-term information storage in the CNS (for a more historical account, see CURRAN and MORGAN 1995; HUGHES and DRAGUNOW 1995; CHIASSON et al. 1997). The mechanism suggested that activation of growth factor/neurotransmitter receptors through their respective intracellular cascades could lead to the rapid induction of cellular genes (many of which are inducible TFs). These early genes would subsequently activate late-effector genes encoding for proteins involved in aspects of cell function capable of endowing long-term storage effects, such as alterations in the level of structural proteins of the synapse (BERRIDGE 1986; GOELET et al. 1986). It should be noted that the general idea that protein synthesis is important in the formation of memory is not a recent belief (for a review, see DAVIS and SQUIRE 1984). Based on this hypothesis, investigators examined whether traditional neurotransmission could alter gene expression. These early studies demonstrated that IEGs, such as c-*fos*, were activated in the brain and spinal cord under various conditions including drug-induced seizures, kindling-induced seizures and noxious stimulation (DRAGUNOW and ROBERTSON 1987; HUNT et al. 1987; MORGAN et al. 1987). Thus, the genetic material of the cell appeared to be accessible to manipulation by what might be thought of as day-to-day events. Therefore, genes may play a role in brain function in a way not previously anticipated by many neuroscientists. Thus, a search for genes which responded rapidly and transiently (i.e. IEG) to stimulation was begun by neuroscientists interested in neuroplasticity, since these genes might represent candidates regulating such phenomena. Although many investigators rapidly focused their attention on the study of IEG TFs, most of these genes were discovered in cell culture systems and often for reasons very different from those envisaged by researchers interested in brain function.

II. The Origin of Immediate-Early Genes

The identification of the c-*fos* gene itself resulted from studies examining the neoplastic properties of the Finkel-Biskis-Jinkins murine osteosarcoma virus (FBJ-MSV; CURRAN and TEICH 1982; CURRAN et al. 1987). The 55-kD phosphoprotein, Fos, was believed to have transforming ability when associated with other proteins. In particular, Jun, a 39-kD protein coded by the cellular proto-oncogene counterpart of avian sarcoma virus 17, was eventually identified as the low molecular-weight partner of Fos (CURRAN 1988; MAKI et al. 1987; MORGAN and CURRAN 1991). Jun and Fos proteins dimerize to form the AP-1 TF complex via a leucine zipper motif (CURRAN and FRANZA 1988; MORGAN and CURRAN 1991). Both the c-*fos* and c-*jun* gene families have other members which display the ability to form AP-1 complexes and alter transcriptional activity.

Members of the Fos family such as *fra*-1 (COHEN and CURRAN 1988), *fos*-B (ZERIAL et al. 1989) and *fra*-2 (NISHINA et al. 1990) and members of the Jun family, *jun*-B (LAU and NATHANS 1987; RYDER et al. 1988) and *jun*-D (HIRAI et al. 1989; RYDER et al. 1989) have been cloned and sequenced. Other IEGs which have received considerable attention were isolated as growth factor-stimulated candidate competence genes (genes involved in cell cycle events) and were obtained by differential cDNA cloning. These genes include *krox* 20 and *krox* 24 (LEMAIRE et al. 1988). The latter is also termed *zif*268 (CHRISTY et al. 1988), *ngfi*-a (MILBRANDT 1987) or *egr*-1 (SUKHATME et al. 1988). As a group these genes are now often termed immediate-early genes. This nomenclature was introduced when it became obvious that these genes were also expressed in conditions not involving the cell cycle. The nomenclature originates in virology in which viral genes are defined as early or late depending on whether their expression occurs before or after replication of the viral genome. Moreover, a set of viral genes is expressed rapidly or "immediately" after infection of a cell even in the presence of protein synthesis inhibitors. Thus, the term "viral immediate-early gene" (e.g. v-*fos*) was adopted and modified to "cellular immediate-early genes" (e.g. c-*fos*) for cellular genes which are rapidly induced in the presence of protein synthesis inhibition.

III. The Immediate-Early Gene as an Inducible Transcription Factor

IEGs belong to an inducible class of TFs. Thus, they are not usually constitutively expressed, but rather they must be transcriptionally activated by external signals impinging on the cell. The IEG c-*fos* is activated by constitutively expressed TFs such as the cyclic AMP (cAMP) response element binding protein (CREB) which is, in turn, activated by phosphorylation following a cascade of second messenger signals coupled to the external stimuli (SHENG et al. 1990; HUGHES and DRAGUNOW 1995). The protein products of these inducible TFs then re-enter the nucleus to stimulate a second bout of transcriptional activity. It is this second phase of transcriptional activity which is thought to endow neurons with long-lasting changes in cellular function. These IEG TFs include the Fos family (c-*fos*, *fra*-1, *fra*-2, *fos*-B (long), *fos*-B (short; also known as Δ *fos*-B), the Jun family (c-*jun*, *jun*-B, *jun*-D) and the zinc finger-containing genes (*krox* 20 and *krox* 24). While some of these genes demonstrate constitutive expression in particular neurons and cell lines (implicating additional functions), they all have the capacity to be upregulate by appropriate stimulation [for an excellent review of IEGs in the CNS see HUGHES and DRAGUNOW (1995)].

IV. Immediate-Early Gene Transcription Factors and Their Targets

It is now well known that Fos/Jun dimers form a complex called the AP-1 TF which transactivates gene expression through the AP-1/TPA responsive element (TRE). Formation of the dimeric complex occurs via a leucine zipper motif which consists of four to five regularly spaced leucine residues found at every seventh position along an α-helical structure (KOUZARIDES and ZIFF 1988; McKNIGHT 1991). Substitution of two or more of these leucine residues completely eliminates dimer formation. The AP-1/TRE site is also known as the dyad symmetry consensus sequence having the following octameric composition, ATGACTCA. When the AP-1 complex binds to this DNA sequence it results in DNA "bending" which promotes the assembly of an initiation complex (KERPPOLA and CURRAN 1991a,b). Fos/Jun binding to the AP-1/TRE site can be prevented by a protein of 30–40 kD known as IP-1. This protein is constitutively expressed in both the cytoplasm and nucleus of cells. Its repressor function at the AP-1/TRE site is modulated by phosphorylation whereby protein kinase A (PKA) or protein kinase C (PKC) activation results in IP-1 inactivation (AUWERX and SASSONE-CORSI 1991, 1992). Thus, it is likely that subtle regulation of AP-1/TRE activity is strictly controlled. The AP-1/TRE consensus sequence is also modulated via the TF CREB which regulates c-*fos* expression. CREB can disrupt Fos/Jun transactivation at the AP-1/TRE by what appears to be a competitive interaction for the consensus sequence (MASQUILIER and SASSONE-CORSI 1992). It is interesting to note that the CRE and AP-1/TRE consensus sequences differ only by one nucleotide and that Fos and Fos-related antigens may also be involved in autoregulation (MORGAN and CURRAN 1991).

While other consensus sequences may also be under the regulatory control of Fos/Jun dimers, the AP-1/TRE sequence is by far the most thoroughly examined to date. Currently, the Fos family contains five members and the Jun family contains three. While Jun family members can dimerize with each other to form active homo- and hetero-dimeric TFs, the members of the Fos family do not appear able to do so. The present findings suggest that heterodimeric assemblies between the two families provide greater transactional ability of the various dimers assembled between Fos and Jun families (see HUGHES and DRAGUNOW 1995 for the transactivating potential of different dimer combinations).

In theory, any of a host of genes that contain the AP-1/TRE-like consensus sequence within their promoter/enhancer regions could be influenced by the presence of the Fos/Jun TF. These potential target genes include tyrosine hydroxylase (TH), thyrotropin-releasing hormone (TRH), cholecystokinin (CCK), glutamic acid decarboxylase (GAD), preproenkephalin (Enk), nerve growth factor (NGF) and others (HUGHES and DRAGUNOW 1995; ROBERTSON 1992a; SHARP 1994). However, recent data would suggest that some of these genes are regulated by different TFs. For example, early studies suggested that Enk was regulated by Fos (SONNERBERG et al. 1989), whereas more recent studies suggest that this particular gene is under the control of the CREB TF (KONRADI et al. 1993).

It is now clear that IEGs which function as TFs are capable of altering the expression of other genes which may have direct consequence on the function of neurons. However, until recently, the proper tools were lacking to explore the idea that specific genes, such as c-*fos*, may be involved in a molecular cascade which lead to altered brain function. The available arsenal of pharmacological agents, which tend to be receptor ligands, falls short of this task since it addresses a pre-transcriptional event. However, in large part due to the work performed by Zamecnik and colleagues, the idea of antisense oligonucleotide technology and the development of synthetic oligodeoxynucleotide (ODN) synthesis in the 1980s established a potential tool for dissecting the function of specific genes (STEPHENSON and ZAMECNIK 1978; ZAMECNIK and STEPHENSON 1978; for a review see GOODCHILD 1989). Today antisense ODNs have become popular research tools and several clinical trials are in progress assessing their therapeutic potential (MATTEUCCI and WAGNER 1996). With the use of antisense ODNs, researchers make certain assumptions about the actions of these compounds: (a) The issue of cellular uptake is important. It is assumed that the ODN will somehow enter the cell. The mechanism(s) by which this occurs remains obscure. (b) ODNs must demonstrate stability under in vitro and in vivo conditions so that an effective quantity reaches the target mRNA. (c) The proposed mechanism of action involves the formation of DNA-RNA duplexes and thus the ODN must undergo hybridization with the target. (d) Ultimately, the ODN must inhibit expression of the targeted gene. (e) The ODN must show selectivity of action and binding so as not to interfere with other cellular processes (GOODCHILD 1989; WAHLESTEDT 1994; YEE et al. 1994). It is clear from this list of criteria that few in vivo studies will be able to rigorously address all these issues (WHITESELL et al. 1993; WOOLF et al. 1992). However, several studies have now used antisense ODNs to investigate a variety of targets within the CNS. One of those targets has been the inducible TF c-*fos*. The remainder of this chapter will focus on the use of antisense ODNs in the CNS with particular emphasis on the study of c-*fos*.

D. Studies Using Antisense Oligodeoxynucleotides in the CNS

I. The Induction of c-*fos* in the Striatum

Systemic administration of dopamine agonists have been demonstrated to activate the IEG c-*fos* within the striatum (G.S. ROBERTSON et al. 1989; GRAYBIEL et al. 1990). Although the significance of this activation is unclear, certain studies have suggested that this IEG may have a role in regulating the output of the basal ganglia system. For example, earlier studies by UNGERSTEDT showing rotational behaviour in rats following ipsilateral lesions of the dopaminergic innervation to the striatum (UNGERSTEDT 1971a,b) have more recently been associated with activation of c-*fos* (G.S. ROBERTSON et al. 1989; PAUL et al. 1992). In the UNGERSTEDT model, a unilateral lesion of the nigrostriatal fibres is made using a selective neurotoxin. Rats bearing a dopaminergic lesion on the right side of the brain rotate to the left (contraversive rotation) when given a dopamine agonist such as apomorphine, and to the right (ipsiversive rotation) when amphetamine or cocaine are administered. The contraversive rotation observed after an agonist results from direct activation of super-sensitive dopamine receptors on the side of the brain which is depleted of dopamine. Ipsiversive turning induced by amphetamine and cocaine are the result of increased transsynaptic dopamine levels on the side of the brain which has an intact dopamine innervation. These drugs achieve this action by elevating dopamine levels by either increasing its release (amphetamine) or blocking its uptake (cocaine).

Rotation induced by D_1 dopamine receptor agonists, L-Dopa or amphetamine also induce expression of IEGs including c-*fos* on the side of the brain opposite to the direction of rotation in rats with unilateral dopamine depletion (G.S. ROBERTSON et al. 1989; H.A. ROBERTSON et al. 1989). In naive (dopamine intact) animals, amphetamine and cocaine induce the expression of c-*fos*, *egr*-1 and other IEGs equally in both striata through a D_1 receptor mechanism (G.S. ROBERTSON et al. 1989; GRAYBIEL et al. 1990; MORATELLA et al. 1992, 1993; BERETTA et al. 1993). In naive animals given amphetamine, the brain receives equal dopaminergic stimulation on both sides and therefore these animals do not demonstrate any preference in rotational direction (HOOPER et al. 1994). Taken together, this information suggests that c-*fos*, *egr*-1 and perhaps other IEGs, may play a role in the rotational behaviour seen in both the lesioned and naive animals. In lesioned animals, D_1 receptor stimulation results in rotation away from the side expressing c-*fos*, while in naive animals no overall preference in rotation is seen when c-*fos* expression is balanced between sides of the brain (striata). Therefore, one would predict that a unilateral decrease in c-*fos* expression in otherwise normal rats would result in a rotational bias with animals turning away from the striatum with the greater levels of c-*fos* activation (see Fig. 1).

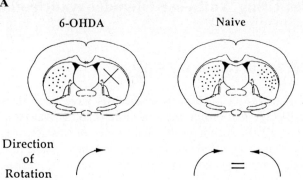

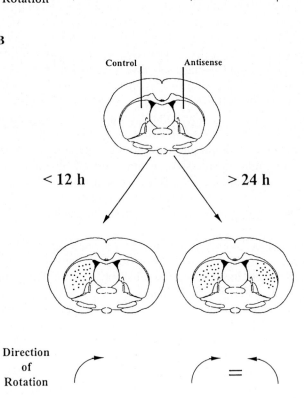

II. Knockdown of Amphetamine-Induced c-*fos* Using Antisense Oligodeoxynucleotides: Studies of the Basal Ganglia Function

To examine the putative role of c-*fos* in the functional output of the striatum, asymmetry in c-*fos* expression could potentially be achieved using antisense ODNs. Initial studies using antisense ODNs to c-*fos* were performed with fully substituted phosphorothioate s-ODNs since reports at that time suggested that non-substituted ODNs would not be likely to remain stable in biological fluids long enough to have any antisense action (CAMPBELL et al. 1990). It was evident then, as it is now, that antisense ODN knockdown of gene expression could be a powerful and relatively simple tool, and it also became apparent that IEG expression in the CNS provided an excellent system in which to study the effects of antisense ODNs in vivo. The reasons are as follows: first, the constitutive expression of c-*fos*, as with many other IEGs, is very low in the absence of stimulation but is rapidly induced upon stimulation. Secondly, due to the bilateral nature of the brain, one hemisphere can receive antisense ODNs and the other serves as a control to which a non-targeting ODN (sense, random or mismatch) can be delivered. Finally, knocking-down IEG expression in the brain can be correlated with changes in behaviour and perhaps provide insight to the role of the IEG in the living brain. This would allow the "true" function of IEGs to be deciphered in freely moving adult animals without the complications associated with the use of null mutations in which developmental deficits and compensatory mechanisms often complicate later analysis of function.

Since little was known about the application of antisense ODNs in brain tissue, it was first necessary to determine whether a gene of interest could be knocked-down using this approach. This was accomplished by infusing an antisense s-ODN to c-*fos* directly into one striatum and a sense control s-ODN into the opposing striatum of the same animal (CHIASSON et al. 1992a,b). Following the s-ODN treatment, Fos production was stimulated by the systemic administration of D-amphetamine (5 mg/kg, i.p.). Studying Fos-like im-

Fig. 1A,B. Representations of coronal sections through the striatum under different conditions in which Fos-LI is induced along with rotational behaviour. **A** Amphetamine-induced Fos-LI. Animals whose right nigrostriatal pathway has been interrupted by a 6-hydroxydopamine (6-OHDA)-lesion (usually produced 3 weeks before testing dopaminergic drugs) show ipsiversive rotation when challenged with D-amphetamine. That is, they rotate toward the lesioned side (to the right) or away from the side expressing Fos-LI. Naive animals challenged with amphetamine show a balanced bilateral Fos-LI in the striata and rotate in both directions equally. **B** Amphetamine-induced Fos-LI following antisense oligodeoxynucleotide (ODN) treatment. Animals which receive an infusion of an antisense ODN to c-*fos* mRNA into the right striatum and a sense (or other control) ODN into the left striatum and then are challenged with amphetamine (within 12 h) show rotation to the right, away from the side expressing the greatest number of Fos-positive cells. However, if the amphetamine challenge is delivered 24 h after the infusion of ODNs, no rotation is observed and Fos-LI is equal on both sides (see text for further explanation)

munoreactivity (Fos-LI) following s-ODN treatment demonstrated that the antisense s-ODN attenuated Fos expression, but that the sense s-ODN did not. However, this reduced expression associated with the antisense s-ODN was only observed at 12 h post-infusion, (10 h following ODNs + 2 h of amphetamine), whereas at 24 h post-infusion (22 h following ODNs + 2 h of amphetamine) no difference in the Fos-LI between hemispheres was observed. These initial studies suggested that the antisense s-ODNs could be delivered in vivo and effectively attenuate, in a reversible fashion, expression of the IEG, c-*fos*. In the same animals, induced *egr*-1 expression was also monitored. Induced Egr-1-LI was studied in order to determine if the s-ODN treatment would affect other IEGs that are activated through the same D_1 receptor mechanism. In other words, Egr-1 expression served as a measure of selectivity of the antisense s-ODN and measured whether the receptor coupling mechanism was intact. Since induced Egr-1-LI remained unaffected in either time group (12 or 24 h) these studies revealed that the D_1 receptor mechanism remained intact and that the effects seen on Fos-LI were selective (CHIASSON et al. 1992a,b; HOOPER et al. 1994).

As mentioned above, the early studies that linked IEG activation and rotation demonstrated that animals with unilateral reductions of striatal Fos-LI rotated away from the side that expressed higher Fos levels. Animals placed in a rotometer following an antisense and sense s-ODN infusion (as described above) and given amphetamine 10 h later, demonstrated a preference to rotate in the direction away from the sense s-ODN-treated side. That is, animals turned (within 15 min following the administration of amphetamine) in the direction away from the side expressing the greater Fos-LI. However, at 22 h following the ODN treatment, animals no longer demonstrated a preference to rotate in either direction (CHIASSON et al. 1992a). A subsequent and more comprehensive study revealed time- and dose-dependency to both the knockdown of Fos-LI and the rotation behaviour (HOOPER et al. 1994). This study revealed that as long as a significant difference in Fos-LI existed between the striata animals would rotate away from the side with the highest Fos expression. Other studies have also reported this relationship (DRAGUNOW et al. 1993; SOMMER et al. 1993). However, the knockdown of c-*fos* expression and rotation is sequence-specific since not all sequences targeting the c-*fos* mRNA were effective at attenuating expression (HOOPER et al. 1994). Thus, use of the antisense ODNs to c-*fos* in the CNS reveals a general phenomenon in the field of antisense technology which is that not all antisense sequences work well or at all (see CHIANG et al. 1991; MATTEUCCI and WAGNER 1996).

The idea of stimulus-transcription coupling predicts that the IEG c-*fos* would couple to a late-effector gene(s) which would have a direct effect(s) on neuronal function. Thus, while the mechanism by which c-*fos* alters the physiology of the basal ganglia is not yet fully elucidated, recent data suggest that IEGs do regulate other genes in the CNS. Using antisense s-ODNs directed against c-*fos* it has been demonstrated that c-*fos* expression can be linked to the levels of neurotensin expression in the striatum (MERCHANT 1994;

ROBERTSON et al. 1995). Lowering neurotensin expression in one hemisphere but not the other could lead to an altered functional output resulting in a behavioural asymmetry. Other AP-1-containing genes expressed within the striatum could also mediate the altered physiology. Proenkephalin, whose gene contains an AP-1 site is, however, unlikely to be directly involved since antisense s-ODNs to c-*fos* did not show an effect on Enk expression (MERCHANT 1994; ROBERTSON et al. 1995). This is consistent with another recent study that demonstrated that Enk is regulated by the constitutively expressed TF CREB and not Fos (KONRADI et al. 1993). Using antisense ODNs applied to the spinal cord, others have demonstrated that Dyn expression is under the control of Fos (HUNTER et al. 1995). An asymmetry in Dyn expression may also play a role in creating the rotational bias observed. These are a few of the possibilities that could account for the behavioural effects seen using s-ODN infusions into the striatum. However, in all of these instances it is difficult to understand why animals rotate so shortly after the amphetamine stimulus (15–30 min), since we would predict that the expression of late-effector genes would not be so rapidly affected. More recent observations may provide insight into this issue.

In a follow-up study SOMMER et al. (1996) have shown that partially substituted ODNs (every alternate internucleotide bond is thio-substituted) may interfere with γ-aminobutyric acid (GABA) transmission in some striatal projection neurons. They demonstrated using in vivo microdialysis that intrastriatal infusion of an antisense ODN to c-*fos* decreases GABA transmission in the striatonigral but not the striatopallidal projection neurons. Some interesting observations were made during this study which relate to antisense function and IEGs. For example, both striatonigral and striatopallidal neurons seem to show uptake of the antisense ODN effectively and quickly, but only nigral projecting neurons show a diminished GABA release. Although GABA transmission was altered, levels of dopamine remained the same since the measurements were made in animals which did not receive a dopamine agonist such as amphetamine. It is clear that an altered level of GABA transmission, as reported by SOMMER et al. (1996), could lead to a bias in the rotational behaviour as they and others have observed (CHIASSON et al. 1992a; DRAGUNOW et al. 1993; HOOPER et al. 1994; SOMMER et al. 1993). However, the timing for the rotational effect in the earlier studies, and the reduced GABA levels in the more recent study, do not easily lend themselves to explanation. The problem is that in the recent article by SOMMER et al. (1996), they report that baseline levels of GABA returned in approximately 3 h following the infusion of the ODN. However, the previous studies reported rotational behaviour at later times (5–12 h following infusion) and in fact no rotation was seen at earlier time points with fully substituted s-ODNs (HOOPER et al. 1994). Perhaps a more recent study using partially substituted antisense ODNs to c-*fos* can provide some insight into this discrepancy. In this study, HEBB and ROBERTSON (1997a), demonstrated that amphetamine-induced rotational behaviour was optimal at 1–3 h following unilateral infusion into the striatum of a partially substituted antisense ODNs to c-*fos*. These are interesting observations because they may allow us to infer that there are considerable differences between ODNs which are fully substituted and those with only partial substitution. In fact it was noted in an earlier study that some of the animals (three of six) which had been treated with partially substituted ODNs only demonstrated rotational behaviour at an earlier time (approximately 2 h following infusion; HOOPER et al. 1994). The animals which showed a rotational bias (three of six) using the partially substituted ODNs also showed a knockdown of Fos-LI at the early time point (2 h), once again strengthening the relationship between Fos expression and functional output of the basal ganglia. However, these data do not convincingly demonstrate that c-*fos* is actually involved in regulating the functional output of the striatum. They do, however, present the intriguing possibility that c-*fos* may regulate, either directly or indirectly, GABAergic transmission in D_1 receptor-positive striatonigral neurons. Given the current evidence, a strong prediction would be that fully substituted antisense s-ODNs would also cause a decrease in GABA transmission in the same striatonigral cells but that this would occur at a later time corresponding to the rotational behaviour (5–12 h following infusion). This remains to be tested but could elucidate whether the effects seen with the antisense ODN (which is the same sequence in all studies) are due to its antisense action on c-*fos* or some other sequence-dependent but nonantisense-mediated action (none of the control ODNs produced any effect on rotation or GABA release). The differences in timing between the partially and fully substituted ODNs could simply reflect the time necessary for uptake, different mechanisms of action, non-specific actions brought about by degradation products and/or non-selective actions which are sequence-dependent. It is well know that phosphorothioate substitutions confer nuclease resistance, but that the degree of substitution may alter rates of uptake and add to non-selective actions (CAMPBELL et al. 1990; GUVAKOVA et al. 1995; KRIEG 1993; THIERRY and DRITSCHILO 1992). Thus, it remains possible that both the partially and fully substituted ODNs achieve the same end measure (rotation away from the side expressing more c-*fos*, albeit at different times) but for different reasons relating not to Fos expression, but to different actions due to the sulphur content and the consequent metabolism of the compounds. Although the above evidence is not conclusive in supporting a role for c-*fos* in the function of the basal ganglia, it does provide relatively strong support for the general idea. However, the role of c-*fos* may not be as a TF involved in stimulus-transcription coupling, but rather as a messenger involved in other aspects of cell function. Alternatively, tonic expression of c-*fos* (at very low levels) may play a role as a transcription factor, but could potentially regulate proteins whose expression are necessary only under basal resting conditions.

SOMMER et al. (1996) have suggested that a short-lived protein under the control of tonic c-*fos* expression may be responsible for the altered level of GABA transmission seen in D_1 receptor-positive striatonigral neurons. They demonstrate that low levels of c-*fos* expression are found in the striatum and that within 90 min following the delivery of the antisense ODN an increased expression of c-*fos*, *jun*-B and *egr*-1 can be observed in non-amphetamine-treated animals. The increased levels of the IEGs, while significant, required an RNase protection assay to be detected. This is unfortunate, since examining whether the increased IEG expression was associated with the D_1/Dyn-positive striatonigral or the D_2/Enk-positive striatopallidal neurons could provide insight to potential mechanisms underlying this phenomenon. Thus, an antisense ODN to c-*fos* can attenuate amphetamine-induced c-*fos* expression without affecting induced *egr*-1 expression (CHIASSON et al. 1992a,b; DRAGUNOW et al. 1993; HOOPER et al. 1994; HEBB and ROBERTSON 1997a). However, a partially substituted ODN injected directly into the striatum of non-amphetamine-treated animals resulted in increased levels of c-*fos*, *jun*-B and *egr*-1 within 90 min (SOMMER et al. 1996). The increased IEG expression could clearly have profound effects on several striatal systems such as the D_1/Dyn-positive striatonigral or the D_2/Enk-positive striatopallidal neurons. However, given previous data (HUNTER et al. 1995; KONRADI et al. 1993), the Dyn-containing striatonigral cells may be more sensitive to the alteration in these IEG TFs. Although it is not clear why the IEGs respond by increasing their mRNA levels, the mechanism may relate to the strong autoregulation that is seen with c-*fos* via the AP-1 regulatory element (SASSONE-CORSI et al. 1988; SOMMER et al. 1996). Indeed, other studies using antisense ODNs also suggest that there is some form of regulation between IEGs (DRAGUNOW et al. 1993, 1994; HEBB and ROBERTSON 1997b).

One possible factor which could effect the striatonigral versus striatopallidal systems is a non-selective and differential action of the antisense ODN on the D_1 and D_2 receptors. However, it would appear that the antisense ODN does not have an effect on dopamine receptors (D_1 and D_2) nor on the adenosine (A_2) and cannabinoid receptors (DRAGUNOW et al. 1994). Other studies also suggest that the dopamine receptors remain functional (HOOPER et al. 1994; HEBB and ROBERTSON 1997a). Additionally, work performed by HEILIG and colleagues (1993) is consistent with the observation that Fos may regulate motor output (i.e. rotational behaviour) from the striatum. They demonstrated that bilateral attenuation of cocaine-induced Fos expression in the nucleus accumbens was associated with a decrease in cocaine-induced locomotory activity. This finding suggests that cocaine-induced c-*fos* expression is also associated with the functional output of the nucleus accumbens. This is particularly interesting in view of the potential role of this mesolimbic structure in the addictive process (GRAYBIEL et al. 1990; NESTLER et al. 1993). This would suggest that c-*fos* may play a role in coupling the short-term exposure to a drug like cocaine to longer-lasting events underlying addiction.

Taken together, initial results using the amphetamine/striatal model of IEG expression demonstrated that some, but not all, fully substituted s-ODNs targeting c-*fos* could effectively attenuate Fos-LI (CHIASSON et al. 1992a,b,

1994; DRAGUNOW et al. 1993; HOOPER et al. 1994; SOMMER et al. 1993). Dose- and time-dependency have been demonstrated and the importance of the degree of sulphur modification within the backbone of the ODN is demonstrably important. Uptake of the ODNs has been examined and they appear to be rapidly and selectively internalized by striatal neurons (SOMMER et al. 1993, 1996). It also appears that this D_1-mediated Fos expression was important in the functional output of the striatum (basal ganglia), since only animals in which Fos-LI differed significantly between striata demonstrated a significant directional preference in rotational behaviour. This behaviour was also consistent with what has been observed in the lesioned model where animals stimulated with D_1 agonist rotate away from the side of the brain expressing the most Fos positive nuclei (PAUL et al. 1992, 1995; G.S. ROBERTSON et al. 1989; H.A. ROBERTSON et al. 1989b). These results indicate that the IEG c-*fos* may play an important role in mediating the physiology of the basal ganglia. Using antisense ODNs in these studies has also begun to reveal the potential interaction between IEGs which may play an important role in stimulus-transcription coupling, leading to altered brain activity and function. In summary, it would appear that c-*fos* regulates at least two temporally and spatially distinct systems within the basal ganglia. Altering low constitutive expression of Fos leads to a change in the system which regulates motor output (e.g. GABA), whereas altering other aspects of the D_1 system (e.g. Dyn) may have pronounced effects on long-term changes relating to addiction (Fig. 2).

III. The Amygdala and Gene Expression

In addition to the studies performed on IEGs and their role in the regulation of striatal function, many additional studies have explored the idea that stimulus-transcription coupling may play a role in the long-term changes in brain function associated with memory and pathophysiological changes such as those seen in kindling. Kindling is the progressive development of behavioural convulsions and epileptiform activity (afterdischarge) that occurs following the repeated administration of an initially sub-convulsant electrical or chemical stimulation to the brain (GODDARD et al. 1969; RACINE 1972). Kindling is perhaps one of the best models of enduring change in brain function since, once kindled, animals remain so for the rest of their lives. Multiple sites in the brain can be kindled, but most researchers use electrical kindling of the amygdala as the model of choice (CAIN 1992). It is interesting that one of the first examples of kindling used repeated exposure to cocaine (DOWNS and EDDY 1932) which, as we have seen, is believed to cause alterations in brain function through activation of IEGs (DRAGUNOW et al. 1989; GRAYBIEL et al. 1990; ROBERTSON 1992a).

Following the general introduction of the concept of stimulus-transcription coupling (BERRIDGE 1986; GOELET et al. 1986), it was demonstrated that a kindling stimulation to the hippocampus activated c-*fos* within the hippocampal formation and thus established the first link between kindling

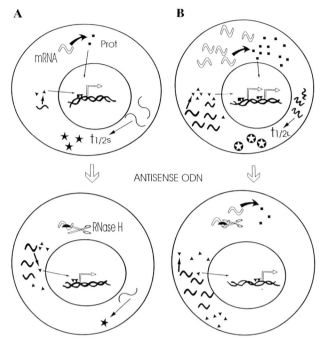

Fig. 2A,B. A model for the role of c-*fos* in both short- and long-term information processing in the brain. **A** Resting condition: a cell in a resting or basal state. Low constitutive levels of Fos protein (*black squares*, indicated as *Prot*) can be seen (c-*fos* mRNA is an *open sinus wave*). Fos protein interacts with other immediate-early genes (IEGs) (*black triangles*) which are expressed under these basal conditions to activate expression of a protein(s) (*black star*) whose half-life ($t_{1/2}s$) is very short. This protein(s) could control basal levels of neurotransmission (e.g. GABAergic transmission) or other important systems necessary for normal neural output. When an antisense oligodeoxynucleotide (ODN) to c-*fos* (*black half circle*) is administered, it binds to the c-*fos* mRNA which is rapidly degraded by RNase H (*scissors*). Consequently, the balance between Fos and other IEGs is changed and regulatory elements that were under the control of Fos are now subject to other transcription factors (TFs). In some cases IEGs may rebound due to the complex interactions which exist among these TFs, shown here as an increase in mRNA (*filled sinus wave*) and the IEG protein (*black triangle*). The new combination of IEGs may now alter transcription of other gene products involved in short-term neural function. Thus, under resting conditions altering low levels of Fos may have effects on short-term function by changing the expression or availability of proteins with short $t_{1/2}$ lives. **B** Stimulated condition: a cell has been stimulated and levels of IEG TF expression have risen dramatically. At this high level of activation, Fos and other TFs can alter expression of late-effector genes (*white star*, with a long half-life, $t_{1/2}L$). It is these late-effector genes which lead to more permanent changes in the cell's function. However, application of antisense ODNs to c-*fos* prohibits activation of some late-effector genes by lowering Fos protein levels to near baseline (these genes may be involved in processes such as kindling or conditioned taste aversion memory). Clearly, in the presence of a stimulus, a new balance of IEG TFs is struck and many downstream genes are differentially regulated (other genes not shown). Thus, under stimulated conditions (e.g. a kindling stimulus), Fos can alter expression of other genes which result in long-term alterations in brain function. Interfering with this process by using antisense ODNs has provided insight into some of the potential candidate genes. [For a discussion of combinatorial interactions amongst TFs see STRUHL (1991)]

and IEG expression (DRAGUNOW and ROBERTSON 1987). Since then numerous studies have demonstrated IEG activation following kindling (CHIASSON et al. 1995; HOSFORD et al. 1995; HUGHES et al. 1994; LABINER et al. 1993; SHIN et al. 1990). Additionally, a link has been established between the afterdischarge (AD), which has been known for years to be necessary for kindling to develop (RACINE 1972), and the activation of IEGs in amygdala kindling (CHIASSON et al. 1995; HUGHES et al. 1994). As mentioned above, kindling requires that repeated stimuli be delivered to the brain (usually one or two stimuli per day) resulting in repeated activation of IEGs (most IEGs are detected for only a few hours following the AD). Therefore, in order to knockdown IEG expression under these conditions, either repeated or continuous delivery of antisense ODNs is required since it has been demonstrated that the antisense actions have dissipated by 24 h in vivo (CHIASSON et al. 1992a,b; HOOPER et al. 1994).

In an initial study it was determined that Fos expression induced by a single AD could be attenuated by an infusion of antisense s-ODN to c-*fos*. This study demonstrated that a single infusion of c-*fos* antisense s-ODNs, but not control s-ODNs, could attenuate Fos-LI when infused 10 h prior to the amygdala kindling stimulus (CHIASSON et al. 1994). This is consistent with the findings of MÖLLER et al. (1994) in their studies on the role of Fos in anxiety, as well as studies on conditioned taste aversion (CTA) memory by LAMPRECHT and DUDAI (1996), both of which demonstrated a knockdown of Fos expression using antisense s-ODNs to c-*fos* in the amygdala. Furthermore, LAMPRECHT and DUDAI (1996) have demonstrated that infusion of the s-ODN into the striatum or near, but not in, the amygdala does not alter Fos-LI induced by the stimulus. These studies, as well as those on the striatal system, demonstrated that the antisense ODNs to c-*fos* are able to attenuate Fos-LI independent of the inducing stimuli (these stimuli vary greatly from the taste of sucrose to an electrical stimulus delivered directly to the amygdala). However, following repeated injections and kindling stimulations, it was discovered that the AD signals normally recorded from the amygdala during kindling were absent after between three and four kindling sessions. Histological examination of the brain tissue (cresyl violet and glial fibrillary acidic protein immunohistochemistry) revealed lesions created by the ODNs (sense and antisense s-ODNs, but not vehicle). Extending the interval between infusions from 1 day to either 3 or 5 days decreased the damage but did not eliminate the lesions (CHIASSON et al. 1994). Thus, it appears that single infusions of fully phosphorothioated ODNs produce no lesions, but that multiple infusions lead to deleterious effects on tissue extending from severe gliosis to large lesions (CHIASSON et al. 1994). Although others have also reported toxicity problems associated with s-ODNs (GUVAKOVA et al. 1995; WOOLF et al. 1990; HEILIG, personal communication), some studies using repeated or continuous infusions of s-ODNs did not reveal any noticeable damage (ZHANG and CREESE 1993; ZHOU et al. 1994). Although the cause of the toxic effects observed following repeated administration are not yet understood, they are clearly influenced by factors such as dose, route of administration and phosphorothioate modification and purity (BRYSCH and SCHLINGENSIEPEN 1994). In general, phosphorothioate substitution has been used to increase nuclease resistance and half-life of ODNs in biological fluids (CAMPBELL et al. 1990). However, the benefits of this increased stability may be counteracted by reduced rates of uptake and increased cytotoxicity (KRIEG 1993; THIERRY and DRITSCHILO 1992). To circumvent some of the toxicity-related problems associated with fully substituted s-ODNs, several groups have used partially substituted or end-capped ODNs in which the sulfur content is greatly reduced . It has been shown that this reduced sulfur content lowers some of the potentially toxic non-selective effects of these compounds (GUVAKOVA et al. 1995; KRIEG et al. 1995). Using end-capped ODNs we have attempted to study the effects of c-*fos* on amygdala kindling. First, we examined whether these chimeric ODNs were useful as antisense molecules in vivo. Secondly, we examined if they could be repeatedly applied to the brain without causing neural damage. Lastly, we determined whether repeated administration of end-capped antisense ODNs to c-*fos* altered kindling.

As indicated above, several groups have now demonstrated that partially substituted phosphorothioate ODNs can be effective at attenuating Fos-LI in vivo (GILLARDON et al. 1994; HEBB and ROBERTSON 1997a; SOMMER et al. 1996). With this knowledge, the toxicity of these compounds was examined by repeated injections into the amygdala. The histological effects of multiple infusions of end-capped ODNs into the amygdala were examined using cresyl violet to detect any gross morphological abnormalities. Animals were infused daily at a dose two to three times greater than necessary for effective knockdown of Fos-LI. Histological examinations revealed only minor local damage which did not extend far (within milimetres) from the site of infusion. Animals which received the end-capped ODNs every third day showed little or no damage even when compared to vehicle infused animals (CHIASSON et al. 1997). Thus, repeated infusion of end-capped ODNs showed no gross histological evidence of toxicity.

As a consequences of these findings, animals were administered kindling stimuli following the infusion of an end-capped antisense ODN. The AD in animals treated with end-capped ODNs appeared normal even after several infusions (CHIASSON et al. 1997). These observation were consistent with the morphological results and suggested that the tissue at the injection site was normal in animals treated with end-capped ODNs (or at least not different from vehicle-treated animals). The effect of a single infusion of the end-capped antisense ODN to c-*fos* on Fos-LI was also examined. Animals received a single infusion and were then given an AD-generating kindling stimulus; we observed a knockdown of Fos-LI similar to that seen in the striatum using end-capped antisense ODNs to c-*fos*. Control animals, treated with sense ODNs, had normal AD-induced Fos-LI following the stimulus (CHIASSON 1995). Taken together, the results suggested that repeated applications of antisense end-capped ODNs could be used to study the role of c-*fos* in amygdala kindling. A subsequent study compared the rate of amygdala kindling between animals receiving an end-capped ODN to c-*fos* and three control groups of animals which received either a random ODN, a sense ODN or a vehicle infusion. All three control groups kindled at the normal rate expected from this site and species [for information on kindling rates and site dependency, see CAIN (1992)]. However, animals infused with the antisense ODN demonstrated generalized seizures within approximately half the number of stimulations (Fig. 3) when compared to the control groups (CHIASSON 1995). These results are consistent with the previous observations implicating IEGs in kindling (CHIASSON et al. 1995; DRAGUNOW and ROBERTSON 1987; DRAGUNOW et al. 1988; HOSFORD et al. 1995; HUGHES et al. 1994; SHIN et al. 1990). However, a recent study examining kindling in mice with a null mutation of c-*fos* demonstrated that knockout mice take longer to achieve generalized seizures than wild-type or heterozygotes (WATANABE et al. 1996).

These observations elucidate an important aspect of using antisense ODNs in vivo as well as using null mutations to study behaviour. It must be

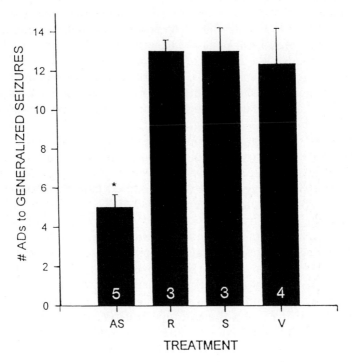

Fig. 3. Histogram displaying the number of afterdischarges (ADs) required to obtain the first generalized seizure [stage 4–5, according to RACINE (1972)] with differing oligodeoxynucleotide (ODN) or vehicle treatments. Animals receiving an antisense (AS) ODN to c-*fos* kindle within approximately half the number of ADs when compared to animals treated with either random (*R*) or sense (*S*) ODNs. Kindling in the presence of the vehicle (*V*) is identical to either of the control ODNs. The numbers within the *bars* represent the number of animals per group

recognized that antisense ODNs generally result in a knockdown of the gene of interest but do not completely eliminate that gene. Therefore, in many ways, using antisense ODNs allows one to examine the role of a given "dose" of a gene within a confined area or neural circuit within the CNS. Antisense strategies also allow one to examine reversible phenomena do to the fact that antisense ODNs have limited half-lives. Thus, it may be difficult to reconcile differences observed between studies using ODN gene knockdown from other studies using transgenic technologies which currently lead to complete gene knockout and usually on a permanent basis. However, the model presented in Fig. 2 can address some of the important issues related to alterations in IEG function. For example, the model would predict that animals which received the antisense ODN to c-*fos* should demonstrate a decreased GABAergic transmission, which is probably due to alterations in the balance of IEGs expressed under basal conditions (as is seen in the striatal studies). It is obvious that this would accelerate the kindling process since GABA is the principle inhibitory transmitter which restrains excitatory drive. It is also known that the GABA transporter protein is decreased by 50% following amygdala kindling (DURING et al. 1995). This protein is believed to provide a significant source of GABA during neuronal excitation leading to seizures. Thus, in the presence of an antisense ODN to c-*fos*, the GABAergic system may be doubly compromised. The model also makes further predictions related to the expression of late-effector genes which would normally be under the control of higher levels of Fos expression seen following a kindling stimulus. In this case, these higher levels of Fos expression are suppressed by the actions of the antisense ODN only locally in the amygdala region, whereas other brain regions undergoing epileptogenic activity would express high levels of Fos. This could create an imbalance between the circuitry involved in the generalization of seizures and, therefore, enhance the rate at which generalized seizures emerge during kindling in the presence of an antisense ODN to c-*fos*.

The circuitry sub-serving opioid function in the brain could potentially alter the rate of seizure development during amygdala kindling . Recent results suggest that induced Fos regulates Dyn expression in vivo. (HUNTER et al. 1995), but not Enk (MERCHANT 1994; ROBERTSON et al. 1995). This finding is particularly interesting in light of the fact that levels of Dyn and Enk have been shown to undergo changes in kindling (reviewed in MORRIS and JOHNSON 1995). Since LTP and kindling share many features (CAIN 1989), one can extrapolate from the LTP studies that a change in the balance of opioids alters the equilibrium between the excitatory and inhibitory components. It is now clear that Enk promotes neuronal excitation in LTP (through μ and δ receptors) while Dyn inhibits this process (through κ receptors; MOORE et al. 1994; MORRIS and JOHNSON 1995; TERMAN et al. 1994; WAGNER et al. 1993). Thus, if induced Fos is involved in activating Dyn expression (HUNTER et al. 1995), then attenuating Fos levels by antisense treatment should reduce the inhibitory effects of this opioid. This could then result in a longer term imbalance between excitatory and inhibitory systems and lead to an altered susceptibility to a kindling stimulus.

It is important to remember that once kindled, an animal remains so for its lifetime. Although alterations in opioid levels have been observed in kindling, it is clear that they do not remain changed long enough to account for the long-term susceptibility to the kindling stimulus. Interestingly, several groups studying kindling have observed structural or molecular changes which last long periods of time and could at least partially explain the long-term nature of kindling (GEINISMAN et al. 1988; NISHIZUKA et al. 1991; PERLIN et al. 1993; SUTULA et al. 1988). This is interesting since WATANABE et al. (1996) used c-*fos* knockout mice to demonstrate that this gene may be necessary for long-term sprouting of dentate granule cells of the hippocampus during kindling. This may suggest that Fos is involved in regulating growth factor expression during kindling since many neurotrophins play an important role in sprouting and it is known that kindling stimuli activate these factors (ERNFORS et al. 1991). This suggests that Fos may have several targets which are necessary during kindling and that altering Fos levels in different regions of the brain during the kindling process could either increase or decrease the rate of kindling. Experimental approaches are necessary to determine if any or all of these systems are affected by the end-capped antisense ODNs to c-*fos*. Using the model presented in Fig. 2, strong testable predictions can be made about the outcome of experiments examining the issues surrounding the role of c-*fos* and other IEGs in CNS function.

Together with the studies on the striatum, nucleus accumbens and the amygdala, the kindling studies would suggest that c-*fos* has a role in both the short- and long-term modulation of brain function. In addition, the kindling studies which used repeated application of s-ODNs demonstrated that there can be potential pitfalls associated with the use of s-ODNs in the CNS (CHIASSON et al. 1994). Some of these problems may be related to the degree of sulfur present within the backbone of the ODN.

IV. Exploring Alternative Modifications to Oligonucleotides

Phosphorothioate ODNs may exert some of their toxic effects through the inactivation of various proteins essential for basic cellular function. Phosphorothioate ODNs bind a variety of heparin-binding growth factors including PDGF and many members of the fibroblast growth factor family (bFGF, aFGF, FGF-4, VEGF) with strong affinity, and inhibit their biological activities (GUVAKOVA et al. 1995). Broad-spectrum inhibition of a variety of neurotrophic growth factors might be expected to cause general neurotoxic trauma since many such factors are essential for cellular subsistence. The s-ODNs may also demonstrate interactions with the immune system which could be detrimental. One recent finding demonstrated that the degree of sulfur substitution in CpG-rich ODNs played a significant role in B-cell activation (KRIEG et al. 1995). While such a side-effect may cause significant problems to experimental scientists, this particular observation may be of use to clinicians. Such a finding could be employed within clinical settings to stimulate immune-mediated responses against tumours. Although it is clear that the issue of toxicity may be related to the thiophosphate modification, the factors involved may be very complicated and involve the route of administration, ODN sequence in combination with the modification and the purity of the ODN (BRYSCH and SCHLINGENSIEPEN 1994; KRIEG et al. 1995). If toxicity is uniquely a consequence of the modified backbone of the ODN, then non-modified ODNs should be tolerated. In fact, several studies have demonstrated that high doses (relative to the doses used with s-ODNs) of non-modified ODN delivered via i.c.v. or i.t. routes produce specific and effective attenuation of receptors such as the neuropeptide Y (NPY)-Y1, N-methyl-D-aspartate (NMDA)-R1 and the δ opioid receptor (WAHLESTEDT et al. 1993a,b; STANDIFER et al. 1994). Presumably, this is because cerebrospinal fluid (CSF) and brain tissue are nuclease-poor (BRYSCH and SCHLINGENSIEPEN 1994). However, non-modified ODNs in vivo have generally been relatively ineffective in producing a knockdown of gene expression due to rapid degradation (THIERRY and DRITSCHILO 1992).

Alternate strategies for design and delivery of antisense ODNs need to be considered to avoid these potential pitfalls of antisense therapy as well as to enhance the utility of antisense strategies in biological research. The polyanionic nature of phosphorothioate ODNs increases the likelihood of non-specific protein interactions, and also decreases the rate of cellular uptake (THIERRY and DRITSCHILO 1992). In addition, the phosphorothioate ODNs appear to exert a biphasic effect on RNase H, the enzyme by which they exert most of their antisense action (GAO et al. 1991; LIU et al. 1994; MATTEUCCI and WAGNER 1996). This particular finding may account for the relatively narrow window of therapeutic effectiveness seen with ODNs in the brain (HOOPER et al. 1994; HEBB and ROBERTSON 1997a).

Methyl-phosphonate-modified ODNs, which carry less charge than phosphorothioate ODNs and appear to be more readily internalized (KRIEG 1993; STEIN and CHENG 1993; WOOLF et al. 1990), seem to exert their effects through translational arrest rather than by RNase H activation (STEIN and CHENG 1993; STEIN et al. 1988). Thus, modifications of ODNs, such as with methyl substitutions, coupled with tissue selective carrier molecules, could prove useful in improving ODN delivery and efficacy. However, reports on methyl phosphonates do not appear to substantiate this possibility since efficiency and solubility with these compounds are problematic (BRYSCH and SCHLINGENSIEPEN 1994). Another alternative is to examine partially modified phosphorothioate ODNs which exert antisense action, support some nuclease resistance and reduce non-specific interactions (GILLARDON et al. 1994; GUVAKOVA et al. 1995; HEBB and ROBERTSON 1997a,b; HOOPER et al. 1994; KRIEG et al. 1995). Other ODN modification strategies have been undertaken largely by industry and second generation compounds are being synthesized and currently evaluated. These modifications are largely out of reach of the average academic neurobiology laboratory and may remain so in the foreseeable future (see WAGNER 1994; MATTEUCCI and WAGNER 1996). The above discussion suggests that we must interpret the results of antisense ODN experiments cautiously.

V. Reconciling Differences

The interpretation of antisense effects within the living brain makes it particularly difficult to apply the usual stringent criteria simply due to the fact that the cells of the brain are interconnected and can consequently alter gene expression rapidly by shifting excitatory or inhibitory tone. Other factors which can further hinder the interpretation of results from antisense experiments are the biphasic activity of RNase H and the complex inter-regulatory signals between IEGs. These factors may contribute to some of the seemingly conflicting results that have been reported in the studies on IEGs. For example, DRAGUNOW et al. (1993) reported that an antisense s-ODN to c-*fos* attenuated both amphetamine-induced Fos and Jun-B-LI, whereas ROBERTSON et al. (1995) reported that haloperidol activation of *egr*-1, *jun*-B and *fos*-B remained unaffected by the antisense s-ODN to c-*fos*. Other studies have shown that amphetamine-induced Egr-1-LI is largely unaffected by an antisense s-ODN to c-*fos* (CHIASSON et al. 1992a,b; HOOPER et al. 1994; HEBB and ROBERTSON 1997a), whereas non-induced levels of this gene appear to be attenuated (DRAGUNOW et al. 1994) or elevated (SOMMER et al. 1996). These differences are difficult to interpret because many factors between the studies varied including the ODN (as well as the source, which is important due to purity), the methods of detection (immunohistochemistry, in situ hybridization and RNase protection assay), the inducing stimulus (no stimulus, haloperidol, which is a D_2 antagonist, and amphetamine) and the time following the delivery of the ODN before the tissue was monitored for gene expression. One possible explanation for these disparate results would be that altering the level of an IEG through an antisense action can have pronounced effects on other IEGs because of interactions which exist between these genes. This, coupled with the fact that RNase H appears to obey dose-dependent modulation of ODN activity could produce different effects on gene expression. The antisense ODN could maintain its same mode of action, which is to hybridize with its target mRNA, but be doing so under very different levels of gene expression and under a different stochiometry with RNase H. These possible interpretations may help to clarify some of the conflicting data and perhaps provide a stimulus for some of the experimental studies required. In fact, LIU et al. (1994) have shown that phosphorothioate antisense ODNs to c-*fos* can selectively hybridize and cleave c-*fos* mRNA in the presence of RNase H. Thus, assuming that autoregulation and interactive modulation of IEG expression occurs in the living brain, these discrepancies, which on the surface appear to be irreconcilable, are merely instructing us about the cellular and molecular nature of IEGs and the mechanism of antisense activity. For example, these sorts of regulatory mechanisms might predict that, under certain circumstances, it may be necessary to disrupt two IEGs seemingly involved in a biological process. Recent reports have demonstrated that the light-induced c-*fos* expression in the SCN (RUSAK et al. 1990) may have functional significance. Attenuating Fos-LI and Jun-B-LI with antisense ODNs targeting both genes resulted in a reduction of the phase shift normally induced by a light pulse when presented to animals in their subjective night (WOLLNIK et al. 1995). These results, as well as the findings discussed previously, suggest that c-*fos* and perhaps *Jun*-B function not only to couple stimuli to long-term changes in gene expression but also function in the short-term regulation of function. Overall, studies using antisense ODNs to examine the role of IEGs in the CNS have provided new insights into the molecular mechanism utilized by the brain. However, these results are largely based on one technology and thus further investigation is certainly warranted to further characterize IEG function in the CNS. On the other hand, using antisense ODNs within the CNS to examine receptor function brings with it the added advantage of the numerous studies which have used more conventional pharmacological approaches and the considerable arsenal of compounds against which the actions of the ODNs can be compared.

VI. Other Studies Using Antisense Oligonucleotides in Nervous Tissue

Some of the first studies to demonstrate that antisense ODNs could knock down receptor levels in vivo examined the NMDA and NPY receptors (HEILIG et al. 1992; WAHLESTEDT et al. 1993a,b). Reducing binding of the NMDA-R1 component of the NMDA receptor showed that ischaemia-induced cell death could be reduced in a fashion similar to applying the NMDA receptor antagonist MK-801 (WAHLESTEDT et al. 1993a). The proposed anti-anxiety role of NPY has also been demonstrated by selective knockdown of the NPY-Y1 receptor (WAHLESTEDT et al. 1993b). Studies examining the role of dopamine receptors in striatal function and the role of opiate receptors in pain modulation have employed antisense technology (SILVIA et al. 1994; STANDIFER et al. 1994; TSENG and COLLINS 1994; WEISS et al. 1993; ZHANG and CREESE 1993; ZHOU et al. 1994). These studies all examined the effects of the antisense ODNs on the target and related genes. Additionally, they demonstrated that antisense ODNs produce the predicted effects based on earlier pharmacological approaches using dopamine or opioid compounds.

Several studies have now correlated behavioural responses with antisense actions, suggesting that this approach provides a very useful tool for researchers interested in dissecting the molecular basis of behaviour. Using antisense ODNs directed at galanin mRNA, it has been demonstrated that this protein may function in self-mutilation behaviour (autotomy) following a sciatic nerve lesion. This behaviour is believed to be associated with increased pain sensitivity, whereas galanin is believed to act as an endogenous antagonist to substance P. In this study, antisense ODNs showed a repression of the injury-induced up-regulation of galanin in dorsal root ganglion cells that was correlated with an increased level of autotomy (Ji et al. 1994). This lends support to the idea that galanin acts as a natural antagonist to substance P.

Other studies have used the knockdown of a given gene by antisense ODNs combined with the over expression of the same gene using a retrovirally transduced constitutively active allele to demonstrate the function of a gene. Studies on the role of *Notch* in the development of retinal ganglion cells demonstrated that antisense ODNs targeting three different regions on the *Notch* mRNA could alter ganglion cell fate. Furthermore, increasing the number of mismatched nucleotides within the antisense molecule demonstrated a corresponding decline in antisense action. Overexpression of *Notch* produced the opposite effect, resulting in a decreased number of retinal ganglion cells (AUSTIN et al. 1995). Another study interested in the role of synaptosomal-associated protein 25 (SNAP-25) in the establishment of retinal cytoarchitecture used antisense ODNs in vivo. In this study, the application of an antisense ODN to SNAP-25 demonstrated an inhibition of neurite elongation of developing chick retinal amacrine cells, whereas control ODNs did not (OSEN-SAND et al. 1993). Many of these studies share the fact that antisense ODNs were used to demonstrate the function of a particular gene and that function was verified by other technical approaches such as with well-established receptor pharmacology or molecular technology using retroviral vectors. In addition, the outcomes of these studies "fit" the anticipated role for the gene in question and thus, taken together, provide support for the antisense approach.

E. Neural Plasticity and the Role of c-*fos* as Demonstrated by Antisense Technology

In the last decade neuroscientists have made considerable progress in understanding some of the underlying mechanisms involved in neuroplasticity. One of these advances is in the conceptual development of stimulus-transcription coupling as a basis for this plasticity. Consequently, TFs have become a major focus of neuroscience research. As we and others have demonstrated, the prototypical inducible TF, Fos, appears to play an important role in basal ganglia and limbic function (CHIASSON et al. 1992a; DRAGUNOW et al. 1993; HEILIG et al. 1993; HOOPER et al. 1994; LAMPRECHT and DUDAI 1996; MÖLLER et al. 1994; SOMMER et al. 1993, 1996). HUGHES and DRAGUNOW (1995) have compiled data suggesting that Fos appears necessary for high-level transcriptional activity in many of the Fos/Jun dimer combinations. Therefore, it is perhaps not too surprising that altering the levels of this protein will have considerable effects on target genes and subsequently on behaviour. However, it may turn out that Fos and other IEGs are involved not only in long-term alterations of function, but also in mechanisms that are important in the short-term processing of information within the cell. Perhaps the various species of Fos that arise from post-translational processing (CURRAN and MORGAN 1995) have roles that have not yet been identified. Interactions with cell proteins involved in neurotransmission/modulation could alter brain function in the short-term and lead to the results discussed above. Future studies will also have to consider the role of mRNA in processes other than translation since it has become increasingly evident that these molecules may have other functions (reviewed in NELLEN and LICHTENSEIN 1993). It would appear that one of the strategies to be exploited in addressing some of these issues will be antisense technology.

Antisense ODNs have gained considerable popularity as research tools. While studies using antisense ODNs are clearly not without problems, they currently offer an approach to studying the role of genes in adult animals. With the growing application of antisense technology, considerable improvements are likely to present themselves.

Acknowledgements. This study was supported by the MRC of Canada, SmithKline Beecham Pharma Inc., the Savoy Foundation, the Parkinson Foundation of Canada and the Huntington's Society.

References

Abraham WC, Christie BR, Logan B, Lawlor P, Dragunow M (1994) Immediate early gene expression associated with the persistence of heterosynaptic long-term depression in the hippocampus. Proc Natl Acad Sci U S A 91:10049–10053

Austin CP, Feldman DE, Ida JA Jr, Cepko CL (1995) Vertebrate retinal ganglion cells are selected from competent progenitors by the action of Notch. Development 121:3637–3650

Auwerx J, Sassone-Corsi P (1991) IP-1: a dominant inhibitor of Fos/Jun whose activity is modulated by phosphorylation. Cell 64:983–993

Auwerx J, Sassone-Corsi P (1992) AP-1 (Fos-Jun) regulation by IP-1 effect of signal transduction pathways and cell growth. Oncogene 7:2271–2280

Beretta S, Robertson HA, Graybiel AM (1993) Neurochemically specialized projection neurons of the striatum respond differently to psychomotor stimulants. Prog Brain Res 99:201–205

Berridge M (1986) Second messenger dualism in neuromodulation and memory. Nature 323:294–295

Brysch W, Schlingensiepen K-H (1994) Design and application of antisense oligonucleotides in cell culture, in vivo, and as therapeutic agents. Cell Mol Neurobiol 14:557–568

Cain DP (1989) Long-term potentiation and kindling: how similar are the mechanisms? Trends Neurosci 12:6–10

Cain DP (1992) Kindling and the amygdala. In: Aggleton JP (ed) The amygdala: neurobiological aspects of emotion, memory, and mental dysfunction. Wiley-Liss, New York, pp 539–560

Campbell JM, Bacon TA, Wickstrom E (1990) Oligodeoxynucleoside phosphorothioate stability in subcellular extracts, culture media, sera and cerebrospinal fluid. J Biochem Biophys Methods 20:259–267

Chiang M-Y, Chan H, Zounes MA, Freier SM, Lima WF, Bennett CF (1991) Antisense oligonucleotides inhibit intercellular adhesion molecule 1 expression by two distinct mechanisms. J Biol Chem 266:18162–18171

Chiasson BJ (1995) Studies on the role of c-fos in the mammalian brain: Application of antisense technology. PhD thesis, Dalhousie University, Halifax, Nova Scotia, Canada

Chiasson BJ, Hooper ML, Robertson HA (1992a) Amphetamine induced rotational behavior in non-lesioned rats: a role for c-fos expression in the striatum. Soc Neurosci Abstr 562:4

Chiasson, BJ, Hooper, M, Murphy, PR and HA Robertson (1992b) Antisense oligonucleotide eliminates vivo expression of c-fos in mammalian brain. Eur J Pharmacol Mol Pharmacol 227:451–453

Chiasson BJ, Armstrong JN, Hooper ML, Murphy PR, Robertson HA (1994) The application of antisense oligonucleotides to the brain: some pitfalls. Cell Mol Neurobiol 14:507–521

Chiasson BJ, Dennison Z, Robertson HA (1995) Amygdala kindling and immediate-early genes. Mol Brain Res 29:191–199

Chiasson BJ, Hong MGL, Robertson HA (1997) Putative roles for the inducible transcription factor c-fos in the central nervous system: studies with antisense oligonucleotides. Neurochem Int 31:459–475

Christy B, Lau LF, Nathans D (1988) A gene activated in mouse 3T3 cells by serum growth factors encodes a protein with "zinc finger" sequences. Proc Natl Acad Sci U S A 85:7857–7861

Cohen DR, Curran T (1988) fra-1 serum inducible, cellular immediate-early gene that encodes a Fos-related antigen. Mol Cell Biol 8:2063–2069

Cole AJ, Saffen DW, Baraban JM, Worley PF (1989) Rapid increase of an immediate early gene messenger RNA in hippocampal neurons by synaptic NMDA receptor activation. Nature 340:474–476

Curran T (1988) The fos oncogene. In: Reddy EP, Skalka AM, Curran T (eds) The oncogene handbook, vol 16. Elsevier Science (Biomedical Division), Amsterdam, pp 307–325

Curran T, Franza BR Jr (1988) Fos and Jun: the AP-1 connection. Cell 55:395–397

Curran T, Morgan JI (1995) Fos: an immediate-early transcription factor in neurons. J Neurobiol 26:403–412

Curran T, Teich NM (1982) Candidate product of the FBJ murine osteosarcoma virus oncogene: characterization of a 55,000-Dalton phosphoprotein. J Virol 42:114–122

Curran T, Gordon MB, Rubino KL, Sambucetti LC (1987) Isolation and characterization of the c-fos (rat) cDNA and analysis of post-translational modification in vitro. Oncogene 2:79–84

Davis HP, Squire LR (1984) Protein synthesis and memory: a review. Psychol Bull 96:518–559

Downs AW, Eddy NB (1932) The effect of repeated doses of cocaine on the rat. J Pharmacol Exp Ther 46:199–200

Dragunow M, Robertson HA (1987) Kindling stimulation induces c-fos protein(s) in granule cells of the rat dentate gyrus. Nature 329:441–442

Dragunow M, Robertson HA, Robertson GS (1988) Effects of kindled seizures on the induction of c-fos protein(s) in mammalian neurons. Exp Neurol 102:261–263

Dragunow M, Currie RW, Faull RLM, Robertson HA, Jansen K (1989) Immediate-early genes, kindling and long-term potentiation. Neurosci Biobehav Rev 13:301–313

Dragunow M, Lawlor PA, Chiasson BJ, Robertson HA (1993) Antisense to c-fos suppresses both Fos and Jun B expression in rat striatum and generates apomorphine- and amphetamine-induced rotation. Neuroreport 5:305–306

Dragunow M, Tse C, Glass M, Lawlor P (1994) c-fos antisense reduces expression of krox 24 in rat caudate and neocortex. Cell Mol Neurobiol 14:395–405

During MJ, Ryder KM, Spencer DD (1995) Hippocampal GABA transporter function in temporal-lobe epilepsy. Nature 376:174–177

Ernfors P, Bengzon J, Kokaia Z, Persson H, Lindvall O (1991) Increased levels of messenger RNA for neurotrophic factors in the brain during kindling epileptogenesis. Neuron 7:165–176

Gao W-Y, Han F-S, Storm C, Egan W, Cheng Y-C (1991) Phosphorothioate oligonucleotides are inhibitors of human DNA polymerases and RNase H: implications for antisense technology. Mol Pharmacol 41:223–229

Geinisman Y, Morrell F, deToledo-Morrell L (1988) Remodelling of synaptic architecture during hippocampal "kindling". Proc Natl Acad Sci U S A 85:3260–3264

Gillardon F, Beck H, Uhlmann E, Herdegen T, Sandkuler J, Peyman A, Zimmermann M (1994) Inhibition of c-fos protein expression in rat spinal cord by antisense oligodeoxynucleotide superfusion. Eur J Neurosci 6:880–884

Goddard GV, McIntyre D, Leech C (1969) A permanent change in brain function resulting from daily electrical stimulation. Exp Neurol 25:295–330

Goelet P, Castellucci VF, Schacher S, Kandel ER (1986) The long and short of long-term memory – a molecular framework. Nature 322:419–422

Goodchild J (1989) Inhibition of gene expression by oligonucleotide. In: Cohen JS (ed) Oligonucleotides: antisense inhibitors of gene expression. CRC Press, Boca Raton, pp 53–77

Graybiel AM, Moratalla R, Robertson HA (1990) Amphetamine and cocaine induce drug-specific activation of the c-fos gene in striosome-matrix compartments and limbic subdivisions of the striatum. Proc Natl Acad Sci U S A 87:6912–6916

Guvakova MA, Yakubov LA, Vlodavsky I, Tonkinson JL, Stein CA (1995) Phosphorothioate oligodeoxynucleotides bind to basic fibroblast growth factor, inhibit its binding to cell surface receptors, and remove it from low affinity binding sites on extracellular matrix. J Biol Chem 270:2620–2627

Hebb MO, Robertson HA (1997a) End-capped antisense oligonucleotides effectively inhibit gene expression in vivo and offer a low-toxicity alternative to fully modified phosphorothioate oligodeoxynucleotides. Mol Brain Res (in press)

Hebb MO, Robertson HA (1997b) Coordinate suppression of striatal ngfi-a and c-fos produces locomotor asymmetry and upregulation of IEGs in the globus pallidus. Mol Brain Res (in press)

Heilig M, Pich EM, Koob GF, Yee F, Wahlestedt C (1992) In vivo down regulation of neuropeptide Y (NPY) Y1 receptors by ICV antisense oligodeoxynucleotide administration is associated with signs of anxiety in rats. Soc Neurosci Abstr 642:18

Heilig M, Engel JA, Söderpalm B (1993) C-fos antisense in the nucleus accumbens blocks the locomotor stimulant action of cocaine. Eur J Pharmacol 236:339–340

Hengerer B, Lindholm D, Heumann R, Ruther U, Wagner EF, Thoenen H (1990) Lesion-induced increase in nerve growth factor mRNA is mediated by c-fos. Proc Natl Acad Sci U S A 87:3899–3903

Hirai S-I, Ryseck R-P, Mechta F (1989) Characterization of jun-D: a new member of the jun proto-oncogene family. EMBO J 8:1433–1439

Hooper ML, Chiasson BJ, Robertson HA (1994) Infusion into the brain of an antisense oligonucleotide to the immediate-early gene c-fos suppresses production of Fos and produces a behavioral effect. Neuroscience 63:917–924

Hosford DA, Simonato M, Cao Z, Garcia-Cairasco N, Silver JM, Butler L, Shin C, McNamara JO (1995) Differences in the anatomic distribution of immediate-early gene expression in amygdala and angular bundle kindling development. J Neurosci 15:2513–2523

Hughes P, Dragunow M (1995) Induction of immediate-early genes and the control of neurotransmitter-regulated gene expression within the nervous system. Pharmacol Rev 47:133–178

Hughes P, Singleton K, Dragunow M (1994) MK-801 does not attenuate immediate-early gene expression following an amygdala afterdischarge. Exp Neurol 128:276–283

Hunt SP, Pini A, Evan G (1987) Induction of c-fos-like protein in spinal cord neurons following sensory stimulation. Nature 328:632–634

Hunter JC, Woodburn VL, Durieux C, Pettersson EKE, Poat JA, Hughes J (1995) C-fos antisense oligodeoxynucleotide increases formalin-induced nociception and regulates preprodynorphin expression. Neuroscience 65:485–492

Ji R-R, Zhang Q, Bedecs K, Arvidsson J, Zhang X, Xu X-J, Wiesenfeld-Hallin Z, Bartfai T, Hökfelt (1994) Galanin antisense oligonucleotides reduce galanin levels in dorsal root ganglia and induce autotomy in rats after axotomy. Proc Natl Acad Sci U S A 91:12540–12543

Kerppola TK, Curran T (1991a) Fos-Jun heterodimers and Jun homodimers bend DNA in opposite directions: implications for transcription factor cooperativity. Cell 66:317–326

Kerppola TK, Curran T (1991b) DNA bending by Fos and Jun: the flexible hinge model. Science 254:1210–1214

Konradi C, Kobierski LA, Nguyen TV, Heckers S, Hyman SE (1993) The c-AMP-response-element-binding-protein interacts but Fos protein does not interact, with the proenkephalin enhancer in rat striatum. Proc Natl Acad Sci U S A 90:7005–7009

Kouzarides T, Ziff E (1988) The role of the leucine zipper in the fos-jun interaction. Nature 336:646–651

Krieg AM (1993) Uptake and efficacy of phosphodiester and modified antisense oligo-nucleotides in primary cell cultures. Clin Chem 39:710–712

Krieg AM, Yi A-K, Matson S, Waldschmidt TJ, Bishop GA, Teasdale R, Koretzky GA, Klinman DM (1995) CpG motifs in bacterial DNA trigger direct B-cell activation. Nature 374:546–548

Labiner DM, Butler LS, Cao Z, Hosford DA, Shin C, McNamara JO (1993) Induction of c-fos mRNA by kindled seizures: complex relationship with neuronal burst firing. J Neurosci 13:744–751

Lamprecht R, Dudai Y (1996) Transient expression of c-fos in rat amygdala during training is required for encoding conditioned taste aversion memory. Learn Mem 3:31–41

Lau LF, Nathans D (1987) Expression of a set of growth-regulated immediate-early genes in BALB/c3T3 cells: coordinate regulation with c-fos or c-myc. Proc Natl Acad Sci U S A 4:1182–1186

Lemaire P, Revelant O, Bravo R, Charnay P (1988) Two mouse genes encoding potential transcription factors with identical DNA-binding domains are activated by growth factors in cultured cells. Proc Natl Acad Sci U S A 85:4691–4695

Liu PK, Salminen A, He YY, Jiang MH, Xue JJ, Liu JS, Hsu CY (1994) Suppression of ischemia-induced Fos expression and AP-1 activity by an antisense oligodeoxynucleotide to c-fos mRNA. Ann Neurol 36:566–576

Maki Y, Bos TJ, Davis C, Starbuck M, Vogt PK (1987) Avian sarcoma virus 17 carries a new oncogene jun. Proc Natl Acad Sci U S A 84:2848–2852

Masquilier D, Sassone-Corsi P (1992) Transcriptional cross-talk: nuclear factors CREM and CREB bind to AP-1 sites and inhibit activation by Jun. J Biol Chem 267:22460–22466

Matteucci MD, Wagner RW (1996) In pursuit of antisense. Nature 384:20–22

McKnight SL (1991) Molecular zippers in gene regulation. Sci Am 264:54–64

Merchant, KM (1994) c-fos antisense oligonucleotide specifically attenuates haloperidol-induced increases in neurotensin/neuromedin N mRNA expression in rat dorsal striatum. Mol Cell Neurosci 5:336–344

Milbrandt J (1987) A nerve growth factor-induced gene encodes a possible transcriptional regulatory factor. Science 238:797–799

Möller C, Bing O, Heilig M (1994) c-fos Expression in the amygdala: in vivo antisense modulation and role in anxiety. Cell Mol Neurobiol 14:415–423

Moore SD, Madamba SG, Schweitzer P, Siggins GR (1994) Voltage-dependent effects of opioid peptides on hippocampal CA3 pyramidal neurons in vitro. J Neurosci 14:809–820

Moratalla R, Robertson HA, Graybiel AM (1992) Dynamic regulation of NGFI-A (zif268, egr1) gene expression in the striatum. J Neurosci 12:2609–2622

Moratalla R, Vickers EA, Robertson HA, Cochran BH, Graybiel AM (1993) Coordinate expression of c-fos and junB is induced in the striatum by cocaine. J Neurosci 13:423–433
Morgan JI, Curran T (1991) Stimulus-transcription coupling in the nervous system: Involvement of the inducible proto-oncogenes fos and jun. Annu Rev Neurosci 14:421–451
Morgan JI, Cohen DR, Hempstead JL, Curran T (1987) Mapping patterns of c-fos expression in the central nervous system after seizure. Science 237:192–197
Morris BJ, Johnston HM (1995) A role for hippocampal Opioids in long-term functional plasticity. Trends Neurosci 18:350–355
Nellen W, Lichtenstein C (1993) What makes an mRNA anti-sense-itive? Trends Biochem Sci 18:419–423
Nestler EJ, Hope BT, Widnell KL (1993) Drug addiction: a model for the molecular basis of neural plasticity. Neuron 11:995–1006
Nishina H, Sato H, Suzuki T, Sato N, Iba H (1990) Isolation and characterisation of Fra-2, an additional member of the fos gene family. Proc Natl Acad Sci U S A 87:3619–3623
Nishizuka M, Okada R, Seki K, Arai Y, Iizuka R (1991) Loss of dendritic synapses in the medial amygdala associated with kindling. Brain Res 522:351–355
Osen-Sand A, Catsicas M, Stapel JK, Jones KA, Ayala G, Knowles J, Grenningloh G, Catsicas S (1993) Inhibition of axonal growth by SNAP-25 antisense oligonucleotides in vitro and in vivo. Nature 364:445–448
Paul ML, Graybiel AM, David J-C, Robertson HA (1992) D_1-like and D_2-like dopamine receptors synergistically activate rotation and c-fos expression in the dopamine-depleted striatum in a rat model of Parkinson's disease. J Neurosci 12:3729–3742
Paul ML, Currie RW, Robertson HA (1995) Priming of a D1 dopamine receptor behavioural response is dissociated from striatal immediate-early gene activity. Neuroscience 66:347–359
Perlin JB, Gerwin CM, Panchision DM, Vicks RS, Jakoi ER, DeLorenzo RJ (1993) Kindling produces long-lasting and selective changes in gene expression of hippocampal neurons. Proc Natl Acad Sci U S A 90:1741–1745
Racine RJ (1972) Modification of seizure activity by electrical stimulation: motor seizure. Electroencephalogr Clin Neurophysiol 38:281–294
Robertson GS, Herrera DG, Dragunow M, Robertson HA (1989) L-Dopa activates c-fos expression in the striatum of 6-hydroxydopamine-lesioned rats. Eur J Pharmacol 159:99–100
Robertson GS, Tetzlaff W, Bedard A, St-Jean M, Wigle N (1995) c-fos mediates antipsychotic-induced neurotensin gene expression in the rodent striatum. Neuroscience 67:325–344
Robertson HA (1992a) Immediate-early genes, neuronal plasticity, and memory. Biochem Cell Biol 70:729–737
Robertson HA (1992b) Dopamine receptor interactions: some implications for the treatment of Parkinson's disease. Trends Neurosci 15:201–206
Robertson HA, Peterson MR, Murphy K, Robertson GS (1989) D_1-dopamine receptor agonists selectively activate striatal c-fos independent of rotational behaviour. Brain Res 503:346–349
Rusak B, Robertson HA, Wisden W, Hunt SP (1990) Light pulses that shift rhythms induce gene expression in the suprachiasmatic nucleus. Science 248:1237–1240
Ryder K, Lau LF, Nathans D (1988) A gene activated by growth factors is related to the oncogene v-jun. Proc Natl Acad Sci U S A 85:1487–1491
Ryder K, Lanahan A, Perez-Albuerne E, Nathans D (1989) Jun-D: a third member of the Jun gene family. Proc Natl Acad Sci U S A 86:1500–1503
Sassone-Corsi P, Sisson JC, Verma IM (1988) Transcriptional autoregulation of the proto-oncogene fos. Nature 334:314–319
Sharp FR (1994) The sense of antisense fos oligonucleotides. Ann Neurol 36:555–556

Sheng M, Greenberg ME (1990) The regulation and function of c-fos and other immediate-early genes in the nervous system. Neuron 4:477–485

Sheng M, McFadden G, Greenberg ME (1990) Membrane depolarization and calcium induce c-fos transcription via phosphorylation of transcription factor CREB. Neuron 4:571–582

Shin C, McNamara JO, Morgan JI, Curran T, Cohen DR (1990) Induction of c-fos mRNA expression by afterdischarge in the hippocampus of naive and kindled rats. J Neurochem 55:1050–1055

Silvia CP, King GR, Lee TH, Xue Z-Y, Caron MG, Ellinwood EH (1994) Intranigral administration of D_2 dopamine receptor antisense oligodeoxynucleotides establishes a role for nigrostriatal D_2 autoreceptors in the motor actions of cocaine. Mol Pharmacol 46:51–57

Sinomato M, Hosford DA, Labiner DM, Shin C, Mansbach HH, McNamara JO (1991) Differential expression of immediate early genes in the hippocampus in the kindling model of epilepsy. Mol Brain Res 11:115–124

Sommer W, Bjelke B, Ganten D, Fuxe K (1993) Antisense oligonucleotide to c-fos induces ipsilateral rotational behavior to d-amphetamine. Neuroreport 5:277–280

Sommer W, Rimondini R, O'Connor W, Hansson AC, Ungerstedt U, Fuxe K (1996) Intrastriatal injected c-fos antisense oligonucleotide interferes with striatonigral but not striatopallidal γ-aminobutyric acid transmission in the conscious rat. Proc Natl Acad Sci U S A 93:14134–14139

Sonnenberg JL, Rauscher JR III, Morgan JI, Curran T (1989) Regulation of proenkephalin by Fos and Jun. Science 246:1622–1625

Standifer KM, Chien C-C, Wahlestedt C, Brown GP, Pasternak GW (1994) Selective loss of d opioid analgesia and binding by antisense oligodeoxynucleotides to a d opioid receptor. Neuron 12:805–810

Stein CA, Cheng Y-C (1993) Antisense oligonucleotides as therapeutic agents – is the bullet really magical? Science 261:1004–1012

Stein CA, Subasinghe C, Shinozuka K, Cohen JS (1988) Physicochemical properties of phosphorothioate oligodeoxynucleotides. Nucleic Acids Res 16:3209–3221

Stephenson ML, Zamecnik PC (1978) Inhibition of Rous sarcoma viral RNA translation by specific oligodeoxynucleotide. Nucleic Acid Res 16:3209–3221

Struhl K (1991) Mechanisms for diversity in gene expression patterns. Neuron 7:177–181

Sukhatme VP, Cao X, Chang LC, Tsai-Morris C-H, Stamenkovich D, Ferreira PCP, Cohen DR, Edwards SA, Shows TB, Curran T, LeBeau MM, Adamson EDA (1988) A zinc finger-encoding gene coregulated with c-fos during growth and differentiation and after cellular depolarization. Cell 53:37–43

Sutula T, He XX, Cavazos J, Scott G (1988) Synaptic reorganization in the hippocampus induced by abnormal functional activity. Science 239:1147–1150

Terman GW, Wagner JJ, Chavkin C (1994) Kappa opioids inhibit induction of long-term potentiation in the dentate gyrus of the Guinea pig hippocampus. J Neurosci 14:4740–4747

Teskey CG, Atkinson BG, Cain DP (1991) Expression of the proto-oncogene c-fos following electrical kindling in the rat. Mol Brain Res 11:1–10

Thierry AR, Dritschilo A (1992) Intracellular availability of unmodified, phosphorothioated, and liposomally encapsulated oligodeoxynucleotides for antisense activity. Nucleic Acid Res 20:5691–5698

Tseng LF, Collins KA (1994) Antisense oligodeoxynucleotide to a d-opioid receptor given intrathecally blocks ICV administered b-endorphin-induced antinociception in the mouse. Life Sci 55:PL127–131

Ungerstedt U (1971a) Striatal dopamine release after amphetamine or nerve degeneration revealed by rotational behaviour. Acta Physiol Scand [Suppl] 367:49–68

Ungerstedt U (1971b) Postsynaptic supersensitivity after 6-hydroxydopamine induced degeneration of the nigro-striatal dopamine system. Acta Physiol Scand [Suppl] 367:69–93

Wagner JJ, Terman GW, Chavkin C (1993) Endogenous dynorphins inhibit excitatory neurotransmission and block LTP induction in the hippocampus. Nature 363:451–454

Wagner RW (1994) Gene inhibition using antisense oligodeoxynucleotides. Nature 372:333–335

Wahlestedt C (1994) Antisense oligonucleotide strategies in neuropharmacology. Trends Pharmacol Sci 15:42–46

Wahlestedt C, Golanov E, Yamamoto S, Yee F, Ericson H, Yoo H, Inturrisi CE, Reis DJ (1993a) Antisense oligonucleotides to NMDA-R1 receptor channel protect cortical neurons from excitotoxicity and reduce focal ischaemic infarctions. Nature 363:260–263

Wahlestedt C, Pich EM, Koob GF, Yee F, Heilig M (1993b) Modulation of anxiety and neuropeptide Y-Y1 receptors by antisense oligodeoxynucleotides. Science 259:528–531

Watanabe Y, Johnson RS, Butler LS, Binder DK, Spiegelman BM, Papaioannou VE, McNamara JO (1996) Null mutation of c-fos impairs structural and functional plasticities in the kindling model of epilepsy. J Neurosci 16:3827–3836

Weiss B, Zhou L-W, Zhang S-P, Qin Z-H (1993) Antisense oligodeoxynucleotide inhibits D_2 Dopamine receptor-mediated behavior and D_2 messenger RNA. Neuroscience 55:607–612

Whitesell L, Geselowitz D, Chavany C, Fahmy B, Walbridge S, Alger JR, Neckers LM (1993) Stability, clearance, and disposition of intraventricularly administered oligodeoxynucleotides: implications for therapeutic application within the central nervous system. Proc Natl Acad Sci U S A 90:4665–4669

Wisden W, Errington ML, Williams S, Dunnett SB, Waters C, Hitchcock D, Evan G, Bliss TVP, Hunt SP (1990) Differential expression of immediate-early genes in the hippocampus and spinal cord. Neuron 4:603–614

Wollnik F, Brysch W, Uhlmann E, Gillardon F, Bravo R, Zimmermann M, Schlingensiepen KH, Herdegen T (1995) Block of c-fos and jun-B expression by antisense-oligonucleotides inhibits light-induced phase shifts of the mammalian circadian clock. Eur J Neurosci 7:388–393

Woolf TM, Jennings GB, Rebagliati M, Melton DA (1990) The stability, toxicity and effectiveness of unmodified and phosphorothioate antisense oligodeoxynucleotides in Xenopus oocytes and embryos. Nucleic Acids Res 18:1763–1769

Woolf TM, Melton DA, Jennings CGB (1992) Specificity of antisense oligonucleotides in vivo (Xenopus oocytes). Proc Natl Acad Sci U S A 89:7305–7309

Yee F, Ericson H, Reis DJ, Wahlestedt C (1994) Cellular uptake of intracerebroventricularly administered biotin- or digoxigenin- labeled antisense oligodeoxynucleotides in the rat. Cell Mol Neurobiol 14:475–486

Zamecnik PC, Stephenson NL (1978) Inhibition of Rous sarcoma virus replication and cell transformation by a specific oligodeoxynucleotide. Proc Natl Acad Sci U S A 75:280–294

Zerial M, Toschi L, Ryseck R-P, Schuermann M, Muller R, Bravo R (1989) The product of a novel growth factor activated gene, fos-B, interacts with JUN proteins enhancing their DNA binding activity. EMBO J 8:805–813

Zhang M, Creese I (1993) Antisense oligonucleotide reduces brain dopamine D2 receptor: behavioral correlates. Neurosci Lett 161:223–226

Zhou LM, Zhang SP, Qin ZH, Weiss B (1994) In vivo administration of an oligonucleotide antisense to the D2 dopamine receptor messenger RNA inhibits D2 dopamine receptor mediated behavior and the expression of D2 dopamine receptors in mouse striatum. J Pharmacol Exp Ther 268:1015–1023

CHAPTER 11
Inhibition of G Proteins by Antisense Drugs

F. KALKBRENNER, B. WITTIG, and G. SCHULTZ

A. Introduction

In a multicellular organism, cells communicate with each other mostly through the cell surface, which is represented by the plasma membrane with its lipids and proteins. A subset of the membrane proteins, i.e., receptors, are specialized to recognize and bind signalling molecules if arriving at sufficient concentrations at the cell surfaces. These signalling molecules, called receptor ligands, are either synthesized and secreted by other cells of the organism or taken up from the outside world. Ligands can be hormones, neurotransmitters, autacoids, growth factors, odorants, or physical signals, such as light. The binding of ligands to their cognate receptors leads to intracellular responses, e.g., changes in intracellular ion concentrations, de novo synthesis of proteins, or secretion of presynthesized proteins or transmitters. In many regulatory processes, the intracellular response is mediated by so-called second messengers, e.g., cyclic adenosine or guanosine monophosphate (cAMP or cGMP) or inositol 1,4,5-trisphosphate (IP_3), which are generated upon receptor activation as a consequence of the ligand signalling. The extracellular signalling components or ligands are called first messengers, and the membrane structures leading to second messenger generation are generally covered by the terms receptors, signal transducers (G proteins), and effectors.

Membrane receptors can be divided into three groups, each consisting of a superfamily of proteins with a common architecture. The first group typically consists of receptors for fast neurotransmitters. These receptors are directly ligand-gated channels, e.g., the nicotinic acetylcholine receptors, the γ-aminobutyric acid $(GABA)_A$ receptors or the ionotropic glutamate receptors. They are formed by oligomeric proteins spanning the membrane several times to function as an ion channel. After binding of the ligand to the extracellular domain, protein conformation changes, allowing for channel opening and subsequent change of intracellular ion concentration. The second class of membrane receptors, i.e., the enzyme-linked receptors represented by the protein tyrosine-kinase-/protein tyrosine phosphatase-/guanylyl-cyclase-linked types, are typical receptors for various hormones, e.g., insulin and growth factors. They consist of monomeric or homodimeric proteins with one transmembrane domain. Binding of the ligand to the extracellular domain

List of Abbreviations

ATG	Adenine, thymine, and guanine (the nucleotides)
BSA	Bovine serum albumin
$[Ca^{2+}]_i$	Free cytosolic calcium concentration
cAMP	Cyclic AMP
FITC	Fluorescein isothiocyanate
fMLP	Formyl-Met-Leu Phe
GABA	γ-Aminobutyric acid
GDP	Guanosine diphosphate
Gpp(NH)p	Guanosine 5'-[β,γ-imido]triphosphate
G Protein	Heterotrimeric GTP binding protein
GTP	Guanosine triphosphate
5-HT	5-Hydroxytryptamine, serotonin
IRS-1	Insulin receptor substrate-1
LTB_4	Leukotriene B_4
MDCK	Madin-Derby canine kidney (*cells*)
M6G	Morphine-6β-glucuronide
P3	inositol 1,4,5-trisphosphate
PLC	Phospholipase C
PTPase	Phosphotyrosine phosphatase
RT-PCR	Reverse transcriptase polymerase chain reaction
SST	Somatostatin
TRH	Thyrotropin-releasing hormone
VIP	Vasoactive intestinal peptide

and accompanied conformational change activate intracellular kinase-, phosphatase-, or cyclase-domains and lead to tyrosine phoshorylation or dephosphorylation of intracellular proteins, contact to adapter proteins to stimulate signalling cascades, or generation of cGMP as second messenger. The third class of membrane receptors, the G-protein-coupled type, typically includes receptors for the majority of hormones, such as adrenoceptors, muscarinic acetylcholine receptors, peptide neurotransmitter receptors, and metabotropic glutamate receptors. The superfamily of G-protein-coupled receptors includes at least 1000 molecularly different polypeptides of which the cDNAs of several hundreds have been cloned and sequenced. Their common structural motifs are an external amino (N) terminus, a cytoplasmatic carboxy (C) terminus, and seven stretches of hydrophobic amino acids which are believed to span seven times the cytoplasmic membrane, thereby generating three extra- and three intracellular loops (for reviews, see BALDWIN 1994; STRADER et al. 1994; GUDERMANN et al. 1995). The seven transmembrane domains are predicted to form α helices; therefore, these receptors are alternatively named heptahelical receptors. The G-protein-coupled receptors do not possess intrinsic ion channel or enzyme activity; instead they interact with heterotrimeric guanine nucleotide-binding proteins (G proteins), which in turn couple to different effector systems.

G proteins are composed of three subunits, α, β, and γ, of which at least 23 (including splice variants) different α forms (Gα), six different β forms (Gβ), and 11 different γ forms (Gγ) exist (for reviews, see SIMON et al. 1991; BIRNBAUMER 1992; OFFERMANNS and SCHULTZ 1994; NEER 1995; MILLIGAN 1995; NÜRNBERG et al. 1995; GUDERMANN et al. 1996). Binding of the ligand to the extracellular binding pocket of the heptahelical receptor induces a conformational change which causes coupling to G proteins, release of guanosine diphosphate (GDP) and guanosine triphosphate (GTP) binding to the G-protein α subunit, dissociation of G$\alpha\beta\gamma$ from the receptor and release of the G$\beta\gamma$ dimer from the Gα-GTP subunit. Both the activated GTP-bound α subunit and free $\beta\gamma$ subunits have the capability of activating or inactivating different effector systems, e.g., adenylyl cyclases, phospholipases C (PLCs)-β, and ion channels, whereby Gα and G$\beta\gamma$ act independently of each other. The intrinsic GTPase activity of the Gα terminates the activation of the α subunit, which then, in its GDP-bound form, reassociates with the G$\beta\gamma$ dimer. Figure 1 illustrates the G-protein effector systems involved in regulation of intracellular calcium homeostasis.

A given receptor uses a limited set of G proteins to interact with and to affect a given effector system. A few years ago, it was assumed that the specificity in receptor–G protein–effector coupling is only determined by the α subunit of the G protein, but it is now well accepted that in addition to the α subunits the $\beta\gamma$ dimers determine specificity of receptor–G protein–effector interaction. This increases the number of hypothetical receptor–G protein–effector combinations from several hundreds to thousands.

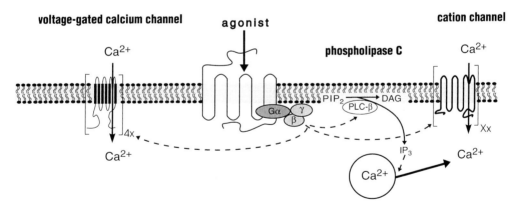

Fig. 1. G-protein-coupled receptor. G-protein-mediated signal transduction pathways leading to an increase in the intracellular calcium concentration. 4× indicates the four repeats of calcium channel α_1 subunits; X× indicates the likely oligomeric structure of nonselective cation channels of the trp family. *PLC*, phospholipase C; *PIP$_2$*, phosphatidyl inositol 4,5-bisphosphate; *DAG*, diacylglycerol; trp, protein of the gene trp (transient receptor potential)

In G-protein-mediated pathways, receptors, effectors, and downstream elements, e.g., calcium channels and phosphodiesterases, are targets for drugs. In the last few years, results from antisense experiments have revealed that a high specificity and selectivity of receptor–G protein–effector coupling arises from the diversity of the G-protein subunit composition. This makes G-protein heterotrimers of interest as pharmacological targets, assuming that compounds can be developed that specifically interfere with individual G proteins.

Our review will focus on the question of how antisense technology can be used to determine the identity of individual subtypes of G-protein subunits involved in signal transduction cascades, and how the specific inhibition of G-protein expression by antisense drugs can be used as a new therapeutic approach.

B. Antisense Approaches to Study Signal Transduction Pathways

In recent years antisense technology has attracted much attention. It enables researchers to selectively inhibit the biosynthesis of one particular gene product and to investigate which cell function is lost (for reviews, see HÉLÈNE and TOULMÉ 1990; CROOKE 1993; MILLIGAN et al. 1993; STEIN and CHENG 1993; CROOKE and BENNETT 1996; for recent books covering this topic, see CROOKE 1995; AGRAWAL 1996; SCHLINGENSIEPEN et al. 1996; RAFFA and PORRECA 1996).

Antisense techniques have been used to inhibit G-protein-mediated signal transduction on each of three levels. The expression of heptahelical receptors was successfully inhibited in cell culture by oligonucleotides and in animals after intracerebroventricular injection of antisense oligonucleotides directed against neuropeptide receptors and dopamine receptors (for reviews, see WAHLESTEDT 1994, 1996; and PASTERNAK and STANDIFER 1995; see also Chap. 9, this volume). Different methods were used to inhibit G-protein expression (for reviews, see BAERTSCHI 1994; ALBERTS and MORRIS 1994; KALKBRENNER et al. 1996; RAFFA 1996). Finally, expression of effector proteins, e.g., calcium channels or PLC, were reached by the antisense approach.

In principle, several ways exist to inhibit the expression of one gene selectively. One is the transcription of the respective cDNA in antisense orientation within the nucleus. For this purpose the target sequence is recombined in reversed orientation with an expression plasmid and the construct transfected or injected into the cells of interest. The length of the transcribed antisense RNA can vary from 40 to 50 up to several hundred nucleotides. Antisense sequences can also be delivered as synthetic oligoribo- or oligodeoxyribonucleotides. Various chemical modifications may be introduced to increase the stability or the cellular uptake of the oligomers as far as modifications do not impair recognition of the complementary sequence.

C. Inhibition of G-Protein Functions by Antisense RNA and Genomic Knockout

For inhibition of expression of the mRNA encoding a particular G-protein subunit, part of the corresponding cDNA can be expressed in reversed orientation. The expressed antisense RNA anneals to the mRNA and interferes in this way with its transport and translation. Table 1 gives an overview of the studies in which this technique has been used to inhibit the expression of G-protein subunits. For stable expression, antisense RNA constructs have to be transfected, together with a resistance marker gene, and cells have to be clonally selected for high level antisense RNA transcription. Liu et at. (1994) used subclones of the pituitary cell line GH_4C_1 stably expressing the long and short forms of the dopamine D_2 receptor. These cells were then transfected for a second time with constructs for transcription of α_o and α_{i2} antisense RNA, and single clones of transfected cells were selected. In clones with the highest transcription of antisense constructs, α_o mRNA was completely eliminated and α_{i2} mRNA was reduced in concentration. These subclones were functionally characterized with respect to inhibition of Ca^{2+} influx induced by the dihydropyridine Ca^{2+} channel opener Bay K 8644. Inhibition of adenylyl cyclase by activation of transfected D_{2L} and D_{2S} receptors, as well as by activation of the endogenous M_4 and somatostatin (SST) receptors, was measured as a second effector system. In different subclones, the inhibition of calcium influx induced by activated D_{2S}, M_4, and SST receptors was abolished by expression

Table 1. G-protein involvements detected by expression of antisense RNA

Receptors	G-protein	Effector system	Cells	Reference
Somatostatin	α_o	VOC↓	GH_4C_1	Liu et al. 1994
Somatostatin	α_{i2}	AC↓	GH_4C_1	Liu et al. 1994
Somatostatin	α_{i2}	AC↓	FTO-2B	Moxham et al. 1993a
A_1 adenosine	α_{i2}	AC↓	FTO-2B	Moxham et al. 1993a
M_4 muscarinic	α_o	VOC↓	GH_4C_1	Liu et al. 1994
M_4 muscarinic	α_{i2}	AC↓	GH_4C_1	Liu et al. 1994
D_{2L}, D_{2S} dopamine	α_{i2}	AC↓	GH_4C_1	Liu et al. 1994
D_{2L}, D_{2S} dopamine	α_o	VOC↓	GH_4C_1	Liu et al. 1994
TRH	α_s	AC↓	GH_3	Paulsen et al. 1992
VIP	α_s	AC↓	GH_3	Paulsen et al. 1992
fMLP	α_o	$[Ca^{2+}]_i$↑	HL-60	Goetzel et al. 1994
LTB_4	α_o	$[Ca^{2+}]_i$↑	HL-60	Goetzel et al. 1994
LTB_4	α_{i2}	AC↓	HL-60	Goetzel et al. 1994
	α_s	Differentiation	HL-60	Meißner et al. 1996
	α_{i2}	Differentiation	F9	Watkin et al. 1992
	α_{i2}	AC, body weight, liver mass, fat mass	Transgenic mice	Moxham et al. 1993a

AC, adenylyl cyclase; $[Ca^{2+}]_i$, cytosolic calcium concentration; fMLP, formyl-Met-Leu-Phe; LTB_4, leukotriene B_4; TRH, thyrotropin-releasing hormone; VIP, vasoactive intestinal polypeptide; VOC, voltage-operated calcium channels; ↓, inhibition; ↑, stimulation.

of the α_o antisense construct. Inhibition by the D_{2L} receptor was only reduced by 70%, even in subclones in which no mRNA coding for α_o or α_{i2} protein could be detected. Inhibition of basal cAMP accumulation from activation of these four receptors was abolished in all subclones in which the α_{i2} antisense RNA was transcribed. LIU and colleagues (1994) also investigated the inhibition of vasoactive intestinal peptide (VIP)-stimulated cAMP accumulation in the α_{i2} antisense RNA transcribing clones. The inhibition of VIP-induced adenylyl cyclase activity mediated by D_{2L} and M_4 receptors was abolished, whereas the inhibition by activated D_{2L} receptors was reduced by only 30%. Paradoxically, the inhibition of VIP-stimulated cAMP accumulation mediated by activation of SST receptors switched to a small stimulation in cells depleted of $G\alpha_{i2}$ mRNA. Serotonin-induced stimulation instead of inhibition of VIP-mediated adenylyl cyclase activation has been shown by the same group in a clone transcribing the $G\alpha_{i2}$ antisense RNA and stably expressing 5-hydroxytryptamine (5-HT)$_{1A}$ receptors (ALBERT and MORRIS 1994). The authors explained these curious results by a potentiation of G_s-activation of adenylyl cyclase by an excess of $\beta\gamma$ subunits released from G_{i1} or G_o in the $G\alpha_{i2}$-depleted cells. Nevertheless, they demonstrated a clear difference in coupling of the closely related D_{2S} and D_{2L} receptors to voltage-gated Ca^{2+} channels, as well as to adenylyl cyclases.

In cells constitutively transcribing antisense RNA, one has to take care that subcloning procedures do not alter the expression of other components in a particular signal transduction cascade. This could change the specificity and degree of coupling between the components under investigation. To overcome these problems, GOETZL et al. (1994) used transient transcription of $G\alpha_{i1}$, $G\alpha_{i2}$, $G\alpha_{i3}$ and $G\alpha_{oA}$ antisense RNA from constructs transfected into HL-60 cells by lipofection. The antisense constructs were identical to those used by LIU et al. (1994). Maximal repression (80%–95%) of target proteins as analyzed by Western blotting was reached 6–7 days after transfection. The authors reported that suppression of $G\alpha_o$, but not of $G\alpha_i$, protein expression partially reduced the responses to the chemotractants formyl-Met-Leu-Phe (fMLP) and leukotriene B_4 (LTB$_4$) in these cells as measured by diminished increases in intracellular Ca^{2+}, β-glucuronidase release, and chemotaxis. Inhibition of adenylyl cyclase was affected in cells transfected with the $G\alpha_{i2}$ antisense RNA-transcribing construct. These results are surprising in functional aspects because $G\alpha_o$ is a G protein mostly expressed in neuronal and neuroendocrine tissue. In addition, coupling of G_o to PLC was not shown before for mammalian systems. However, coupling of different heptahelical receptors to PLC-β via G_o has been show in *Xenopus* oocytes (CHEN et al. 1995; KASAHARA and SUGIYAMA 1994; QUICK et al. 1994).

PAULSSEN et al. (1992) used a similar approach to target the expression of $G\alpha_s$ in the rat pituitary cell line GH$_3$. They recombined a part of the coding sequence followed by the known part of the 3'-noncoding sequence in reverse orientation with a plasmid driving transcription by an inducible promoter. This plasmid was transfected by electroporation into GH$_3$ cells. Twenty-four hours after induction of the promoter, a concentration-dependent reduction of $G\alpha_s$ mRNA and protein by up to 95% was measured. The VIP-induced increase in cAMP was reduced by up to 30% in transfected cells and the thyrotropin-releasing hormone (TRH)-induced increase in cAMP was reduced by up to 85%. The guanosine 5'-[β,γ-imido]triphosphate [Gpp(NH)p]-induced increase in cAMP was unaffected by transcription of the $G\alpha_s$ antisense RNA. Results indicate that even after a 95% reduction of $G\alpha_s$ protein and mRNA, the remaining $G\alpha_s$ molecules are sufficient to mediate a coupling to the effector system in these cells. One can also speculate that different receptors need different amounts of G proteins to accomplish full functional coupling to an effector system.

The studies described so far demonstrate that the transcription of antisense RNA effectively suppresses gene expression. One limitation of this method arises from the length of the antisense RNA. If antisense RNAs of several hundred nucleotides are used, the selective knockout of one member in gene families, like in those for β and γ subunits of G proteins, cannot be achieved. The untranslated regions, which in most of the known genes contain a sufficiently high sequence variability, are not known for most of the β and γ subunit genes of G proteins and can, therefore, not be utilized for long antisense transcripts.

The antisense constructs used by WATKINS et al. (1992) provide the best solution for this problem. This group constructed very short antisense RNA constructs which transcribe only 39 bases of the 5'-untranslated region including the ATG translation start codon. Sequences for $G\alpha_{i1}$, $G\alpha_{i2}$, $G\alpha_{i3}$, and $G\alpha_o$ were cloned in reverse orientation into a retroviral expression vector. These constructs were used to create recombinant viruses for efficient infection of cell lines and subsequent stable integration of the vector into the cellular genome. Subclones with constitutive high level transcription of short antisense RNA can easily be selected. A clone transcribing the 39-nucleotide long α_{i2} antisense RNA at high level reduced the cellular content of $G\alpha_{i2}$ protein by up to 85%. When in F9 stem cells $G\alpha_{i2}$ expression was inhibited by antisense transcription from this construct, the ability of thrombin to inhibit forskolin-stimulated adenylyl cyclase was decreased by 70%. In addition, these cells changed their morphology following retinoic acid-induced differentiation. They displayed an endoderm-like morphology. The authors confirmed the results by stable transfection of F9 cells with an expressing construct for a constitutively active mutant of $G\alpha_{i2}$. The transfected cells did not differentiate when retinoic acid was applied. These findings revealed an important role for $G\alpha_{i2}$ in differentiation. The same $G\alpha_{i2}$ antisense construct was used by this group to show that in F9 teratocarcinoma cells and in rat osteosarcoma 17/2.8 cells, in which $G\alpha_{i2}$ was depleted, the basal accumulation of inositol phosphates was increased. Stimulation of PLC by different hormones was potentiated (WATKINS et al. 1994).

By using even shorter antisense RNA constructs, i.e., by cloning only 20 bp of the $G\alpha_s$ and the $G\alpha_{i1}$ genes in reversed orientation into an eukaryotic expression plasmid, the involvement of $G\alpha_s$ in retinoic induced differentiation of HL-60 cells has been demonstrated (MEISSNER et al. 1996). The antisense RNA constructs were cotransfected with a selection plasmid into HL-60 cells and stably expressing clones were selected. In cells expressing the $G\alpha_s$ antisense RNA the amount of $G\alpha_s$ protein was reduced by more than 90% when compared to $G\alpha_{i1}$ antisense RNA-expressing cells or untransfected cells. The cells depleted of $G\alpha_s$ displayed a granulocyte-like morphology similar to retinoic-induced HL-60 cells and were resistant to further differentiation, indicating that $G\alpha_s$ expression is essential for keeping HL-60 cells in an undifferentiated state. This finding was supported by experiments showing that in untransfected cells the amount of $G\alpha_s$ protein declined during retinoid-induced differentiation.

A similar strategy, i.e., expression of short antisense RNAs, was used to target the $G\alpha_{i2}$ mRNA in transgenic mice. A 39-bp segment from the 5'-untranslated region was recombined in antisense orientation with a promoter active only in fat cells, liver, and kidney after birth (MOXHAM et al. 1993a,b). The major characteristics of transgenic mice transcribing the short antisense RNA in these tissues was their failure to thrive. There was a remarkable reduction of body weight and blunted skeletal growth. In adipocytes isolated from transgenic animals, as well as in rat hepatoma cells transfected with the same antisense construct, the basal cAMP accumulation was increased, and the hormonal-induced inhibition of adenylyl cyclase was reduced. These findings confirmed the important role of $G\alpha_{i2}$ as a transducer for the inhibition of adenylyl cyclase (MOXAM et al. 1993b). In a subsequent paper (MOXAM and MALBON 1996) the authors showed that the mice deficient in $G\alpha_{i2}$ displayed a phenotype of insulin resistance characteristic of non-insulin-dependent diabetes mellitus, e.g., runted phenotype, fasting hypoinsulinaemia, and impaired glucose and insulin tolerance. In adipocytes derived from transgenic animals, insulin stimulated phosphorylation of insulin receptor substrate-1 (IRS-1) is reduced compared to adipocytes from control animals, a finding which could be explained by increased phosphotyrosine phosphatase (PTPase) activity caused by high expression rates of the enzyme subtype PTP1B and release of this enzyme from its membrane-associated endoplasmic reticular localization to the cytosol. High expression of PTP1B could be found in adipocytes, liver cells and skeletal muscle cells of antisense RNA-expressing animals. These results suggest $G\alpha_{i2}$ as a link between insulin receptors and phosphotyrosine phosphatase 1B. A phenotype similar to that described for $G\alpha_{i2}$-deficient transgenic mice has been described in mice with disrupted IRS-1 gene (ARAKI et al. 1994; TAMEMOTO et al. 1994).

The approach to use inducible and tissue-specific transcription of short antisense RNA in transgenic animals has advantages and disadvantages compared to knockouts created by homologous recombination (MORTENSEN and SEIDEMAN 1994). Most noticeable is the advantage that lethal effects in utero can be avoided since antisense transcription starts at birth. In addition, transcription restricted to certain tissues opens the opportunity to investigate the effects of gene suppression in tissues and interesting subsets. Compared to genomic knockouts, it is impossible to achieve a complete loss of the target mRNA; by now the extent to which the targeted protein has to be reduced to get the intended loss of function is not clear. The first described genomic knockout of a G protein was the disruption of $G\alpha_{i2}$ by homologous recombination (RUDOLPH et al. 1995). Transgenic mice with a loss of $G\alpha_{i2}$ function developed ulcerative colitis and adenocarcinoma of the colon aside from retardation of growth. Prior to clinical symptoms, the mice showed alteration in thymocyte maturation and function with similarities to transgenic mice expressing the S1 subunit of pertussis toxin. However, from this study it is not clear whether the immunological defect is a prerequisite for the defects in gut epithelium. In a subsequent paper the authors described that the hormone-induced inhibition of adenylyl cyclase in cardiac and fat cells were partially, but not completely, lost (RUDOLPH et al. 1996). In 50% of the α_{i2} -/- mice the inhibition of adenylyl cyclase by three different hormones was even unaltered. The authors explained the failure to detect an effect of the α_{i2} knockout in these animals by compensation of the $G\alpha_{i2}$ function by $G\alpha_{i3}$. Indeed, in various tissues and cells derived from α_{i2} -/- mice, a compensatory upregulation of $G\alpha_{i3}$ could be detected, which may functionally substitute for $G\alpha_{i2}$.

VOISIN et al. (1996a) used the same target sequence in the 5' untranslated region of the subtypes of $G\alpha_i$ for expression of antisense RNA as WATKINS and colleagues (1992) used to inhibit $G\alpha_i$ expression in renal proximal tubule cells. In stable clones of renal proximal tubule cells expressing these antisense constructs, different levels of inhibition of $G\alpha_{i2}$ expression were observed. Two clones were characterized, one reaching 90% and the other 60% inhibition of $G\alpha_{i2}$ expression. No change in the expression of $G\alpha_{i3}$ was observed (VOISIN et al. 1996a,b). In both clones the coupling of peptide YY receptor to adenylyl cyclase was abolished, and no substitution of $G\alpha_{i3}$ for the depleted $G\alpha_{i2}$ function could be observed. These results indicate an exclusive coupling of peptide YY receptor to $G\alpha_{i2}$. In contrast to the cells derived from $G\alpha_{i2}$ -/- mice, in the cell line used by VOISIN and colleagues (1996a) no substitution of $G\alpha_{i2}$ function by $G\alpha_{i3}$ was found, although the cell clones depleted of $G\alpha_{i2}$ underwent several rounds of selection. These results suggest that there is either a difference in the specificity in coupling to $G\alpha_{i2}$ between the receptors investigated by RUDOLPH and colleagues (1996), i.e., muscarinic, adenosine A_1, prostaglandin EP_3 and nicotinic receptors, and the receptor investigated by VOISIN and colleagues (1996a), i.e., peptide YY receptor, or there is a difference in the capacity of one $G\alpha_i$ subtype to substitute for another in the knockout mouse model used by RUDOLPH et al. (1996), and the $G\alpha_i$ protein depletion by antisense RNA expression used by VOISIN et al. (1996a). A selective knockout of one $G\alpha_i$ with no substitution by another $G\alpha_i$ was also reached by antisense oligodeoxynucleotides in other cell systems (see Sect. D.VI).

Up to now, only one additional knockout mouse for a G protein subunit has been published – the knockout mouse for $G\alpha_{13}$ (OFFERMANNS et al. 1997). No homozygous α_{13} -/- mouse was born, indicating a recessive lethal phenotype. Homozygous α_{13} -/- embryos died between embryonic days 9.5–11. Embryos at embryonic day 9.5 displayed defects in the process of angiogenesis leading to a lack of blood vessel formation, and fibroblasts derived from embryos at embryonic day 8.5 showed an inhibition of thrombin-induced cell migration. The thrombin receptor had been shown before to couple to G_{13} in platelet membranes (OFFERMANNS et al. 1994). Obviously, the related G protein G_{12}, also coupling to the thrombin receptor, is not able to substitute for $G\alpha_{13}$.

D. Inhibition of G-Protein Functions by Antisense Oligodeoxynucleotides in Cell Culture

Antisense sequences can also be delivered to cells as short oligodeoxyribonucleotides, i.e., 15–35 residues long. The minimum length to target antisense oligodeoxynucleotides specifically to only one kind of transcript among the entire mRNA population is 15 bp. This length range facilitates the search for a unique target sequence of a particular mRNA out of the mRNAs coding for a highly related gene family like that for β subunits of G proteins. Since the hybridization temperature is fixed to 37°C, oligodeoxynucleotides longer than 25–30 nucleotides will not necessarily increase specific annealing of the oligodeoxynucleotide to the targeted RNA. Table 2 summarizes the studies to date in which antisense oligodeoxynucleotides directed against G proteins were used.

I. Nuclear Microinjection of G-Protein Antisense Oligodeoxynucleotides

KLEUSS et al. (1991) established the repression of biosynthesis of G-protein subunits by microinjection of antisense oligodeoxynucleotides into the nuclei of single cells (for experimental details, see KLEUSS et al. 1994; KALKBRENNER et al. 1997; DEGTIAR et al. 1997). First, cells were grown on marked cover slips, then antisense oligodeoxynucleotides, generally 20–30 nucleotides long and designed to anneal specifically to the mRNAs of subtypes of α, β and γ subunits of G proteins, were injected into the nuclei of single cells. To measure microinjection efficiency, cells were injected with fluorescein isothiocyanate (FITC)-marked oligodeoxynucleotides; the fluorescence signal was detected in the nuclei of about 90% of injected cells. After 2–4 days the G-protein function was determined, e.g., by measuring hormone-induced inhibition of voltage-gated Ca^{2+} channels, release of calcium from intracellular stores or influx of calcium from extracellular medium.

Table 2. G-protein involvements detected by antisense oligodeoxynucleotides in cell culture

Receptors	G-protein	Effector system	Cells	Reference
Somatostatin	$\alpha_{o2}\beta_1\gamma_3$	VOC$\downarrow$	GH$_3$, RINm5F	Kleuss et al. 1993, Degtiar et al. 1996
M$_4$-muscarinic	$\alpha_{o1}\beta_3\gamma_4$	VOC$\downarrow$	GH$_3$	Kleuss et al. 1993
M$_{2/4}$-muscarinic	$\alpha_{o1}\beta_3\gamma_4$	VOC$\downarrow$	PC-12	Kalkbrenner et al. unpublished data
Galanin	$\alpha_{o1}\beta_3\gamma_4$	VOC$\downarrow$	GH$_3$, RINm5F	Kalkbrenner et al. 1995
Galanin	$\alpha_{o1}\beta_2\gamma_2$	VOG$\downarrow$	GH$_3$, RINm5F	Kalkbrenner et al. 1995
GABA$_B$	α_o	VOC$\downarrow$	DRG	Campell et al. 1993
D$_2$ dopamine	α_o	VOC$\downarrow$	Lactotropes	Baertschi et al. 1992
α_2-Adrenergic	α_o	VOC$\downarrow$	SCG	Buckley et al. 1995
5-HT$_{2C}$	α_o, α_{i1}	PLC$\uparrow$	*Xenopus* oocytes	Chen et al. 1995
5-HT$_{1C}$	α_o	PLC$\uparrow$	*Xenopus* oocytes	Quick et al. 1994
mGluR$_1$	α_o	PLC$\uparrow$	*Xenopus* oocytes	Kasahara and Sugiyama 1994
mGluR$_5$	α_o	PLC$\uparrow$	*Xenopus* oocytes	Kasahara and Sugiyama 1994
M$_1$-muscarinic	α_o, α_{i1}	PLC$\uparrow$	*Xenopus* oocytes	Kasahara and Sugiyama 1994
TRH	$\alpha_{i2/i3}$	VOC$\uparrow$	GH$_3$	Gollasch et al. 1993
D$_2$ dopamine	α_{i3}	K$^+$ channel$\uparrow$	Lactotropes	Baertschi et al. 1992
Galanin	α_{i3}	AC$\downarrow$	RINm5F	deMazancourt et al. 1994
δ-Opioid	α_{i2}	[Ca^{2+}]$_i\uparrow$	ND8-47	Tang et al. 1995
M$_2$-muscarinic	α_{i1}	GIRK1$\downarrow$	*Xenopus* oocytes	Schreibmayer et al. 1996
(No receptor)	α_s	Differentiation	3T3-L1	Wang et al. 1992
TRH	$\alpha_{q/11}$	VOC$\uparrow$	GH$_3$	Gollasch et al. 1993
TRH	α_q	PLC$\uparrow$	*Xenopus* oocytes	Quick et al. 1994
M-muscarinic	α_{11}	K$^+$ channel$\uparrow$	VMH	French-Mullen et al. 1994
M$_1$-muscarinic	$\alpha_q/\alpha_{11}\beta_1/\beta_4\gamma_4$	[Ca^{2+}]$_i\uparrow$	RBL-2H3-hml	Dippel et al. 1996
A$_3$ adenosine	$\alpha_{i3}\beta_2\gamma_2$	[Ca^{2+}]$_i\uparrow$	RBL-2H3-hml	Dippel et al. submitted
α_{1A}-Adrenergic	$\alpha_q\beta_1\gamma_3$	[Ca^{2+}]$_i\uparrow$	Venous myocytes	Macrez-Leprêtre et al. 1997a
α_{1A}-Adrenergic	$\alpha_{11}\beta_3\gamma_2$	Calcium influx$\uparrow$	Venous myocytes	Macrez-Leprêtre et al. 1997a
Angiotensin AT$_{1A}$	$\alpha_{i3}\beta_1\gamma_3$	[Ca^{2+}]$_i\uparrow$	Venous myocytes	Macrez-Leprêtre et al. 1997b
Neuromedin B	α_q	PLC$\uparrow$	*Xenopus* oocytes	Shapira et al. 1994
M$_3$-muscarinic	$\alpha_{q/11}$	PLC$\uparrow$	*Xenopus* oocytes	Stehno-Bittel et al. 1995

AC, adenylyl cyclase; [Ca^{2+}]$_i$, cytosolic calcium concentration; DRG, dorsal root ganglion neurones; PLC, phospholipase C; TRH, thyrotropin-releasing hormone; VMH, ventromedial hypothalamic neurones; SCG, superior cervical ganglia neurones; VOC, voltage-operated calcium channels; GIRK1, inwardly rectifying K$^+$ channel of cardiac/brain type; $\downarrow$, inhibition; $\uparrow$, stimulation.

II. Signal Transduction Pathways Leading to G-Protein-Mediated Inhibition of Voltage-Gated Calcium Channels in Neuroendocrine Cells

To determine which type of G proteins couple a given hormone receptor to voltage-gated Ca^{2+} channels in GH_3 cells, KLEUSS et al. (1991) microinjected antisense oligodeoxynucleotides into the nuclei of GH_3 cells; the oligodeoxynucleotides were designed to hybridize to the mRNA of $G\alpha_{o1}$ and $G\alpha_{o2}$ subunits. To check whether the oligodeoxynucleotides are able to suppress biosynthesis of the respective proteins, the G-protein α subunits were visualized by using an affinity-purified α_{common} antibody recognizing the α_s, α_i, α_o, α_z and α_t subunits of G proteins, and the cells were monitored by immunofluorescence microscopy. Antisense oligodeoxynucleotides complementary to the mRNAs of all pertussis toxin-sensitive G proteins and G_z (α-com) suppressed the corresponding proteins one day after injection of oligodeoxynucleotides, and G proteins started to reappear on the third day after injection. The function of the G_o protein, measured as inhibition of voltage-gated Ca^{2+} channels, was reduced in a similar time-dependent manner. The largest reduction was seen 48 h following injection of antisense oligodeoxynucleotides. After 72 h, channel function had completely recovered. By nuclear injection of antisense oligodeoxynucleotides selectively targeted to the mRNA coding for the $G\alpha_o$ subtypes $G\alpha_{o1}$ and $G\alpha_{o2}$, the authors were able to demonstrate that the SST receptor couples to G_{o2} and that the muscarinic M_4 receptor couples to G_{o1} to inhibit the voltage-gated Ca^{2+} channel in these cells.

This study was extended to the β and γ subunits of G proteins. In two subsequent papers KLEUSS et al. (1992, 1993) demonstrated that the G proteins coupling the SST receptors to the voltage-gated Ca^{2+} channels are composed of the subunits $\alpha_{o2}\beta_1\gamma_3$ and that the G proteins coupling the muscarinic M_4 receptor to the same effector system are composed of the subunits $\alpha_{o1}\beta_3\gamma_4$. The specificity in coupling of a given receptor to a certain heterotrimeric G protein is identical in other cell models. In PC-12 cells, the muscarinic $M_{2/4}$ receptor uses the same triplet as in GH_3 cells to inhibit voltage-gated Ca^{2+} channels; the same is true in RINm5F cells for the SST receptor (DEGTIAR et al. 1996; Fig. 2).

Studying the identity of the G protein coupling the galanin receptors to voltage-gated Ca^{2+} channels, KALKBRENNER et al. (1995) demonstrated that this one receptor couples via two different G-protein heterotrimers to the same effector system. From the quantity of antisense effects on function and from coexpression data of $\beta\gamma$ dimers, they deduced that galanin receptors in GH_3 and RINm5F cells couple preferentially to G-protein trimers consisting of $\alpha_{o1}\beta_2\gamma_2$ and less efficiently to trimers consisting of $\alpha_{o1}\beta_3\gamma_4$, a heterotrimer which is also used by the M_4 muscarinic receptor (KALKBRENNER et al. 1995; see Fig. 2).

CAMPBELL et al. (1993) used the technique of microinjection of antisense oligodeoxynucleotides to analyze which G protein couples the $GABA_B$ recep-

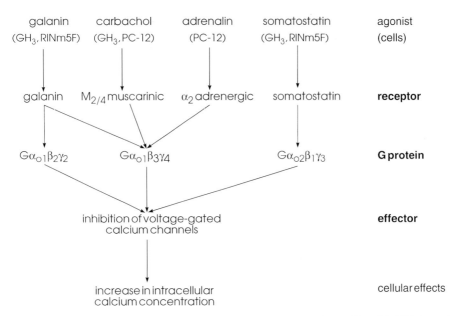

Fig. 2. The signal transduction pathways leading to G-protein-mediated inhibition of voltage-gated calcium channels in neuroendocrine cell lines

tor to voltage-operated Ca^{2+} channels in root dorsal ganglion cells. Injection of an anti-α_o antisense oligodeoxynucleotide common for either splice variant of $G\alpha_o$ reduced baclofen-induced inhibition of voltage-operated Ca^{2+} channels from 34%–38% in control, non-sense or anti-α_i oligodeoxynucleotide-injected cells to 21% in cells injected with anti-α_o antisense oligodeoxynucleotides. Staining of $G\alpha_o$ with $G\alpha_{o\text{-common}}$ antibodies was reduced by 76% in cells injected with anti-α_o antisense oligodeoxynucleotides, whereas staining of $G\alpha_i$ with $G\alpha_{i\text{-common}}$ antibodies was not reduced. However, injection of anti-α_i antisense oligodeoxynucleotides, designed to anneal to the mRNAs of all α_i subunits, reduced the staining of $G\alpha_i$ by 68% but did not influence $G\alpha_o$ staining. These data suggest that the level of G proteins in these cells has to be reduced by two thirds before observing measurable functional effects by injected antisense oligodeoxynucleotides. In studies of RINm5F cells concerning the coupling of G_o to voltage-gated Ca^{2+} channels, the reduction in immunofluorescence paralleled the reduction in G-protein function, i.e., a 50% reduction in the protein level of $G\alpha_{o1}$ resulted in a 50% reduction in inhibition of voltage-gated Ca^{2+} channels (KALKBRENNER et al. 1995).

By injecting antisense oligodeoxynucleotides which target mRNA sequences of $G\alpha_q$, $G\alpha_{11}$, $G\alpha_z$, $G\alpha_{i2}$ and $G\alpha_{i3}$, GOLLASCH et al. (1993) showed that two pathways are involved in TRH-induced stimulation of voltage-gated Ca^{2+} channels in GH_3 cells. One leads mainly via $G\alpha_{i2}$ (and to a minor extent via $G\alpha_{i3}$) to a probably direct stimulation of Ca^{2+} channels. The other pathway leads via $G_{q/11}$ to a stimulation of protein kinase C and thereby to a necessary phosphorylation of signal transduction components (e.g., Ca^{2+} channels or G proteins), which finally results in stimulation of voltage-gated Ca^{2+} channels. Figure 2 summarizes the specificity in G-protein usage in coupling of hormonal receptors to voltage calcium channels in neuroendocrine cells; only G proteins for which the heterotrimer composition have been determined are shown.

III. Signal Transduction Pathways Leading to a G-Protein-Mediated Increase in the Intracellular Calcium Concentration

Recently, DIPPEL et al. (1996) extended the antisense studies to another effector system, namely PLC-β. As a model system they used rat basophilic leukemia cells stably transfected with the m_1 muscarinic receptor (RBL-2H3-hm1; JONES et al. 1991). To answer the question of which specific G-protein subunits are involved in the coupling of the muscarinic m_1 and adenosine A_3 receptor to PLC-β, they inhibited the expression of G-protein subunits by intranuclear microinjection of antisense oligodeoxynucleotides complementary to sequences of mRNAs of G-protein subunits into RBL-2H3-hm1 cells. As an indicator of PLC-β activity, the increase in the free cytoplasmatic Ca^{2+} concentration was measured by single cell imaging of the injected cells loaded with Fura-2. The results indicate that the m_1 receptor selectively couples via a G-protein complex composed of $G\alpha_{q/11}\beta_{1/4}\gamma_4$ to PLC-β, whereas the endogenously expressed A_3 receptor selectively couples via G-protein heterotrimers composed of $G\alpha_{i3}\beta_2\gamma_2$ to PLC-β (DIPPEL et al. 1996; Fig. 3). In cells injected with anti-α_q and anti-α_{11} antisense oligodeoxynucleotides, the amounts of the corresponding immunochemically detected proteins were reduced by 85% after 2 days; the expression of $G\alpha_q$ and $G\alpha_{11}$ completely recovered after 4 days, demonstrating the reversibility of the antisense effect. Figure 3 summarizes the composition of the specific G-protein heterotrimers involved in the activation of PLC-β in RBL-2H3-hm1 cells.

By using the same methods, e.g., nuclear microinjection of antisense oligodeoxynucleotides corresponding to the mRNA sequences coding for G-protein α, β and γ subunits in combination with fluorometric determination of increase in cytosolic calcium concentration $[Ca^{2+}]_i$, MACREZ-LEPRÊTRE et al. (1997a) showed that the norepinephrine-induced transient increase in $[Ca^{2+}]_i$ depending on Ca^{2+} release from the intracellular store in primary rat portal vein myocytes is mediated by the G protein $G\alpha_q\beta_1\gamma_3$, whereas the Ca^{2+} entry evoked by norepinephrine in these cells is mediated by the G protein $G\alpha_{11}\beta_3\gamma_5$. In a subsequent paper, MACREZ-LEPRÊTRE and colleagues (1997b) showed that the AT_{1A} receptor in rat portal venous myocytes specifically couples to the G protein $G\alpha_{13}\beta_2\gamma_3$ to induce an increase in $[Ca^{2+}]_i$. The data obtained with the α-antisense oligodeoxynucleotides were confirmed by microinjection of $G\alpha_{13}$-specific antibodies, as well as by microinjection of a peptide corresponding to the last 13 amino acids of $G\alpha_{13}$, both inhibiting the angiotensin-induced

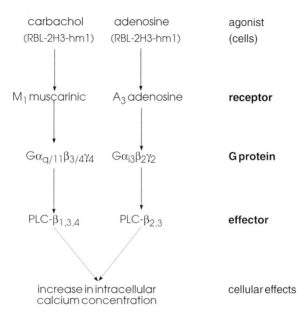

Fig. 3. The signal transduction pathways leading to G-protein-mediated stimulation of phospholipase C-β (*PLC-β*) with subsequent increase in intracellular calcium concentration in the rat basophilic cells RBL-2H3-hm1

increase in $[Ca^{2+}]_i$ with a similar efficiency compared to the antisense oligodeoxynucleotide. A corresponding reduction of the G-protein subunit expression in cells injected with antisense oligodeoxynucleotides was immunochemically shown by using α-, β- and γ-specific antibodies. The immunofluorescence signal specific for $G\alpha_{13}$ was not significantly reduced in cells injected with anti-$β_1$ and anti-$γ_3$ antisense oligodeoxynucleotides. These results indicate that the specific inhibition of the expression of one subunit is sufficient to disturb the function of one specific heterotrimer without changing the expression rate of the other subunits involved in the composition of the corresponding heterotrimer (Fig. 4).

Heptahelical receptors expressed in *Xenopus* oocytes can use endogenous G_o to couple to PLC-β. CHEN et al. (1995) demonstrated coupling of the 5-HT_{2C} receptor to PLC via G_o, and KASAHARA and SUGIYAMA (1994) demonstrated coupling of metabotropic glutamate receptor subtypes 1 and 5 and m_1 muscarinic receptors to G_o and, in addition, coupling of m_1 muscarinic receptors to G_{11} in *Xenopus* oocytes by using injection of antisense oligodeoxynucleotides specific for G_o and the subtypes of G_i. Using the same method, i.e., injection of antisense oligodeoxynucleotides directed against G-protein mRNA sequences, coupling of the serotonin 1c (5-HT_{1c}) receptor to G_o and coupling of the TRH receptor to G_q in *Xenopus* oocytes have been shown (QUICK et al. 1994).

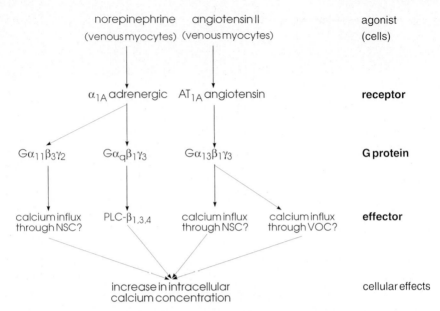

Fig. 4. The signal transduction pathways leading to G-protein-mediated increase in intracellular calcium concentration in primary rat portal vein myocytes. *VOC*, voltage-operated calcium channels; *NSC*, nonselective cation channels; *PLC*, phospholipase C

IV. Specificity of Antisense Effects

KALKBRENNER et al. (1995) showed subtype-specific reduction in protein expression of $G\alpha_{o1}$ and $G\alpha_{o2}$ when RINm5F cells were injected with anti-α_{o1} and anti-α_{o2} antisense oligodeoxynucleotides, respectively. In these experiments, 48 h after injection the immunofluorescence signal of $G\alpha_{o1}$ was reduced in cells injected with α_{o1} antisense oligodeoxynucleotides compared to cells injected with α_{o2} antisense oligodeoxynucleotides and vice versa; in cells injected with α_{o2} antisense oligodeoxynucleotides, the immunofluorescence signal of $G\alpha_{o2}$ was reduced compared to cells injected with α_{o1} antisense oligodeoxynucleotides. Similarly, the specificity for the antisense oligodeoxynucleotides directed against $G\alpha_{i2}$, $G\alpha_{i3}$, $G\alpha_q$, $G\alpha_{11}$, $G\alpha_{12}$, $G\alpha_{13}$, $G\beta_1$, $G\beta_3$, $G\gamma_2$ and $G\gamma_3$ on protein expression of the corresponding G-protein subunit was shown by DIPPEL et al. (1996) and MACREZ-LEPRÊTRE et al. (1997a,b).

Another important proof of specificity of the antisense effects comes from the specificity in loss of function. In one and the same GH_3 cell in which antisense oligodeoxynucleotides annealing to the β_1 subunit were injected, the inhibition of voltage-gated Ca^{2+} channels by SST was diminished, but the inhibition induced by carbachol and galanin was still present. Vice versa the inhibition of Ca^{2+} channels by galanin was diminished in cells injected with β_2 antisense oligodeoxynucleotides, but the inhibition induced by SST and carbachol was still present in one and the same cell. (For composition of heterotrimers coupling M_4 muscarinic, SST and galanin receptors to voltage-gated Ca^{2+} channels, see Fig 2.) Similar results were obtained by injection of antisense oligodeoxynucleotides annealing to the mRNAs of G-protein α and G-protein γ subunits (DEGTIAR et al. 1997). Similar control experiments were performed in RBL-2H3-hm1 cells by comparing adenosine- to carbachol-induced increases in $[Ca^{2+}]_i$ (DIPPEL et al. 1996; DIPPEL et al., submitted) and in rat portal vein myocytes by comparing norepinephrine- to angiotensin-induced increases in $[Ca^{2+}]_i$ (MACREZ-LEPRÊTRE et al. 1997a,b).

Thus, the most convincing demonstration for the high degree of selectivity by which one particular heterotrimer is used for receptor–effector coupling comes from the subsequent application of several hormones. Activation of at least two different receptors by their cognate hormones and measurements conducted on one and the same cell would obviously reveal nonspecific effects of the injected antisense oligodeoxynucleotides. It is well known that oligodeoxynucleotides containing phosphorothioate modifications in their backbone bind in a more or less unspecific manner to proteins (e.g., see GUVAKOVA et al. 1995; SRINIVASAN et al. 1995). This may explain effects of p53, c-*myb* and c-*myc* antisense oligodeoxynucleotides (BARTON and LEMOINE 1995; BURGESS et al. 1995) on cell proliferation which were recently identified as not being related to antisense-mediated gene suppression, but rather to effects caused by certain sequences, i.e., clusters of three or four guanosines in the case of the c-*myb* and the c-*myc* antisense oligodeoxynucleotides (BURGESS et al. 1995). In microinjection studies performed in our laboratory, nonspecific effects were observed only at concentrations of higher than 70 μM (concentration of injection solution) using oligodeoxynucleotides completely protected by phosphorothioate linkages.

Therefore, antisense experiments to explore specificity in signal transduction pathways should generally include the comparison of two related pathways in one and the same cell, allowing for internal controls similar to those described above.

V. Application of G-Protein Antisense Oligodeoxynucleotides Through the Patch-Clamp Pipette

BAERTSCHI et al. (1992) used a further method to introduce antisense oligodeoxynucleotides into single cells. They patch-clamped bovine serum albumin (BSA) gradient-enriched cultured lactotrophes from lactating rats twice. During the first round of patch-clamping, the cells were dialyzed for 160 s on average with antisense oligodeoxynucleotides which were able to anneal to the mRNA of $G\alpha_o$, $G\alpha_{i1}$, $G\alpha_{i2}$, $G\alpha_{i3}$ and $G\alpha_s$. After 48 h, the same cells were patch-clamped a second time, and the dopamine-induced inhibition of voltage-gated calcium current and increase in voltage-activated potassium current were measured. Inhibition of calcium current was reduced by dialyzing the cells with anti-α_o antisense oligodeoxynucleotides, whereas the increase in potassium current was abolished by dialyzing the cells with anti-α_{i3} antisense oligodeoxynucleotides. Compared to microinjection of antisense oligodeoxynucleotides, the greatest advantage of the dialyzing method is that the cells can be chosen by functional parameters before oligonucleotides will be applied. However, patch-clamping is much more damaging to cells than microinjection. In addition, oligodeoxynucleotides applied to the cytoplasm will be exposed to degradation by unspecific endo- and exonucleases, which reduces the amount of antisense oligodeoxynucleotides available for annealing to their target mRNA in the nucleus.

VI. Application of G-Protein Antisense Oligodeoxynucleotides by Addition to the Cell Culture Medium

Oligodeoxynucleotides applied at high concentration to the cell culture medium will be taken up by passive diffusion or by active transport mechanisms. An active transport was demonstrated in HL-60 cells (LOKE et al. 1989), and similar mechanisms are postulated for other cells.

In cultured ventromedial hypothalamic neurons from rat, carbachol-induced suppression of the outward delayed rectifying potassium current was reduced after treating the cells with antisense phosphorothioate oligodeoxynucleotides, designed to anneal to α-subunit mRNA of G_{11} (FRENCH-MULLEN et al. 1994). The oligodeoxynucleotides were used at a concentration of 10 μM, and the time period of treatment was 24–48 h. Other antisense oligodeoxynucleotides complementary to sequences of $G\alpha_q$ or sense oligodeoxynucleotides resembling sequences of $G\alpha_{11}$ were without effect on potassium current.

WANG et al. (1992) applied antisense oligodeoxynucleotides annealing to the mRNA of $G\alpha_s$, $G\alpha_{i1}$ and $G\alpha_{i3}$ into the medium of 3T3-L1 fibroblasts at 30 μM. Dexamethasone/methylisobutylxanthine-induced differentiation of these fibroblasts into adipocytes occurred after 3 days in cells treated with $G\alpha_s$ antisense oligodeoxynucleotide compared to 7–10 days in control cells. In addition, treatment of fibroblasts with $G\alpha_s$ antisense oligodeoxynucleotides by itself induced differentiation of the cells within 7 days. Other oligonucleotides used in this study had no effect on differentiation. After incubation of cells with the $G\alpha_s$ antisense oligodeoxynucleotides for 48 h, the expression of the respective proteins was reduced by more than 90%, as indicated by Western blotting. The ability of 3T3-L1 fibroblasts to differentiate after addition of dexamethasone/methylisobutylxanthine was blocked by expression of constitutive GTPase active mutants of $G\alpha_s$, confirming the antisense data (WANG et al. 1996). The domain responsible for regulation of differentiation in $G\alpha_s$ was mapped by mutants to amino acids 146–220 (WANG et al. 1996). The results of these works raise interesting questions about the role of G_s proteins in such complex biological processes as cellular differentiation.

DE MAZANCOURT et al. (1994) determined the α subunit of G_i proteins mediating galanin-induced inhibition of adenylyl cyclase in RINm5F cells by applying subtype-specific phosphorothioate antisense oligodeoxynucleotides repeatedly for 12 h to cells in serum-free medium at concentrations of 1–10 μM. By applying oligodeoxynucleotides at 5-μM concentrations, the protein expression of various α subunits was specifically decreased by 41%–47%, as shown by Western blots with specific antibodies. Treating cells with α_{i3} antisense oligodeoxynucleotides totally blocked the galanin-induced inhibition of forskolin-stimulated adenylyl cyclase activity in RINm5F cells. This means that reduction in protein expression by half led to a complete decline of protein function. The authors explained this by the fact that a large part of the Gα_i proteins is associated with Golgi membranes and only a limited population of G$_{i3}$ proteins is localized at the plasma membrane and is available there for galanin receptors. Analyses of fractionated membranes of RINm5F cells by Western-blotting indeed showed a significant amount of G$_{i3}$ protein associated with microsomal membranes.

TANG et al. (1995) used antisense oligodeoxynucleotides directed against G proteins to elucidate which G proteins mediate the increase in [Ca^{2+}]$_i$ induced by δ-opioid receptors in neuroblastoma × dorsal root ganglion hybrid cells (ND8-47). They treated the cells for up to 6 days with partially protected phosphorothioate oligodeoxynucleotides at a concentration of 10 μM. The cell culture medium contained 0.5% serum and was replenished, together with fresh oligodeoxynucleotides, every 48 h. The amount of the respective G proteins declined after 4 days of treatment of ND8-47 cells and was significantly reduced after 6 days of oligodeoxynucleotide treatment, as shown by Western blotting experiments. The opioid-induced increase in [Ca^{2+}]$_i$ was reduced in cells treated with antisense oligodeoxynucleotides directed against Gα_{i2}. In time courses the reduction of the functional effect went parallel to the decrease in the respective protein amount in these cells. From these experiments, and from experiments with inhibitors of voltage-gated calcium channels, the authors concluded that the δ-opioid receptor in ND8-47 cells stimulates voltage-gated calcium channels via Gα_{i2} and in this way raises [Ca^{2+}]$_i$.

Application of antisense oligodeoxynucleotides to the medium is an attractive way to administer oligonucleotides to cells because very high numbers of cells can be treated with oligodeoxynucleotides. This allows for quantification of effector systems not measurable in single cell assays, e.g., of adenylyl cyclase, phospholipases C, D and A$_2$ or of more complex processes, e.g., differentiation, growth, secretion and chemotaxis. On the other hand, high concentrations of oligodeoxynucleotides in the medium are required to overcome degradation and to obtain reasonable effects. This may cause unspecific effects, although successful examples of G-protein-mediated effects have been published, as summarized above. We performed similar experiments and used up to 100 μM of either unprotected or protected oligodeoxynucleotides and did not obtain reproducible effects of G-protein antisense oligodeoxynucleotides: neither by studying functions of effector systems nor by determining RNA expression as measured by reverse transcriptase polymerase chain reaction (RT-PCR) experiments. The addition of cationic liposomes (BENNETT et al. 1992) did not improve our results in three cell lines studied, i.e., GH_3, RINm5F and RBL-2H3. However, the failure to detect significant oligodeoxynucleotide uptake in our experiments may depend on the cell lines used.

Recently, CROOKE et al. (1995) showed that the pharmacokinetics of phosphorothioate antisense oligodeoxynucleotides in tissue culture can vary as a function of cellular assay conditions and analytical methods used (e.g., time, temperature, concentration, sequence of the oligodeoxynucleotides, cell lines and methods for labelling and detection of oligodeoxynucleotides). To overcome all these problems, we developed another method to introduce antisense oligodeoxynucleotides into a large number of cells at one time. For this purpose we combined the earlier described method of ballistic transfer of DNA to cells with magnetic separation of cells to which paramagnetic beads were attached (KALKBRENNER et al. 1997). The oligodeoxynucleotides were attached to gold particles covered with magnetic beads. The particles were then accelerated by applying high pressure and transferred into the nuclei of cells. By separating hit cells from missed cells, the amount of cells successfully transfected by antisense oligodeoxynucleotides can be enriched by up to more than 90%. We compared microinjection and ballistomagnetic transfer in RBL-2H3-hm1 cells and found both methods to be equal in respect to inhibition of G-protein expression by antisense oligodeoxynucleotides (DIPPEL et al., submitted). While only a few hundred up to maximally a few thousand cells can be used for one experiment with microinjection, up to 1×10^8 cells can be successfully transfected at the same time with ballistomagnetic transfer.

E. Inhibition of G-Protein Functions by Antisense Oligodeoxynucleotides in Animals

Several studies have been published in which antisense oligodeoxynucleotides against G proteins were used in vivo in animal studies (Table 3). RAFFA et al. (1994) intracerebroventricularly injected phosphorothioate antisense oligodeoxynucleotides in a single dosage of 6 nmol directed against different G-protein α subunits to inhibit antinociception induced by supraspinal μ-opioid receptors. Injection of α_{i2} antisense oligodeoxynucleotides but not the injection of α_{i1}, α_{i3}, and α_s antisense oligodeoxynucleotides into the ventricles of rats reduced the morphine-induced antinociception (mediated by μ-subtype opioid receptors). Antinociception was measured as latency in heat tail flick assay. In a subsequent study the same group showed that injection of anti-α_{i2} antisense oligodeoxynucleotides had no influence on morphine-related side effects such as constipation and dependence liability. The authors found no differences between the mice injected with anti-α_{i2} and mice injected with other antisense oligodeoxynucleotides (RAFFA et al. 1996a). These results suggest that either these side effects are less sensitive to application of antisense oligodeoxynucleotides directed against $G\alpha_{i2}$, or the signal transduc-

Table 3. G-protein involvements detected by antisense oligodeoxynucleotides applied in vivo in animals

Receptors	G-protein	Effector system	Species	References
μ-Opioid	α_{i2}	Supraspinal antinociception	Mouse	RAFFA et al. 1994
μ-Opioid	α_{i2}	Supraspinal antinociception	Rat	ROSSI et al. 1995
μ-Opioid	α_{i2}, α_z	Supraspinal antinociception	Mouse	SÁNCHEZ-BLÁZQUEZ et al. 1995
μ-Opioid	α_{i2}, α_o, α_s	Supraspinal antinociception	Mouse	STANDIFER et al. 1996
μ-Opioid	α_{i2}, α_z	Spinal antinociception	Mouse	STANDIFER et al. 1996
δ-Opioid	α_{i2}, α_{i3}	Supraspinal antinociception	Mouse	SÁNCHEZ-BLÁZQUEZ et al. 1995
δ-Opioid	α_{i1}, α_{i2}, α_{i3}, α_o, α_s, α_q, α_z	Spinal antinociception	Mouse	STANDIFER et al. 1996
κ_1-Opioid	α_q	Spinal antinociception	Mouse	STANDIFER et al. 1996
κ_3-Opioid	α_{i1}, α_{i3}, α_s, α_q, α_z	Supraspinal antinociception	Mouse	STANDIFER et al. 1996
M6G receptor	α_{i1}	Supraspinal antinociception	Rat	ROSSI et al. 1995
M6G receptor	α_{i1}, α_Z, α_s	Supraspinal antinociception	Mouse	STANDIFER et al. 1996
M6G receptor	α_{i1}, α_{i3}, α_o, α_z	Spinal antinociception	Mouse	STANDIFER et al. 1996
α_2-Adrenergic	$\alpha_{i3} \gg \alpha_{i2}$	Supraspinal antinociception	Mouse	RAFFA et al. 1996b
	α_{o1}	Decrease in food intake	Rat	PLATA-SALAMÁN et al. 1995
MA-receptor	α_s	VOC	Rat	COSTA et al. 1995

M6G receptor, receptor for morphine-6β-glucoronide; MA-receptor, receptor for megestrol acetate; VOC, voltage-operated calcium channel; $\alpha_{i3} \gg \alpha_{i2}$, α_{i3}, antisense oligonucleotides were more effective than α_{i2} antisense oligonucleotides.

tion pathways leading to antinociception differ on the level of G proteins from the pathways leading to acute dependence and constipation; this observation is of possible interest for the development of new analgetic compounds with less side effects. In the same system the α_2-adrenoreceptor-mediated antinociception was also inhibited by $G\alpha_i$ antisense oligodeoxynucleotide, namely by anti-$\alpha_{i3} \gg$ anti-α_{i2} (i.e., the α_2-adrenoreceptor-mediated antinociception was more strongly inhibited in animals injected with anti-α_{i3} antisense oligonucleotides than in animals injected with anti-α_{i2} antisense oligonucleotides); oligodeoxynucleotides directed against $G\alpha_{i1}$ and $G\alpha_s$ were without effect (RAFFA et al. 1996b). Single intracerebroventricular injection of phosphodiester oligodeoxynucleotides was sufficient to inhibit opioid-induced supraspinal antinociception; the effect started 1.5 h after injection and lasted up to 48 h, indicating high stability of phosphodiester oligodeoxynucleotide in spinal fluid (RAFFA 1996). SANCHEZ-BLAZQUEZ et al. (1995) compared the effect of single intracerebroventricular administration of partially phosphorothioate-stabilized oligodeoxynucleotides to chronic administration for 5 days with increasing concentrations on supraspinal analgesia evoked by μ- and δ-opioid agonists. The reduction of G-protein levels was confirmed in Western blots. The authors found that a single dose lowered the supraspinal analgesia on the second day and the pattern of antisense effect was equal to multiple doses, but the effect of multiple administrations for 5 days (day 1–2, 0.2 nmol; day 3–4, 0.4 nmol; day 5, 0.8 nmol) resulted in higher inhibition of

G-protein expression and suppression of supraspinal analgesia. Expression of G protein was decreased by 50%–70% in periaqueductal gray matter and hypothalamus compared to a 20%–40% decrease in striatum and thalamus. Analgesia evoked by agonists of μ-opioid receptors was inhibited by injection of antisense oligodeoxynucleotides directed against $G\alpha_{i2}$ and $G\alpha_z$, whereas analgesia evoked by δ-opioid receptor agonists was inhibited by injection of oligodeoxynucleotides directed against $G\alpha_{i2}$ and $G\alpha_{i3}$ (SANCHEZ-BLAZQUEZ et al. 1995). The results obtained with injection of antisense oligodeoxynucleotides were confirmed by intracerebroventricular injection of G protein α subunit-specific antibodies. There were no discrepancies between the results obtained with either method. The finding that $G\alpha_{i2}$ is involved in supraspinal analgesia mediated by the μ-subtype of the opioid receptors was confirmed by comparing antisense oligodeoxynucleotides directed against $G\alpha_{i2}$ and MOR-1, the cDNA coding for a μ receptor, in the same study (ROSSI et al. 1995). A total of 25 μg of phosphodiester oligodeoxynucleotides were injected in a volume of 1 μl within 30 s through a cannula implanted into the periaqueductal grey and the effect measured after 48 h. The analgesic effect of the potent morphine metabolite, morphine-6β-glucuronide (M6G), was not suppressed by injection of $G\alpha_{i2}$ antisense oligodeoxynucleotides, but was suppressed by injection of $G\alpha_{i1}$ antisense oligodeoxynucleotides, indicating that another opioid-receptor subtype is responsible for the effects of M6G (ROSSI et al. 1995). Using the same methods, e.g., administration of G-protein antisense oligodeoxynucleotides in vivo and measurement of opioid-induced analgesia in tail flickering assay, this group extended the classification of interaction of opioid receptors to G-protein α subunits (STANDIFER et al. 1996). In this study only 5 μg of phosphodiester oligonucleotides were administered. By using both intracerebroventricular and intrathecal injection of oligodeoxynucleotides, the authors discriminated between supraspinal and spinal antinociception evoked by different opioid agonists. Injection of oligodeoxynucleotides directed against $G\alpha_{i2}$, $G\alpha_o$ and $G\alpha_s$ blocked supraspinal μ-opioid analgesia, whereas $G\alpha_{i2}$ and $G\alpha_z$ antisense oligodeoxynucleotides blocked spinal μ-opioid analgesia. The analgesia evoked by M6G showed a different sensitivity profile. $G\alpha_{i1}$ and $G\alpha_z$ antisense oligodeoxynucleotides blocked supraspinal M6G analgesia, whereas $G\alpha_{i1}$, $G\alpha_{i3}$, $G\alpha_o$ and $G\alpha_z$ blocked spinal M6G analgesia. The spinal analgesia mediated by the κ_1-opioid receptor was only sensitive to $G\alpha_q$ antisense oligodeoxynucleotides, whereas spinal analgesia mediated through the δ-opioid receptor was sensitive to all antisense probes tested, e.g., $G\alpha_{i1}$, $G\alpha_{i2}$, $G\alpha_{i3}$, $G\alpha_o$, $G\alpha_s$, $G\alpha_q$ and $G\alpha_z$. The analgesia evoked by an opioid specific for the κ_3-subtype receptor was suppressed by injection of antisense oligodeoxynucleotides directed against $G\alpha_{i1}$, $G\alpha_{i3}$, $G\alpha_s$, $G\alpha_q$ and $G\alpha_z$. The results of this study indicate a high complexity for analgesia mediated by certain receptors, i.e., δ-opioid receptors, whereas analgesia mediated to other receptors, i.e., the κ_1-opioid receptor, seems to mediated only by one G protein. However, the variation in these experiments was high, and for all agonists tested nearly all oligodeoxynucleotides displayed antisense effects, but only the oligodeoxynucleotides cited above reached statistic differences compared to saline injection. In addition, the coupling of opioid receptors has been restricted to G_i and G_o subtypes of the G-protein family. Therefore, the results of this study should be verified by using more than one antisense sequence for each G-protein subunit and by using various concentrations of oligonucleotides, or by the injection of specific antibodies as in the study of SANCHEZ-BLAZQUEZ et al. (1995). COSTA and colleagues (1995) used continuous intraventricular infusion of completely protected phosphorothioate antisenses oligodeoxynucleotides at a rate of 14 μg/24 h for 48 h to suppress the expression of $G\alpha_s$ in freshly dissociated rat ventromedial hypothalamic nucleus neurones in mice. In cells dissociated from animals treated with antisense oligonucleotides directed against the mRNA of $G\alpha_s$, the inhibition of voltage-gated calcium channels mediated by the progesterone derivative megestrol acetate was reduced by about 50% compared to cells dissociated from animals treated with the corresponding sense oligodeoxynucleotides or from untreated animals. These are the first experiments indicating the involvement of G proteins in non-genomic effects of steroids by using G-protein antisense oligodeoxynucleotides. By using a comparable protocol of continuous intraventricular infusion of phosphorothioate antisense oligodeoxynucleotides, the same group showed that $G\alpha_{o1}$ is involved in the modulation of feeding behaviour in rats (PLATA-SALAMAN et al. 1995). Table 3 summarizes the studies in which G-protein antisense oligodeoxynucleotides have been used to examine signal transduction pathways in animals.

F. Perspectives

The question arises whether the specificity in heterotrimers is due to a specific coupling of activated receptors to particular G-protein heterotrimers or to localization of G-protein heterotrimers and receptors in distinct areas of the plasma membrane, and thereby limiting the available combinations of receptors and G proteins. Indeed, there are indications that compartmentalization may be an important principle in G-protein-mediated signal transduction (for review, see NEUBIG 1994). In S49 lymphoma cells, $\beta\gamma$ subunits become sequestered in Triton-X-100 insoluble fractions together with cytoskeletal fragments (SARGIACOMO et al. 1993). In smooth muscle cells, G-protein α and β subunits copurify with caveolin, an important structural protein of plasmalemma caveolae (CHANG et al. 1994; LISANTI et al. 1994). Furthermore, γ_5 subunits in neonatal cardiac fibroblasts colocalize in focal adhesions with vinculin (HANSEN et al. 1994). Recently, G-protein α subunits have been shown to interact with caveolin, an integral membrane component of caveolar membranes. G-protein α subunits expressed in Madin-Derby canine kidney (MDCK) cells interacted with caveolin expressed in Sf9 insect cells; furthermore, caveolin and G-protein α subunits recombinantly expressed in

Escherichia coli, purified and reconstituted in phospholipid membranes, bound specifically to each other (LI et al. 1995, 1996). Caveolin interacts with the G-protein α subunit in its inactive GDP-bound form and the interaction does not require coexpression or co-reconstitution with G$\beta\gamma$ (LI et al. 1995, 1996). In addition, the mobility of G-protein $\beta\gamma$ subunits, if exogenously added to intact cells, is limited in NG108-15 cells (KWON et al. 1994). Thus, compartmentalization of signalling components may prevent free movement of receptors and G proteins in membranes, which may play an important role in the specific interactions between receptor and effectors. Taking all results into account, one may speculate that receptors interact locally with distinct pools of preformed specific G-protein heterotrimers, presumably in caveolae of cell membranes. For G_q and G_{11}, data indicate that preformed or precoupled complexes can include more than one heterotrimer (DIPPEL et al. 1996; MACREZ-LEPRÊTRE et al. 1997a). Inhibition of the expression of only one of the involved subunits leads to abrogation of the functional coupling, which means that all subunits involved contribute equally to functional coupling (DIPPEL et al. 1996). This is in accordance with the studies which showed that even partial inhibition of one particular G-protein subunit abrogates a signal transduction pathway completely; however, other studies showed that inhibition of one subunit can be at least partially substituted by other related G-protein subunits. The mechanisms behind this have to be worked out before G proteins are accessible as possible targets for future therapeutics.

Targeting the individual G-protein subunits by antisense knockout has led to the identification of specificity in receptor–G protein and G protein–effector coupling in the genuine environment of the transmembrane signal transduction system. Various methods have been used, e.g., stable and transient expression of antisense RNA in cells, as well as in animals, and different ways of applying antisense oligodeoxynucleotides with different modifications have been used. All these methods have their restrictions, but all together provide useful tools to specifically target individual members in a highly-related and widely-expressed family of proteins such as G proteins.

Acknowledgements. The authors' own studies were supported by the Deutsche Forschungsgemeinschaft and the Fonds der Chemischen Industrie.

References

Agrawal S (1996) Antisense therapeutics. In: Walker JM (ed) Methods in molecular medicine. Humana, Totowa, NJ

Albert PR, Morris SJ (1994) Antisense knockouts: molecular scalpels for the dissection of signal transduction. Trends Pharmacol Sci 15:250–254

Araki E, Lipes MA, Patti M-E, Brüning JC, Haag III B, Johnson RS, Kahn CR (1994) Alternative pathway of insulin signalling in mice with targeted disruption of IRS-1 gene. Nature 372:186–190

Baertschi AJ (1994) Antisense oligonucleotide strategies in physiology. Mol Cell Endocrinol 101:R15–R24

Baertschi AJ, Audigier Y, Lledo P-M, Israel J-M, Bockaert J, Vincent J-D (1992) Dialysis of lactotropes with antisense oligonucleotides assigns guanine nucleotide binding protein subtypes to their channel effectors. Mol Endocrinol 6:2257–2265

Baldwin JM (1994) Structure and function of receptors coupled to G proteins. Curr Opin Cell Biol 6:180–190

Barton CM, Lemoine NR (1995) Antisense oligonucleotides directed against p53 have antiproliferative effects unrelated to effects on p53 expression. Br J Cancer 71:429–437

Bennett FC, Chiang M-Y, Chan H, Shoemaker JEE, Mirabelli CK (1992) Cationic lipids enhance cellular uptake and activity of phosphorothioate antisense oligonucleotides. Mol Pharmacol 41:1023–1033

Birnbaumer L (1992) Receptor-to-effector signaling through G proteins: roles for $\beta\gamma$ dimers as well as α subunits. Cell 71:1069–1072

Buckley NJ, French-Mullen J, Caulfield M (1995) Use of antisense oligonucleotides and monospecific antisera to inhibit G-protein gene expression in cultured neurons. Biochem Soc Trans 23:137–141

Burgess TL, Fisher EF, Ross SL, Gready JV, Qian Y-X, Bayewitch LA, Cohen AM, Herrera CJ, Hu SS-F, Kramer TB, Lott FD, Martin FH, Pierce GF, Simonet L, Farrell CF (1995) The antiproliferative activity of c-myb and c-myc antisense oligonucleotides in smooth muscle cells is caused by a nonantisense mechanism. Proc Natl Acad Sci USA 92:4051–4055

Campbell V, Berrow N, Dolphin AC (1993) $GABA_B$ receptor modulation of Ca^{2+} currents in rat sensory neurones by the G protein G_o: antisense oligonucleotide studies. J Physiol (Lond) 470:1–11

Chang W-J, Ying Y-S, Rothberg KG, Hooper NM, Turner AJ, Gambiel HA, DeGunzburg J, Mumby SM, Gilman AG, Anderson RGW (1994) Purification and characterization of smooth muscle cell caveolae. J Cell Biol 126:127–138

Chen Y, Baez M, Yu L (1995) Functional coupling of the $5-HT_{2C}$ serotonin receptor to G proteins in Xenopus oocytes. Neurosci Lett 179:100–102

Costa A-MN, Spence KT, Plata-Salamán CR, French-Mullen JMH (1995) Residual Ca^{2+} channel current modulation by megestrol acetate via a G-protein α_s-subunit in rat hypothalamic neurones. J Physiol (Lond) 487:291–303

Crooke RS, Graham MJ, Cooke ME, Crooke ST (1995) In vitro pharmacokinetics of phosphorothioate antisense oligonucleotides. J Pharmacol Exp Ther 275:462–473

Crooke ST (1993) Progress toward oligonucleotide therapeutics: pharmacodynamic properties. FASEB J 7:533–539

Crooke ST (1995) Therapeutic applications of oligonucleotides. Medical intelligence unit. Landes, Austin, Texas, USA

Crooke ST, Bennett CF (1996) Progress in antisense oligonucleotide therapeutics. Annu Rev Pharmacol Toxicol 36:107–129

Degtiar VE, Wittig B, Schultz G, Kalkbrenner F (1996) A specific heterotrimer couples somatostatin receptors to voltage-gated calcium channels in RINm5F cells. FEBS Lett 380:137–141

Degtiar VE, Wittig B, Schultz G, Kalkbrenner F (1997) Microinjection of antisense oligonucleotides and electrophysiological recording of whole cell currents as tool to identify specific G-protein subtypes coupling hormone receptors to voltage-gated calcium channels. In: Bar-Sagi D (ed) Transmembrane signaling protocols. Methods in molecular biology, vol 84. Humana, Totowa, NJ (in press)

De Mazancourt P, Goldsmith PK, Weinstein LS (1994) Inhibition of adenylate cyclase activity by galanin in rat insulinoma cells is mediated by the G-protein G_{i3}. Biochem J 303:369–375

Dippel E, Kalkbrenner F, Wittig B, Schultz G (1996) The muscarinic m1 receptor couples to specific G-protein heterotrimers to increase the cytosolic calcium concentration. Proc Natl Acad Sci USA 93:1391–1396

French-Mullen JMH, Plata-Salamán CR, Buckley NJ, Danks P (1994) Muscarinic modulation by a G-protein alpha-subunit of delayed rectifier K^+ current in rat ventromedial hypothalamic neurones. J Physiol (Lond) 474:21–26

Goetzl EJ, Shames RS, Yang J, Birke FW, Liu YF Albert PR, An S (1994) Inhibition of human HL-60 cell responses to chemotactic factors by antisense messenger RNA depletion of G proteins. J Biol Chem 269:809–812

Gollasch M, Kleuss C, Hescheler J, Wittig B, Schultz G (1993) G_{12} and protein kinase C are required for thyrotropin-releasing hormone-induced stimulation of voltage-dependent Ca^{2+} channels in rat pituitary GH_3 cells. Proc Natl Acad Sci USA 90:6265–6269

Gudermann T, Nürnberg B, Schultz G (1995) Receptors and G proteins as primary components of transmembrane signal transduction. 1. G-protein-coupled receptors: structure and function. J Mol Med 73:51–63

Gudermann T, Kalkbrenner F, Schultz G (1996) Diversity and selectivity of receptor-G protein interaction. Annu Rev Pharmacol Toxicol 36:429–459

Guvakova MA, Yakubov LA, Vlodavsky I, Tonkinson JL, Stein CA (1995) Phosphorothioate oligodeoxynucleotides bind to basic fibroblast factor, inhibits its binding to cell surface receptors, and remove it from low affinity binding sites on extracellular matrix. J Biol Chem 270:2620–2627

Hansen CA, Schoering AG, Carey DJ, Robishaw JD (1994) Localisation of a heterotrimeric G protein γ subunit to focal adhesions and associated stress fibers. J Cell Biol 126:811–819

Hélène C, Toulmé J-J (1990) Specific regulation of gene expression by antisense, sense and antigene nucleic acids. Biochim Biophys Acta 1049:99–125

Jones SV, Choi OH, Beaven MA (1991) Carbachol induces secretion in a mast cell line (RBL-2H3) transfected with the m1 muscarinic receptor. FEBS Lett 289:47–50

Kalkbrenner F, Degtiar VE, Schenker M, Brendel S, Zobel A, Hescheler J, Wittig B, Schultz G (1995) Subunit composition of G_o proteins functionally coupling galanin receptors to voltage-gated calcium channels. EMBO J 14:4728–4737

Kalkbrenner F, Dippel E, Wittig B, Schultz G (1996) Specificity of the receptor–G-protein interaction: using antisense techniques to identify the function of G protein-subunits. Biochim Biophys Acta 1314:125–139

Kalkbrenner F, Dippel E, Schroff M, Wittig B, Schultz G (1997) Intranuclear microinjection and ballistomagnetic transfer of antisense oligonucleotides as tools to determine receptor interaction of G-protein heterotrimers causing phospholipase C-β activation. In Challis RAJ (ed) Receptor signal transduction protocols. Methods in molecular biology, vol 83. Humana, Totowa, NJ, pp 203–216

Kasahara J, Sugiyama H (1994) Inositol phospholipid metabolism in Xenopus oocytes mediated by endogenous G_o and G_i protein. FEBS Lett 355:41–44

Kleuss C, Hescheler J, Ewel C, Rosenthal W, Schultz G, Wittig B (1991) Assignment of G-protein subtypes to specific receptors inducing inhibition of calcium currents. Nature 353:43–48

Kleuss C, Scherübel H, Hescheler J, Schultz G, Wittig B (1992) Different β-subunits determine G-protein interaction with transmembrane receptors. Nature 358:424–426

Kleuss C, Scherübel H, Hescheler J, Schultz G, Wittig B (1993) Selectivity in signal transduction determined by γ subunits of heterotrimeric G proteins. Science 259:832–834

Kleuss C, Schultz G, Wittig B (1994) Microinjection of antisense oligonucleotides to assess G-protein subunit function. Methods Enzymol 237:345–355

Kwon G, Axelrod D, Neubig RR (1994) Lateral mobility of tetramethylrhodamine (TMR) labelled G protein α and $\beta\gamma$ subunits in NG 108–15 cells. Cell Signal 6:663–679

Li S, Okamoto T, Chun M, Sargiacomo M, Casanova JE, Hansen SH, Nishimoto I, Lisanti MP (1995) Evidence for a regulated interaction between heterotrimeric G proteins and caveolin. J Biol Chem 270:15693–15701

Li S, Song K, Lisanti MP (1996) Expression and characterization of recombinant caveolin. J Biol Chem 271:568–573

Lisanti MP, Scherer PE, Vidugiriene J, Tang ZL, Hermanowski-Vosatka A, Tu Y-H, Cook RF, Sargiacomo M (1994) Characterization of caveolin-rich membrane domains isolated from an endothelin-rich source: implications for human disease. J Cell Biol 126:111–126

Liu YF, Jakobs KH, Rasenick MM, Albert PR (1994) G protein specificity in receptor-effector coupling. J Biol Chem 269:13880–13886

Loke SL, Stein CA, Zhang XH, Mori K, Nakanishi M, Subashinghe C, Cohen JS, Neckers LM (1989) Characterization of oligonucleotide transport into living cells. Proc Natl Acad Sci USA 86:3474–3478

Macrez-Leprêtre N, Kalkbrenner F, Schultz G, Mironneau J (1997a) Distinct functions of G_q and G_{11} proteins in coupling α_{1A}-adrenoceptors to Ca^{2+} release and Ca^{2+} entry in rat portal vein myocyte. J Biol Chem 272:5261–5268

Macrez-Leprêtre N, Kalkbrenner F, Morel J-L, Schultz G, Mironneau J (1997b) G protein heterotrimer $G\alpha_{13}\beta_1\gamma_3$ couples angiotensin AT_{1A} receptor to increase in cytosolic Ca^{2+} in rat portal vein myocytes. J Biol Chem 272:10095–10102

Meissner JD, Brown GA, Mueller WH, Scheibe RJ (1996) Retinoic acid-mediated decrease of $G\alpha_s$ protein expression: involvement of $G\alpha_s$ in the differentiation of HL-60 myeloid cells. Exp Cell Res 225:112–121

Milligan G (1995) Signal sorting by G-protein-linked receptors. Adv Pharmacol 32:1–29

Milligan JF, Matteucci MD, Martin JC (1993) Current concepts in antisense drug design. J Med Chem 36:1923–1937

Mortensen RM, Seidman JG (1994) Inactivation of G-protein genes: double knockout in cell lines. Methods Enzymol 237:356–386

Moxam CM, Malbon CC (1996) Insulin action impaired by deficiency of the G-protein subunit $G_{i\alpha 2}$. Nature 379:840–844

Moxam CM, Hod Y, Malbon CC (1993a) Induction of $G\alpha_{i2}$-specific antisense RNA in vivo inhibits neonatal growth. Science 260:991–995

Moxam CM, Hod Y, Malbon CC (1993b) $Gi\alpha_2$ mediates the inhibitory regulation of adenylylcyclase in vivo: analysis in transgenic mice with $Gi\alpha_2$ suppressed by inducible antisense RNA. Dev Genet 14:266–273

Neer EJ (1995) Heterotrimeric G proteins: organizer of transmembrane signals. Cell 80:249–257

Neubig RR (1994) Membrane organization in G-protein mechanisms. FASEB J 8:939–946

Nürnberg B, Gudermann T, Schultz G (1995) Receptors and G proteins as primary components of transmembrane signal transduction, part 2. G proteins: structure and function. J Mol Med 73:123–132, correction: 73:379

Offermanns S, Schultz G (1994) Complex information processing by the transmembrane signaling system involving G proteins. Naunyn Schmiedebergs Arch Pharmacol 350:329–338

Offermanns S, Laugwitz K-L, Spicher K, Schultz G (1994) G proteins of the G_{12} family are activated via thromboxane A_2 and thrombin receptors in human platelets. Proc Natl Acad Sci USA 31:504–508

Offermanns S, Mancino V, Revel, J-P, Simon MI (1997) Vascular system defects and impaired cell chemokinesis as a result of $G\alpha_{13}$ deficiency. Science 275:533–536

Pasternak GW, Standifer KM (1995) Mapping of opioid receptors using antisense oligodeoxynucleotides: correlating their molecular biology and pharmacology. Trends Pharmacol Sci 16:344–350

Paulssen RH, Paulssen EJ, Gautvik KM, Gordeladze JO (1992) The thyroliberin receptor interacts directly with a stimulatory guanine-nucleotide-binding protein in the activation of adenylyl cyclase in GH_3 rat pituitary tumour cells. Eur J Biochem 204:413–418

Plata-Salamán CR, Wilson CD, Sonti G, Borkowski JP, French-Mullen JM (1995) Antisense oligodeoxynucleotides to G-protein α-subunit subclasses identify a transductional requirement for the modulation of normal feeding dependent on $G\alpha_{oA}$ subunit. Brain Res Mol Brain Res 33:72–78

Quick MW, Simon MI, Davidson N, Lester HA, Aragay A (1994) Differential coupling of G protein α subunits to seven-helix receptors expressed in Xenopus oocytes. J Biol Chem 269:30164–30172

Raffa RB (1996) In vivo antisense strategy for the study of second-messengers: application to G-proteins. In: Raffa RB, Porecca F (eds) Antisense strategies for the study of receptor mechanisms. Landes, Austin, Texas, USA, pp 53–69

Raffa RB, Porecca F (1996) Antisense strategies for the study of receptor mechanisms. Landes, Austin, Texas, USA

Raffa RB, Martinez RP, Connelly CD (1994) G-protein antisense oligodeoxyribonucleotides and μ-opioid supraspinal antinociception. Eur J Pharmacol 258:R5–R7

Raffa RB, Goode TL, Martinez RP, Jacoby HI (1996a) A G_{i2}a antisense oligonucleotide differentiates morphine antinociception, constipation and acute dependence in mice. Life Sci 58:73–76

Raffa RB, Conelly CD, Chambers JR, Stone DJ (1996b) α-subunits G-protein antisense oligonucleotide effects on supraspinal (i.c.v.) α_2-adrenoceptor antinociception in mice. Life Sci 58:77–80

Rossi GC, Standifer KM, Pasternak GW (1995) Differential blockade of morphine and morphine-6β-glucuronide analgesia by antisense oligodeoxynucleotides directed against MOR-1 and G-protein α subunits in rats. Neurosci Lett 198:99–102

Rudolph U, Finegold MJ, Rich SS, Harriman GR, Srinivansan Y, Brabet P, Boulay G, Bradley A, Birnbaumer L (1995) Ulcerative colitis and adenocarcinoma of the colon in $G\alpha_{i2}$-deficient mice. Nature Genet 10:143–150

Rudolph U, Spicher K, Birnbaumer L (1996) Adenylyl cyclase inhibition and altered G protein subunit expression and ADP-ribosylation patterns in tissues and cells from $G_{i2}\alpha$ -/- mice. Proc Natl Acad Sci USA 93:3209–3214

Sargiacomo M, Sudol M, Tang ZL, Lisanti MP (1993) Signal transduction molecules and glycosyl-phosphatidylinositol-linked proteins form caveolin-rich insoluble complex in MDCK cells. J Cell Biol 122:789–807

Sánchez-Blázquez P, García-España A, Garzón J (1995) In vivo injection of antisense oligodeoxynucleotides to Gα subunits and supraspinal analgesia evoked by mu and delta opioid agonists. J Pharmacol Exp Ther 275:1590–1596

Shapira H, Way J, Lipinsky D, Oron Y, Battey JF (1994) Neuromedin B receptor, expressed in Xenopus laevis oocytes, selectively couples to $G\alpha_q$ and not $G\alpha_{11}$. FEBS Lett 348:89–92

Schreibmayer W, Dessauer CW, Vorobiov D, Gilman AG, Lester HA, Davidson N, Dascal N (1996) Inhibition of an inwardly rectifying K^+ channel by G-protein α-subunits. Nature 380:624–627

Schlingensiepen KH, Schlingensiepen R, Brysch W (1996) Antisense oligodeoxynucleotides: from technology to therapy. Blackwell International/Blackwell Wissenschaft, Berlin

Simon MI, Strathmann MP, Gautam N (1991) Diversity of G proteins in signal transduction. Science 232:802–808

Srinivasan SK, Tewary HK, Iversen PL (1995) Characterization of binding sites, extent of binding, and drug interaction of oligonucleotides with albumin. Antisense Res Dev 5:131–139

Standifer KM, Rossi GC, Pasternak GW (1996) Differential blockade of opioid analgesia by antisense oligodeoxynucleotides directed against various G protein α subunits. Mol Pharmacol 50:293–298

Stehno-Bittel L, Krapivinsky G, Krapivinsky L, Perez-Terzic C, Clapham D (1995) The G protein $\beta\gamma$ subunit transduces the muscarinic receptor signal for Ca^{2+} release in Xenopus oocytes. J Biol Chem 270:30068–30074

Stein CA, Cheng Y-C (1993) Antisense oligonucleotides as therapeutic agents – is the bullet really magical? Science 261:1004–1012

Strader CD, Fong TM, Tota MR, Underwood D, Dixon RAF (1994) Structure and function of G protein-coupled receptors. Annu Rev Biochem 63:101–132

Tamemoto H, Kadowaki T, Tobe K, Yagi T, Sakura H, Hayakawa T, Terauchi Y, Ueki K, Kaburagi Y, Satoh S, Sekihara H, Yoshioka S, Horikoshi H, Furuta Y, Ikawa Y, Kasuga M, Yazaki Y, Aizawa S (1994) Insulin resistance and growth retardation in mice lacking insulin receptor substrate-1. Nature 372:182–186

Tang T, Kiang, JG, Côte TE, Cox BM (1995) Antisense oligodeoxynucleotide to the G_{i2} protein α subunit sequence inhibits an opioid-induced increase in the intracellular free calcium concentration in ND8-47 neuroblastoma × dorsal root ganglion hybrid cells. Mol Pharmacol 48:189–193

Voisin T, Lorinet A-M, Maoret J-J, Couvineau A, Laburthe M (1996a) $G\alpha_i$ RNA antisense expression demonstrates the exclusive coupling of peptide YY receptors to G_{i2} proteins in renal proximal tubule cells. J Biol Chem 271:574–580

Voisin T, Lorinet A-M, Laburthe M (1996a) Partial knockdown of $G\alpha_{i2}$ protein is sufficient to abolish the coupling of PYY receptors to biological response in renal proximal tubule cells. Biochem Biophys Res Commun 225:16–21

Wahlestedt C (1994) Antisense oligonucleotide strategies in neuropharmacology. Trends Pharmacol Sci 15:42–46

Wahlestedt C (1996) Antisense "knockdown" strategies in neurotransmitter receptor research. In: Raffa RB, Porecca F (eds) Antisense strategies for the study of receptor mechanisms. Landes, Austin, Texas, USA, pp 1–10

Wang H-Y, Watkins DC, Malbon CC (1992) Antisense oligodeoxynucleotides to G_s protein α-subunit sequence accelerate differentiation of fibroblasts to adipocytes. Nature 358:334–337

Wang H-Y, Johnson GL, Liu X, Malbon CC (1996) Repression of adipogenesis by adenylyl cyclase stimulatory G-protein α subunit is expressed within region 146–220. J Biol Chem 271:22022–22029

Watkins DC, Johnson GL, Malbon CC (1992) Regulation of differentiation of teratocarcinoma cells into primitive endoderm by $G\alpha_{i2}$. Science 258:1373–1375

Watkins DC, Moxham CM, Morris AJ, Malbon CC (1994) Suppression of $G_{i\alpha 2}$ enhances phospholipase C signalling. Biochem J 299:593–596

CHAPTER 12
Use of Antisense Oligonucleotides to Modify Inflammatory Processes

C.F. BENNETT and T.P. CONDON

A. Introduction

The inflammatory process is vital for the survival of higher eukaryotic organisms. Despite being a tightly regulated system, failures in its checks and balances can occur, resulting in disease. With the exception of rare genetic deficiencies, such as leukocyte adhesion disorders due to defects in $\beta 2$ integrin (ANDERSON and SPRINGER 1987), X-linked hyper-immunoglobulin (Ig)M syndrome due to defects in CD40L (ARUFFO et al. 1993) or an autosomal recessive form of severe combined immunodeficiency due to a mutation in zap-70 (CHAN et al. 1994), the underlying causes of disease are unknown. Historical approaches to the regulation of the immune system utilized either general cytotoxic compounds or glucocorticoids. While these agents clearly provide benefit to the patient, they also expose him or her to undesirable risks due to their nonspecific nature of activity. The identification of cyclosporin A, and more recently FK-506 and rapamycin, demonstrated that it was possible to attenuate an immune response without causing generalized myelosuppression, thus identifying methods for selectively modulating an immune response. These findings, combined with our vastly increased understanding of how the immune system functions, has opened up tremendous opportunities for the treatment of inflammatory disorders. Unfortunately, the identification of new chemical entities which selectively inhibit specific pathways in immune cell function has been difficult, with few selective inhibitors forthcoming. Many investigators and companies have relied on monoclonal antibodies or expressed protein products to validate targets and also to serve as therapeutic approaches. In fact, several of these products are on the market, such as OKT3, granulocyte macrophage colony-stimulating factor (GM-CSF), and various interferons. There is still a need for alternative strategies to identify selective inhibitors of proteins which are thought to play important roles in regulating an immune response.

Antisense oligonucleotides represent one such alternative approach to inhibiting the function of proteins thought to be important in regulating immune cell function. In contrast to more conventional approaches the target of antisense oligonucleotides is the RNA which codes for the protein, rather than the protein product itself. Identification of a lead antisense oligonucleotide can be a rapid process, enabling validation of a molecular target within a couple of

List of Abbreviations

ICAM-1	Intercellular adhesion molecule 1
IFN	Interferon
Ig	Immunoglobulin
IL	Interleukin
GM-CSF	Granulocyte macrophage colony-stimulating factor
LFA	Lymphocyte function associated antigen
Mac-1	Macrophage 1
MadCAM	Mucosal addressin cell adhesion molecule
NF	Nuclear factor
NK	Natural killer (*cells*)
PECAM	Platelet/endothelial cell adhesion molecule
SCID	Severe combined immunodeficiency
TNBS	2,4,6 trinitrobenzene sulfonic acid
TNF	Tumor necrosis factor
VCAM	Vascular cell adhesion molecule
VLA	Very late activation (*antigen*)

months. One of the advantages of antisense oligonucleotides is that the subcellular location of the protein product does not matter. The RNA product which codes for these proteins is all, in theory, equally accessible to the antisense oligonucleotide. Therefore, intracellular as well as extracellular targets may be selected. In contrast, monoclonal antibodies or other protein products are primarily useful for extracellular targets. Another advantage of antisense oligonucleotides is that they do not appear to be antigenic, a characteristic which limits the long-term application of monoclonal antibody products.

This review will focus primarily on intercellular adhesion molecule 1 (ICAM-1) as one example of how antisense oligonucleotides can be used to target a mRNA encoding a protein, which plays a central role in immune regulation. Other examples of the use of antisense oligonucleotides to suppress immune responses will be highlighted. Previous reviews have discussed the in vitro application of antisense oligonucleotides (BENNETT 1993; BENNETT and CROOKE 1996; CROOKE 1992; CROOKE and BENNETT 1996); therefore, the focus of this review will be primarily on recent in vivo applications.

Before proceeding, it should be pointed out, as discussed in more detail in Chap. 8, that some types of oligodeoxynucleotides are effective stimulators of the immune system through non-antisense effects. Phosphodiester oligodeoxynucleotides can stimulate a polyclonal B cell proliferative response or activation of natural killer (NK) cells if they contain the appropriate sequence motif (BALLAS et al. 1996; KRIEG et al. 1995; PISETSKY and REICH 1993). In our experience, all phosphorothioate-modified oligodeoxynucleotides will promote some degree of B cell proliferation (BENNETT et al. 1997; HENRY et al. 1997a,b; MONTEITH et al. 1997). Incorporation of the appropriate sequence motifs into phosphorothioate oligodeoxynucleotide (BOGGS et al. 1997; KRIEG et al. 1995) significantly increases the potency for B cell proliferation. Because of the potential to nonspecifically modulate an immune response with oligonucleotides, it is important to incorporate controls. With the use of appropriate controls it is possible to effectively utilize oligonucleotides as tools for modulating immune response and, potentially, as therapeutic agents for the treatment of inflammatory diseases or diseases with an inflammatory component.

B. Cell Adhesion Molecules

Extravasation of leukocytes from the circulation through postcapillary venules into tissue is a carefully orchestrated process involving the production of soluble chemotactic factors at sites of inflammation, adhesion of leukocytes to vascular endothelium, and diapedesis between endothelial cells (BUTCHER 1991; EBNET et al. 1996; SPRINGER 1990a, 1994). At least three distinct steps can be identified in leukocyte emigration: reversible adhesion or rolling on vascular endothelial cells, activation of leukocytes resulting in firm adhesion, and diapedesis, each mediated by cell–cell adhesion. The initial rolling steps appear to be mediated by selectins (E-, L- and P-selectin; Table 1) expressed on either vascular endothelial cells or on leukocytes interacting with specific carbohydrate structures expressed on the cognate cell type (BEVILACQUA 1993; KANSAS 1996). Recently, several papers have been published suggesting that vascular cell adhesion molecule 1 (VCAM-1) interacting with VLA4 can also mediate rolling on endothelial cells (ALON et al. 1995; BERLIN et al. 1995).

Leukocytes can become activated by a variety of chemotactic factors either soluble or cell associated (EBNET et al. 1996; ZIMMERMAN et al. 1996). Best characterized are neutrophils which undergo marked shape changes upon activation, translocate macrophage 1 (MAC-1) from intracellular granules to the cell surface, shed L-selectin, and increase affinity of integrins for their ligands, among other changes. Firm adhesion is mediated by the $\beta 2$ integrins lymphocyte function associated antigen (LFA)-1 and MAC-1 binding to ICAM-1 and ICAM-2 on endothelial cells and the $\beta 1$ integrin very late activation (*antigen*) (VLA)-4 binding to VCAM-1. The activation of $\beta 2$ integrins LFA-1 and MAC-1 on neutrophils is thought to be required for firm adhesion and transmigration (BUTCHER 1991; SPRINGER 1990b, 1994). The passage of leukocytes through endothelial monolayers is predominately mediated by ICAM-1–LFA-1 interactions (FURIE et al. 1991; OPPENHEIMER-MARKS et al. 1991), although contributions by platelet/endothelial cell adhesion molecules (PECAM)-1 have also been suggested (NEWMAN 1997).

In addition to playing a role in the migration of leukocytes to sites of inflammation, most adhesion molecules are also capable of signaling leukocytes and endothelial cells (ALTMANN et al. 1989; SPRINGER 1990b; DAMLE et al. 1992, 1994; KUHLMAN et al. 1991; LO et al. 1991; WADDELL et al. 1995). In the

Table 1. Endothelial cell–leukocyte adhesion molecules

Cell adhesion molecule	Other names	Gene family	Expression pattern	Counter-receptor
E-Selectin	ELAM-1, CD62E	Selectin	Induced on endothelial cells	Sialyl Lewis X
L-Selectin	CD62L	Selectin	Constitutively expressed on most leukocytes	Sialyl Lewis X expressed on GlyCAM-1, MadCAM-1, PSGL-1
P-Selectin	PADGEM-1, CD62P	Selectin	Stored in Weibel-Palade bodies of endothelial cells and α-granules of platelets. Expressed on cell surface after cell activation	Sialyl Lewis X expressed on PSGL-1
ICAM-1	CD54	Immunoglobulin	Activated endothelial cells, keratinocytes, fibroblasts, B-lymphocytes, monocytes, etc.	LFA-1, MAC-1, fibrinogen, hylarounic acid, rhinovirus
ICAM-2	CD102	Immunoglobulin	Constitutively expressed on endothelial cells induced on activated lymphocytes, platelets	LFA-1, MAC-1
ICAM-3	ICAM-R, CD50	Immunoglobulin	Constitutively expressed on lymphocytes	LFA-1, $\alpha d\beta 2$ integrin
MadCAM-1	Mucosal addressin, MECA-367, MECA-89	Immunoglobulin and mucin	Expressed on high endothelial venules in Peyers patches, mesenteric lymph nodes, endothelial cells of lamina propria and spleen	$\alpha 4\beta 7$ Integrin, L-selectin

PECAM-1	CD31	Immunoglobulin	Constitutively expressed on endothelial cells, platelets, monocytes and neutrophils	PECAM-1, $\alpha v\beta 3$ integrin
VCAM-1	CD106	Immunoglobulin	Induced on endothelial cells, dendritic cells, smooth muscle cells	VLA4 ($\alpha 4\beta 1$), and $\alpha 4\beta 7$
LFA-1 ($\alpha L\beta 2$)	CD11a/CD18	Integrin	Constitutively expressed on most leukocyte populations	ICAM-1, ICAM-2, ICAM-3, ICAM-4
MAC-1 ($\alpha M\beta 2$)	CD11b/CD18	Integrin	Expressed on monocytes, NK cells and neutrophils (stored in secondary granules)	ICAM-1, ICAM-2, fibrinogen, iC3b, factor X
P150/95 ($\alpha x\beta 2$)	CD11c/CD18	Integrin	Constitutively expressed on most leukocytes, higher level of expression on monocytes	iC3b, fibrinogen
$\alpha d\beta 2$	CD11d/CD18	Integrin	Constitutively expressed on most leukocytes	ICAM-1, ICAM-3
VLA4 ($\alpha 4\beta 1$)	CD49D/CD29	Integrin	Constitutively expressed on lymphocytes and monocytes	VCAM-1, fibronectin
$\alpha 4\beta 7$	LPAM-1	Integrin	Constitutively expressed on lymphocytes	MadCAM-1, VCAM-1, fibronectin

ICAM, intercellular adhesion molecule; VCAM, vascular cell adhesion molecule; PECAM, platelet/endothelial cell adhesion molecule; LFA, lymphocyte function associated antigen; VLA, very late activation (*antigen*); NK, natural killer.

case of lymphocytes, both LFA-1 and VLA-4 provide co-stimulatory signals required for a productive response to antigens. Inhibiting these co-stimulatory signals has been shown to attenuate the response of the lymphocyte. Binding of leukocytes to endothelial cell has also been shown to activate the endothelial cells, facilitating the emigration process (DOUKAS and POBER 1990; DURIEU-TRAUTMANN et al. 1994; KARMANN et al. 1996; PFAU et al. 1995).

Because of the fundamental role adhesion molecules play in initiating and propagating an immune response, there has been much interest in identifying inhibitors. Monoclonal antibodies have been used to demonstrate proof of concept in a variety of preclinical pharmacological models, as well as in early clinical studies (DOERSCHUK et al. 1990; HAUG et al. 1993; ISOBE et al. 1992; KAVANAUGH et al. 1994; OROSZ et al. 1992; PODOLSKY et al. 1994; WEGNER et al. 1990; WINN and HARLAN 1993; WINN et al. 1993). The identification of selective low-molecular-weight inhibitors of cell adhesion is highly desirable; unfortunately, efforts to identify such compounds have not met with great success. There is still a need, therefore, for alternative approaches. We have used antisense oligonucleotides to inhibit the expression of a variety of endothelial cell adhesion molecules and are currently evaluating them for use in preclinical pharmacology models as well as in the clinic. ICAM-1 will serve as an example of how antisense oligonucleotides can be used to regulate the expression of adhesion molecules, while results obtained with other adhesion molecules will serve to exemplify additional observations.

I. Intercellular Adhesion Molecule 1

ICAM-1 is expressed at low levels on resting endothelial cells and can be markedly upregulated in response to inflammatory mediators such as tumor necrosis factor (TNF)-α, interleukin (IL)-1, and interferon (IFN)-γ. We have tested over 100 different oligonucleotides of various chemistries for effects on ICAM-1 expression. The most effective first generation phosphorothioate oligodeoxynucleotides identified targeted, specific sequences in the 3'-untranslated region of the human ICAM-1 mRNA, ISIS 1939 and ISIS 2302 (BENNETT et al. 1994; CHIANG et al. 1991). Both ISIS 1939 and ISIS 2302 inhibit ICAM-1 expression by a RNase H-dependent mechanism of action (BENNETT et al. 1994). Because ISIS 1939 was very pyrimidine rich (90% C and T), there was concern that this oligonucleotide might produce undesirable effects through interaction with other molecules (both RNAs and non-RNAs). Because of these concerns, ISIS 2302 was selected for further additional studies.

ISIS 2302 selectively inhibits ICAM-1 expression in a variety of cell types (BENNETT et al. 1994; MIELE et al. 1994; NESTLE et al. 1994). Both sense and a variety of scrambled control oligonucleotides fail to inhibit ICAM-1 expression, including a two-base mismatch control (BENNETT et al. 1994; MIELE et al. 1994; NESTLE et al. 1994). Treatment of endothelial cells with ISIS 2302 blocked adhesion of leukocytes, demonstrating that blocking the expression of

ICAM-1 will attenuate adhesion of leukocytes to activated endothelial cells (BENNETT et al. 1994). ISIS 2302 also blocked a one-way mixed lymphocyte reaction when the antigen-presenting cell was pretreated with ISIS 2302 to down regulate ICAM-1 expression prior to exposure to the lymphocyte (VICKERS et al., manuscript submitted). Thus, ISIS 2302 is capable of blocking both leukocyte adhesion to activated endothelial cells and co-stimulatory signals to T lymphocytes; both activities were predicted based on previous studies with monoclonal antibodies to ICAM-1.

1. Pharmacology of ICAM-1 Antisense Oligonucleotides

a) Proof of Mechanism

Much has been written concerning the non-antisense effects of phosphorothioate oligodeoxynucleotides (STEIN 1995, 1996; WAGNER 1994). With this in mind, can the pharmacological activity observed with an oligonucleotide be ascribed to an antisense effect, rather than non-antisense effects? Although it is difficult to unequivocally conclude that all pharmacological activity reported for antisense oligonucleotides is due to antisense effects, with proper controls and experimental design it has been possible to build a strong case that the ICAM-1 antisense oligonucleotides are producing the described pharmacological activity by an antisense mechanism of action. There are multiple lines of evidence which support this conclusion:

1. In cell culture, the oligonucleotides were identified after screening multiple oligonucleotides all capable of hybridizing to ICAM-1 mRNA. The oligonucleotides used for pharmacological evaluation were identified as being the most effective at inhibiting ICAM-1 expression in cell culture-based assays (BENNETT et al. 1994; CHIANG et al. 1991). Optimizing oligonucleotides for increased potency by screening multiple target sites on an mRNA results in compounds with a greater signal/noise ratio. Given that phosphorothioate oligodeoxynucleotides will produce non-antisense effects at higher concentrations or doses, identification of potent compounds enables the use of oligonucleotides at doses which do not produce non-antisense effects.
2. The ICAM-1 antisense oligonucleotides will selectively reduce ICAM-1 proteins in multiple cell types with a wide range of stimuli, including IL-1α and β, TNF-α, IFN-γ, bacterial endotoxin, and phorbol esters (BAKER et al. 1997; BENNETT et al. 1992, 1993, 1994; CHIANG et al. 1991; MIELE et al. 1994; NESTLE et al. 1994; STEPKOWSKI et al. 1994). It is unlikely that the oligonucleotides are interfering with a central signaling pathway or receptor–ligand interaction as these agents induce ICAM-1 expression by different signaling mechanisms (CORNELIUS et al. 1993; OHH et al. 1994; READ et al. 1995; STRASSMAN et al. 1994; WERTHEIMER et al. 1992).
3. For the ICAM-1 antisense oligonucleotides which target the 3'-untranslated region of ICAM-1 mRNA, it is possible to demonstrate a

selective reduction in mRNA which appears to be due to RNase H (BENNETT et al. 1994; CHIANG et al. 1991).
4. The antisense oligonucleotides are species-specific, which is what would be predicted based on the poor conservation of sequence between different species.
5. In several in vivo experiments, it has been possible to demonstrate a reduction in ICAM-1 mRNA or protein following treatment with the ICAM-1 antisense oligonucleotide (BENNETT et al. 1996, 1997; CHRISTOFIDOU-SOLOMIDOU et al. 1997; KUMASAKA et al. 1996). For chronic models, it is difficult to unequivocally conclude that reductions in ICAM-1 by the oligonucleotides are a direct antisense effect, as the oligonucleotides could also affect expression of cytokines which induce ICAM-1 expression or activation of cells which release the cytokines. However, we have demonstrated selective reductions in ICAM-1 expression in acute models when ICAM-1 is directly induced by bacterial endotoxins (KUMASAKA et al. 1996). It is highly unlikely that these effects would be due to a non-antisense mechanism as the reductions in ICAM-1 expression were detected 2–4h after stimulation.
6. In several in vivo models the effects of the ICAM-1 antisense oligonucleotide were similar to the effects produced with ICAM-1 monoclonal antibodies (KATZ et al. 1995; KUMASAKA et al. 1996; STEPKOWSKI et al. 1994).
7. The ICAM-1 antisense oligonucleotide produces the expected pharmacology in vivo for an agent inhibiting ICAM-1 expression.

b) Human Xenografts

ISIS 2302 is selective for human ICAM-1 mRNA, limiting its application for in vivo pharmacology studies. To test the pharmacology of the human-specific antisense oligonucleotide, we have resorted to experimental models in which human tissue is xenografted to immuno-compromised mice. One model examined the role of ICAM-1 in metastasis of human melanoma cells to the lung of mice. ICAM-1 is expressed at high levels in advanced primary melanomas and melanoma metastasis (JOHNSON et al. 1989; NATALI et al. 1990). Treating human melanoma cells with either TNF-α or IFN-γ prior to injection into nude mice results in a significant increase in the number of lung metastasis and ICAM-1 expression (MIELE et al. 1994). To address the question of whether ICAM-1 played a role in the increased number of lung metastasis, the melanoma cells were pretreated with ICAM-1 antisense oligonucleotides prior to treatment with cytokines. Pretreatment of the melanoma cells with the ICAM-1 antisense oligonucleotides reduced the number of lung metastases, while an irrelevant control oligonucleotide failed to decrease the number of metastases (MIELE et al. 1994). The rank order potency for inhibition of ICAM-1 expression correlated with the rank order potency for inhibition of ICAM-1 expression. The mechanism by which ICAM-1 contributes to the development of lung metastasis is not clear. One possibility is that the adhesion of leukocytes to circulating melanoma cells results in embolism in the microvasculature of the lung. Also, ICAM-1 is known to activate leukocytes which could release a variety of proteases and other mediators, enhancing colonization of the melanoma cells in lung tissue.

A second study addressed the role of ICAM-1 in an experimental model of cytotoxic dermatitis (lichen planus). In this model human skin is xenografted onto severe combined immunodeficiency (SCID) mice (MURRAY et al. 1994; YAN et al. 1993). When the human tissue becomes engrafted, heterologous lymphocytes are injected into the graft which migrate into the epidermis (epidermal tropism) and produce a cytotoxic interaction between effector lymphocytes and epidermal cells (CHRISTOFIDOU-SOLOMIDOU et al. 1997). Migration of the lymphocytes from the dermis into the epidermis was correlated with expression of ICAM-1 in the epidermis. Treatment with ISIS 2302 inhibited ICAM-1 expression in the human graft, decreased migration of lymphocytes into the epidermis and subsequently decreased lesion formation. A scrambled control oligonucleotide failed to attenuate the responses. These data demonstrate that an ICAM-1 antisense oligonucleotide administered systemically can attenuate an inflammatory response in the skin.

c) Rodent Allografts

Because of the lack of conservation between human, mouse, and rat ICAM-1 mRNAs in the 3'-untranslated region where ISIS 2302 hybridizes, it was necessary to identify rat- and mouse-specific antisense oligonucleotides. ISIS 3082 and ISIS 9125 are 20-base phosphorothioate oligodeoxynucleotides which hybridize to an analogous region in the 3'-untranslated region of murine and rat ICAM-1 mRNA, respectively. Similar to ISIS 2302, ISIS 3082 and ISIS 9125 selectively inhibit ICAM-1 expression in mouse or rat cells by an RNase H-dependent mechanism (STEPKOWSKI et al. 1994).

Previous studies have demonstrated that monoclonal antibodies to ICAM-1 prolong heterotopic cardiac allograft survival in mice (ISOBE et al. 1992). ISIS 3082 was tested in the same model to determine if an ICAM-1 antisense oligonucleotide would prolong cardiac allografts (STEPKOWSKI et al. 1994). Treatment of recipient C3H mice with ISIS 3082 for 7 or 14 days by continuous intravenous infusion resulted in a dose-dependent prolongation of C57BL/10 cardiac allograft survival. Maximal effects occurred at 5–10 mg/kg per day. Treatment of recipient mice with 5 mg/kg per day for 14 days increased cardiac allograft survival from 7.7 ± 1.4 days to 23.0 ± 7.5 days. Similar results were obtained with two additional strain combinations. Two control phosphorothioate oligodeoxynucleotides failed to prolong cardiac allograft survival. ISIS 3082 was either additive or synergistic with anti-lymphocyte serum, brequinar, or rapamycin in prolonging cardiac allograft survival. Similar to previous reports using monoclonal antibodies to ICAM-1 and LFA-1 (ISOBE et al. 1992), the combination of ISIS 3082 plus a monoclonal antibody to LFA-1 increased survival of the cardiac allograft to greater than 150 days.

These results suggest that the combination of a LFA-1 monoclonal antibody and an inhibitor of ICAM-1 (either an antibody or antisense oligonucleotide) induces donor-specific transplantation tolerance.

In the mouse model of cardiac allograft rejection, the combination of ISIS 3082 plus cyclosporin A attenuated the effect of each agent when given alone. This apparent antagonism between ISIS 3082 and cyclosporin A was unique to the mouse heterotopic heart model. ISIS 9125 (the rat-specific oligonucleotide) is synergistic with cyclosporin A in rat kidney and heart allograft models (STEPKOWSKI et al., manuscript submitted). ISIS 2302 does not attenuate the effects of cyclosporin A in a primate kidney transplant model (S. STEPKOWSKI, unpublished data). Finally, in a two-way mixed lymphocyte reaction, ISIS 2302 did not reverse the inhibitory effects of cyclosporin A (Fig. 1). These data suggest that the apparent lack of synergy between cyclosporin A and ISIS 3082 is unique to the mouse cardiac allograft model. Given that the effects of cyclosporin A in mice are variable, these results are not unexpected.

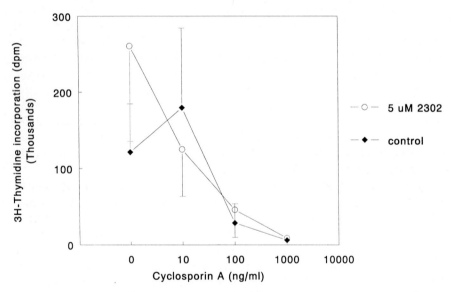

Fig. 1. Effect of ISIS 2302 on cyclosporin A inhibition of a two-way mixed lymphocyte reaction. B cell depleted (Immunotech CD19 magnetic beads) peripheral blood mononuclear cells were isolated from two different normal volunteers. Cells from both donors (10^5 cells each/well) were mixed in a 96-well plate and incubated for 5 days in the presence of the indicated concentration of cyclosporin A and ISIS 2302. ^{3}H-Thymidine (1 μCi) was added to each well 18 h prior to harvesting the cells to label proliferating cells. Cells were lysed and DNA collected on Whatman GF/B filters. Data are expressed as the mean ± standard deviation of ^{3}H-thymidine incorporation into the cells. Results demonstrate that cyclosporin A inhibits a two-mixed lymphocyte reaction in a concentration-dependent manner and that ISIS 2302, at a concentration of 5 μM, does not attenuate the effects of cyclosporin A

ISIS 3082 also prolonged survival of mouse islet cell allografts, demonstrating that the effects are not restricted to the heart (KATZ et al. 1995). In both the cardiac allograft model and the islet cell allograft model, the effects of ISIS 3082 were as good or better than an ICAM-1 monoclonal antibody. The rat ICAM-1 antisense oligonucleotide, ISIS 9125, prolongs survival of rat cardiac and kidney allografts in a dose-dependent manner (STEPKOWSKI et al., manuscript submitted). The effects of the oligonucleotide were more pronounced in the kidney allograft model, which is consistent with the pharmacokinetics of phosphorothioate oligodeoxynucleotides in that the kidney is the major organ of disposition (AGRAWAL et al. 1991; COSSUM et al. 1993; CROOKE et al. 1996).

There are several mechanisms by which the ICAM-1 antisense oligonucleotides may prolong survival of allografts such as: (a) inhibition of ICAM-1 expression on endothelial cells of the graft, preventing leukocyte infiltration into the graft tissue, (b) inhibition of ICAM-1 expression on either professional or nonprofessional antigen-presenting cells in the graft tissue, or (c) inhibition of ICAM-1 expression on recipient lymphocytes or NK cells. We have performed several experiments to address this question. The most revealing data was generated in the rat renal allograft model, in which treatment of the donor animal or perfusion of the graft at the time of harvest with ISIS 9125 resulted in prolongation of the kidney allograft survival, suggesting that the effects of the oligonucleotide are on the donor tissue rather than the recipient.

d) Colitis

Increased expression of ICAM-1 has been detected in both ulcerative colitis and Crohn's disease (KOIZUMI et al. 1992; SCHUERMANN et al. 1993). The murine-specific ICAM-1 antisense oligonucleotide, ISIS 3082, was evaluated in a dextran sulfate model of colitis in mice (BENNETT et al. 1997). Administration of 5% dextran sulfate in the drinking water of mice for 5–7 days produced a colitis, which persisted for up to 6 weeks from discontinuation of treatment (OKAYASU et al. 1990). Mice treated with dextran sulfate for 7 days exhibited increased ICAM-1 expression on endothelial cells in the submucosa and lymphoid structures, demonstrating that ICAM-1 was expressed in inflamed colon tissue. ICAM-1 was also detected on mucosal leukocytes infiltrating the tissue. The localization of the ICAM-1 antisense oligonucleotide in normal and diseased colon tissue was determined using a rhodamine-labeled oligonucleotide. In normal tissue the ICAM-1 oligonucleotide was localized in the lamina propria and, to a lesser extent, in epithelial cells in normal mice. The distribution of rhodamine-labeled ISIS 3082 in mice with colitis changed in that epithelial cells accumulated significantly more of the oligonucleotide compared to normal animals (BENNETT et al. 1997). Treatment of mice with ISIS 3082 decreased ICAM-1 expression and leukocyte infiltration into the colon of dextran sulfate-treated mice. ISIS 3082 was effective in preventing the development of colitis when administered prophylactically and also in attenuating existing colitis. The optimal dose for preventing development of colitis was 0.3–1.0 mg/kg per day, while approximately ten-fold higher concentrations were required to attenuate existing disease. Several control oligonucleotides were also evaluated in the model and found to have minimal effects. Thus, ICAM-1 antisense oligonucleotides could be of value in treating inflammatory bowel disease.

e) Renal Ischemia

HALLER et al. used an ICAM-1 antisense oligonucleotide to decrease acute renal injury following ischemia in rats (HALLER et al. 1996). They identified a 20-base phosphorothioate oligodeoxynucleotide, targeting the 3′-untranslated region of ICAM-1 mRNA. This oligonucleotide was shown to inhibit ICAM-1 expression in rat cells in a sequence-specific manner. Using a cationic lipid formulation of the oligonucleotide infused into the femoral vein, they demonstrated decreased ICAM-1 protein expression following ischemic injury and decreased leukocyte infiltrates (HALLER et al. 1996). The ICAM-1 antisense oligonucleotide also preserved renal function since blood urea nitrogen and serum creatine were reduced in the antisense oligonucleotide-treated group 12–24 h following injury, compared to the saline- or control oligonucleotide-treated group. These data suggest that inhibition of ICAM-1 expression or function protects against ischemia-reperfusion injury in kidney.

2. Toxicology of ICAM-1 Antisense Oligonucleotides

The human ICAM-1 antisense oligonucleotide, ISIS 2302, has been evaluated for both acute and chronic toxicities in primates and mice (HENRY et al. 1997a,b). In addition, the murine-specific antisense oligonucleotide, ISIS 3082, has been evaluated for exaggerated pharmacological toxicities in mice (BENNETT et al. 1997; HENRY et al. 1997b). ISIS 2302 contains a one-base mismatch for the same region in cynomolgus monkey ICAM-1 mRNA, and is accordingly approximately two-fold less effective in inhibiting ICAM-1 expression in cynomolgus monkey cells compared to human cells (unpublished data). Therefore, ISIS 2302 is capable of inhibiting ICAM-1 expression in cynomolgus monkey tissue, albeit at higher doses than would be effective for humans. In both the mouse and monkey studies there was no evidence of toxicities which could be attributed to exaggerated pharmacology. These results were not unexpected, given that genetic deletion of the ICAM-1 gene in mice does not result in any marked phenotypic changes (SLIGH et al. 1993; XU et al. 1994). The observed toxicities were common to other phosphorothioate oligodeoxynucleotides discussed in Chap. 5. Briefly, in the monkey studies dose-dependent prolongation in activated partial thromboplastin time and evidence for oligonucleotide accumulation in proximal tubular epithelial cells were observed. In the mice studies, both ISIS 2302 and the murine analog, 3082, caused dose-dependent increases in spleen and liver weights and mononuclear cell infiltrates in several organs, although the magnitude of changes were smaller than observed for some phosphorothioate oligodeoxynucleotides (MONTEITH et al. 1997). Other changes noted included increases in circulating monocytes in the 100 mg/kg-dose groups and increases in liver enzymes in serum at the same dosage level (BENNETT et al. 1997). These data suggest that at pharmacological relevant doses, the ICAM-1 antisense oligodeoxynucleotides are well tolerated with repeat administrations.

3. Clinical Studies with ISIS 2302

A phase-1 safety assessment of ISIS 2302 in normal volunteers was recently completed (GLOVER et al. 1997). As discussed in Chap. 18, the results of this study demonstrated that ISIS 2302 was well tolerated in normal volunteers, with no adverse events reported. The pharmacokinetics of ISIS 2302 in man was similar to the pharmacokinetics in cynomolgus monkeys. ISIS 2302 is currently being evaluated for efficacy in five different clinical indications: rheumatoid arthritis, psoriasis, acute renal transplant rejection, Crohn's disease, and ulcerative colitis. Preliminary results in Crohn's disease are encouraging and suggest that ISIS 2032 may provide some benefit to this patient group (YACYSHYN et al., manuscript submitted).

4. Second and Third Generation Chemistry

Four major objectives for the application of medicinal chemistry in antisense oligonucleotide-based therapeutics are to increase potency, decrease toxicity, alter the pharmacokinetics, and reduce costs. It is gratifying to see that there have been tremendous advances in all four areas.

A large number of different chemically modified oligonucleotides and derivatives have been evaluated for activity against ICAM-1. Two types of approach have been taken: maintaining an oligodeoxynucleotide segment or "gap" to support RNase activity (MONIA et al. 1993), or utilization of uniformly modified oligonucleotides which do not support RNase H activity (BAKER et al. 1997; CHIANG et al. 1991). Surprisingly, among the most potent oligonucleotides identified to date are oligonucleotides uniformly modified on the 2'- position of sugars such as 2'-fluoro or 2'-methoxyethyl. These oligonucleotides do not support RNase H activity, yet are 10- to 20-fold more potent than the best phosphorothioate oligodeoxynucleotide which supports RNase H (BAKER et al. 1997). These results demonstrate that it is not necessary to induce RNA turnover to obtain potent antisense oligonucleotides. In general, we have observed a correlation between increased hybridization affinity for the target RNA and antisense activity in cell culture. However, it should be kept in mind that methods used to introduce oligonucleotides into cells, such as cationic lipids or electroporation, may skew the data towards one type of chemistry. As an example we have observed that oligonucleotides with reduced or no charge interact poorly with cationic lipids, yet are very effective at inhibiting ICAM-1 expression when electroporated or microinjected into cells. Thus, direct comparison of different chemistries using cationic lipids as the only means for enhancing cellular delivery may be biased towards highly charged species.

In addition to increasing potency, we have found that several modifications will decrease the class-specific toxicity of phosphorothioate oligodeoxynucleotides. For example, both 5-substituted pyrimidines and 2'-sugar-modified oligonucleotides produce less polyclonal B cell proliferation compared to unmodified oligodeoxynucleotides (BOGGS et al. 1997; KRIEG et al. 1995; ZHAO et al. 1995). The 2'-methoxyethyl modification also appears to decrease the potential for acute toxicities in primates, namely increases in aPTT and complement activation (MONTEITH et al., manuscript submitted).

The in vivo pharmacokinetics for several oligonucleotide modifications have been described (AGRAWAL et al. 1995; CROOKE et al. 1996; PARDRIDGE et al. 1995; ZHANG et al. 1995b, 1996). Results from these studies demonstrate that it is possible to change the tissue distribution of oligonucleotides with different chemical modifications. In addition, more stable oligonucleotide analogs have been identified which would allow for less frequent administration of the drug.

II. Other Endothelial Cell–Leukocyte Adhesion Molecules

We have taken a similar approach as described for ICAM-1 to identify antisense oligonucleotides targeting VCAM-1, PECAM-1 and E-selectin (BENNETT et al. 1994). In each case, targeting several sites on the respective mRNAs with antisense oligonucleotides resulted in the identification of phosphorothioate oligodeoxynucleotides which would selectively inhibit expression of the respective adhesion molecules in different species. One of the antisense oligonucleotides targeting human E-selectin, ISIS 4730, provides some interesting insights into the molecular mechanism of action of phosphorothioate antisense oligodeoxynucleotides (CONDON and BENNETT 1996). Treatment of human umbilical vein endothelial cells with ISIS 4730 inhibits the synthesis of E-selectin in a dose- and sequence-specific manner. Following treatment of cells with ISIS 4730, a novel, lower molecular weight transcript was induced due to RNase H cleavage of the pre-mRNA trapping the last intron in the cleavage product. The resulting transcript was stable and appeared to remain in the cell nucleus. These results demonstrate that phosphorothioate oligodeoxynucleotides are capable of binding to the pre-mRNA in the cell nucleus prior to RNA processing. Furthermore, they open up interesting opportunities for the use of oligonucleotides to regulate RNA maturation.

Recently, the effects of a porcine E-selectin antisense oligodeoxynucleotide have been evaluated in a septic shock model (GOLDFARB et al., to be published). Pretreatment of pigs with 10 mg/kg of the E-selectin antisense oligonucleotide attenuated the drop in cardiac output and increased peripheral vascular resistance due to endotoxin administration. In addition, the E-selectin antisense oligodeoxynucleotide prevented endotoxin-induced neutropenia, presumably due to the inhibition of neutrophil margination. A control oligonucleotide had no effect on these parameters. These results demonstrate that an E-selectin antisense oligonucleotide can attenuate acute inflammatory changes.

We have found that human VCAM-1 is particularly sensitive to the non-antisense effects of oligonucleotides. In cell culture-based experiments, specific VCAM-1 oligonucleotides are approximately five-fold more effective at inhibiting VCAM-1 expression compared to control oligonucleotides. Other adhesion molecules exhibit a greater selectivity than this. The non-antisense effects are not limited to phosphorothioate oligodeoxynucleotides since more potent phosphodiester- and heterocyclic-modified oligonucleotides also inhibit VCAM-1 expression with similar selectivity. The reason why VCAM-1 appears to be more prone to the non-antisense effects of oligonucleotides compared to other molecules is currently not known. VCAM-1 antisense oligodeoxynucleotides have been evaluated for pharmacological activity in several animal models. Although they showed activity in the models, they are either equal to or less effective than the ICAM-1 antisense oligonucleotide.

C. Nuclear Factor-κB

Nuclear factor (NF)-κB is a member of a family of transcription factors which regulate expression of a large number of gene products including Ig κ, IL-2 receptor, GM-CSF, E-selectin, ICAM-1, VCAM-1, IL-1, IL-6, IL-8, TNF, etc. (BAEUERLE and HENKEL 1994). There are several members of the NF-κB, or Rel, family which form either homo- or heterodimers. NF-κB is a p50-p65 heterodimer which forms a trimeric complex with IκB (inhibitor of κB) in resting cells. Upon activation, IκB becomes phosphorylated and subsequently degrades, releasing NF-κB from the complex. NF-κB translocates into the nucleus where it activates transcription of a variety of gene products. Two study groups have independently utilized antisense oligonucleotides targeting the p65 subunit of NF-κB to inhibit the growth of tumor cells in mice (HIGGINS et al. 1993; KITAJIMA et al. 1992; NARAYANAN et al. 1993). More recently, NEURATH et al. (1996) demonstrated that antisense oligonucleotides to the p65 subunit reverse established colitis in mice. A single intravenous injection or intracolonic application of a p65 antisense oligodeoxynucleotide reversed clinical symptoms in mice with 2,4,6 trinitrobenzene sulfonic acid (TNBS)-induced colitis and reversed ongoing intestinal inflammation determined by histology. Macrophages isolated from the intestine of antisense oligonucleotide-treated mice produced significantly lower amounts of IL-1, IL-6, and TNF-α. The p65 antisense oligodeoxynucleotide was also found to be effective in reversing clinical and histological scores in IL-10-deficient mice which develop chronic intestinal inflammation. In each case the mismatched phosphorothioate oligodeoxynucleotides failed to exhibit activity. Although the authors did not directly show an effect induced by systemically or locally administered oligonucleotides on p65 levels within cells in the tissue, the data are compelling in that the oligonucleotide was producing these dramatic effects by an antisense mechanism of action. It has been previously demonstrated that some of the earlier pharmacological activity described for p65 antisense oligonucleotides may be due to non-antisense effects (Maltese et al. 1995); the oligonucleotides used in the colitis study did not contain the four consecutive guanines which contributed in part to the non-antisense effects. These studies also demonstrate that short-term suppression of an ongoing inflammatory response may provide long term benefit. Similar observations have been made for ICAM-1 and TNF-α monoclonal antibodies (Elliott et al. 1993, 1994; Kavanaugh et al. 1994).

D. Interleukin-1 Receptors

Perhaps one of the first demonstrations of the in vivo efficacy of an antisense oligonucleotide was a study performed by Burch and Mahan in which they identified an oligodeoxynucleotide targeting human and murine type 1 IL-1 receptors (Burch and Mahan 1991). They demonstrated that subcutaneous injection of a phosphorothioate oligodeoxynucleotide targeting murine IL-1 receptor inhibited IL-1-induced neutrophil influx into skin by 37%. To obtain this response, it was necessary to treat the mice at least 48 h prior to injection with IL-1, presumably to downregulate existing receptors (Burch and Mahan 1991). The human-specific antisense oligonucleotide which contains five mismatched bases to the murine antisense oligonucleotide did not modify neutrophil influx. The authors did not demonstrate an effect of the antisense oligonucleotide on receptor expression in murine tissue to determine if the degree of receptor inhibition correlated with a decrease in neutrophil influx. As there is some evidence for spare IL-1 receptors (Dinarello 1994), it is possible that the antisense oligonucleotide was more effective than would appear based upon neutrophil influx.

E. Conclusions

With an increased understanding of how the immune system functions under normal conditions and a greater appreciation of how dysregulation of immune responses contributes to a variety of inflammatory diseases, a large number of potentially therapeutically useful molecular targets are being identified. For intracellular targets, antisense oligonucleotides represent one of the most direct routes to address the role a given protein plays in normal and abnormal immune responses. In addition, antisense oligonucleotides have a utility for inhibiting the expression of extracellular targets and offer some advantages over other approaches. However, it should be kept in mind that not all cell types within a tissue will be equally sensitive to the effects of the antisense oligonucleotide due to the pharmacokinetic behavior of specific oligonucleotide chemistries (AGRAWAL et al. 1991; BUTLER et al. 1996; COSSUM et al. 1993; PLENAT et al. 1995; RIFAJ et al. 1996; ZHANG et al. 1995). Therefore, like any other technology, careful experimental design and interpretation of the results are required.

We and others have demonstrated that phosphorothioate oligodeoxynucleotides targeting ICAM-1 exhibit broad anti-inflammatory activity in a variety of animal models (BENNETT et al. 1996, 1997; CHRISTOFIDOU-SOLOMIDOU et al. 1997; HALLER et al. 1996; KATZ et al. 1995; KUMASAKA et al. 1996; STEPKOWSKI et al. 1994). The effects of the oligonucleotides were sequence-specific and in many instances direct effects of the oligonucleotide on tissue expression of ICAM-1 have been demonstrated. These findings support the conclusion that the oligonucleotides were acting by an antisense mechanism of action. We are currently developing a parenteral formulation of the ICAM-1 antisense oligodeoxynucleotide for several inflammatory diseases on the basis of pharmacological activity, safety profile, pharmacokinetic behavior, and medical needs.. Data from a Crohn's disease study, in which an ICAM-1 antisense oligonucleotide was given by intravenous infusion, are very encouraging, suggesting that the oligonucleotide has beneficial effects in this group of patients. To obtain a broad utility for chronic inflammatory diseases, it is clear that more convenient dosage forms are needed. Both second generation chemistries and advanced formulations appear to meet this need. Although still in its infancy, the outlook for the application of antisense technologies in the treatment of human diseases is very promising.

References

Agrawal S, Temsamani J, Tang JY (1991) Pharmacokinetics, biodistribution, and stability of oligodeoxynucleotide phosphorothioates in mice. Proc Natl Acad Sci U S A 88:7595–7599

Agrawal S, Zhang X, Lu Z, Zhao H, Tamburin JM, Yan J, Cai H, Diasio RB, Habus I, Jiang Z, Iyer RP, Yu D, Zhang R (1995) Absorption, tissue distribution and vivo stability in rats of a hybrid antisense oligonucleotide following oral administration. Biochem Pharmacol 50:571–576

Alon R, Kassner PD, Carr MW, Finger EB, Hemler ME, Springer TA (1995) The integrin VLA-4 supports tethering and rolling in flow on VCAM-1. J Cell Biol 128:1243–1251

Altmann DM, Hogg N, Trowsdale J, Wilkinson D (1989) Cotransfection of ICAM-1 and HLA-DR reconstitutes human antigen-presenting cell function in mouse L cells. Nature 338:512–514

Anderson DC, Springer TA (1987) Leukocyte adhesion deficiency: an inherited defect in the Mac-1, LFA-1, and p150,95 glycoproteins. Annu Rev Med 38:175–194

Aruffo A, Farrington M, Hollenbaugh D, Li X, Milatovich A, Nonoyama S, Bajorath J, Grosmaire LS, Stenkamp R, Neubauer M, Roberts RL, Noelle RJ, Ledbetter JA, Francke U, Ochs H D (1993) The CD40 ligand, gp39, is defective in activated T cells from patients with X-linked hyper-IgM syndrome. Cell 72:291–300

Baeuerle PA, Henkel T (1994) Function and activation of the NF-κB in the immune system. Annu Rev Immunol 12:141–179

Baker BF, Lot SF, Condon TP, Cheng-Flournoy S, Lesnik E, Sasmor HM, Bennett CF (1997) 2′-O-(2-methoxy)ethyl modified anti-ICAM-1 oligonucleotides selectively increase the ICAM-1 mRNA level and inhibit formation of the ICAM-1 translation initiation complex in HUVECs. J Biol Chem 272:11994–12000

Ballas ZK, Rasmussen WL, Krieg A M (1996) Induction of NK activity in murine and human cells by CpG motifs in oligodeoxynucleotides and bacterial DNA. J Immunol 157:1840–1845

Bennett CF (1993) Antisense oligonucleotides in inflammation research and therapeutics. In: Crooke ST, Lebleu B (eds) Antisense research and applications. CRC Press, Boca Raton, pp 547–562

Bennett CF, Crooke ST (1996) Oligonucleotide-based inhibitors of cytokine expression and function. In: Henderson B, Bodmer MW (eds) Therapeutic modulation of cytokines. CRC Press, Boca Raton, pp 171–193

Bennett CF, Chiang M-Y, Chan H, Shoemaker JEE, Mirabelli CK (1992) Cationic lipids enhance cellular uptake and activity of phosphorothioate antisense oligonucleotides. Mol Pharmacol 41:1023–1033

Bennett CF, Chiang M-Y, Chan H, Grimm S (1993) Use of cationic lipids to enhance the biological activity of antisense oligonucleotides. J Liposome Res 3:85–102

Bennett CF, Condon T, Grimm S, Chan H, Chiang M-Y (1994) Inhibition of endothelial cell–leukocyte adhesion molecule expression with antisense oligonucleotides. J Immunol 152:3530–3540

Bennett CF, Dean N, Ecker DJ, Monia BP (1996) Pharmacology of antisense therapeutic agents: cancer and inflammation. In: Agrawal S (ed) Antisense therapeutics. Humana, Totowa, NJ, pp 13–46 (Methods in molecular medicine)

Bennett CF, Kornbrust D, Henry S, Stecker K, Howard R, Cooper S, Dutson S, Hall W, Jacoby HI (1997) An ICAM-1 antisense oligonucleotide prevents and reverses dextran sulfate sodium-induced colitis in mice. J Pharmacol Exp Ther 280:988–1000

Berlin C, Bargatze RF, Campbell JJ, von Adrian UH, Szabo MC, Hasslen SR, Nelson RD, Berg EL, Erlandsen SL, Butcher EC (1995) Alpha 4 integrins mediate lymphocyte attachment and rolling under physiologic flow. Cell 80:413–420

Bevilacqua MP (1993) Endothelial-leukocyte adhesion molecules. Annu Rev Immunol 11:767–804

Boggs RT, McGraw K, Condon T, Flournoy S, Villiet P, Bennett CF, Monia BP (1997) Characterization and modulation of immune stimulation by modified oligonucleotides. Antisense Nucleic Acid Drug Dev (in press)

Burch RM, Mahan LC (1991) Oligonucleotides antisense to the interleukin 1 receptor mRNA block the effects of interleukin 1 in cultured murine and human fibroblasts and in mice. J Clin Invest 88:1190–1196

Butcher EC (1991) Leukocyte-endothelial cell recognition: three (or more) steps to specificity and diversity. Cell 67:1033–1036

Butler M, Stecker K, Bennett CF (1996) Histological localization of phosphorothioate oligodeoxynucleotides in normal rodent tissue. Lab Invest (in press)

Chan AC, Kadlecek TA, Elder ME, Filipovich AH, Kuo W-L, Iwashima M, Parslow TG, Weiss A (1994) ZAP-70 deficiency in an autosomal recessive form of severe combined immunodeficiency. Science 264:1599–1601

Chiang M-Y, Chan H, Zounes MA, Freier SM, Lima WF, Bennett CF (1991) Antisense oligonucleotides inhibit intercellular adhesion molecule 1 expression by two distinct mechanisms. J Biol Chem 266:18162–18171

Christofidou-Solomidou M, Albelda SM, Bennett FC, Murphy GF (1997) Experimental production and modulation of human cytotoxic dermatitis in human-murine chimeras. Am J Pathol 150:631–639

Condon TP, Bennett CF (1996) Altered mRNA splicing and inhibition of human E-selectin expression by an antisense oligonucleotide in human umbilical vein endothelial cells. J Biol Chem 271:30398–30403

Cornelius LA, Taylor JT, Degitz K, Li L-J, Lawley TJ, Caughman SW (1993) A 5′ portion of the ICAM-1 gene confers tissue-specific differential expression levels and cytokine responsiveness. J Invest Dermatol 100:753–758

Cossum PA, Sasmor H, Dellinger D, Truong L, Cummins L, Owens SR, Markham PM, Shea JP, Crooke S (1993) Disposition of the ^{14}C-labeled phosphorothioate oligonucleotide ISIS 2105 after intravenous administration to rats. J Pharmacol Exp Ther 267:1181–1190

Crooke ST (1992) Therapeutic applications of oligonucleotides. Annu Rev Pharmacol Toxicol 32:329–376

Crooke ST, Bennett CF (1996) Progress in antisense oligonucleotide therapeutics. Annu Rev Pharmacol Toxicol 36:107–129

Crooke ST, Graham MJ, Zuckerman JE, Brooks D, Conklin BS, Cummins LL, Greig MJ, Guinosso CJ, Kornbrust D, Manoharan M, Sasmor HM, Schleich T, Tivel KL, Griffey RH (1996) Pharmacokinetic properties of several novel oligonucleotide analogs in mice. J Pharmacol Exp Therap 277:923–937

Damle NK, Klussman K, Linsley PS, Aruffo A (1992) Differential costimulatory effects of adhesion molecules B7, ICAM-1, LFA-3, and VCAM-1 on resting and antigen-primed CD4+ T lymphocytes. J Immunol 148:1985–1992

Damle NK, Klussman K, Leytze G, Myrdal S, Aruffo A, Ledbetter JA, Linsley PS (1994) Costimulation of T lymphocytes with integrin ligands intercellular adhesion molecule-1 or vascular cell adhesion molecule-1 induces functional expression of CTLA-4, a second receptor for B7. J Immunol 152:2686–2697

Dinarello CA (1994) The interleukin-1 family: 10 years of discovery. FASEB J 8:1314–1325

Doerschuk CM, Winn RK, Coxson HO, Harlan JM (1990) CD18-dependent and -independent mechanisms of neutrophil emigration in the pulmonary and systemic microcirculation of rabbits. J Immunol 144(6):2327–2333

Doukas J, Pober JS (1990) Lymphocyte-mediated activation of cultured endothelial cells (EC). J Immunol 145:1088–1098

Durieu-Trautmann O, Chaverot N, Cazaubon S, Strosberg AD, Couraud P-O (1994) Intercellular adhesion molecule 1 activation induces tyrosine phosphorylation of the cytoskeleton-associated protein cortactin in brain microvessel endothelial cells. J Biol Chem 269:12536–12540

Ebnet K, Kaldjian EP, Anderson AO, Shaw S (1996) Orchestrated information transfer underlying leukocyte endothelial interactions. Annu Rev Immunol 14:155–177

Elliott MJ, Maini RN, Feldmann M, Long-Fox A, Charles P, Katsikis P, Brennan FM, Walker J, Bijl H, Ghrayeb J, Woody JN (1993) Treatment of rheumatoid arthritis with chimeric monoclonal antibodies to tumor necrosis factor. Arthritis Rheum 36:1681–1690

Elliott MJ, Maini RN, Feldmann M, Kalden JR, Antoni C, Smolen JS, Leeb B, Breedveld FC, Macfarlane JD, Bijl H, Woody JN (1994) Randomized double-blind comparison of chimeric monoclonal antibody to tumor necrosis factor alpha (cA2) versus placebo in rheumatoid arthritis. Lancet 344:1105–1110

Furie MB, Tancinco MCA, Smith CW (1991) Monoclonal antibodies to leukocyte integrins CD11a/CD18 and CD11b/CD18 or intercellular adhesion molecule-1 inhibit chemoattractant-stimulated neutrophil transendothelial migration in vitro. Blood 78:2089–2097

Glover JM, Leeds JM, Mant TGK, Kisner DL, Zuckerman J, Levin AA, Shanahan WR (1997) Phase I safety and pharmacokinetic profile of an ICAM-1 antisense oligodeoxynucleotide (ISIS 2302). J Pharmacol Exp Ther (in press)

Haller H, Dragun D, Miethke A, Park JK, Weis A, Lippoldt A, Grob V, Luft FC (1996) Antisense oligonucleotides for ICAM-1 attenuate reperfusion injury and renal failure in the rat. Kidney Int 50:473–480

Haug CE, Colvin RB, Delnonico FL, Auchincloss H, Tolkoff-Rubin N, Preffer FI, Rothlein R, Norris S, Scharschmidt L, Cosimi AB (1993) A phase I trial of immunosuppression with anti-ICAM-1 (CD54) mAb in renal allograft recipients. Transplantation 55:766–773

Henry SP, Bolte H, Auletta C, Kornbrust DJ (1997a) Evaluation of the toxicity of ISIS 2302, a phosphorothioate oligonucleotide: 2) 4-week study in cynomolgus monkeys. Toxicology 120:145–155

Henry SP, Taylor J, Midgley L, Levin AA, Kornbrust DJ (1997b) Evaluation of the toxicity profile of ISIS 2302, a phosphorothioate oligonucleotide, following repeated intravenous administration: 1) 4-week study in CD-1 mice. Arch Toxicol (in press)

Higgins KA, Perez JR, Coleman TA, Dorshkind K, McComas WA, Sarmiento UM, Rosen CA, Narayanan R (1993) Antisense inhibition of the p65 subunit of NF-kappaB blocks tumorigenicity and causes tumor regression. Proc Natl Acad Sci U S A 90:9901–9905

Isobe M, Yagita H, Okumura K, Ihara A (1992) Specific acceptance of cardiac allograft after treatment with antibodies to ICAM-1 and LFA-1. Science 255:1125–1127

Johnson JP, Stade BG, Holzmann B, Schwable W, Riethmuller G (1989) De novo expression of intercellular adhesion molecule-1 in melanoma correlates with increased risk of metastasis. Proc Natl Acad Sci U S A 86:641–644

Kansas GS (1996) Selectin and their ligands: current concepts and controversies. Blood 88:3259–3287

Karmann K, Min W, Fanslow WC, Pober JS (1996) Activation and homologous desensitization of human endothelial cells by CD40 ligand, tumor necrosis factor, and interleukin 1. J Exp Med 184:173–182

Katz SM, Browne B, Pham T, Wang ME, Bennett CF, Stepkowski SM, Kahan BD (1995) Efficacy of ICAM-1 antisense oligonucleotide in pancreatic islet transplantation. Transplant Proc 27:3214

Kavanaugh AF, Davis LS, Nichols LA, Norris SH, Rothlein R, Scharschmidt LA, Lipsky PE (1994) Treatment of refractory rheumatoid arthritis with a monoclonal antibody to intercellular adhesion molecule 1. Arthritis Rheum 37:992–999

Kitajima I, Shinohara T, Bilakovics J, Brown DA, Xiao X, Nerenberg M (1992) Ablation of transplanted HTLV-1 tax-transformed tumors in mice by antisense inhibition of NF-κB. Science 258:1792–1795

Koizumi M, King N, Lobb R, Benjamin C, Podolsky DK (1992) Expression of vascular adhesion molecules in inflammatory bowel disease. Gastroenterology 103:840–847

Krieg AM, Yi A-K, Matson S, Waldschmidt TJ, Bishop GA, Teasdale R, Koretzky GA, Klinman DM (1995) CpG motifs in bacterial DNA trigger direct B-cell activation. Nature 374:546–549

Kuhlman P, Moy VT, Lollo BA, and Brian AA (1991) The accessory function of murine intercellular adhesion molecule-1 in T lymphocyte activation. J Immunol 146:1773–1782

Kumasaka T, Quinlan WM, Doyle NA, Condon TP, Sligh J, Takei F, Beaudet AL, Bennett CF, Doerschuk CM (1996) The role of the Intercellular Adhesion Molecule-1 (ICAM-1) in endotoxin-induced pneumonia evaluated using ICAM-1 antisense oligonucleotides, anti-ICAM-1 monoclonal antibodies, and ICAM-1 mutant mice. J Clin Invest 97:2362–2369

Lo SK, Lee S, Ramos RA, Lobb R, Rosa M, Chi-Rosso G, Wright SD (1991) Endothelial-leukocyte adhesion molecule 1 stimulates the adhesive activity of leukocyte integrin CR3 (CD11b/CD18, Mac-1, amB2) on human neutrophils. J Exp Med 173:1493–1500

Maltese J-Y, Sharma HW, Vassilev L, Narayanan R (1995) Sequence context of antisense RelA/NF-κB phosphorothioates determines specificity. Nucleic Acids Res 23:1146–1151

Miele ME, Bennett CF, Miller BE, Welch DR (1994) Enhanced metastatic ability of TNF-α-treated malignant melanoma cells is reduced by intercellular adhesion molecule-1 (ICAM-1, CD54) antisense oligonucleotides. Exp Cell Res 214:231–241

Monia BP, Lesnik EA, Gonzalez C, Lima WF, McGee D, Guinosso CJ, Kawasaki AM, Cook PD, Freier SM (1993) Evaluation of 2′ modified oligonucleotides containing deoxy gaps as antisense inhibitors of gene expression. J Biol Chem 268:14514–14522

Monteith DK, Henry SP, Howard RB, Flournoy S, Levin AA, Bennett CF, Crooke ST (1997) Immune stimulation – a class effect of phosphorothioate oligonucleotides in rodents. Anti Cancer Drug Design 12:421–432

Murray AG, Petzelbauer P, Hughes CC, Costa J, Askenase P, Pober JS (1994) Human T-cell-mediated destruction of allogeneic dermal microvessels in a severe combined immunodeficient mouse. Proc Natl Acad Sci U S A 91:9146–9150

Narayanan R, Higgins KA, Perez JR, Coleman TA, Rosen CA (1993) Evidence for differential functions of the p50 and p65 subunits of NF-κB with a cell adhesion model. Mol Cell Biol 13:3802–3810

Natali P, Nicotra MR, Cavaliere R, Bigotti A, Romano G, Temponi M, Ferrone S (1990) Differential expression of intercellular adhesion molecule 1 in primary and metastatic melanoma lesions. Cancer Res 50:1271–1278

Nestle FO, Mitra RS, Bennett CF, Chan H, Nickoloff BJ (1994) Cationic lipid is not required for uptake and selective inhibitory activity of ICAM-1 phosphorothioate antisense oligonucleotides in keratinocytes. J Invest Dermatol 103:569–575

Neurath MF, Pettersson S, Buschenfelde MK-H, Strober W (1996) Local administration of antisense phosphorothioate oligonucleotides to the p65 subunit of NF-κB abrogates established experimental colitis in mice. Nature Med 2:998–1004

Newman PJ (1997) The biology of PECAM-1. J Clin Invest 99:3–8

Ohh M, Smith CA, Carpenito C, Takei F (1994) Regulation of intercellular adhesion molecule-1 gene expression involves multiple mRNA stabilization mechanisms: effects of interferon-gamma and phorbol myristate acetate. Blood 84:2632–2639

Okayasu I, Hatakeyama S, Yamada M, Ohkusa T, Inagaki Y, Nakaya R (1990) A novel method in the induction of reliable experimental acute and chronic ulcerative colitis in mice. Gastroenterology 98:694–702

Oppenheimer-Marks N, Davis LS, Bogue DT, Ramberg J, Lipsky PE (1991) Differential utilization of ICAM-1 and VCAM-1 during the adhesion and transendothelial migration of human T lymphocytes. J Immunol 147:2913–2921

Orosz CG, van Buskirk A, Sedmak DD, Kincade P, Miyake K, Pelletier RP (1992) Role of the endothelial adhesion molecule VCAM-1 in murine cardiac allograft rejection. Immunol Lett 32:7–12

Pardridge WM, Boado RJ, Kang Y-S (1995) Vector-mediated delivery of a polyamide ("peptide") nucleic acid analog through the blood–brain barrier in vivo. Proc Natl Acad Sci U S A 92:5592–5596

Pfau S, Leitenberg D, Rinder H, Smith BR, Pardi R, Bender JR (1995) Lymphocyte adhesion-dependent calcium signaling in human endothelial cells. J Cell Biol 128:969–978

Pisetsky DS, Reich C (1993) Stimulation of vitro proliferation of murine lymphocytes by synthetic oligodeoxynucleotides. Mol Biol Rep 18:217–221

Plenat F, Klein-Monhoven N, Marie B, Vignaud J-M, Duprez A (1995) Cell and tissue distribution of synthetic oligonucleotides in healthy and tumor-bearing nude mice. Am J Pathol 147:124–135

Podolsky DK, Lobb R, King N, Benjamin CD, Pepinsky B, Sehgal P, deBeaumont M (1994) Attenuation of colitis in the Cotton-top Tamarin by anti-alpha 4 integrin monoclonal antibody. J Clin Invest 92:372–380

Read MA, Neish AS, Luscinskas FW, Palombella VJ, Maniatis T, Collins T (1995) The proteasome pathway is required for cytokine-induced endothelial-leukocyte adhesion molecule expression. Immunity 2:493–506

Rifaj A, Brysch W, Fadden K, Clark J, Schlingensiepen K-H (1996) Clearance kinetics, biodistribution, and organ saturability of phosphorothioate oligodeoxynucleotides in mice. Am J Pathol 149:717–725

Schuermann GM, Aber-Bishop AE, Facer P, Lee JC, Rampton DS, Dor CJ, Polak JM (1993) Altered expression of cell adhesion molecules in uninvolved gut in inflammatory bowel disease. Clin Exp Immunol 94:341–347

Sligh JE, Ballentyne CM, Rich SS, Hawkins HK, Smith CW, Bradley A, Beaudet AL (1993) Inflammatory and immune responses are impaired in mice deficient in intercellular adhesion molecule 1. Proc Natl Acad Sci U S A 90:8529–8533

Springer TA (1990a) Adhesion receptors of the immune system. Nature 346:425–434

Springer TA (1990b) The sensation and regulation of interactions with the extracellular environment: the cell biology of lymphocyte adhesion receptors. Annu Rev Cell Biol 6:359–402

Springer TA (1994) Traffic signals for lymphocyte recirculation and leukocyte emigration: the multistep paradigm. Cell 76:301–314

Stein CA (1995) Does antisense exist? Nature Med 1:1119–1121

Stein CA (1996) Phosphorothioate antisense oligodeoxynucleotides: questions of specificity. Tibtech 14:147–149

Stepkowski SM, Tu Y, Condon TP, Bennett CF (1994) Blocking of heart allograft rejection by intercellular adhesion molecule-1 antisense oligonucleotides alone or in combination with other immunosuppressive modalities. J Immunol 153:5336–5346

Strassman G, Graber N, Goyert SM, Fong M, McCullers S, Rong G-W, Beall LD (1994) Inhibition of lipopolysaccharide and IL-1 but not of TNF-induced activation of human endothelial cells by suramin. J Immunol 153:2239–2247

Waddell TK, Fialkow L, Chan CK, Kishimoto TK, Downey GP (1995) Signaling functions of L-selectin enhancement of tyrosine phosphorylation and activation of MAP kinase. J Biol Chem 270:15403–15409

Wagner RW (1994) Gene inhibition using antisense oligodeoxynucleotides. Nature 372:333–335

Wegner CD, Gundel RH, Reilly P, Haynes N, Letts LG, Rothlein R (1990) Intercellular adhesion molecule-1 (ICAM-1) in the pathogenesis of asthma. Science 247:456–459

Wertheimer SJ, Myers CL, Wallace RW, Parks TP (1992) Intercellular adhesion molecule-1 gene expression in human endothelial cells. Differential regulation by tumor necrosis factor-alpha and phorbol myristate acetate. J Biol Chem 267:12030–12035

Winn RK, Harlan JM (1993) CD18-independent neutrophil and mononuclear leukocyte emigration into the peritoneum of rabbits. J Clin Invest 92:1168–1173

Winn RK, Liggitt D, Vedder NB, Paulson JC, Harlan JM (1993) Anti-P-selectin monoclonal antibody attenuates reperfusion injury to the rabbit ear. J Clin Invest 92:2042–2047

Xu H, Gonzalo JA, St Pierre Y, Williams IR, Kupper TS, Cotran RS, Springer TA, Gutierrez-Ramos J-C (1994) Leukocytosis and resistance to septic shock in intercellular adhesion molecule 1-deficient mice. J Exp Med 180:95–109

Yan H-C, Juhasz I, Pilewski JM, Murphy GF, Herlyn M, Albelda SM (1993) Human/severe combined immunodeficient mouse chimeras: an experimental in vivo model system to study the regulation of human endothelial cell–leukocyte adhesion molecules. J Clin Invest 91:986–996

Zhang R, Diasio RB, Lu Z, Liu T, Jiang Z, Galbraith WM, Agrawal S (1995a) Pharmacokinetics and tissue distribution in rats of an oligodeoxynucleotide phosphorothioate (GEM 91) developed as a therapeutic agent for human immunodeficiency virus type-1. Biochem Pharmacol 49:929–939

Zhang R, Lu Z, Zhao H, Zhang X, Diasio RB, Habus I, Jiang Z, Iyer RP, Yu D, Agrawal S (1995b) In vivo stability, disposition and metabolism of a "hybrid" oligonucleotide phosphorothioate in rats. Biochem Pharmacol 50:545–556

Zhao Q, Temsamani J, Iadarola PL, Jiang Z, Agrawal S (1995b) Effect of different chemically modified oligodeoxynucleotides on immune stimulation. Biochem Pharmacol 51:173–182

Zhang R, Iyer RP, Yu D, Tan W, Zhang X, Lu Z, Zhao H, Agrawal S (1996) Pharmacokinetics and tissue distribution of a chimeric oligodeoxynucleoside phosphorothioate in rats after intravenous administration. J Pharmacol Exp Ther 278: 971–979

Zimmerman GA, McIntyre TM, Prescott SM (1996) Adhesion and signalling in vascular cell–cell interactions. J Clin Invest 98:1699–1702

CHAPTER 13
Antisense Oligonucleotides and Their Anticancer Activities

D. Fabbro, M. Müller, and T. Geiger

A. Potential Cancer Targets for Pharmaceutical Intervention

Available cancer therapies still have only limited success against most solid cancer types and inevitably result in the development of multidrug-resistant tumors. There is an urgent need for therapeutic alternatives aiming at compounds with better tolerability at efficacious doses that are directed at defined, disease-relevant molecular targets. The progress made in understanding the molecular basis of mammalian cell transformation has led to the unifying concept of growth regulation and its disorders in cancer cells. It is now well recognized that many products of "cancer genes" encode for proteins that regulate normal mitogenesis and apoptosis. Taken together, these indicate that the carcinogenic process may be viewed as a progressive disorder of signal transduction (Croce 1987; Alitalo et al. 1988; Bos 1989; Bishop 1991; Rabbitts 1994; Weinberg 1994). In fact, many of the genes that are mutated or lost in cancer cells, including both the oncogenes and tumor suppressers, encode proteins that are crucial regulators for both intra- and intercellular signal transduction (Croce 1987; Alitalo et al. 1988; Bos 1989; Bishop 1991; Rabbitts 1994; Weinberg 1994). This conceptual framework has provided a basis for the development of novel anticancer strategies and therapeutic modalities aimed at inhibiting cancer growth either by blocking mitogenic signal transduction or specifically inducing apoptosis of cancer cells. Although the various approaches have not been clinically validated, these strategies are likely to identify compounds with fewer side effects compared to standard chemotherapeutic agents.

Specific inhibition of cancer-causing gene products can in principle be accomplished by appropriately designed small molecules, provided that the chosen targets display reasonable enzymatic functions (e.g., inhibitors for protein kinases, extracellular matrix-degrading proteases, farnesyltransferases). However, a large proportion of putative cancer-causing or cancer-associated oncoproteins either do not have intrinsic enzymatic functions, such as the various transcription factors (e.g., myc, jun, fos) and cell death suppressors (Bcl-2 or Bcl-X), or their enzymatic functions are complex (e.g., multigene families of proteins such as protein kinases, GTP-binding proteins) and/or are not readily amenable to a conventional high-throughput random screening (e.g., proteasome, raf).

It is against this background that therapeutic agents directed at cancer targets with poorly defined enzymatic functions may develop from oligonucleotide-based strategies (CROOKE 1992; see Chap. 1, this volume). These approaches are based on the inhibition of protein expression by binding of synthetic oligonucleotides (ODN) to single-stranded RNA (*antisense approach*) or double-stranded genomic DNA (*antigene approach*) and have been extensively reviewed in the recent literature (see also Chap. 1, this volume). Antisense-based approaches offer the opportunity to inhibit the expression of specific disease-causing target genes. It should be remembered that the most challenging issue for all of these novel anticancer approaches continues to be the preclinical and clinical validation of the particular anticancer target under investigation (see Chaps. 1, 14, 16, 17, this volume).

B. Antisense-Based Approaches and Cancer

The increasing interest in the antisense technology and in antisense-based approaches for the treatment of cancer, hyperproliferative disorders, and other diseases is reflected by a growing list of review articles (HÉLÈNE and TOULMÉ 1990; CALABRETTA et al. 1993; CARTER and LEMOINE 1993; CHO-CHUNG 1993; BAYEVER and IVERSEN 1994; MAGRATH 1994; PIERGA and MAGDALENAT 1994). Various nuclear factors, oncogenes, growth factors, and their receptors have been targeted by antisense ODN (AS-ODN) or antisense vector constructs (PIERGA and MAGDALENAT 1994). Progress has also been made in the synthetic capabilities of AS-ODN which has not only allowed tests to be carried out in animals models to investigate their disease-modulating activities, but has also permitted the initiation of clinical trials (see Chap. 16, this volume). The studies listed in Tables 1–3 are a selection rather than a comprehensive summary of all the studies that have been published so far. In the majority of these studies, downregulation of the targeted mRNA and/or inhibition of protein expression have been observed (Table 1), although most of these studies lack appropriate control experiments. Only a limited number of studies have included in vitro antiproliferative effects and antitumor activity in vivo and of these studies only a few showed adequate control experiments (Tables 2, 3). The most relevant in vivo anticancer activities of AS-ODN are summarized in Table 3 and are discussed in the following sections.

C. Antisense Targets Related to Solid Tumors

I. *ras* and *raf*

Mutations in *ras* genes have been found to occur frequently in human malignancies (Bos 1989). Consequently, various AS-ODN have been targeted to this oncogene. Treatment of Ha-*ras*-transformed NIH3T3 cells with AS-ODN

Table 1. In vitro studies of antisense oligonucleotides (AS-ODN) targeted against nuclear transcription factors, oncogenes, growth factors, and growth factor receptors

Target	Approach	Cell systems	Result	Reference
c-*myc*	Antisense RNA	F9 Terato Ca	Inhibition of proliferation	Nishikura et al. 1990
	PO, poly(L-lysine)	L929, HeLa cells	Reduction of *myc* and growth	Degols et al. 1991
	PS	MCF-7, MDA-MB	Reduction of *myc* and growth	Watson et al. 1991
	PS	Human SMC	Reduction of *myc*; inhibition of proliferation or colony formation	Ebbecke et al. 1992
	PS	Rat SMC		Shi et al. 1993
	Phosphoramidates	HL-60 cells		Bennett et al. 1993
				Gryaznov et al. 1996
c-*myb*	PS	Human T cells	Inhibition of T cell proliferation and DNA Pol.α or PCNA expression	Venturelli et al. 1990
	PO	AML, CML cells	Inhibition of clonal growth	Calabretta et al. 1991
	PS	Lovo, Colo 205, HT29	Reduction of *myb* and growth	Melani et al. 1991
	PS	Rat vascular SMC	Inhibition of *myb*-induced cytosolic calcium rise	Simons et al. 1993
	PS	Rat/dog/human SMC	Inhibition of growth by a non-antisense mechanism (G quartet)	Burgess et al. 1995
	Phosphoramidates	HL-60 cells	Reduction of *myb*, growth, and colonies	Gryaznov et al. 1996
c-*fos*	Antisense RNA	F9 terato Ca	Reduction of *fos* induction by interferon and phorbol ester	Levi and Ozato 1988
	PO	Murine 7TD1 and B9 lymphoid cells	Inhibition of *fos* expression and PCD in lymphoid cells	Colotta et al. 1992
	PO	Adipocytes (OB1771)	Reduction of *fos* and lipoprotein lipase	Barcellini-Couget et al. 1993
	Ribozyme	A2780 ovarian Ca cells	Inhibition of *fos* expression; reversion of MDR	Scanlon et al. 1994

Table 1. *Continued*

Target	Approach	Cell systems	Result	Reference
c-jun	Antisense RNA	NIH3T3 fibroblast	Reduction of *jun* and *fos*	Schönthal et al. 1989
	PO	Murine 7TD1 and B9 lymphoid cells	Reduction of *jun* expression; induction of PCD in lymphoid cells	Colotta et al. 1992
NF-κB (p65)	PS, antisense RNA	3T3 cells, breast and colon Ca (SW480, MCF-7)	Inhibition of soft agar colonies, adhesion to ECM, and tumors	Higgins et al. 1993
	PS, antisense RNA	Murine embryonic stem cells	Inhibition of p65 expression and cellular adhesion	Narayanan et al. 1993
	PS	HL-60 cells	Inhibition of p65, CD11b expression, and adhesion	Sokoloski et al. 1993
	Antisense RNA	Balb/c fibroblasts, B16 melanoma	Inhibition of p65 expression and proliferation	Perez et al. 1994
	PS	Vascular SMC	Inhibition of p65 expression, proliferation, and adhesion	Autieri et al. 1995
p120	PO	Human PBL	Inhibition of p120 expression, proliferation, and S-phase entry	Fonagy et al. 1992
	PS	Melanoma, renal Ca and HeLa cells	Inhibition of p120 expression and growth; induction of PCD	Perlaky et al. 1993
	PS	HeLa cells	Inhibition of p120 expression; induction of PCD	Busch et al. 1994
	PS	Human melanoma	Inhibition of p120 expression	Saijo et al. 1993
	Antisense RNA	NIH3T3 fibroblast	Growth and transformation reduced	Perlaky et al. 1992
PCNA	PO	Balb/c 3T3 cells	Reduction of PCNA and growth	Jaskulski et al. 1988
	PO methylethylamine	Rat vascular SMC	Inhibition of proliferation	Speir and Epstein 1992
	PS	Gastric Ca cells	Reduction of PCNA and growth	Sakakura et al. 1994, 1995
EGFR	Antisense RNA	Epidermoid and colon Ca cells	Reduction of EGFR, colony formation, and invasion	Moroni et al. 1992
				Chakrabarty et al. 1995
IGF-IR	PS	Balb/c 3T3 cells and PBMC	Inhibition of 3T3 and PBMC growth	Baserga et al. 1992

Gene	Type	Cell line	Effect	Reference
	PO	Balb/c 3T3 cells	Inhibition of EGF-stimulated 3T3 cell growth	PIETROZKOWSKI et al. 1992
	Antisense RNA	Rhabdomyosarcoma (Rh30)	Inhibition of IGF-IR expression, growth, and colony in soft agar	SHAPIRO et al. 1994
	Antisense RNA	Glioblastoma (T98G)	Inhibition of IGF-IR expression and IGF-I-dependent growth	AMBROSE et al. 1994
	PS	Colo-357, HL-60	Inhibition of proliferation	BERGMANN et al. 1995
	Antisense RNA	MCF-7 breast Ca	Reduction of IGF-IR and growth	NEUENSCHWANDER et al. 1995
PDGF-A	PS	Melanoma cells	Inhibition of proliferation	BEHL et al. 1993
	PO	Rat vascular SMC	Inhibition of proliferation	ITOH et al. 1993
	PS	Rat vascular SMC	Inhibition of SMC growth	CALARA et al. 1996
	PS	Mesangial cells	Reduction of TGF-β-dependent synthesis of collagen IV; mesangial sclerosis	HANSCHE et al. 1995
TGF-β	PO	Rat muscular SMC	Reduction of TGF-β and growth	ITOH et al. 1993
	PS, mixed PS/PO	C3 fibrosarcoma	Inhibition of TGF-β expression, invasion, and metastatic potential	SPEARMAN et al. 1994
	PS	Glioblastoma (U138MG)	Inhibition of TGF-β expression, cellular adhesion, and invasion	PAULUS et al. 1995
	Antisense RNA	Rat 9L gliosarcoma	Inhibition of TGF-β expression	FAKHRAI et al. 1996
	PS	Human gliomas	Reduction of TGF-β and growth	JACHIMCZAK et al. 1996
c-raf	PS, antisense RNA	Hematopoietic cells	Reduction of c-raf and growth	SKORSKI et al. 1995
	PS	Hematopoietic cells (FDCP1, 32D)	Inhibition of proliferation and colony formation	MUSZYNSKI et al. 1995
	PS	T24 and A549 cells	Reduction of c-raf and growth	MONIA et al. 1996
PKC-α	Antisense RNA	MCF-7 breast Ca	Reduction of PKC-α and MDR	AHMAD et al. 1993
	Antisense RNA	U-87 glioblastoma	Reduction of PKC-α and growth	AHMAD et al. 1994
	PO	Rat hepatoma	Reduction of PKC-α and growth	PERLETTI et al. 1994
	PS	A549 lung Ca	Reduction of PKC-α and induction of ICAM-1	DEAN et al. 1994

Ca, carcinoma; PS, phosphorothioate ODN; PO, wild-type unmodified ODN; PCD, programmed cell death; ECM, extracellular matrix; PBM, peripheral blood monocytes; PBL, peripheral blood lymphocytes; MDR, multidrug resistance; PCNA, proliferating cell nuclear antigen; EGF, epithelial growth factor; EGFR, EGF receptor; IGF, insulin-like growth factor; IGF-IR, IGF-I receptor; SMC, smooth muscle cell; TGF, tumor growth factor; PKC, protein kinase C; ICAM, intercellular adhesion molecule.

Table 2. In vitro studies of antisense oligonucleotides (AS-ODN) targeted against hematological malignancies

Target	Approach	Cell systems	Result	Reference
bcr-abl	PS, antisense RNA	BaF3 and K562 cells	Reduction of p210 bcr-abl; induction of PCD	Lewalle et al. 1993
	Antisense RNA	B10 and K562 cells	Reduction of bcr-abl and growth; induction of PCD	Martiat et al. 1993
	PO, mafosfamide	CML (BV173 cells)	Reduction of bcr-abl and growth in SCID	Skorski et al. 1993
	PS	CML (BV173), NB4, HL-60 cells	Reduction of bcr-abl, growth, and ras activation	Skorski et al. 1994
	PO, PS	CML (BV173, LAMA 84, KY01)	Reduction of bcr-abl (not sequence dependent) and growth	O'Brien et al. 1994
	Ribozyme	CML (K562)	Cleavage of bcr-abl mRNA; reduction of bcr-abl	Lange et al. 1994
	PS	CML (BV173), HL-60	Reduction of bcr-abl and growth; induction of PCD	Smeters et al. 1994
	PS PO (5′ and 3′ capping)	K562, BV173	Reduction of bcr-abl and growth	Thomas et al. 1994
	PS	CML (K562, BV173, Lama-84)	Induction of PCD without bcr-abl reduction	Smeters et al. 1995
	PO	CML (BV173, KCL22, Ky01, Lama-84, K562)	Inhibition of CML cell growth; sequence-specific but non-antisense mechanism	Vaerman et al. 1995
c-myb	PO	CML cells	Reduction of CFU and bcr-abl	Ratajczak et al. 1992
	PS, PO (5′ and 3′ capping)	K562, BV173 cells	Reduction bcr-abl and growth (myb less active than bcr-abl ODN); only capped PO is sequence specific	Thomas et al. 1994
c-myc	PO	HL60 cells	Reduction of c-myc and S-phase entry	Wickstrom et al. 1988
bcl-2	PS	DoHH2 lymphoma cells injected into SCID	Reduction of bcl-2 and growth of lymphoma in vivo	Cotter et al. 1994

PCD, programmed cell death (apoptosis); PS, phosphorothioate ODN; PO, phosphodiester, wild-type unmodified ODN; SCID, severely compromised immunodeficient mice; CML, chronic myelogenous leukemia; CFU, colony-forming unit.

Table 3. Antitumor activities of antisense oligonucleotides (AS-ODN) in animal models

Target	Approach (application)	Cell system/Animal model	Result	Reference
Ki-*ras*	Antisense RNA (intra-tracheal)	Large cell lung Ca (H460); orthotopic	Inhibition of tumor growth (orthotopic)	Georges et al. 1993
	Antisense RNA	Pancreatic Ca (AsPC-1, BxPC-3); i.p. nu/nu	Reduction of tumor growth and metastasis	Aoki et al. 1995
Ha-*ras*	PS (systemic)	NIH3T3 Ha-*ras* cells; s.c. nu/nu	Reduction of tumor size	Gray et al. 1993
	PO (nanoparticles, local)	Mammary Ca (HBL100); s.c. nu/nu	Reduction of Ha-*ras* and tumor growth	Schwab et al. 1994
TGF-β1, -β2	Antisense RNA	Mouse/human mesothelioma (AC29, DeHm); s.c. nu/nu	Reduction of TGF-β and tumor growth	Fitzpatrick et al. 1994
TGF-α	Antisense RNA	Rat liver epithelial cells (LE2)	Inhibition of tumor growth	Laird et al. 1994
	PS (i.v. by Alzet, local)	PC-3 prostate Ca; s.c. nu/nu	Tumor necrosis by local application	Rubenstein et al. 1995
VEGF	Antisense RNA	Rat C6 glioma cells; s.c. nu/nu	Reduction of blood vessels and tumor size; increased necrosis	Saleh et al. 1996
bFGF	PS	Kaposi sarcoma (KS); s.c. nu/nu	Peduction of bFGF, KS-like lesions, and angiogenesis	Ensoli et al. 1994
IGF-IR	Antisense RNA	Melanoma cells (FO-1); nu/nu	Reduction of IGF-IR and tumor growth	Resnicoff et al. 1994
	Antisense RNA	Alveolar rhabdomyosarcoma; nu/nu	Reduction of IGF-IR, proliferation, and tumor growth	Shapiro et al. 1994
	Antisense RNA	Lewis lung Ca (H-69)	Reduction of IGF-IR and tumor growth/metastosis	Long et al. 1995
EGFR	PS (i.v. by Alzet, local)	Human PC-3 prostate Ca; nu/nu	Necrosis after local application and tumor regression	Rubenstein et al. 1996

Table 3. *Continued*

Target	Approach (application)	Cell system/Animal model	Result	Reference
NF-κB, p65	PS (i.p.)	Fibroblastic tumors induced with HTLV-1 *Tax* gene; nu/nu	Inhibition of fibrosar-coma	Kitajima et al. 1992
	PS, antisense RNA (s.c.)	Fibrosarcoma (3T3); nu/nu	Reduction of p65 and of tumorigenicity	Higgins et al. 1993
Urokinase-type PA	Antisense RNA	Epidermoid carcinoma (Hep3); nu/nu	Inhibition of tumor growth	Kook et al. 1994
	PS (i.v.)	Ovarian cancer (OV-MZ-6); nu/nu	Inhibition of invasive capacity and metastasis	Wilhelm et al. 1995
Matrilysin	Antisense RNA	Colon carcinoma (SW620); nu/nu	Inhibition of tumorigenicity and metastasis	Witty et al. 1994
c-myb	PS (minipump, s.c.)	CML line (K562); SCID	Inhibition of leukemic disease	Ratajczak et al. 1992
	PS (s.c.)	Melanoma; SCID	Reduction of *c-myb* and tumor growth	Huiya et al. 1994
c-myc	Antisense RNA	Esophageal cancer (EC8712); nu/nu	Reduction of tumor growth	Ye and Wu 1992
EWS-fli, -erg	Antisense RNA	Ewing sarcoma; nu/nu	Inhibition of EWS fusion and tumors	Ouchida et al. 1995
hsp 70	Antisense RNA	Fibrosarcoma (WEHI-S); nu/nu	Reduction of hsp 70, tumorigenicity, and resistance to TNF	Jäättelä et al. 1995
Cyclin D_1	Antisense RNA	Esophageal Ca (HCE7); nu/nu	Inhibition of anchorage indepebnt growth and tumor growth	Zhou et al. 1995

Target	Method	Model	Effect	Reference
CRIPTO	Antisense RNA	Colon Ca (GEO, CGS, WiDR); nu/nu	Reduction of CRIPTO, soft agar colonies, and tumor growth	Ciardiello et al. 1994
PKA (RIα)	PS (s.c.)	Colon Ca (LS-174T); nu/nu	Reduction of RIα and tumor growth	Nesterova et al. 1995
PKC-α	PS (i.v.)	A549 lung, T24 bladder, MDA MB231 breast CA; nu/nu	Reduction of PKC-α and tumor growth	Dean et al. 1996
c-raf	PS (i.v.)	A549 lung, T24 bladder, MDA MB231 breast Ca; nu/nu	Reduction of raf and tumor growth	Monia et al. 1996
Osteopontin	Antisense RNA	Murine PAP2 cells (oncogenic ras)	Reduction of metastasis and osteopontin	Behrend et al. 1994
bcl-2	PO (in vitro)	DoHH2 lymphoma; SCID	Reduction of bcl-2; inhibition of lymphoma	Cotter et al. 1994
bcr-abl	PS (i.v.)	CML blast crisis cell line (BV173); nu/nu	Reduction of leukemia and organ infiltration	Skorski et al. 1994
p120	PS (i.p. with lipids)	Melanoma (LOX); nu/nu	Inhibition of tumor growth	Perlaky et al. 1993
	Antisense RNA	Human breast carcinoma (MCF-7); nu/nu	Inhibition of tumor growth	Saijo et al. 1993

Ca, carcinoma; i.v., intravenously; s.c., subcutaneously; i.p., intraperitoneally, nu/nu, nude mice; SCID, severely compromised immunodeficient mice; PCD, programmed cell death (apoptosis); PS, phosphorothioate ODN; PO, phosphodiester, wild-type unmodified ODN; TGF, tumor growth factor; VEGF, vascular endothelial growth factor; bFGF, basic fibroblast growth factor; IGF-IR, insulin-like growth factor I receptor; EGFR, epithelial growth factor receptor; PA, plasminogen activator; PKA, protein kinase A; PKC, protein kinase C; CML, chronic myelogenous leukemia; TNF, tumor necrosis factor; HTLV, human T-lymphotropic virus.

to the 5'-flanking region of the *ras* mRNA resulted in drastic reduction of Ha-*ras* levels; when these cells were implanted into nude mice, tumor growth of AS-ODN-treated cells was dramatically reduced compared to cells treated with nonspecific control ODN (GRAY et al. 1994). In another study, phosphodiester AS-ODN directed at codon 12 mutation of Ha-*ras* selectively inhibited the proliferation of cells expressing the mutated Ha-*ras* (SCHWAB et al. 1994). In this study, phosphodiester AS-ODN that were absorbed onto polyalkylcyanoacrylate nanoparticles displayed both a remarkable potency in vitro and antitumor activity in vivo (SCHWAB et al. 1994). As the central effector system for *ras*, c-*raf* kinase, which regulates one of the mitogen-activated protein (MAP) kinase signaling pathways, was also targeted in various studies by antisense-based approaches (KASID et al. 1989; KOLCH et al. 1991; MONIA et al. 1996). The *raf* antisense constructs or AS-ODNs were able to revert the *ras*-induced malignant phenotype (KOLCH et al. 1991), to increase the radiation sensitivity of cells in vitro (KASID et al. 1989), and to inhibit cellular proliferation as well as tumor growth in nude mice (KASID et al. 1989; MONIA et al. 1996), suggesting that reduced expression of *raf* kinase is sufficient to modulate the tumorigenicity and the radiation sensitivity of carcinoma cells.

II. Growth Factors and Growth Factor Receptors

Various studies have targeted growth factors or growth factor receptors with AS-ODN. In one study, the growth factor receptor c-*erbB2* (or HER-2), whose gene amplification has been correlated with nodal status, early relapse, and shortened survival (SLAMON et al. 1987), was targeted with AS-ODN to inhibit growth and DNA synthesis of breast cancer cell lines (COLOMER et al. 1994). In this study, the *erbB2*-targeted AS-ODN had no effect on breast cancer cell lines that had no amplification of the *erbB2*, indicating that overexpression of this growth factor receptor is important for the proliferation of breast cancer cells (COLOMER et al. 1994). In another study, an antisense construct to the insulin-like growth factor receptor type I (IGF-IR) was expressed in FO-1 human melanoma cells, resulting in both a marked reduction in the number of IGF-IR and strong inhibition of the growth of cells implanted into nude mice (RESNICOFF et al. 1994). A similar inhibition of tumorigenesis was also observed when FO-1 cells were treated in vitro with synthetic AS-ODN prior to the injection into nude mice (RESNICOFF et al. 1994).

Malignant melanomas, unlike normal melanocytes, can proliferate in the absence of exogenous basic fibroblast growth factor (bFGF) (BECKER et al. 1989, 1992). Exposure of primary melanomas and metastatic melanomas to unmodified AS-ODN targeted to bFGF mRNA inhibited cell proliferation and colony formation in soft agar (Becker et al. 1989, 1992). However, the expression of antisense RNA to transforming growth factor β growth factor (TGF)-β reduced the anchorage-independent growth of murine mesothelioma cells in vitro and their tumorigenesis in vivo, but did not influence anchorage-dependent growth (FITZPATRICK et al. 1994). Reduction of endogenous TGF-β mRNA was in the order of 60%–80%, suggesting that very strong promoter systems have to be used for vector-based antisense inhibition of target genes to achieve efficient inhibition of gene expression.

III. Proteases

The metastatic process is usually associated with an increase in extracellular matrix (ECM)-degrading proteases (FIDLER 1995). For example, the urokinase-type plasminogen activator (uPA) and its surface receptor (uPAR) have been shown to correlate with an invasive tumor cell phenotype (FIDLER 1995). An antisense construct to uPAR transfected in a human malignant epidermoid carcinoma cell line reduced the uPAR mRNA and protein by 50%–75%. Cells expressing the lowest level of uPAR showed a significantly lower level of invasion and a reduced tumorigenicity as well as local invasion in vivo (KOOK et al. 1994; QUATTRONE et al. 1995). Similarly, the reduction of matrilysin expression by AS-ODN decreased tumorigenicity in vivo and inhibited subsequent metastasis to the liver, indicating that matrilysin is involved in tumorigenicity and metastasis of colon carcinoma cell lines. Expression of this particular matrix metalloproteinase correlates with the ability to invade artificial basement membrane in vitro and to metastasize to the liver following orthotopic injection into the cecum of nude mice in vivo (WITTY et al. 1994). AS-ODN have also been targeted to the p65 subunit of NF-κB, which is implicated in the induction of adhesion molecule expression; they were shown to inhibit the tumorigenicity of a fibrosarcoma cell line in vivo (HIGGINS et al. 1993). In this particular study, the mechanism of action of the AS-ODN was explicitly verified by using control ODN as well as by a conditional antisense construct. The effects of this inducible antisense construct in inhibiting in vivo tumor growth were comparable to the effect of systemically administered AS-ODN (HIGGINS et al. 1993).

IV. Multidrug Resistance

A persistent problem in the treatment of cancer is the emergence of drug-resistant tumors after chemotherapy. The most common and serious manifestation of this problem is the phenomenon of acquired pleiotropic drug resistance or multidrug resistance (MDR). MDR consists of a complex system of genetic and epigenetic mechanisms (VAN KALKEN et al. 1991; DIETEL 1993; KAYE 1993; SHUSTIK et al. 1995). The type of MDR that is presently best understood is the one mediated by P-glycoprotein (Pgp), which is mediated by the overexpression of a membrane-spanning glycoprotein of 170 kDa that is able to deplete cancer cells of the various natural-product-based anticancer agents (GOTTESMAN 1993; PASTAN and GOTTESMAN 1991). In fact, increased levels of Pgp expression after chemotherapy have been reported in various cancer types, and correlation between Pgp expression in tumor specimens and treatment outcome have also been described (VAN KALKEN et al. 1991; KAYE 1993; SHUSTIK et al. 1995). Although Pgp-mediated MDR may not be the only mechanism involved in human cancers, the inhibition of Pgp function by modulators has been considered an important aspect of postoperative cancer treatment. AS-ODN to Pgp in human mesangial cells have been shown to decrease the efflux of drugs that are actively transported by Pgp to a similar extent as cyclosporin A and verapamil (BELLO and ERNEST 1994). In another study, the effect of AS-ODN on Pgp was studied in human colon carcinoma cell lines that were either resistant or sensitive to doxorubicin (RIVOLTINI et al. 1990). Interestingly, this study showed that the extent of verapamil-induced reversal of MDR was similar to the antisense-mediated downregulation of Pgp (RIVOLTINI et al. 1990). Similarly, a 15-mer phosphorothioate-capped ODN directed at the 5' end of the coding region of *mdr*-1 mRNA reduced P-glycoprotein expression and doxorubicin resistance in human ovarian cancer cells (THIERRY et al. 1993). Reversal of Pgp-mediated MDR has also been achieved using hammerhead ribozymes directed at various regions of Pgp mRNA (KIEHNTOPF et al. 1994; KOBAYASHI et al. 1994; SCANLON et al. 1994). Ribozymes are catalytically active RNA molecules which, after binding to an mRNA target, are capable of cleaving the bound RNA in the vicinity of a GUC sequence motif. This irreversible mechanism of action makes ribozymes very attractive antisense agents, but their practical applicability is limited due to their pronounced susceptibility to degradation by RNAses. Ribozyme-mediated cleavage of the Pgp transcript in drug-resistant human cells significantly reduced Pgp expression and restored the sensitivity of the cells to chemotherapeutic drugs (KIEHNTOPF et al. 1994; KOBAYASHI et al. 1994; SCANLON et al. 1994). In one study, it was clearly shown that the effects observed were the result of ribozyme action and were not due to a conventional antisense mechanism, as mutated analogues incapable of RNA cleavage were devoid of any activity (KOBAYASHI et al. 1994). Since expression of Pgp is dependent on c-*fos* (the promoter of the *mdr*-1 gene contains an AP-1 binding site), a hammerhead ribozyme targeting c-*fos* was expressed in cells displaying Pgp-mediated MDR (SCANLON et al. 1994). The hammerhead ribozyme reduced both c-*fos* and Pgp expression and restored sensitivity to chemotherapeutic agents. Surprisingly, the ribozyme directed at c-*fos* was more potent in the reversal of the MDR phenotype than the ribozyme targeted to Pgp mRNA (SCANLON et al. 1994).

V. Miscellaneous

Antisense-based approaches have commonly been used to validate the anticancer targets under investigation in vitro and in vivo. In a study representing the first report showing enhancement of antisense effects by cationic lipids in vivo in an animal model, AS-ODN targeted to the nucleolar antigen p120 were shown to inhibit growth of human melanoma cells in nude mice (PERLAKY et al. 1993). In addition, an AS-ODN targeted to osteopontin reduced the tumorigenic and metastatic properties of *ras*-transformed NIH3T3 tumors in vivo, thus demonstrating the involvement of osteopontin (a secreted, calcium-binding phosphoprotein) in tumorigenesis (BEHREND et al. 1994). Similarly, exposure of cells to a 21-mer phosphorothioate ODN targeted to the human regulatory subunit (RIα) of the cyclic adenosine monophosphate (cAMP)-dependent protein kinase A (PKA) exhibited growth inhibition both in vitro and in vivo (YOKOZAKI et al. 1993), indicating the potential involvement of this regulatory subunit of PKA in tumor growth (CHO-CHUNG and CLAIR 1993; CHO-CHUNG 1993).

D. Antisense Targets Related to Hematological Malignancies

I. Chronic Myelogenous Leukemia and bcr-abl

In many malignant hematological diseases, the disease-causing gene products are preferentially expressed in the malignant cells and are either not present or only present in low levels as nonmutated counterparts in normal cells. Prototype examples are the bcr-abl fusion proteins, which contain a constitutively activated protein tyrosine kinase activity; as a result of a chromosomal translocation, this activity is the hallmark of chronic myelogenous leukemia (CML) (NOWELL and HUNGERDORF 1960; ROWLEY 1982). While the chronic phase of CML is characterized by an increase in immature and mature hematopoietic stem cells in the peripheral blood and bone marrow, the CML blast crisis involves a marked degree of differentiation arrest of leukemic cells in the peripheral blood and bone marrow. This latter stage ultimately leads to death, since the blast phase is refractory to any established therapeutic regimens. AS-ODN targeted to the bcr-abl fusion region showed that the tendency to form blast colonies in vitro was inhibited (80%–90%) compared to cells treated with control ODN (SZCZYLIK et al. 1991). A mismatched control ODN inhibited the specific effect of the antisense sequence on blast cell colony formation (SKORSKI et al. 1993). In another study, systemic treatment of severely compromised immunodeficient (SCID) mice (injected with the CML blast cell line BV173) with a 26-mer AS-ODN targeting bcr-abl reduced both the bcr-abl transcripts and the number of leukemic cells and prolonged the survival of mice two- to threefold (SKORSKI et al. 1994). In addition, bcr-abl AS-ODN in combination with a low dose of mafosfamide drastically reduced clonogenicity and growth of BV173 cells in nude mice (SKORSKI et al. 1993). Thus AS-ODN may potentially be used as effective agents for ex vivo bone marrow purging in autologous bone marrow transplantation to specifically eliminate CML cells.

II. *myc*, *myb*, and *bcl*-2

Overexpression of *bcl*-2 seems to be responsible for follicular B cell lymphoma due to a chromosomal translocation of the *bcl*-2 gene to the immunoglobulin heavy chain sequences. Since the *bcl*-2 gene product is known to inhibit programmed cell death, the elimination of its aberrant expression in these lymphomas should lead to tumor regression. In vitro treatment of a B cell lymphoma cell line with an AS-ODN to the translational start codons of the *bcl*-2 mRNA before injection into SCID mice resulted in a reduction in *bcl*-2 expression and prevented the establishment of lymphoma disease in the mice, whereas sense and nonsense control ODN had no effect (COTTER et al. 1994).

Aberrant gene expression can also occur through gene amplification. For example, a 20- to 40-fold amplification of the c-*myc* gene is found in HL-60 cells (COLLINS et al. 1982). Exposure of HL-60 cells to a phosphodiester AS-ODN targeted to the translational start codon of c-*myc* inhibited c-*myc* expression and proliferation of HL-60 cells by inhibiting S-phase entry, whereas various control ODN with unrelated sequences had no effects (WICKSTROM et al. 1988). Similarly, an AS-ODN complementary to c-*myb* mRNA inhibited c-*myb* RNA and protein expression in myelogenous leukemia cell lines and primary leukemia cells obtained from patients (CALABRETTA et al. 1991). Downregulation of c-*myb* expression was accompanied by inhibition of proliferation and of clonogenic growth of the leukemic cells (CALABRETTA et al. 1991). However, in contrast to c-*myc*, inhibition of c-*myb* expression in HL-60 cells was not accompanied by the induction of terminal differentiation, suggesting that c-*myc* and c-*myb* regulate leukemia cell proliferation and differentiation at different levels. In vivo treatment of SCID mice (inoculated with human K562 leukemia cells) with c-*myb* AS-ODN (5 mg/kg) prolonged their survival three- to eightfold compared to untreated mice (RATAJCZAK et al. 1992). In addition, c-*myb* AS-ODN-treated mice had significantly less leukemic cell infiltration in the central nervous system and in the ovary (the two sites that most frequently show manifestation of disease).

E. Protein Kinase C-α and *raf* as Cancer Targets for Antisense

I. In Vivo Activity of Phosphorothioate Antisense Oligonucleotides Targeting Protein Kinase C-α and c-*raf*

Almost all mitogenic signals are amplified and transduced inside cells by protein kinase cascades either by receptor-activated tyrosine phosphorylation or by receptor coupling to GTP-binding proteins (BISHOP 1991; CANTLEY et al. 1991; HUNTER 1991; BLENIS 1993). Most mitogenic pathways utilize unique and/or overlapping parts of these protein kinase cascades (BISHOP 1991;

CANTLEY et al. 1991; HUNTER 1991; BLENIS 1993). Accordingly, mutant alleles of these protein kinase genes (e.g., PKC or *raf*) or of other oncogenes (e.g., *ras*) that signal through these protein kinases are able to perturb entire signaling networks, leading to the deregulation of cell differentiation, division, and apoptosis (BISHOP 1991; CANTLEY et al. 1991; HUNTER 1991; BLENIS 1993). On the basis of these data, it can be assumed that blocking deregulated mitogenic signal transduction at the level of protein kinases should inhibit cancer growth. The reasons for choosing PKC (in particular PKC-α) and *raf* (in particular c-raf-1) as anticancer targets for our antisense-based approach are presented in Chap. 14.

Inhibitors of c-*raf* and PKC-α expression were identified in "gene walks" using 20-mer phosphorothioate ODN (CGP 69846A targeting the 3'-UTR of human c-*raf* mRNA and CGP 64128A targeting the 3'-UTR of human PKC-α mRNA; see Chap. 14, this volume). Both As-ODN inhibited the expression of their target gene in cells with an IC_{50} of 100–150 nM (in the presence of cationic lipids) by a RNase H-dependent mechanism (DEAN et al. 1994, 1996; MONIA et al. 1996). The inhibition of c-*raf* and PKC-α expression by CGP 69846A and CGP 64128A is sequence specific and gradually lost upon introduction of increasing numbers of mismatches (MONIA et al. 1997). As a consequence of the reduction of c-*raf* and PKC-α protein levels, proliferation of cells, activation of MAP kinase, and phorbol ester-dependent intercellular adhesion molecule (ICAM) expression are inhibited or strongly reduced (DEAN et al. 1994; see also Chap. 14, this volume).

CGP 69846A and CGP 64128A showed sequence-specific antitumor activity in the dose range of 0.006–6.0 mg/kg (i.v. daily doses) against various human tumors, including A549 lung, T24 bladder, MDA-MB231 breast, and Colo205 colon tumor (DEAN et al. 1996; MONIA et al. 1996). CGP 69846A downregulated c-*raf* mRNA levels in the tumors, and the introduction of increasing numbers of mismatches into CGP 69846A resulted in a gradual loss of the antitumor activity (MONIA et al. 1996, 1997). Similarly, CGP 64128A, but not its scrambled version, was able to downregulate PKC-α in treated tumors, as demonstrated by immunohistochemistry (DEAN et al 1996). All these data indicate that CGP 69846A and CGP 64128A exerted their antitumor activity in nude mice by a sequence-specific mechanism of action.

The antitumor activity of CGP 69846A and CGP 64128A was studied in combination with standard chemotherapeutic agents (cisplatin, mitomycin C, vinblastine, tamoxifen, estracyt, 5-fluorouracil, adriamycin, or ifosfamide) against a variety of human tumors transplanted subcutaneously into nude mice (MCF-7, BT 20, and MDA-MB 231 breast carcinomas; PC3 and Du145 prostate carcinomas; Colo205, WiDr, and HCT 116 colon carcinomas; NCI-H69 small cell lung carcinoma; NCI-H460 large cell lung carcinoma; NCI-H520 squamous lung carcinoma; SK-mel 1 and SK-mel 3 melanomas; NIH-Ovcar 3 ovarian cancer). For most of the combinations studied, additive antitumor effects were found. Antagonistic effects were not seen in any of the combinations (GEIGER et al. 1997). In some of the combinations (mitomycin C, cisplatin, and vinblastine), synergistic antitumor effects with complete regression of tumor were observed (GEIGER et al. 1997).

II. In Vivo Activity of Modified and Formulated Antisense Oligonucleotides Targeting c-*raf*

We have tried to improve the in vivo activity of the AS-ODN targeting PKC-α and c-*raf* either by introducing modifications into the AS-ODN or by improving and/or modifying their pharmacokinetic and/or pharmacodynamic behavior using various formulations and/or modifications. The efforts that have been undertaken to modify AS-ODN targeting c-*raf* are summarized in the following sections.

1. 2′-Methoxy-ethoxy-Modified Oligonucleotides

The activity of ODN modifications of CGP 69846A were studied both in vitro and in vivo. Of all the modifications introduced in either the CGP 69846A or CGP 64128A sequences, the 2′-methoxy-ethoxy (2′-MoE) modification have been shown to display the most potent activities in vivo. Whether this is due to an increase in the hybridization efficiency, an increase in nuclease resistance, favorable pharmacokinetics, or a combination thereof still remains to be demonstrated. With 2′-MoE modification, an unmodified phosphodiester backbone can be used without loss in nuclease resistance. Phosphorothioate AS-ODN may create problems related to their protein binding and toxicology which appear to be due to their sulphur content (see Chaps. 3–7, this volume). The reduction of the sulphur of phosphorothioate AS-ODN may therefore have beneficial effects with respect to toxicology and protein binding. However, in order not to lose the RNAse H-dependent mechanism, a chimeric approach was chosen in which at least six phosphorothioates are retained in the central portion of the molecule, while the phosphodiesters are located in the wings. The 2′-MoE ODN modified from the CGP 69846A targeting c-*raf* and used for the in vivo antitumor studies are shown in Fig. 1. Note that CGP 69845A is a chimeric 2′-MoE ODN with phosphodiester wings, while CGP 71849A has a full phosphorothioate backbone.

The 2′-MoE modified ODN CGP 69845A (phosphodiester wings) and CGP 71849A (full phosphorothioate backbone) are more potent in inhibiting c-*raf* mRNA expression than CGP 69846A (unmodified phosphorothioate) 24 h after treatment (Fig. 2A). CGP 71849A is even more potent than CGP 69845A in inhibiting c-*raf* mRNA expression 75 h after treatment (Fig. 2B). Their antitumor activities were compared in A549 human lung carcinomas transplanted subcutaneously into nude mice. Compared to 69846A (Fig. 3) and CGP 71849A (Fig. 4), CGP 69845A displayed a more potent antitumor activity. The finding that CGP 71849A was less potent than CGP 69845A in vivo is in contrast to the inhibition of c-*raf* mRNA expression in vitro (Fig. 2B).

	"wing"　　　　　"gap"　　　　　"wing"	
CGP 69845A (20-mer):	5'-T C C C G C^{S}C^{S}T^{S}G^{S}T^{S}G^{S}A^{S}C^{S}A^{S}T G C A T T-3'	
CGP 69846A (20mer):	5'-T^{S}C^{S}C^{S}C^{S}G^{S}C^{S}C^{S}T^{S}G^{S}T^{S}G^{S}A^{S}C^{S}A^{S}T^{S}G^{S}C^{S}A^{S}T^{S}T-3'	
CGP 71849A (20-mer):	5'-T$\underline{^S}$C$\underline{^S}$C$\underline{^S}$-C$\underline{^S}$G$\underline{^S}$C$\underline{^S}$C^{S}T^{S}G^{S}T^{S}G^{S}A^{S}C^{S}A^{S}T$\underline{^S}$G^{S}C^{S}A$\underline{^S}$T$\underline{^S}$T-3'	

CGP 69845A	CGP 71849A	CGP 69846A
(phosphodiester, 2'-MoE)	(phosphorothioate, 2'-MoE)	(phosphorothioate, 2'-deoxy)

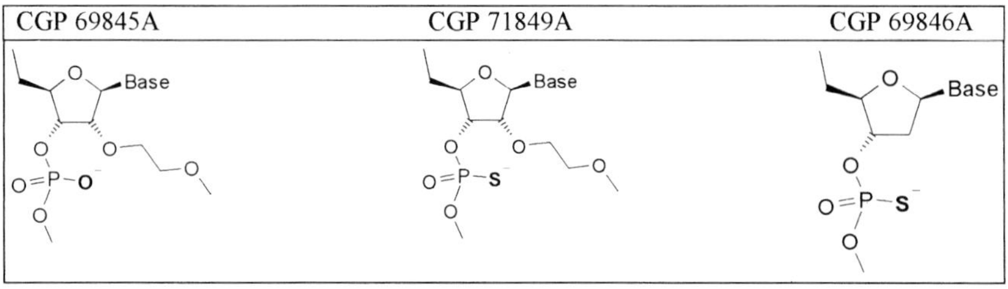

Fig. 1. 2'-Methoxy-ethoxy-modified chimeric backbones of CGP 69846A

2. Antitumor Activity of CGP 69846A in Stealth Liposomes

It was investigated whether ODN formulations could improve the antitumor activity of CGP 69846A, especially whether it would be possible to treat mice less frequently without a reduction in antitumor activity. A stealth liposomal formulation of CGP 69846A was compared to the ODN given i.v. as a saline solution. As shown in Fig. 5A, the antitumor activity of CGP 69846A in saline solution is rapidly lost when treatment intervals exceed 2–3 days. In fact, antitumor activity is almost completely lost when the mice are only treated once weekly. In contrast, the stealth liposome formulation of CGP 69846A allows reduction of treatment frequency to every second and third day without a significant loss of antitumor activity (Fig. 5B).

3. Antitumor Activity of CGP 69846A with Dextran Sulfate or as Cholesterol Conjugates of CGP 69846A

CGP 69846A was applied either i.v. or p.o. in combination with dextran sulfate. It was hoped that coadminstration with dextran sulfate would increase the antitumor activity of CGP 69846A after oral application. Figure 6A shows that dextran sulfate alone (60 mg/kg p.o. once daily) did not influence antitumor activity. While CGP 69846A in combination with dextran sulfate (20 mg/kg p.o.) was less potent than CGP 69846A given once daily i.v. (0.6 mg/kg without dextran sulfate), the combination of CGP 69846A with dextran sulfate clearly potentiated antitumor activity against A549 human lung adenocarcinomas (Fig. 6A). The conjugation of CGP 69846A with cholesterol had a lower antitumor activity than the nonconjugated ODN (Fig. 6B). CGP 69845A (2'-MoE, phosphodiester wings) is considerably more potent than CGP 69846A (Fig. 6B).

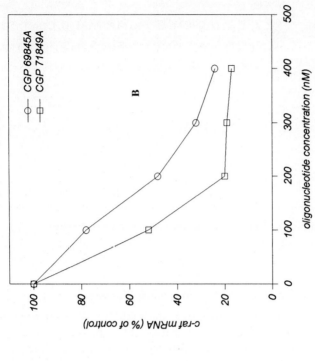
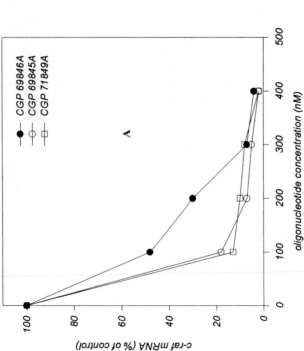

Fig. 2A,B. Inhibition of c-*raf* mRNA expression by 2′-methoxy-ethoxy (MoE)-modified CGP 69846A in A549 cells. A549 lung carcinoma cells were treated for 4 h with CGP 69845A, CGP 69846A, and CGP 71849A in the presence of cationic lipids. mRNA was isolated and c-*raf* mRNA expression was measured by northern blotting; c-*raf* mRNA levels were quantified using a PhosphorImager as described by MONIA et al. (1996). Results are expressed as a percentage of control (without oligonucleotides, ODN). **A** c-*raf* mRNA expression 24 h after treatment. *Closed circles,* CGP 69846A; *open circles,* CGP 69845A; *open squares,* CGP 71849A. **B** c-*raf* mRNA expression 75 h after treatment with CGP 69845A (*open circles*) and CGP 71849A (*open squares*)

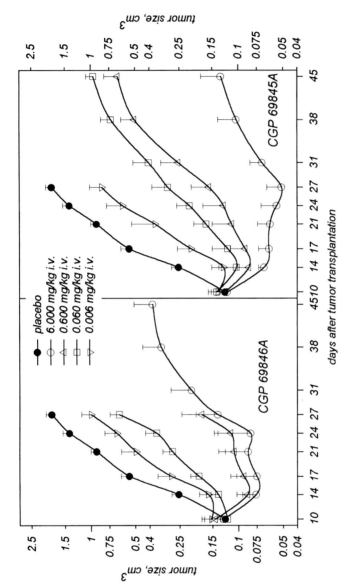

Fig. 3. Antitumor activity of CGP 69845A and CGP 69846A. A549 lung carcinoma was transplanted s.c. into nude mice as described (GEIGER et al. 1997; MONIA et al. 1996). CGP 69845A and CGP 69846A were administered once daily i.v. from day 10 to day 43 at doses of 6 (*open circles*), 0.6 (*upward triangles*), 0.06 (*squares*), and 0.006 mg/kg (*downward triangles*). Tumor growth was monitored twice weekly, and tumor volumes were calculated as described by GEIGER et al. (1997) and MONIA et al. (1996). *Closed circles*, placebo-treated controls

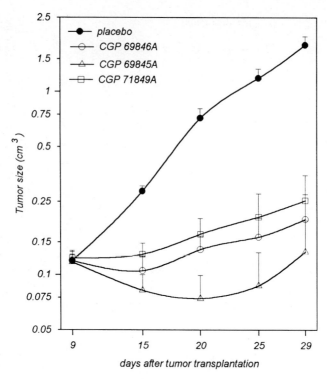

Fig. 4. Antitumor activity of CGP 698456A (*open circles*), CGP 69846A (*open triangles*), and CGP 71849A (*open squares*) (daily doses of 0.6 mg/kg from day 9 to day 28) administered i.v. after s.c. transplantation of A549 human lung carcinoma. *Closed circles*, placebo-treated controls

F. Concluding Remarks

Enormous progress has been made in antisense research over the last few years, and it has now become an established fact that AS-ODN can be highly potent and very specific inhibitors of gene expression in vitro. Data are now starting to emerge indicating that the concept may also be workable in vivo; indeed, several antisense compounds are presently undergoing phase I or even phase II clinical trials (see Chaps. 15–19, this volume). However, much remains to be done before the potential of this technology can be fully appreciated, especially with respect to understanding and ultimately altering the uptake and the pharmacokinetic and pharamacodynamic properties of AS-ODN. It should be borne in mind that, although some of these molecules are currently undergoing clinical trials, phosphorothioates (and methylphosphonates) may only represent a first generation of antisense molecules. Modified ODN have been identified in recent years that are clearly superior to full 2′-deoxy phosphorothioates in terms of RNA-binding affinity and in vitro antisense activity. Thus far, of all the modifications introduced into

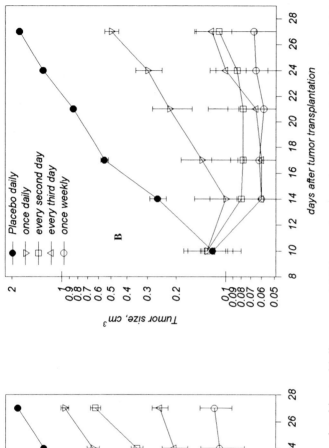
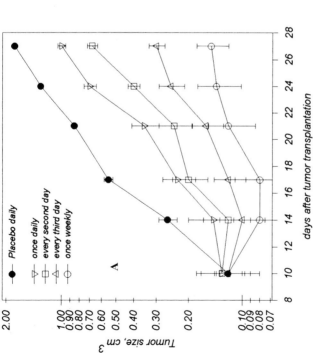

Fig. 5A,B. Antitumor activity of CGP 69846A formulated in stealth liposomes. A549 human lung carcinomas were transplanted s.c. into nude mice. The mice were treated with CGP 69846A at a dose of 6 mg/kg i.v. once daily (*downward triangles*), every second day (*squares*), every third day (*upwards triangles*), or once weekly (*open circles*) either **A** as a saline solution or **B** as a stealth liposome formulation. Sterically stabilized liposomes with CGP 69846A were prepared as described (see Chap. 4, this volume). *Closed circles*, placebo-treated controls

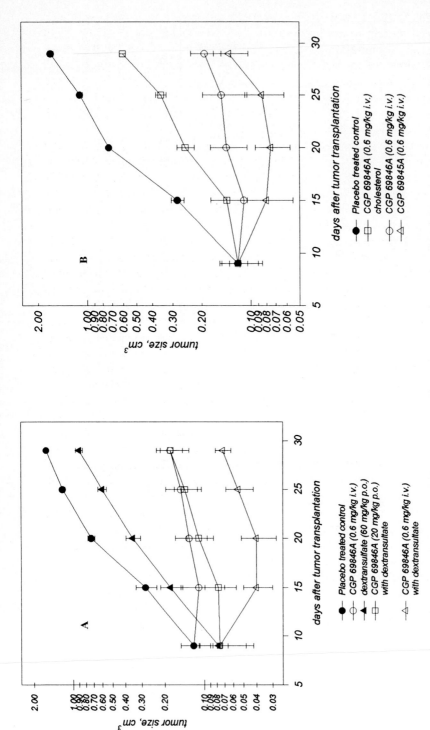

Fig. 6A,B. Antitumor activity of CGP 69846A with dextran sulfate and as a cholesterol conjugate. A549 human lung carcinomas were transplanted s.c. into nude mice. **A** Dextran sulfate was given p.o. (at a dose of 60 mg/kg) alone (*closed triangles*) or in combination with CGP 69846A either i.v. or p.o. at a dose of 0.6 mg/kg (*open circles, open triangles*) or at a dose of 20 mg/kg p.o. (*open squares*). Treatment was from day 7 to day 29. *Closed circles*, placebo-treated controls. **B** CGP 69846A (*open circles*), CGP 69845A (*open triangles*), and a cholesterol conjugate of CGP 69846A (*open squares*) were given at a dose of 0.6 mg/kg i.v. from day 7 to day 29. *Closed circles*, placebo-treated controls

CGP 69846A, the 2'-MoE modifications have been found to be the most potent with respect to antitumor activity in vivo (ALTMANN et al. 1996). In addition, all attempts to improve the antitumor activity of the phosphorothioate AS-ODN targeted to PKC-α or c-*raf* have not yielded the expected results. Although the stealth liposomal formulation did not increase the overall antitumor activity, less frequent dosing was possible with almost complete conservation of the activity. Similarly, conjugation of AS-ODN with cholesterol or coadministration with dextran sulfate did not reveal any improvement in antitumor activity, although in the case of dextran sulfate oral application of AS-ODN resulted in significant antitumor activity. More experiments are required to verify whether the coadministration of AS-ODN and dextran sulfate will make these molecules orally bioavailable. Similar results using various formulations of AS-ODN have been reported that demonstrated an increased antitumor activity (see Table 3).

It should also be remembered that, so far, the in vivo activity of ODN drugs has never been explicitly demonstrated to be mediated by an antisense mechanism (e.g., by demonstrating downregulation of the target mRNA and/or protein in treated tumors), although indirect evidence exists in several cases that strongly supports an antisense hypothesis (HIGGINS et al. 1993; DEAN et al. 1996; MONIA et al. 1996). However, in many cases, reported activities of AS-ODN in vivo and in vitro may in fact depend on non-antisense mechanisms, especially for phosphorothioates (although the effect may still be specific for a particular base sequence). Therefore, appropriate concept validation (e.g., proof of downregulation of mRNA and/or protein in the tumor) is mandatory. Should such direct proof be difficult to obtain in particular cases (especially in vivo), the use of adequate control ODN (to demonstrate reduced activity or complete absence thereof) may serve as an acceptable substitute strategy. The use of mismatched ODN, differing from the active compound at only few base positions, appears to be more appropriate for this purpose than the (frequently encountered) use of scrambled or sense controls (MONIA et al. 1997).

Although it is too early at this stage to make a final assessment of the viability of antisense-based treatment strategies in human disease, the data that have accumulated so far concerning in vitro and, more recently, in vivo experiments are very encouraging and indicate that antisense-based therapeutic approaches may become clinical reality in the not too distant future. In this context, it should also be noted that production costs for ODN have decreased tremendously over the last few years, suggesting that ODN-based therapies can be affordable if an adequate level of efficacy can be achieved, particularly when life-threatening diseases are involved.

References

Ahmad S, Glazer RI (1993) Expression of the antisense cDNA for protein kinase Cα attenuates resistance in doxorubicin-resistant MCF-7 breast carcinoma cells. Mol Pharmacol 43:858

Ahmad S, Mineta T, Martuza RL, Glazer RI (1994) Antisense expression of protein kinase Cα inhibits the growth and tumorigenicity of human glioblastoma cells. Neurosurgery 35:904

Alitalo K, Schwab M (1988) Oncogene amplification in tumor cells. Adv Cancer Res 46:235

Altmann KH, Dean NM, Fabbro D, Freier SM, Geiger T, Häner R, Hüsken D, Martin P, Monia BP, Müller M, Natt F, Nicklin P, Phillips J, Pieles U, Sasmor H, Moser HE (1996) Second generation of antisense oligonucleotides: From nuclease resistance to biological efficacy in animals. Chimia 50:168–176

Ambrose D, Resnicoff M, Coppola D, Sell C, Miura M, Jameson S, Baserga R, Rubin R (1994) Growth regulation of human gioblastoma T98G cells by insulin-like growth factor-1 and its receptor. J Cell Physiol 159:92

Aoki K, Yoshida T, Sugimura T, Terada M (1995) Liposome-mediated in vivo gene transfer of antisense K-ras construct inhibits pancreatic tumor dissemination in the murine pertoneal cavity. Cancer Res 55:3810

Autieri MV, Yue TL, Ferstein GZ, Ohlstein E (1995) Antisense oligonucleotides to the p65 subunit of NF-kB inhibit human vascular smooth muscle cell adherence and proliferation and prevent neointima formation in rat carotid arteries. Biochem Biophys Res Commun 213:827

Barcellini-Couget S, Pradines-Figuere A, Roux P, Dani C, Iilhaud G (1993) The regulation of growth hormone of lipoprotein lipase gene expression is mediated by c-fos protooncogene. Endocrinology 132:53

Baserga R, Reiss K, Alder H, Pietrozkowski Z, Surmacz E (1992) Inhibition of cell cycle progression by antisense oligodeoxynucleotides. Ann NY Acad Sci 660:64

Bayever E, Iversen P (1994) Oligonucleotides in the treatment of leukemia. Hematol Oncol 12:9

Becker D, Meier CB, Herlyn M (1989) Proliferation of human malignant melanomas is inhibited by antisense oligodeoxynucleotides targeted against basic fibroblast growth factor. EMBO J 8:3685

Becker D, Lee PL, Rodeck U, Herlyn M (1992) Inhibition of the fibroblast growth factor receptor 1 gene in human melanocytes and malignant melanomas leads to inhibition of proliferation and signs indicative of differentiation. Oncogene 7:2303

Behl C, Bogdahn U, Winkler J, Apfel R, Brysch WI, Schlingensiepen KH (1993) Antiinduction of platelet-derived growth factor A-chain mRNA expression in a human malignant melanoma cell line and growth-inhibitory effects of PDGF A-chain mRNA-specific antisense molecules. Biochem Biophys Res Commun 193:744

Behrend E I, Craig AM, Wilson SM, Denhardt DT, Chambers AF (1994) Reduced malignancy of ras-transformed NIH 3T3 cells expressing antisense osteopontin RNA. Cancer Res 54:832

Bello-Rehss E, Ernest S (1994) Expression and function of P-glycoprotein in human mesangial cells. Am J Physiol 267:C1351

Bennett MR, Anglin S, McEwan FR, Jagoe R, Newby AC, Evan GI (1994) Inhibition of vascular smooth muscle cell proliferation in vitro and in vivo by c-myc antisense oligodeoxynucleotides. J Clin Invest 93:820

Bergmann R, Funatomi H, Yokoyama M, Berger HG, Korc M (1995) Insulin-line growth factor-1 overexpression in human pancreatic cancer: evidence for autocrine and paracrine roles. Cancer Res 55:2007

Bishop JM (1991) Molecular themes in oncogenesis. Cell 64:235

Blenis J (1993) Signal transduction via the MAP kinases: proceed at your own RSK. Proc Natl Acad Sci U S A 90:5889

Bos JL (1989) Ras oncogene in human cancer: a review. Cancer Res 49:4682

Burgess TL, Fisher EF, Ross SL, Bready JV, Qian YX, Bayewitch LA, Cohen A, Herrera CJ, Hu SS, Kramer TB (1995) The antiproliferative activity of c-myb and c-myc antisense oligonucleotides in smooth muscle cells is caused by a non-antisense mechanism. Proc Natl Acad Sci U S A 92:4051

Busch RK, Perlaky L, Valdez BC, Henning D, Busch H (1994) Apoptosis in human tumor cells following treatment with p120 antisense oligodeoxynucleotide ISIS 3466. Cancer Lett 86:151

Calabretta B, Sims RB, Valtieri M, Caracciolo D, Szczylik C, Venturelli D, Ratajczak M, Beran M, Gewirtz AM (1991) Normal and leukemic hematopoietic cells manifest differential sensitivity to inhibitory effects of c-myb antisense oligonucleotides: an in vitro study relevant to bone marrow purging. Proc Natl Acad Sci U S A 88:2351

Calabretta B, Skorski T, Szczylik C, Zon G (1993) Prospects for gene-directed therapy with antisense oligodeoxynucleotides. Cancer Treat Rev 19:169

Calara F, Ameli S, Hultgardh-Nilsson J (1996) Autocrine induction of DNA synthesis by mechanical injury of cultured smooth muscle cells. Potential role of FGF and PDGF. Arterioscler Thromb Vasc Biol 16:187

Cantley LC, Auger KR, Carpenter C, Duckworth B, Graziani A, Kapeller R, Soltoff S (1991) Oncogenes and signal transduction. Cell 64:281

Carter G, Lemoine NR (1993) Antisense technology for cancer therapy: does it make sense? Br J Cancer 67:869

Chakrabarty S, Rajagopal S, Huang S (1995) Expression of antisense epidermal growth factor receptor RNA down modulates the malignant behaviour of human colon cancer cells. Clin Exp Metastasis 13:191

Cho-Chung Y-S (1993) Antisense oligonucleotides for the treatment of cancer. Curr Opin Ther Pat 3:1737

Cho-Chung Y-S, Clair T (1993) The regulatory subunit of cAMP-dependent protein kinase as a target for chemotherapy and cancer and other cellular dysfunctional-related disease. Pharmacol Ther 60:265

Ciardiello F, Tortora G, Bianco C, Selvam MP, Basolo F, Fontanini G, Pacifico F, Normanno N, Brandt R, Persico, MG (1994) Inhibition of CRIPTO expression and tumorigenicity in human colon cancer cells by antisense RNA and oligodeoxynucleotides. Oncogene 9:291

Collins S, Groudine M (1982) Amplification of endogenous myc-related DNA sequences in a human myeloid leukemia cell line. Nature 299:679

Colomer R, Lupu R, Bacus SS, Gelman EP (1994) erbB-2 antisense oligonucleotides inhibit the proliferation of breast carcinoma cells with erbB-2 oncogene amplification. Br J Cancer 70:819

Colotta F, Polentarutti N, Sironi M, Mantovani A (1992) Expression and involvement of c-fos and c-jun protooncogenes in programmed cell death induced by growth factor deprivation in lymphoid cell lines. J Biol Chem 267:18278

Cotter FE, Johnson P, Hall P, Pocock C, As Mahdi NA, Cowell JK, Morgan G (1994) Antisense oligonucleotides suppress B-cell lymphoma growth in a SCID-hu mouse model. Oncogene, 9:3049

Croce CM (1987) Role of chromosome translocations in human neoplasia. Cell 49:155

Crooke ST (1992) Therapeutic application of oligonucleotides. Annu Rev Pharmacol Toxicol 32:329

Dean NM, McKay R, Condon TP, Bennett CF (1994) Inhibition of protein kinase C-α expression in human A549 cells by antisense oligonucleotides inhibits induction of intercellular adhesion molecule 1 (ICAM-1) mRNA by phorbol esters. J Biol Chem 269:1

Dean NM, McKay R, Miraglia L, Howard R, Meister L, Ziel R, Geiger T, Müller M, Fabbro D (1996) Inhibition of growth of human tumor cell lines in nude mice by an antisense oligonucleotide inhibitor of PKC-α expression. Cancer Res 56: 3499

Degols G, Leonetti JP, Mechti N, Lebleu B (1991) Antiproliferative effects of antisense oligonucleotides directed to the RNA of c-myc oncogene. Nucleic Acids Res 19:945

Dietel M (1993) Meeting report: second international symposium on cytostatic drug resistance. Cancer Res 53:2683

Ebbecke M, Unterberg C, Buchwald A, Stohr S, Wiegand V (1992) Antiproliferative effects of a c-myc antisense oligonucleotide on human arterial smooth muscle cells. Basic Res Cardiol 87:585

Ensoli JB, Markham P, Kao V, Barillari G, Fiorelli V, Gendelman R, Raffeld M, Zon G, Galo RC (1994) Block of AIDS-Kaposi's sarcoma cell growth, angiogenesis and lesion formation in nude mice by antisense oligonucleotide targeting basic fibroblast growth factor. A novel strategy for the therapy of Kaposi sarcoma. J Clin Invest 94:1736

Fakhrai H, Dorigo O, Shawler DL, Lin HN, Mercola D, Black KL, Royston I, Sobol RE (1996) Eradication of established intracranial rat gliomas by transforming growth factor beta antisense gene therapy. Proc Natl Acad Sci U S A 93:2902

Fidler IJ (1995) Invasion and metastasis. In: Abeloff MD, Armitage JO, Lichter AS, Niederhuber JE (eds) Clinical oncology. Churchill Livingstone, Edinburgh, pp 45–76

Fitzpatrick DR, Bielefeldt-Ohmann H, Himbeck RP, Jarnicki AG, Marzo AL, Robinson RWS (1994) Transforming growth factor-beta: antisense RNA-mediated inhibition affects anchorage-independent growth, tumorigenicity and tumor infiltrating T-cells in malignant mesothelioma. Growth Factors 11:29

Fonagy A, Swiderski C, Dunn M, Freeman JW (1992) Antisense-mediated specific inhibition of p120 protein expression prevents G1 to S-phase transition. Cancer Res 52:5250

Geiger T, Müller M, Monia BP, Fabbro D (1997) Antitumor activity of a c-raf antisense oligonucleotide in combination with standard chemotherapeutic agents against various human tumors transplanted subcutaneously into nude mice. Clin Cancer Res 3:1179–1185

Georges RN, Mukhopadhyay T, Zhang Y, Yen N, Roth JA (1993) Prevention of orthotopic human lung cancer growth by intratracheal instillation of a retroviral antisense K-ras construct. Cancer Res 53:1743

Gottesman MM (1993) How cancer cells evade chemotherapy: sixteenth Richard and Hinda Rosenthal Foundation Award lecture. Cancer Res 53:747

Gray GD, Hernandez OM, Hebel D, Root M, Pow-Sang JM, Wickstrom E (1993) Antisense DNA inhibition of tumor growth induced by c-Ha-ras oncogene in nude mice. Cancer Res 53:577

Gryaznov S, Skorski T, Cucco C, Nieborowska-Skorska M, Chiu CY, Lloyd D, Chen JK, Koziolkiewicz M, Calabretta B (1996) Oligonucleotide N3'-P5' phosphoroamidates as antisense agents. Nucleic Acids Res 24:1508

Hansche GM, Wagner C, Burger JA, Dong W, Staehler G, Stoeck M (1995) Matrix protein synthesis by glomerular mesangial cells in culture: effects of transforming growth factor beta and platelet-derived growth factor on fibronectin and collagen type IV mRNA. J Cell Physiol 163:451

Hélène C, Toulmé JJ (1990) Specific regulation of gene expression by antisense, sense and antigene nucleic acids. Biochim Biophys Acta 1049:99–125

Higgins KA, Perez JR, Coleman TA, Dorshkind K, McComas WA, Sarmiento UM, Rosen CA, Narayanan R (1993) Antisense inhibition of the p65 subunit of NF-κB blocks tumorigenicity and causes tumor regression. Proc Natl Acad Sci U S A 90:9901

Hijiya N, Zhang J, Ratajczak MZ, Kant JA, DeRiel K, Herlyn M, Zon G, Gewirtz AM (1994) Biologic and therapeutic significance of myb expression in human melanoma. Proc Natl Acad Sci U S A 91:4499

Hunter T (1991) Cooperation between oncogenes. Cell 64:249

Itoh H, Mukoyama M, Pratt RE, Dzau VJ (1992) Specific blockade of basic fibroblast growth factor gene expression in endothelial cells by antisense oligonucleotide. Biochem Biophys Res Commun 188:1205

Jaattelä M (1995) Overexpression of hsp70 confers tumorigenicity to mouse fibrosarcoma cells. Int J Cancer 60:689

Jachimczak P, Hessdorfer B, Fabel-Schulte K, Wismeth C, Brysch W, Schlingensiepen KH, Bauer A, Blesch A, Bogdahn U (1996) Transforming growth factor-beta-mediated autocrine growth regulation of gliomas as detected with phosphorothioate antisense oligonucleotides. Int J Cancer 65:332

Jaskulski D, deRiel JK, Mercer WE, Calabretta B, Baserga R (1988) Inhibition of cellular proliferation by antisense oligodeoxynucleotides to PCNa cyclin. Science 240:1544

Kasid U, Pfeifer A, Brennan T, Beckett M, Weichselbaum RR, Dristchilo A, Mark GE (1989) Effect of antisense c-raf-1 on tumorigenicity and radiation sensitivity of human squamous carcinoma. Science 243:1354

Kaye SB (1993) P-glycoprotein (Pgp) and drug resistance-time for reappraisal? Br J Cancer 67:641

Kiehntopf M, Brach MA, Licht T, Petschauer S, Karawajew L, Kirschning C, Herrmann F (1994) Ribozyme-mediated cleavage of the MDR-1 transcript restores chemosensitivity in previously resistant cancer cells. Embo J 13:4645,

Kitajima I, Shinohara T, Biladovics J, Brown DA, Yu X, Nerenberg M (1992) Ablation of transplanted HTLV-1 Tax-transformed tumors in mice by antisense inhibition of NF-κB. Science 258:1792

Kobayashi H, Dorai T, Holland JF, Ohnuma T (1994) Reversal of drug sensitivity in multidrug-resistant tumor cells by a MDR1 (PGY1) ribozyme. Cancer Res 54:1271

Kolch W, Heidecker G, Lloyd P, Rapp UR (1991) Raf-1 protein kinase is required for growth of induced NIH/3T3 cells. Nature 349:426

Kook YH, Adamski J, Zelen A, Ossowski L (1994) The effect of antisense inhibition of urokinase receptor in human squamous cell carcinoma on malignancy. EMBO J 13:3983

Laird AD, Brown PI, Fausto N (1994) Inhibition of tumor growth in liver epithelial cells transfected with a transforming growth factor-α antisense gene. Cancer Res 54:4224

Lange W, Daskalakis M, Finke J, Dölken G (1994) Comparison of different ribozymes for efficient and specific cleavage of bcr/abl related mRNAs. FEBS Lett 338:175

Levi BZ, Ozato K (1988) Constitutive expression of c-fos antisense RNA blocks c-fos gene induction by interferon and by phorbol ester and reduces c-myc expression in F9 embryonal carcinoma cells. Genes Dev 2:554

Lewalle P, Martiat P (1993) Inhibition of p210 expression in chronic myeloid leukaemia: oligonucleotides and/or transduced antisense sequences. Leuk Lymphoma 11:139

Long L, Rubin R, Baserga R, Brodt P (1995) Loss of the metastatic phenotype in murine carcinoma cells expressing an antisense RNA to the insulin-like growth factor receptor. Cancer Res 55:1006

Magrath IT (1994) Prospects for the therapeutic use of antisense oligonucleotides in malignant lymphomas. Ann Oncol 5:S67

Martiat P, Lewalle P, Taj AS, Philippe M, Larondelle Y, Vaerman JL, Wildmann C, Goldman JM, Michaux JL (1993) Retrovirally transduced antisense sequences stably suppress p210 expression and inhibit the proliferation of bcr/abl-containing cell lines. Blood 81:502

Melani C, Rivoltini L, Parmiani G, Calabretta B, Colombo MP (1991) Inhibition of proliferation of c-myb antisense oligodeoxynucleotides in colon adenocarcinoma cell lines that express c-myb. Cancer Res 51:2897

Monia BP, Johnston JF, Geiger T, Müller M, Fabbro D (1996) Antitumor activity of a phosphorothioate antisense oligodeoxynucleotide targeted against c-raf kinase. Nat Med 2:668

Monia BP, Samsor H, Johnston JF, Freier SM, Lesnik EA, Müller M, Geiger T, Altmann KH, Moser H, Fabbro D (1997) Sequence-specific antitumor activity of a phosphorothioate oligodeoxynucleotide targeted to human c-raf kinase supports an antisense mechanism of action in vivo. Proc Natl Acad Sci U S A 93:15481–15484

Moroni MC, Willingham MC, Beguinot L (1992) EGF receptor antisense RNA blocks expression of the epidermal growth factor receptor and suppresses the transforming phenotype of a human carcinoma cell line. J Biol Chem 267:2714

Muszynski KW, Ruscetti FW, Heidecker G, Rapp U, Troppmair J, Gooya JM, Keller JR (1995) Raf-1 protein is required for growth factor-induced proliferation of hematopoietic cells. J Exp Med 181:2189

Narayanan R, Higgins KA, Perez JR, Coleman TA, Rosen CA (1993) Evidence for differential functions of the 50 and p65 subunits of NF-κB with a cell adhesion model. Mol Cell Biol 13:3802

Nesterova M, Cho-Chung Y-S (1995) A single injection protein kinase A-directed antisense treatment to inhibit tumor growth. Nat Med 1:528

Neuenschwander S, Roberts CT, LeRoith D (1995) Growth inhibition of MCF-7 breast cancer cells by stable expression of an insulin-like growth factor-1 receptor antisense ribonucleic acid. Endocrinology 136:4298

Nishikura K, Kim U, Murray JM (1990) Differentiation of F9 cells is independent of c-myc expression. Oncogene 5:981

Nowell PC, Hungerford DA (1960) A minute chromosome in human chronic granulocytic leukemia. Science 132:1491

O'Brien SG, Kirkland MA, Goldman JM (1994) Gene therapy – a future for cancer management. Eur J Cancer 30A:1160

Ouchida M, Ohno T, Fujimura Y, Rao VN, Reddy ES (1995) Loss of tumorigenicity of Ewing's sarcoma cells expressing antisense RNA to EWS-fusion transcripts. Oncogene 11:1049

Pastan I, Gottesman MM (1991) Multidrug resistance. Annu Rev Med 41:277

Paulus W, Baur I, Huettner C, Schmausser B, Roggendorf W, Schlingensiepen KH, Brysch W (1995) Effects of transforming growth factor-beta-1 on collagen synthesis, integrin expression, adhesion and invasion of glioma cells. J Neuropathol Exp Neurol 54:236

Perez JR, Higgins-Sochaski KA, Maltese JY, Narayanan R (1994) Regulation of adhesion and growth of fibrosarcoma cells by NF-κB RelA involves transforming growth factor β. Mol Cell Biol 14:5326

Perlaky L, Valdez BC, Busch RK, Larson RG, Jhiang SM, Zhang WW, Brattain M, Busch H (1992) Increased growth of NIH/3T3 cells by transfection with human p120 complementary DNA and inhibition by a p120 antisense construct. Cancer Res 52:428

Perlaky L, Saijo Y, Busch RK, Bennett CF, Mirabelli CK, Crooke ST, Busch H (1993) Growth inhibition of human tumor cell lines by antisense oligonucleotides designed to inhibit p120 expression. Anticancer Drug Des 8:3

Perletti GP, Smeraldi C, Porro D, Piccinini F (1994) Involvement of the alpha isoenzyme of protein kinase C in the growth inhibition induced by phorbol esters in MH1C1 hepatoma cells. Biochem Biophys Res Commun 205:1589

Pierga J-Y, Magdelenat H (1994) Applications of antisense oligonucleotides in oncology, Cell Mol Biol 40:237

Pietrozkowski Z, Sell C, Lammers R, Ullrich A, Baserga R (1992) Roles of insulin-like growth factor-1 and the IGF-1 receptor in epidermal growth factor-stimulated growth of 3T3 cells. Mol Cell Biol 12:3883

Quattrone A, Fibbi G, Anichini E, Zamperini A, Cappacioli S, del Rosso M (1995) Reversion of the invasive phenotype of transformed human fibroblasts by inhibition of urokinase receptor gene expression. Cancer Res 55:90

Rabbitts TH (1994) Chromosomal translocations in human cancer. Nature 372:143

Ratajczak MZ, Hijiya N, Catani L, deRiel K, Luger SM, McGlave P, Gewirtz AM (1992) Acute- and chronic-phase chronic myelogenous leukemia colony-forming units are highly sensitive to the growth inhibitory effects of c-myb antisense oligodeoxynucleotides. Blood 79:1956

Resnicoff M, Coppola D, Sell C, Rubin R, Ferrone S, Baserga R (1994) Growth inhibition of human melanoma cells in nude mice by antisense strategies to the type-1 insulin-like growth factor receptor. Cancer Res 54:4848

Rivoltini L, Colombo MP, Supino R, Ballinari D, Tsuruo T, Parmiani G (1990) Modulation of multidrug resistance by verapamil or mdr1 antisense oligodeoxynucleotides does not change the high susceptibility to lymphokine-activated killers in mdr-resistant human carcinoma (LoVo) line. Int J Cancer 46:727

Rowley JD (1982) Identification of the constant chromosome regions involved in human hematologic malignant diseases. Science 216:749

Rubenstein M, Mirochnik Y, Choi P, Guinan P (1996) Antisense oligonucleotide intralesional therapy for human PC-3 prostate tumors carried in athymic nude mice. J Surg Oncol 62:194–200

Saijo Y, Perlaky L, Valdez BC, Busch RK, Henning D, Zhang WW, Busch H (1993a) The effect of antisense p120 construct on p120 expression and cell proliferation in human breast cancer MCF-7 cells. Cancer Lett 68:95

Saijo Y, Perlaky L, Valdez BC, Wang H, Henning D, Busch H (1993b) Cellular pharmacology of p120 antisense oligodeoxynucleotide phosphorothioate ISIS 3466. Oncol Res 5:283

Sakakura C, Hagiwara A, Tsujimoto H, Ozaki K, Sakakibara T, Oyama T, Ogaki M, Takahashi T (1994) Inhibition of gastric cancer cell proliferation by antisense oligonucleotides targeting the messenger RNA encoding proliferating cell nuclear antigen. Br J Cancer 10:1060

Sakakura C, Hagiwara A, Tsujimoto H, Ozaki K, Sakakibara T, Oyama T, Ogaki M, Takahashi T (1995) The anti-proliferative effect of proliferating cell nuclear antigen-specific antisense oligonucleotides on human gastric cancer cell lines. Surg Today 25:184

Saleh M, Stacker SA, Wilks AF (1996) Inhibition of growth of C6 glioma cells in vivo by expression of antisense vascular endothelial growth factor sequence. Cancer Res 56:393

Scanlon KJ, Ishida H, Kashani-Sabet M (1994) Ribozyme-mediated reversal of multidrug-resistant phenotype. Proc Natl Acad Sci U S A 91:11123

Schönthal A, Büscher M, Angel P, Rahmsdorf HJ, Ponta H, Hattori K, Chiu R, Karin M, Herrlich P (1989) The Fos and Jun/AP-1 proteins are involved in the downregulation of Fos transcription. Oncogene 4:629

Schwab G, Chavany C, Duroux I, Goubin G, Lebeau J, Hélène C, Saison-Behmoaras T (1994) Antisense oligonucleotides absorbed to polyalkyl-cyanoacrylate nanoparticles specifically inhibit mutated Ha-ras-mediated cell proliferation and tumorigenicity in nude mice. Proc Natl Acad Sci 91:10460

Shapiro DN, Jones BG, Shapiro LH, Dias P, Houghton PJ (1994) Antisense-mediated reduction in insulin-like growth factor-1 receptor expression suppresses the malignant phenotype of a human alveolar rhabdomyosarcoma. J Clin Invest 94:1235

Shi Y, Hutchinson HG, Gall DJ, Zalewski A (1993) Downregulation of c-myc expression by antisense oligonucleotides inhibits proliferation of human smooth muscle cells. Circulation 88:1190

Shustik C, Dalton W, Gros P (1995) P-glycoprotein-mediated multidrug resistance in tumor cells: biochemistry, clinical relevance and modulation. Mol Aspects Med 16:1

Simons M, Morgan KG, Parker C, Collins E, Rosenberg RD (1993) The proto-oncogene c-myb mediates an intracellular calcium rise during the late G1 phase of the cell cycle. J Biol Chem 286:627

Skorski TT, Nieborowska-Skorska M, Barletta C, Malaguarnera L, Szczylik C, Chen ST, Lange B, Calabretta B (1993) Highly efficient elimination of Philadelphia leukemic cells by exposure to bcr/abl antisense oligodeoxynucleotides combined with mafosfamide. J Clin Invest 92:194

Skorski T, Nieborowska-Skorska M, Nicolaides NC, Szcylik C, Iversen P, Iozzo RV, Zon G, Calabretta B (1994) Suppression of Philadelphia leukemia cell growth in mice by bcr-abl antisense oligodeoxynucleotides. Proc Natl Acad Sci U S A 91:4504

Skorski T, Nieborowska-Skorska M, Szczylik C, Kanakaraj P, Perrotti D, Zon G, Gewirtz, A, Perussia B, Calabretta B (1995) c-RAF-1 serine/threonine kinase is required in BCR/ABL-dependent and normal hematopoiesis. Cancer Res 55: 2275

Slamon DJ, Clark GM, Wong SG, Levin MJ, Ullrich A, McGuire WL (1987) Human breast carcinoma: correlation of relapse and survival with amplification of the Her-2/neu oncogene. Science 235:177

Smeters TFC, Skorski T, van de Locht LTF, Wessels HM, Pennings AH, de Witte T, Calabretta B, Mensink EJ (1994) Antisense BCR-ABL oligonucleotides induce apoptosis in the Philadelphia chromosome-positive cell line BV173. Leukemia 8:129

Smeters TFC, van de Locht LTF, Pennings AHM, Wessels HMC, de Witte TM, Mensink EJB (1995) Phosphorothioate Bcr-abl antisense oligonucleotides induce cell death, but fail to reduce cellular bcr-abl protein levels. Leukemia 9:118

Sokoloski JA, Sartorelli AC, Rosen CA, Narayanan R (1993) Antisense oligonucleotides to the p65 subunit of NF-κB block CD11b expression and alter adhesion properties of differentiated HL-60 granulocytes. Blood 82:625

Spearman M, Taylor WR, Greenberg AH, Wright JA (1994) Antisense oligodeoxynucleotide inhibition of TGF-β1 gene expression and alterations in the growth and malignant properties of mouse fibrosarcoma cells. Gene 149:25

Speir E, Epstein SE (1992) Inhibition of smooth muscle cell proliferation by an antisense oligodeoxynucleotide targeting the messenger RNA encoding proliferating cell nuclear antigene. Circulation 86:538

Szczylik C, Skorski T, Nicolaides NC, Manzella L, Malaguarnera L, Venturelli D, Gewirtz A, Calabretta B (1991) Selective inhibition of leukemia cell proliferation by bcr-abl antisense oligonucleotides. Science 253:562

Thierry AR, Rahman A, Dritschilo A (1993) Overcoming multidrug resistance in human cells using free and liposomally encapsulated antisense oligonucleotides. Biochem Biophys Res Commun 190:952

Thomas M, Kosciolek B, Wang N, Rowley P (1994) Capping of bcr-abl antisense oligonucleotides enhances antiproliferative activity against chronic myeloid leukemia cell lines. Leuk Res 18:401

Vaerman JL, Lammineur C, Moureau P, Lewall P, Deldime M, Blumenfeld M, Martiat P (1995) BCR-ABL antisense oligodeoxyribonucleotides suppress the growth of leukemic and normal hematopoietic cells by a sequence-specific but nonantisense mechanism. Blood 10:3891

van Kalken, CK, Pinedo HM, Giaccone G (1991) Multidrug resistance from the clinical point of view. Eur J Cancer 27:1481

Venturelli D, Travali S, Calabretta B (1990) Inhibition of T-cell proliferation by a MYB antisense oligomer is accompanied by selective down-regulation of DNA polymerase α expression. Proc Natl Acad Sci U S A 87:5963

Watson PH, Pon RT, Shiu RPC (1991) Inhibition of c-myc expression by phosphorothioate antisense oligonucleotide identifies a critical role for c-myc in the growth of human breast cancer. Cancer Res 51:3996

Weinberg RA (1994) Oncogenes and tumor suppressor genes. CA Cancer J Clin 44: 160

Wickstrom EL, Bacon TA, Gonzalez A, Freeman DL, Lyman GH, Wickstrom E (1988) Human promyelocytic leukemia HL-60 cell proliferation and c-myc protein expression are inhibited by an antisense pentadecadeoxynucleotide targeted against c-myc mRNA. Proc Natl Acad Sci U S A 85:1028

Wilhelm O, Schmitt M, Höhl S, Senekowitsch R, Graeff H (1995) Antisense inhibition of urokinase reduces spread of human ovarian cancer in mice. Clin Exp Metastasis 13:296

Witty JP, McDonnell S, Newell KJ, Cannon P, Navre M, Tressler RJ, Matrisian L (1994) Modulation of matrilysin levels in colon carcinoma cell lines affects tumorigenicity in vivo. Cancer Res 54:4805

Ye X, Wu M (1992) Retrovirus mediated transfer of antisense human c-myc gene into human esophageal cancer cells suppressed cell proliferation and malignancy. Sci China B 35:76

Yokozaki H, Budillon A, Tortora G, Meissner S, Beaucage SL, Miki K, Cho-Chung YS (1993) An antisense oligodeoxynucleotide that depletes RIα subunit of cyclic AMP-dependent protein kinase induces growth inhibition in human cancer cells. Cancer Res 53:868

Zhou P, Jiang W, Zhang YJ, Kahn SM, Schieren I, Santella RM, Weinstein IB (1995) Antisense to cyclin D1 inhibits growth and reverses the transformed phenotype of human esophageal cancer cells. Oncogene 11:571

CHAPTER 14
Pharmacological Activity of Antisense Oligonucleotides in Animal Models of Disease

B.P. MONIA and N.M. DEAN

A. Introduction

Over the last 20 years, antisense oligonucleotide approaches have proven to be extremely valuable methods for characterizing the function of specific gene products and for addressing and optimizing antisense mechanisms of action. Numerous reviews have been published describing some of the accomplishments in this area, some of which are cited here (CROOKE 1992; BENNETT et al 1995; FIELD and GOODCHILD 1995; CROOKE and BENNETT 1996). However, it has only been within the last 5–7 years that antisense technology has progressed such that the therapeutic value of these compounds can be evaluated in animal models of human disease. Over this period, reports demonstrating pharmacological activities of antisense oligonucleotides in animal models have accelerated at an impressive rate in a number of therapeutic areas, most commonly cancer, restenosis, and inflammation. In many of these studies, the pharmacological activity reported for the antisense oligonucleotides was very controlled such that it is difficult to ascribe a mechanism of action other than antisense. Many of these studies have been included in prior reviews of the antisense field (BENNETT et al. 1995; FIELD and GOODCHILD 1995; CROOKE and BENNETT 1996).

The focus of this review is to summarize the findings reported using antisense oligonucleotides in animal models of vascular, skin, and peripheral organ disease. Reviews covering the pharmacological activity of antisense oligonucleotides in disease models of cancer, central nervous system (CNS), and inflammation will be covered elsewhere in this volume.

B. Vascular System

I. Restenosis

Excessive proliferation of vascular smooth muscle cells (SMC) is a critical process underlying the development of atherosclerosis and restenosis after balloon angioplasty. Growth factors and cytokines regulate proliferation by binding to receptors, often containing intrinsic protein tyrosine kinase activity, resulting in the activation of intracellular signaling cascades (SCHLESSINGER 1994). To date, the precise molecular mechanisms that underlie excessive proliferation of SMC remain poorly defined, although it is likely that the signals induced by diverse mitogenic stimuli will converge to a certain extent at common downstream effectors. Antisense approaches have proven to be extremely valuable for the identification of signaling and other regulatory molecules that are required for SMC proliferation. Inhibition of in vitro proliferation of vascular SMC has been demonstrated following treatment with antisense oligonucleotides targeted against a variety of extracellular and intracellular regulatory molecules, including proliferating cell nuclear antigen (PCNA; PICKERING et al. 1992; SPEIR and EPSTEIN 1992; MORISHITA et al. 1993; PICKERING et al. 1996), c-*myb* (SIMONS and ROSENBERG 1992), c-*myc* (SHI et al. 1993; BENNETT et al. 1994), cdc2 kinase (MORISHITA et al. 1993), cdk2 kinase (MORISHITA et al. 1994a), the receptors for thrombin (CHAIKOF et al. 1995) and insulin-like growth factor I (DELAFONTAINE et al. 1995), NF-κB (AUTIERI et al. 1995), and C-*raf* kinase (CIOFFI et al. 1997; SCHUMACHER et al. 1997). This topic has been the subject of previous reviews and editorials (EPSTEIN et al. 1993; BENNETT and SCHWARTZ 1995; FIELD and GOODCHILD 1995).

Reports demonstrating in vivo efficacy of antisense oligonucleotides in animal models of restenosis have been essentially limited to slight variations of a rat carotid artery model in which neointimal thickening is induced following balloon angioplasty or other forms of injury that result in deendothelialization (SIMONS et al. 1992; MORISHITA et al. 1993,1994a; ABE et al. 1994; BENNETT et al. 1994; SIMONS et al. 1994; AUTIERI et al. 1995; Table 1). In addition, a rabbit artery model has been employed to characterize the effects of antisense oligonucleotides on SMC proliferation in artery walls ex vivo (BURGESS et al. 1995), and a porcine model of coronary artery balloon injury has been examined to a limited extent (SHI et al. 1994). In all cases in which pharmacology was examined in vivo, oligonucleotides were administered locally to the injured artery either using a transcatheter device (SHI et al. 1994), formulation with a liposome administered into the lumen of the injured artery (MORISHITA et al. 1993, 1994a,b; PICKERING et al. 1996), or by delivering the oligonucleotides periadventially using a pluronic gel or by a polyethylene cuff with a osmotic minipump (SIMONS et al. 1992,1994; ABE et al. 1994; BENNETT et al. 1994; AUTIERI et al. 1995).

The first report published describing in vivo efficacy of an antisense oligonucleotide in a model of restenosis was by SIMONS et al. (1992), who utilized an 18-base phosphorothioate antisense oligonucleotide targeted to position +4 to +22 of the mouse C-*myb* mRNA sequence. In this study, the investigators administered the oligonucleotide in a pluronic gel with a single application to the injured site around the artery. Two weeks following treatment, c-*myb* mRNA levels were measured and were found to be substantially reduced, whereas a sense control phosphorothioate analogue had no effect. Moreover, minimal intimal accumulation of SMC was observed within the injured artery in animals receiving the antisense oligonucleotide, whereas extensive accumulation was observed in the sense-treated and untreated animal groups.

Table 1. In vivo efficacy studies employing antisense oligonucleotides in animal models of vascular, skin, and peripheral organ disease

Therapeutic area	Molecular target	Animal model	Method of oligonucleotide administration	References
Restenosis	c-myb	Rat	Local/pluronic gel	Simons et al. (1992)
	c-myc	Rat	Local/pluronic gel	Bennett et al. (1994)
	c-myc	Pig	Local/transcatheter	Shi et al. (1994)
	cdc2 kinase	Rat	Local/HVJ liposomes	Morishita et al. (1993, 1994a,b)
	cdc2 kinase	Rat	Local/pluronic gel	Abe et al. (1994)
	cdk2 kinase	Rat	Local/pluronic gel	Abe et al. (1994)
	cdk2 kinase	Rat	Local/HVJ liposomes	Morishita et al. (1994a,b)
	PCNA	Rat	Local/HVJ liposomes	Morishita et al. (1993, 1994a,b)
	PCNA	Rat	Local/pluronic gel	Simons et al. (1994)
	Cyclin B_1	Rat	Local/HVJ liposomes	Morishita et al. (1994a,b)
	NF-κB	Rat	Local/polyethylene cuff	Autieri et al. (1995)
Atherosclerosis	TGF-β1	Rat (organ culture)	Local	Merrilees and Scott (1994)
	cdc2 kinase	Rabbit	Local/HVJ liposomes	Mann et al. (1995)
	PCNA	Rabbit	Local/HVJ liposomes	Mann et al. (1995)
	CETP	Rabbit	Systemic	Sugano and Makina (1996)
Hypertension	Angiotensinogen	Rat	Local/catheter (CNS)	Gyurko et al. (1993)
				Wielbo et al. (1995)
Ocular neovascularization	VEGF	Mouse	Local/injection (intraocular)	Smith et al. (1995)
				Aiello et al. (1995)
				Robinson et al. (1996)
Scarring/wounding	TGF-β1	Mouse	Local/topical	Choi et al. (1996)
Kidney ischemia	INOS	Rat	Systemic	Noiri et al. (1996)
Asthma	Adenosine A_1 receptor	Rabbit	Local/inhalation	Nyce and Metzger (1997)

PCNA, proliferating cell nuclear antigen; TGF, tumor growth factor; CETP, cholesteryl ester transfer protein; VEGF, vascular endothelial growth factor; INOS, inducible nitric oxide synthase; HVJ, hemagglutinating virus of Japan.

Similar to the findings reported for c-*myb*, BENNETT et al. (1994), also using a pluronic gel formulation, reported that a single application of a 15-mer phosphorothioate antisense oligonucleotide targeted to the first five codons of human c-*myc* mRNA resulted in a substantial reduction of c-*myc* mRNA levels in injured arterial wall tissue and significantly reduced neointimal formation following balloon injury. Neither a pluronic gel-formulated sense control phosphorothioate analogue of the c-*myc* antisense sequence nor the gel by itself had any effect on c-*myc* mRNA levels or neointimal formation. c-*myc* antisense oligonucleotides have also been evaluated for their ability to inhibit neointimal formation following a single transcatheter delivery into denuded porcine coronary arteries (SHI et al. 1994). Despite rapid clearance following local delivery, oligonucleotides were detectable at the site of injection for at least 3 days, and significant inhibition of neointimal thickening was observed up to 1 month following application of the c-*myc* antisense. Based on these experiments, the authors concluded that the c-*myc* proto-oncogene is involved in the process of vascular remodeling, regulation of SMC proliferation, and extracellular matrix formation.

The in vivo results described above for the c-*myb* and c-*myc* antisense oligonucleotides were attributed to an antisense mechanism based on two observations: (1) sequence specificity was demonstrated by the fact that sense control oligonucleotides had no effect on target RNA levels or neointimal formation, and (2) inhibition of target RNA levels was demonstrated. Despite this, however, the in vivo (and in vitro) mechanism of action of these oligonucleotides was brought into question by BURGESS et al. (1995), who reported that their activity in vitro and in an ex vivo model was due to a sequence-specific non-antisense mechanism in which the presence of a guanosine (G) quartet was responsible for the pharmacological activity of the c-*myb* and c-*myc* oligonucleotides. However, although the data presented in this report supporting a non-antisense mechanism to explain the effects of the oligonucleotides on SMC proliferation in vitro are very convincing, a number of factors in this study make it difficult to completely discount the possibility that the in vivo effects of these oligonucleotides are the result, at least in part, of an antisense mechanism of action. First, Burgess et al. employed a very different model to examine SMC proliferation in injured arteries; second, the pluronic gel formulation was not employed and thus the possibility remains that this formulation provides a means of optimally transfecting the oligonucleotides into arterial SMC, whereby they can inhibit expression of their target mRNA through an antisense mechanism. The authors also failed to explain how a contiguous stretch of guanosines could inhibit c-*myb* and c-*myc* mRNA levels through non-antisense mechanisms. Nevertheless, non-antisense pharmacological effects of contiguous stretches of guanosines have been well documented (ECKER et al. 1993; YASWEN et al. 1993; BENNETT et al. 1994; WYATT et al. 1994) and the possibility therefore remains that the in vivo effects of these oligonucleotides in animal restenosis models are the result of sequence-specific non-antisense mechanisms, a true antisense effect, or a combination of mechanisms.

Cell cycle-regulatory proteins have also received a great deal of attention in the evaluation of antisense approaches for the treatment of restenosis. In particular, phosphorothioate antisense inhibitors of cell division cycle 2 (cdc2) kinase, cyclin-dependent kinase 2 (cdk2) kinase, PCNA, and cyclin B_1 have been shown to be effective inhibitors of neointimal thickening in rat restenosis models following a single application (MORISHITA et al. 1993, 1994a,b, 1995; ABE et al. 1994).

ABE et al. (1994) employed the pluronic gel formulation that was originally reported by SIMONS et al. (1992) to examine the efficacy of 18-mer phosphorothioate antisense oligonucleotides targeted to cdc2 kinase and cdk2 kinase (complementary to codons −3 to +3) in preventing SMC accumulation and neointimal thickening in rat carotid arteries following balloon angioplasty. Again, a single application of oligonucleotide around the injured artery was tested for efficacy. Both of the antisense cdc2 and cdk2 oligonucleotides were effective in inhibiting cdc2 and cdk2 enzyme activity in a target-specific manner. Moreover, both antisense oligonucleotides substantially reduced neointimal formation following injury, whereas application of sense control phosphorothioate oligonucleotides did not affect target enzyme activity or neointimal formation. Furthermore, neither the cdc2 or cdk2 antisense sequences contained contiguous stretches of guanosines.

SIMONS et al. (1994) also reported positive results in the rat restenosis model using the pluronic gel approach with phosphorothioate antisense inhibitors targeted against rat PCNA mRNA sequences +4 to +21 and +22 to +39 (SIMONS et al. 1994). Similar to the findings reported above for c-*myb*, c-*myc*, cdc2 kinase, and cdk2 kinase, short-term extraluminal delivery of antisense oligonucleotides immediately following arterial injury produced significant suppression of target (PCNA) mRNA levels, a significant decrease in the frequency of proliferating medial SMC, and a significant decrease in neointimal thickening for up to 2 weeks following injury. Again, these sequences were devoid of contiguous stretches of guanosine, and control oligonucleotides (both sense and scrambled) had no effect on the biological end points mentioned above.

Taking a different approach, MORISHITA et al. (1993) reported the development of a strategy utilizing an intraluminal molecular delivery method that employs the protein coat of a Sendai virus complexed with liposomes (hemagglutinating virus of Japan or HVJ liposomes); this method apparently enhances tissue penetration by oligonucleotides. Employing this strategy, administration of a combination of phosphorothioate antisense oligonucleotides targeted to cdc2 kinase mRNA (mouse sequence, positions −9 to +9) and PCNA mRNA (rat sequence, positions +4 to +22) resulted in a marked decrease in cdc2 and PCNA mRNA levels in artery tissue and completely inhibited neointimal formation in a dose-dependent manner. Moreover, the inhibitory effect of the antisense treatment on neointimal formation persisted for up to 8 weeks after a single administration. Further support for an antisense mechanism of action was provided by the fact that treatment with sense control oligonucleotides or the PCNA antisense oligonucleotide alone had no effect on neointimal formation. Active oligonucleotides were also devoid of contiguous stretches of guanosine.

In a follow-up set of studies in which the same delivery method was employed (MORISHITA et al. 1994a,b), antisense phosphorothioate oligonucleotides targeted against cdc2 kinase, cdk2 kinase, and cyclin B_1 were evaluated for activity in the rat balloon injury model. Again, neointimal formation was inhibited in a potent and dose-dependent manner by application of antisense combinations (cdc2–cdk2 and cdc2–cyclin B1), and these combinations were more effective than administration of any single oligonucleotide alone. Oligonucleotide administration also resulted in reduced target mRNA levels, and sense control oligonucleotides were without effect. Furthermore, using fluorescein-conjugated (FITC) oligonucleotides, it was shown that oligonucleotides administered by the HVJ liposome method localized to the medial layer of the injured artery and accumulated within nuclei, and this localization pattern persisted for up to 2 weeks following a single application. Administration of unformulated FITC-labeled oligonucleotide also resulted in localization to the medial layer, but no nuclear accumulation was observed and fluorescence disappeared within 1 day following administration. Accordingly, sustained localization of oligonucleotide within the tissue using the liposome procedure correlated with sustained activity in preventing neointimal formation.

The pleiotropic transactivator NF-κB has also been targeted with phosphorothioate antisense oligonucleotides for the prevention of neointimal formation in rat carotid arteries following balloon injury (AUTIERI et al. 1995). In this study, antisense oligonucleotides targeted to the p65 subunit of NF-κB (positions –3 to +18) were administered directly to the injured artery following angioplasty by employing a polyethylene cuff around the artery to deliver oligonucleotides chronically (50 µg/day for 14 days) and periadventially with osmotic minipumps. Administration of NF-κB antisense oligonucleotides was found to significantly inhibit neointimal formation 14 days following balloon injury, whereas administration of a sense control oligonucleotide had no significant effect. However, the mechanism of action for this oligonucleotide must be viewed with caution, since it does contain a contiguous stretch of guanosines and target gene expression was not examined in vivo. Nevertheless, this is an important study, because it is the first and only report in which an oligonucleotide was found to have pharmacological activity in a rat restenosis model in which a pluronic gel or a liposome formulation was not employed for oligonucleotide delivery.

II. Atherosclerosis

Atherosclerosis is characterized by the formation of intimal fibrofatty lesions called atherosclerotic plaques, which narrow the vascular lumen and are associated with degenerative changes in media and adventitia. Some of these plaques are prone to undergo further complications that worsen the luminal narrowing or cause total occlusion. The most common and serious consequences of atherosclerotic plaque deposition in humans are found when associated with coronary arteries or arteries supplying the brain or when contained within the aorta, common manifestations of which are myocardial infarction, stroke, and aneurysm. Although much remains to be learned about the molecular mechanisms underlying atherosclerosis in humans, this disease has clearly been associated with certain risk factors, including hypercholesterolemia, hypertension, cigarette smoking, and diabetes mellitus.

Surprisingly, few reports have been published in which antisense oligonucleotides were employed to investigate the underlying mechanisms of atherosclerosis or for the treatment of the disease in animal models. MERRILEES and SCOTT (1994) reported on the use of phosphorothioate antisense oligonucleotides designed against transforming growth factor (TGF)-β1 to elucidate the role of this cytokine, secreted from endothelial cells, in the stimulation of proteoglycan synthesis in adjacent SMC. Since atherosclerotic lesions are characterized by increased amounts of chondroitin sulfate-containing proteoglycans, inhibitors of proteoglycan synthesis may be of value in the prevention of atherosclerosis. In this study, a rat carotid artery organ culture model was employed to demonstrate that TBF-β1 is upregulated in concert with proteoglycan synthesis. Moreover, application of a 20-mer phosphorothioate TBF-β1 antisense oligonucleotide (positions +1 to +20) to endothelial cells in culture blocked TBF-β1 production in vitro and reduced proteoglycan biosynthesis in both endothelial cells and SMC in the artery organ culture model in a dose-dependent manner after 24 h of treatment. Sense and scrambled phosphorothioate oligonucleotide controls displayed minimal effects on proteoglycan biosynthesis in the organ culture model. Although this report employed an ex vivo model and additional experiments are needed to prove an antisense mechanism, the results are very encouraging and suggest that TBF-β1 may be an excellent molecular target for the prevention of atherosclerotic plaque formation.

In another report in which antisense oligonucleotides were employed to prevent the formation of atherosclerotic plaques, MANN et al. (1995) sought to prevent the accelerated atherosclerosis that is associated with vein grafts following surgery for the treatment of occlusive arterial disease. Since this process remains a major limiting factor in the successful treatment of this disease, therapeutic strategies that retard or prevent accelerated atherosclerosis in vein grafts would be of considerable value. In this study, the HVJ liposome procedure (using phosphorothioate antisense oligonucleotides targeted to cdc2 kinase and PCNA) was employed that was previously utilized to prevent neointimal thickening in rat restenosis models (MORISHITA et al. 1993, 1994a,b). The system employed in this study was a diet-induced atherosclerotic rabbit vein graft model. The results of the study demonstrated that a single intraoperative application of PCNA and cdc2 kinase antisense oligonucleotides in combination prevents neointimal hyperplasia in response to the acute injury of surgery and to chronic hemodynamic stress and that this response dramatically inhibits accelerated vein graft atherosclerosis. The authors also demonstrated that the antisense oligonucleotides inhibited PCNA and cdc2 kinase gene expression in vivo, whereas control (sense and scrambled) oligonucleotides had no effect on either target gene expression or accelerated atherosclerosis.

In a more recent study, SUGANO and MAKINO (1996) reasoned that the inhibition of plasma cholesteryl ester transfer protein (CETP), an enzyme that facilitates the transfer of cholesteryl ester from high-density lipoprotein (HDL) to apoB-containing lipoproteins and regulates low-density lipoprotein (LDL) metabolism, should result a reduction in plasma LDL levels and cholesterol levels while increasing HDL levels. Since a high LDL to HDL ratio is strongly correlated with the development of atherosclerosis, insights into the mechanisms that regulate cholesterol metabolism and therapeutic approaches to modulate this process would be of great value. Since the liver is a major tissue responsible for the production of CETP, the authors postulated that a reduction in CETP levels in liver by antisense oligonucleotides may cause a reduction in plasma LDL cholesterol concentrations. To test this, a phosphodiester antisense oligonucleotide was employed that targets positions +148 to +168 of the rabbit CETP mRNA sequence. This oligonucleotide was conjugated to asialglycoprotein-poly-L-lysine, administered systemically to cholesterol-fed rabbits in a single dose (0.03 mg/kg), and monitored over time. The purpose of oligonucleotide conjugation was that the asialglycoprotein-poly-L-lysine moiety enhances uptake of oligonucleotides in liver and protects oligonucleotides from degradation in plasma (CHIOU et al. 1994; LU et al. 1994). Interestingly, administration of the antisense oligonucleotide complex resulted in a time-dependent reduction in liver CETP enzyme activity and mRNA levels. Moreover, this inhibition correlated well with a reduction in total plasma cholesterol levels and an increase in plasma HDL cholesterol levels. A sense control oligonucleotide (asialglycoprotein-poly-L-lysine conjugated) displayed no significant effects on these biological end points. These results suggest that inhibition of liver CETP by antisense oligonucleotides may be beneficial for reducing plasma levels of LDL cholesterol and increasing plasma levels of HDL cholesterol, possibly by enhancing LDL catabolism and decreasing the transfer of cholesteryl ester from HDL to apoB-containing lipoproteins.

III. Hypertension

Hypertension has been identified as the single most important risk factor in vascular disease. Approximately 90% of hypertension is idiopathic and apparently primary (essential hypertension). Of the remaining 10%, most is secondary to renal disease. Although the cause of most cases of hypertension is unknown, the renin–angiotensin system (RAS) is known to play a major role in regulating normal blood pressure through the elaboration of renin and the subsequent formation of angiotensinogen II. Despite the fact that the kidney

RAS has been clearly shown to play an extremely important role in blood pressure regulation, evidence has accumulated supporting the existence of a central (CNS) RAS that functions separately from the classic RAS. In fact, an overactive brain RAS has been implicated in the development and maintenance of blood pressure in the spontaneously hypertensive rat (SHR), the animal model of essential hypertension (BERECK et al. 1987; PHILLIPS and KIMURA 1988).

To investigate the role of the central RAS in the SHR model and to determine the potential of antisense approaches for the treatment of primary hypertension, GYURKO et al. (1993) designed phosphodiester antisense oligonucleotides against angiotensinogen mRNA (positions −5 to +13) and examined their effects on angiotensinogen mRNA levels and various blood pressure parameters following a single central injection. Administration of the antisense compound was found to cause a transient decrease in angiotensinogen mRNA, and this effect correlated with a transient drop in blood pressure such that hypertensive animals became normotensive. Sense oligonucleotides had no effect on either of these biological end points. The transient nature of the pharmacological response was attributed to the sensitivity of phosphodiester oligonucleotides to nuclease degradation. To test this hypothesis, WIELBO et al. (1995) reported in a follow-up study experiments in which phosphorothioate antisense oligonucleotides targeted against angiotensinogen mRNA were examined in the SHR model. As was found in their earlier study with phosphodiester oligonucleotides, administration of the phosphorothioate antisense oligonucleotides resulted in a reduction in blood pressure from hypertensive to normotensive levels. Moreover, the maximal decrease in mean arterial pressure (MAP) was greater and the duration of action was far longer for the phosphorothioate antisense in comparison with phosphodiesters. Antisense treatment also resulted in reduced angiotensinogen in the brain stem and hypothalamus, whereas sense control phosphorothioates had no effect on either angiotensinogen expression or MAP. The authors also examined the effects of peripheral administration of the antisense inhibitors and, consistent with the inability of phosphorothioate oligonucleotides to penetrate the blood–brain barrier, no effects on brain angiotensinogen levels or on MAP were observed. These studies provide strong evidence for a centrally acting RAS in the regulation of blood pressure and suggest that antisense approaches may be of considerable value as a therapeutic approach for the control of hypertension.

IV. Ocular Neovascularization

Ocular diseases of neovascularization, including macular degeneration, diabetic retinopathy, retinopathy of prematurity, central vein occlusion, and corneal neovascularization, are among the principle causes of loss of vision worldwide. Loss of vision in these pathological conditions is characterized by the extensive proliferation of new blood vessels in the retina. Current therapies aimed at controlling this aberrant angiogenesis are only partially effective and are often destructive to the retina.

In a series of studies reported by Lois E. Smith and colleagues (AIELLO et al. 1995; SMITH et al. 1995; ROBINSON et al. 1996), the role of vascular endothelial growth factor (VEGF) in a mouse retinopathy of prematurity model was examined by employing phosphorothioate antisense oligonucleotides. In vitro selection of a series of antisense oligonucleotides was carried out to identify the most potent inhibitor of VEGF mRNA and protein expression. The oligonucleotides determined to be most active were targeted to positions +648 to +668 and +37 to +56 of the murine VEGF sequence. Intravitreal injection of VEGF antisense oligonucleotides in neonatal mice prior to the onset of proliferative retinopathy resulted in a statistically significant reduction in retinal VEGF levels and in the level of retinal neovascularization. Administration of sense control phosphorothioate oligonucleotides elicited no significant effects on either biological end point. However, as pointed out by the investigators, inhibition of neovascularization was incomplete, implying that other angiogenic factors may be involved in retinal neovascularization in this model. Nevertheless, the results of this study are very encouraging and demonstrate the value of antisense as a method to delineate the role of specific gene products in the etiology of disease.

C. Skin

Uptake and internalization of phosphorothioate oligodeoxynucleotides has been demonstrated in keratinocytes both in vitro and in vivo (NESTLE et al. 1994; CHRISTOFIDOU-SOLOMIDOU et al. 1997). In human skin xenografted onto the backs of severely compromised immunodeficient (SCID) mice, intradermally injected oligonucleotides become localized throughout the lower epidermis and superficial dermis (CHRISTOFIDOU-SOLOMIDOU et al. 1997). These findings suggest that the intradermal route of oligonucleotide administration may be applicable for inhibiting the expression of genes normally expressed in skin and may be useful to treat various skin disorders. In spite of this, relatively few studies have been conducted on the ability of phosphorothioate oligodeoxynucleotides to inhibit gene expression in skin.

Of those studies published, most have focused on targeting proteins involved in mediating signal transduction. GILLARDON and colleagues (1994) have investigated ultraviolet (UV) effects on transcription factor expression in mouse skin. A number of components of the AP-1 transcription factor complex (c-*fos*, *fos*B, c-*jun*, and *jun*B) are upregulated by UV treatment, and the authors found that superfusion of rat skin with a c-*fos* targeting antisense oligonucleotide was capable of inhibiting the upregulation of c-*fos* expression by UV irradiation. A control oligonucleotide had no effect on c-*fos* upregulation. Oligonucleotides were applied to the skin at a dose of 500 μM for 3 h prior to irradiation, suggesting that some compound is able to penetrate through the upper stratified cornified epidermis, and the oligonucleotides used were 17-mer phosphodiester molecules with two phosphorothioate linkages at each of the 5′ prime and 3′ prime ends. These additional phosphorothioate modifications were included to increase the nuclease resistance of the oligonucleotides; however, the phosphodiester linkages present at the center of the oligonucleotide are likely to be fairly rapidly degraded by endonucleases (MONIA et al. 1996). The authors also did not demonstrate the ability of their c-*fos* oligonucleotide to inhibit c-*fos* expression in tissue culture. The oligonucleotide is targeted to the AUG initiation of protein synthesis codon, a target sequence which does not always yield active oligonucleotides (DEAN et al. 1996). It is therefore likely that, although some activity was clearly seen with this oligonucleotide in this animal model, the activity observed may be greatly improved by optimization of both oligonucleotide chemistry and sequence. In a related subsequent study (GILLARDON et al. 1995), the same oligonucleotide was found to suppress UV activation of PCNA, presumably by inhibiting the c-*fos*-mediated upregulation of genes necessary for cell proliferation. In this latter study, the authors also examined the ability of their oligonucleotide to penetrate into skin after topical application. Using fluorescein-labeled compounds, they found that oligonucleotide was localized in the nucleus of skin keratinocytes, consistent with the observed inhibition of c-*jun* expression in these cells.

The ability of a 25-mer phosphorothioate oligonucleotide targeting tumor growth factor beta-1 (TGF-β1) to control scarring in adult mouse skin after wounding has also been investigated (CHOI et al. 1996). In this study, the authors found that scarring could be inhibited by their active compound, but not by a sense control oligonucleotide, when oligonucleotides were administered topically at a dose of 100 μg. The expression of TGF-β1 mRNA was also determined in skin and was found to be suppressed by the antisense compound, but not the sense control. Although this was a well-controlled study, the conclusions should be viewed with some caution, since the oligonucleotide sequence used contains a stretch of four contiguous guanine residues. Therefore, confirming that additional TGF-β1-targeting oligonucleotides without four contiguous guanine residues are capable of inhibiting wound healing would help validate that these compounds are functioning through a true antisense mechanism of action and that TGF-β1 plays a role in this process.

D. Peripheral Organs

The pharmacokinetics of oligonucleotides has been well documented (see Chap. 6, this volume), and phosphorothioate oligonucleotides following systemic administration are found to accumulate in all peripheral organs, the major organs of deposition being liver and kidney. Notwithstanding these findings, relatively few in vivo studies on the pharmacological activities of phosphorothioate oligodeoxynucleotides in these organs have been reported.

I. Liver

In an early study, we examined whether systemic administration of a phosphorothioate oligonucleotide could reduce expression of a constitutively expressed gene in liver. The target investigated was protein kinase C (PKC)-α, a gene whose mRNA is expressed in all major organs of the body. We initially identified a 20-mer phosphorothioate oligonucleotide sequence (ISIS 4189), which potently and specifically was able to inhibit the expression of both PKC-α mRNA and protein in tissue culture cells (DEAN and MCKAY 1994). This oligonucleotide was found to induce the degradation of PKC-α mRNA through an RNase-H-mediated mechanism (DEAN and MCKAY 1994). When ISIS 4189 was administered intraperitoneally in mice, we found that the compound induced a dose-dependent, oligonucleotide sequence-dependent reduction of PKC-α mRNA expression in liver. The IC_{50} for this effect was 30–50 mg/kg body weight. A reduction in PKC-α mRNA expression of 64 ± 11% was achieved after a single 50 mg/kg dose. The expression of other PKC family members (PKC-δ, -ε, and -ζ) was unaffected by this treatment. In addition, the ability of the oligonucleotide to inhibit PKC-α mRNA expression did not require the presence of cationic liposomes, although in vitro the oligonucleotide required cationic liposomes for activity. This study demonstrated for the first time the utility of phosphorothioate oligonucleotides as specific inhibitors of gene expression in vivo after systemic administration.

DESJARDINS and IVERSEN (1995) have used a phosphorothioate oligonucleotide to inhibit expression of cytochrome P450 3A2 (CYP3A2) in rat liver after intraperitoneal administration. A 22-mer compound designed to hybridize to the translation start site of the CYP32A mRNA was administered at doses of 0.25, 0.5, and 1.0 mg per animal (equivalent to approximately 1.25, 2.5, and 5.0 mg/kg body weight). The authors were able to demonstrate a specific reduction in both protein expression of CYP3A2 and the activity of liver microsomal erythromycin demethylase, a marker for CYP3A2 activity. The authors conclude by suggesting that phosphorothioate oligonucleotide-mediated inhibition of CYP3A2 activity may be valuable as a mechanism to modulate drug pharmacokinetics in patients. In a second series of studies, the same authors (DESJARDINS et al. 1995) investigated the ability of both a phosphorothioate oligonucleotide and a cholesterol-conjugated phosphorothioate oligonucleotide targeted against another cytochrome P450, CYP2B1. Compounds were administered intraperitoneally at doses up to approximately 5 mg/kg and 0.5 mg/kg, respectively. CYP2B1 expression was measured indirectly by determining pentoxyresorufin O-dealkylase activity. Rather surprisingly, when compared to the previous study, no effect of the phosphorothioate oligonucleotide was found at this dose. In contrast, the cholesterol-conjugated phosphorothioate oligonucleotide inhibited pentoxyresorufin O-dealkylase activity at a dose of approximately 0.5 mg/kg.

Scrambled control compounds had no effect. Although intriguing, interpretation of these results would benefit from a direct measurement of CYP2B1 expression, either by northern or western blotting.

II. Kidney

To our knowledge, only a single study has been reported showing an antisense-mediated reduction in expression of a targeted gene in kidney (NOIRI et al. 1996). The authors administered phosphorothioate oligonucleotides targeting inducible nitric oxide synthase (INOS) intracardially at a dose of 1 mg/kg and examined their effect on ischemia in rat kidneys. Active oligonucleotides were 20-mers designed to hybridize to open reading frame sequences present in both mouse and rat INOS isoform sequences. Several control (sense and scrambled) oligonucleotides were also employed. Initial kinetic studies using a 3'-biotinylated oligonucleotide found a maximal accumulation of oligonucleotide in kidneys 4–8 h after administration. Ischemia was therefore induced 8 h after the oligonucleotides were given. The oligonucleotides targeting INOS gave a dramatic functional protection of kidney from acute ischemia; in contrast, the control oligonucleotides gave no protection. This protection was accompanied by a virtual abolition in the usual upregulation in INOS protein expression in this tissue. The authors conclude that these data provide direct evidence for the cytotoxic effects of INOS in the course of ischemic acute renal failure and offer a novel method to selectively prevent the induction of this enzyme.

III. Lung

Antisense oligonucleotides targeting the adenosine A_1 receptor have recently been shown to be effective inhibitors of asthma in a rabbit model of the disease (NYCE and METZGER 1997). Asthma is an inflammatory disease characterized by bronchial hyperresponsiveness which may be mediated by adenosine in asthmatic individuals. The oligonucleotide used was a 21-mer phosphorothioate oligodeoxynucleotide designed to hybridize to the initiation of protein synthesis codon of both the human and rabbit adenosine A_1 receptor. Various control oligonucleotides were also employed. Oligonucleotides were administered via aerosols directly into the lungs through an intratracheal tube. The adenosine A_1 receptor-targeting oligonucleotides were able to inhibit allergen-induced airway obstruction and bronchial hyperresponsiveness in the rabbits in a sequence-specific manner. Supporting the hypothesis that these pharmacologic effects were mediated by an antisense mechanism of action, adenosine A_1 receptor binding sites in airway smooth muscle tissue were decreased by oligonucleotide treatment (NYCE and METZGER 1997). Furthermore, an oligonucleotide inhibitor of bradykinin B_2 receptor expression had no effect on either the induction of asthma or on adenosine A_1 receptor expression. These results indicate that the delivery of oligonucleotides directly to the lung may be of value therapeutically for the treatment of diseases of this tissue.

E. Conclusions and Future Directions

The wealth of data supporting the conclusion that antisense technology is a valuable approach for delineating the roles of specific gene products and as a mode of therapy in animal models of disease is overwhelming. As summarized in this review, as well as in numerous other reviews within this volume, most of the studies that have been reported investigating the pharmacology of antisense oligonucleotides in animal models have been well controlled and strongly support an antisense mechanism as the basis of their findings. Certainly, some issues concerning the mechanism of action remain to be clarified in some models (e.g., the role of contiguous stretches of guanosines in certain restenosis models), but these effects will most certainly be sorted out in future studies. What is needed for the future is a much more thorough and systematic examination of antisense inhibitors in a wide variety of animal models using inhibitors that are designed against both previously investigated and novel molecular targets. Furthermore, the wealth of new pharmacokinetic information on oligonucleotides in vivo must be exploited to take full advantage of antisense technology in animal disease models. Finally, all of the in vivo studies reported to date with antisense oligonucleotides in animal models of disease have essentially been limited to the application of phosphorothioate oligonucleotides. Since many novel oligonucleotide chemistries have been well documented to possess a variety of attractive and potentially beneficial pharmacological and toxicological properties (as indicated in other reviews in this volume), determining the potential of antisense approaches in animal disease models, and ultimately in humans, will require thorough examination of these novel chemistries.

References

Abe J-I, Zhou W, Taguchi J-I, Takuwa N, Miki K, Okazaki H, Kurokawa K, Kumada M, Takuwa Y (1994) Suppression of neointimal smooth muscle cell accumulation in vivo by antisense cdc2 and cdk2 oligonucleotides in rat carotid artery. Biochem Biophys Res Comm 198(1):16–24

Aiello LP, Pierce EA, Robinson GS, Ferrara N, Smith LEH (1995a) Vascular endothelial growth factor (VEGF) antagonists inhibit retinal neovascularization. Diabetes 44(1):53

Aiello LP, Pierce EA, Robinson GS, Ferrara N, Smith LEH (1995b) Vascular endothelial growth factor (VEGF) antagonists inhibit retinal neovascularization. Diabetes 44(1):53

Autieri MV, Yue T-L, Ferstein GZ, Ohlstein E (1995) Antisense oligonucleotides to the P65 subunit of NF-kB inhibit human vascular smooth muscle cell adherence and proliferation and prevent neointima formation in rat carotid arteries. Biochem Biophys Res Commun 213(3):827–836

Bennett MR, Schwartz SM (1995) Antisense therapy for angioplasty restenosis. Circulation 92(7):1981–1993

Bennett CF, Chiang M-Y, Wilson-Lingardo L, Wyatt JR (1994) Sequence specific inhibition of human type II phospholipase A_2 enzyme activity by phosphorothioate oligonucleotides. Nucleic Acids Res 22(15): 3202–3209

Bennett CF, Dean N, Ecker DJ, Monia BP (1995) Pharmacology of antisense therapeutic agents: cancer and inflammation. In: Agrawal S, Totowa N (eds) Methods in molecular medicine: antisense therapeutic. Humana, Clifton

Bereck KH, Kirk KA, Nagahama S, Oparil S (1987) Sympathetic function in spontaneously hypertensive rats after chronic administration of captopril. Am J Physiol 252: H796–H806

Burgess TL, Fisher EF, Ross SL, Bready JV, Qian Y-X, Bayewitch LA, Cohen AM, Herrera CJ, Hu SS-F, Kramer TB, Lott FD, Martin FH, Pierce GF, Simonet L, Farrell CL (1995) The antiproliferative activity of c-*myb* and c-*myc* antisense oligonucleotides in smooth muscle cells is caused by a nonantisense mechanism. Proc Natl Acad Sci U S A 92:4051–4055

Chaikof EL, Caban R, Yan G-N, Rao GN, Runge MS (1995) Growth-related responses in arterial smooth muscle cells are arrested by thrombin receptor antisense sequences. J Biol Chem 270:7481–7486

Chiou HC, Tangco MV, Levine SM, Robertson D, Kormis K, Wu CH, Wu GY (1994) Enhanced resistance to nuclease degradation of nucleic acids complexed to asialoglycoprotein-polylysine carriers. Nucleic Acids Res 22(24):5439–5446

Choi B-M, Kwak H-J, Jun C-D, Park S-D, Kim K-Y, Kim H-R, Chung H-T (1996) Control of scarring adult wounds using antisense transforming growth factor-β1 oligodeoxynucleotides. Immunol Cell Biol 74:144–150

Christofidou-Solomidou M, Albelda SM, Bennett CF, Murphy GF (1997) Experimental production and modulation of human cytotoxic dermatitis in human-murine chimeras. Am J Pathol 150(2):631–639

Cioffi CL, Garay M, Johnston JF, McGraw K, Boggs RT, Hreniuk D, Monia BP (1997) Selective inhibition of A-raf and C-raf mRNA expression by antisense oligodeoxynucleotides in rat vascular smooth muscle cells: role of A-raf and C-raf in serum-induced proliferation. Mol Pharmacol 51:383–389

Crooke ST (1992) Therapeutic applications of oligonucleotides. Annu Rev Pharmacol Toxicol 32:329–76

Crooke ST, Bennett CF (1996) Progress in antisense oligonucleotide therapeutics. Annu Rev Pharmacol. Toxicol 36:107–29

Dean NM, McKay R. (1994) Inhibition of protein kinase C-α expression in mice after systemic administration of phosphorothioate antisense oligodeoxynucleotides. Proc Natl Acad Sci 91:11762–11766

Delafontaine P, Meng XP, Ku L, Du J (1995) Regulation of vascular smooth muscle cell insulin-like growth factor 1 receptors by phosphorothioate oligonucleotides. Effects on cell growth and evidence that sense targeting at the ATG site increases receptor expression. J Biol Chem 270:14383–14388

Desjardins JP, Iversen PL (1995) Inhibition of rat cytochrome P450 3A2 by an antisense phosphorothioate oligodeoxynucleotide in vivo. J Pharmacol Exp Ther 275:1608–1613

Desjardins J, Mata J, Brown T, Graham D, Zon G, Iversen P (1995) Cholesterylconjugated phosphorothioate oligodeoxynucleotides modulate CYP2B1 expression in vivo. J Drug Target 2:477–485

Ecker DJ, Vickers TA, Hanecak R, Driver F, Anderson K (1993) Rational screening of oligonucleotide combinatorial libraries for drug discovery. Nucleic Acids Res 21(8):1853–1856

Epstein SE, Speir E, Finkel T (1993) Do antisense approaches to the problem of restenosis make sense? Circulation 88:1351–1353

Field AK, Goodchild J (1995) Antisense oligonucleotides: rational drug design for genetic pharmacology. Exp Opin Invest Drugs 4(9):799–821

Gillardon F, Eschenfelder C, Uhlmann E, Hartschuh W, Zimmermann M (1994) Differential regulation of c-*fos*, *fos*B, c-*jun*, *jun*B, bcl-2 and *bax* expression in rat skin following single or chronic ultraviolet irradiation and in vivo modulation by antisense oligodeoxynucleotide superfusion. Oncogene 9:3219–3225

Gillardon F, Moll I, Uhlmann E (1995) Inhibition of c-*fos* expression in the UV-irradiated epidermis by topical application of antisense oligodeoxynucleotides suppresses activation of proliferating cell nuclear antigen. Carcinogenesis 16(8):1853–1856

Gyurko R, Wielbo D, Phillips MI (1993) Antisense inhibition of AT_1 receptor mRNA and angiotensinogen mRNA in the brain of spontaneously hypertensive rats reduces hypertension of neurogenic origin. Regul Pept 49:167–174

Lu X, Fischman AJ, Jyawook SL, Hendricks K, Tompkins RG, Yarmush ML (1994) Antisense DNA delivery in vivo: liver targeting by receptor-mediated uptake J Nucl Med 35:269–275

Mann MJ, Gibbons GH, Kernoff RS, Diet FP, Tsao PS, Cooke JP, Kaneda Y, Dzau VJ (1995) Genetic engineering of vein grafts resistant to atherosclerosis. Proc Natl Acad Sci U S A 92:4502–4506

Merrilees MJ, Scott L (1994) Antisense S-oligonucleotide against transforming growth factor-β_1 inhibits proteoglycan synthesis in arterial wall. J Vasc Res 31:322–329

Monia BP, Johnston JF, Sasmor H, Cummins LL (1996) Nuclease resistance and antisense activity of modified oligonucleotides targeted to ha-*ras*. J Biol Chem 271(24):14533–14540

Morishita R, Gibbons GH, Ellison KE, Nakajima M, Zhang L, Kaneda Y, Ogihara T, Dzau VJ (1993) Single intraluminal delivery of antisense cdc2 kinase and proliferating-cell nuclear antigen oligonucleotides results in chronic inhibition of neointimal hyperplasia. Proc Natl Acad Sci U S A 90:8474–8478

Morishita R, Gibbons GH, Ellison KE, Nakajima M, Leyen HVD, Zhang L, Kaneda Y, Ogihara T, Dzau VJ (1994a) Intimal hyperplasia after vascular injury is inhibited by antisense cdk2 kinase oligonucleotides. J Clin Invest 93:1458–1464

Morishita R, Gibbons GH, Kaneda Y, Ogihara T, Dzau VJ (1994b) Pharmacokinetics of antisense oligodeoxyribonucleotides (cyclin B_1 and CDC2 kinase) in the vessel wall in vivo: enhanced therapeutic utility for restenosis by HVJ-liposome delivery. Gene 149:13–19

Nestle FO, Mitra RS, Bennett CF, Chan H, Nickoloff BJ (1994) Cationic lipid is not required for uptake and selective inhibitory activity of ICAM-1 phosphorothioate antisense oligonucleotides in keratinocytes. J Invest Dermatol 103(4):569–575

Noiri E, Peresieni T, Miller F, Goligorsky MS (1996) In vivo targeting of inducible NO synthase with oligodeoxynucleotides protects rat kidney against ischemia. J Clin Invest 97(10):2377–2383

Nyce JW, Metzger WJ (1997) DNA antisense therapy for asthma in an animal model. Nature 385(20):721

Phillips MI, Kimura BC (1988) Brain angiotensin in the developing spontaneously hypertensive rat. J Hypertens 6:607–612

Pickering G, Weir I, Jekanowski J, Isner JM (1992) Inhibition of proliferation of human vascular smooth muscle cells using antisense oligodeoxynucleotides to PCNA. J Am Coll Cardiol 19:165 (abstr)

Pickering JG, Isner JM, Ford CM, Weir L, Lazarovits A, Rocnik EF, Chow LH (1996) Processing of chimeric antisense oligonucleotides by human vascular smooth muscle cells and human atherosclerotic plaque. Circulation 93(4):772–780

Robinson GS, Pierce EA, Rook SL, Foley E, Webb R, Smith LEH (1996) Oligodeoxynucleotides inhibit retinal neovascularization in a murine model of proliferative retinopathy. Proc Natl Acad Sci U S A 93:4851–4856

Schlessinger J (1994) SH2/SH8 signaling proteins. Curr Opin Genet Dev 4:25–30

Schumacher C, Sharif H, Haston W, Wennogle L, Monia BP, Cioffi CL (1997) Exposure of human vascular smooth muscle cells to *raf*-1 antisense oligodeoxynucleotides: cellular responses and pharmacodynamic implications. Mol Pharmacol (in press)

Shi Y, Hutchinson HG, Hall DJ, Zalewaski A (1993) Downregulation of c-*myc* expression by antisense oligonucleotides inhibits proliferation of human smooth muscle cells. Circulation 88:1190–1195

Shi Y, Fard A, Galeo A, Hutchinson HG, Vermani P, Dodge GR, Hall DJ, Shaheen F, Zalewski A (1994) Transcatheter delivery of c-*myc* antisense oligomers reduces neointimal formation in a porcine model of coronary artery balloon injury. Circulation 90:944–951

Simons M, Rosenberg RD (1992) Antisense nonmuscle myosin heavy chain and c-*myb* oligonucleotides suppress smooth muscle cell proliferation in vitro. Circ Res 70(4):835–843

Simons M, Edelman ER, DeKeyser JL, Langer R, Rosenberg RD (1992) Antisense c-myb oligonucleotides inhibit intimal arterial smooth muscle cell accumulation in vivo. Nature 359:67–70

Simons M, Edelman ER, Rosenberg RD (1994) Antisense proliferating cell nuclear antigen oligonucleotides inhibit intimal hyperplasia in a rat carotid artery injury model. J Clin Invest 93:2351–2356

Smith LEH, Pierce EA, Aiello LP, Foley E, Sullivan R, Rook SL, Robinson GS (1995) Inhibition of proliferative retinopathy using antisense phosphorothioate oligonucleotides against vascular endothelial growth factor (VEGF/VPF). Invest Ophthalmol Vis Sci 36(4):871

Speir E, Epstein SE (1992) Inhibition of smooth muscle cell proliferation by an antisense oligodeoxynucleotide targeting the messenger RNA encoding proliferating cell nuclear antigen. Circulation 86:588–547

Sugano M, Makino N (1996) Changes in plasma lipoprotein cholesterol levels by antisense oligodeoxynucleotides against cholesteryl ester transfer protein in cholesterol-fed rabbits. J Biol Chem 271(32):19080–19083

Wielbo D, Sernia C, Gyurko R, Phillips MI (1995) Antisense inhibition of hypertension in the spontaneously hypertensive rat. Hypertension 25(3):314–319

Wyatt JR, Vickers TA, Roberson JL, Buckheit RW Jr, Klimkait T, DeBaets E, Davis PW, Rayner B, Imbach JL, Ecker DJ (1994) Combinatorially selected guanosine-quartet structure is a potent inhibitor of human immunodeficiency virus envelope-mediated cell fusion. Proc Natl Acad Sci U S A 91(4):1356–1360

Yaswen P, Stampfer MR, Ghosh K, Cohen JS (1993) Effects of sequence of thioated oligonucleotides on cultured human mammary epithelial cells. Antisense Res Dev 3:67–77

CHAPTER 15
Clinical Antiviral Activities

S.L. HUTCHERSON

A. Introduction

Development of new antiviral drugs is needed and challenging. Antiviral drug discovery has focused on viral inhibition strategies principally directed at viral penetration and uncoating or viral genome replication, and the era of human immunodeficiency virus (HIV) infection and acquired immunodeficiency syndrome (AIDS) has helped regenerate interest in antiviral drug discovery and development (Table 1). Discovery of many of the active antivirals has been through random drug screening rather than rational drug design (DE CLERCQ et al. 1983; DE CLERCQ 1991). The cellular toxicity due to lack of viral target specificity and the typically short therapeutic half-lives limit the usefulness of many antiviral drugs, including nucleoside analogues. More pharmacologically selective and specific antiviral drugs are needed to help limit toxicity to the host while inhibiting viral replication.

To theoretically improve therapeutic index and to design antiviral drugs rationally, several laboratories have advanced the initial observations of Zamecnik (ZAMECNIK and STEPHENSON 1978) that short strands of oligodeoxynucleotides can bind targeted mRNA for specific inhibition of gene expression. These advancements led to a number of different approaches to oligonucleotide (ODN) therapeutics, with antisense therapeutics being chief among these.

The principal theoretical advantages of antisense ODN therapeutics are affinity, selectivity, and specificity for its target. MIRABELLI and CROOKE (1993) thoroughly described the potential affinity, selectivity, and specificity for antisense ODN therapeutics. COHEN (1991), AGRAWAL (1992), and COWSERT (1993) reviewed antisense ODN drug discovery strategies and progress as antiviral drugs. Some skeptics demand proof that nuclease-stabilized ODN work by an antisense mechanism (STEIN 1992, 1995, 1996; STEIN et al. 1991; STEIN and CHENG 1993; STEIN and KRIEG 1994), thus delivering the theoretical advantages of this approach to antiviral drug therapy. CROOKE (1996a,b) and CROOKE and BENNETT (1996) clearly and carefully defined the general principles for demonstration of an antisense drug mechanism and answered some of the questions raised by skeptics. ANDERSON et al. (1996b) provided evidence for both a nucleic acid base sequence-specific effect and a sequence-independent effect of a phosphorothioate ODN (sODN) targeted against

Table 1. Antiviral drugs approved in the United States (1997)

Antiviral	Manufacturer	Mechanism	Indication
Amantidine	DuPont	Inhibitor of viral penetration and uncoating	Influenza, chemoprophylaxis
Rimantadine	Forest	Inhibitor of viral penetration and uncoating	Influenza, chemoprophylaxis
Vidarabine	Parke-Davis	Unknown	HSV keratitis
Trifluridine	GlaxoWellcome	Thymidilate synthetase inhibitor	HSV keratitis
Acyclovir	GlaxoWellcome	DNA polymerase inhibitor	HSV keratitis, labialis, genitalis, VZV
Ganciclovir	Roche	DNA polymerase inhibitor	CMV retinitis
Famciclovir	SmithKline	DNA polymerase inhibitor	HSV labialis, genitalis, VZV
Foscarnet	Astra	Pyrophosphate analogue DNA polymerase inhibitor	CMV retinitis
Ribavirin	ICN	Unknown	Broad-spectrum antiviral
Nevirapine	Roxane	Non-nucleoside RT inhibitor	HIV/AIDS
Zidovudine	GlaxoWellcome	Nucleoside RT inhibitor	HIV/AIDS
Lamivudine	GlaxoWellcome	Nucleoside RT inhibitor	HIV/AIDS
Zalcitabine	Roche	Nucleoside RT inhibitor	HIV/AIDS
Didanosine	Bristol Myers Squibb	Nucleoside RT inhibitor	HIV/AIDS
Stavudine	Bristol Myers Squibb	Nucleoside RT inhibitor	HIV/AIDS
Indinavir	Merck	Protease inhibitor	HIV/AIDS
Saquinavir	Roche	Protease inhibitor	HIV/AIDS
Ritonavir	Abbott	Protease inhibitor	HIV/AIDS

RT, reverse transcriptase; HSV, herpes simplex virus; VZV, varicella-zoster virus; CMV, cytomegalovirus; HIV, human immunodeficiency virus; AIDS, acquired immunodeficiency syndrome.

immediate-early gene products essential for cytomegalovirus (CMV) replication, and this drug (fomivirsen sodium, ISIS 2922) is now being evaluated in phase III controlled clinical trials. The strategy and tactics for development of fomivirsen sodium will be reviewed as an example of how one might rationally address these challenges.

B. Antisense Oligonucleotide Antiviral Pharmacology

During drug discovery, scientists struggle with targeted selection of viral replication inhibition opportunities to optimize therapeutic index and minimize selection for viral strains resistant to the antiviral. Knowledge of viral replication mechanisms is enhanced by permissive cell lines that allow in vitro study of viral pathogenesis. The biological function of viral-encoded proteins required for replication can be more thoroughly studied when one can use laboratory-adapted human viruses or clinical isolates in cell culture. New therapeutic targets can then be identified, and with antisense ODN technology nuclease-stabilized ODN can be synthesized quickly and efficiently to test hypotheses about therapeutic inhibition of one or more viral-encoded proteins essential for replication (ZAMECNIK et al. 1986; DE BENEDETTI et al. 1987; AGRAWAL et al. 1988, 1989, 1992; GOODCHILD et al. 1988; SARIN et al. 1988; ZAIA et al. 1988; KULKA et al. 1989; LETSINGER et al. 1989; LEITER et al. 1990; COHEN 1991; HOKE et al. 1991; AGRAWAL 1992; COULSON et al. 1992; AZAD et al. 1993; SANGHVI and COOK 1993; OJWANG et al. 1994; AZAD et al. 1995). Thus much of the early antisense antiviral discovery research focused on human herpesviruses, principally herpes simplex virus HSV-1 and HSV-2 (Table 2; SMITH et al. 1986; GAO et al. 1989, 1990a; KULKA et al. 1989, 1993; GAO et al. 1990b; BRANDT 1991; HOKE et al. 1991; CANTIN et al. 1992; CROOKE et al. 1992a; JACOB et al. 1993; VINOGRADOV et al. 1994; SHOJI et al. 1996).

Focus on viral inhibition mechanisms that are not already being exploited with other antiviral drugs is important so that antiviral agents without cross-resistance will be available for clinical use in patients who are not responsive to other antivirals (DESATNIK et al. 1996; DODDS et al. 1996; JABS et al. 1996). For example, CMV DNA polymerase could be a viable antisense ODN target, but other antivirals (ganciclovir, foscarnet, and cidofovir) already inhibit this viral-encoded enzyme. When CMV mutates to become resistant to one of the DNA polymerase antivirals, all of them become less effective for treatment of CMV disease (DUNN et al. 1995). As viral and clinical resistance becomes more prevalent with DNA polymerase inhibitors, antiviral drugs with novel mechanisms of action become important for development. Fomivirsen sodium is a potent and selective inhibitor of human CMV (AZAD et al. 1993). Its antiviral

Table 2. Antisense phosphorothioate oligonucleotides (sODN) in clinical trials

ODN identification	Targeted virus
ISIS 2105 (afovirsen sodium)	HPV-6 and -11
GEM 91	HIV
ISIS 2922 (fomivirsen sodium)	CMV

HPV, human papillomavirus; HIV, human immunodeficiency virus; CMV, cytomegalovirus.

activity and antisense mechanism of action have been previously described (ANDERSON et al. 1996b). In vitro, fomivirsen inhibits replication of clinical CMV isolates, including those resistant to DNA polymerase inhibitors (ANDERSON et al. 1996a). Traditional approaches to investigative pharmacology suggest antiviral evaluation in animal models prior to clinical trials (HSIUNG and CHAN 1989). In the absence of an animal pharmacology model or with one that is poorly suited to study the activity of a new antiviral drug (virus specificity for host), the drug developer must move forward to assess the pharmacokinetics and toxicology of the new drug without animal pharmacology.

C. Pharmacokinetic Assessment Strategies

Evaluation of intravenously administered ODN is sensible to establish the distribution, metabolism, and elimination of the drug in animals (AGRAWAL et al. 1991, 1995; CROOKE 1993; COSSUM et al. 1994a,b; CROOKE et al. 1994, 1996; IVERSEN et al. 1994; SAIJO et al. 1994; SRINIVASAN and IVERSEN 1995; ZHANG et al. 1995a,b, 1996; AGRAWAL and TEMSAMANI 1996; WALLACE et al. 1996; PHILLIPS et al., in press). Because phosphodiester oligonucleotides are rapidly metabolized in vivo by exo- and endonucleases (CROOKE 1991; CROOKE et al. 1995), chemical modifications were made to stabilize ODN from nuclease degradation. sODN are the most widely studied first-generation ODN chemical modifications, and their pharmacokinetic profile after intravenous administration is similar irrespective of base sequence. These drugs are rapidly distributed to peripheral tissues from the bloodstream, and they concentrate in liver, kidney, spleen, bone marrow, and lung tissues (CROOKE et al. 1996; GEARY et al. 1996). Medicinal chemistry modifications to an ODN or ODN liposomal formulations can alter the distribution pattern and metabolism, allowing an ODN to be effectively delivered to many other tissues (AKHTAR et al. 1991; JULIANO and AKHTAR 1992; BENNETT et al. 1993, 1997; BRIGHAM and SCHREIER 1993; LAPPALAINEN et al. 1994; ZELPHATI et al. 1994; AOKI et al. 1995; BENNETT 1995; LITZINGER et al. 1996; ZELPHATI and SZOKA 1996a,b; ZELPHATI and SZOKA 1996).

Measurement of parent ODN requires skillful application of technology and careful interpretation of results. Prior to the development of an assay for parent ODN in biological tissue following administration of non-radiolabeled or fluorescent-labeled drug, our understanding of absorption, distribution, metabolism, and excretion of ODN was clouded by different interpretation of the results. Metabolism of ODN where radiolabel tags were placed on multiple bases or phosphate linkages resulted in measurement of a mixture of parent ODN and its metabolites (DEWANJEE et al. 1993; CROOKE et al. 1994; PIWNICA-WORMS 1994; SAIJO et al. 1994; SANDS et al. 1994; HUGHES et al. 1995; BENNETT et al. 1996; GLOVER et al. 1997). Fluorescently tagged oligonucleotides were conveniently visible with light microscopy in tissues and cells, but questions concerning alteration in biodistribution of the parent ODN because of the biochemical properties of the fluorescent label were a cause of concern (URDEA et al. 1988; FISHER et al. 1993; PROUDNIKOV and MIRZABEKOV 1996). Capillary gel electrophoretic assay and high-pressure liquid chromatography have provided reasonable and reliable methods for the measurement of parent ODN and their metabolic shorter versions (CUMMINS et al. 1996; LEEDS et al. 1996). Using these assay methods, one can be much more confident of the biological distribution of ODN, so that targeted drug delivery can be better achieved.

Strategically, the drug developer needs to understand how and where the targeted virus replicates to produce pathogenic disease in human tissue so that the route of drug administration or drug delivery methods can be optimized. For example, CMV retinitis is the most common disease produced by CMV infection (Jabs 1995). Viral replication occurs within retinal tissue, with many characteristic viral inclusion bodies observed histopathologically in retinal pigment epithelium. Systemic administration of any drug, including an ODN, for retinal bioavailability is almost always very inefficient, and the patient is systemically and needlessly exposed to the drug. A simple aqueous solution of ODN can be intravitreally injected to achieve effective retinal concentrations of fomivirsen sodium for at least 14 days (LEEDS et al. 1994). Using this rational approach to select the route of administration and drug delivery technology can reduce overall costs of drug development, principally the cost of the ODN (much smaller doses can be used than would be required systemically) and the cost of toxicology studies. If one can demonstrate little to no systemic bioavailability of a locally administered ODN, one can argue that the drug should be evaluated under guidelines for topical rather than systemic administration. One could expect to avoid such costly and time-consuming toxicology studies as carcinogenicity assessment or reproduction/fertility studies.

D. Toxicology for Antisense Oligonucleotide Antivirals

Toxicokinetic studies of any new drug are only useful if they help predict doses, schedules, or routes of administration that will minimize untoward effects in humans. Some viral infections are of short duration with viral latency periods. Others are chronic infections. Some viral infections can be present for extended periods of time prior to development of end organ disease. The intended use of an antiviral ODN will help define the types and duration of toxicology studies that must be completed to enable clinical trials to be initiated and to meet regulatory agency expectations for prescription drug labeling of a drug to be marketed.

Although there are generally accepted guidelines for the types of toxicology studies required prior to clinical trials and prior to market approval, the drug developer can benefit from thoughtful consideration of the intended use of the ODN before starting any toxicology studies (LUSTER et al. 1988, 1992; BLACK et al. 1993, 1994; TAKAYAMA 1993). The route of drug administration in toxicology studies should mimic the intended route for humans. Intravitreally administered fomivirsen sodium for treatment of CMV retinitis allowed toxicologic evaluation according to guidelines for topical agents because of limited systemic bioavailability due to the quantity of drug injected (maximum, 330 µg), the route of administration, and the metabolism of the parent drug in retinal cells. Furthermore, animal models proved to be poor predictors of toxic effects that paralleled the principal adverse experiences observed in clinical trials. Lack of an animal model with both a compromised immune system and retinal tissue with vascular leakage (due to HIV microvasculopathy) meant that one could not study fomivirsen sodium in animal models that would help predict toxicity for humans. This practical and sensible approach to a toxicology development strategy is often missed in drug development.

E. Chemical Synthesis Scale-Up and Chemical Development

Scaling up solid support chemistry to produce sufficient quantity and quality of an ODN for clinical trials and marketing required advancement of methods for synthesis and analytical characterization of the drug (ANDRADE et al. 1994; RAVIKUMAR et al. 1995). Analytical methods for identification and characterization of synthetic ODN provided new challenges to demonstrate to regulatory authorities that these compounds can be consistently manufactured to meet good manufacturing practice (GMP) standards. Methods for sequencing ODN (WYRZYKIEWICZ and COLE 1994), nucleic acid base composition determination (SCHUETTE et al. 1994), and quantitation of the parent drug substance in pharmaceutical formulations (SRIVATSA et al. 1994) required extensive research and development. Improved efficiency of chemical reactions to synthesized sODN provided better yields of fully thioated ODN (CHERUVALLATH et al. 1996). Methods for increasing the purity of ODN have also been developed (KROTZ et al. 1997). All of these chemical development advancements, while essential for development of any new drug, have been critical to the advancement of ODN therapeutics. Continued improvements to oligonucleotide synthesis will result in even lower costs for this new approach to antiviral drug discovery and treatment.

F. Clinical Research of Antisense Oligonucleotides

Table 2 shows the antisense ODN that have advanced to clinical trials, and the most advanced of these is fomivirsen. After successful completion of intravitreal fomivirsen dose and schedule evaluation in AIDS patients with recalcitrant CMV retinitis (R. LIEBERMAN et al., manuscript in preparation), controlled clinical trials began in 1996 and are ongoing. In phase III trials, doses of 150 μg per injection are being evaluated for newly diagnosed CMV retinitis patients, while a higher dose of 330 μg per injection is being studied for patients with recalcitrant CMV retinitis. Fomivirsen has demonstrated clinically acceptable responses for control of CMV retinitis in patients who were not responding to other antivirals (ganciclovir, foscarnet, and cidofovir), and because this sODN is the first to demonstrate therapeutic utility in clinical trials, results from an open-label clinical trial for "salvage therapy" of CMV retinitis are presented. This study accompanies the ongoing well-controlled studies.

CMV retinitis occurs in 25%–40% of AIDS patients. Progressive, necrotizing CMV retinitis can lead to blindness in less than 1 month if not treated with an effective antiviral drug. All antiviral drugs for CMV are virostatic, and the aim of therapy can therefore only be to delay progressive loss of vision due viral replication that causes retinal necrosis. Ganciclovir, foscarnet, and cidofovir all inhibit CMV replication by a common mechanism, i.e., DNA polymerase inhibition presumably leading to viral DNA chain elongation termination.

Patients with active CMV retinitis that was progressing despite current treatment were eligible for fomivirsen sodium intravitreal injections. CMV retinitis diagnosis was based upon characteristic white, necrotic, granular lesions. Lesions were in zone 1 (area of macula and optic disc) or zones 2 or 3 (at least 1500 μm from the margin of the optic disc or 3000 μm from the fovea). Continuation of systemic antiviral agents after initiation of fomivirsen was not prohibited by the protocol, but prior treatment failure of systemic antivirals indicates that these drugs were not effectively controlling CMV retinitis for the patients who entered this open-label study of fomivirsen. After topical application of anesthetic, 0.05 cc fomivirsen was intravitreally injected at a dose of 330 μg once weekly for 3 weeks (induction period). Following induction and control of an advancing CMV retinitis lesion border, fomivirsen dosing was decreased to one dose every 14 days. Patients were clinically evaluated using indirect ophthalmoscopy for CMV retinitis activity and CMV lesion border advancement immediately prior to each dose of fomivirsen. A complete ophthalmic examination and a clinical assessment of treatment side effects were also performed at the same time. In contrast to other clinical studies, in which the efficacy end point was defined as active CMV lesion border advancement (greater than or equal to half a disc diameter or 750 μm), additional criteria were used in this study to assess therapeutic utility of fomivirsen. These functional criteria of vision loss included retinal detachment in an area of active CMV retinitis and CMV optic neuritis with a decrease in best corrected visual acuity to worse than 20/400. Patients were allowed to continue receiving fomivirsen injections and evaluated for antiviral effectiveness and drug safety for as long as they received therapeutic benefit from fomivirsen.

Ninety-two eyes (68 patients) were evaluated. The number of prior drug treatments, including combination treatment regimens, was complex; Table 3

Table 3. Cytomegalovirus (CMV) retinitis treatment history prior to fomivirsen treatment

Other antiviral treatment	Patients with therapeutic failure (%)
intravenous ganciclovir	>95
intravenous foscarnet	64
intravitreal ganciclovir	26
intravitreal foscarnet	37
ganciclovir implant	12
cidofovir	11

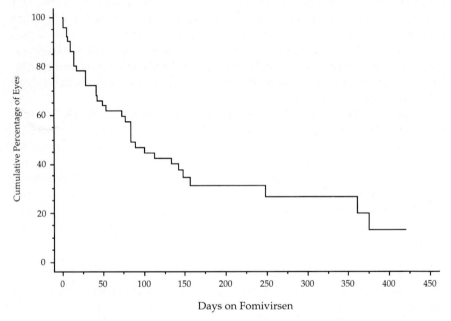

Fig. 1. Survival analysis of eyes that derived therapeutic benefit from fomivirsen treatment (eyes treated with 330 µg fomivirsen sodium, $n = 51$)

illustrates the range of antiviral treatments that these patients received prior to fomivirsen. Ninety-one percent of these patients received multiple antiviral drugs prior to receiving fomivirsen, and all received multiple courses of other drugs (median of two drugs administered prior to fomivirsen treatment).

Figure 1 (Kaplan-Meier plot) illustrates the cumulative percentage of eyes that derived therapeutic benefit from fomivirsen treatment among 40 patients. Patients who were removed from the study for reasons other than advancement of their active CMV retinitis lesion border or an adverse experience that required discontinuation of fomivirsen (including death due to AIDS-related illness) were censored from this analysis at the time of removal from the study without reaching an end point for evaluation of therapeutic utility. This stringent analysis helps assessment of the practical utility of fomivirsen as a treatment for CMV retinitis among patients who either no longer respond to other antivirals or who cannot receive these treatments because of adverse reactions to the agents. Figure 2 provides a more traditional Kaplan-Meier analysis of days to first observation of CMV retinitis lesion border advancement (CMV retinitis progression). Nine patients were removed from the study because of insufficient effectiveness of fomivirsen.

Both analyses demonstrate prolonged and durable response to fomivirsen treatment in eyes that failed to respond to any other antiviral treatments (including ganciclovir implants and the newly available antiviral cidofovir). Because some of the eyes included in these analyses were still receiving fomivirsen at the time of the analyses, calculation of a median time to lesion border advancement would be misleading and is not presented. However, more than half of the patients received therapeutic benefit from fomivirsen for at least 75 days when all other therapies failed. The effectiveness of fomivirsen in this study illustrates the importance of having an antiviral mechanism of action distinctive from ganciclovir, foscarnet, and cidofovir.

It is not surprising that there were no fomivirsen-related, systemic adverse events noted in this study. Given the size of the dose and the intravitreal route

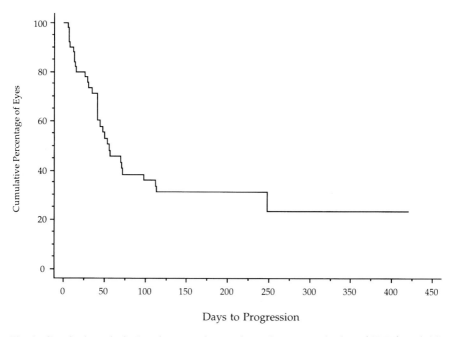

Fig. 2. Survival analysis for time to observation of cytomegalovirus (CMV) retinitis lesion border advancement of at least 750 μm (eyes treated with 330 mg fomivirsen sodium, $n = 92$)

of administration, systemic exposure to parent fomivirsen would be below the limits of detection in plasma. Not unexpectedly, there was no evidence of drug–drug interaction or interference, an important consideration in AIDS patients, who have to take multiple drugs to combat their primary disease.

Table 4 shows the incidence of ocular side effects that exceeded 5%, a commonly used reference mark for clinically important and possibly drug-related side effects. Intravitreally administered fomivirsen appeared to be well tolerated, as evidenced by the relatively small percentage of patients (8.8%) who discontinued treatment because of a fomivirsen-related side effect: three patients with elevated intraocular pressure (IOP), one with retinal edema, one with decreased peripheral vision, and one with epiretinal membrane.

Transiently increased IOP was the most commonly observed side effect, and when hypertony was observed, topical treatment with antiglaucoma drugs was effective in bringing the pressure back within normal limits in most patients. Anterior chamber paracentesis was performed on eyes with transient IOP above 50 mmHg, and following a rapid return of IOP to within normal limits, patients usually resumed their fomivirsen injections. Patients who entered the study with baseline IOP near the upper end of the normal range (approximately 18 mmHg) were the patients who experienced increased IOP while on fomivirsen, suggesting some predisposing, as yet unknown factor that correlated with transient post-treatment hypertony. Anterior chamber inflammation (cells and proteaneous flare) was positively correlated with transient hypertony, suggesting the possibility that aqueous outflow passage through the trabecular network might have been impaired by inflammatory cells and protein.

Table 4. Incidence of ocular side effects observed with fomivirsen in acquired immunodeficiency syndrom (AIDS) patients with advanced cytomegalovirus (CMV) retinitis

Ocular side effect	Incidence (% of patients)
Increased IOP	19
Transient decreased VA	12
Retinal detachment	10
Eye pain	9
Vitritis	9
Retinal edema	9
Anterior chamber inflammation	7
Conjunctival hemorrhage	7
Visual field defect	7
Retinal hemorrhage	6
Retinal pigment epithelium hyperpigmentation	6
Transient desaturation of color vision	6

Patients treated, $n = 68$; eyes dosed, $n = 92$.
IOP, intraocular pressure; VA, visual acuity.

Transient decreased visual acuity (incidence, 12%) can be explained by increased IOP, anterior or posterior chamber inflammation, retinal hemorrhage, or retinal edema.

Intraocular inflammation was also observed. Krieg and colleagues have shown that ODN with CpG base sequences, especially at the terminal ends of the ODN, can produce immunostimulation (KRIEG et al. 1995; BALLAS et al. 1996; KLINMAN et al. 1996). Since fomivirsen contains these sequences at both terminal ends, it is not surprising that local inflammatory reactions were observed. Both anterior chamber inflammation and posterior chamber (vitritis, uveitis) inflammation were noted. Topical steroids or nonsteroidal antiinflammatories helped decrease the anterior chamber inflammation, and in most cases the posterior chamber inflammation subsided with maintenance dosing of fomivirsen over time.

Despite repeated intravitreal injections, the cumulative retinal detachment incidence (10%) among these fomivirsen-treated patients was low in comparison to the cumulative incidence reported for patients with CMV retinitis who have been treated with other therapies (SANDY et al. 1995). This finding underscores the importance of effective antiviral treatment that halts further retinal necrosis and the safety of repeated intravitreal injections. No cases of bacterial endophthalmitis were reported. Both retinal detachment and endophthalmitis are important sight-threatening events that can occur following intraocular surgery or intravitreal injection.

The rationale for intraocular fomivirsen treatment of CMV retinitis and the potential advantages of this antisense ODN are shown below:

- Treats a sight-threatening disease with locally sustained therapeutic concentrations of an effective antiviral
- Lack of systemic bioavailability avoids systemic side effects and drug–drug interactions
- Improves quality of life for patients through avoidance of an indwelling catheter for antiviral administration
- Permits use of systemic antivirals for suppression of CMV viremia with potential for additive effects to intraocular fomivirsen
- Potent anti-CMV activity in vitro (mean ED_{50}, 0.5 μM)
- Novel mechanism of action
- Additive antiviral activity with DNA polymerase inhibitors
- Effective in "clinically resistant" CMV retinitis
- Prolonged therapeutic benefit and utility for patients with no treatment alternatives

G. Future Antisense Antiviral Development

Positive results from well-controlled clinical trials will be necessary for the approval of new drugs, and results from well-controlled studies of fomivirsen sodium are eagerly anticipated. Once these trials have been completed, fomivirsen could be the first antisense ODN to receive regulatory review for marketing approval. Its approval will usher in a new era in antiviral research using a rational drug discovery process that can be optimized for treatment of other viral diseases.

Continued improvement in chemical manufacturing methods will help lower the costs of ODN production and, with that change, medicinal chemists will develop alternative ODN to improve nuclease stability and change the biodistribution of the drugs so that other diseased end organs can be effectively treated with antisense ODN. Pharmaceutics will provide additional options for ODN delivery, e.g., liposomes and liposomal-like vehicles. Oral bioavailability will allow development of new ODN that can be used more conveniently to treat patients on an outpatient basis.

Antisense ODN provide a new and important alternative to nonselective and often more toxic traditional nucleoside antivirals. One can envision new molecular strategies for synergistic antiviral activity where ODN are designed to inhibit viral replication by sequential blockage of proteins essential for viral replication. As a possible result, less frequent clinical resistance and an even better safety profile could be expected as the overall therapeutic index is raised.

References

Agrawal S (1992) Antisense oligonucleotides as antiviral agents. Trends Biotechnol 10:152–158

Agrawal S, Temsamani J (1996) Comparative pharmacokinetics of antisense oligonucleotides. In: Agrawal S (ed) Antisense Therapeutics. Humana, Totowa, NJ, pp 247–270

Agrawal S, Goodchild J, Civeira MP (1988) Oligodeoxynucleoside phosphoramidates and phosphorothioates as inhibitors of human immunodeficiency virus. Biochemistry 85:7079–7083

Agrawal S, Ikeuchi T, Sun D, Sarin PS, Konopka A, Maizel J, Zamecnik PC (1989) Inhibition of human immunodeficiency virus in early infected and chronically infected cells by antisense oligodeoxynucleotides and their phosphorothioate analogues. Proc Natl Acad Sci USA 86:7790–7794

Agrawal S, Temsamani J, Tang JY (1991) Pharmacokinetics, biodistribution, and stability of oligodeoxynucleotide phosphorothioates in mice. Proc Natl Acad Sci USA 88:7595–7599

Agrawal S, Temsamani J, Galbraith W, Tang J (1995) Pharmacokinetics of antisense oligonucleotides. Clin Pharmacokinet 28:7–16

Akhtar S, Basu S, Wickstrom E, Juliano RL (1991) Interactions of antisense DNA oligonucleotide analogs with phospholipid membranes (liposomes). Nucleic Acids Res 19:5551–5559

Anderson KP, Azad RF, Zhang H, Humman J, Miner R, Drew WL (1996a) Growth inhibition of clinical and drug-resistant isolates of human cytomegalovirus by the antisense phosphorothioate oligonucleotide, ISIS 2922. Antiviral Res (submitted)

Anderson KP, Fox MC, Brown-Driver V, Martin MJ, Azad RF (1996b) Inhibition of human cytomegalovirus immediate-early gene expression by an antisense oligonucleotide complementary to immediate-early RNA. Antimicrob Agents Chemother 40(9):2004–2011

Andrade M, Scozzari A, Cole DL, Ravikumar VT (1994) Efficient synthesis of antisense oligodeoxyribonucleotide phosphorothioates. Bioorg Med Chem Lett 4(16):2017–2022

Aoki K, Yoshida T, Sugimura T, Terada M (1995) Liposome-mediated in vivo gene transfer of antisense K-ras construct inhibits pancreatic tumor dissemination in the murine peritoneal cavity. Cancer Res 55:3810–3816

Azad RF, Driver VB, Tanaka K, Crooke RM, Anderson KP (1993) Antiviral activity of a phosphorothioate oligonucleotide complementary to RNA of the human cytomegalovirus major immediate-early region. Antimicrob Agents Chemother 37(9):1945–1954

Azad RF, Brown-Driver VB, Buckheit RW, Anderson KP (1995) Antiviral activity of a phosphorothioate oligonucleotide complementary to human cytomegalovirus RNA when used in combination with antiviral nucleoside analogs. Antiviral Res 28(2):101–111

Ballas ZK, Rasmussen WL, Krieg AM (1996) Induction of NK activity in murine and human cells by CpG motifs in oligodeoxynucleotides and bacterial DNA. J Immunol 157:1840–1845

Bennett CF (1995) Intracellular delivery of oligonucleotides with cationic liposomes. In: Akhtar S (ed) Delivery strategies for antisense oligonucleotide therapeutics. CRC Press, Boca Raton, pp 223–232

Bennett CF, Chiang M-Y, Chan H, Grimm S (1993) Use of cationic lipids to enhance the biological activity of antisense oligonucleotides. J Liposome Res 3:85–102

Bennett CF, Zuckerman JE et al (1996) Pharmacokinetics in mice of a [^{3}H]-labeled phosphorothioate oligonucleotide formulated in the presence and absence of a cationic lipid. J Control Release 41(1–2):121–130

Bennett CF, Mirejovsky D, Crooke RM, Tsai YJ, Felgner J, Sridhar CN, Wheeler CJ, Felgner PL (1997) Structural requirements for cationic lipid mediated phosphorothioate oligonucleotide delivery to cells in culture. J Drug Target (in press)

Black LE, DeGeorge JJ, Cavagnaro JA, Jordan A, Ahn AW (1993) Regulatory considerations for evaluating the pharmacology and toxicology of antisense drugs. Antisense Res Dev 3:399–404

Black LE, Farrelly JG, Cavagnaro JA, Ahn CH, DeGeorge JJ, Taylor AS, Defelice AF, Jordan A (1994) Regulatory considerations of oligonucleotide drugs: updated recommendations for pharmacology and toxicology studies. Antisense Res Dev 4:299–301

Brandt CR (1991) Evaluation of the efficacy of ISIS 1082 in a murine model of herpes simplex virus ocular infection. Internal report. University of Wisconsin-Madison

Brigham KL Schreier H (1993) Cationic liposomes and DNA delivery. J Liposome Res 3:31–49

Cantin EM, Podsakoff G, Willey DE, Openshaw H (1992) Antiviral effects of herpes simplex virus specific anti- sense nucleic acids. Adv Exp Med Biol 312:139–150

Cheruvallath ZS, Cole DL, Ravikumar VT (1996). Sulfurization efficiency in the solution phase synthesis of deoxyribonucleoside phosphorothioates – comparison of sulfur triethylamine with various sulfurizing agents. Nucleosides Nucleotides 15(9):1441–1445

Cohen JS (1991) Antisense oligodeoxynucleotides as antiviral agents. Antiviral Res 16:121–133

Cossum PA, Sasmor H, Dellinger D, Truong L, Cummins L, Owens SR, Markham PM, Shea PJ, Crooke S (1994a) Disposition of the ^{14}C-labeled phosphorothioate oligonucleotide ISIS 2105 after intravenous administration to rats. J Pharmacol Exp Ther 267(3):1181–1190

Cossum PA, Truong L, Owen SR, Markham PM, Shea JP, Crooke ST (1994b) Pharmacokinetics of a ^{14}C-labeled phosphorothioate oligonucleotide, ISIS 2105, after intradermal administration to rats. J Pharmacol Exp Ther 269(1):89–94

Coulson JM, Blake NW, Archard LC, Malcolm AD (1992) Antisense oligodeoxynucleotides as antiviral agents. Biochem Soc Trans 20:321S

Cowsert LM (1993) Antiviral activities of antisense oligonucleotides. In: Crooke ST, Lebleu B (eds) Antisense research and applications. CRC Press, Boca Raton, pp 521–533

Crooke RM (1991) In vitro toxicology and pharmacokinetics of antisense oligonucleotides. Anticancer Drug Des 6:609–646

Crooke RM (1993) Cellular uptake, distribution and metabolism of phosphorothioate, phosphodiester, and methylphosphonate oligonucleotides. In: Crooke ST, Lebleu B (eds) Antisense research and applications. CRC Press, Boca Raton, pp 427–449

Crooke RM, Hoke GD, Shoemaker JE (1992) In vitro toxicological evaluation of ISIS 1082, a phosphorothioate oligonucleotide inhibitor of herpes simplex virus. Antimicrob Agents Chemother 36:527–532

Crooke RM, Graham MJ, Cooke ME, Crooke ST (1995) In vitro pharmacokinetics of phosphorothioate antisense oligonucleotides. J Pharm Exp Ther 275(1):462–473

Crooke ST (1996a) Advances in understanding the pharmacological properties of antisense oligonucleotides. Adv Pharmacol 40:1–49

Crooke ST (1996b) Proof of mechanism of antisense drugs. Antisense Nucleic Acid Drug Dev 6(2):145–147

Crooke ST, Bennett CF (1996) Progress in antisense oligonucleotide therapeutics. Annu Rev Pharmacol Toxicol 36:107–129

Crooke ST, Grillone LR, Tendolkar A, Garrett A, Fratkin MJ, Leeds J, Barr WH (1994) A pharmacokinetic evaluation of ^{14}C-labeled afovirsen sodium in patients with genital warts. Clin Pharmacol Ther 56:641–646

Crooke ST, Graham MJ, Zuckerman JE, Brooks D, Conklin BS, Cummins LL, Greig MJ, Guinosso CJ, Kornbrust D, Manoharan M, Sasmor HM, Schleich T, Tivel KL, Griffey RH (1996) Pharmacokinetic properties of several novel oligonucleotide analogs in mice. J Pharmacol Exp Ther 277(2):923–937

Cummins LL, Leeds JM, Greig M, Griffey RH, Graham MJ, Crooke R, Gaus HJ (1996) Capillary gel electrophoresis and mass spectrometry: powerful tools for the analysis of antisense oligonucleotides and their metabolites. XII International roundtable proceedings: nucleosides, nucleotides, and their biological applications ("Making drugs out of nucleosides and oligonucleotides"). La Jolla, CA, 15-19 September 1996

De Benedetti AB, Pytel A, Baglioni C (1987) Loss of (2'-5')oligoadenylate synthetase activity by production of antisense RNA results in lack of protection by interferon from viral infections. Proc Natl Acad Sci USA 84:658–662

De Clercq E (1991) Targets and strategies for the antiviral chemotherapy of AIDS. TIPS 11:198–205

De Clercq E, Descamps J, Balzarini J, Giziewicz J, Barr PJ, Robins MJ (1983) Nucleic acid related compounds. 40. Synthesis and biological activities of 5-alkynyluracil nucleosides. J Med Chem 26:661–666

Desatnik R, Foster RE, Lowder CY (1996) Treatment of clinically resistant cytomegalovirus retinitis with combined intravitreal injections of ganciclovir and foscarnet. Am J Ophthalmol 122(1):121–123

Dewanjee MK, Kapadvanjwala M, Krishan A, Serafini AN, Ghafouripour AK, Oates EL, Lopez DM, Sfakianakis GN (1993) Radiolabeled antisense oligodeoxynucleotides: In vitro and in vivo applications. J Clin Immunoassay 16:276–289

Baglioni CY, Foster RE, Avery RK, Prayson RA (1996) Serous retinal detachments in a patient with clinically resistant cytomegalovirus retinitis. Arch Ophthalmol 114(7): 896–897

Dunn JP, MacCumber MW, Forman MS, Charache P, Apuzzo L, Jabs DA (1995) Viral sensitivity testing in patients with cytomegalovirus retinitis clinically resistant to foscarnet or ganciclovir. Am J Ophthalmol 119(5):587–596

Fisher TL, Terhorst T, Cao X, Wagner RW (1993) Intracellular disposition and metabolism of fluorescently-labeled unmodified and modified oligonucleotides miroinjected into mammalian cells. Nucleic Acids Res 21(16):3857–3865

Gao W-Y, Stein CA, Cohen JS, Dutschman GE, Cheng YC (1989) Effect of phosphorothioate homo-oligodeoxynucleotides on herpes simplex virus type 2-induced DNA polymerase. J Biol Chem 264:11521–11526

Gao W-Y, Hanes RN, Vasquez-Padua MA, Stein CA, Cohen JS, Cheng YC (1990a) Inhibition of herpes simplex virus type 2 growth by phosphorothioate oligodeoxynucleotides. Antimicrob Agents Chemother 34:808–812

Gao W-Y, Jaroszewski JW, Cohen JS, Cheng YC (1990b) Mechanisms of inhibition of herpes simplex virus type 2 growth by 28-mer phosphorothioate oligodeoxycytidine. J Biol Chem 265(33):20172

Geary RS, Leeds JM, Shanahan W, Glover J, Pribble J, Truong L, Fitchett J, Burckin T, Nicklin P, Philips J, Levin AA (1996) Sequence independent plasma and tissue kinetics for three antisense phosphorothioate oligonucleotides: mouse to man. American association of pharmaceutical sciences meeting, Seattle, WA, 27-31 October 1996

Glover JM, Leeds JM, Mant TGK, Amin D, Kisner DL, Zuckerman J, Geary RS, Levin AA, Shanahan WR (1997) Phase 1 safety and pharmacokinetic profile of an ICAM-1 antisense oligodeoxynucleotide (ISIS 2302). J Pharmacol Exp Ther (in press)

Goodchild J, Agrawal S, Civeira MP, Sarin PS, Sun D, Zamecnik PC (1988) Inhibition of human immunodeficiency virus replication by antisense oligodeoxynucleotides [published erratum appears in Proc Natl Acad Sci USA 1989 Mar; 86(5):1504]. Proc Natl Acad Sci U S A 85:5507–5511

Hoke GD, Draper K, Freier S, Gonzalez C, Driver VB, Zounes MC, Ecker DJ (1991) Effects of phosphorothioate capping on antisense oligonucleotide stability, hybridization and antiviral efficacy versus herpes simplex virus infection. Nucleic Acids Res 19(20):5743–5748

Hsiung GD, Chan VF (1989) Evaluation of new antiviral agents. II. The use of animal models. Antiviral Res 12:239–258

Hughes JA, Avrutskaya AV, Brouwer KL, Wickstrom E, Juliana RL (1995) Radiolabeling of methylphosphonate and phosphorothioate oligonucleotides and evaluation of their transport in everted rat jejunum sacs. Pharm Res 12:817–824

Iversen PL, Mata J, Tracewell WG, Zon G (1994) Pharmacokinetics of an antisense phosphorothioate oligodeoxynucleotide against rev from human immunodeficiency virus type 1 in the adult male rat following single injections and continuous infusion. Antisense Res Dev 4:43–52

Jabs DA (1995) Ocular manifestations of HIV infection. Trans Am Ophthalmol Soc 93(83):623–683

Jabs DA, Dunn JP, Enger C, Forman M, Bressler N, Charache P (1996) Cytomegalovirus retinitis and viral resistance. Prevalence of resistance at diagnosis, 1994. Cytomegalovirus Retinitis and Viral Resistance Study Group. Arch Ophthalmol 114(7):809–814

Jacob A, Duval-Valentin G, Ingrand D, Thuong NT, Helene C (1993) Inhibition of viral growth by an α-oligonucleotide directed to the splice junction of herpes simplex virus type-1 immediate-early pre-mRNA species 22 and 47. Eur J Biochem 216:19–24

Juliano RL, Akhtar S (1992) Liposomes as a drug delivery system for antisense oligonucleotides. Antisense Res Dev 2:165–176

Klinman DM, AE-Kyung Y, Yi AK, Beaucage SL, Conover J, Krieg AM (1996) CpG motifs present in bacterial DNA rapidly induce lymphocytes to secrete interleukin 6, interleukin 12, and interferon χ. Proc Natl Acad Sci USA 93:2879–2883

Krieg AM, Yi A-K, Matson S, Waldschmidt TJ, Koretzky GA, Klinman DM (1995) CpG motifs in bacterial DNA trigger direct B-cell activation. Nature 374:546–549

Krotz AH, Klopchin P, Cole DL, Ravikumar VT (1997) Phosphorothioate oligonucleotides: largely reduced (n-1)-mer and phosphodiester content through the use of dimeric phosphoramidite synthons. Bioorg Med Chem Lett 7(1):73–78

Kulka M, Smith CC, Aurelian L, Fishelevich R, Meade K, Miller P, Ts'o POP (1989) Site specificity of the inhibitory effects of oligo(nucleoside methylphosphonate)s complementary to the acceptor splice junction of herpes simplex virus type 1 immediate early mRNA 4. Proc Natl Acad Sci USA 86:6868–6872

Kulka M, Wachsman M, Miura S, Fishelevich R, Miller PS, Ts'o POP, Aurelian L (1993) Antiviral effect of oligo(nucleoside methylphosphonates) complementary to the herpes simplex virus type 1 immediate early mRNAs 4 and 5. Antiviral Res 20:115–130

Lappalainen K, Urtti A, Jaaskelainen I, Syrjanen K, Syrjanen S (1994) Cationic liposomes mediated delivery of antisense oligonucleotides targeted to HPV 16 E7 mRNA in CaSki cells. Antiviral Res 23:119–130

Leeds JM, Graham M, Truong L, Cummins LL (1996) Quantitation of phosphorothioate oligonucleotides in human plasma. Anal Biochem 235:36–43

Leeds JM, Kornbrust D, Truong L, Henry S (1994) Metabolism and pharmacokinetic analysis of a phosphorothioate oligonucleotide after intravitreal injection. American association of pharmaceutical sciences meeting, 9th annual meeting, San Diego, CA, 6-10 November 1994 [Abstr in Pharm Res 11(10):S353]

Leiter JME, Agrawal S, Palesc P, Zamecnik PC (1990) Inhibition of influenza virus replication by phosphorothioate oligodeoxynucleotides. Proc Natl Acad Sci USA 87:3430–3434

Letsinger RL, Zhang GR, Sun DK, Ikeuchi T, Sarin PS (1989) Cholesteryl-conjugated oligonucleotides: synthesis, properties, and activity as inhibitors of replication of human immunodeficiency virus in cell culture. Proc Natl Acad Sci USA 86:6553–6556

Litzinger DC, Brown JM, Wala I, Kaufman SA, Van GY, Farrell CL, Collins D (1996) Fate of cationic liposomes and their complex with oligonucleotide in vivo. Biochem Biophy Acta 1281:139–149

Luster MI, Portier C, Pait DG, White KL Jr, Gennings C, Munson AE, Rosenthal GJ (1992) Risk assessment in immunotoxicology. Fundam Appl Toxicol 18:200–210

Luster MI, Munson AE, Thomas PT, Holsapple MP, Fenters JD, White KL Jr, Lauer LD, Germolec DR, Rosenthal GJ, Dean JH (1988) Development of a testing battery to assess chemical-induced immunotoxicity: national toxicology program's guidelines for immunotoxicity evaluation in mice. Fundam Appl Toxicol 10:2–19

Mirabelli CK, Crooke ST (1993) Antisense oligonucleotides in the context of modern molecular drug discovery and development. In: Crooke ST, Lebleu B (eds) Antisense research and applications. CRC Press, Boca Raton, pp 7–35

Ojwang J, Okleberry KM, Marshall HB, Vu HM, Huffman JH, Rando RF (1994) Inhibition of Friend murine leukemia virus activity by guanosine/thymidine oligonucleotides. Antiviral Res 25:27–41

Phillips JA, Craig SJ, Bayley D, Christian RA, Geary R, Nicklin PL (in press) Pharmacokinetics, metabolism and elimination of a 20-mer phosphorothioate oligodeoxynucleotide (CGP 69846A) after intravenous and subcutaneous administration. Biochem Pharmacol

Piwnica-Worms D (1994) Making sense out of anti-sense: challenges of imaging gene translation with radiolabeled oligonucleotides. J Nucl Med 35:1064–1066

Proudnikov D, Mirzabekov A (1996). Chemical methods of DNA and RNA fluorescent labeling. Nucleic Acids Res 24:4535–4532

Ravikumar VT, Andrade M, Wyrzykiewicz, Scozzari A, Cole DL (1995) Large-scale synthesis of oligodeoxyribonucleotide phosphorothioate using controlled-pore glass as support. Nucleosides Nucleotides 14(6):1219

Saijo Y, Perlaky L, Wang H, Busch H (1994) Pharmacokinetics, tissue distribution, and stability of antisense oligodeoxynucleotide phosphorothioate ISIS 3466 in mice. Oncol Res 6:243–249

Sands H, Gorey-Feret LJ, Cocuzza AJ, Hobbs FW, Chidester D, Trainor GL (1994) Biodistribution and metabolism of internally 3H-labeled oligonucleotides. I. Comparison of a phosphodiester and phosphorothioate. Mol Pharmacol 45:932–943

Sandy CJ, Bloom PA, Graham EM, Ferris JD, Shah SM, Schulenburg WE, Migdal CS (1995) Retinal detachment in AIDS-related cytomegalovirus retinitis. Eye 9 (3):277–281

Sanghvi YS, Cook PD (1993) Towards second-generation synthetic backbones for antisense oligonucleosides. In: Chu CK, Baker DC (eds) Nucleosides, nucleotides as antiviral and antitumor agents. Plenum, New York, pp 309–322

Sarin PS, Agrawal S, Civeira MP, Goodchild J, Ikeuchi T, Zamecnik PC (1988) Inhibition of acquired immunodeficiency syndrome virus by oligodeoxynucleoside methylphosphonates. Proc Natl Acad Sci USA 85:7448–7451

Schuette JM, Cole DL, Srivatsa GS (1994) Development and validation of a method for routine base composition analysis of phosphorothioate oligonucleotides. J Pharm Biomed Anal 12:1345–1353

Shoji Y, Shimada J, Mizushima Y, Iwasawa A, Nakamura Y, Inouye K, Azuma T, Sakurai M, Nishimura T (1996) Cellular uptake and biological effects of antisense oligodeoxynucleotide analogs targeted to herpes simplex virus. Antimicrob Agents Chemother 40(7):1670–1675

Smith CC, Aurelian L, Reddy MP, Miller PS, Ts'o POP (1986) Antiviral effect of an oligo(nucleoside methylphosphonate) complementary to the splice junction of herpes simplex virus type 1 immediate early pre-mRNAs 4 and 5. Proc Natl Acad Sci USA 83:2787–2791

Srinivasan SK, Iversen P (1995) Review of in vivo pharmacokinetics and toxicology of phosphorothioate oligonucleotides. J Clin Lab Anal 9:129–137

Srivatsa GS, Batt M, Schuette J, Carlson RH, Fitchett J, Lee C, Cole DL (1994) Quantitative capillary gel electrophoresis (QCGE) assay of phosphorothioate oligonucleotide in pharmaceutical formulations. J Chromatogr 680:469–477

Stein CA (1992) Anti-sense oligodeoxynucleotides – promises and pitfalls. Leukemia 6:967–974

Stein CA (1995) Does antisense exist? Nat Med 1(11):1119–1121

Stein CA (1996) Phosphorothioate antisense oligodeoxynucleotides: questions of specificity. TIBTECH 14:147–149

Stein CA, Cheng YC (1993) Antisense oligonucleotides as therapeutic agents – is the bullet really magical? Science 261(5124):1004–1012

Stein CA, Krieg AM (1994) Problems in interpretation of data derived from in vitro and in vivo use of antisense oligodeoxynucleotides. Antisense Res Dev 4:67–69

Stein CA, Tonkinson JL, Yakubov L (1991) Phosphorothioate oligodeoxynucleotides – antisense inhibitors of gene expression? Pharmacol Ther 52:365–384

Takayama S, Hayashi Y (1993) The establishment of international guidelines for reproductive an developmental toxicity studies. Appl Clin Trials 2(9):28–35

Urdea MS, Warner BD, Running JA, Stempien M, Clyne J, Horn T (1988) A comparison of non-radioisotopic hybridization assay methods using fluorescent, chemiluminescent and enzyme labeled synthetic oligodeoxyribonucleotide probes. Nucleic Acids Res 16: 4937–4956

Vinogradov SV, Suzdaltseva Y, Alakhov VYu, Kabanov AV (1994) Inhibition of herpes simplex virus 1 reproduction with hydrophobized antisense oligonucleotides. Biochem Biophys Res Commun 203:959–966

Wallace TL, Bazemore SA, Kornbrust DJ, Cossum PA (1996) Repeat-dose toxicity and pharmacokinetics of a partial phosphorothioate anti-HIV oligonucleotide (AR177) after bolus intravenous administration to cynomolgus monkeys. J Pharmacol Exp Ther 278(3):1313–1317

Wyrzykiewicz TK, Cole DL (1994) Sequencing of oligonucleotide phosphorothioates based on solid-supported desulfurization. Nucleic Acids Res 22:2667–2669

Zaia JA, Rossi JJ, Murakawa GJ, Spallone PA, Stephens DA, Kaplan BE, Eritja R, Wallace RB, Cantin EM (1988) Inhibition of human immunodeficiency virus by

using an oligonucleoside methylphosphonate targeted to the tat-3 gene. J Virol 62:3914–3917
Zamecnik PC, Stephenson ML (1978) Inhibition of Rous sarcoma virus replication and cell transformation by a specific oligodeoxynucleotide. Proc Natl Acad Sci USA 75:289–294
Zamecnik PC, Goodchild J, Taguchi Y, Sarin PS (1986) Inhibition of replication and expression of human T-cell lymphotropic virus type III in cultured cells by exogenous synthetic oligonucleotides complementary to viral RNA. Proc Natl Acad Sci USA 83:4143–4146
Zelphati O, Szoka FC (1996a) Liposomes as a carrier for intracellular delivery of antisense oligonucleotides: A real or magic bullet? J Control Release 41(1/2):99–119
Zelphati O, Szoka FC (1996b) Mechanism of oligonucleotide release from cationic liposomes. Proc Natl Acad Sci USA 93(21):11493–11498
Zelphati O, Imbach JL, Signoret N, Zon G, Rayner B, Leserman L (1994) Antisense oligonucleotides in solution or encapsulated in immunoliposomes inhibit replication of HIV-1 by several different mechanisms. NAR 22:4307–4314
Zhang R, Diasio RB, Lu Z, Liu T, Jiang Z, Galbraith WM, Agrawal S (1995a) Pharmacokinetics and tissue distribution in rats of an oligodeoxynucleotide phosphorothioate (GEM 91) developed as a therapeutic agent for human immunodeficiency virus type-1. Biochem Pharmacol 49:929–939
Zhang R, Yan J, Shahinian H, Amin G, Lu Z, Liu T, Saag MS, Jiang Z, Temsamani J, Martin RR et al (1995b) Pharmacokinetics of an anti-human immunodeficiency virus antisense oligodeoxynucleotide phosphorothioate (GEM 91) in HIV-infected subjects. Clin Pharmacol Ther 58:44–53
Zhang R, Iyer RP, Yu D, Tan W, Zhang X, Lu Z, Zhoa H, Agrawal S (1996) Pharmacokinetics and tissue distribution of a chimeric oligodeoxynucleoside phosphorothioate in rats after intravenous administration. J Pharmacol Exp Ther 278:971–979

CHAPTER 16
Antisense Oligonucleotides to Protein Kinase C-α and C-*raf* Kinase: Rationale and Clinical Experience in Patients with Solid Tumors

F.A. DORR and D.L. KISNER

A. Introduction

The mechanisms of action of classical small-molecule cytotoxic chemotherapy are varied, but each depends in part on excessive proliferation of tumor cells relative to that of normal tissue for their therapeutic indices. Small-molecule chemotherapeutic agents have narrow therapeutic indices for a variety of reasons. The most global reason is simply that small molecules do not specifically target cancer cells. However, in addition cytotoxic agents lack target specificity. While designed with specific mechanisms or targets in mind, many small molecules inhibit additional proteins, whether intracellular, membrane associated, or at the cell surface. Historically, the screening process for identifying new anticancer agents has selected agents for their ability to inhibit tumor growth in various animal models rather than with significant consideration for target selectivity.

Antisense oligonucleotides offer the potential of highly target-specific inhibitors of protein synthesis. In contrast to most anticancer drugs which modulate the activity by binding directly to protein or to cell surface receptors, the molecular target for antisense oligodeoxynucleotides is mRNA, which encodes for a specific protein. By hybridizing to mRNA coding for a specific protein, antisense oligodeoxynucleotides reduce or inhibit the expression of the protein. Interaction with unintended mRNA sequences is avoided by appropriate-length oligonucleotides with high binding affinity to the targeted sequence. Potent inhibition of a specific protein could potentially result in toxicity if other salvage pathways in normal tissues are not present. Isis Pharmaceuticals, along with its corporate partner, Novartis, has chosen to initiate clinical development of two antisense oligodeoxynucleotides which target genes demonstrated to have a significant role in tumorigenesis, tumor growth, and metastasis and are members of multigene families. These are inhibitors of protein kinase C (PKC)-α and C-*raf* kinase (DEAN et al. 1996; MONIA et al. 1996).

B. Rationale for the Development of an Inhibitor of Protein Kinase C (ISIS 3521)

Interference with growth factor-activated signal transduction pathways represents a promising strategy for the development of novel anticancer modalities. PKC is a phospholipid-dependent, cytoplasmic serine/threonine kinase responsible for signal transduction in response to growth factors, hormones, and neurotransmitters. Gene-cloning studies have demonstrated that PKC exists as a family of proteins consisting of multiple discrete isozymes (Nishizuka 1988). These isozymes are products of three distinct genes and differ in their biochemical properties, tissue-specific expression, and intracellular localization (Nishizuka 1988; Basu 1993). The PKC isozymes are classified into three groups: group A (classical), which consists of PKC-α, -βI, -βII, and -γ; group B (new), which consists of PKC-δ, -ε, -η, -θ, and -μ; and group C (atypical), which consists of PKC-λ and -ζ (Gescher 1992). The classical and new PKC are believed to be activated by 1,2-diacylglycerol (1,2-DAG), which is generated by phospholipase cleavage of membrane phospholipids. Phospholipases are regulated by many growth factors and hormones, and PKC is therefore believed to play an important role in regulating normal cell differentiation and proliferation (Basu 1993; Blobe et al. 1994).

Considerable experimental evidence also exists implicating PKC in abnormal proliferation which occurs during the processes of tumor promotion and carcinogenesis. In mice, chronic activation of PKC in skin by phorbol esters leads to the selective outgrowth of initiated keratinocytes and the growth of multiple squamous tumors (Yuspa 1994). Functional PKC appears to be required for the development of a transformed phenotype after transfection of keratinocytes with activated v-*ras*-Ha, suggesting that some of the effects of *ras* are mediated through the PKC signaling pathway (Dlugosz et al. 1994). Stable overexpression of different isoforms of PKC in cell lines also results in deregulated growth, both positive and negative. For example, transfection of PKC-βI and PKC-γ into fibroblasts results in a phenotype consistent with transformation (i.e., increased growth rate and anchorage-independent growth; Housey et al. 1988; Persons et al. 1988), whereas overexpression of PKC-βI in colon cells causes growth inhibition (Choi et al. 1990). The overexpression of PKC-α in MCF-7 breast cancer cells leads to a more aggressive phenotype, which exhibits enhanced proliferation, anchorage-independent growth, and increased tumorigenicity in athymic mice (Ways et al. 1995). These studies suggest that abnormal PKC regulation can have a profound effect on growth and transformation of a variety of cell types. In patient-derived tumor cells, the concentration of the various PKC isoforms has been demonstrated to be elevated or decreased in different types of cancer (O'Brien et. al. 1989; Aflalo et al. 1992; Alvaro et al. 1992; Benzil et al. 1992). One hypothesis associating dietary fat with the development of colon cancer through activation of PKC by 1,2-DAG has been proposed. Bacteria in the intestinal lumen are able to convert lipids to 1,2-DAG, which may enter colonic epithelium and activate PKC, resulting in induction of cellular proliferation. This would cause a sustained state of increased proliferation of the colonic epithelium, a condition which is observed in populations at high risk of developing colon cancer (WEINSTEIN 1991).

The role of PKC in regulating tumor growth makes it an attractive target for investigational cancer therapy. Small molecules have been identified which bind to either the catalytic site or regulatory site of the protein (BASU 1993; O'BRIEN and KUO 1994). Although these small molecules are sometimes specific for PKC, isotype selectivity has not been achieved. Although the roles of the individual PKC isozymes in carcinogenesis and growth control have only recently begun to be explored, it is unlikely that this process will be regulated by all isozymes. Nonspecific inhibitors of PKC are likely to have unwanted biologic effects by inhibiting PKC isozymes which do not participate in the process or maintenance of transformation.

ISIS 3521 (also referred to as CGP 64128A) is a 20-base phosphorothioate oligodeoxynucleotide antisense inhibitor of human PKC-α mRNA expression in vitro and in vivo. It selectively hybridizes to a sequence in the 3'-untranslated region of the PKC-α mRNA. It has been shown to specifically and potently reduce the quantity of PKC-α mRNA in multiple human cell lines. It has also been demonstrated to inhibit the growth of multiple human tumor xenografts in athymic mice. The reduction in tumor growth has been shown to be associated with both a sequence- and dose-dependent reduction in PKC-α protein expression in the tumors evaluated.

When ISIS 3521 is combined with standard cytotoxic chemotherapy in human tumor xenografts grown in athymic nude mice, additive and supra-additive antitumor activity has been observed in several different tumor models (MONIA et al. 1996).

C. Rationale for the Development of an Inhibitor of C-*raf* Kinase (ISIS 5132)

The *raf* kinase gene family is comprised of three highly conserved genes, A-*raf*, B-*raf*, and C-*raf*. *raf* genes code for serine/threonine-specific protein kinases that play pivotal roles in intracellular signaling pathways. *raf* kinases serve as central regulators of mitogenic signaling pathways by their role in mediating upstream growth factor-mediated tyrosine kinase stimulation and downstream activation of serine threonine kinases (DAUM et al. 1994). In this signaling cascade, receptor activation leads to the activation of *ras* (ROZAKIS-ADCOCK et al. 1992; SKOLNIK et al. 1993; SASAOKA et al. 1994). Activated *ras* interacts directly with the NH_2-terminal regulatory domain of *raf* kinase (VOJTEK et al. 1993; WARNE et al. 1993; ZHANG et al. 1993; FABIAN et al. 1994), resulting in the recruitment of *raf* to the plasma membrane, where it associates with 14-3-3 (cytoskeletal) proteins (FU et al. 1994; LEEVERS et al. 1994; STOKOE et al. 1994). Once activated, *raf* initiates a cascade of reactions by direct activation of mitogen-activated protein kinase, a dual-specificity kinase that phosphorylates mitogen-activated protein kinase, causing its activation (CREWS et al. 1992; DENT et al. 1992; KOSAKO et al. 1992; KYRIAKIS et al. 1992; ZHENG et al. 1994). Mitogen-activated protein kinase has a number of substrates, including transcription factors (PULVERER et al. 1991; GILLE et al. 1992), phospholipase A_2 (LIN et al. 1993; NEMENOFF et al. 1993), and other kinases (STURGILL et al. 1988). This signaling cascade is often referred to as the mitogen-activated protein kinase signaling pathway (DAUM et al. 1994).

Substantial evidence exists supporting a direct role for *raf* kinases in the development and maintenance of human malignancies. The mitogen-activated protein kinase signaling cascade has been shown to be essential for cellular proliferation and mediation of cellular transformation by most oncogenes (STURGILL et al. 1988; NISHIDA and GOTOH 1993; DAUM et al. 1994). Raf proteins have been shown to be direct effectors of ras protein function within the mitogen-activated protein kinase signaling pathway (VOJTEK et al. 1993; WARNE et al. 1993; ZHANG et al. 1993; FABIAN et al. 1994). Since *ras* mutations are present in a high proportion of human cancers (BOS 1988, 1989), novel therapies directed against *raf* kinases may prove useful in the treatment of *ras*-dependent tumors. Mutations in *raf* genes have been shown to transform cells in vitro (RAPP et al. 1983; STANTON and COOPER 1988; HEIDECKER ct al. 1990; FABIAN et al. 1993) and have been associated with certain human tumors. Finally, expression of unusually high levels of C-*raf* kinase mRNA and protein have been reported in patients with small cell lung cancer and breast cancer (RAPP et al. 1988).

ISIS 5132 (also referred to as CGP 69846A) is a 20-base phosphorothioate oligodeoxynucleotide antisense inhibitor of human C-*raf* mRNA expression in vitro and in vivo. It was designed to hybridize to a sequence in the 3'-untranslated region of C-*raf* kinase mRNA.

It has been demonstrated to specifically and potently reduce the expression of C-*raf* kinase mRNA in multiple human tumor cell lines. Inhibition of several different human tumor xenografts in athymic nude mice has been demonstrated. The reduction in tumor growth has been shown to be accompanied by an oligonucleotide dose- and sequence-dependent reduction in C-*raf* kinase mRNA expression in the tumors studied (MONIA et al. 1996).

D. Preclinical Toxicology and Pharmacokinetics

The nonclinical toxicity of phosphorothioate oligodeoxynucleotide is fairly consistent from one sequence to another among those tested thus far. Repeated intravenous administration in mice results in reversible and dose-dependent immunostimulation characterized by lymphoid hyperplasia, splenomegaly, and multiorgan monocytic infiltration. There appear to be no target-specific toxicities with either c-*raf* kinase or PKC-α, as murine-specific antisense oligodeoxynucleotides do not produce distinct clinical or histopathologic toxicities compared to the human construct (HENRY et al. 1997).

In monkeys, the interaction of phosphorothioate oligodeoxynucleotides with plasma proteins has resulted in two specific acute reactions following either intravenous (IV) or subcutaneous (SC) administration. One acute toxicity is a transient prolongation of activated partial thromboplastin time (aPTT), which is maximal at maximum concentration in plasma (C_{max}) and is rapidly reversible as the compound is cleared from plasma. At doses less than 10 mg/kg administered as a 2-h IV infusion, no prolongation of aPTT is observed. In addition, there is no apparent cumulative effect with repeat dosing at those doses which do prolong aPTT. This effect appears to be mediated through binding to factor VIII or to other factors in the intrinsic coagulation pathway (J.P. SHEEHAN, unpublished data). Reversal of the anticoagulant effect is not accomplished by administration of protamine or platelet factor 4, but because prolongation of aPTT is associated with plasma concentration, changes in aPTT rapidly normalize following drug discontinuation (J.P. SHEEHAN, unpublished data).

The second acute toxicity observed in monkeys is activation of the alternative complement pathway. Complement activation appears to be related to threshold drug concentration. Plasma concentrations of intact oligodeoxynucleotide in excess of 40–50 μg/ml or of total oligodeoxynucleotide in excess of approximately 80 μg/ml result in complement activation, as measured by increases in Bb split product. Factor B is a component unique to the alternative complement pathway, which, when activated, results in the release of the Bb fragment. No activation of the classical complement pathway has been observed in animals. Further activation of the alternative complement pathway can lead to release of the anaphylotoxins C3a and C5a. Complement activation can lead to a number of secondary effects, including alterations in neutrophil activation and vascular integrity. At high doses (10 mg/kg and 20 mg/kg as 2-h IV infusions), vascular instability has been observed in monkeys, including two deaths. Complement activation was observed in both monkeys (HENRY et al. 1997).

Administration of antisense oligodeoxynucleotides by IV infusion results in rapid clearance from plasma into tissue. Tissues with the greatest drug distribution include liver, kidney, spleen, and bone. Repeated administration does not result in altered pharmacokinetics. Peak plasma concentrations (C_{max}), AUC, and terminal plasma elimination half-life are dose dependent and predictable, though nonlinear. Distribution of drug to the kidney is associated with atrophic and regenerative changes in the proximal renal tubular epithelium in monkeys. Discontinuation of therapy resulted in reversal of these changes in the kidney, although at the highest dose delivered (80 mg/kg SC every other day for four doses) tubular degeneration and renal dysfunction were observed (HENRY et al. 1997).

E. Status of Clinical Development

I. ISIS 3521

On the basis of the preclinical human tumor xenograft activity, the correlation of this activity with inhibition of PKC-α, and the potential for a high therapeutic index, ISIS 3521 entered phase I clinical trials in early 1996. Because of the association of complement activation and coagulation (aPTT) abnormalities associated with high peak plasma levels, two schedules of IV administration are being investigated, both designed to avoid high peak plasma concentrations. In the first study, Clinical Study 1, ISIS 3521 is being given by 2-h infusion three times per week for 3 weeks followed by 1 week off treatment. This cycle is repeated every 4 weeks. In Clinical Study 2, administration of ISIS 3521 is accomplished by 21-day continuous infusion, followed by 1 week of rest, again repeated every 4 weeks. The development strategy for phase I is to attempt to identify the maximally tolerated dose of ISIS 3521 on both schedules. This is being accomplished with a fairly classical phase I design, but with conservative dose escalation steps, given the modest amount of experience with this class of drugs in cancer patients.

The objectives of the phase I studies currently in progress are to identify the maximally tolerated dose of ISIS 3521 and to evaluate pharmacokinetic behavior of the drug, to identify antitumor activity, and to begin to characterize the safety of this compound in a heterogeneous group of patients with advanced cancer for whom alternative therapy is considered to be of limited utility. Routine clinical safety monitoring is being conducted. In addition, the potential for complement activation and coagulation abnormalities is being monitored with frequent assessment of C3a, C5a, aPTT, and prothrombin time.

1. Clinical Study 1

The first patient was treated with ISIS 3521 on 25 March 1996. As of 1 January 1997, a total of 21 patients have been treated, with three remaining on the study. Table 1 summarizes the characteristics of patients entered in this study. The amount and type of prior therapy is not yet available for these patients.

Fourteen of the 21 patients discontinued treatment because of objective evidence of disease progression. Three patients voluntarily withdrew from the study, and one patient withdrew because of an unrelated adverse event. Three patients have been treated at each of the following dose levels: 0.15, 0.30, 0.60, 1.0, 1.5, 2.0, and 2.5 mg/kg per day of dosing. There has been no evidence of activation of the alternative complement pathway or clinically significant prolongation of aPTT, although transient (<2h) increases in aPTT by approximately 25%–50% have been observed. There has been no evidence of dose-limiting (grade 3 or 4) toxicity. One 62-year-old man with colon cancer treated at the 1.0 mg/kg dose level withdrew from the study because of a

Table 1. ISIS 3521 clinical study 1: patient characteristics

Characteristic ($n = 21$)	Value
Median age (years)	62[a]
Sex	
Male	11
Female	10
Tumor type (n)	
Melanoma	5
Colon	5
Renal	3
Breast	1
Lymphoma	1
Neuroendocrine	1
NSCLC	1
Ovarian	1
Pancreatic	1
Sarcoma	1
SCLC	1

NSCLC, non-small-cell lung carcinoma; SCLC, small cell lung carcinoma.
[a] Range, 24–73 years.

myocardial infarction at the end of the second week of his third treatment cycle. This patient had documented coronary artery disease, and the acute event was felt to be unrelated to drug administration. A patient with low-grade lymphoma has received four cycles of therapy at 0.6 mg/kg per day and maintained stable disease with no evidence for cumulative toxicity. He is continuing therapy.

Pharmacokinetic evaluations have demonstrated that the plasma concentrations of ISIS 3521 at the end of infusion (C_{max}) have increased proportional to dose with both intact oligodeoxynucleotide and metabolites rapidly cleared from plasma with a half-life of approximately 30–45 min (Table 2). Metabolites were composed of chain-shortened oligodeoxynucleotides, consistent with nuclease degradation of the parent nucleotide strand. There has been no evidence of accumulation of drug with repeated dosing, nor of enhanced metabolism.

2. Clinical Study 2

As mentioned, Clinical Study 2 is evaluating a 21-day continuous infusion schedule given every 4 weeks. Because of the potential for renal accumulation of oligonucleotide and the association of renal proximal tubule degeneration with dose, renal tubule function is being monitored by serial determinations of urinary retinol-binding protein and N-acetyl-β-glucosaminidase. The first patient was treated on 13 August 1996, and as of 1 January 1997, there have been nine patients treated in the study. Five patients have remained on the study, while four have discontinued therapy because of disease progression. Table 3 lists the characteristics of the nine patients entered on the trial.

Three patients have been treated at each of the following dose levels: 0.5, 1.0, and 1.5 mg/kg per day for 21 days. Dose escalation will continue with planned dose levels of 2.0, 3.0, 4.0, and 5.0 mg/kg per day. There has been no evidence of dose-limiting (grade 3 or 4) toxicity. No patients have experienced evidence of activation of the alternative complement pathway, prolongation of aPTT, changes in renal function, or laboratory evidence of renal tubule toxicity. Grade 2 toxicities have included thrombocytopenia in one patient and leukopenia in one patient. These toxicities resolved in the presence of continued therapy. A patient with ovarian cancer has received five cycles of therapy at 0.5 mg/kg per day with less than 50% reduction in measurable tumor. Her

Table 2. Pharmacokinetics of ISIS 3521 administered as a 2-h infusion at three dose levels

Dose (mg/kg)	Day of dosing	EOI plasma concentration (μg/ml)[a]	Intact (%)	$t_{1/2}$ (min)
0.15	1	0.40 ± 0.04	nm	nm
	5	0.30 ± 0.06	nm	ND
	19	0.27 ± 0.02	nm	ND
0.30	1	0.99 ± 0.55	81	35
	5	0.84[b]	80	ND
0.60	1	1.81 ± 0.70	77	36
	5	2.14 ± 1.07	68	ND

EOI, end of infusion; nm, not measurable; ND, not done.
[a] Mean ± SD.
[b] $n = 2$.

Table 3. ISIS 3521 clinical study 2: patient characteristics

Characteristic ($n = 9$)	Value
Median age (years)	63[a]
Sex	
Male	6
Female	3
Tumor type (n)	
Pancreatic	3
Colon	2
Breast	1
Gastric	1
NSCLC	1
Ovarian	1

NSCLC, non-small-cell lung carcinoma.
[a] Range, 43–67 years.

CA-125 levels initially rose, but have subsequently declined as her measurable tumor has decreased in size and has developed extensive central necrosis.

In patients treated at 0.5 mg/kg per day for 21 days, plasma concentrations at the end of the 21-day infusion have been 0.08–0.09 µg/ml. The plasma concentration was variable during the infusion, with a mean concentration of 0.15 ± 12 µg/ml. Following the end of infusion, the plasma concentration declined rapidly to nondctectable levels. At these low plasma levels, only intact oligonucleotide was quantifiable. Chain-shortened metabolites, though detectable, were below the lower limit of quantitation.

II. ISIS 5132

On the basis of the preclinical activity and similar toxicology to ISIS 3521, two phase I trials using identical designs and end points were begun in 1996. ISIS 5132 Clinical Study 1 is evaluating a 2-h infusion three times per week for 3 weeks followed by 1 week of rest with cycles repeated every 4 weeks. Standard safety monitoring of clinical, organ, and hematological parameters is being conducted. In addition, aPTT and complement split products (C3a, C5a, and Bb) are being monitored with the anticipation that these will represent adequate surrogate markers for the potential interaction with the intrinsic coagulation pathway and the alternative complement pathway. In ISIS 5132 Clinical Study 2, the drug is given by 21-day continuous infusion followed by 1 week of rest with the cycle repeated every 4 weeks. As with the continuous infusion study for ISIS 3521, urinary retinol-binding protein and N-acetyl-β-glucosaminidase is being monitored to look for subclinical proximal renal tubule toxicity.

1. Clinical Study 1

The first patient in Clinical Study 1 was treated on 11 April 1996. As of 1 January 1997, 15 patients have been treated, three each at 0.5, 1.0, 1.5, 2.0, and 2.5 mg/kg. All 15 patients have withdrawn from the study because of progressive disease after one or two cycles of therapy. Patient characteristics are summarized in Table 4.

There has been no evidence of treatment-related dose-limiting toxicity, complement activation (C3a or C5a), or clinically significant prolongation of aPTT. Changes in aPTT have included transient increases during the infusion with rapid normalization within 2 h of the end of the 2-h infusion. The C_{max} in plasma was observed at the end of infusion and was proportional to dose. Oligonucleotide was rapidly cleared from plasma with a half-life of 30–50 min. Metabolites were composed of chain-shortened oligodeoxynucleotides, consistent with exonuclease degradation. There has been no evidence of drug accumulation or enhanced metabolism with repetitive dosing. The maximally tolerated dose has not yet been reached and accrual continues. Table 5 displays a brief summary of the pharmacokinetics of ISIS 5132 by dose level.

Table 4. ISIS 5132 clinical study 1: patient characteristics

Characteristic ($n = 15$)	Value
Median age (years)	57[a]
Sex	
Male	11
Female	10
Tumor type (n)	
Colon	4
Bowel	1
Mesothelioma	1
NSCLC	6
Ovarian	1
SCLC	2

NSCLC, non-small-cell lung carcinoma; SCLC, small cell lung carcinoma.
[a] Range, 35–78 years

Table 5. Pharmacokinetics of ISIS 5132 administered as a 2-h infusion in three patients

Dose (mg/kg)	Day of dosing	EOI plasma concentration (μg/ml)[a]	Intact (%)	$t_{1/2}$ (min)
0.5	1	2.79 ± 1.42	60	45
	19	1.27 ± 0.28	53	ND
1.0	1	4.62 ± 0.81	52	37
	5	3.63 ± 0.07	56	ND

EOI, end of infusion; ND, not done.
[a] Mean ± SD.

2. Clinical Study 2

Clinical Study 2 accrued its first patient on 24 September 1996. As of 1 January 1997, 12 patients have been entered. Three patients have been treated at 0.5 mg/kg per day, six at 1.0 mg/kg per day, and three at 1.5 mg/kg per day. The cumulative dose administered in a 70-kg patient receiving 1.5 mg/kg per day is 31.5 mg/kg or 2205 mg. Comparable cumulative dose on the 2-h infusion schedule three times per week would be achieved at a dose of 3.5 mg/kg per day of dosing. Patients have received from less than one up to three cycles of therapy. Five of 12 patients remain on the study. Two patients discontinued therapy because of concurrent medical problems and five because of progressive disease. Patient characteristics are summarized in Table 6.

The three patients treated at the 0.5 mg/kg dose level experienced no toxicity. One of the first three patients treated at 1.0 mg/kg developed an *Escherichai coli* sepsis in the presence of a normal total white blood count and granulocyte count. However, during the septic event, he experienced grade 4 thrombocytopenia. He continued to have rigors until his indwelling central catheter was removed. His platelet count returned to normal over 5–7 days.

Table 6. ISIS 5132 clinical study 2: patient characteristics

Characteristic ($n = 11$)	Value
Median age (years)	63[a]
Sex	
Male	7
Female	4
Tumor type (n)	
NSCLC	2
Breast	1
Colon	1
Hepatocellular	1
Lymphoma	1
Prostate	1
Rectal	1
Renal cell	1
SCLC	1
Sarcoma	1

NSCLC, non-small-cell lung carcinoma; SCLC, small cell lung carcinoma.
[a] Range, 43–67 years.

The other five patients at this dose level and those at 1.5 mg/kg per day have had no evidence of myelosuppression. There have been four episodes of sepsis, bacteremia, or fungemia related to indwelling catheters in three different patients on this study. Thrombocytopenia has not been associated with any of these other episodes. In addition, there has been no evidence of acute or cumulative toxicity in this trial. Specifically, there have been no interpretable or clinically significant changes in C3a, C5a, aPTT, urinary retinol-binding protein, or N-acetyl-β-glucosaminidase. At the time of writing, pharmacokinetic parameters are not yet available for this trial.

F. Future Clinical Development

It is anticipated that the phase I trials with ISIS 3521 and ISIS 5132 will be completed during 1997 and phase II trials will begin. Phase II single-agent trials for ISIS 3521 are planned for the following indications: glioma, melanoma, metastatic colon carcinoma, non-small-cell lung carcinoma, breast cancer, and prostate cancer. ISIS 5132 will be studied as a single agent in the following tumor types: pancreas, small cell lung, non-small-cell lung, colon, breast, and prostate cancers. Preclinical human tumor xenograft studies have demonstrated additive and supra-additive antitumor activity. Thus combination phase I studies are planned for both drugs with carboplatin, taxol, and 5-fluorouracil with the anticipation of conducting randomized trials in patients with colon and non-small-cell lung cancer. Additional trials may include dose–response studies once a sensitive tumor type is identified.

References

Aflalo E, Wolfson M, Ofir R, Weinstein Y (1992) Elevated activities of protein kinase C and tyrosine kinase correlate to leukemic cell aggressiveness. Int J Cancer 50:136

Alvaro V, Touraine P, Raisman Vozari R, Bai-Grenier F, Birman P, Joubert D (1992) Protein kinase C activity and expression in normal and adenomatous human pituitaries. Int J Cancer 50:724

Basu A (1993) The potential of protein kinase C as a target for anticancer treatment. Pharmacol Ther 59:257

Benzil DB, Finkelstein SD, Epstein MH, Finch PW (1992) Expression pattern of alpha-protein kinase C in human astrocytomas indicates a role in malignant progression. Cancer Res 52:2951

Blobe GC, Obeid LM, Hannun YA (1994) Regulation of protein kinase C and role in cancer biology. Cancer Metastasis Rev 13:411–431

Bos JL (1988) The ras gene family and human carcinogenesis. Mutat Res 195:255

Bos JL (1989) ras oncogenes in human cancer: a review. Cancer Res 49:4682

Castagna M, Takai Y, Kaibuchi K, Sano K, Kikkawa U, Nishizuka Y (1982) Direct activation of calcium-activated, phospholipid-dependent protein kinase by tumor-promoting phorbol esters. J Biol Chem 257:7847

Choi PL, Tchou-Wong KM, Weinstein IB (1990) Overexpression of protein kinase C in Ht29 colon cancer cells causes growth inhibition and tumor suppression. Mol Cell Biol 10:4650

Crews CM, Alessandrini A, Erikson RL (1992) The primary structure of MEK, a protein kinase kinase that phosphorylates the ERK gene product. Science 258:478

Daum G, Eisenmann-Tappe I, Fries H-W, Troppmair J, Rapp UR (1994) The ins and outs of raf kinases. TIBS 19:474

Dean NM, McKay R, Miraglia L, Geiger T, Muller M, Fabbro D, Bennett CF (1996) Antisense oligonucleotides as inhibitors of signal transduction: development from research tools to therapeutic agents. Biochem Soc Trans 24:623

Dent P, Haser W, Haystead TA, Vincent LA, Roberts TM, Sturgill TW (1992) Activation of mitogen-activated protein kinase kinase by v-raf in NIH 3T3 cells and in vitro. Science 257:1404

Dlugosz A, Cheng C, Williams EK, Dharia AG, Denning MF, Yuspa SH (1994) Alterations in murine keratinocyte differentiation induced by activated Ha-ras genes are mediated by protein kinase C-alpha. Cancer Res 54:6413

Fabian JR, Daar IO, Morrison DK (1993) Critical tyrosine residues regulate the enzymatic and biological activity of Raf-1 kinase. Mol Cell Biol 13:7170

Fabian JR, Vojtek AB, Cooper JA, Morrison DK (1994) A single amino acid change in Raf-1 inhibits ras binding and alters Raf-1 function. Proc Natl Acad Sci USA 91:5982

Fu H, Xia K, Pallas DC, Cui C, Conroy K, Narsimhan RP, Mamon H, Collier RJ, Roberts TM (1994) Interaction of the protein kinase Raf-1 with 14-3-3 proteins. Science 266:126

Gescher A (1992) Towards selective pharmacological modulation of protein kinase C – opportunities for the development of novel antineoplastic agents. Br J Cancer 66:10

Gille H, Sharrocks AD, Shaw PE (1992) Phosphorylation of transcription factor p62TCF by MAP kinase stimulates ternary complex formation at c-fos promoter. Nature 358:414

Heidecker G, Huleihel M, Cleveland JL, Kolch W, Beck TW, Lloyd P, Pawson T, Rapp UR (1990) Mutational activation of c-raf-1 and definition of the minimal transforming sequence. Mol Cell Biol 10:2503

Henry SP, Monteith D, Levin AA (1997) Antisense oligonucleotide inhibitors for the treatment of cancer. 2. Toxicological properties of phosphorothioate oligodeoxynucleotides. Anticancer Drug Des 12:395–408

Housey GM, Johnson MD, Hsiao WLW, O'Brian CA, Murphy JP, Kirshmeier P, Weinstein IB (1988) Overproduction of protein kinase C causes disordered growth control in rat fibroblasts. Cell 52:343

Kosako H, Gotoh Y, Matsuda S, Ishikawa M, Nishida E (1992) Xenopus MAP kinase activator is a serine/threonine/tyrosine kinase activated by threonine phosphorylation. Embo J 11:2903

Kyriakis JM, App H, Zhang XF, Banerjee P, Brautigan DL, Rapp UR, Avruch J (1992) Raf-1 activates MAP kinase-kinase. Nature 358:417

Leevers SJ, Paterson HF, Marshall CJ (1994) Requirement for ras in raf activation is overcome by targeting raf to the plasma membrane. Nature 369:411

Lin L-L, Wartmann M, Lin AY, Knopf JL, Seth A, Davis RJ (1993) cPLA2 is phosphorylated and activated by MAP kinase. Cell 72:269

Monia BP, Johnston JF, Geiger T, Muller M, Fabbro D (1995) Antitumor activity of a phosphorothioate oligodeoxynucleotide targeted against C-raf kinase. Nat Med 2:668

Monia BP, Johnston JF, Geiger T, Muller M, Altmann K-H, Fabbro D, Smyth J (1996) CGP69846A, a phosphorothioate antisense oligodeoxynucleotide targeted to human c-raf-1 displays potent antitumor activity, vol 7. Kluwer, Vienna, p 123

Nemenoff RA, Winitz S, Qian NX, Van Putten V, Johnson GL, Heasley LE (1993) Phosphorylation and activation of a high molecular weight form of phospholipase A2 by p42 microtubule-associated protein 2 kinase and protein kinase C. J Biol Chem 268:1960

Nishida E, Gotoh Y (1993) The MAP kinase cascade is essential for diverse signal transduction pathways. Trends Biochem Sci 18:128

Nishizuka Y (1988) The molecular heterogeneity of protein kinase C and its implications for cellular regulation. Nature 334:661

Nishizuka Y (1992) Intracellular signaling by hydrolysis of phospholipids and activation of protein kinase C. Science 258:607

O'Brian C, Vogel VG, Singletary SE, Ward NE (1989) Elevated protein kinase C expression in human breast tumor biopsies relative to normal breast tissue. Cancer Res 49:3215

O'Brian CA, Kuo JF (1994) Protein kinase C inhibitors. In: Kuo JF (ed) Protein kinase C. Oxford University Press, Oxford, p 96

Persons DA, Wilkison WO, Bell RM, Finn OJ (1988) Altered growth regulation and enhanced tumorigenicity of NIH 3T3 fibroblasts transfected with protein kinase C-1 cDNA. Cell 52:447

Pulverer BJ, Kyriakis JM, Avruch J, Nikolakaki E, Woodgett JR (1991) Phosphorylation of c-jun mediated by MAP kinases. Nature 353:670

Rapp UR, Goldsborough MD, Mark GE, Bonner TI, Groffen J, Reynolds F Jr, Stephenson JR (1983) Structure and biological activity of v-raf, a unique oncogene transduced by a retrovirus. Proc Natl Acad Sci USA 80:4218

Rapp UR, Cleveland JL, Bonner TI, Storm SM (1988) The raf oncogene. In: Reddy EP (ed) The oncogene handbook. Elsevier, Amsterdam, p 213

Rozakis-Adcock M, McGlade J, Mbamalu G, Pelicci G, Daly R, Li W, Batzer A, Thomas S, Brugge J, Pelicci PG (1992) Association of the Shc and Grb2/Sem5 SH2-containing proteins is implicated in activation of the ras pathway by tyrosine kinases. Nature 360:689

Sasaoka T, Draznin B, Leitner JW, Langlois WJ, Olefsky JM (1994) Shc is the predominant signaling molecule coupling insulin receptors to activation of guanine nucleotide releasing factor and p21ras-GTP formation. J Biol Chem 269:10734

Skolnik EY, Batzer A, Li N, Lee CH, Lowenstein E, Mohammadi M, Margolis B, Schlessinger J (1993) The function of GRB2 in linking the insulin receptor to ras signaling pathways. Science 260:1953

Stanton V Jr, Cooper GM (1988) Activation of human raf transforming genes by deletion of normal amino-terminal coding sequences. Mol Cell Biol 7:1171

Stokoe D, Macdonald SG, Cadwallader K, Symons M, Hancock JF (1994) Activation of raf as a result of recruitment to the plasma membrane [published erratum appears in Science 1994 Dec 16; 266(5192):1792–1793]. Science 264:1463

Sturgill TW, Ray LB, Erikson E, Maller JL (1988) Insulin-stimulated MAP-2 kinase phosphorylates and activates ribosomal protein S6 kinase II. Nature 334:715

Vojtek AB, Hollenberg SM, Cooper JA (1993) Mammalian Ras interacts directly with the serine/threonine kinase raf. Cell 74:205

Warne PH, Viciana PR, Downward J (1993) Direct interaction of ras and the amino-terminal region of Raf-1 in vitro. Nature 364:352

Ways DK, Kukoly CA, deVente J, Hooker JL, Bryant WO, Posekany KJ, Fletcher DJ, Cook PP, Parker PP (1995) MCF-7 breast cancer cells transfected with PKC-alpha exhibit altered expression of other protein kinase C isoforms and display a more aggressive neoplastic phenotype. J Clin Invest 95:1906

Weinstein BI (1991) Cancer prevention: recent progress and future opportunities. Cancer Res [Suppl] 51:5080s

Yuspa SH (1994) The pathogenesis of squamous cell cancer: lessons learned from studies of skin carcinogenesis. Thirty-third G.H.A. Clowes Memorial Award Lecture. Cancer Res 54:1178

Zhang XF, Settleman J, Kyriakis JM, Takeuchi-Suzuki E, Elledge SJ, Marshall MS, Bruder JT, Rapp UR, Avruch J (1993) Normal and oncogenic p21ras proteins bind to the amino-terminal regulatory domain of c-Raf-1. Nature 364:308

Zheng CF, Ohmichi M, Saltiel AR, Guan KL (1994) Growth factor induced MEK activation is primarily mediated by an activator different from c-raf. Biochemistry 33:5595

CHAPTER 17
Nucleic Acid Therapeutics for Human Leukemia: Development and Early Clinical Experience with Oligodeoxynucleotides Directed at c-*myb*

A.M. GEWIRTZ

A. Introduction

For the past several years, we have been engaged in trying to develop an effective strategy of disrupting specific gene function with antisense oligodeoxynucleotides (ODN). We have also been actively engaged in attempting to utilize this strategy in the clinic. This latter pursuit has focused on finding appropriate gene targets that can be successfully targeted using an antisense approach and then developing "scale-up" methods so that techniques developed in the laboratory can be applied in the clinic. It was our opinion that human leukemias would be particularly amenable to this therapeutic strategy. They can be successfully manipulated ex vivo, the tumor is "liquid" in vivo and therefore more likely to successfully take up ODN, and a great deal is known about their cell and molecular biology. The latter in particular facilitates the choice of a gene target. Accordingly, if ODN were going to be developed as therapeutics, the hematopoietic system seemed an ideal model system.

B. c-*myb* Proto-oncogene

Of the genes that we have targeted for disruption using the antisense ODN strategy (GEWIRTZ and CALABRETTA 1988; LUGER et al. 1996; RATAJCZAK et al. 1992a,b; SMALL et al. 1994; TAKESHITA et al. 1993), one that has been of particular interest to our laboratory, and one where therapeutically motivated disruptions are now in clinical trial, is the c-*myb* gene (LYON et al. 1994). c-*myb* is the normal cellular homologue of v-*myb*, the transforming oncogene of the avian myeloblastosis virus (AMV) and avian leukemia virus E26. It is a member of a family composed of at least two other highly homologous genes designated A-*myb* and B-*myb* (NOMURA et al. 1993). Located on chromosome 6q in humans, the predominant transcript of c-*myb* encodes an approximately 75-kDa nuclear binding protein (Myb) which recognizes the core consensus sequence 5′-PyAAC(G/Py)G-3′ (BIEDENKAPP et al. 1988). Myb consists of three primary functional regions (SAKURA et al. 1989; Fig. 1).

At the NH2 terminus, the DNA-binding domain is situated. This region consists of three imperfect tandem repeats (R1, R2, R3), each consisting of 51

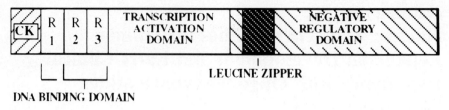

Fig. 1. Functional map of the c-Myb protein. See text for details

or 52 amino acids. Three perfectly conserved tryptophan residues are found in each repeat. Together they form a cluster in the hydrophobic core of the protein which maintains the DNA-binding helix–turn–helix structure. The midportion of the protein contains an acidic transcriptional activating domain. The DNA-binding portion of the protein is required for these transcriptional effects to be observed. The protein also contains a negative regulatory domain which has been localized to the carboxy terminus. Interestingly, the carboxy terminus is deleted in v-*myb*, and this has been thought to contribute to its transforming ability. Recently reported experiments have confirmed this hypothesis and have further demonstrated that amino-terminal deletions give rise to a protein with even more potent transforming ability (Dini et al. 1995). Deletions of both the amino and carboxy termini create a protein with the greatest transforming ability and one which induces the formation of hematopoietic cells that are more primitive than those produced by amino-terminal deletions alone (Dini et al. 1995). These data suggest that the simultaneous loss of Myb's ability to bind DNA and to interact with as yet unidentified proteins is a potent transforming stimulus. Nevertheless, this simple hypothesis is complicated by the observation that overexpression of the C-terminal portion of c-*myb* can also be oncogenic (Press et al. 1994), whereas overexpression of the whole protein is not (Dini et al. 1995). At the least, one may conclude that sequestration of certain potential Myb-binding proteins may also be an oncogenic event.

The above discussion suggests that c-*myb* might play a role in leukemogenesis. Additional, albeit indirect, evidence also supports this contention. For example, c-*myb* amplification in acute myelogenous leukemia (AML) and overexpression in 6q syndrome has been reported (Barletta et al. 1987). The mechanism whereby overexpressed Myb might be leukemogenic is uncertain, but points out the important difference of working with primary cells as opposed to cell lines. As noted above, it has been reported that overexpression of Myb is not by itself leukemogenic (Dini et al. 1995), but this work was carried out in cell lines, which may give results that are valid only for the lines tested. As was also noted above, one could reasonably postulate that Myb-driven transformation might be regulated by the binding of additional protein partners in the leucine zipper domain (Kanei-Ishii et al. 1992). Recent evidence demonstrating that Myb interacts with other nuclear binding proteins, and that Myb's carboxy terminus may interact with a cellular inhibitor of transcription, supports this hypothesis (BURK et al. 1993; VORBRUEGGEN et al. 1994). Other potential mechanisms might relate to Myb's ability to regulate hematopoietic cell proliferation (GEWIRTZ et al. 1989), perhaps by its effects on important cell cycle genes, including c-*myc* (COGSWELL et al. 1993) and cdc2 (KU et al. 1993). Finally, Myb also plays a role in regulating hematopoietic cell differentiation (WEBER et al. 1990). It functions as a transcription factor for several cellular genes, including the neutrophil granule protein mim-1 (NESS et al. 1989), CD4 (NAKAYAMA et al. 1993), insulin-like growth factor (IGF)-1 (TRAVALI et al. 1991), and CD34 (MELOTTI et al. 1994), and possibly other growth factors (SZCZYLIK et al. 1993), including c-*kit* (RATAJCZAK et al. 1992c,d). The latter is of particular interest, since it has been shown that when hematopoietic cells are deprived of c-*kit* ligand (steel factor), they undergo apoptosis (YU et al. 1993). Accordingly, Myb is clearly an important hematopoietic cell gene which may, directly or indirectly, contribute to the pathogenesis or maintenance of human leukemias. For this reason, it is a rational target for therapeutically motivated disruption strategies.

I. Targeting the c-*myb* Gene

Our investigations were initially designed to elucidate the role of Myb protein in regulating hematopoietic cell development. Because the results obtained from these studies had obvious clinical relevance, more translationally oriented studies were also undertaken. These have now culminated in clinical trials which are presently ongoing at the Hospital of the University of Pennsylvania. The steps carried out in the clinical development of the c-*myb*-targeted antisense ODN are summarized in the following. In addition, brief mention of our initial clinical experience with the myb-targeted ODN will also be made.

II. In Vitro Experience in the Hematopoietic Cell System

1. Role of c-*myb*-Encoded Protein in Normal Human Hematopoiesis

Attempts to exploit the c-*myb* gene as a therapeutic target for antisense ODN began as an outgrowth of studies which were seeking to define the role of Myb protein in regulating normal human hematopoiesis (GEWIRTZ et al. 1989; GEWIRTZ and CALABRETTA 1988). During the course of these studies, it was determined that exposing normal bone marrow mononuclear cells (MNC) to c-*myb* antisense ODN resulted in a decrease in cloning efficiency and progenitor cell proliferation. The effect was independent of lineage, since c-*myb* antisense DNA inhibited granulocyte-macrophage colony-forming units (GM-CFU), CFU-e (erythroid), and Meg-CFU (megakaryocyte). In contrast, c-*myb* ODN with the corresponding sense sequence had no consistent effect on hematopoietic colony formation when compared to growth in control cultures. Finally, inhibition of colony formation was also related to dose. Inhibition of the targeted mRNA was also demonstrated. Sequence-specific, dose-related biologic effects accompanied by a specific decrease or total elimination of the targeted mRNA constituted strong evidence to suggest that the effects we were observing were due to an "antisense" mechanism. It should be added that the effects we observed were largely confirmed using homologous recombination (MUCENSKI et al. 1991). In other investigations, it was also determined that hematopoietic progenitor cells appeared to require Myb protein during specific stages of development, in particular when actively cycling (CARACCIOLO et al. 1990), as might be expected given the above functional description of Myb protein.

2. Myb Protein Requirement for Leukemic Hematopoiesis

Since c-*myb* antisense ODN inhibited normal cell growth, we were also interested in determining their effect on leukemic cell growth. While one could reasonably postulate that aberrant c-*myb* expression or Myb function might play a role in carcinogenesis, demonstrating this was another matter. To address this question, we employed a variety of leukemic cell lines, including those of myeloid and lymphoid origin. In addition, we also employed primary patient material. We first determined the effect of *myb* sense and antisense ODN on the growth of HL-60, K562, KG-1, and KG-a myeloid cell lines (ANFOSSI et al. 1989). The antisense ODN inhibited the proliferation of each leukemia cell line, although the effect was most pronounced on HL-60 cells. Specificity of this inhibition was demonstrated by the fact that the sense ODN had no effect on cell proliferation; "antisense" sequences with two or four nucleotide mismatches also had no effect. To determine whether the treatment with *myb* antisense ODN-modified cell cycle distribution of HL-60 cells, we measured the DNA content in exponentially growing cells exposed to either sense or antisense *myb* ODN. Control cells and cells treated with c-*myb* sense ODN had twice the DNA content of HL-60 cells exposed to the antisense ODN. The majority of these cells appeared to reside either in the G_1 compartment or were blocked at the G_1/S boundary. To examine the effect of the c-*myb* ODN on lymphoid cell growth, we employed a lymphoid leukemia cell line, CCRF-CEM. As noted in the case of normal lymphocytes (GEWIRTZ et al. 1989), CCRF-CEM cells were extremely sensitive to the antiproliferative effects of the c-*myb* antisense ODN. When exposed to the sense ODN, we found negligible effects on CEM cell growth in short-term suspension cultures. In contrast, exposure to c-*myb* antisense DNA resulted in a daily decline in cell numbers. Compared to untreated controls, antisense DNA inhibited growth by approximately 2 logs. Growth reduction was not a cytostatic effect, since cell viability was reduced by only approximately 70% after exposure to the antisense ODN, and CEM cell growth did not recover when cells were left in culture for an additional 9 days.

Results obtained from primary patient material were equally encouraging (Table 1; CALABRETTA et al. 1991). We began by attempting to determine whether colony forming unit-leukemia (CFU-L) from AML patients could be inhibited by exposure to c-*myb* antisense ODN. Of the 28 patients we initially studied, colony and cluster data were available in 16 and 23 cases, respectively. After exposure to relatively low doses of c-*myb* antisense ODN (60 μg/ml), colony formation was inhibited in a statistically significant manner in 12 out of 16 patients (approximately 75%). Inhibition of cluster formation fell in a similar range. Of equal importance, the numbers of residual colonies in the antisense treated dishes was approximately 10%. An obvious problem with interpreting these results, however, was determining the nature of the residual

Table 1. Effect of c-*myb* oligomers on primary acute myelogenous leukemia (AML) cell colony/cluster formation

Case	Colonies (% control)			Clusters (% control)		
	Sense	Antisense	p^a	Sense	Antisense	p^a
1	86	18	0.058	60	37	0.080
4	NG	NG		90	28	0.036
5	NG	NG		70	22	0.101
6	NG	NG		79	22	0.026
7	170	100	0.423	76	128	0.502
8	92	11	0.008	96	46	0.020
10	NG	NG		190	216	0.034
11	45	14	0.021	58	21	0.084
14	68	01	0.152	90	53	0.071
15	66	81	0.736	100	100	0.896
16	NG	NG		66	24	0.001
17	NG	NG		16	8	0.023
18	NG	NG		110	77	0.164
19	113	116	0.717	91	91	0.763
20	92	09	0.051	100	50	0.009
21	94	00	0.006	90	06	0.004
22	80	13	0.001	103	11	0.015
23	63	06	0.001	74	27	0.004
24	87	17	0.002	91	26	0.018
25	100	00	0.019	107	38	0.364
26	76	00	0.009	89	00	0.001
27	79	21	0.014	59	18	0.043
28	88	20	0.009	94	152	0.096

Blast cells were isolated from the peripheral blood of AML patients and exposed to sense or antisense oligomers. Colonies and clusters were enumerated, and values were compared with growth in control cultures containing two oligomers. For each case, the number of colonies or clusters arising in the untreated control dishes was assumed to represent maximal (100%) growth for that patient. The numbers of colonies or clusters arising in the oligomer-treated dishes are expressed as a percentage of this number. NG, No growth.
[a] The statistical significance (determined by Student's *t* test for unpaired samples) of the change observed in the antisense-treated dishes relative to the untreated control is given as a *p* value.

cells, i.e., whether they were the progeny of residual normal CFU or CFU-L. To try to answer this question in a rigorous manner, we turned out attention to chronic myelogenous leukemia (CML), where the presence of the t(9:22) or bcr-abl neogene provided an unequivocal marker of the malignant cells (RATAJCZAK et al. 1992a). Exposure of CML cells to c-*myb* antisense ODN resulted in inhibition of GM-CFU-derived colony formation in 50% of patients evaluated, and thus far we have studied in excess of 40 patients. Representative data are shown in Fig. 2 and are presented as a function of

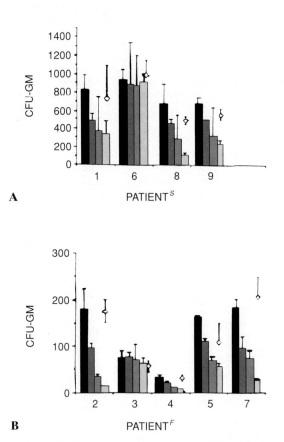

Fig. 2A,B. Effect of c-*myb* oligomers on chronic myeloid leukemia cell colony formation by cells with **A** "high" and **B** "low" cloning efficiency. Colony-forming cells were enriched from patient peripheral blood or bone marrow and exposed to oligomers as detailed in the text. At 24 h, cells were plated and resulting colonies were enumerated in plates containing untreated control cells (*black bars*), antisense-treated (20 μg/ml then 10 μg/ml, *dark hatched bars*; 40 μg/ml then 20 μg/ml, *tan bars*; 100 μg/ml then 50 μg/ml, *light hatched bars*), and sense-treated (100 μg/ml then 50 μg/ml, *diamonds*) cells. Values plotted are mean ± SD of actual colony counts compared to growth in control cultures that contained no oligomers

oligomer effect on cells with "greater" cloning efficiency (control colonies, 250 per plate; Fig. 2A) versus "lesser" cloning efficiency (control colonies, less than 250 per plate; Fig. 2B). In this particular study, colony formation was observed in eight of 11 patients evaluated and was statistically significant ($p \leq 0.03$) in seven. The amount of inhibition seen was dose dependent and ranged between 58% and 93%. In two patients, the effect of the c-*myb* oligomers on granulocyte erythrocyte macrophage monocyte (GEMM)-CFU colony formation was also determined to assess the effect of the oligomers on progenitors more primitive than GM-CFU. In each case, significant inhibition of GEMM-CFU-derived colony formation was noted. It is also important to note that colony inhibition was sequence specific. Finally, c-*myb* sense sequence ODN fails to significantly inhibit colony formation when employed at the highest antisense doses utilized.

3. Differential Reliance of Normal and Leukemic Progenitor Cells on c-*myb* Function

In order to be useful as a therapeutic target, leukemic cells would have to be more dependent on Myb protein than their normal counterparts. To examine this critical issue, we incubated phagocyte and T cell-depleted normal human MNC, human T lymphocyte leukemia cell line blasts (CCRF-CEM), or 1:1 mixtures of these cells with sense or antisense ODN to codons 2–7 of human c-*myb* mRNA (CALABRETTA et al. 1991). ODN were added to liquid suspension cultures at 0 h and 18 h. Control cultures were untreated. In controls or in cultures to which "high" doses of sense ODN were added, CCRF-CEM proliferated rapidly, whereas MNC numbers and viability decreased by less than 10%. In contrast, when CCRF-CEM were incubated for 4 days in c-*myb* antisense DNA, cultures contained $4.7 \pm 0.8 \times 10^4$ cells/ml (mean $\pm$ SD; n = 4) compared to $285 \pm 17 \times 10^4$ cells/ml in controls. At the effective antisense dose, MNC were largely unaffected. After four days in culture, remaining cells were transferred to methylcellulose supplemented with recombinant hematopoietic growth factors. Myeloid colonies/clusters were enumerated at day ten of culture inception. Depending on cell number plated, control MNC formed from 31 ± 4 to 274 ± 18 colonies. In dishes containing equivalent numbers of untreated or sense ODN exposed CCRF-CEM, colonies were too numerous to count. When MNC were mixed 1:1 with CCRF-CEM in antisense oligomer concentrations ≤ 5 µg/ml, only leukemic colonies could be identified by morphologic, histochemical and immunochemical analysis. However, when antisense oligomer exposure was intensified, normal myeloid colonies could now be found in the culture while leukemic colonies could no longer be identified with certainty using the same analytic methods. Finally, at antisense DNA doses used in the above studies, AML blasts from eighteen of twenty-three patients exhibited ~75% decrease in colony and cluster formation compared to untreated or sense oligomer treated controls. When 1:1 mixing experiments were carried out with primary AML blasts and normal MNC, we were again able to preferentially eliminate AML blast colony formation while normal myeloid colonies continued to form.

4. Use of c-*myb* Oligodeoxynucleotides as Bone Marrow-Purging Agents

The above experiments suggested that leukemic cell growth could be preferentially inhibited after exposure to c-*myb* antisense ODN. In contemplating a clinical use for our findings, application in the area of bone marrow transplantation seemed compelling. In this application, exposure conditions are entirely under the control of the investigator. In addition, the patient's exposure to the antisense DNA is minimal. This circumstance would also make approval by regulatory agencies less difficult. We therefore determined whether the antisense ODN could be utilized as ex vivo bone marrow-purging agents.

To examine this issue, normal MNC were mixed (1:1) with primary AML or CML blast cells and then exposed to the ODN using a slightly modified protocol designed to test the feasibility of a more intensive antisense exposure.

With this in mind, an additional ODN dose ($20\,\mu g/ml$) was given just prior to plating the cells in methylcellulose. In control growth factor-stimulated cultures, leukemic cells formed 25.5 ± 3.5 (mean $\pm$ SD) colonies and 157 ± 8.5 clusters (per 2×10^5 cells plated). Exposure to c-*myb* sense ODN did not significantly alter these numbers (19.5 ± 0.7 colonies and 140.5 ± 7.8 clusters, $p > 0.1$). In contrast, equivalent concentrations of antisense ODN totally inhibited colony and cluster formation by the leukemic blasts. Colony formation was also inhibited in the plates containing normal MNC, but only by approximately 50% in comparison to untreated control plates (control colony formation, 296 ± 40 per 2×10^5 cells plated; treated colony formation, 149 ± 15.5 per 2×10^5 cells).

To assess the potential effectiveness of an antisense purge, we carried out co-culture studies with cells obtained from CML patients in blast crisis (BC) and in the chronic phase (CP) of their disease (RATAJCZAK et al. 1992a). CML was a particularly useful model, because cells from the malignant clone carry a tumor-specific chromosomal translocation which can be easily identified in tissue culture by looking for bcr-abl, the mRNA product of the gene produced by the translocation (WITTE 1993). RNA was therefore extracted from cells cloned in methylcellulose cultures after exposure to the highest c-*myb* antisense ODN dose. The RNA was then reverse-transcribed and the resulting cDNA amplified. For each patient studied, mRNA was also extracted from a comparable number of cells derived from untreated control colonies using the same technique. Eight cases were evaluated, and in each case bcr-abl expression as detected by reverse transcription-polymerase chain reaction (RT-PCR) correlated with colony growth in cell culture. In cases which were inhibited by exposure to c-*myb* antisense ODNs (seven out of 11 patients), bcr-abl expression was also greatly decreased or not detectable (Fig. 3A). These results suggested that bcr-abl-expressing CFU might be substantially or entirely eliminated from a population of blood or marrow MNC by exposure

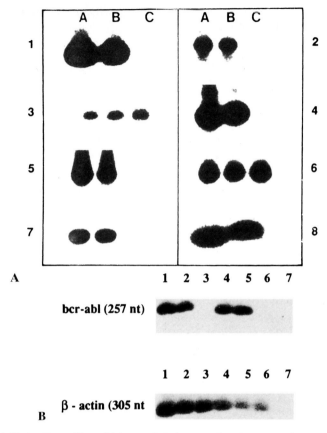

Fig. 3. A Detection of bcr-abl transcripts in granulocyte-macrophage colony-forming unit (GM-CFU)-derived colonies from marrow of eight patients (nos. 1–8) whose marrow was not exposed to c-*myb* antisense oligodeoxynucleotides (AS ODN, *A*), to c-*myb* sense ODN (*B*), or to c-*myb* antisense ODN (*C*). All colonies present in the variously treated methylcellulose cultures were harvested and subjected to analysis. Colony selection bias was therefore avoided. **B** Detection of bcr-abl transcripts in GM-CFU-derived (*lanes 1–3*) and GEMM-CFU-derived (*lanes 4–6*) colonies obtained from reseeded primary colonies of patient no. 8. Note that while β-actin transcripts are clearly detected in all colony samples, bcr-abl is only detectable in colonies derived from untreated control colonies (lanes 1, 4) and colonies previously exposed to Myb sense ODN (lanes 2, 5). Cells derived from colonies originally exposed to c-*myb* AS ODN do not have detectable bcr-abl-expressing cells. *Lane 7* is a control lane for the polymerase chain reaction (PCR) reactions and is appropriately empty

to the antisense ODN. To explore this possibility further, replating experiments were carried out on samples from two patients (Fig. 3B). We hypothesized that if CFU belonging to the malignant clone were present at the end of the original 12-day culture period, but not detectable due to their failure to express bcr-abl, they might reexpress the message upon regrowth in fresh cultures. Accordingly, cells from these patients were exposed to ODN and then plated into methylcellulose cultures formulated to favor growth of either GM-CFU or GEMM-CFU. As was found with the original specimens, untreated control cells and cells exposed to sense ODN had RT-PCR-detectable bcr-abl transcripts. Those exposed to the c-*myb* antisense ODN had none. One of the paired dishes from these cultures was then solubilized with fresh medium, and all cells contained therein were washed, disaggregated, and replated into fresh methylcellulose cultures without reexposing the cells to ODN. After 14 days, GM-CFU and GEMM-CFU colony cells were again probed for bcr-abl expression. Control and sense-treated cells had RT-PCR-detectable mRNA, but none was found in the antisense-treated colonies. These results suggest that elimination of bcr-abl-expressing cells and CFU was highly efficient and perhaps permanent.

5. Efficacy of c-*myb* Oligodeoxynucleotides In Vivo: Development of Animal Models

The studies described above were carried out primarily with unmodified DNA. Such molecules are subject to endo- and exonuclease attack at the phosphodiester bonds and are therefore of little utility in vivo. We therefore needed to address two questions at this point. First, we needed to know whether a more stable, chemically modified ODN would give similar results. Second, we needed to know whether these materials would be effective in an in vivo system against human leukemia cells. Since we could not give this material to patients, we established a human leukemia/severely compromised immunodeficient (SCID) mouse model system which would allow us to address both questions simultaneously (RATAJCZAK et al. 1992). To carry out these experiments, SCID mice were injected IV with K562 chronic myeloid leukemia cells after cyclophosphamide conditioning. K562 cells express c-*myb*, the antisense oligodeoxynucleotide target, and the tumor-specific bcr-abl oncogene, which was utilized for tracking the human leukemia cells in the mouse host. After tumor cell injection, animals developed blasts in the peripheral blood within 4–6 weeks. After peripheral blood blast cells appeared, the mean ($\pm$ SD) survival of untreated mice ($n = 20$) was 6 ± 3 days. Dying animals had prominent central nervous system (CNS) infiltration, marked infiltration of the ovary, and scattered abdominal granulocytic sarcomas. Infusion of either sense or scrambled-sequence c-*myb* phosphorothioate ODN (24 bp; codons 2–9) for 3, 7, or 14 days had no statistically significant effect on sites of disease involvement or on animal survival in comparison to control animals. In contrast, animals treated for 7 or 14 days with c-*myb* AS ODN survived 3.5 to eight times longer ($p < 0.001$) than the various control animals ($n = 60$; Fig. 4). In addition, animals receiving c-*myb* AS DNA had either rare microscopic foci or no obviously detectable CNS disease (Fig. 3) and a 50% reduction of ovarian involvement. A 3-day infusion of *myb* AS ODN (100 μg per day) was without effect. Infusing mice ($n = 12$) with AS ODN (200 μg per day for 14 days) complementary to the c-*kit* proto-oncogene, which K562 cells do not

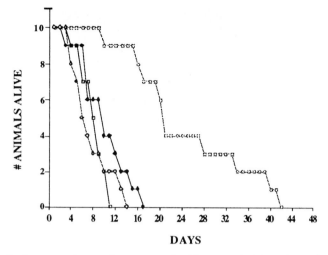

Fig. 4. Survival curves of severely compromised immunodeficient (SCID)–human chimeric animals transplanted with K562 chronic myelogenous leukemia (CML) cells. Animals received a 14-day infusion of oligomers at a dose of 100 µg/day. *Solid line, squares,* control; *solid line, diamonds,* sense; *zigzag line, squares,* antisense; *zigzag line, diamonds,* scrambled

express, also had no effect on disease burden or survival. These results suggested that phosphorothioate-modified c-*myb* antisense DNA might be efficacious for the treatment of human leukemia in vivo.

III. Why Downregulating *myb* Kills Leukemic Cells Preferentially: A Hypothesis

Our initial studies on the function of the c-*kit* receptor in hematopoietic cells suggested that c-*kit* might be a Myb-regulated gene (RATAJCZAK et al. 1992c,d). Since c-*kit* encodes a critical hematopoietic cell tyrosine kinase receptor (RATAJCZAK et al. 1992), we hypothesized that dysregulation of c-*kit* expression may be an important mechanism of action of Myb AS ODN. In support of this hypothesis, it has been shown that, when hematopoietic cells are deprived of c-*kit* receptor ligand (steel factor), they undergo apoptosis (YU et al. 1993). It has also recently been shown that when CD56bright NK cells, which express c-*kit*, are deprived of their ligand (steel factor), they too undergo apoptosis, perhaps because *bcl-*2 is downregulated (CARSON et al. 1994). Malignant myeloid hematopoietic cells, in particular CML cells, also express c-*kit* and respond to steel factor. Accordingly, we postulate that perturbation of Myb expression in malignant hematopoietic cells may force them to enter an apoptotic pathway by downregulating c-*kit*. Preliminary studies of K562 cells exposed to c-*myb* AS ODN demonstrate that such cells do in fact undergo nuclear degenerative changes characteristic of apoptosis (Fig. 5). Of necessity,

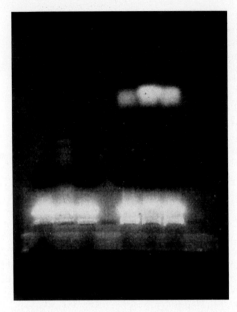

Fig. 5. c-*myb* antisense oligodeoxynucleotides (ODN) cause K562 cells to undergo apoptosis. K562 cells were exposed to c-*myb* sense or antisense ODN (approximately 20 µM) for 36 h. After this time, DNA was extracted from the cells, electrophoresed, and stained with ethidium bromide. *Lane 1*, untreated control (nuclear DNA is intact); *lane 2*, cells exposed to c-*myb* antisense ODN (note characteristic "laddering" of DNA); *lane 3*, cells exposed to c-*myb* sense ODN

we must also postulate that normal progenitor cells, at least at some level of development, are more tolerant of this transient disturbance. Since neither steel nor white spotting (W) mice (which lack the Kit receptor and its ligand, respectively) are aplastic, this is a tenable hypothesis (RATAJCZAK et al. 1992c).

IV. Pharmacodynamic Studies with an *myb*-Targeted Oligodeoxynucleotide

As we have noted in the past, understanding uptake mechanisms and intracellular handling might allow ODN to be used with enhanced biologic effectiveness. We have examined these issues at the ultrastructural level in the hope of gaining information that will be useful for oligonucleotide design and rational administration in the clinic (BELTINGER et al. 1995).

We first sought to identify ODN-binding proteins on hematopoietic cell surfaces (Fig. 6). We then attempted to follow ODN trafficking once inside the cell. To conduct these studies, K562 cells were incubated at 4°C with biotin-labeled ODN alone or with excess unlabeled ODN of identical sequence. Binding proteins were identified by cross-linking them to bound ODN, extracting the complexes with 2% NP-40, resolving them on sodium dodecyl

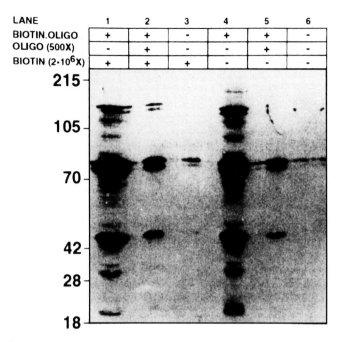

Fig. 6. K562 cells were incubated at 4°C with biotin-labeled phosphorothioate (PS)-oligodeoxynucleotides (ODN). Proteins to which the PS-ODN bound were identified by chemical cross-linking with BS3, followed by extraction of the complexes with 2% NP-40. These were subsequently resolved on sodium dodecyl sulfate polyacrylamide gel electrophoresis (SDS-PAGE) gels and identified colorometrically by western blotting using chemiluminescence. We identified five doublet bands representing ODN-BP ranging in size from approximately 20 kDa to approximately 143 kDa (*lanes 1, 4*). Excess unlabeled PS-ODN (*lanes 2, 5*), but not free biotin (*lane 1*), inhibited PS-ODN binding, suggesting specificity of BP interactions. D-Biotin not conjugated to ODN could not be detected bound to ODN-BP (*lane 2*). Molecular weight markers are depicted on the left. *Lane 6 is empty*

sulfate polyacrylamide gel electrophoresis (SDS-PAGE) gels, followed by colorometric detection. In contrast to previous reports (LOKE et al. 1989), we identified at least five major ODN binding proteins (BP), ranging in size from approximately 20 to 143 kDa. Excess unlabeled PS-ODN (500-fold), but not free biotin, inhibited PS-ODN binding, suggesting specificity of BP interactions. Neuraminidase treatment of the cells prior to incubation with ODN decreased binding to most ODN-BP, indicating that sugar moieties may play a role in binding. In other experiments, ODN-binding proteins were examined after metabolic labeling with [^{35}S]methionine. Binding proteins of similar number, migration pattern, and relative abundance were again identified after gel resolution under nondenaturing conditions. Thus the ODN-binding proteins on hematopoietic cells are not composed of subunits.

To visualize receptor-mediated endocytosis, K562 cells were again incubated at 4°C with biotinylated ODN. After cross-linking, cells were incubated with gold-streptavidin particles, warmed to 37°C, fixed, and then processed for electron microscopy (EM). ODN were clearly identified in clathrin-coated pits, consistent with receptor-mediated endocytosis. EM studies employing gold-streptavidin particles and sbAS ODN (biotinylated AS ODN) were also carried out to visualize intracellular trafficking. ODN were seen in endosomes, lysosome-like bodies, and throughout the cytoplasm. A significant amount of labeled material was also observed in the nucleus.

Finally, it is clearly important that, in order to document an "antisense" mechanism, it is necessary to correlate mRNA levels of the targeted gene with biologic effects observed. We have demonstrated that this is feasible and that one can adequately follow target gene mRNA levels in tissues of animals receiving ODN (Fig. 7; Hijiya et al. 1994).

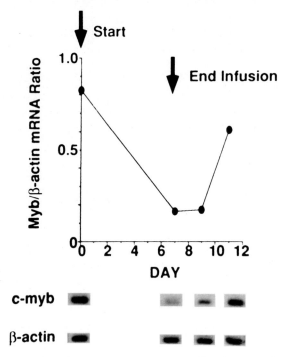

Fig. 7. Effect of c-*myb* antisense oligodeoxynucleotides (AS ODN) on *myb* mRNA expression in tumor tissue obtained from human melanoma-bearing severely compromised immunodeficient (SCID) mice. Mice were infused with the AS ODN (500 µg/day for 7 days) and tumors were excised on days 7, 9, and 11 after infusions were begun. c-*myb* and β-actin mRNA was detected in the same tumor tissue sample (approximately 1 g) by semiquantitative reverse transcriptase-polymerase chain reaction (RT-PCR) and quantitated by scanning densitometry. The relative amount of c-*myb* mRNA in each sample was estimated by normalization to the actin mRNA present in each sample

C. Use of Antisense Oligonucleotides in a Clinical Setting

For the purpose of developing an antisense oligonucleotide therapeutic, CML seemed to us to be an excellent disease model. As mentioned above, CML is relatively common and has a convenient marker chromosome and gene for objectively following potential therapeutic efficacy of a test compound (GALE et al. 1993). In addition to these considerations, CML is uniformly fatal except for in individuals who are fortunate enough to have an allogeneic bone marrow donor. Picking a gene target in CML was actually somewhat problematic. An obvious target was the bcr-abl gene-encoded mRNA [Fig. 8; MELO (1996)]. However, because bcr-abl is not expressed in primitive hematopoietic stem cells (BEDI et al. 1993) and because it is uncertain whether transient interruption of bcr-abl signaling actually results in the death of CML cells, we felt that an alternative target might be of greater use in treating this disease. Based on the type of data presented above, a favorable therapeutic index in toxicology testing, and more detailed knowledge of the pharmacokinetics of oligonucleotides, we have begun to evaluate *myb*-targeted AS ODN in the clinic (GEWIRTZ et al. 1996a).

To this end, we initiated clinical trials to evaluate the effectiveness of phosphorothioate-modified ODN antisense to the c-*myb* gene as marrow-purging agents for CP or accelerated phase (AP) CML patients; in addition, a phase I intravenous infusion study for BC patients and patients with other refractory leukemias was performed. ODN purging was carried out for 24h on $CD34^+$ marrow cells. Patients received busulfan and cytoxan, followed by reinfusion of previously cryopreserved PS-ODN-purged MNC. In the pilot

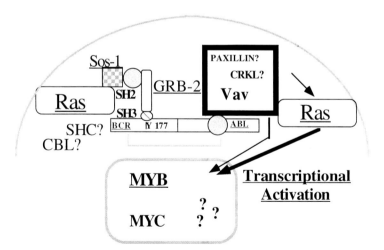

Fig. 8. Molecular pathogenesis of chronic myelogenous leukemia (CML). Bcr-abl, product of the *bcr-abl* gene created by the t(9:22) cooperates with a number of signaling partners to activate Ras. Ras activation is postulated to play a critical role in hematopoietic cell transformation. *bcr-abl*, breakpoint cluster region-abelson oncogene

marrow-purging study, seven CP and one AP CML patients have been treated. Seven out of eight engrafted. In four out of six evaluable CP patients, metaphases were 85%–100% of normal 3 months after engraftment, suggesting that a significant purge had taken place in the marrow graft. Five CP patients have demonstrated marked, sustained, hematologic improvement with essential normalization of their blood counts. Follow-up ranges from 6 months to approximately 2 years. In an attempt to further increase purging efficiency, we incubated patient MNC for 72 h in P-ODN. Although PCR and long-term culture-initiating cell (LTCIC) studies suggested that a very efficient purge had occurred, engraftment in five patients was poor. In the phase I systemic infusion study, there were 18 refractory leukemia patients (two patients were treated at two different dose levels; 13 had AP or BC CML). Myb AS ODN was delivered by continuous infusion at dose levels ranging between 0.3mg/kg to 2.0mg/kg per day for 7 days. No recurrent dose-related toxicity has been noted, although idiosyncratic toxicities, not clearly drug related, were observed (one transient renal insufficiency, one pericarditis). One BC patient survived approximately 14 months with transient restoration of CP disease. These studies show that ODN may be administered safely to leukemic patients. Whether patients treated on either study derived clinical benefit is uncertain, but the results of these studies suggest to us that ODN may eventually demonstrate therapeutic utility in the treatment of human leukemias.

D. Future Outlook

The power of the antisense approach has been demonstrated in experiments in which critical biological information has been gathered using antisense technology and has been subsequently verified by other laboratories using other methodologies (Gewirtz and Calabretta 1988; Metcalf 1994; Mucenski et al. 1991). However, this technology, in spite of its successes, has been found to be highly variable in its efficiency. To the extent that many have tried to employ ODN and more than a few have been perplexed and frustrated by results that were at best noninformative or, even worse, misleading or nonreproducible, it is easy to understand why this approach has become somewhat controversial. We believe that progress on two fronts would help address this problem.

First, in order for an ODN to hybridize with its mRNA target, it must find an accessible sequence. Sequence accessibility is at least in part a function of mRNA physical structure, which in turn is dictated by internal base composition and associated proteins in the living cell. Attempts to describe the in vivo structure of RNA, in contrast to DNA, have been fraught with difficulty (Baskerville and Ellington 1995). Accordingly, mRNA targeting is largely a hit or miss process, accounting for many experiments in which the addition of an ODN yields no effect on expression. Hence, the ability to determine which regions of a given mRNA molecule are accessible for ODN targeting is a significant impediment to the application of this technique in many cell systems. We have begun to approach this issue by developing a footprinting assay to determine which physical areas of an RNA are accessible to the oligonucleotide. We have proceeded under the assumption that sequence which remains accessible to single-stranded RNases in a more physiologic environment may also remain accessible for hybridization with an ODN. Preliminary experiments performed in our laboratory in which a labeled RNA transcript is allowed to hybridize with an oligonucleotide in the presence or absence of nuclear extracts from the cells of interest along with RNase T1 suggest that footprinting of this type is feasible (Fig. 9). Of greater interest, our preliminary results suggest that this approach may be of use in designing oligonucleotides.

Second, the ability to deliver ODN into cells and to have them reach their target in a bioavailable form also remains problematic (GEWIRTZ et al. 1996b). Without this ability, it is clear that even an appropriately targeted sequence is not likely to be efficient. Native phosphodiester ODN, and the widely used phosphorothioate-modified ODN, which contain a single sulfur substituting for oxygen at a nonbridging position at each phosphorus atom, are polyanions. Accordingly, they diffuse across cell membranes poorly and are only taken up by cells through energy-dependent mechanisms. This appears to be accomplished primarily through a combination of adsorptive endocytosis and fluid-phase endocytosis, which may be triggered in part by the binding of the ODN to receptor-like proteins present on the surface of a wide variety of cells (BELTINGER et al. 1995; LOKE et al. 1989). After internalization, confocal and EM studies have indicated that the bulk of the ODN enter the endosome/lysosome compartment. These vesicular structures may become acidified and

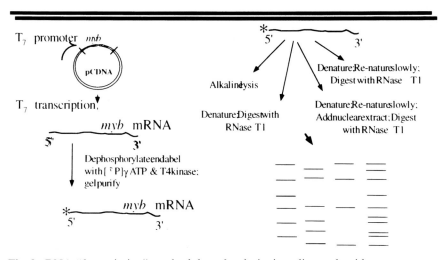

Fig. 9. RNA "footprinting" methodology for designing oligonucleotides

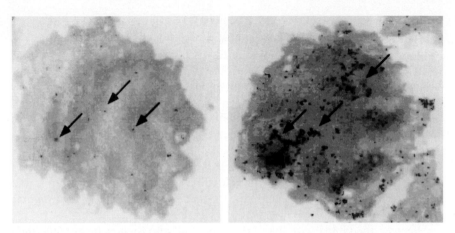

Fig. 10. Uptake of phosphorothioate oligodeoxynucleotides delivered by streptolysin permeabilization (*left*) or cationic lipid (*right*). Oligonucleotides were synthesized with biotin tags so that they could be decorated with avidin-gold particles which appear as *black dots* in the photomicrographs. Most of the material appears to be localized to the nucleus of the cells

acquire other enzymes which degrade the ODN. Biologic inactivity is the predictable result of this process. Recently described strategies for introducing ODN into cells, including various cationic lipid formulations, may address this problem (Fig. 10; BERGAN et al. 1996; LEWIS et al. 1996; SPILLER and TIDD 1995).

E. Conclusions

The ability to block gene function with AS ODN has become an important tool in many research laboratories. Since activation and aberrant expression of proto-oncogenes appears to be an important mechanism in malignant transformation, targeted disruption of these genes and other molecular targets with ODN could have significant therapeutic utility as well. In this regard, the potential therapeutic usefulness of ODN has been demonstrated in many systems and against a number of different targets, including viruses, oncogenes, proto-oncogenes, and an increasing array of cellular genes. These studies in aggregate suggest that synthetic ODN have the potential to become an important new therapeutic agent for the treatment of human cancer. Nevertheless, it is clear that considerable optimization will be required before AS ODN will emerge as an effective agent for treating human disease. Progress will need to occur on several fronts. These include issues related to the chemistry of the molecules employed, e.g., how chemical modification impacts on uptake, stability, and hybridization efficiency of the synthetic DNA molecule. A clearer understanding of the mechanism of antisense-mediated inhibition, including where such inhibition takes place, will also be required. Finally, cellular "defense" mechanisms, such as increasing transcription of the targeted message, may also be factors to consider in planning effective treatment strategies with these agents. Choice of target is clearly also an important issue. Nevertheless, while many issues remain to be resolved, we remain optimistic that this approach will one day prove useful for the treatment of patients with a variety of hematologic malignancies.

Acknowledgements. This work was supported by grants from the National Institutes of Health (NIH) and the Leukemia Society of America.

References

Anfossi G, Gewirtz AM, Calabretta B (1989) An oligomer complementary to c-myb-encoded mRNA inhibits proliferation of human myeloid leukemia cell lines. Proc Natl Acad Sci USA 86:3379–3383

Barletta C, Pelicci PG, Kenyon LC, Smith SD, Dalla-Favera R (1987) Relationship between the c-myb locus and the 6q-chromosomal aberration in leukemias and lymphomas. Science 235:1064–1067

Baskerville S, Ellington AD (1995) RNA structure. Describing the elephant. Curr Biol 5:120–123

Bedi A, Zehnbauer BA, Collector MI, Barber JP, Zicha MS, Sharkis SJ, Jones RJ (1993) BCR-ABL gene rearrangement and expression of primitive hematopoietic progenitors in chronic myeloid leukemia. Blood 81:2898–2902

Beltinger C, Saragovi HU, Smith RM, LeSauteur L, Shah N, DeDionisio L, Christensen L, Raible A, Jarett L, Gewirtz AM (1995) Binding, uptake, and intracellular trafficking of phosphorothioate-modified oligodeoxynucleotides. J Clin Invest 95:1814–1823

Bergan R, Hakim F, Schwartz GN, Kyle E, Cepada R, Szabo JM, Fowler D, Gress R, Neckers L (1996) Electroporation of synthetic oligodeoxynucleotides: a novel technique for ex vivo bone marrow purging. Blood 88:731–741

Biedenkapp H, Borgmeyer U, Sippel AE, Klempnauer KH (1988) Viral myb oncogene encodes a sequence-specific DNA-binding activity. Nature 335:835–837

Burk O, Mink S, Ringwald M, Klempnauer KH (1993) Synergistic activation of the chicken mim-1 gene by v-myb and C/EBP transcription factors. Embo J 12:2027–2038

Calabretta B, Sims RB, Valtieri M, Caracciolo D, Szczylik C, Venturelli D, Ratajczak M, Beran M, Gewirtz AM (1991) Normal and leukemic hematopoietic cells manifest differential sensitivity to inhibitory effects of c-myb antisense oligodeoxynucleotides: an in vitro study relevant to bone marrow purging. Proc Natl Acad Sci USA 88:2351–2355

Caracciolo D, Venturelli D, Valtieri M, Peschle C, Gewirtz AM, Calabretta B (1990) Stage –related proliferative activity determines c-myb functional requirements during normal human hematopoiesis. J Clin Invest 85:55–61

Carson WE, Halder S, Baiocchi RA, Croce CM, Caligiuri MA (1994) The c-kit ligand suppresses apoptosis of human natural killer cells through the upregulation of bcl-2. Proc Natl Acad Sci USA 91:7553

Cogswell JP, Cogswell PC, Kuehl WM, Cuddihy AM, Bender TM, Engelke U, Marcu KB, Ting JP (1993) Mechanism of c-myc regulation by c-Myb in different cell lineages. Mol Cell Biol 13:2858–2869

Dini PW, Eltman JT, Lipsick JS (1995) Mutations in the DNA-binding and transcriptional activation domains of v-Myb cooperate in transformation. J Virol 69:2515–2524

Gale RP, Grosveld G, Canaani E, Goldman JM (1993) Chronic myelogenous leukemia: biology and therapy. Leukemia 7:653–658

Gewirtz AM, Calabretta B (1988) A c-myb antisense oligodeoxynucleotide inhibits normal human hematopoiesis in vitro. Science 242:303–306

Gewirtz AM, Anfossi G, Venturelli D, Valpreda S, Sims R, Calabretta B (1989) G1/S transition in normal human T-lymphocytes requires the nuclear protein encoded by c-myb. Science 245:180–183

Gewirtz AM, Luger S, Sokol D, Gowdin B, Stadtmauer E, Reccio A, Ratajczak MZ (1996a) Oligodeoxynucleotide therapeutics for human myelogenous leukemia: interim results. Blood 88 [Suppl 1]:270a

Gewirtz AM, Stein CA, Glazer PM (1996b) Facilitating oligonucleotide delivery: helping antisense deliver on its promise. Proc Natl Acad Sci USA 93:3161–3163

Hijiya N, Zhang J, Ratajczak MZ, Kant JA, DeRiel K, Herlyn M, Zon G, Gewirtz AM (1994) Biologic and therapeutic significance of MYB expression in human melanoma. Proc Natl Acad Sci USA 91:4499–4503

Kanei-Ishii C, MacMillan EM, Nomura T, Sarai A, Ramsay RG, Aimoto S, Ishii S, Gonda TJ (1992) Transactivation and transformation by Myb are negatively regulated by a leucine-zipper structure. Proc Natl Acad Sci USA 89:3088–3092

Ku DH, Wen SC, Engelhard A, Nicolaides NC, Lipson KE, Marino TA, Calabretta B (1993) c-myb transactivates cdc2 expression via Myb binding sites in the 5'-flanking region of the human cdc2 gene [published erratum appears in J Biol Chem 1993 Jun 15; 268(17):13010]. J Biol Chem 268:2255–2259

Lewis JG, Lin KY, Kothavale A, Flanagan WM, Matteucci MD, DePrince RB, Mook RA Jr, Hendren RW, Wagner RW (1996) A serum-resistant cytofectin for cellular delivery of antisense oligodeoxynucleotides and plasmid DNA. Proc Natl Acad Sci USA 93:3176–3181

Loke SL, Stein CA, Zhang XH, Mori K, Nakanishi M, Subasinghe C, Cohen JS, Neckers LM (1989) Characterization of oligonucleotide transport into living cells. Proc Natl Acad Sci USA 86:3474–3478

Luger SM, Ratajczak J, Ratajczak MZ, Kuczynski WI, DiPaola RS, Ngo W, Clevenger CV, Gewirtz M (1996) A functional analysis of protooncogene Vav's role in adult human hematopoiesis. Blood 87:1326–1334

Lyon J, Robinson C, Watson R (1994) The role of Myb proteins in normal and neoplastic cell proliferation. Crit Rev Oncog 5:373–388

Melo JV (1996) The molecular biology of chronic myeloid leukaemia. Leukemia 10:751–756

Melotti P, Ku DH, Calabretta B (1994) Regulation of the expression of the hematopoietic stem cell antigen CD34: role of c-myb. J Exp Med 179:1023–1028

Metcalf D (1994) Blood. Thrombopoietin – at last. Nature 369:519–520

Mucenski ML, McLain K, Kier AB, Swerdlow SH, Schreiner CM, Miller TA, Pietryga DW, Scott WJ Jr, Potter SS (1991) A functional c-myb gene is required for normal murine fetal hepatic hematopoiesis. Cell 65:677–689

Nakayama K, Yamamoto R, Ishii S, Nakauchi H (1993) Binding of c-Myb to the core sequence of the CD4 promoter. Int Immunol 5:817–824

Ness SA, Marknell A, Graf T (1989) The v-myb oncogene product binds to and activates the promyelocyte-specific mim-1 gene. Cell 59:1115–1125

Nomura N, Zu YL, Maekawa T, Tabata S, Akiyama T, Ishii S (1993) Isolation and characterization of a novel member of the gene family encoding the cAMP response element-binding protein CRE-BP1. J Biol Chem 268:4259–4266

Press RD, Reddy EP, Ewert DL (1994) Overexpression of C-terminally but not N-terminally truncated Myb induces fibrosarcomas: a novel nonhematopoietic target cell for the myb oncogene. Mol Cell Biol 14:2278–2290

Ratajczak MZ, Hijiya N, Catani L, DeRiel K, Luger SM, McGlave P, Gewirtz AM (1992a) Acute-and chronic-phase chronic myelogenous leukemia colony-forming units are highly sensitive to the growth inhibitory effects of c-myb antisense oligodeoxynucleotides. Blood 79:1956–1961

Ratajczak MZ, Kant JA, Luger SM, Hijiya N, Zhang J, Zon G, Gewirtz AM (1992b) In vivo treatment of human leukemia in a scid mouse model with c-myb antisense oligodeoxynucleotides. Proc Natl Acad Sci USA 89:11823–11827

Ratajczak MZ, Luger SM, DeRiel K, Abrahm J, Calabretta B, Gewirtz AM (1992c) Role of the KIT protooncogene in normal and malignant human hematopoiesis. Proc Natl Acad Sci USA 89:1710–1714

Ratajczak MZ, Luger SM, Gewirtz AM (1992d) The c-kit proto-oncogene in normal and malignant human hematopoiesis. Int J Cell Cloning 10:205–214

Sakura H, Kanei-Ishii C, Nagase T, Nakagoshi H, Gonda TJ, Ishii S (1989) Delineation of three functional domains of the transcriptional activator encoded by the c-myb protooncogene. Proc Natl Acad Sci USA 86:5758–5762

Small D, Levenstein M, Kim, E, Carow C, Amin S, Rockwell P, Witte L, Burrow C, Ratajczak MZ, Gewirtz AM et al (1994) STK-1, the human homolog of Flk-2/Flt-3, is selectively expressed in CD34+ human bone marrow cells and is involved in the proliferation of early progenitor/stem cells. Proc Natl Acad Sci USA 91:459–463

Spiller DG, Tidd DM (1995) Nuclear delivery of antisense oligodeoxynucleotides through reversible permeabilization of human leukemia cells with streptolysin O. Antisense Res Dev 5:13–21

Szczylik C, Skorski T, Ku DH, Nicolaides NC, Wen SC, Rudnicka L, Bonati A, Malaguarnera L, Calabretta B (1993) Regulation of proliferation and cytokine expression of bone marrow fibroblasts: role of c-myb. J Exp Med 178:997–1005

Takeshita K, Bollekens JA, Hijiya N, Ratajczak M, Ruddle FH, Gewirtz A M (1993) A homeobox gene of the Antennapedia class is required for human adult erythropoiesis. Proc Natl Acad Sci USA 90:535–538

Travali S, Reiss K, Ferber A, Petralia S, Mercer WE, Calabretta B, Baserga R (1991) Constitutively expressed c-myb abrogates the requirement for insulinlike growth factor 1 in 3T3 fibroblasts. Mol Cell Biol 11:731–736

Vorbrueggen G, Kalkbrenner F, Guehmann S, Moelling K (1994) The carboxyterminus of human c-myb protein stimulates activated transcription in trans. Nucleic Acids Res 22:2466–2475

Weber BL, Westin EH, Clarke MF (1990) Differentiation of mouse erythroleukemia cells enhanced by alternatively spliced c-myb mRNA. Science 249:1291–1293

Witte ON (1993) Role of the BCR-ABL oncogene in human leukemia: fifteenth Richard and Hinda Rosenthal Foundation Award Lecture. Cancer Res 53:485–489

Yu H, Bauer B, Lipke GK, Phillips RL, Van Zant G (1993) Apoptosis and hematopoiesis in murine fetal liver. Blood 81:373–384

CHAPTER 18
Properties of ISIS 2302, an Inhibitor of Intercellular Adhesion Molecule-1, in Humans

W.R. SHANAHAN JR.

A. Introduction

ISIS 2302 is a 20-base phosphorothioate oligodeoxynucleotide designed to specifically hybridize to a sequence in the 3′-untranslated region of the human intercellular adhesion molecule (ICAM)-1 message. ICAM-1, a member of the immunoglobulin superfamily, is an inducible transmembrane glycoprotein constitutively expressed at low levels on vascular endothelial cells and on a subset of leukocytes (DUSTIN et al. 1986; ROTHLEIN et al. 1986; SIMMONS et al. 1988). In response to proinflammatory mediators, many cell types upregulate expression of ICAM-1 on their surface. The primary counterligands for ICAM-1 are the β_2-integrins lymphocytic function-associated antigen (LFA)-1 and macrophage differentiation antigen (Mac)-1, expressed on leukocytes (MARLIN and SPRINGER 1987; DIAMOND et al. 1990).

ICAM-1 serves multiple functions in the propagation of inflammatory processes, the best characterized being facilitation of leukocyte emigration from the intravascular space in response to inflammatory stimuli (BUTCHER 1991; FURIE et al. 1991; OPPENHEIMER-MARKS et al. 1991). ICAM-1 also appears to provide an important secondary signal during antigen presentation (ALTMANN et al. 1989; VAN SEVENTER et al. 1990; KUHLMAN et al. 1991) and to play an important facilitatory role in cytotoxic T cell (MAKGOBA et al. 1988), natural killer (NK) cell (ALLAVENA et al. 1991), and neutrophil-mediated (ENTMAN et al. 1992) damage to target cells. Any or all of the above functions makes ICAM-1 a theoretically attractive therapeutic target for a broad spectrum of inflammatory and autoimmune diseases.

Numerous studies have demonstrated an increase in ICAM-1 expression within involved tissues from patients suffering from a wide range of diseases with an inflammatory component, including inflammatory dermatoses, inflammatory bowel disease, rheumatoid arthritis, glomerulonephritis, systemic lupus, vasculitis, atherosclerosis, organ allograft rejection, multiple sclerosis, and reperfusion injury.

ICAM-1 monoclonal antibodies have been used to demonstrate beneficial effects in a variety of animal models of disease, including pulmonary inflammation and asthma (BARTON et al. 1989; WEGNER et al. 1990), prevention of allograft rejection (COSIMI et al. 1990; ISOBE et al. 1992), nephritis (HARNING et al. 1992; KAWASAKI et al. 1993), ischemic injury (MA et al. 1992; KELLY et al.

1994), arthritis (IIGO et al. 1991), and contact dermatitis (SCHEYNIUS et al. 1993), further suggesting that inhibitors of ICAM-1 function or expression could have broad therapeutic utility.

ISIS 2302 selectively inhibits cytokine-induced ICAM-1 expression in a wide variety of human cells in vitro (BENNETT et al. 1994; MIELO et al. 1994; NESTLE et al. 1994). As discussed in Chap. 12, a murine analogue, ISIS 3082, has been shown to be active in multiple models of inflammation, including prolongation of cardiac allograft survival (STEPKOWSKI et al. 1994), carrageenan-induced neutrophil infiltration (Isis Pharmaceuticals, unpublished data), dextran sulfate-induced colitis (BENNETT et al., in press), endotoxin-induced neutrophil migration (KUMASAKA et al. 1996), and collagen-induced arthritis (C.F. Bennett et al., unpublished observation). In each study, control oligonucleotides failed to demonstrate pharmacological activity, suggesting that the anti-inflammatory activity of ISIS 3082 was due to inhibition of ICAM-1 expression. In the endotoxin pneumonitis and the dextran colitis models, downregulation of either ICAM-1 message or protein in the involved tissue was demonstrated.

As discussed in Chap. 5, doses of ISIS 2302 up to 100mg/kg and of ISIS 3082, the murine analogue of ISIS 2302, have been administered i.v. every other day for 4 weeks to mice (HENRY et al. 1997a), and doses of ISIS 2302 up to 50mg/kg i.v. every other day for 4 weeks to monkeys (HENRY et al. 1997b). The simian ICAM-1 message differs by only one of 20 bases in the targeted region of the message, and ISIS 2302 therefore has substantial pharmacological activity (IC_{50}, approximately twice that of the simian analogue) in vitro in simian endothelial cells. As discussed in Chaps. 5 and 8, the major toxicological finding in mice, as with other phosphorothioate oligodeoxynucleotides, was a dose-related, nonspecific immune stimulation consisting of splenic and lymph node hypertrophy, polyclonal B cell hyperplasia, and a mixed mononuclear cell infiltrate in various organs. In monkeys, nonspecific immune stimulation was not observed. However, dose-related and transient prolongation of the activated partial thromboplastin time (aPTT) and alternative pathway complement activation were demonstrated. Both laboratory effects were clearly related to peak plasma drug levels, with a threshold of approximately 50µg/ml for complement activation and approximately 30µg/ml for prolongation of aPTT.

In rodents and primates, pharmacokinetic studies demonstrated similar behavior, with a plasma distribution half-life on the order of 30–45min but with "tissue half-lives" of 1–3 days. These data suggested that an every-other-day dose regimen would be appropriate for humans, with minimal tissue accumulation.

In planning initial exposure in humans, it was felt that a peak plasma level of 10–15µg/ml, or approximately 20%–30% of the threshold for complement activation in monkeys, would provide an adequate margin of safety for initial phase I studies. Extrapolating from monkey data, it was projected that a 2-mg/kg 2-h infusion would produce such peak drug levels in humans. Furthermore,

2 mg/kg was within the therapeutic range in animal models (see Chap. 12). A dose of 2 mg/kg by 2-h i.v. infusion was therefore selected as an appropriate target regimen for the initial phase I trial.

ISIS 2302 has successfully completed a phase I trial in volunteers in whom multiple doses of 0.5–2 mg/kg administered i.v. every other day by 2-h infusion were well tolerated. The compound is presently being evaluated in small (20–40 patients) phase II i.v. trials in five indications: Crohn's disease, rheumatoid arthritis, ulcerative colitis, psoriasis, and prophylaxis of acute renal allograft rejection. All but the psoriasis study are placebo controlled and double-blinded. In these trials, ISIS 2302 in doses of 0.5–2 mg/kg is being administered three times weekly for 2–4 weeks by 2-h i.v. infusions. In addition, a phase I subcutaneous trial of the intravenous formulation (phosphate-buffered saline, PBS) has just been completed. Data from the phase I trials and the status of the phase II trials are presented.

B. Phase I Intravenous Trial

The purpose of the first clinical trial with ISIS 2302 was to assess the safety and pharmacokinetics of intravenous administration of an anti-ICAM-1 oligodeoxynucleotide in healthy subjects before commencing pilot therapeutic trials in target disease states.

I. Methods and Materials

The phase I i.v. trial was a double-blinded, placebo-controlled, randomized (ISIS 2302 to placebo ratio, 3:1) study (GLOVER et al. 1997). Four healthy male volunteers were enrolled in each of seven single-dose (0.06, 0.12, 0.24, 0.5, 1.0, 1.5, and 2.0 mg/kg) and multiple-dose groups (0.2, 0.5, 1.0, and 2.0 mg/kg every other day for four doses). Groups were studied in a rising-dose fashion, and multiple dosing commenced after the first five single-dose groups had completed the trial.

ISIS 2302 (or placebo) was administered by i.v. infusion in a volume of 80 ml over 2 h. Sterile normal saline was used as placebo.

Subjects remained recumbent, with continuous electrocardiographic (ECG) monitoring for 4 h after the beginning of each infusion. The following were measured before and at intervals after each infusion: supine blood pressure and pulse, clotting screen (aPTT; thrombin time, TT; prothrombin time, PT), serum complement split products (C3a, C5a), neutrophil count (single-dose groups only), urine microproteins (retinol-binding protein, N-acetyl glucosaminidase, microalbumin; single-dose groups only), and standard laboratory safety screen (hematology, blood biochemistry, and urinalysis). Serum samples were collected from multiple-dose groups at 14 and 21 days after the last infusion to be analyzed for the presence of antibodies to ISIS 2302.

Blood samples were taken for plasma ISIS 2302 concentration before and up to 24 h after the beginning of infusion from all dose groups. More complete pharmacokinetic profiling was performed and urine was collected up to 12 h after the beginning of infusion in the 0.50- and 2.0-mg/kg single- and multiple-dose groups.

Complement split products were measured by commercially available C3a and C5a desArg ^{125}I assay kits (Amersham).

Plasma was examined for the presence of antidrug antibodies using a modification of a previously described enzyme-linked immunosorbent assay (ELISA) methodology (Lacy and Voss 1989). Uncoated areas on plates were blocked by incubating with 2% nonfat, dried Carnation milk powder. Medium from a hybridoma cell culture line producing monoclonal antibodies which recognize ISIS 2302 served as a positive control. ISIS 2302 (and other phosphorothioate oligonucleotides) does not appear to be antigenic; consequently, monoclonal antibodies to ISIS 2302 were raised by immunizing mice with ISIS 2302 conjugated to keyhole limpet hemocyanin.

Drug analysis was performed by capillary gel electrophoresis (CGE), as previously described (Leeds et al. 1996), on triplicate aliquots from each sample of plasma and urine. A phosphorothioate oligonucleotide composed of 27 thymidine nucleotides (T_{27}) was added to both plasma and urine as an internal standard. The linear range of concentrations of oligonucleotides detectable in plasma using this method is $10\,nM$ to $20\,\mu M$ (approximately 0.07–140 µg/ml).

II. Results

1. Safety

All single- and multiple-dose regimens of ISIS 2302 were well tolerated. ECG, supine blood pressure, and pulse were unaffected by dosing. No serious adverse events were reported, and no adverse event appeared to occur with disproportionate frequency in the drug-treated groups, with the exception of headache (seven ISIS 2302, one placebo; expected ratio, 3:1). However, the incidence of headache was not dose related. With the exceptions of transient increases in aPTT and C3a levels, there were no dose-related differences between placebo- and drug-treated patients in the median percentage change from baseline in laboratory safety variables.

A consistent and dose-related increase in aPTT was seen in subjects who received single or multiple doses of 0.5 mg ISIS 2302 per kg and above (Fig. 1). The maximum increase in aPTT occurred between 1 and 2 h after the beginning of infusion, and values returned to baseline (or below) within 2–4 h after the end of infusion. The greatest median increase in aPTT (expressed as a percentage of control value) was 75% (51.8 s) in the 2.0-mg/kg single-dose group. There was no evidence of exaggeration or attenuation of this effect with multiple dosing.

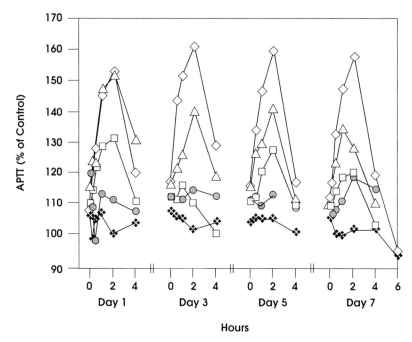

Fig. 1. Median activated partial thrombin time (*APTT*) expressed as a percentage of control following each of four doses of ISIS 2302 ($n = 3$) or placebo ($n = 4$). Placebo (*cross*) or ISIS 2302 (0.2 mg/kg, *circle*; 0.5 mg/kg, *square*; 1.0 mg/kg, *triangle*; 2.0 mg/kg, *diamond*) was administered by 2-h intravenous infusion on days 1, 3, 5, and 7 (GLOVER et al. 1997)

Much smaller, clinically insignificant, and less clearly dose-related increases in PT and TT occurred in subjects who received ISIS 2302. TT rose slightly above the reference range in some cases, but PT did not. The maximum median increases in these variables (approximately 6% for PT and 10% for TT) occurred between 30 and 120 min after the beginning of a single infusion and recovery occurred within 2 h from this time. Bleeding time was not changed by ISIS 2302.

Complement C5a split product concentrations remained unchanged throughout the study in all subjects, and single-dose groups showed no evidence of drug-related increases in complement C3a concentrations. In subjects in the multiple-dose groups, a small increase in median C3a concentration was seen after the fourth dose of 1.0 mg ISIS 2302 per kg and the third and fourth doses of 2.0 mg/kg (Fig. 2). These increases were not of clinical significance and were not associated with symptoms or changes in vital signs. The maximum change in this variable (up to 130% median increase from baseline) was seen at 60 or 120 min after the beginning of infusion, with almost complete recovery occurring by 240 min.

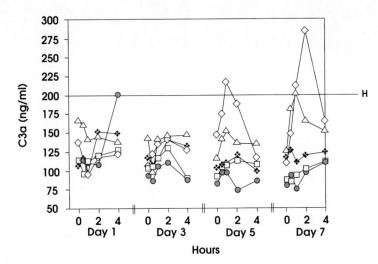

Fig. 2. Median complement C3a split product concentrations following each of four doses of ISIS 2302 ($n = 3$) or placebo ($n = 4$). The *horizontal line* (*H*) indicates the upper limit of the normal range. Placebo (*cross*) or ISIS 2302 (0.2 mg/kg, *circle*; 0.5 mg/kg, *square*; 1.0 mg/kg, *triangle*; 2.0 mg/kg, *diamond*) was administered by 2-h intravenous infusion on days 1, 3, 5, and 7 (GLOVER et al. 1997)

Plasma samples were taken from each multiple-dose subject before dosing and on day 22; these were negative for the presence of IgG or IgM antibodies to ISIS 2302.

Urinary excretion of microproteins was examined in the single-dose subjects, and no changes were observed.

2. Pharmacokinetics

Results are expressed as both the amount of parent compound and of total oligonucleotide present (defined as the sum of intact ISIS 2302 plus apparent n-1, n-2, and n-3 chain-shortened metabolites). In plasma, metabolites apparently shorter than 17-mer were not generally detected (Fig. 3). Although very little intact drug or metabolites were excreted in urine (less than 0.5% of the total drug administered was excreted in the first 6 h), detectable amounts of material in urine comigrated with shorter oligonucleotide standards (Fig. 4). These apparently shorter metabolites were not quantitated because baseline separation was not reliably achieved.

Plasma concentration–time curves for subjects who received single infusions of 0.5 and 2.0 mg/kg are shown in Fig. 5. Even though the plasma concentration–time curves appeared biphasic, an open, one-compartmental model produced the best fit to the data by both statistical and visual criteria, and pharmacokinetic parameters were calculated for each subject in the 0.5- and 2.0-mg/kg single-dose groups (Table 1) using this model. Plasma half-life was unchanged by dose. However, although end-of-infusion plasma concen-

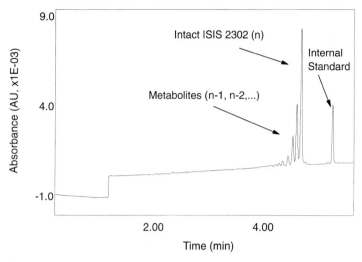

Fig. 3. Representative electropherogram of ISIS 2302 and chain-shortened metabolites extracted from plasma obtained from a subject 1 h after the end of the fourth of four every-other-day, 2-h infusions of ISIS 2302 (2 mg/kg). Capillary gel electrophoresis (CGE) provides resolution between intact ISIS 2302 (*N*) and shortmer metabolites (*N-1*, *N-2*, *N-3*, etc.) (GLOVER et al. 1997)

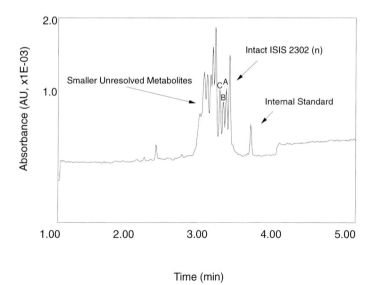

Fig. 4. Representative electropherogram of ISIS 2302 and shortmer metabolites extracted from urine collected from 0 to 6 h from a subject during and after the third of four every-other-day 2-h infusions of ISIS 2302 (2 mg/kg). Capillary gel electrophoresis (CGE) provides resolution between intact ISIS 2302 (*N*) and shortmer metabolites (N-1, *A*; N-2, *B*; N-3, *C*, etc.) (GLOVER et al. 1997)

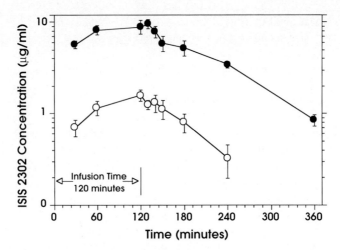

Fig. 5. Logarithmic mean plasma concentration–time profiles during and after 2-h hour infusions of 0.5 mg (*open symbols*) and 2.0 mg ISIS 2302 (*closed symbols*) per kg (GLOVER et al. 1997)

Table 1. Mean plasma pharmacokinetic paramerers derived for ISIS 2302 and total oligonucleotide from those groups ($n = 3$) that received a single 2-h infusion of 0.5 or 2.0 mg ISIS 2302 per kg (GLOVER et al. 1997)

Parameter	0.5-mg/kg dose group		2.0-mg/kg dose group	
	ISIS 2302	Total oligonucleotide	ISIS 2302	Total Oligonucleotide
V_{ss} (ml/kg)	155.4 ± 13.6	121.9 ± 5.6	97.5 ± 7.1*	77.5 ± 5.3*
$AUC_{(0-\infty)}$ (μg min/ml)	249.7 ± 43.8	392.7 ± 92.5	1824.6 ± 111.1	2765.0 ± 30.0
$K_{10}t_{1/2}$ (min)	54.4 ± 16.1	66.7 ± 18.2	52.9 ± 6.0	74.1 ± 7.7
C_{max} (μg/ml)	1.60 ± 0.16	2.30 ± 0.26	10.3 ± 0.06	15.5 ± 1.10
C_{max}/dose (μg/ml)	3.21 ± 0.33	4.60 ± 0.52	5.17 ± 0.34*	7.75 ± 0.65*
Clearance (ml/min kg)	2.07 ± 0.48	1.33 ± 0.34	1.28 ± 0.12*	0.73 ± 0.08*

V_{ss}, steady state volume of distribution; $AUC_{(0-\infty)}$, area under the plasma concentration – time curve extrapolated to infinity; $K_{10}t_{1/2}$, plasma half-life; C_{max}, Maximum plasma concentration.
*$p < 0.05$, as compared to the 0.5-mg/kg dose group.

trations of ISIS 2302 increased linearly over the tested dose range of 0.06–2 mg/kg, with increasing doses of drug (Fig. 6), area under the curve (AUC) values increased disproportionately. A fourfold increase in single dose produced a sevenfold increase in AUC of both intact drug and total oligonucleotide, suggesting a saturable component to disposition. Pharmacokinetic modeling, demonstrating a reduction in volume of distribution and plasma clearance as dose increases (Table 1), further supports this suggestion.

During 2-h, single infusions of ISIS 2302, metabolites comigrating with synthesized n-1, n-2, and n-3 chain-shortened forms (shortmers) of the intact drug appeared rapidly in plasma, constituting 20% of total oligonucleotide after 30 min of infusion. Interestingly, the relative proportion of total oligonucleotide constituted by full-length oligonucleotide and n-1, n-2, and n-3 shortmers remained relatively constant during the 2 h of study drug infusion and for at least the 4 h post-infusion during which metabolites could be measured. Intact drug therefore constituted the majority of oligonucleotide present at all times at which drug or metabolites were detectable (Fig. 7).

There was no evidence of accumulation of oligonucleotide in plasma with alternate-day dosing by infusion over 2 h (Fig. 8). Furthermore, there were no clear differences in the proportions or concentrations of intact drug and metabolites over time between the first and fourth infusions, suggesting that ISIS 2302 does not induce or inhibit its own metabolism.

Urine samples from the 1.0- and 2.0-mg/kg multiple-dose groups were analyzed for concentrations of intact drug and metabolites. Although very low concentrations of ISIS 2302 and metabolites were present in urine, intact drug and n-1, n-2, and n-3 shortmers could be measured, and the quantity of shorter

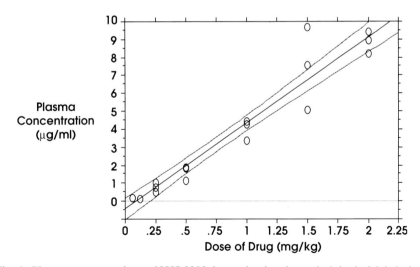

Fig. 6. Plasma concentrations of ISIS 2302 determined at the end of single 2-h infusions of ISIS 2302 (GLOVER et al. 1997)

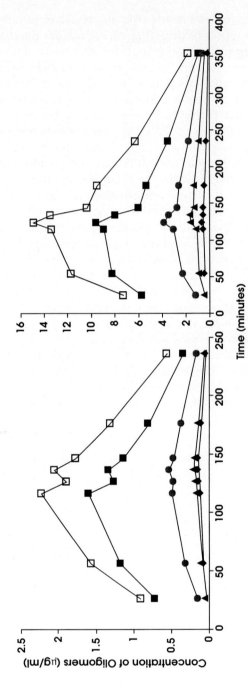

Fig. 7. Mean concentrations of intact ISIS 2302 and metabolites in plasma during and after 2-h intravenous infusions of 0.5 (*left*) or 2.0 mg ISIS 2302 (*right*) per kg. *Closed squares*, n; *circles*, n-1; *triangles*, n-2; *diamonds*, n-3; *open squares*, total oligomers (GLOVER et al. 1997)

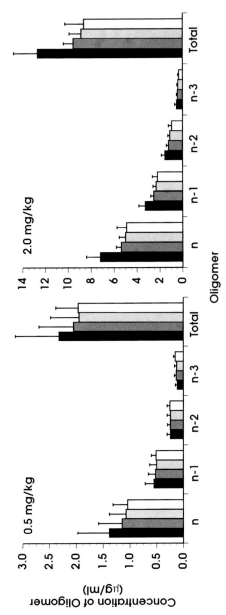

Fig. 8. Plasma concentrations of ISIS 2302 and metabolites at the end of 2-h infusions during multiple dosing. *Bars* show concentrations of ISIS 2302 (n), chain-shortened metabolites (n-1, n-2, n-3) and total oligonucleotide (sum of n, n-1, n-2, and n-3) on days 1 (*black*), 3 (*dark gray*), 5 (*light gray*), and 7 (*white*). Each bar represents the mean concentration from three subjects, with *vertical lines* showing standard deviation (GLOVER et al. 1997)

metabolites, although not well resolved, could be visually estimated from electropherograms. The amount of intact drug excreted over 6 h after the beginning of infusion averaged approximately 0.05% of the administered dose, and the estimated total excretion of parent drug and metabolites in this time period was less than 0.5% of the total dose.

III. Discussion

The primary purpose of this trial was to establish the safety of a range of single and multiple doses of ISIS 2302, a drug which had not previously been administered to humans. All dosages administered, up to 2 mg/kg every other day for four doses, were well tolerated. The only drug-related effects were transient, dose-related increases in aPTT and C3a observed at the higher doses.

In cynomolgus monkeys, ISIS 2302 has consistently caused transient increases in aPTT and alternative pathway complement activation at plasma concentrations in excess of 30 and 50 μg/ml, respectively, effects which are related in magnitude to peak plasma drug concentration (HENRY et al. 1994; HENRY et al., submitted). In this trial, single- and multiple-dose groups demonstrated increases in aPTT at doses of 0.5 mg/kg (peak plasma levels, approximately 1.5 μg/ml) of ISIS 2302 and above. These increases were very clearly dose related in magnitude and unaffected by multiple dosing. Peak prolongations in aPTT were seen at the time of peak plasma concentration of both intact ISIS 2302 and total oligonucleotide (at the end of infusion) and reversed spontaneously within 2–4 h after the end of infusion, the same period over which oligonucleotide was detectable in plasma after higher doses.

The threshold for prolongation of aPTT appears to be approximately 20-fold lower in humans than in monkeys; the reason for this is not known at present. The mechanism of the aPTT effect is not yet fully understood, but it is known that ISIS 2302, like other phosphorothioate oligonucleotides, is a polyanionic molecule which binds reversibly to a number of proteins, including thrombin (HENRY et al. 1994; HENRY et al., submitted), and other coagulation proteins may also be involved. Levels of blood proteins are similar in humans and monkeys, but it is possible that phosphorothioate oligonucleotides bind more avidly to human than to monkey thrombin or other proteins in the aPTT cascade or to different sites on these proteins. The physiological relevance of the prolongation in aPTT is not known, since the threshold for in vitro effects on human clotting, as assessed by thromboelastography, is approximately 50 μg/ml. At this drug concentration, an effect on time to clot initiation becomes evident, but effects on the physical properties of the clot are not seen until drug concentrations of approximately 70 μg/ml or greater. Regardless of the physiological importance of this finding, the reliable relationship between plasma drug concentration and prolongation of aPTT allows for the calculation of conservative ratios of maximum doses to rates of infusion for clinical trials on the basis of maximum acceptable increase in aPTT.

No effect on the C5a complement split product was seen, but small, brief increases in C3a were noted after repeated infusions of higher doses of ISIS 2302. The mechanism of complement activation or conversion by ISIS 2302 observed in this study is also not known, although this is also thought to be related to protein binding or nonspecific enzymatic degradation by leukocyte proteases. Nonclinical experiments are underway to investigate these possibilities.

Rapid fluctuations in neutrophil count and hemodynamic changes, thought to be related to complement activation, have been observed in monkeys at higher dose levels and infusion rates with ISIS 2302 and other phosphorothioate oligonucleotides (GALBRAITH et al. 1994; HENRY et al., submitted). These variables were also measured at frequent intervals after the beginning of infusion in the single-dose subjects in this trial, and no significant changes nor any trends to change were observed.

It is likely that all phosphorothioate oligonucleotides of sufficient chain length will cause prolongation of aPTT and alternative pathway complement activation in animals and humans if administered in such a way as to exceed threshold peak plasma drug concentrations.

There was no evidence of antibody formation to ISIS 2302, a finding consistent with the experience at Isis in other animal and clinical trials with similar oligonucleotides.

Other authors have described the human pharmacokinetics of intravenously injected ^{35}S-labeled GEM 91 (a 25-mer; ZHANG et al. 1995), intradermal ^{14}C-labeled afovirsen sodium (a 20-mer; CROOKE et al. 1994), and unlabeled OL(1)p53 (a 20-mer; BAYEVER et al. 1993), all phosphorothioate oligodeoxynucleotides, but this was the first human study to employ CGE to measure nonradiolabeled, systemically administered phosphorothioate oligonucleotide. The results of pharmacokinetic analysis in this trial are generally consistent with those obtained with other systemically administered phosphorothioate oligonucleotides in animals (COSSUM et al. 1993, 1994; AGRAWAL et al. 1995) and humans (BAYEVER et al. 1993; CROOKE et al. 1994; AGRAWAL et al. 1995; ZHANG et al. 1995), but there are some important differences.

The rapid distribution phase and the volume of distribution observed in this study appear to be consistent with previous animal and human data. However, this study differs from previous studies in that the terminal half-life described in studies using radiolabeled drug was not observed. This difference may be related to differences in assay sensitivity and the ability to measure single nucleotide metabolites by radiochemical analysis. Irrespective of the analytic method, however, the plasma half-life of intact drug is long compared to unmodified oligonucleotides with phosphodiester backbones.

An additional difference between reports lies in the assessment of the importance of urinary excretion. In this study and those conducted with ^{14}C-labeled afovirsen sodium (COSSUM et al. 1993, 1994; CROOKE et al. 1994), urine was found to be a minor route of excretion of oligonucleotides (or of radioactivity derived from drug), whereas studies with ^{35}S-labeled 20- and 25-mers showed substantial urinary excretion of drug-derived radioactive material, mostly in the form of metabolites (AGRAWAL et al. 1995; ZHANG et al. 1995). These differences could be due to differences in the route of excretion of carbon- and sulfur-related moieties resulting from metabolism of phosphorothioate oligodeoxynucleotides. ZHANG et al. (1995) described urinary excretion of approximately 70% of GEM 91-derived ^{35}S over 96 h after a single dose in humans, while COSSUM et al. (1993, 1994) and CROOKE et al. (1994) found approximately 50% and 30% of afovirsen sodium-derived ^{14}C in expired air after single doses in rats and humans, respectively.

In a human study, BAYEVER et al. (1993) used high-performance electrophoretic chromatography (HPEC) to measure cumulative urinary excretion of an unlabeled 20-mer phosphorothioate oligodeoxynucleotide (OL(1)p53) given by continuous i.v. infusion at a rate of 0.05 mg/kg per h (1.2 mg/kg per day) for 10 days to five patients with acute myelogenous leukemia or myelodysplastic syndrome. Cumulative urinary excretion of apparently undifferentiated oligonucleotide ranged from 42% to 63% of administered drug. Although plasma drug levels were not directly measured, peak plasma concentrations (2.1–6.4 mg/ml) and half-life (4.9–14.7 days) were calculated from the rate of urinary excretion.

Among the potential explanations for discrepancies in plasma half-life and the relative importance of urinary excretion found between the ISIS 2302 study and the study by Bayever et al. are differences in analytic methodology and directness of determination of plasma levels. Substantial methodological differences exist between the HPEC technique used by Bayever et al. and the CGE technique (LEEDS et al. 1996) used in this study to measure drug concentrations, including methods of extraction, scale of analytic method, and ability to identify intact drug and metabolites. Furthermore, plasma drug concentrations were not directly measured by Bayever et al, but calculated from urinary excretion. Differences in plasma half-life and the relative importance of urinary excretion found in these two studies could also be due to differences in dosing regimen (continuous infusion for 10 days versus every-other-day 2-h infusions). Continuous infusion might saturate plasma binding capacity, cellular uptake, and/or renal proximal tubular reabsorption.

There are no other published accounts of human pharmacokinetic behavior after multiple dosing of phosphorothioate oligonucleotides. In this trial, no accumulation of drug or metabolites was seen in plasma, nor was there any apparent change in the kinetics or metabolism of ISIS 2302 in plasma with repeated administration. Similar plasma pharmacokinetic behavior has been seen in monkeys (Isis Pharmaceuticals, unpublished data). Although tissue levels cannot be determined in clinical trials, we can speculate by extrapolation from monkey data demonstrating tissue half-lives on the order of 24–72 h (Isis Pharmaceuticals, unpublished data) that concentrations of ISIS 2302 and its metabolites in target tissues are maintained with an alternate-day regimen.

This study showed the peak plasma concentration and AUC for total oligonucleotide to be approximately 50% higher than for ISIS 2302 alone. Nonclinical studies have indicated that synthesized n-1, n-2, and n-3 chain-shortened metabolites are capable of inhibiting ICAM-1 expression and of causing the toxicities typical of phosphorothioate oligonucleotides (Isis Pharmaceuticals, unpublished data). The observation that the relative proportion of total oligonucleotide constituted by full-length oligonucleotide and n-1, n-2, and n-3 shortmers remained fairly constant throughout the 2-h infusion period and for at least 4 h post-infusion suggests that ISIS 2302 is shortened one base at a time by exonucleases that compete at a constant rate for substrate irrespective of chain length (at least n-3 mers, 17-mers) and base sequence. The absence of significant urinary excretion and the failure to detect a buildup of metabolites, either long (17- to 19-mers) or short (less than 17-mers) suggests that ISIS 2302 is principally cleared from plasma by extravascular distribution and subsequent cellular uptake in tissues.

The major plasma metabolites detected in this study comigrated with standards shortened by one, two, or three nucleotides from the 3' end of the ISIS 2302 molecule, consistent with the hypothesis that phosphorothioate oligonucleotides undergo metabolism by exonucleases which remove single bases in a sequential manner from the end of the molecule. Alternatively, nuclease activity might remove pairs or triplets of bases from the end or the interior of the molecule in order to produce n-2 and n-3 metabolites. Although CGE cannot determine the sequence of the apparent 19-, 18-, and 17-mers seen, the latter hypothesis is difficult to reconcile with the observed metabolic profile. Furthermore, phosphodiester oligodeoxynucleotides in plasma undergo exonuclease digestion from their 3' end (EDER et al. 1991).

It is likely that the resulting shortmers are metabolized in much the same way as endogenous nucleotides. The finding (described above) that approximately 50% of ^{14}C radiolabel derived from a similar phosphorothioate oligonucleotide, labeled at the C_2 position of thymidine, was eliminated from rats as carbon dioxide in expired air (COSSUM et al. 1993, 1994) supports this hypothesis. It is unlikely that biliary excretion plays any significant role in disposition, as fecal elimination of labeled, i.v. administered phosphorothioates has been minimal in rodents (COSSUM et al. 1993; AGRAWAL et al. 1995; Isis Pharmaceuticals, unpublished data).

The predictability of the clinical profile and pharmacokinetics of repeated infusions of ISIS 2302 in this study enabled the commencement of pilot therapeutic trials in a spectrum of inflammatory and autoimmune conditions.

C. Phase IIa Trials

Phase II trials in five indications are presently underway: Crohn's disease, rheumatoid arthritis, ulcerative colitis, psoriasis, and prophylaxis of acute renal allograft rejection. All 20 patients in the Crohn's disease study have completed the treatment phase and 3 months of follow-up, and a preliminary discussion of these data will be presented.

The goals of these studies are to obtain preliminary evidence of efficacy for a systemically administered antisense compound, to validate the relevance of ICAM-1 in multiple inflammatory diseases, and to obtain some preliminary dose–response and dose regimen data. In addition, these studies will provide data on the time to induction and duration of response and preliminary data on disease-specific tolerability and will help to select diseases for further development. These studies by design lack sufficient numbers of patients to provide statistically significant results.

To increase the opportunity of demonstrating a therapeutic effect, drug exposure in these trials was maximized within the constraints of available 1-month toxicology data and experience in normal volunteers. The animal data were sufficient to support 1 month of exposure in North America, and threshold effects on complement activation (C3a generation) and 50% increases in aPTT with a 2-h infusion of 2mg/kg in normal volunteers argued that initial exposure in patients should be limited to this dose and regimen. Such an increase in aPTT was a concern in inflammatory bowel disease and in the perioperative state, and it was unknown whether patients with autoimmune and inflammatory conditions might be more or less sensitive to alternative pathway complement activation. Furthermore, 2mg/kg is within the therapeutic range in mice, and with similar pharmacokinetics in humans and mice, including drug exposure at a given dose, it was reasonable to expect that this dose level would be therapeutic in humans. The phase IIa trials were therefore designed as fixed-dose-within-patient, dose-escalation studies, beginning at 0.5 mg/kg and escalating to 1 and 2mg/kg every other day i.v. in successive cohorts.

These studies are enrolling refractory patients, with efficacy being assessed by standard, well-accepted outcome measures. All but the psoriasis trial are double-blinded, placebo controlled, and randomized (study drug to placebo ratio, 3:1), and all but the renal transplant study involve a 4-week treatment period (treatment in the renal transplant study is for 2 weeks). In addition, the renal transplant study differs from the other studies in that two additional lower-dose groups of 0.05 and 0.1 mg/kg are specified and a phase I and a phase II segment are provided for. In each study, 20–40 patients are being enrolled at one to four centers per trial. Patients are followed for up to 6 months after the treatment period or until disease relapses or fails to respond to study drug.

I. Crohn's Disease

The Crohn's disease study is the most mature of the phase IIa ISIS 2302 studies (YACYSHYN et al. 1997). All patients have completed the treatment phase and have had the opportunity to complete at least 5 months of follow-up. Final data are being collected and analyzed, but a preliminary analysis can be summarized here.

This study was a double-blinded, placebo-controlled, randomized (study drug to placebo ratio, 3:1) study that enrolled steroid-dependent patients with moderately active Crohn's disease despite background corticosteroids (less than 40 mg prednisone or equivalent per day). The study was conducted at a single center by one investigator by B.R. Yacyshyn at the University of Alberta (Edmonton, Alberta, Canada). "Moderately active" was defined as a Crohn's disease activity index (CDAI) greater than 200 and less than 350. Background 5-aminosalicylic acid (5-ASA) drugs in stable dosage were also permitted. Corticosteroids and 5-ASA drugs were to remain stable during the treatment period and for 1 month of follow-up. The primary efficacy measure was the CDAI, and a secondary measure was the endoscopic index of severity (EIS).

The CDAI is a validated clinical instrument (BEST et al. 1976, 1979) that is widely accepted in the evaluation of clinical disease activity and response. The CDAI is calculated on the basis of a patient diary over the previous 7 days and objective criteria. It includes eight measures: the number of liquid or very soft stools, abdominal pain, general well-being, extraintestinal manifestations of Crohn's disease, the use of opiates to treat diarrhea, abdominal mass on physical examination, hematocrit, and body weight. The composite score can range from 0 to approximately 600. A CDAI of less than 150 is widely accepted as defining disease remission.

EIS (MARY and MODIGLIANI 1989) is calculated by assessing the extent and severity of intestinal inflammation averaged over the four or five regions (unless surgically removed) of the colon and ileum examined: rectum and descending colon, left colon, transverse colon, right colon, and ileum (if visualizable). All evaluations were performed by the same, blinded investigator. Previous studies have failed to demonstrate a strong correlation between endoscopy scores and simultaneous CDAI (GOMES et al. 1986; MODIGLIANI et al. 1990; CELLIER et al. 1994), but the predictive value of changes in EIS has not been evaluated.

CDAI was measured weekly through to day 40, at Day 60, and monthly thereafter. Colonoscopy was performed and EIS calculated at baseline and at days 26, 60, and 120. A clinical remission was defined as a CDAI of less than 150 and a clinical response as a decrease from baseline of more than 100 points.

Over the investigated dose range of 0.5–2 mg/kg, there was no consistent evidence of a dose response for CDAI or EIS. Combining the drug-treated patients into a single group, trends favoring ISIS 2302 over placebo were seen for the proportion of remitters (47% vs. 0%), and percentage change in CDAI (−38% vs. −13%) and EIS (−56% vs. −47%) at the end of therapy (day 26 EIS, day 33 CDAI). Highly significantly lower ($p = 0.001$) mean daily corticosteroid requirements were apparent in the ISIS 2302 group than in the placebo group by the end of the first month, a significant difference that has persisted through the last time point analyzed to date (month 5). Although decreases in mean CDAI could be seen at day 7, for patients responding to ISIS 2302, the mean time to clinical remission was approximately 25 days. As some patients remain in remission, the mean duration of remission has not yet been defined.

The proportions of remitters and the percentage improvement in EIS compare favorably with recent trials with active biologicals (antitumor necrosis factor antibody, anti-TNF Ab; interleukin-10, IL-10) and methotrexate in similar disease populations (remission rates, 30%–50%; 72% decrease in EIS). In a recent dose–response trial of single i.v. infusions of anti-TNF Ab (chimeric) in 108 patients (approximately 25 patients per group), no dose response was evident over a fourfold dose range (5–20 mg/kg), and the 4-week remission rates (CDAI < 150) were 33% for the combined drug treatment group and 15% for the placebo-treated patients (TARGAN et al. 1996). Response rates (decrease in CDAI > 70) were 65% and 17%, respectively.

In another recent trial of recombinant IL-10 given as an i.v. infusion daily for 7 days to 32 patients in doses from 0.5 to $25\,\mu g/kg$, no dose response was observed, and rates of remission (requiring both a CDAI of less than 150 and a decrease of more than 100 points) within 4 weeks for the combined active drug ($n = 32$) and placebo ($n = 13$) groups were 50% and 23%, respectively (VAN DEVENTER et al. 1996).

Anti-TNF Ab (chimeric) was administered as a single i.v. infusion in doses of 1, 5, 10, and 25 mg/kg to cohorts of five patients in an open-label, uncontrolled trial (MCCABE et al. 1996). A flat dose–response curve was noted for CDAI across all doses (average of 40% remitters) and for EIS across the three highest doses (average of approximately 72% decrease).

Methotrexate (25 mg i.m. weekly) was compared to placebo in 141 steroid-dependent Crohn's disease patients in a recent double-blinded, randomized study (study drug to placebo ratio, 2:1; FEAGAN et al. 1995). Patients were stratified according to high (≥20 mg per day) or low (<20 mg per day) prednisone dosage at baseline and received 20 mg prednisone per day at baseline to be tapered to 0 over 10 weeks. At the end of 16 weeks, 39.4% of the methotrexate patients and 19.1% of the placebo patients ($p = 0.025$) were in remission (CDAI < 150 and off corticosteroids).

ISIS 2302 was well tolerated. Apart from one patient who repeatedly experienced transient facial flushing at the end of drug infusions, and the expected transient increases in aPTT (approximately 50% in the 2-mg/kg dose group), no pattern of drug-related adverse events was observed.

In summary, ISIS 2302 appears to be a well-tolerated and effective therapy for patients with active, steroid-dependent Crohn's disease. At the end of 1 month of therapy, substantial trends favoring ISIS 2302 over placebo treatment were observed in proportions of remitters and responders and in endoscopy scores, and these responses in CDAI and EIS compare favorably to improvements observed with active biological agents in similar patient populations. The likelihood that these trends represent a therapeutic response is strengthened by the finding of persistently and highly statistically significantly lower steroid requirements in the ISIS 2302 group than in the placebo group. For ISIS 2302-treated responding patients, the mean time to clinical response was approximately 25 days.

These safety and efficacy data, though limited by very small patient numbers, are seen as encouraging preliminary evidence of the safety and efficacy of ISIS 2302 in Crohn's disease, and a pivotal-quality, placebo-controlled efficacy trial is being planned.

II. Rheumatoid Arthritis

The rheumatoid arthritis trial is being conducted at two centers in patients with active disease, i.e., more than ten swollen joints plus at least two of the following: more than 12 tender joints; morning stiffness for longer than 1 h; an erythrocyte sedimentation rate (ESR) higher than 25 for men and 35 for women. This is also a placebo-controlled, double-blinded trial that is recruiting 40 patients, four each at the 0.5- and 1-mg/kg level and the remainder at the 2-mg/kg level. The treatment regimen, duration of treatment, and follow-up period are the same as for the Crohn's disease study. Efficacy is being assessed by changes in swollen joint count, tender joint count, patient global assessment, physician global assessment, and pain by a visual analogue scale. Composite scoring (Paulus criteria) will also be performed (PAULUS et al. 1990).

III. Ulcerative Colitis

The ulcerative colitis study is identical in design to the Crohn's disease study, except for the patient population and the efficacy assessments. Steroid-dependent patients with a clinical activity index (CAI) of at least 10 of a possible 21 points despite corticosteroids (less than 40 mg prednisone per day or equivalent) plus 5-ASA drugs are being enrolled. The CAI is a variation of the Truelove-Witts criteria (TRUELOVE and WITTS 1959), incorporating diarrhea, visible blood in the stool, abdominal pain, general well-being, abdominal tenderness, and need for antidiarrheal medication (LICHTIGER et al. 1994). Endoscopic disease is also being assessed by the EIS. This study is being carried out at three centers.

IV. Psoriasis

Due to anticipated difficulty with patient acceptance of a placebo-controlled, unproven i.v. therapy in this disease, the psoriasis study is the only one of the phase IIa studies that is open-labeled and uncontrolled. Twenty patients with moderately active (5%–40% body surface area involvement), plaque-type psoriasis vulgaris resistant to medium-potency topical corticosteroids are being enrolled. Patients are washed out from all present antipsoriatic medication prior to baseline. All patients receive 13 infusions of ISIS 2302 over

26 days: three patients each are assigned to the 0.5- and 1-mg/kg dose, and the remaining 14 patients to the 2-mg/kg dose group.

Efficacy is being primarily assessed by change in the psoriasis area and severity index (PASI), which gives a single number value to the extent and severity of a patient's disease (FREDRIKSSON and PETTERSSON 1978). The body surface area is divided into four segments (head, trunk, upper limbs, and lower limbs), the severity of disease (erythema, scaling, and thickness), and percentage involvement of each segment assigned a numerical value; the scores for each segment, multiplied by a correction factor for the percentage of the total body surface area that the segment represents, are added. Mathematically possible scores range from 0 to 72 (severe involvement of the entire body surface area).

V. Renal Transplantation

The prophylaxis of acute renal cadaveric allograft rejection study is being conducted at a single center as a double-blinded, placebo-controlled, randomized (3:1) study. In addition to indication, it differs from the other phase IIa studies in four respects: (1) since most patients come from a distance, therapy is limited to 2 weeks for logistical reasons; (2) the study is divided into a phase I and a phase II segment; (3) two lower-dose groups (0.05 and 0.1 mg/kg) are included; and (4) due to the unknown potential for complement activation and bleeding in the immediate perioperative period, the first two infusions are administered over 6 h, the second two infusions over 4 h, and the remaining three infusions over 2 h in order to reduce maximum plasma concentrations of drug during the immediate perioperative and postoperative periods.

In the phase I segment, which has now been completed, patients with stable cadaveric renal allografts at least 6 months postoperatively received ISIS 2302 in a single 6-h infusion followed 1 week later by an every-other-day regimen for 2 weeks against a stable background of prednisone and cyclosporin A (CsA). Four patients each were assigned to the 0.05-, 0.5-, 1-, and 2-mg/kg dose groups. All doses were well tolerated. As predicted, dose-related, transient increases in aPTT similar in magnitude to increases experienced by normal volunteers at equivalent doses were observed.

The phase II segment is about to begin. Four patients each will be assigned to the 0.05-, 0.1-, 0.5-, and 1-mg/kg dose groups, and 12 patients to the 2-mg/kg dose group. De novo renal transplant patients, recipients of cadaveric allografts, will receive seven infusions of ISIS 2302 administered every other day, beginning during the transplant procedure, in addition to their usual regimen of prednisone and CsA. Patients will be followed for 6 months post-transplantation to assess the incidence of acute rejection in this time period.

The incidence of acute rejection has been validated as an early surrogate for graft survival (WILLIAMS 1988; CECKA et al. 1992; BURKE et al. 1994; KAHAN

1993) and served as the basis for the recent United States Food and Drug Administration (FDA) approval of mycophenolate mofetil for the prophylaxis of acute renal allograft rejection. With standard therapy, consisting of prednisone and CsA with or without azathioprine, the incidence of acute rejection within the first 3–6 months post-transplantation is 40%–50%. It is hoped that the addition of ISIS 2302 to the standard regimen will have a significant impact on the incidence of rejection, and hence graft survival, without the addition of medically significant toxicities.

D. Exploration of Subcutaneous Dosing

In order to explore the feasibility of subcutaneous self-injection, initial pharmacokinetic studies with this route of administration were undertaken in the rat and monkey, after first assessing in vitro release characteristics, if appropriate, of a variety of formulations in addition to the PBS i.v. formulation. A spectrum of alternative salts and lipid formulations were investigated, and none with sufficient loading capacity for a 1- to 2-mg/kg subcutaneous injection in a volume of 1 ml or less appeared to offer significant advantages over the saline formulation in terms of bioavailability or sustained release. The PBS formulation subcutaneously produced about 50% plasma bioavailability as compared to i.v. administration, with a time to maximal plasma concentration of 1–3 hours.

Every-other-day subcutaneous toxicity studies lasting 1 month with the PBS formulation were therefore undertaken in the mouse and the monkey. Tolerability and toxicity paralleled that in the i.v. studies, and 20 mg/kg at a concentration of 80 mg/ml, the highest monkey dose level, was well tolerated without significant local (injection site), regional (lymph nodes), or laboratory findings. Plasma bioavailability was dependent on dose and drug concentration, but concentrations of 50 mg/ml and above and doses within or greater than the anticipated therapeutic range in humans (0.5–2 mg/kg) produced approximately 50% bioavailability.

With these data, a phase I subcutaneous study was initiated in normal volunteers. In the first phase of the study, the tolerability and pharmacokinetics of single 1-ml subcutaneous injections of concentrations of ISIS 2302 ranging from 50 to 200 mg/ml were administered in double-blinded, placebo-controlled, randomized (study drug to placebo ratio, 3:1) fashion to cohorts of four subjects. Without an apparent concentration response, all concentrations in the tested range produced mild but clinically noticeable injection site erythema, edema, and induration that lasted for several days. This response was more of a clinical observation than a subject complaint, and it appeared that all concentrations were adequately tolerated. Data are now being reviewed, but plasma bioavailability appears to be about 50%, with a time to maximal concentration (t_{max}) of 1–3 h. It was originally planned to test the maximum well-tolerated concentration in cohorts of four subjects at doses of 1, 2, and 4 mg/kg every other day for four doses and 1 and 2 mg/kg daily for 7 days. Single injections of at least 1 and 2 mg/kg had been tested during the concentration tolerability evaluations, and a single exposure of 4 mg/kg in two injections was also tested in this initial phase.

The 1-mg/kg every-other-day regimen was fairly well tolerated, although there was evidence of very low grade complement activation (C3a only, and not C5a) and low-grade lymphadenopathy in addition to the mild injection site reactions described above. The 2-mg/kg every-other-day regimen, however, was very poorly tolerated, with definite increases in C3a, marked and tender inguinal lymphadenopathy, and low-grade constitutional symptoms, including pyrexia, myalgias, and malaise.

With these observations, the 4-mg/kg cohort was canceled, and the initial daily regimen was decreased from 1 to 0.5 mg/kg. This regimen was well tolerated, with only the expected injection site reactions being experienced. The next regimen tested was 1 mg/kg once weekly for 4 weeks. As with the 1-mg/kg regimen every other day for 4 doses, there was evidence of low-grade C3 conversion and low-grade lymphadenopathy in at least one subject.

In an attempt to increase tolerated drug exposure by the subcutaneous route, we next investigated 6-day continuous subcutaneous administration via an external insulin pump and indwelling subcutaneous catheter, starting with a dose of 0.5 mg/kg per day. This regimen was not well tolerated, with considerable injection site reaction and regional lymphadenopathy.

In summary, a subcutaneous regimen of 0.5 mg/kg daily for 7 days was well tolerated, and regimens of 1 mg/kg every other day for four doses or weekly for 4 weeks were marginally tolerated. A higher-dose (2 mg/kg) bolus regimen and low-dose (0.5 mg/kg per day) continuous infusion regimens were poorly tolerated, presumably due to exceeding locally the threshold for complement activation and/or to polyclonal B cell proliferation and activation, as observed with phosphorothioates in human cells in vitro (LIANG et al. 1996) and in rodents in vivo (MONTEITH et al. 1997).

E. Future Plans

In steroid-dependent, active Crohn's disease, a randomized, double-blinded, placebo-controlled, 6-month study with 300 patients is being planned. A 4- and a 2-week regimen of 2 mg/kg i.v. three times weekly during months 1 and 3 will be compared to placebo. A positive interim analysis could trigger the initiation of additional pivotal-quality and dose–response trials. In addition, since at least in Crohn's disease, 0.5 mg/kg every other day i.v. appears to be well within the therapeutic range, it is anticipated that subcutaneous ISIS 2302 injections of 1 mg/kg once or twice weekly or daily injections of 0.5 mg/kg might be effective, and a dose regimen study investigating these regimens and a daily i.v. infusion of 2 mg/kg for 5 days is being planned to parallel the efficacy study in Crohn's disease.

A similar approach will likely be taken in other indications if phase II data are promising.

References

Agrawal S, Temsamani J, Galbraith W, Tang J (1995) Pharmacokinetics of antisense oligonucleotides. Clin Pharmacokinet 28(1):7–16

Allavena P, Paganin C, Martin-Padura I, Peri G, Gaboli M, Dejana E, Marchisio PC, Mantovani A (1991) Molecules and structures involved in the adhesion of natural killer cells to vascular endothelium. J Exp Med 173:439–448

Altmann DM, Hogg N, Trowsdale J, Wilkinson D (1989) Cotransfection of ICAM-1 and HLA-DR reconstitutes human antigen-presenting cell function in mouse L cells. Nature 338:512–514

Barton RW, Rothlein R, Ksiazek J, Kennedy C (1989) The effect of anti-intercellular adhesion molecule-1 on phorbol-ester-induced rabbit lung inflammation. J Immunol 143:1278–1282

Bayever E, Iversen PL. Bishop MR, Sharp JG, Tewary HK, Arneson MA, Pirruccello SJ, Ruddon RW, Kessinger GZ, Armitage JO (1993) Systemic administration of phosphorothioate oligonucleotide with a sequence complementary to p53 for acute myeloblastic leukemia and myelodysplastic syndrome: initial results of a phase I trial. Antisense Res Dev 3:383–390

Bennett CF, Condon TP, Grimm S, Chan H, Chiang MY (1994) Inhibition of endothelial cell leukocyte adhesion molecule expression with antisense oligonucleotides. J Immunol 152:3530–3541

Bennett CF, Kornbrust D, Henry S, Stecker K, Howard R, Cooper S, Dutson S, Hall W, Jacoby HI (1997) An ICAM-1 antisense oligonucleotide prevents and reverses dextran sulfate sodium-induced colitis in mice. J Pharmacol Exp Ther 280:988–1000

Best WR, Bectel JM, Singleton JW, Kern F (1976) Development of a Crohn's disease activity index: National Cooperative Crohn's Disease Study. Gastroenterology 70:439–444

Best WR, Bectel JM, Singleton JW (1979) Rederived values of the eight coefficients of the Crohn's Disease Activity Index. Gastroenterology 77:483–486

Burke JF, Pirsch JD, Ramos EL, Salomon DR, Stablein DM, Van Buren DH, West JC (1994) Long-term efficacy and safety of cycloporine in renal-transplant patients. New Engl J Med 331:358–363

Butcher EC (1991) Leukocyte-endothelial cell recognition: three (or more) steps to specificity and diversity. Cell 67:1033–1036

Cecka JM, Cho YW, Terasaki PI (1992) Analysis of the UNOS scientific transplant registry at three years – early events affecting transplant success. Transplantation 53:59–64

Cellier C, Sahmoud T, Froguel E, Adenis A, Belaiche J, Bregagner JF, Florent C, Bouvry M, Mary JY, Modigliana R (1994) Correlation between clinical activity, endoscopic severity, and biological parameters in colonic or ileocolonic Crohn's disease: a prospective multicenter study of 121 cases. Gut 35:231–235

Cosimi AB, Conti D, Delmonico FL, Preffer FI, Wee SL, Rothlein R, Faanes R, Colvin RB (1990) In vivo effects of monoclonal antibody to ICAM-1 (CD54) in non-human primates with renal allografts. J Immunol 144:4604–4612

Cossum PA, Sasmor H, Dellinger D, Truong L, Cummins L, Owens SR, Markham PM, Shea JP, Crooke S (1993) Disposition of the ^{14}C-labeled phosphorothioate oligonucleotide ISIS 2105 after intravenous administration to rats. J Pharmacol Exp Ther 267(3):1181–1190

Cossum PA, Truong L, Owens SR, Markham PM, Shea JP, Crooke ST (1994) Pharmacokinetics of a ^{14}C-labeled phosphorothioate oligonucleotide, ISIS 2105, after intradermal administration to rats. J Pharmacol Exp Ther 269:89–94

Crooke ST, Grillone LG, Tendolkar A, Garrett A, Fratkin MJ, Leeds J, Barr WH (1994) A pharmacokinetic evaluation of ^{14}C-labeled afovirsen sodium in patients with genital warts. Clin Pharmacol Ther 56:641–646

Diamond MS, Staunton DE, deFougerolles AR, Stacker SA, Garcia-Aguilar J, Hibbs ML, Springer TA (1990) ICAM-1 (CD54): a counter-receptor for Mac-1 (CD11b/CD18). J Cell Biol 111:3129–3139

Dustin ML, Rothlein R, Bhan AK, Dinarello CA, Springer TA (1986) Induction by IL-1 and interferon: tissue distribution, biochemistry, and function of a natural adherence molecule (ICAM-1). J Immunol 137:245–254

Eder PS, DeVine RJ, Dagle JM, Walder JA (1991) Substrate specificity and kinetics of degradation of antisense oligonucleotides by a 3' exonuclease in plasma. Antisense Res Dev 1:141–151

Entman ML, Youker K, Shoji T, Kukielka G, Shappell SB, Taylor AA, Smith CW (1992) Neutrophil induced oxidative injury of cardiac myocytes. J Clin Invest 90:1335–1345

Feagan BG, Rochon J, Fedorak RN, Irvine EJ, Wild G, Sutherland L, Steinhart AH, Greenberg GR, Gillies R, Hopkins M, Hanauer SB, McDonald JWD (1995) Methotrexate for the treatment of Crohn's disease. New Engl J Med 332:292–297

Fredriksson T, Pettersson U (1978) Severe psoriasis – oral therapy with a new retinoid. Dermatologica 157:238–244

Furie MB, Tancinco MCA, Smith CW (1991) Monoclonal antibodies to leukocyte chemoattractant-stimulated neutrophil transendothelial migration in vitro. Blood 78:2089–2097

Galbraith WM, Hobson, WC, Giclas PC, Schecter, PJ, Agrawal S (1994) Complement activation and hemodynamic changes following administration of phosphorothioate oligonucleotides in the monkey. Antisense Res Dev 4:201–206

Glover JM, Leeds JM, Mant TGK, Amin D, Kisner DL, Zuckerman JE, Geary RS, Levin AA, Shanahan WR (1997) Phase I safety and pharmacokinetic profile of an ICAM-1 antisense oligonucleotide. J Pharmacol Exp Ther 282:1173–1180

Gomes P, Du Boulay C, Smith CL, Holdstock G (1986) Relationship between disease activity indices and colonoscopic findings. Gut 27:92–95

Harning R, Pelletier J, Van G, Takel F, Merluzzi VJ (1992) Monoclonal antibody to MALA-2 (ICAM-1) reduces acute autoimmune nephritis in kdkd mice. Clin Immunol Immunopathol 64:129–134

Henry SP, Larkin R, Novotny WF, Kornbrust DJ (1994) Effects of ISIS 2302, a phosphorothioate oligonucleotide, on in vitro and in vivo coagulation parameters. Pharmaceutical Res 11:S-353

Henry SP, Taylor J, Midgley L, Levin AA, Kornbrust DJ (1997a) Evaluation of the toxicity of ISIS 2302, a phosphorothioate oligonucleotide, in a 4-week study in CD-1 mice. Antisense Nucleic Acid Drug Dev (in press)

Henry SP, Bolte H, Auletta C, Kornbrust DJ (1997b) Evaluation of the toxicity of ISIS 2302, a phosphorothioate oligonucleotide, in a 4-week study in cynomolgus monkeys. Toxicology 120:145–155

Iigo Y, Takashi T, Tamatani T, Miyasaka M, Higashida T, Yagita H, Okumura K, Tsukada W (1991) ICAM-1 dependent pathway is critically involved in the pathogenesis of adjuvant arthritis in rats. J Immunol 147(12):4167–4171

Isobe M, Yagita H, Okumura K, Ihara A (1992) Specific acceptance of cardiac allograft after treatment with antibodies to ICAM-1 and LFA-1. Science 255:1125–1127

Kahan BD (1993) Toward a rationale design of clinical trials of immunosuppressive agents in transplantation. Immunol Rev 136:29–49

Kawasaki K, Yaoita E, Yamamoto T, Tamatani T, Miyasaka M, Kihara I (1993) Antibodies against intercellular adhesion molecule-1 and lymphocyte function-

associated antigen-1 prevent glomerular injury in rat experimental crescentic glomerulonephritis. J Immunol 150(3):1074–1083
Kelly KJ, Williams WW, Colvin RB, Bonventre JB (1994) Antibody to intercellular adhesion molecule 1 protects the kidney against ischemic injury. Proc Natl Acad Sci USA 91:812–816
Kuhlman P, Moy VT, Lollo BA, Brian AA (1991) The accessory function of murine intercellular adhesion molecule-1 in T lymphocyte activation. J Immunol 146:1773–1782
Kumasaka T, Quinlan WM, Doyle NA, Condon TP, Sligh J, Takei F, Beaudet AL, Bennett CF, Doerschuk CM (1996) The role of ICAM-1 in endotoxin-induced pneumonia evaluated using ICAM-1 antisense oligonucleotides, anti-ICAM-1 monoclonal antibodies, and ICAM-1 mutant mice. J Clin Invest 97:2362–2369
Lacy MJ, Voss EW (1989) Direct adsorption of ssDNA to polystyrene for characterisation of the DNA/anti-DNA interaction and immunoassay for anti-DNA autoantibody in New Zealand White mice. J Immunol Methods 116:87–98
Leeds JM, Graham MJ, Truong L, Cummins LL (1996) Quantitation of phosphorothioate oligonucleotides in human plasma. Anal Biochem 235:36–43
Liang H, Nishioka Y, Reich CF, Pisetsky DS, Lipsky PE (1996) Activation of human B cells by phosphorothioate oligodeoxynucleotides. J Clin Invest 98:1119–1129
Lichtiger S, Present DH, Kornbluth A, Gelernt I, Bauer J, Galler G, Michelassi F, Hanauer S (1994) Cyclosporine in severe ulcerative colitis refractory to steroid therapy. New Engl J Med 330:1841–1845
Ma XI, Lefer DJ, Lefer AM, Rothlein R (1992) Coronary endothelial and cardiac protective effects of a monoclonal antibody to intercellular adhesion molecule-1 in myocardial ischemia and reperfusion. Circulation 86:937–946
Makgoba MW, Sanders ME, Luce GEG, Gugel EA, Dustin TL, Springer TA, Shaw S (1988) Functional evidence that intercellular adhesion molecule-1 (ICAM-1) is a ligand for LFA-1-dependent adhesion in T cell-mediated cytotoxicity. Eur J Immunol 18:637–640
Marlin SD, Springer TA (1987) Purified intercellular adhesion molecule-1 (ICAM-1) is a ligand for lymphocyte function-associated antigen-1 (LFA-1). Cell 51:813–819
Mary JY, Modigliani R (1989) Development and validation of a Crohn's disease endoscopic index: a prospective multicentric study. Gut 30:983–989
McCabe RP, Woody S, Van Deventer S, Targan SR, Mayer L, Van Hogezand R, Rutgeerts P, Hanauer SB, Podolsky D, Elson CO (1996) A multicenter trial of cA2 anti-TNF chimeric monoclonal antibody in patients with active Crohn's disease. Gastroenterology 110 [Suppl 4]:A962
Mielo ME, Bennett CF, Miller BE, Welch DR (1994) Enhanced metastatic ability of TNF-a treated malignant melanoma cells is reduced by intercellular adhesion molecule-1 (CD54) antisense oligonucleotides. Exp Cell Res 214:231–241
Modigliani R, Mary JY, Simon JF, Cortot A, Soule JC, Gendre JP, Rene E (1990) Clinical, biological, and endoscopic picture of attacks of Crohn's disease: evolution on prednisolone. Gastroenterology 98:811–818
Monteith DK, Henry SP, Howard RB, Flournoy S, Levin AA, Bennett CF, Crooke ST (1997) Immune stimulation – a class effect of phosphorothioate oligodeoxynucleotides in rodents. Anticancer Drug Des 12:421–432
Nestle F, Mitra RS, Bennett CF, Nickoloff BJ (1994) Cationic lipid is not required for uptake and inhibitory activity of ICAM-1 phosphorothioate antisense oligonucleotide in keratinocytes. J Invest Dermatol 103:569–575
Oppenheimer-Marks N, Davis LS, Bogue DT, Ramberg J, Lipsky PE (1991) Differential utilization of ICAM-1 and VCAM-1 during the adhesion and transendothelial migration of human T lymphocytes. J Immunol 147:2913–2921
Paulus HE, Egger MJ, Ward JR, Williams HJ (1990) Analysis of improvement in individual rheumatoid arthritis patients treated with disease-modifying antirheumatic drugs, based on the findings in patients treated with placebo. Arthritis Rheum 33:477–484

Rothlein R, Dustin ML, Marlin SD, Springer TA (1986) A human intercellular adhesion molecule (ICAM-1) distinct from LFA-1. J Immunol 137:1270–1274

Scheynius A, Camp RL, Pure E (1993) Reduced contact sensitivity reactions in mice treated with monoclonal antibodies to leukocyte function-associated molecule-1 and intercellular adhesion molecule-1. J Immunol 150(2):655–663

Simmons D, Makgoba MW, Seed B (1988) ICAM, an adhesion ligand of LFA-1, is homologous to the neural cell adhesion molecule NCAM. Nature 331:624–627

Springer TA (1990) The sensation and regulation of interactions with the extracellular environment: the cell biology of lymphocyte adhesion receptors. Annu Rev Cell Biol 6:359–401

Statistical Consultants Inc (1986) PCNONLIN and NONLIN84: software for the statistical analysis of nonlinear models. Am Stat 40:52

Stepkowski SM, Tu Y, Condon TP, Bennett CF (1994) Blocking of heart allograft rejection by ICAM-1 antisense oligonucleotides alone or in combination with other immunosuppressive modalities. J Immunol 153:5336–5346

Targan SR, Rutgeerts P, Hanauer SB, Van Deventer SJH, Mayer L, Present DH, Braakman TAJ, Woody JN (1996) A multicenter trial of anti-tumor necrosis factor (TNF) antibody (cA2) for treatment of patients with active Crohn's disease. Gastroenterology 110 [Suppl 4]:A1026

Truelove SC, Witts LJ (1959) Cortisone and corticotrophin in ulcerative colitis. Br Med J i:387–394

Van Deventer SJH, Elson CO, Fedorak RN (1996) Safety, tolerance, pharmacokinetics and pharmacodynamics of recombinant interleukin-10 (SCH 52000) in patients with steroid refractory Crohn's disease. Gastroenterology 110 [Suppl 4]:A1034

Wegner CD, Gundel RH, Reilly P, Haynes N, Letts LG, Rothlein R (1990) Intercellular adhesion molecule-1 (ICAM-1) in the pathogenesis of asthma. Science 247:456–459

Williams CM (1988) Transplant rejection: an overview from the clinical perspective. In: Meryman HT (ed) Transplantation: approaches to graft rejection. Proceedings of the 18th annual session of the American Red Cross. Liss, New York

Yacyshyn B, Woloschuk B, Yacyshyn MB, Martini D, Doan K, Tami J, Bennett F, Kisner D, Shanahan W (1997) Efficacy and safety of ISIS 2302 (ICAM-1 antisense oligonucleotide) treatment of steroid-dependent Crohn's disease. Gastroenterology 112:A1123

Zhang R, Yan J, Shahinian H, Amin G, Lu Z, Liu T, Saag MS, Jiang Z, Temsamani J, Martin RR, Schechter PJ, Agrawal S (1995) Pharmacokinetics of an anti-human immunodeficiency virus antisense oligodeoxynucleotide phosphorothioate (GEM 91) in HIV-infected subjects. Clin Pharmacol Ther 58:44–53

CHAPTER 19
Pharmacokinetics and Bioavailability of Antisense Oligonucleotides Following Oral and Colorectal Administrations in Experimental Animals

S. AGRAWAL and R. ZHANG

A. Introduction

Antisense oligonucleotides are a novel class of therapeutic agents being developed for the treatment of various diseases. Antisense activity has been demonstrated with several oligodeoxynucleotides both in vitro and in vivo (AGRAWAL 1996). As the first generation of antisense oligonucleotides, several phosphorothioate (PS)-oligonucleotides have entered human clinical trials (CROOKE et al. 1994; ZHANG et al. 1995a). Pharmacokinetic studies have shown that PS-oligonucleotides have a short distribution half-life and a longer elimination half-life in plasma and are distributed widely into all major tissues following intravenous (i.v.), intraperitoneal, or subcutaneous administration (AGRAWAL et al. 1991, 1995a; AGRAWAL and TANG 1992; COSSUM et al. 1993, 1994; IVERSEN et al. 1994; SANDS et al. 1994; ZHANG et al. 1995b). Analysis of PS-oligonucleotides extracted from various tissues following administration shows that PS-oligonucleotides are degraded primarily from the 3' end in a tissue- and time-dependent fashion; however, in kidney and liver, degradation from the 5' end has been observed as well. In the development of the second generation of antisense oligonucleotides, therefore, major efforts have been devoted to stabilizing PS-oligonucleotides in vivo by various modification of their structure (MONIA et al. 1993; METELEV et al. 1994; AGRAWAL and IYER 1995; AGRAWAL et al. 1995c; ZHANG et al. 1995b,c,d, 1996; ZHAO et al. 1996). More stable oligonucleotides have at least two advantages: (1) an intact antisense oligonucleotide provides a longer duration of action and would therefore require less frequent dosing, and (2) fewer degradation products generated or increased clearance of degradation products would avoid potential unwanted side effects from these metabolites.

In continuing our effort to design second-generation antisense oligonucleotides with better in vivo stability, we have studied various end-capped, self-stabilized, and mixed-backbone oligonucleotides (MBO) (METELEV et al. 1994; AGRAWAL and IYER 1995; AGRAWAL et al. 1995c, 1997; ZHANG et al. 1995b,c,d, 1996; ZHAO et al. 1996). Of particular interest for this chapter, we have recently reported the in vivo stability, disposition, metabolism, and excretion of a 25-mer PS-oligonucleotide, and its two end-protected analogues – one containing segments of 2'-O-methyloligoribonucleotide phosphorothioates at both the 3' and 5' ends (MBO-1) and another containing methylphosphonate linkages at both the 3′ and 5′ ends (MBO-2) – were determined in rats after i.v. bolus administration of [^{35}S]-radiolabeled oligonucleotides (ZHANG et al. 1995b,c, 1996). The two MBOs exhibited a wide tissue distribution with a significantly better in vivo stability than the PS-oligonucleotide (ZHANG et al. 1995c, 1996).

Interestingly, following i.v. injection of PS-oligonucleotide and the two MBOs to rats, relatively high concentrations of oligonucleotide-derived radioactivity were detected in the liver, bile, and gastrointestinal tissues and contents, but minimal fecal excretion of oligonucleotide-derived radioactivity was observed (10-day cumulative excretion was less than 7% of the administered dose; ZHANG et al. 1995b,c, 1996), indicating that there is an enterohepatic circulation of these oligonucleotides.

Biliary excretion of PS-oligonucleotide and MBO-1 was examined in rats with biliary fistula (ZHAO et al. 1995). Following i.v. administration of [^{35}S]-labeled PS-oligonucleotide or MBO-1 at various doses (10–50 mg/kg), dose-dependent biliary excretion of oligonucleotide-derived radioactivity was observed. High-performance liquid chromatography (HPLC) analysis of the extracted oligonucleotides revealed both the intact form of the oligonucleotides and their metabolites in the bile. These results demonstrated that oligonucleotides were excreted through bile.

The results suggest the possibility of gastrointestinal drug administration as an alternative method of delivery of antisense oligonucleotides as therapeutic agents in the future. In this chapter, we will summarize recent studies carried out in our laboratories to determine the pharmacokinetics and bioavailability of antisense oligonucleotides following oral and colorectal administration.

B. Oral Administration

I. Experimental Design

1. Synthesis of Unlabeled and [^{35}S]-Labeled Oligonucleotides

To carry out the pharmacokinetic studies, we used [^{35}S]-labeled oligonucleotides. The sequences and chemical structure of the oligonucleotides studied (PS-oligonucleotide, MBO-1, and MBO-2) are illustrated in Fig. 1. The methods for synthesis, purification, and radiolabeling of these oligonucleotides were the same as we previously reported (AGRAWAL and TANG 1992; PADMAPRIYA et al. 1994; ZHANG et al. 1995c, 1996).

2. Animals and Drug Administration

Male Sprague-Dawley rats (110 ± 10g, Harlan Laboratories, Indianapolis, IN) and male CD/F2 mice (25 ± 3g, Charles River Laboratory) were utilized. The animals were given commercial diet and water ad libitum for 1 week prior to

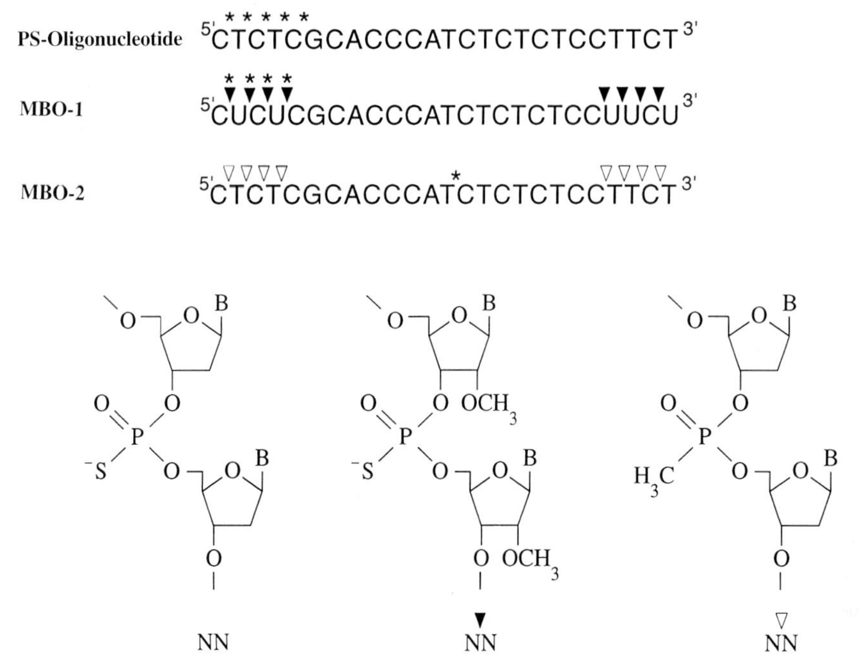

```
                        * * * * *
PS-Oligonucleotide   5'CTCTCGCACCCATCTCTCTCCTTCT 3'
                       * * * *                    ▼▼▼▼
MBO-1                5'CUCUCGCACCCATCTCTCTCCUUCU 3'
                      ▽▽▽▽              *        ▽▽▽▽
MBO-2                5'CTCTCGCACCCATCTCTCTCCTTCT 3'
```

Fig. 1. Chemical structure of phosphorothioate (*PS*)-oligonucleotide and end-modified mixed-backbone oligonucleotides (MBO)-1 and -2

the study. Unlabeled and [^{35}S]-labeled oligonucleotides were dissolved in physiological saline (0.9% NaCl) to a concentration of 25 mg/ml and were administered to the animals via gavage at a designated dose. Doses were based on the pretreatment body weight and rounded to the nearest 0.01 ml. After dosing, each animal was placed in a metabolism cage and fed a commercial diet and water ad libitum. Total voided urine was collected, and each metabolism cage was washed following the collection intervals. Total excreted feces was collected from each animal at various time points, and feces samples were homogenized prior to quantitation of radioactivity. Blood samples were collected and the animals were killed at various time points (i.e., 1, 3, 6, 12, 24, and 48h; two to three animals per time point). Plasma was separated by centrifugation. Tissues/organs were removed, trimmed of extraneous fat or connective tissue, emptied and cleaned of all contents, and individually weighed prior to homogenization.

3. Absolute Bioavailability

To quantify the total absorption of oligonucleotide, additional groups of animals (three per group) for each test oligonucleotides were treated using the same procedure as above. Animals were killed at 6 or 12h after dosing, and the gastrointestinal tract was removed. Radioactivity in the gastrointestinal tract, feces, urine, plasma, and the remainder of the body was determined separately. The percentage of the absorbed hybrid oligonucleotide-derived radioactivity was determined by the following calculation:

$$\frac{\text{Total radioactivity in the remainder of the body and plasma} + \text{total radioactivity in urine}}{\text{Total radioactivity in the gastrointestinal tract, feces, urine, plasma, and the remainder of the body}}$$

4. Sample Preparation and Total Radioactivity Measurements

The total radioactivity in tissues and body fluids was determined by liquid scintillation spectrometry using a method described previously (AGRAWAL et al. 1995d; ZHANG et al. 1995b,c, 1996). In brief, biological fluids (plasma, 50–100 µl; urine, 50–100 µl) were mixed with 6 ml scintillation solvent. Feces were ground and weighed prior to being homogenized in a ninefold volume of 0.9% NaCl saline. Following their removal, tissues were blotted immediately on Whatman no. 1 filter paper and weighed prior to being homogenized in 0.9% NaCl saline (3–5 ml/g wet weight). An aliquot of the homogenate (100 µl) was mixed with tissue solubilizer and then with scintillation solvent (6 ml) to permit quantification of total radioactivity.

5. Gel Electrophoresis

Polyacrylamide gel electrophoresis (PAGE) of the extracted oligonucleotides was carried out using methods previously described (AGRAWAL et al. 1995d; ZHANG et al. 1995b,c, 1996). Plasma and tissue homogenates were incubated with proteinase K (2 mg/ml) in extraction buffer – 0.5% sodium dodecyl sulfate (SDS)/10 mM NaCl/20 mM Tris-HCl, pH 7.6/10 mM ethylenediaminetetraacetate (EDTA) – for 1 h at 60°C. The samples were then extracted twice with phenol/chloroform (1:1, v/v) and once with chloroform. After ethanol precipitation, the extracts were analyzed by electrophoresis in 20% polyacrylamide gels containing 7M urea. Urine samples were filtered, desalted, and then analyzed by PAGE. The gels were fixed in 10% acetic acid/10% methanol solution and then dried before autoradiography.

6. High-Performance Liquid Chromatography Analysis

Oligonucleotide-derived radioactivity in plasma and tissue samples was extracted and analyzed by ion-paired HPLC using a modification of the method described previously (AGRAWAL et al. 1995d; ZHANG et al. 1995c). A Microsorb MV-C4 column (Rainin Instruments, Woburn, MA) was employed in a Hewlett Packard 1050 HPLC with a quaternary pump for gradient making. The mobile phase included two buffers: buffer A was 5 mM PIC-A reagent (Waters Co., Bedford, MA) in water, and buffer B was 4:1 (v/v) acetonitrile (Fisher)/water. The column was eluted at a flow rate of 1.5 ml/min using the following gradient: (a) 0–5 min, 0% buffer B; (b) 5–15 min, 0%–35% buffer B; and (c) 15–70 min, 35%–80% buffer B. The column was equilibrated with buffer A for at least 30 min prior to the next run. By using a RediFrac fraction collector, 1-min fractions (1.5 ml) were collected in 7-ml scintillation vials and mixed with 5 ml scintillation solvent to determine radioactivity in each fraction.

II. Results and Discussion

1. Phosphorothioate Oligonucleotide

Oral administration of a 25-mer PS-oligonucleotide at a dose of 30 mg/kg via gavage in rats showed that the PS-oligonucleotide remained stable in the contents of the stomach up to 3 h after administration (Fig. 2A). Analysis of PS-oligonucleotide extracted from the contents of the small intestine showed extensive degradation, and a ladder of shorter lengths of PS-oligonucleotide was observed at 3 and 6 h after administration. Analysis of PS-oligonucleotide extracted from the contents of the large intestine also showed extensive degradation of PS-oligonucleotide. In rats receiving PS-oligonucleotide via gavage, the total radioactivity observed from the gastrointestinal tract in the body was $17.3 \pm 5.5\%$ over 6 h and $35.5 \pm 6.0\%$ over 12 h; however, radioactivity was associated to largely degraded forms of PS-oligonucleotide. Therefore, the bioavailability of PS-oligonucleotide following oral administration is limited and can be improved either by administering the prodrug form of PS-oligonucleotide or by minimizing degradation by administering the PS-oligonucleotide directly into large intestines (see below, Sect. C.II.2).

2. Mixed-Backbone Oligonucleotide-1

Initially, stability of the end-modified oligonucleotide was determined in rats following oral administration of [^{35}S]-labeled oligonucleotide at a single dose of 50 mg/kg (AGRAWAL et al. 1995d). MBO-1 was stable in the stomach and in the small intestines, as analyzed by HPLC, up to 6 h following administration. Extensive degradation of this oligonucleotide was observed in the large intestine 6 h after dosing (AGRAWAL et al. 1995d). Radioactivity was detectable in various tissues following oral administration of MBO-1. The chemical forms of radioactivity in portal venous plasma, systemic plasma, liver, and kidneys were further examined by HPLC; both intact and degraded forms of the MBO-1 were found in these samples. Radioactivity was also detected in the urine within 1 h following oral administration. Following complete urine collection, the mean cumulative excretion of urinary radioactivity was determined to be 2.5% of the administered dose in the 24-h period and 3.8% in the 48-h period after dosing. The majority of radioactivity in the urine was associated with the degraded forms of the oligonucleotide, but trace amounts of the intact MBO-1 were also detected. Based on the quantitation of total radioactivity in the

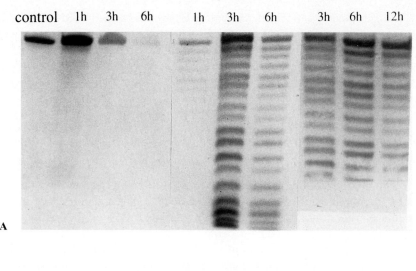

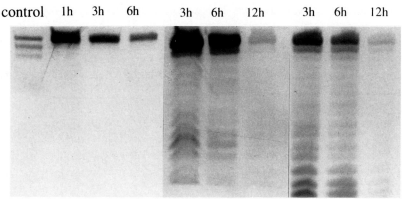

Fig. 2A,B. Gel electrophoresis analysis of radioactivity in contents of stomach (*left*), small intestine (*center*), and large intestine (*right*) of rats following oral administration of **A** phosphorothioate (PS)-oligonucleotide and **B** mixed-backbone oligonucleotide (MBO)-1

gastrointestinal tract, feces, urine, plasma, and the remainder of the body in two groups of animals (three animals per group), the total absorption of the MBO-1-derived radioactivity was determined to be $10.2 \pm 2.5\%$ over 6 h and $25.9 \pm 4.7\%$ over 12 h following oral administration. Total recovery of radioactivity in the study was $95 \pm 6\%$.

Further demonstration of oral bioavailability of MBO-1 was carried out in lower doses in both rats and mice. As illustrated in Fig. 2B and Fig. 3A, MBO-1 was shown by PAGE to be stable in the stomach and small intestines up to 6 h following oral administration. The majority of the radioactivity in the contents of the small intestine was associated with the intact MBO-1, but significant degradation was observed in the large intestine.

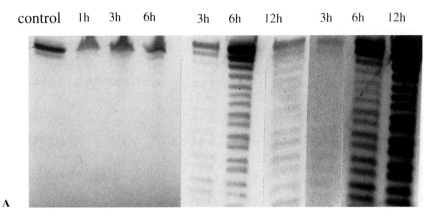

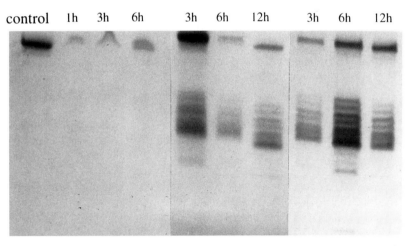

Fig. 3A,B. Gel electrophoresis analysis of radioactivity in contents of stomach (*left*), small intestine (*center*), and large intestine (*right*) of mice following oral administration of **A** mixed-backbone oligonucleotide (MBO)-1 and **B** MBO-2

Following oral administration of the radiolabeled MBO-1 to mice at a dose of 10 mg/kg, radioactivity was detectable in various tissues, including liver, kidneys, spleen, heart, lungs, bone marrow, and lymph nodes. The intact oligonucleotide was detected in various tissues by gel electrophoresis (Fig. 4A).

Oral absorption of oligonucleotides in fasting animals was also determined with PS-oligonucleotide and MBO-1, and decreased absorption rates

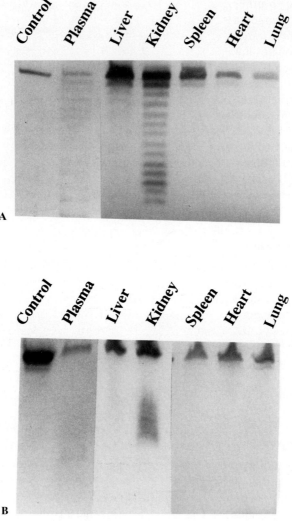

Fig. 4A,B. Gel electrophoresis analysis of tissue radioactivity in mice following oral administration of **A** mixed-backbone oligonucleotide (MBO)-1 and **B** MBO-2. Samples taken from animals killed at 6h after dosing. Representative gels from one animal for each oligonucleotides. Similar profiles were observed with other animals

were found (about 50% of that observed in nonfasting animals). The retention time of oligonucleotides in the gastrointestinal tract in fasting animals may be shorter than that in nonfasting animals, resulting in decreased absorption.

3. Mixed-Backbone Oligonucleotide-2

From the studies with MBO-1, it becomes clear that oligonucleotide that remains stable in the gastrointestinal tract is absorbed into systemic circulation and distributed to other tissues. We then investigated whether there is any impact of the charges associated with oligonucleotides on oral absorption. In MBO-2, the charges were reduced by about 30% by incorporating methylphosphonate linkages. The stability of MBO-2 in the contents of the stomach and small and large intestines of mice following oral gavage (30 mg/kg) showed the presence of mainly intact MBO-2 in the contents of the small and large intestine (Fig. 3B).

The mean percentages of absorption of MBO-2 from the gastrointestinal tract were 23.6 ± 2.8% over 6 h and 39.3 ± 2.4% over 12 h following gavage. Analysis of the extracted radioactivity from plasma and various tissues by PAGE and HPLC showed the presence of both intact and degraded forms of MBO-2 (Fig. 4B). The results from experiments with MBO-2 suggest that increased oral absorption is possible if oligonucleotides are resistant to degradation and are less polyanionic or have reduced phosphorothioate linkages.

4. General Discussion

Using an in vitro model (everted rat jejunum sacs), HUGHES et al. (1995) demonstrated that up to 15% of [^{14}C]-labeled PS-oligonucleotides was transported to the sacs during the 1-h experimental period, with no significant differences in transport rates between methylphosphonate oligonucleotides and PS-oligonucleotides. They also demonstrated no significant degradation during the transport of oligonucleotides across the intestinal wall (HUGHES et al. 1995). It is not clear, however, how these in vitro observations can be extrapolated to in vivo conditions, particularly in terms of bioavailability. Our data showed that only 10% of the oligonucleotides was absorbed over a 6-h period in rats.

Our studies establish four major points regarding the oral bioavailability of antisense oligonucleotides:

1. PS-oligonucleotides remain stable in the stomach, but undergo extensive degradation in the intestinal tract and liver; the use of PS-oligonucleotides as oral drugs may therefore be limited. With appropriate formulation or prodrugs of PS-oligonucleotides that protect the oligonucleotide from digestion in the intestinal tract and in the liver, however, PS-oligonucleotides may still be used by oral delivery.
2. Compared to PS-oligonucleotide, two end-modified MBOs show better oral bioavailability, which is associated with their stability in gastrointestinal tract, liver, and other tissues.
3. Oral absorption of oligonucleotides is dependent on the nature of the internucleotide linkages and charge as well as the stability of the oligonucleotides in the gastrointestinal tract; the significantly more stable 2'-O-methyloligoribonucleotide phosphorothioates had minimal absorption (S. Agrawal and R. Zhang, unpublished data).
4. Since oligonucleotides, in general, have short plasma half-lives and are retained in most tissues, traditional approaches to pharmacokinetic analy-

sis, using area under the curve (AUC) in plasma or urinary excretion as indices to determine oral bioavailability, may be not appropriate.

The mechanisms of the transport of oligonucleotides across the gastrointestinal tract and the metabolism of absorbed oligonucleotides in the liver and other tissues were not determined in the present study, but additional observations made during this study may be important to the design of future oral antisense therapeutics. First, oligonucleotides in fasting animals were absorbed less than in nonfasting animals, indicating that (a) oligonucleotides may be absorbed through stomach and (b) the absorption of oligonucleotide across the gastrointestinal tract is a slow process, and the retention time of intact oligonucleotides in the gastrointestinal tract is therefore critical. Second, unmodified PS-oligonucleotide had a relatively high oral absorption rate based on radioactivity quantification, but limited intact compound was detectable in the liver and extrahepatic tissues. Two modified PS-oligonucleotides that have been shown to have greater in vivo stability after i.v administration (ZHANG et al. 1995c, 1996) not only had good oral absorption, but a high degree of integrity in vivo following oral administration. The results indicate that the first-pass effect of the liver on the metabolism of oligonucleotides is a key factor in determining the bioavailability and ultimately the therapeutic efficacy of oral delivery of these agents. Finally, no significant differences were observed in the present study between rats and mice, indicating that both animals are useful in screening oral absorption of antisense oligonucleotides.

C. Colorectal Administration

I. Special Considerations in Experimental Design

As discussed above, the oral bioavailability of antisense oligonucleotides may be affected by several factors, including their stability and retention in the gastrointestinal tract and their degradation in the liver. Following intestinal absorption, PS-oligonucleotides underwent extensive degradation in the liver, while modified analogues were relatively stable. These results suggest the possibility of alternative delivery of antisense oligonucleotides as therapeutic agents in the future for the treatment of chronic diseases such as acquired immunodeficiency syndrome (AIDS) and cancers, which require long-term therapy. The study described below was designed to explore the possibility of rectal administration as a novel drug delivery means for antisense oligonucleotides. The major advantage of rectal administration is to avoid extensive degradation of oligonucleotides in the liver, which would provide more intact oligonucleotides at the target sites with a longer period of activity.

Male Sprague-Dawley rats (150–200 g; Harlan Laboratories, Indianapolis, IN) were utilized in this study. The animals were given commercial diet and water ad libitum for 1 week prior to study. After each animal had been anesthetized using pentobarbital, an incision was made on the lower part of the abdomen to expose the large intestine. The colon was cut open 0.5 cm from the caecum. The contents of the large intestine were washed out using 30 ml physiological saline (0.9% NaCl) at 37°C. After the anus had been ligated, unlabeled and [^{35}S]-labeled oligonucleotides dissolved in physiological saline (0.9% NaCl) at designated concentrations were injected into the large intestine through the cut, which was ligated after drug administration. The abdomen was then closed, and body temperature was maintained at 38 ± 0.5°C by means of a heat lamp. Oligonucleotides were administered to rats at four dose levels, i.e., 3.3, 10, 30, and 90 mg/kg (three rats per dose level). Blood samples were collected in heparinized tubes from animals at the various times, i.e., 1, 2, 3, and 4 h. Plasma was separated by centrifugation. At 4 h after drug administration, animals were euthanized by exsanguination under sodium pentobarbital anesthesia. All tissues/organs were subsequently collected, immediately blotted on Whatman no. 1 filter paper, trimmed of extraneous fat or connective tissue, emptied and cleaned of all contents, and individually weighed prior to quantitation of oligonucleotide-derived radioactivity. Biological samples were analyzed by determination of total radioactivity, HPLC, and PAGE analysis using the methods described above.

II. Results and Discussion

1. Mixed-Backbone Oligonucleotide-1

MBO-1 was stable in the large intestine as analyzed by HPLC and PAGE for up to 4 h following administration, with minimal degradation observed (Fig. 5). Gel electrophoresis revealed that the majority of extracted radioactivity in the large intestine and its contents was intact oligonucleotide (Fig. 6). Absorption of MBO-1 was examined at various doses, i.e., 3.3, 10, 30, and 90 mg/kg. Oligonucleotide-derived radioactivity was detectable in various tissues following large intestine administration of the radiolabeled MBO-1. Figure 7 illustrates the concentration of the MBO-1 equivalents in plasma, indicating that the oligonucleotide was absorbed in a time- and concentration-dependent fashion. Significant accumulation of oligonucleotide-derived radioactivity was observed in various tissues. Figure 8 illustrates the concentration of MBO-1 equivalents in selected tissues, including kidneys, liver, spleen, bone marrow, lymph nodes, and brain, 4 h after administration. As can be seen in Fig. 9, HPLC analysis revealed both intact and degraded forms of MBO-1 in these tissue samples. The majority of the radioactivity in the liver and kidneys was associated with the intact form of MBO-1. Gel electrophoresis also revealed that the majority of the extracted radioactivity in these samples was associated with the intact form of MBO-1 (Fig. 6). No significant degraded products were detected in the large intestine for up to 4 h after administration. Approximately 4%–14% of administered MBO-1 was absorbed within 4 h in the anesthetized animals, depending on the dose levels.

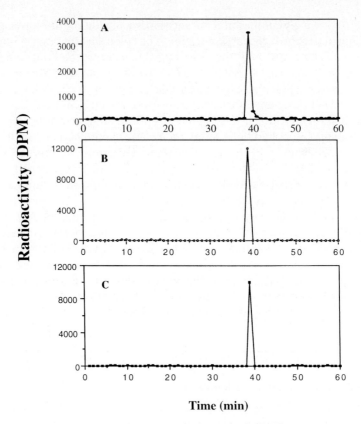

Fig. 5A–C. High-performance liquid chromatography (HPLC) profile of radioactivity in the large intestine and its contents following large intestine administration of radio-labeled mixed-backbone oligonucleotide (MBO)-1. Samples taken from animals killed at 4 h after dosing. **A** Standard. **B** Large intestine contents. **C** Large intestine

2. Phosphorothioate Oligonucleotide and End-Modified Mixed-Backbone Oligonucleotide-2

In separate studies, similar results were obtained following rectal administration of PS-oligonucleotide and MBO-2. At 10 mg/kg, PS-oligonucleotide had a 4-h absorption rate of 8.74% of the administered dose, and MBO-1 had an absorption rate of 6.6% of the administered dose. Our previous studies demonstrated that, following oral administration, PS-oligonucleotides could be well absorbed through the gastrointestinal wall, but were extensively degraded in the liver; few intact PS-oligonucleotides were available, therefore, in the systemic tissues. Rectal delivery can avoid the first-pass effect in the liver. Following large intestine administration, PS-oligonucleotide was well absorbed largely in the intact form and was less extensively degraded in other tissues.

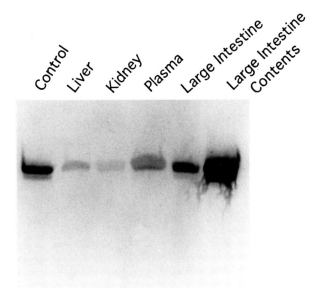

Fig. 6. Polyacrylamide gel electrophoresis (PAGE) profile of radioactivity in the large intestine and its contents as well as various tissues following large intestine administration of radiolabeled mixed-backbone oligonucleotide (MBO)-1. Samples taken from animals killed at 4 h after dosing

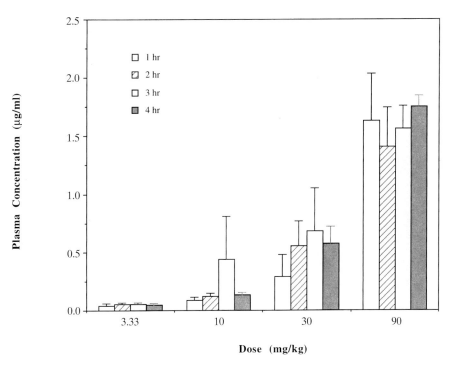

Fig. 7. Concentration of mixed-backbone oligonucleotide (MBO)-1 equivalents in plasma, indicating that the oligonucleotide was absorbed in a time- and concentration-dependent fashion. Samples taken from animals killed at 4 h after dosing. Time points indicated (1–4 hr) are those after drug administration

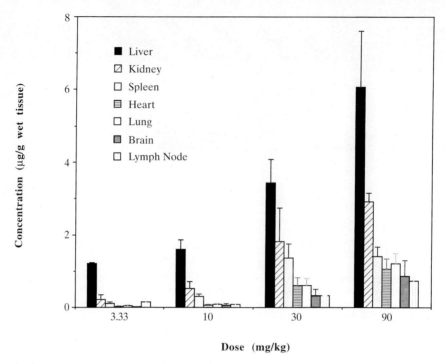

Fig. 8. Concentration of mixed-backbone oligonucleotide (MBO)-1 equivalents in selected tissues 4 h following large intestine administration of MBO-1 at various doses. Samples taken from animals killed at 4 h after dosing

3. General Discussion

These studies represent the first report on the bioavailability of antisense oligonucleotides following colorectal administration in experimental animals. The rationale of the present study was to demonstrate the rectal bioavailability of antisense oligonucleotides, which may facilitate the development of this class of compound as therapeutic agents for chronic diseases such as human immunodeficiency virus (HIV) infection/AIDS and cancers, which require long-term therapy. We have now shown that, following large intestine administration, the following is true:

1. PS-oligonucleotide and end-modified oligonucleotides were stable in the lumen of the large intestine.
2. These oligonucleotides were absorbed through the wall of the large intestine.
3. The absorbed oligonucleotide-derived radioactivity was widely distributed to various tissues with a pattern similar to that seen following i.v. administration.

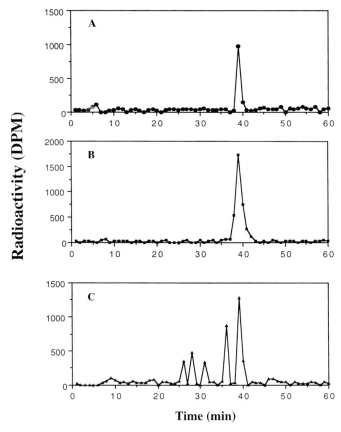

Fig. 9A–C. High-performance liquid chromatography (HPLC) profile of radioactivity in **A** plasma, **B** liver, and **C** kidneys following large intestine administration of radiolabeled mixed-backbone oligonucleotide (MBO)-1. Samples taken from animals killed at 4 h after dosing

4. Radioactivity in tissues such as liver and kidneys was associated with intact oligonucleotide as well as metabolites.

The present study therefore provides initial evidence for the possibility of colorectal administration as an alternate delivery means for antisense oligonucleotides as therapeutic agents. It should be noted that the absorption rates were estimated in anesthetized rats; thus the actual bioavailability of colorectal oligonucleotides may be underestimated. The mechanisms responsible for absorption of the oligonucleotides have not been defined in the present study. Further studies examining the mechanisms of transport of oligonucleotides in the gastrointestinal tract and liver, first-pass effects of liver, enterohepatic circulation, and formulation of oligonucleotides are needed. Studies in large animals and humans may be indicated.

D. Conclusions

In this chapter, we have briefly summarized the bioavailability of antisense oligonucleotides administered through the gastrointestinal tract, namely oral and colorectal administration. The pharmacokinetics and factors affecting gastrointestinal absorption of oligonucleotides are summarized in Fig. 10. Briefly, when oligonucleotides are administered orally, they may be stable in the stomach contents, and whether they are absorbed through the stomach wall is not clear. When the administered oligonucleotides move into small intestines, extensive degradation of PS-oligonucleotides and some degradation of MBO may occur. Intact oligonucleotides (and perhaps also degradative forms) are absorbed through portal venous blood and enter the liver. The absorbed oligonucleotides may undergo metabolism in the liver (the first-pass effect) and enter the systemic circulation. Oligonucleotides and their metabolites are excreted into bile, enter the intestinal lumen, and reenter the enterohepatic circulation. Oligonucleotides in the systemic circulation are distributed into various tissues and excreted into urine as seen following i.v. administration. When orally administered oligonucleotides move into the large intestine, most PS-oligonucleotides and MBO may be present as degradation products. In general, oligonucleotides absorbed through the upper portion of the large intestine enter the liver, and oligonucleotides absorbed

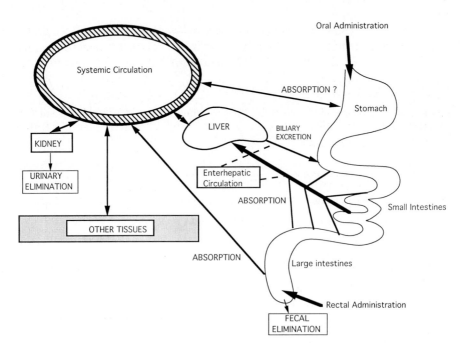

Fig. 10. Enterohepatic circulation of oligonucleotides and the drug delivery of oligonucleotides through the gastrointestinal tract

through the lower portion of the large intestine directly enter the systemic circulation. The latter are not metabolized in the liver, and the first-pass effect of the liver is avoided. Colorectal administration of oligonucleotides takes the advantage of this opportunity. When oligonucleotides are administered into the rectum, most absorbed oligonucleotides enter the systemic circulation. In general, the following factors will be important in the development of oral or rectal oligonucleotides therapeutics: (a) stability of oligonucleotides in the gastrointestinal tract, (b) duration of the retention of oligonucleotides in the gastrointestinal tract, (c) the structure and physical and biochemical properties of oligonucleotides, e.g., charges, (d) the first-pass effect of the liver, (e) diet and host status of the gastrointestinal and hepatic functions, and (f) formulations.

The advantages of delivery of oligonucleotides through oral or rectal administration are obvious. The slow but continuous release of oligonucleotides into the systemic circulation may increase the uptake of target tissues. In addition, it will avoid the high plasma concentrations associated with i.v. injection and reduces the risk of side effects resulting from these high concentrations. In this chapter, we have only summarized the single-dose studies. In fact, multiple dosing (once a day) significantly increased the plasma and tissue concentrations of MBO (R. Zhang and S. Agrawal, unpublished data). More recently, we have observed the biological efficacy (antitumor activity) of MBO targeted at protein kinase A following oral administration (ZHANG et al. 1997).

Since the mechanisms responsible for the absorption of oligonucleotides through the gastrointestinal tract are largely unknown, further studies are needed in this area. Our data provide the basis for future studies of oligonucleotides as orally or rectally available therapeutic agents. Additional ongoing studies are designed to further elucidate the mechanisms of transport and metabolism of oligonucleotides. Most importantly, the therapeutic efficacy of antisense oligonucleotides with different targets is to be further demonstrated following oral or rectal administration.

Acknowledgement. We would like to thank X. Zhang, H. Zhao, Z. Lu, Q. Cai, J.M. Tamburin, J. Yan, H. Cai, X. Wu, L. High, Y. Li, W. Tan, and Z. Jiang for their excellent technical assistance, and Dr. R.B. Diasio for helpful discussions.

References

Agrawal S (1996) Antisense oligonucleotides: towards clinical trial. Trends Biotechnol 14:376–387
Agrawal S, Iyer RP (1995) Modified oligonucleotides as therapeutic and diagnostic agents. Curr Opin Biotechnol 6:112–119
Agrawal S, Tang JY (1992) GEM 91 – an antisense oligonucleotide phosphorothioate as a therapeutic agent for AIDS. Antisense Res Dev 2:261–66
Agrawal S, Temsamani J, Tang JY (1991) Pharmacokinetics, biodistribution and stability of oligodeoxynucleotide phosphorothioates in mice. Proc Natl Acad Sci USA 88:7595–7599

Agrawal S, Temsamani J, Galbraith W, Tang J (1995a) Pharmacokinetics of antisense oligonucleotides. Clin Pharmacokinet 28:7–16

Agrawal S, Temsamani J, Tang JY (1995b) Self-stabilized oligonucleotides as novel antisense agents. In: Akhtar S (ed) Delivery strategies: antisense oligonucleotide therapeutics. CRC Press, Boca Raton, pp 105–21

Agrawal S, Rustagi PK, Shaw DR (1995c) Novel enzymatic and immunological responses to oligonucleotides. Toxicol Lett 82/83:431–434

Agrawal S, Zhang X, Zhao H, Lu Z, Yan J, Cai H, Diasio RB, Habus I, Jiang Z, Iyer RP, Yu D, Zhang R (1995d) Absorption, tissue distribution and in vivo stability in rats of a hybrid antisense oligonucleotide following oral administration. Biochem Pharmacol 50:571–576

Agrawal S, Jiang Z, Zhao Q, Shaw D, Cai Q, Roskey A, Channavajjala L, Saxinger C, Zhang R (1997) Mixed-backbone oligonucleotides as second generation antisense oligonucleotides: in vitro and in vivo studies. Proc Natl Acad Sci USA 94:2620–2625

Cossum PA, Sasmor H, Dellinger D, Truong L, Cummins L, Owens SR, Markham PM, Shea JP, Crooke S (1993) Disposition of the ^{14}C-labeled phosphorothioate oligonucleotide ISIS 2105 after intravenous administration to rats. J Pharmacol Exp Ther 267:1181–1190

Cossum PA, Truong L, Owens SR, Markham PM, Shea JP, Crooke S (1994) Pharmacokinetics of a ^{14}C-labeled phosphorothioate oligonucleotide, ISIS 2105, after intradermal administration to rats. J Pharmacol Exp Ther 269:89–94

Crooke ST, Grillone LR, Tendolkar A, Garrett A, Fratkin MJ, Leeds J, Barr WH (1994) A pharmacokinetic evaluation of ^{14}C-labeled afovirsen sodium in patients with genital warts. Clin Pharmacol Ther 56:641–646

Hughes JA, Avrutskaya AV, Brouwer KLR, Wickstrom E, Juliano RL (1995) Radiolabeling of methylphosphonate and phosphorothioate oligonucleotides and evaluation of their transport in everted rat jejunum sacs. Pharm Res 12:817–824

Iversen PL, Mata J, Tracewell WG, Zon G (1994) Pharmacokinetics of an antisense phosphorothioate oligodeoxynucleotide against *rev* from human immunodeficiency virus type 1 in the adult male rat following single injection and continuous infusion. Antisense Res Dev 4:43–52

Metelev V, Lisziewicz J, Agrawal S (1994) Study of antisense oligonucleotide phosphorothioates containing segments of oligodeoxynucleotides and 2'-*O*-methyloligoribonucleotides. Bioorg Med Chem Lett 4:2929–2934

Monia BP, Lesnik EA, Gonzalez C, Lima WF, McGee D, Guinosso CJ, Kawasaki AM, Cook PD, Freier SM (1993) Evaluation of 2'-modifies oligonucleotides containing 2'-deoxy gaps as antisense inhibitors of gene expression. J Biol Chem 268:14514–14522

Padmapriya AP, Tang JY, Agrawal S (1994) Large-scale synthesis, purification and analysis of oligodeoxynucleotide phosphorothioates. Antisense Res Dev 4:185–199

Sands H, Gorey-Feret LJ, Cocuzza AJ, Hobbs FW, Chidester D, Trainor GL (1994) Biodistribution and metabolism of internally ^{3}H-labeled oligonucleotides. I. Comparison of a phosphodiester and a phosphorothioate. Mol Pharmacol 45:932–943

Zhang R, Yan J, Shahinian H, Amin G, Lu Z, Liu T, Saag MS, Jiang Z, Temsamani J, Martin RR, Schechter P, Agrawal S, Diasio RB (1995a) Pharmacokinetics of an oligodeoxynucleotide phosphorothioate (GEM 91) in HIV-infected subjects. Clin Pharmacol Ther 58:44–53

Zhang R, Diasio RB, Lu Z, Liu TP, Jiang Z, Galbraith WM, Agrawal S (1995b) Pharmacokinetics and tissue disposition in rats of an oligodeoxynucleotide phosphorothioate (GEM 91) developed as a therapeutic agent for human immunodeficiency virus type-1. Biochem Pharmacol 49:929–939

Zhang R, Lu Z, Zhao H, Zhang X, Diasio RB, Habus I, Jiang Z, Iyer RP, Yu D, Agrawal S (1995c) In vivo stability, disposition, and metabolism of a "hybrid" oligonucleotide phosphorothioate in rats. Biochem Pharmacol 50:545–556

Zhang R, Lu Z, Zhang X, Zhao H, Diasio RB, Jiang Z, Agrawal S (1995d) In vivo stability and disposition of a self-stabilized oligodeoxynucleotide phosphorothioate in rats. Clin Chem 41:836–843

Zhang R, Iyer P, Yu D, Zhang X, Lu Z, Zhao H, Agrawal S (1996) Pharmacokinetics and tissue disposition of a chimeric oligodeoxynucleotide phosphorothioate in rats following intravenous administration. J Pharm Exp Ther 278:971–979

Zhang R, Cai Q, Li Y, Xu J, Xie X, Tan W, Agrawal S (1997) Novel mixed-backbone oligonucleotides (MBO) targeted at protein kinase A with improved in vivo antitumor activities against human cancer xenografts. Proc Am Assoc Cancer Res 38:316 (abstr)

Zhao H, Lu Z, Diasio RB, Agrawal S, Zhang R (1995) Biliary excretion of oligonucleotides: Previously unrecognized pathway in metabolism of antisense oligonucleotides. FASEB J 9(3):A410

Zhao Q, Temsamani J, Iadarola PL, Jiang Z, Agrawal S (1996) Effect of different modified oligodeoxynucleotides on immune stimulation. Biochem Pharmacol 51:173–182

CHAPTER 20
Antisense Properties of Peptide Nucleic Acid

P.E. NIELSEN

A. Introduction

The hybridization properties of peptide nucleic acid (PNA) combined with its ease of synthesis and high chemical and biological stability rapidly made this molecule a very attractive lead compound for the development of antisense gene therapeutic drugs.

In the past 6 years, much has been learned about the chemical and biological properties of PNA, and the present chapter will discuss these developments in terms of antisense technology. The reader is also referred to several recent reviews dealing with the chemistry, biology, and biophysical chemistry of PNA (DUEHOLM and NIELSEN 1996; ERIKSSON and NIELSEN 1996b; HYRUP and NIELSEN 1996; KNUDSEN and NIELSEN 1997; LARSEN and NIELSEN 1998).

B. Chemistry

Since PNA (Fig. 1) is a polyamide (or a pseudopeptide) (NIELSEN et al. 1991, 1994a; EGHOLM et al. 1992a,b), its chemistry is very closely related to that of peptides rather than to that of oligonucleotides. In fact, PNA can be synthesized from amino acid monomers (Fig. 2) on (slightly modified) peptide synthesizers using either the tBoc (DUEHOLM et al. 1994b; CHRISTENSEN et al. 1995) or the Fmoc protection strategy (THOMPSON et al. 1995; WILL et al. 1995; BREIPOHL et al. 1996), and both types of monomers are now commercially available.

C. Hybridization Properties

Sequence-specific hybridization of PNA to single-stranded RNA or DNA appears to follow the Watson-Crick-Hoogsteen base-pairing rules found for natural nucleic acids (EGHOLM et al. 1993). However, both antiparallel (PNA amino-terminal facing the 3′ end of the oligonucleotide) and parallel PNA–nucleic acid duplexes are stable, although the antiparallel duplex is the more stable by 1–1.5°C per base pair (EGHOLM et al. 1993). In contrast to oligonucleotides, PNA are electrostatically neutral, and thus the stability of PNA–nucleic acid complexes is not significantly influenced by counter ions (EGHOLM

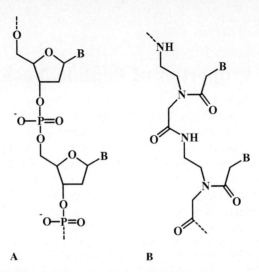

Fig. 1A,B. Chemical structures of **A** DNA and **B** peptide nucleic acid (PNA). *B* is a nucleobase, e.g., adenine cytosine, guanine, thymine. Note that the PNA backbone is uncharged and achiral. Furthermore, both structures have six bonds in the backbone between each unit and three bonds between the backbone and the nucleobase ("6+3 rule")

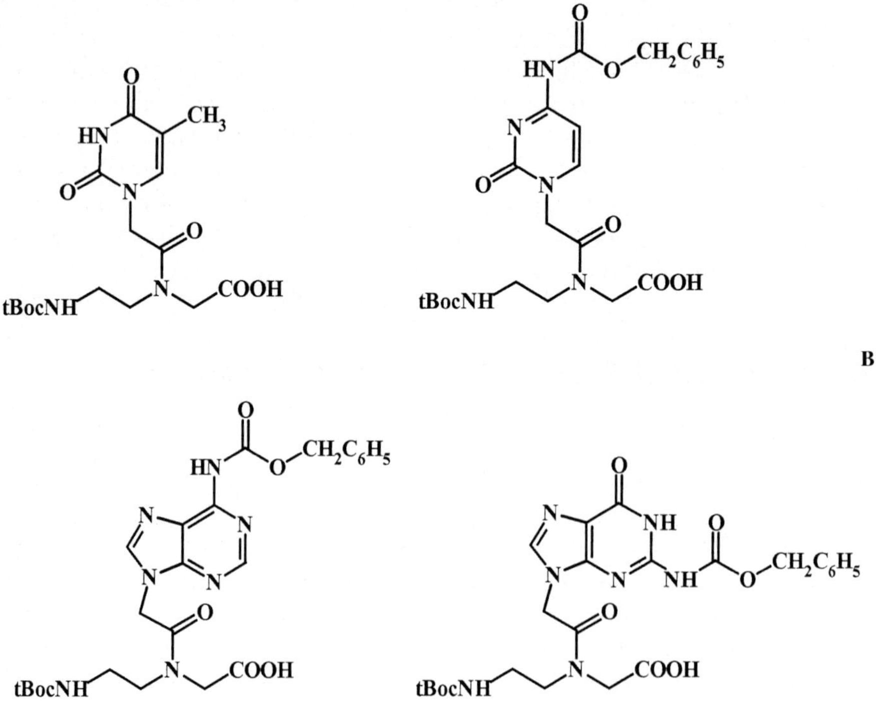

Fig. 2A–D. Examples of peptide nucleic acid (PNA) monomers for Boc-type peptide oligomerization. **A** Thymine. **B** Cytosine. **C** Adenine. **D** Guanine

et al. 1993; TOMAC et al. 1996). In general, it has been found that PNA–DNA (or –RNA) duplexes are approximately 1°C more stable than the corresponding DNA–DNA (or –RNA) complexes at 140 mM Na$^+$, whereas equal thermal stability is observed at 1 M Na$^+$. Thermal stability studies have also indicated that the sequence discrimination of PNA hybridization is at least as good as that of DNA or RNA, since ΔTm (change in melting transition) values upon introduction of mismatches in PNA-DNA or PNA-RNA duplexes are equal to or greater than those observed for DNA–DNA or DNA–RNA duplexes (EGHOLM et al. 1993; JENSEN et al. 1997). Thorough kinetic studies of PNA hybridization are still lacking, but the results obtained so far (EGHOLM et al. 1993), including recent "Biacore" measurements (JENSEN et al. 1997), indicate that PNA hybridization is at least as fast as DNA or RNA hybridization. Most interestingly, PNA$_2$–nucleic acid triplexes of extraordinarily high stability are formed between homopyrimidine PNA and homopurine oligonucleotide targets. The most stable PNA$_2$–DNA triplexes have the PNA Watson-Crick strand antiparallel and the Hoogsteen PNA strand parallel to the oligonucleotide target (EGHOLM et al. 1995). For instance, the complex between PNA-T$_{10}$ and dA$_{10}$ has a T$_m$ of 71°C (EGHOLM et al. 1992a). As expected, sequence discrimination by homopyrimidine PNA is even greater due to the dual recognition of triplexes (EGHOLM et al. 1992a,b, 1995). Even disregarding the propensity of pyrimidine-rich PNA to form stable triplexes, the PNA–nucleic acid hybrid sequence stability correlation is more complex than that governing DNA duplex formation. It appears that, in addition to the influence of the GC to AT ratio, the purine content of the PNA strand is also of major importance. For instance, it was found that the thermal stability of the duplex between a decamer PNA (A$_4$G$_2$AGAG) and the complementary antiparallel oligonucleotide is 70°C, i.e., as high as that of a PNA$_2$–DNA decamer triplex (NIELSEN and CHRISTENSEN 1996).

D. Structure of Peptide Nucleic Acid Complexes

Structure determinations of PNA–RNA (BROWN et al. 1994) and PNA–DNA duplexes (ERIKSSON and NIELSEN 1996a) by nuclear magnetic resonance methods indicated that the PNA strand would adapt to its nucleic acid partner to a great extent, since in both complexes the nucleic acid adopts close to its natural conformation: A form for RNA and B form for DNA. However, the PNA–DNA duplex shows A-type characteristics in terms of base-pair helical displacement. Recent X-ray determinations of the structures of a PNA$_2$–DNA triplex (BETTS et al. 1995) and not least a pure PNA duplex (RASMUSSEN et al. 1997) have revealed that the PNA in fact prefers a helical conformation, the P form, which is distinctly different from both the A and the B form. The P form is characterized by a large diameter (28Å) and a very large pitch (18bp). Nonetheless, consecutive base-pair stacking overlaps are very close to those seen in the A-form helix. Thus, despite the fact that PNA is indeed a very good mimic of DNA, it prefers a conformation that is suboptimal for DNA or RNA hybridization, although (in accordance with the higher stability of the PNA–RNA duplex) it is closer to the RNA A form than to the DNA B form.

E. Antisense Activity (Mechanism of Action)

Antisense activity of PNA oligomers has been demonstrated in in vitro translation systems (HANVEY et al. 1992; BONHAM et al. 1995; GAMBACORTI et al. 1996; KNUDSEN and NIELSEN 1996) and by cellular microinjection (HANVEY et al. 1992; BONHAM et al. 1995). However, since PNA–RNA hybrids are not substrates for RNase H (KNUDSEN and NIELSEN 1996), a mechanism based on physical obstruction of the ribosomes is believed to operate. Apparently, PNA_2–RNA triplexes (homopurine targets) of at least decamer length are stable enough to arrest elongating ribosomes, whereas even 20-mer PNA–RNA duplexes (mixed purine–pyrimidine targets) are not (KNUDSEN and NIELSEN 1996). However, when targeted to sequences around the AUG initiation sit, even mixed purine–pyrimidine duplex-forming, 15-mer PNA oligomers are efficient antisense agents in in vitro translation assays (GAMBACORTI et al. 1996; KNUDSEN and NIELSEN 1996; Fig. 3).

Presumably, "gene-walks" will be required to identify the most sensitive regions towards PNA targeting in any particular gene. This procedure is also necessary when employing other antisense agents, such as phosphorothioates (MONIA et al. 1996).

Recently, it was shown that the cancer-related enzyme telomerase, which maintains the telomeres of chromosomes, can be efficiently inhibited by PNA

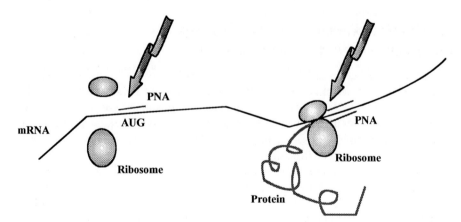

Fig. 3. Antisense mechanisms proposed for peptide nucleic acid (*PNA*). Translation initiation and elongation arrest. Targeting around the AUG initiation codon is usually efficient, with both triplex forming homopyrimidine PNA (>10-mers) and duplex forming mixed purine/pyrimidine sequence PNA (>15 mers), whereas targeting downstream from the initiation requires triplex forming PNA (>10-mers)

targeted to the RNA moiety of this enzyme (NORTON et al. 1996). Thus RNA targets other than the mRNA should be considered as "antisense" targets.

F. Cellular Biology

Very little information is so far available regarding the cellular biology and pharmacokinetic properties of PNA, except that cellular uptake and membrane penetration in vitro is at least as poor for PNA as it is for oligonucleotides (BONHAM et al. 1995; WITTUNG et al. 1995). Cationic lipids are usually required when performing antisense experiments on cells in culture with oligonucleotides (BONHAM et al. 1995). Likewise, it should be possible to develop analogous mediator systems for PNA. However, since PNA is synthesized by peptide chemistry and oligopeptides are easily attached to the PNA as part of the solid support synthesis, an obvious step would be to exploit various peptide motifs now recognized as serving as cellular and nuclear uptake signals (e.g., PROCHIANTZ 1996). More complex carrier systems such as ferritin conjugates have also been suggested for the transportation of PNA over the blood–brain barrier (PARDRIGE et al. 1995).

Regarding medicinal applications, the poor in vitro cellular uptake of PNA may prove a pseudoproblem, since animal studies with C-*raf* kinase-targeted phosphorothioates, which are equally poorly taken up by cells in culture, have shown very promising anticancer effects (MONIA et al. 1996). Thus pharmacokinetic data and animal studies using PNA are eagerly awaited, especially since PNA shows extremely high biological stability (DEMIDOV et al. 1994).

G. Peptide Nucleic Acid Derivatives (Structure–Activity Relationships)

Since the introduction of PNA, a large number of modifications of the original aminoethylglycine backbone have been synthesized and tested for hybridization efficiency in a PNA context. These derivatives have explored some of the structural space for PNA in general and also provide starting structures for second-generation PNA molecules that could aid optimization of pharmacokinetic properties of PNA drug candidates.

So far, it can in general be concluded that any modification that alters the original backbone structure, such as extensions of any of the three "linker portions" of the backbone (HYRUP et al. 1994; Table 1, compounds 2, 3 and 4) results in PNA with highly reduced hybridization efficiency, indicating that the "6 + 3" bond geometry, also found in DNA (and RNA), is crucial (see Fig. 1). Likewise, the restricted conformational flexibility imposed by the two amide moieties in PNA seems essential, since reduction of one of these to a flexible amine (HYRUP et al. 1996; Table 1, compound 5) is detrimental. The "retro-

Table 1. Effects on thermal stability (ΔT_m, °C per monomer) for structurally modified PNA T monomers when incorporated into the oligomer sequence H-GTA GAT CAC T-NH$_2$

Compound	Structure	Backbone/linker	ΔT_m DNA	ΔT_m RNA	Reference
1		Ethylglycine	0	0	–
2		Propylglycine	–8.0	–6.5	Hyrup et al. 1995
3		Ethyl-β-alanine	–10	–7.5	Hyrup et al. 1995

4	Propionyl linker	−20	−16	Hyrup et al. 1994
5	Ethyl linker	−22	−18	Hyrup et al. 1996
6	Retro-inverso	−6.5	ND	Krotz et al. 1995a,b

ND, not determined.

inverso" isomer of PNA (Table 1, compound 6) is not a very good DNA mimic either (KROTZ et al. 1995a,b, 1998).

In agreement with the available structural data on PNA complexes, substitutions at the α-position of the glycine moiety of the backbone do not severely interfere with duplex (or triplex) formation (DUEHOLM et al. 1994a; HAAIMA et al. 1996; Table 2). Naturally, the hybridization efficiency is dependent on the substituent, since other factors (in addition to simple steric hindrance), such as changes in hydration, flexibility, and structure of the single-stranded PNA, also affect the change in free energy upon hybrid formation. Nonetheless, by employing α-amino acid building blocks other than glycine for the PNA backbone, PNA with a wide variety of functionalities can be constructed. This should give ample freedom to exploit PNA as a genuine gene therapeutic drug lead. For instance, simply employing a few lysine backbone units in a PNA oligomer appears to generate PNA with greatly improved aqueous solubility without sacrificing hybridization efficiency or specificity (HAAIMA et al. 1996). Thus any solubility problems that may be experienced with certain PNA oligomers (NOBLE et al. 1995) should be solvable in this way.

Only a few alternative nucleobases have so far been used in a PNA context. Hoogsteen recognition of guanine by cytosine is compromised by neutral or alkaline pH, since protonation of cytosine-N3 is required and the pKa of this is approximately 4.5 (for free cytosine). Employment of

Table 2. Effects on thermal stability (ΔT_m, °C per monomer) for the PNA sequence H-GTA GAT CAC T-NH$_2$ incorporating three chiral monomers as compared to an unmodified PNA (HAAIMA et al. 1996)

Compound	R	Chirality	ΔT_m DNA	ΔT_m RNA
7	CH$_3$	L	−1.8	ND
8	CH$_3$	D	−0.7	ND
9	sec-Bu	L	−2.6	−3.0
10	CH$_2$OH	L	−1.0	−1.0
11	CH$_2$OH	D	−0.6	−1.0
12	CH$_2$CO$_2$H	L	−3.3	ND
13	CH$_2$CH$_2$CO$_2$H	D	−2.3	ND
14	(CH$_2$)$_4$NH$_2$	L	−1.0	−1.3
15	(CH$_2$)$_4$NH$_2$	D	+1.0	0

R, substituent at the α-position of glycine in the backbone; ND, not determined.

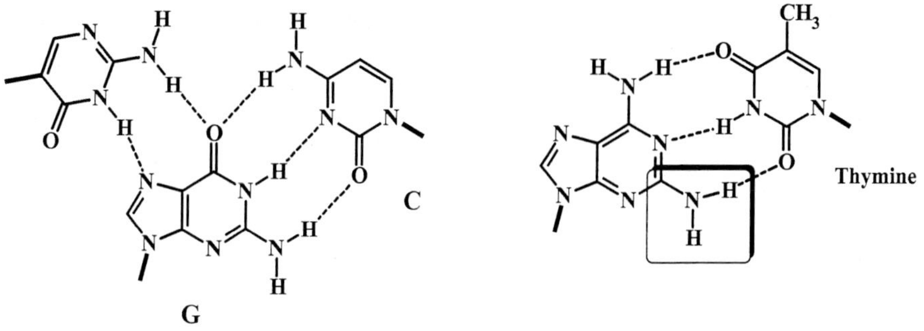

Fig. 4A,B. Examples of modified nucleobases. **A** Pseudoisocytosine (ψiC) is used for efficient, pH-independent third-strand Hoogsteen recognition of guanine, while **B** diaminopurine can replace adenine in Watson-Crick recognition of thymine, forming a more stable base pair

pseudoisocytosine instead of cytosine solves this problem, since the one isomer of this nucleobase does not need protonation to engage in the proper Hoogsteen base pairing with guanine (Fig. 4A). Although pseudoisocytosine can function as a substitute both for cytosine and for protonated cytosine via either of its tautomeric forms, this alternative nucleobase is especially useful in connection with linked (bis-)PNA in which an (antiparallel) Watson-Crick strand recognizing PNA oligomer that contains cytosine is covalently linked to a (parallel) Hoogsteen strand recognizing PNA oligomer that contains pseudoisocytosine (Fig. 5; EGHOLM et al. 1995).

Significant stabilization (2–3°C per base pair) of PNA–DNA duplexes can be accomplished by employing diaminopurine (DAP) instead of adenine for recognition of thymine (HAAIMA et al., 1997), since the DAP base pair is held together by three hydrogen bonds (as is the guanine–cytosine base pair) instead of only two for the adenine–thymine base pair (Fig. 4B).

H. Peptide Nucleic Acid–DNA Chimeras

Several synthetic strategies have been presented for the preparation of PNA–DNA chimeric molecules connected via amino-3′ linkages, carboxy-5′ linkages, or both, creating carboxy-PNA–DNA-5′, amino-PNA–DNA-3′, or amino-PNA–DNA–PNA-carboxy chimeras (Fig. 6; BERGMANN et al. 1995; PETERSEN et al. 1995; BREIPOHL et al. 1996; FINN et al. 1996; UHLMANN et al. 1996). Such chimeras should provide RNase H activation via the DNA moiety, by analogy to the use of 2′-substituted DNA analogues (e.g., ALTMAN et al. 1996). However, it is too early to tell whether this will turn out to be an advantage in antisense technology, since no biological studies have yet been reported for these molecules.

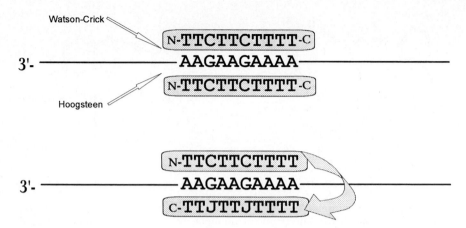

Fig. 5. Peptide nucleic acid (PNA)$_2$–DNA triplexes formed by either two single, identical PNA oligomers or by a chemically linked bis-PNA

UHLMANN et al. (1996) argue that such PNA–DNA chimeric molecules could provide good cellular uptake properties for the PNA, but since DNA oligonucleotides are taken up very poorly themselves, this is not necessarily a viable route to address this issue.

I. Antigene Activity

Eventually, PNA may also be developed into an antigene agent. Homopyrimidine PNA oligomers bind very strongly and with high sequence discrimination to complementary homopurine targets in double-stranded DNA (NIELSEN et al. 1991; DEMIDOV et al. 1995). This binding takes place by strand invasion via formation of a P loop, which is composed of a Watson-Crick-Hoogsteeen PNA$_2$–DNA triplex and a single DNA strand (CHERNY et al. 1993; NIELSEN et al. 1994b; Fig. 7). Such PNA triplex strand displacement complexes efficiently occlude binding of proteins, such as transcription factors (VICKERS et al. 1996, PRASEUTH et al. 1997), restriction enzymes (NIELSEN et al. 1993a; PEFFER et al. 1993), or methylases (VESELKOV et al. 1996), that recognize overlapping or adjacent sequences and also efficiently obstruct the passage of elongating RNA polymerases (HANVEY et al. 1992; NIELSEN et al. 1993b, 1994c; PEFFER et al. 1993). Thus PNA shows many of the properties required of an antigene agent.

Strand invasion of PNA in double-stranded DNA in intact nuclei or cells has not yet been demonstrated, and low efficiency could be anticipated since the binding rate is greatly diminished at elevated ionic strength. In fact, no binding of simple PNA to double-stranded DNA is usually observable at 140 mM K$^+$. However, positively charged, linked (bis-)PNA exhibit greatly enhanced binding rates (EGHOLM et al. 1994) and do indeed bind to their

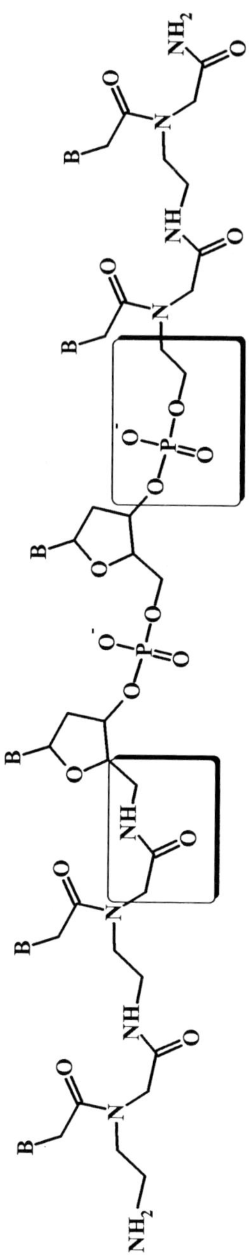

Fig. 6. Chemical structure of a peptide nucleic acid (PNA)–DNA–PNA chimera. Connection linkages are shown in *boxes*

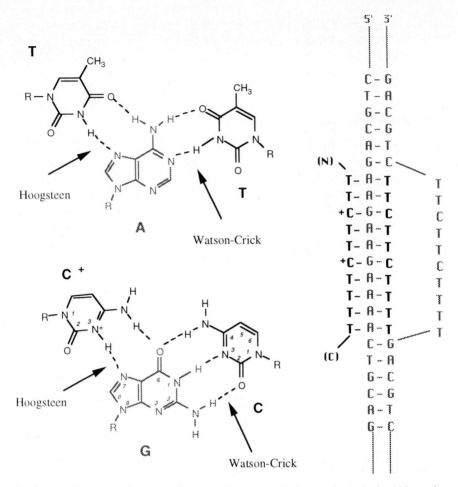

Fig. 7. Peptide nucleic acid (PNA)$_2$–DNA strand displacement complex. Note that the Watson-Crick PNA strand is preferentially antiparallel to the target, whereas the Hoogsteen PNA strand is parallel to the target. The base triplets responsible for the recognition are shown to the *left*. PNA is drawn in *black*, while DNA is drawn in *gray*

double-stranded DNA targets in 140 mM K$^+$ at micromolar PNA concentrations (GRIFFITH et al. 1995). Furthermore, recent results have shown that negative DNA supercoiling, which is a hallmark of actively transcribing genes, increases the PNA binding rate by as much as 200-fold (BENTIN and NIELSEN 1996), and even the transcription process itself, presumably by virtue of the transient DNA opening of the transcription bubble, catalyzes the binding of PNA (LARSEN and NIELSEN 1996). Thus it may well be that, within the living cell, the PNA is indeed directed towards its target – an actively transcribed gene – by these biological effects.

Quite surprisingly, it has also been shown that, under some circumstances, P loops function as efficient promoters for RNA polymerases and that PNA can therefore function as artificial transcription initiation factors for gene activation (MØLLEGAARD et al. 1994).

J. Future Prospects

Although crucial information about the pharmacokinetic properties of PNA is still lacking and no animal studies in general have yet been reported, the properties unveiled so far by PNA seem to give reason for optimism regarding the successful development of PNA-based gene therapeutic agents. The fact that functional modification of the PNA backbone is feasible is particularly encouraging, since such modifications should allow optimization of the physical and pharmacological properties of a given PNA oligomer without compromising its favorable hybridization properties.

Acknowledgements. This work was supported by the Danish National Research Foundation.

References

Altmann K-H, Dean NM, Fabbro D, Freier SM, Geiger T, Häner R, Hüsken D, Martin P, Monia BP, Müller M, Natt F, Nicklin P, Phillips J, Pieles U, Sasmor H, Moser HE (1996) Second generation of antisense oligonucleotides: from nuclease resistance to biological efficacy in animals. Chimia 50:168–176

Bentin T, Nielsen PE (1996) Enhanced peptide nucleic acid (PNA) binding to supercoiled DNA: Possible implications for DNA "breathing" dynamics. Biochemistry 35:8863–8869

Bergmann F, Bannwarth W, Tam S (1995) Solid phase synthesis of directly linked PNA-DNA-hybrids. Tetrahedron Lett 36:6823–6826

Betts L, Josey JA, Veal JM, Jordan SR (1995) A nucleic acid triple helix formed by a peptide nucleic acid-DNA complex. Science 270:1838–1841

Bonham MA, Brown S, Boyd AL, Brown PH, Bruckenstein DA, Hanvey JC, Thomson SA, Pipe A, Hassman F, Bisi JE, Froehler BC, Matteucci MD, Wagner RW, Noble SA, Babiss LE (1995) An assessment of the antisense properties of RNase H-competent and steric-blocking oligomers. Nucleic Acids Res 23:1197–1203

Breipohl G, Knolle J, Langner D, O'Malley G, Uhlmann E (1996) Synthesis of polyamide nucleic acids (PNAs) using a novel Fmoc/Mmt protecting-group combination. Bioorg Med Chem Lett 6:665–670

Cherny DY, Belotserkovskii BP, Frank-Kamenetskii MD, Egholm M, Buchardt O, Berg RH, Nielsen PE (1993) DNA unwinding upon strand displacement of binding of PNA to double stranded DNA. Proc Natl Acad Sci USA 90:1667–1670

Christensen L, Fitzpatrick R, Gildea B, Petersen KH, Hansen HF, Koch T, Egholm M, Buchardt O, Nielsen PE, Coull J, Berg RH (1995) Solid-phase synthesis of peptide nucleic acids (PNA) J Peptide Sci 3:175–183

Demidov V, Potaman VN, Frank-Kamenetskii MD, Buchardt O, Egholm M, Nielsen PE (1994) Stability of peptide nucleic acids in human serum and cellular extracts. Biochem Pharmacol 48:1309–1313

Dueholm K, Nielsen PE (1996) Chemical aspects of peptide nucleic acid. New J Chem 21:19–31

Dueholm K, Petersen KH, Jensen DK, Egholm M, Nielsen PE, Buchardt O (1994a) Peptide nucleic acid (PNA) with a chiral backbone based on alanine. Bioorg Med Chem Lett 4:1077–1080

Dueholm KL, Egholm M, Behrens C, Christensen L, Hansen HF, Vulpius T, Petersen K, Berg RH, Nielsen PE, Buchardt O (1994b) Synthesis of peptide nucleic acid monomers containing the four natural nucleobases: thymine, cytosine, adenine and guanine, and their oligomerization. J Org Chem 59:5767–5773

Egholm M, Buchardt O, Nielsen PE, Berg RH (1992a) Peptide nucleic acids (PNA). Oligonucleotide analogues with an achiral peptide backbone. J Am Chem Soc 114:1895–1897

Egholm M, Buchardt O, Nielsen PE, Berg RH (1992b) Recognition of guanine and adenine in DNA by cytosine and thymine containing peptide nucleic acids (PNA). J Am Chem Soc 114:9677–9678

Egholm M, Buchardt O, Christensen L, Behrens C, Freier SM, Driver DA, Berg RH, Kim SK, Nordén B, Nielsen PE (1993) PNA hybridizes to complementary oligonucleotides obeying the Watson-Crick hydrogen bonding rules. Nature 365:556–568

Egholm M, Christensen L, Dueholm K, Buchardt O, Coull J, Nielsen PE (1995) Efficient pH independent sequence specific DNA binding by pseudoisocytosine-containing bis-PNA. Nucleic Acids Res 23:217–222

Eriksson M, Nielsen PE (1996a) Solution structure of a peptide nucleic acid-DNA duplex. Nat Struct Biol 3:410–413

Eriksson M, Nielsen P E (1996b) PNA-nucleic acid complexes. Structure, stability and dynamics Q Rev Biophys 29:369–394

Finn PJ, Gibson NJ, Fallon R, Hamilton A, Brown T (1996) Synthesis and properties of PNA/DNA chimeric oligomers. Nucleic Acids Res 24:3357–3363

Gambacorti-Passerini C, Mologni L, Bertazzoli C, Marchesi E, Grignani F, Nielsen PE (1996) In vitro transcription and translation inhibition by anti-PML/RARα and -PML peptide nucleic acid (PNA) Blood 88:1411–1417

Griffith MC, Risen LM, Greig MJ, Lesnik EA, Sprangle KG, Griffey RH, Kiely JS, Freier SM (1995) Single and bis peptide nucleic acids as triplexing agents: binding and stoichiometry. J Am Chem Soc 117:831–832

Haaima G, Lohse A, Buchardt O, Nielsen PE (1996) Peptide nucleic acids (PNA) containing thymine monomers derived from chiral amino acids: hybridization and solubility properties of d-lysine PNA. Angew Chem 35:1939–1941

Haaima G, Hansen HF, Christensen L, Dahl O, Nielsen PE (1997) Increased DNA binding and sequence discrimination of PNA upon incorporation of diaminopurine. Nucleic Acids Res 25:4639–4643

Hanvey JC, Peffer NC, Bis, JE, Thomson SA, Cadilla R, Josey JA, Ricca DJ, Hassman CF, Bonham MA, Au KG, Carter SG, Bruckenstein DA, Boyd AL, Noble SA, Babiss LE (1992) Antisense and antigene properties of peptide nucleic acids. Science 258:1481–1485

Hyrup B, Nielsen PE (1996) Peptide nucleic acids (PNA). Synthesis, properties and potential applications. Bioorg Biomed Chem 4:5–23

Hyrup B, Egholm M, Nielsen PE, Wittung P, Nordén B, Buchardt O (1994) Structure-activity studies of the binding of modified peptide nucleic acids (PNA) to DNA. J Am Chem Soc 116:7964–7970

Hyrup B, Egholm M, Buchardt O, Nielsen PE (1996) A flexible and positively charged PNA analogue with an ethylene-linker to the nucleobase: synthesis and hybridization properties. Bioorg Med Chem Lett 6:1083–1088

Jensen KK, Ørum H, Nielsen PE, Norden B (1997) Hybridization kinetics of peptide nucleic acids (PNA) with DNA and RNA studied with BIAcore technique. Biochemistry 36:5072–5077

Knudsen H, Nielsen PE (1996) Antisense properties of duplex and triplex forming PNA. Nucleic Acids Res 24:494–500

Knudsen H, Nielsen PE (1997) Application of PNA in cancer therapy. Anticancer Drugs 8:113–118

Krotz AH, Buchardt O, Nielsen PE (1995a) Synthesis of "retro-inverso" peptide nucleic acids. 1. Characterization of the monomers. Tetrahedron Lett 36:6937

Krotz AH, Buchardt O, Nielsen PE (1995b) Synthesis of "retro-inverso" peptide nucleic acids. 2. Oligomerization and stability. Tetrahedron Lett 36:6941

Krotz AH, Larsen S, Buchardt O, Nielsen PE (1998) "Retro-inverso" PNA: structural implications for DNA binding (submitted)

Larsen HJ, Nielsen PE (1996) Transcription-mediated binding of peptide nucleic acid (PNA) to double stranded DNA: sequence-specific suicide transcription. Nucleic Acids Res 24:458–463

Larsen HJ, Nielsen PE (1998) The potential use of peptide nucleic acid (PNA) for modulation of gene expression. In: Cohen AS, Smisek DL (eds) Analysis of antisense and related compounds. CRC Press, Boca Raton

Lioy E, Kessler H (1996) Synthesis of a new chiral peptide analogue of DNA using ornithine subunits and solid-phase synthesis methodologies. Liebigs Ann 2:201–204

Møllegaard NE, Buchardt O, Egholm M, Nielsen PE (1994) PNA-DNA Strand displacement loops as artificial transcription promoters. Proc Natl Acad Sci USA 91:3892–3895

Monia BP, Johnston JF, Geiger T, Muller M, Fabbro D (1996) Antitumor activity of a phosphorothioate antisense oligonucleotide targeted against C-raf kinase. Nat Med 6:668–675

Nielsen PE, Christensen L (1996) Strand displacement binding of a duplex forming homopurine PNA to a homopyrimidine duplex DNA target. J Am Chem Soc 118:2287–2288

Nielsen PE, Egholm M, Berg RH, Buchardt O (1991) Sequence selective recognition of DNA by strand displacement with a thymine-substituted polyamide. Science 254:1497–1500

Nielsen PE, Egholm M, Berg RH, Buchardt O (1993a) Sequence specific inhibition of restriction enzyme cleavage by PNA. Nucleic Acids Res 21:197–200

Nielsen PE, Egholm M, Berg RH, Buchardt O (1993b) Peptide nucleic acids (PNA). Potential antisense and anti-gene agents. Anticancer Drug Des 8:53–63

Nielsen PE, Egholm M, Buchardt O (1994a) Peptide nucleic acids (PNA). A DNA mimic with a peptide backbone. Bioconjug Chem 5:3–7

Nielsen PE, Egholm M, Buchardt O (1994b) Evidence for $(PNA)_2$/DNA triplex structure upon binding of PNA to dsDNA by strand displacement. J Mol Recognit 7:165–170

Nielsen PE, Egholm M, Buchardt O (1994c) Sequence specific transcription arrest by PNA bound to the template strand. Gene 149:139–145

Noble SA, Bonham MA, Bisi JE, Bruckenstein DA, Brown PH, Brown SC, Cadilla R, Gaul MD, Hanvey JC (1995) Impact of biophysical parameters on the biological assessment of peptide nucleic acids, antisense inhibitors of gene expression. Drug Dev Res 34:184–195

Norton JC, Piatyczek MA, Wright WE, Shay JW, Corey DR (1996) Inhibition of human telomerase activity by peptide nucleic acid. Nat Biotechnol 14:615–619

Pardrige WM, Boado RJ, Kang Y-S (1995) Vector-mediated delivery of a polyamide ("peptide") nucleic acid analogue through the blood-brain barrier in vivo. Proc Natl Acad Sci USA 92:5592–5596

Peffer NJ, Hanvey JC, Bisi JE, Thomson SA, Hassman FC, Noble SA, Babiss LE (1993) Strand-invasion of duplex DNA by peptide nucleic acid oligomers. Proc Natl Acad Sci USA 90:10648–10652

Petersen K, Jensen DK, Egholm M, Nielsen PE, Buchardt O (1995) A PNA-DNA linker synthesis of N-((4,4-dimethoxytrityloxy)ethyl)-N-(thymin-1-ylacetyl)glycine. Bioorg Med Chem Lett 11:1119–1124

Praseuth D, Grigoriev M, Guieysse AL, Pritchard LL, Harel-Bellan A, Nielsen PE, Helene C (1997) Peptide nucleic acids directed to the promoter of the α-chain of the interleukin-2 receptor. Biochim Biophys Acta 1309:226–238

Prochiantz A (1996) Getting hydrophilic compounds into cells: lessons from homeopeptides. Curr Opin Neurobiol 6:629–634

Rasmussen H, Kastrup JS, Nielsen JN, Nielsen JM, Nielsen PE (1997) Crystal structure of a peptide nucleic acid (PNA) duplex at 1.7Å resolution. Nat Struct Biol 4:98–101

Thomson SA, Josey JA, Cadilla R, Gaul MD, Hassman FC, Luzzio MJ, Pipe AJ, Reed KL, Ricca DJ, Wiethe RW, Noble SA (1995) Fmoc mediated synthesis of peptide nucleic acids. Tetrahedron 51:6179–6194

Tomac S, Sarkar M, Ratilainen T, Wittung P, Nielsen PE, Nordén B, Gräslund A (1996) Ionic effects on the stability and conformation of peptide nucleic acid (PNA) complexes. J Am Chem Soc 118:5544–52

Uhlman E, Will DW, Breipohl G, Langner D, Ryte A (1996) Synthesis and properties of PNA/DNA chimeras. Angew Chem 35:2632–2635

Veselkov AG, Demidov VV, Nielsen PE, Frank-Kamenetskii M (1996) A new class of genome rare cutters. Nucleic Acids Res 24:2483–2487

Vickers TA, Griffith MC, Ramasamy K, Risen LM, Freier SM (1995) Inhibition of NF-kappa B specific transcriptional activation by PNA strand invasion. Nucleic Acids Res 23:3003–3008

Will DW, Breipohl G, Langner D, Knolle J, Uhlman E (1995) The synthesis of polyamide nucleic acids using a novel monomethoxytrityl protecting-group strategy. Tetrahedron Lett 51:12069–12082

Wittung P, Kajanus J, Edwards K, Nielsen PE, Nordén B, Malmström BG (1995) Phospholipid membrane permeability of peptide nucleic acid. FEBS Lett 365:27–29

CHAPTER 21
Triple Helix Strategies and Progress

T. AKIYAMA and M. HOGAN

A. Introduction

The binding of single-stranded nucleic acids to a DNA duplex is a phenomenon which has been known for at least 40 years. Recently, it has been recognized that the formation of such triple helices could serve as the basis for the design of site specific duplex DNA binding agents. During the past 7 years, several laboratories have worked to understand the physical chemistry of triple helix formation, for the purposes of exploiting and enhancing this mode of biomolecular recognition. Some of the most advanced approaches to nucleic acid chemistry, drug delivery formulation, and structure-based molecular design have been applied to this effort. Here, we review those design efforts and discuss current progress in the application of triple helix forming oligonucleotides (TFOs) as the basis for duplex DNA specific drug design.

B. Discovery of the Triple Helix Structure

The first evidence of triple helix DNA structure was obtained from binding stoichiometry experiments exploiting UV hyperchromism. Early studies on poly(U) and poly(A) suggested, under high salt conditions, a mixture of poly(U) and poly(A) formed a bound complex which consisted of two strands of poly U and one strand of poly(A) (FELSENFELD et al. 1957). Similar 2:1 complexes were found to be formed by DNA and RNA polynucleotides (LIPSETT et al. 1963, 1964; BROITMAN et al. 1987). The structure of one such 2:1 complex was investigated by X-ray fiber diffraction (ARNOTT and SELSING 1974), in which poly(dT) lays in the major groove of poly(dT):poly(dA) duplex through Hoogsteen hydrogen bonding. In 1980s, elegant techniques for synthesis of short oligonucleotides were developed, which allowed oligonucleotide triple helices to be studied by biotechnological techniques. In 1987, Dervan's group (MOSER and DERVAN 1987) and Helene's group (LE DOAN et al. 1987) independently demonstrated that a short pyrimidine oligonucleotide could be used as a DNA duplex binding ligand by triple helix formation. Subsequently, a number of reports have documented the application of triple helix forming oligonucleotides (TFOs) for various purposes.

C. Detailed Structure of the Triple Helix

To date, the existence of at least three types of triple helix structure have been proposed. In all three structural motifs, the third strand resides in the major groove of the underlying duplex. These motifs are classified by the hydrogen bonding pattern, the base triplet, and the polarity of the third strand relative to that of the Watson-Crick duplex. The first class of triple helix forms by the binding of a pyrimidine (thymine and cytosine)-rich single strand to the duplex through Hoogsteen hydrogen base pairing, T:AT and C+:GC (motif I in Fig. 1; MOSER and DERVAN 1987). To form Hoogsteen hydrogen bonding at GC base pairs, protonation of cytosine base is required to yield the C+GC base triplet. The polarity of the third pyrimidine-rich strand is parallel to the purine-containing strand of the underlying duplex. The second triple helix motif is achieved by the binding of purine bases (guanine and adenine) or a mixture of guanine and thymine bases (motif II in Fig. 2; COONEY et al. 1988; BEAL and DERVAN 1991). In this motif, G:GC and A:AT or T:AT base triplets are formed through reversed Hoogsteen base pairing, and the polarity of the third strand is antiparallel to the purine sequence in the duplex. A third motif also results from the binding of a third strand consisting of guanine and thymine sequences. But, in this kind of triple helix, the third strand binds *parallel* to the purine sequence of the duplex (motif III; GIOVANNANGELI et al. 1992a). The preferred polarity of a guanine and thymine-rich third strand (motif II or motif III) appears to be dependent on the sequence of the duplex binding site. However, the microscopic structure of motif III remains to be determined.

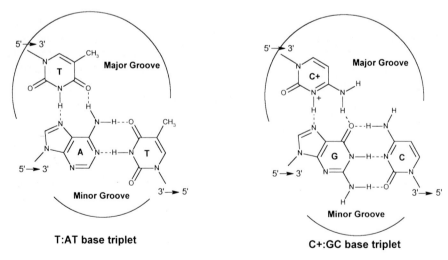

Fig. 1. Base triplet pattern of parallel PuPyPy-type triple helix. The base triplets, T:AT and C+:GC are illustrated. *Arrows* indicate the polarity of backbone. Hydrogen bonding is shown by *dotted lines*

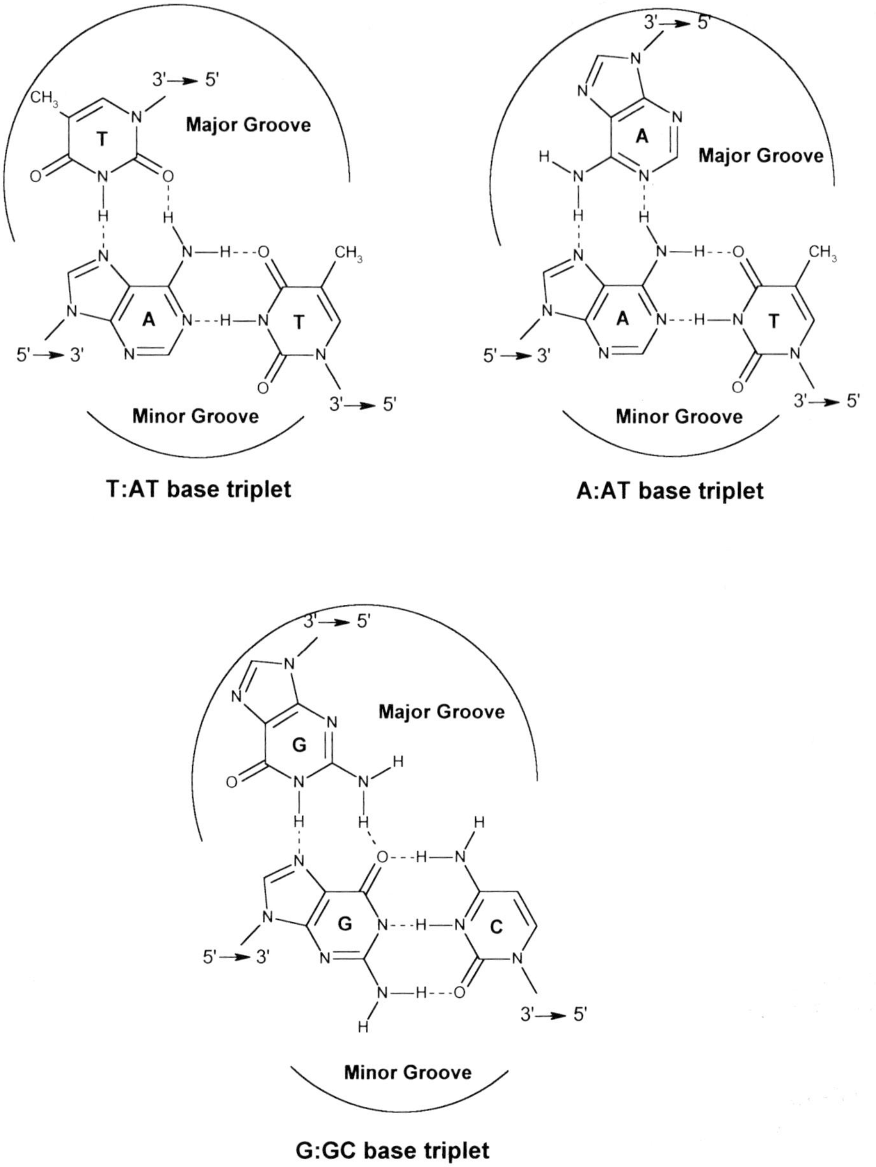

Fig. 2. Base triplet pattern of antiparallel PuPuPy-type triple helix. The base triplets, A:AT, T:AT, and G:GC are illustrated. *Arrows* indicate the polarity of backbone. Hydrogen bonding is shown by *dotted lines*

I. Nuclear Magnetic Resonance Studies

1. Hydrogen Bonding

Nuclear magnetic resonance (NMR) is one of the most powerful methods with which to elucidate the structure of biomolecules as well as that of small organic compounds. Since the first report by Feigon's group (RAJAGOPAL and FEIGON 1989a), a number of NMR studies have been published to determine the detailed structure of motif I, the parallel PyPuPy triple helix (DE LOS SANTOS et al. 1989; RAJAGOPAL and FEIGON 1989b; MOOREN et al. 1990; PILCH et al. 1990a; SKLENAR and FEIGON 1990; LIVE et al. 1991; RADHKRISHNAN et al. 1991b; MACAYA et al. 1992a,b) and of motif II, the antiparallel PuPuPy triple helix (PILCH et al. 1991; RADHAKRISHNAN et al. 1991c, 1993). NMR data for triple helix formation have been measured in the presence of inorganic cations such as Na^+ and/or Mg^{2+}. For parallel triple helix formation, low pH was employed (pH 4–6) so as to ensure cytosine protonation. Intermolecular triple helices were used for early studies. In recent years, single-stranded oligonucleotides which can form intramolecular triple helices have been exploited (SKLENAR and FEIGON 1990). Since Hoogsteen or reverse Hoogsteen hydrogen bonding is involved in complex formation, the analysis of the imino proton region is essential in proton NMR studies of these triplexes. Besides the sharp imino proton signals arising from the well-characterized Watson-Crick base pairing, hydrogen-bonded imino resonances resulting from Hoogsteen base pairing or reversed Hoogsteen base pairing were observed around 12.5–16 ppm for the parallel triple helix and around 12–14 ppm for the antiparallel triple helix (DE LOS SANTOS et al. 1989; RAJAGOPAL and FEIGON 1989a,b; MOOREN et al. 1990; PILCH et al. 1990a, 1991; SKLENAR and FEIGON 1990; RADHAKRISHNAN et al. 1991c, 1993; MACAYA et al. 1992b; SCARIA et al. 1995). In a parallel triple helix motif containing cytosine in the third strand, the imino proton signals display pH-dependent behavior, which is consistent with other spectroscopic methods. Hoogsteen hydrogen bonding in the parallel triple helix has also been probed by measuring ^{15}N-NMR of the triple helix containing ^{15}N-enriched deoxyguanosine and deoxyadenosine in the purine rich strand of the duplex (GAFFNEY et al. 1995).

2. Base Orientation

It is well known that the ribose ring in a nucleoside can adopt a number of conformations by pseudorotation. Sugar conformation within the DNA duplex has been thoroughly investigated by X-ray crystallography, suggesting that S-type (C2'-endo) sugar pucker dominates in the B-type DNA duplex. In the case of triple helix formation, an early study by Arnott et al. proposed an N-type (C3'-endo) conformation of the ribose moiety (ARNOTT and SELSING 1974). However, nuclear Overhauser effect (NOE) intensities (RAJAGOPAL and FEIGON 1989a,b) and detailed analysis on COSY crosspeak patterns and coupling constants (MACAYA et al. 1992a,b) have suggested that the majority of sugars in the motif I triple helix are in an S-type (C2'-endo) conformation. NOE intensities of base-H1' and base-H2',H2" have suggested the *anti* orientation of the glycosyl bond in motif I (parallel PyPuPy) (MACAYA et al. 1992b) and motif II (antiparallel PuPuPy) triple helices (RADHAKRISHNAN et al. 1991c, 1993). An NMR study of an RNA triple helix (KLINCK et al. 1994, 1995; HOLLAND and HOFFMAN 1996) and an DNA-RNA triple helix (VON DONGEN et al. 1996) has also been reported for motif I, yielding a similar indication of nucleoside conformation.

II. Other Methods to Investigate Triple Helix Structure

1. Infrared Spectroscopy

In the infrared (IR) absorbance spectrum of nucleic acids, the region from $800\,cm^{-1}$ to $1000\,cm^{-1}$ and that from $1250\,cm^{-1}$ to $1500\,cm^{-1}$ represent vibrations of the sugar ring bonds and vibrations of the glycosidic bond, respectively (HIGUCHI et al. 1969; TSUBOI 1969) . The IR absorbance wavelength in this region is dependent on furanose sugar conformation. Thus, the IR spectrum of $(dT)_n(dA)_n(dT)_n$ has suggested that sugar conformation for this motif I triple helix resembles that of B-form DNA (HOWARD et al. 1992; OUALI et al. 1993b), which is consistent with the sugar conformation of motif I as determined by NMR analysis. On the other hand, the IR spectra of $(dG)_n(dG)_n(dC)_n$ (OUALI et al. 1993a) and $(dG)_{20}(dG)_{20}(dC)_{20}$ (WHITE and POWELL 1995), which assume motif II, showed the existence of both S- and N-type sugar puckering. Additionally, the spectrum from the 1250 to $1500\,cm^{-1}$ region of these dG:dG:dC type triple helixes suggested an *anti* conformation of the glycosidic bonds (OUALI et al 1993a), again consistent with NMR.

2. Phase-Sensitive Electrophoresis

A unique study to estimate the helical periodicity of triple helix was reported, based on the principle of phasing analysis (ZINKEL and CROTHERS 1987). This study suggests a helical pitch of 11.2 bp/turn for the PuPuPy antiparallel triple helix (motif II) and 11.1 bp/turn for the PyPuPy type triple helix (motif I) (SHIN and KOO 1996). These numbers were very close to the helical pitch derived from modeling based on NMR data (RADHAKRISHNAN and PATEL 1994b) or molecular mechanics calculations (LAUGHTON and NEIDLE 1992b; WEERASINGHE et al. 1995).

3. Electron Microscopy

Electron microscopy has been used to visualized the binding of a triple helix forming oligonucleotide (motif I) , which is biotinylated for detection, to a restriction fragment of plasmid DNA (CHERNY et al. 1993; BOUZIANE et al. 1996).

4. X-ray Crystallography

Since the early X-ray fiber diffraction studies by Arnott et al. (ARNOTT et al. 1972; ARNOTT and BOND 1973; ARNOTT and SELSING 1974), effort has been focused on obtaining an X-ray crystal structure for a triple helix. However, at present, only a partial structure has been obtained, and only for a pair of adjacent GGC triplets (VAN MEERVELT et al. 1995; VLIEGHE et al. 1996).

5. Molecular Dynamics Simulation

Modeling can complement experimental data obtained from NMR or from IR or CD spectroscopy in the understanding of the details of triple helix formation. Several careful molecular dynamics simulation studies on parallel (LAUGHTON and NEIDLE 1992b; OUALI et al. 1993b; CHENG and PETTITT 1995) or antiparallel triple helices (CHENG and PETTITT 1992, 1995; Laughton and Neidle 1992a; WEERASINGHE et al. 1995) have been performed from the starting structure originally derived by X-ray fiber diffraction (ARNOTT and SELSING 1974). Using molecular dynamics simulation, Pettitt's group has studied the behavior of the solvent and conterions around the triple helix (WEERASINGHE et al. 1995), which could be important for predicting details of triple helix structure and also for understanding ligand-induced stabilization of triple helix domains.

The molecular mechanics calculations have generated structural models which appear to be consistent with available NMR, IR, and phasing data. Predictions made from the modeling concerning hydration and ion binding await confirmation by crystallography and more advanced spectroscopic methods.

D. Stability of the Triple Helix

In order to introduce chemical improvements into triple helix forming oligonucleotides, it is essential to understand the thermodynamics of triple helix formation. Thus far, a number of studies have been carried out via physical and biochemical techniques.

I. Thermodynamic Analysis by Spectroscopic and Calorimetric Methods

By analogy with double helix studies, ultraviolet (UV) spectroscopy has frequently been used to evaluate the stability of triple helix formation. When a triple helix formation occurs in solution, UV absorption is diminished relative to the simple sum of the absorbance of the third strand plus that of the duplex. This hypochromic effect has been exploited to monitor triple helix formation, especially as a function of temperature, to yield melting curves. By adjusting solution conditions such as ionic strength or pH, a melting profile of a triple helix to three random coiled single strands can generally be adjusted to reveal a biphasic transition, corresponding to the dissociation of triplex to a duplex and a single strand, followed at a higher temperature by the dissociation of the duplex to two random coiled single strands.

The thermodynamic parameter obtained directly from a melting curve is Tm, which is defined as the temperature at which half of a triple helix is dissociated. Tm data have proven to be useful for comparison of triple helices differing in one or two base triplets (MACAYA et al. 1991; MERGNY et al. 1991a; FOSSELLA et al. 1993; MILLER and CUSHMAN 1993). Several groups have extracted more detailed thermodynamic parameters (ΔG^0, ΔH^0, and ΔS^0) from triple helix melting data, by detailed curve fitting methods (MANZINI et al. 1990; PILCH et al. 1990b, 1991; PLUM et al. 1990; PLUM and BRESLAUER 1995; XODO et al. 1990; ROBERTS and CROTHERS 1991, 1996; DURAND et al. 1992b; VOLKER et al. 1993; SCARIA et al. 1995). Two methods are used. One is to fit the melting curve to an all-or-none two-state model, in which the dissociation of a triple helix to duplex and single strand is considered to occur in a one-step mechanism (MANZINI et al. 1990; XODO et al. 1990; DURAND et al. 1992b; VOLKER et al. 1993; PLUM and BRESLAUER 1995). From the shape of a melting curve, the change in the enthalpy (ΔH^0) and entropy (ΔS^0) accompanying the triple helix dissociation can be derived. Another method is based on the assumption that triple helix formation occurs as a bimolecular association process. Thus, the parameter ΔH^0 and ΔS^0 can be obtained from the plot of 1/T against the logarithm of the strand concentration (MANZINI et al. 1990; PILCH et al. 1990b, 1991; XODO et al. 1990; ROBERTS and CROTHERS 1991, 1996; SCARIA et al. 1995). Therefore, this method is useful only for intermolecular triple helix formation. Both methods are based on the assumptions that triple helix dissociation obeys a two-state model and that ΔH^0 (the van't Hoff enthalpy) is independent of temperature.

Scanning calorimetry has also allowed determination of the ΔH^0 of triple helix formation (PLUM et al. 1990, 1995; XODO et al. 1990; VOLKER et al. 1993; KAMIYA et al. 1996; SCARIA and SHAFER 1996). A heat capacity versus temperature profile is obtained as experimental data in such a differential calorimetric scan. Integration of the heat capacity versus temperature curve gives ΔH^0 directly, which is independent of the equilibrium model employed.

For parallel PyPuPy type triple helix formation, the ΔH^0 value determined by several groups from melting and calorimetric data have been distributed over the range from 2 to 6 kcal/mol per base. This range of values may be ascribed to the difference in the nucleotide sequence used, pH, and/or buffer species (WILSON et al. 1994a).

For antiparallel PuPuPy triple helix formation, fewer melting and calorimetric data are available as compared to the data of parallel PyPuPy type triple helix (PILCH et al. 1991; HOWARD et al. 1995; SVINARCHUK et al. 1995a,b; SCARCIA and SHAFER 1996). Because this type of triple helix is very stable, third strand dissociation appears to be highly cooperative with duplex dissociation under most experimental conditions of salt and pH. Consequently, dissociation transitions appear monophasic and the interpretation of data is complicated by the convolution of triplex and duplex transitions (PILCH et al. 1991; SVINARCHUK et al. 1995b; SCARCIA and SHAFER 1996).

Recently, from detailed van't Hoff analysis of a triple helix set, Roberts and Crothers have developed rules that can successfully predict the ΔG^0 and ΔH^0 of parallel PyPuPy triple helix formation (ROBERTS and CROTHERS 1996). Values derived from this model appeared to be consistent with data obtained previously.

An interesting study was carried out by Shafer's group (SCARIA and SHAFER 1996), in which they compared thermodynamic parameters for the three types of triple helix in the same sequence. Although the data should be interpreted very carefully, the stability estimated from ΔH^0 values per base triplet, which were determined by calorimetric measurement, were as follows: antiparallel PuPuPy triple helix (motif II) > parallel PyPyPu triple helix (motif I) > *parallel* GT type triple helix (motif III).

II. Thermodynamic Analysis by Biochemical Methods

Since a triple helix forming oligonucleotide (TFO) can be considered as a DNA binding ligand, biochemical gel-based methods used in protein studies have been exploited for the analysis of DNA-TFO interaction. There are some advantages in these methods over spectroscopic or calorimetric methods: (a) the binding affinity under physiological conditions can be estimated directly; and (b) only a trace amount of DNA or oligonucleotide is required for the measurement. However, they have the disadvantage that the analysis of experimental data could be complicated by the presence of other equibria in addition to triple helix formation.

The DNase I footprinting method, which has been widely employed for analysis of DNA–protein interaction, has been used not only to confirm the binding specificity of TFOs but also to quantify the binding affinity (SINGLETON and DERVAN 1992b; CHANDLER and FOX 1993, 1995; CHENG and VAN DYKE 1994; GREENBERG and DERVAN 1995; SVINARCHUK et al. 1995b). The method is based on the fact that the region where the triple helix forms is protected from the cleavage by DNase I. By quantitation of cleavage efficiency as a function of TFO concentration added to the duplex solution, the binding affinity of the TFO was determined.

In our laboratory (COONEY et al. 1988; DURLAND et al. 1991; GEE et al. 1994; AKIYAMA and HOGAN 1996b) and those of others (YOON et al. 1992; DURLAND et al. 1994; VASQUEZ et al. 1995; OLIVAS and MAHER 1996; WANG et al. 1996a), a gel titration assay has been frequently used to estimate TFO binding affinity. This assay is based on the retardation of gel mobility by the binding of a TFO to a duplex. The method is limited to the study of relatively long triple helices, since the complex must remain intact during the course of electrophoresis (1-2h) in order to be detected. However, the method is easy, rapid, and a small amount of oligonucleotide is sufficient to provide reliable data. Therefore, this method is extremely useful for screening in search of TFOs that can bind to a duplex of interest. Besides the detection of binding, band quantification allows estimation of TFO binding affinity (COONEY et al. 1988; DURLAND et al. 1991, 1994; AKIYAMA and HOGAN 1996b). Application of this method in our recent work (AKIYAMA and HOGAN 1996a,b) is described below.

Dervan's group have developed an affinity cleaving method to estimate TFO binding affinity. The equilibrium binding constant is determined from the quantification of the cleavage reaction by a cleaving marker molecule attached to TFO (SINGLETON and DERVAN 1992b, 1994; BEST and DERVAN 1995; COLOCCI and DERVAN 1995). An outstanding characteristic of this method is that it allows information on the polarity of TFO binding to be obtained.

Other methods for analysis of TFO binding, such as endonuclease protection (MAHER et al. 1990; WARD 1996), fluorescence energy transfer (YANG et al. 1994), filter binding (SHINDO et al. 1993), or triple helix blotting (NOONBERG et al. 1994b) have also been reported.

III. Factors Influencing the Stability of the Triple Helix

1. Mismatching at Inversion Sites

As described above, the binding of third strand is stabilized by Hoogsteen or reversed Hoogsteen base pairing. This pattern of binding restricts triple helix formation to regions with consecutive purine bases in the duplex. The effect of mismatching or sequence inversion (where one pyrimidine base is inserted in a consecutive purine sequence) on the stability of the triple helix has been studied using a number of the techniques described above (GRIFFIN and DERVAN 1989; HORNE and DERVAN 1991; ROBERTS and CROTHERS 1991; MERGNY et al. 1991a; MACAYA et al. 1991; BEAL and DERVAN 1992a; ROUGEE et al. 1992; FOSSELLA et al. 1993; MILLER and CUSHMAN 1993; XODO et al. 1993; HARDENBOL and VAN DYKE 1996; KAMIYA et al. 1996). The free energy penalty for such a single mismatch or inversion site was approximately 3 kcal/mol per base triplet for both parallel (ROBERTS and CROTHERS 1991; BEST and DERVAN 1995) and antiparallel (GREENBERG and DERVAN 1995) motifs. Interestingly, this is greater selectivity than is seen for most duplex mismatches, suggesting that triple helix formation is generally more selective than duplex formation under the same conditions.

The location of a mismatch site was found to affect the stability of the triple helix. Thus, a higher discrimination was observed when the mismatch site is located at the center of the duplex as compared to at its end (MERGNY et al. 1991a). By systematic experiments by several groups, semistable triplet pairs have been found. In the PyPuPy type parallel triple helix, the G:TA triplet (GRIFFIN and DERVAN 1989; MERGNY et al. 1991a; YOON et al. 1992; FOSSELLA et al. 1993; BEST and DERVAN 1995) and T:CG triplet (YOON et al. 1992) showed higher stability than other triplets in the inversion site. Detailed structural analysis by NMR of these semistable base triplets has suggested the formation of a single hydrogen bond between the third base and the pyrimidine base of the duplex in the invasion site (RADHKRISHNAN et al. 1991a; RADHKRISHNAN and PATEL 1992a,b, 1994; WANG et al. 1992). In the antiparallel motif, the TCG triplet was found to be stable in the inversion site (BEAL and DERVAN 1992a; GREENBERG and DERVAN 1995). NMR analysis by our laboratory (DITTRICH et al. 1994) suggested that the stability of the TCG triplet resulted from a combination of base stacking and H-bonding interaction.

2. Ionic Strength

Ionic strength is an important factor influencing the stability of triple helix formation. In the parallel triple helix motif (PLUM et al. 1990; DURAND et al. 1992b; HOPKINS et al. 1993; SINGLETON and DERVAN 1993; VOLKER et al. 1993; PLUM and BRESLAUER 1995) analysis of the effect of ionic strength on Tm, δTm/δln(M), suggested that the number of cations released during triple helix dissociation was greater than for the duplex (PLUM et al. 1990; DURAND et al. 1992b; VOLKER et al. 1993; WILSON et al. 1994a; PLUM and BRESLAUER 1995). That observation is reasonable because the triple helix has a higher density of negative charge. However, for the parallel triple helix, the number of released cations appears to be largely dependent on the sequence (WILSON et al. 1994a; PLUM and BRESLAUER 1995). Apparently, the number of released cations is smaller in a triple helix formed by a cytidine-rich third strand binding to a GC-rich duplex target. This observation indicates that the protonation on the cytosine base might reduce the cation requirement for shielding of negative phosphate charge.

In the case of the antiparallel PuPuPy triple helix, in which a guanine-rich oligonucleotide is usually used as the third strand, the salt-dependent behavior is complicated by the existence of other equibria, most likely aggregation through guanine tetrad formation. Tetraplex formation by guanine-rich oligonucleotides is known to be stimulated by alkaline metal ions, especially by K^+ cation (JIN et al. 1990; SEN and GILBERT 1990; JIN et al. 1992; SCARIA et al. 1992; LUE et al. 1993). Therefore, the presence of a higher concentration of K^+ would prevent antiparallel triple helix formation. Inhibition of triple helix formation via competition with a tetraplex equilibrium has been observed in several instances (CHENG and VAN DYKE 1993; GEE et al. 1995; KANDIMALLA and AGRAWAL 1995b; NOONBERG et al. 1995; OLIVAS and MAHER 1995b; VASQUEZ et al. 1995). On the other hand, bivalent cations, especially Mg^{2+}, greatly stabilized the antiparallel triple helix (MALKOV et al. 1993; AKIYAMA and HOGAN 1996b) as well as the parallel triple helix (PILCH et al. 1990b; SINGLETON and DERVAN 1993).

3. Effect of pH

In the parallel triple helix motif, protonation of the cytidine base is required for stable C+:GC triplet formation. Because the pKa at N^3 of cytosine is approximately 4.5 (SANGER 1984), parallel triple helixes containing cytidine in the third strand become unstable under neutral pH. Several systematic studies on the pH dependence of the parallel triple helix using various techniques such as UV melting (PLUM et al. 1990; SHINDO et al. 1993; WILSON et al. 1994a; PLUM and BRESLAUER 1995), affinity cleaving (SINGLETON and DERVAN 1992a), and NMR (SKLENAR and FEIGON 1990) have been reported. In addition, a formalism to predict pH effects on the stability of the antiparallel triple helix (HUSLER and KLUMP 1995) and a semi-empirical model to predict pH, temperature, and Na+ dependence of the thermal transitions have been developed (PLUM and BRESLAUER 1995).

IV. Kinetics of Triple Helix Formation

It is generally known that the association and dissociation step of triple helix is a relatively slow process. Several methods including restriction endonuclease protection (MAHER et al. 1990), DNase I footprinting (FOX 1995b), a filter binding assay (SHINDO et al. 1993), fluorescence energy transfer (YANG et al. 1994), UV absorbance (ROUGEE et al. 1992; XODO 1995), bimolecular interaction analysis (BATES et al. 1995), and the gel shift assay (VASQUEZ et al. 1995; AKIYAMA and HOGAN 1996b) have been used to monitor the kinetics of triple helix formation. Except in one instance (SHINDO et al. 1993), a two-state model of a triple helix formation from a duplex and a TFO has been found to fit the data. Both motif I and II have yielded a second order association constant in the 10^3–10^4 M^{-1} s^{-1} range. This kinetic constant is influenced by several factors, for example, temperature, ionic strength, the length of TFO, etc. However, as a class, the data require that triple helix formation is 10^2 to 10^3 fold slower than the diffusion limit, suggesting that conformational change, possibly of the duplex, may be rate limiting.

E. Nucleoside Modification to Improve Triple Helix Stability

I. Modified Nucleosides to Overcome the pH Dependency of the Parallel Triple Helix

The pH dependence of the parallel PyPuPy triple helix greatly limits its application to gene regulation or biotechnological tools. Several chemically modified nucleoside derivatives have been synthesized to overcome this difficulty.

1. Cytidine Derivatives

Stabilization of PyPuPy triple helix at neutral pH by substitution of cytidine with 5-methylcytidine in the third strand was originally found in a polymer system (LEE et al. 1984). Following that study, formation of short triple helices at neutral pH was obtained from the design of oligonucleotides containing 5-methylcytidines (POVSIC and DERVAN 1989; COLLIER et al. 1991b). Since the pKa of cytidine and 5-methylcytidine are almost identical, the driving force behind the stabilization by 5-methylcytidine must be an increase of hydrophobicity or stacking interaction (POVSIC and DERVAN 1989; COLLIER et al. 1991). Molecular mechanics and molecular dynamics simulation have supported the idea that a change in the solvation energy might also play a key role in the stabilization by 5-methylcytidine substitution (HAUSHEER et al. 1992).

In order to circumvent the pH dependency of parallel PyPuPy triple helix, cytidine in parallel TFOs has also been replaced by an N^3-protonated cytidine derivative: pseudoisocytidine (ONO et al. 1991b, 1992), pyrazine (VON KROSIGK and BENNER 1995), or a 6-oxocytidine derivative (XIANG et al. 1994, 1996; BERRESSEM and ENGELS 1995). All of these compounds have two proton donors in the base, which are capable of forming two hydrogen bonds at neutral pH. As expected from their design, Tm data suggested that the binding of the TFO modified by these bases was practically independent of pH in the solution. However, substitution with these compounds seemed to destabilize the triple helix. A cytidine derivative-bearing a spermine at the N^4-exocyclic amine was also found to diminish the pH dependency of parallel triple helix when incorporated in TFO (BARAWKAR et al. 1996); however it, too, appeared to be generally destabilizing.

2. Adenosine Derivatives

8-Oxoadenine derivatives (or more precisely, "7,8-dihydro-8-oxoadenine derivatives") including 8-oxoadenosine (MILLER et al. 1992; JETTER and HOBBS 1993; WANG Q et al. 1994; MILLER et al. 1996) and N^6-methyl-8-oxoadenosine (YOUNG et al. 1991; KRAWCZYK et al. 1992) have also been reported to replace cytidine as N^3-protonated cytidine analogues. An 8-oxoadenosine derivative may be able to adopt two tautomers, a keto form and an enol form. Fortunately, the keto form appears to predominate (CHO and EVANS 1991), making the amidoproton at N^7 available for hydrogen bonding with N^7 of the guanine base of the Watson-Crick duplex in the parallel triple helix motif. In addition, the glycosidic bond of 8-oxoadenosine seems to exist in the *syn* conformation preferentially (GUSCHLBAUER et al. 1991), which might be required for 8-oxoadenosine to form Hoogsteen hydrogen bonding without disturbing backbone positioning of the third strand. Footprinting analysis (KRAWCZYK et al. 1992) and UV melting analysis (MILLER et al. 1992; WANG et al. 1994) have shown that by the replacement of cytidine in the third stand by 8-oxoadenine derivatives, the stability of triple helix becomes independent of buffer pH, suggesting the validity of the predicted structure of the triple helix containing 8-oxoadenine in the third strand.

3. Guanosine Derivatives

N^7-glycosylated guanosine (HUNZIKER et al. 1995; BRUNAR and DERVAN 1996) and its analogues (1-(2-deoxy-β-D-robofuranosyl)-3-methyl-5-amino-1H-pyrazolo[4, 3-d]pyrimidin-7-one: (the P base) (KOH and DERVAN 1992; PRIESTLEY and DERVAN 1995) have also been designed to form Hoogsteen base pairs with the GC Watson-Crick doublet at neutral pH. When cytidine in the third strand is replaced by this guanine derivative in a parallel triple helix, modeling has suggested that the glycosidic bond of the guanosine analogue would be directed toward the position at which the backbone of the third strand of parallel triple helix is located. pH-independent TFO binding affinity was demonstrated for oligomers containing these G analogues by quantitative DNase I footprinting analysis (HUNZIKER et al. 1995; PRIESTLEY and DERVAN 1995; BRUNAR and DERVAN 1996). The G substitutions were found to stabilize the parallel triple helix motif in duplex regions possessing consecutive GC base pairs, which is a sequence which is ordinarily unstable for an unmodified parallel triple helix (PRIESTLEY and DERVAN 1995). The proposed hydrogen bonding pattern of the P:GC base triplet was also confirmed by a NMR study (RADHAKRISHNAN et al. 1993).

II. New Nucleosides to Recognize Invasion Sites

In order to expand the recognition motif at invasion sites in the triple helix, several nonnatural nucleosides were designed, synthesized, and incorporated into TFO for parallel motif binding (GRIFFIN et al. 1992; KIESSLING et al. 1992; HUANG and MILLER 1993; KOSHLAP et al. 1993; HUANG et al. 1996; MICHEL et al. 1996; WANG et al. 1996) or antiparallel motif binding (STILZ and DERVAN 1993; DURLAND et al. 1994, 1995; MICHEL et al. 1996). Several of these have shown a small enhancement of triple helix stability at sites containing inversion sites in polypurine runs (GRIFFIN et al. 1992; DURLAND et al. 1995). However, the development in this area is still in progress and further detailed design may be required.

III. Modification to Overcome Alkali Metal Mediated Inhibition of Antiparallel Triple Helix Formation

The stability of the antiparallel PuPuPy type triple helix is not pH dependent, which makes it well suited for pharmaceutical applications. However, as mentioned above, the binding of antiparallel TFOs can be inhibited by the addition of alkali metal ions, especially K^+, which is most likely due to K^+ mediated formation of guanosine tetrads. This phenomenon may be problematic when this type of triple helix is applied to gene regulation in living cells because the physiological concentration of K⁺ in the cell is near 180 mM (SANGER 1984). In order to circumvent this shortcoming, 6-thio-2'-dexoyguanosine (GEE et al. 1995; OLIVAS and MAHER 1995a; RAO et al. 1995) or 7-deaza-2'-deoxyxanthosine (MILLIGAN et al. 1993) has been incorporated into the TFO. Both modified bases were expected to inhibit the formation of guanine tetrads. Our laboratory and others have found that limited substitution of deoxyguanosine by 6-thio-2-deoxyguanosine in an antiparallel TFO facilitates triple helix formation in K⁺ containing buffer (GEE et al. 1995; OLIVAS and MAHER 1995a; RAO et al. 1995). Dimethysulfate protection analysis (RAO et al. 1995) and a modeling study (OLIVAS and MAHER 1995a) have suggested that the substitution by 6-thio-2'-deoxyguanosine might abolish the tetrad structure by guanine due to the steric hindrance and/or reduced coordination resulting from the substitution of O^6 by sulfur. However, the detailed mechanism of the suppression of K⁺ mediated inhibitory effect remains to be ascertained. Furthermore, the biological application of 6-thioguanosine substituted oligonucleotides may be questionable due to the toxicity of 6-thioguanosine and the fact that the sulfur atom at C^6 of 6-thio-2-deoxyguanosine is gradually replaced by oxygen, probably due to hydrolysis.

F. Triple Helix Binding Ligands

I. The Triple Helix as a Target for Small Molecule Recognition

The equilibrium leading to triple helix formation from a duplex and a single strand is influenced by a number of factors. Simple equilibrium considerations indicate that, if a ligand binds to triple helix more tightly than to a corresponding duplex or single strand, the ligand will stabilize the equilibrium leading to triple helix formation. This observation has lead to the search for ligands whose binding would lead to stabilization via a coupled binding equilibrium of that kind.

II. Ethidium Bromide

Ethidium bromide is known to bind tightly to duplex in intercalative fashion. The binding interaction of ethidium bromide to the triple helix was investigated. Early studies using a polyribonucleotide system demonstrated that the binding of ethidium dye inhibited triple helix formation (WARING 1974a; LEHRMAN and CROTHERS 1977; LEE et al. 1979). Recent Tm data of oligonucleotides has confirmed that ethidium bromide binds more weakly to parallel type triple helix consisting of both C+:GC and T:AT triplets than to duplex DNA (MERGNY et al. 1991a). On the other hand, strong ethidium bromide binding has been reported to a poly(dT:dA:dT) triple helix, relative to a poly(dT-dA) duplex (SCARIA and SHAFER 1991). This apparent discrepancy seems to be explained by the inherently weak binding of ethidium dye to the poly(dT-dA) duplex, which is known to form a variant of the canonical B-form duplex.

III. Benzopyridoindole Derivatives

Helene's group has reported that benzopyridoindole derivatives (MERGNY et al. 1992; PILCH et al. 1993; DUVAL-VALENTIN et al. 1995; ESCUDE et al. 1995, 1996; de BIZEMONT et al. 1996) and a related compound (MARCHAND et al. 1996) specifically stabilize parallel triple helix formation. The stabilizing effect of these ligands was monitored by the resulting increase in the Tm of triple helix formation. Studies of the structure–activity relationship suggested that the positioning of the benzene ring, the substituent group, and structure of the alkylamine side chain were important for the binding to the triple helix (ESCUDE et al. 1995). As expected from the high planarity of these compounds, a fluorescence study and viscosity measurement revealed the intercalation of these derivatives into the triple helix (PILCH et al. 1993b). The binding appears to occur preferentially at runs of consecutive T:AT triplets in the parallel triple helix motif, which is probably due to electrostatic repulsion between the cationic charge of the compound and that of the C+:GC triplet (MERGNY et al. 1992).

Stabilization of an antiparallel triple helix bearing the G:GC and T:AT triplet motif was also demonstrated by a benzopyridoindole (BePI) derivative (ESCUDE et al. 1996b). Interestingly, antiparallel triple helix bearing G:GC and A:AT triplet was not stabilized by the addition of BePI (ESCUDE et al. 1996b). A gel shift experiment in that study implied that the binding kinetics of BePI seems to be relatively slow.

IV. Naphtyl-Quinoline Derivative

A Tm study (WILSON et al. 1993) and DNase I footprinting (CHANDLER et al. 1995; BROWN et al. 1996) have indicated that naphtyl-quinoline derivatives stabilize the parallel triple helix. It is interesting to note that these derivatives have a structure similar to the benzopyridoindole derivatives. The naphtyl-quinoline derivatives have unfused ring systems, but are similar to the benzopyridoindoles in that they both have a cationic ring system and an alkyl side chain. A modeling study suggested the intercalative binding of these compounds (WILSON et al. 1993), which is also a common property of benzopyridoindole derivatives.

V. Other Intercalators

Other compounds such as 2,6-disubstituted amidoanthraquinone (FOX et al. 1995a), coralyne (LEE et al. 1993), quinacrine (WILSON et al. 1994b), and methylene blue (TUITE and NORDEN 1995), all of which are known intercalators, have each been shown to stabilize triple helix significantly and selectively.

VI. Minor Groove Binding Ligands

Since the third strand in triple helix occupies the major groove of the underlying duplex, it is interesting to determine if the minor groove in a triple helix is available to bind ligands. Several studies with known minor groove binders have been reported (Durand et al. 1992a; Park and Breslauer 1992; Chalikian et al. 1994; Durand and Maurizot 1996; Vigneswaran et al. 1996). In these studies, a parallel triple helix motif bearing consecutive T:AT triplets was employed since minor groove binders such as distamycin A or netropsin are known to bind preferentially to the minor helix groove of AT-rich DNA. CD spectroscopy is useful for monitoring the binding of netropsin (Durand et al. 1992a; Park and Breslauer 1992) and distamycin (Durand and Maurizot 1996). Since these compounds are not chiral, a measurable cotton effect appears in the CD spectrum only upon binding to the (chiral) triple helix. CD titration studies have suggested that these compounds can bind to the triple helix, which is consistent with the results derived from NMR studies of a distamycin analogue (Umemoto et al. 1990). However, UV melting studies have shown a destabilization of the triple helix structure by the binding of either netropsin (Durand et al. 1992a; Park and Breslauer 1992) or distamycin (Durand and Maurizot 1996), suggesting the existence of crosstalk between minor groove and major groove binding in a DNA duplex. It has been demonstrated that binding of the minor groove agents berenil and 4′,6-diamidino-2-phenylindole (DAPI) enhanced (rather than inhibited) formation of the poly(rA)poly(rA)poly(dT) triple helix as well as the poly(dT)poly(rA)poly(dT) triple helix (Pilch and Breslauer 1994). This apparent discrepancy between minor groove binders suggests that details of minor groove association may play an important role in the coupling to triple helix formation in the major groove, possibly mediated by conformational changes in the duplex binding site.

VII. Polyamines and Basic Oligopeptides

Based upon the high negative charge of a triple helix, it is expected that polyamines including putrescine, spermidine, and spermine would bind tightly to a triple helix. From a biological perspective, it is important to investigate the effect of spermidine and spermine on triple helix formation, because these compounds are ubiquitous in the nucleus of living cells (Tabor and Tabor 1984) and may play an important role in cell cycle regulation (Pegg 1988). Several studies have revealed the stabilizing effect of polyamines (Hampel et al. 1991; Thomas and Thomas 1993; Musso and Van Dyke 1995). Polyamines have hydrophobic alkyl chains as well as an organic cation moiety. This serves to distinguish them from the spherical, uniformly hydrated inorganic cations.

X-ray crystallography of a duplex-polyamine complex has revealed binding in the major helix groove (EGLI et al. 1991). In a triple helix, it is plausible that polyamines could bind not only through simple electrostatics, but also through a complicated mode including salt bridge and hydrophobic interaction between closely spaced strands. This prediction awaits experimental verification.

Basic oligopeptides consisting of lysine residues were found to stabilize PyPuPy type parallel triple helix formation (POTAMAN and SINDEN 1995). It is expected that studies with other basic peptides and protein are forthcoming.

G. Backbone Modification and Strand Switching

I. DNA and RNA Backbone

The effect of RNA or DNA backbone composition on parallel triple helix formation has been investigated independently via UV melting analysis (ROBERTS and CROTHERS 1992) or affinity cleaving (HAN and DERVAN 1993). These two studies have not yielded a consensus result as to the effect of 2'OH modification, which may be due to the difference in the method, condition used, and/or length of the triple helices. It was reported that the antiparallel triple helix containing an RNA strand could not be detected in gel retardation (SEMERAD and MAHER 1994). Triple helix formation by a TFO with the α nucleotide anomer was investigated (LE DOAN et al. 1987; SUN et al. 1991; SUN and LAVERY 1992). The preferred polarity of the third strand consisting of α-nucleotides in PyPuPy type triple helix was found to depend on the nucleotide sequence (SUN and LAVERY 1992). Triple helix formation with oligonucleotides containing the 2'-5' internucleotide linkage was also demonstrated (JIN et al. 1993).

II. Chemical Modification of Backbones

When the triple helix technology is used for the purpose of gene regulation, it important to consider the stability of TFOs against nuclease mediated degradation in the cell, a problem shared by antisense oligonucleotide technology. A number of chemical modifications to provide nuclease resistance have been reported in the antisense field (UHLMANN and PEYMAN 1990).

The phosphorothioate is one of the best-known modifications to confer nuclease resistance, in which a oxygen of phosphodiester bond is replaced by sulfur. However, melting analysis has demonstrated that phosphorothioate substitution in a TFO decreased the Tm of the parallel triple helix (KIM et al. 1992; XODO et al. 1994). On the other hand, a phosphorothioate modified antiparallel TFO formed a triple helix with an affinity similar to a corresponding unmodified TFO (HACIA et al. 1994; MUSSO and VAN DYKE 1995; TU et al. 1995). Interestingly, triple helix formation by an antiparallel TFO with a phosphorothioate backbone was not inhibited in the presence of physiological concentration of K^+ (Tu et al. 1995) under conditions where the parent unmodified TFO failed to form the triple helix due to G-quartet based aggregation of guanine-rich oligonucleotide. This observation may be very important for the application of antiparallel TFO to gene regulation in living cells.

Ts'o and Miller's group has extensively studied the properties of oligonucleotides possessing a methylphosphonate linkage, which is an uncharged and nuclease resistant analogue. The ability of a methylphosphonate modified TFO to form the triple helix has been investigated by experimental (MILLER et al. 1980, 1981; KIBLER-HERZOG et al. 1990; CALLAHAN et al. 1991 TRAPANE et al. 1996) and theoretical (HAUSHEER et al. 1990) approaches.

Because of the high negative charge of the triple helix, TFO consisting of a positively charged backbone might be of practical utility. Recently, a gel titration experiment showed that an antiparallel TFO with the N,N-diethylethylenediamine phosphoramidate internucleotide linkage displays enhanced binding affinity as compared to an unmodified TFO (DAGLE and WEEKS 1996). In addition, antiparallel triple helix formation by this positively charged TFO can occur even in the presence of 130 mM KCl (DAGLE and WEEKS 1996), under conditions where aggregation of a guanine-rich oligonucleotide would have inhibited triple helix formation in the parent TFO. This observation is reasonable since the formation of guanine tetraplex requires the binding of an alkali metal cation such as K^+ at the center of the guanine tetrad, which could be inhibited by electrostatic repulsion in the positively charged TFO.

A TFO analogue with the N3'-5' phosphoramidate linkage was shown to form a more stable parallel triple helix than TFO with natural internucleotide linkage (GRYAZNOV et al. 1995; ESCUDE et al. 1996). The elegance of this modification is that there is no new chirality at the phosphorus, which could be a problem in other types of modified oligonucleotides due to the existence of a huge number of diasteromers.

It has been found that TFO bearing formacetal (MATTEUCCI et al. 1991) or riboacetetal internucleotide (JONES et al. 1993) linkages formed a more stable triple helix than the TFO with the natural phosphorodiester linkage. However, the difficulty in the synthesis of these linkages might encumber the practical application of these modified TFOs.

III. Alternate Strand Recognition and Minor Helix Groove Bridging

Because of the strict recognition limits imposed by Hoogsteen and reverse Hoogsteen H bonding, sequences capable of stable triple helix formation have been limited to relatively long stretches of homopurine-homopyrimidine duplex sequence. To relieve that limitation, approaches to recognize two adjacent purine tracts on alternate strands have been developed. In the original approach, two TFOs with an identical motif were connected by a 3'-3' linkage (HORNE and DERVAN 1990; ONO et al. 1991a) or 5'-5' linkage (ONO et al. 1991a) to preserve the required polarity for the triple helix formation. The requirement of that polarity switch in TFO was overcome by using both parallel and antiparallel triple helix motifs (BEAL and DERVAN 1992b; JAYASENA and JOHNSTON 1992a,b, 1993; WASHBROOK and FOX 1994; OLIVAS and MAHER 1994; BALATSKAYA et al. 1996). Affinity cleavage experiments suggested that alternate strand recognition considerably enhanced affinity as compared to that of each unlinked TFO (BEAL and DERVAN 1992b). Although there are small discrepancies among labs as to the requirement for a linker base (BEAL and DERVAN 1992b; JAYASENA and JOHNSTON 1992a), it seems to be true that the preferred structure of the junction for motif change differs between 5'-(Pu)m(Py)n-3' and 5'-(Py)m(Pu)n-3' sequences (BEAL and DERVAN 1992b). The application of the alternate strand triple helix formation in a target site in the human p53 gene, which has three alternate purine blocks, was also reported (OLIVAS and MAHER 1994).

A study in our laboratory has demonstrated the cooperative recognition of two noncontiguous purine rich sequences which are separated by one helical turn of duplex, by using two TFOs connected by an inert polymeric linker (KESSLER et al. 1993). Interestingly, the linker connecting the two triple helices passed over the minor groove, in the bound complex. This binding mode presents a new approach to the design of TFO recognizing separated purine tracts. Thus, rather than using a long TFO with several consecutive mismatched bases, high binding affinity can be conferred via short TFOs connected by a linker, which is designed to pass over the minor helix groove.

H. Conjugation of Small Compounds Enhance Triple Helix Stability

I. Conjugation of Intercalators

Conjugation of a duplex specific compounds to a TFO has been shown to enhance TFO binding affinity. An acridine derivative, 2-methoxy-6-chloro-9-aminoacridine, was covalently attached to a parallel TFO (SUN et al. 1989; COLLIER et al. 1991b, 1994; ORSON et al. 1994; STONEHOUSE and FOX 1994; FOX 1995; ZHOU et al. 1995) or antiparallel TFO (FOX 1994) via an aliphatic linker. Tm analysis (SUN et al. 1989; COLLIER et al. 1991b) and DNase I footprinting (STONEHOUSE and FOX 1994) have demonstrated significant stabilization of the triple helix by the conjugation of the acridine derivative. Fluorescence analysis (SUN et al. 1989) and chemical footprinting of the acridine conjugate (COLLIER et al. 1991a) have suggested that intercalation occurred at the triple helix-duplex junction, possibly as a result of a conformational change which occurs there. Interestingly, the conjugate appeared to reduce nonspecific intercalation of the acridine moiety (SUN et al. 1989), which has been ascribed to the electrostatic repulsion between the TFO moiety and the DNA duplex. Length and structure of the linker connecting acridine to the TFO seems to be important for stabilization (ORSON et al. 1994). Incorporation of an acridine into the middle of TFO and a study of its effect on stabilization of triple helix at sites with a single or double base pair inversion have also been reported (ZHOU et al. 1995).

The oxazolopyridocarbazole chromophore, which is also a nonspecific intercalator, has been linked to both parallel and antiparallel TFOs which were targeted to the LTR region of HIV-I (BAZILE et al. 1989; MOUSCADET et al. 1994b). Footprinting data have revealed that the conjugated TFO showed enhanced binding affinity relative to an unmodified TFO (MOUSCADET et al. 1994b).

II. Conjugation of Other Small Compounds

A minor groove binder, a Hoechst 33258 analogue, has been conjugated to the 5'-end of a parallel TFO via a polyethylene glycol linker (ROBLES et al. 1996). A UV melting study has shown that tethering of the Hoechst dye resulted in a large increase of triple helix stability (ROBLES et al. 1996). Polyamines are known to enhance the stability of triple helix structure as mentioned above. Spermine conjugates with a TFO at the 5' end (TUNG et al. 1993) or at the N^4-exocyclic amine of cytidine (BARAWKAR et al. 1996) have been synthesized. Both type of conjugates showed enhanced stability of the triple helix. Based on a similar idea, a parallel TFO tethered to a basic peptide consisting of arginines, lysines, and ornithines was synthesized and showed the enhancement in the stability of the triple helix (TUNG et al. 1996).

Dervan's group has demonstrated the cooperative binding of two TFO domains via connection through a dimerization domain consisting of duplex (DISTEFANO et al. 1991, 1993) or via a distamycin analogue (SZEWCZYK et al. 1996), so as to mimic the symmetry of proteins that bind cooperatively to the duplex as a homo- or heterodimer.

I. Inhibition of Transcription by Triple Helix Technology

I. Prospects for Triple Helix Mediated Gene Regulation

Several mechanisms to inhibit eukaryotic transcription by intermolecular triple helix formation may be possible. The first mechanism is based upon inhibition of transcription factor binding so as to affect the initiation of mRNA expression. Transcription factors (TF) can be classified into two categories. There are the general transcription factors, which are necessary for the basal level of mRNA production. These factors include RNA polymerase II, TFIIB, TFIID, etc. Another class is the gene-specific transcription factors. These factors are required to maintain a high level of mRNA induction. They are referred to as activators and are known to bind to promoter elements of a gene. Both types of transcription factor have been discussed as targets for TFO mediated intervention, where the binding of a TFO could obstruct the binding of these transcription factors. The second possible mechanism is to arrest transcription elongation via triple helix formation at sites downstream of the mRNA start site, for the purpose of inhibiting RNA polymerase at the region where the triple helix is formed.

A third mechanism involves disruption of nucleosome structure. It is now believed that the nucleosome fold can be phased relative to regulatory protein binding sites in gene promotor-enhancer regions. Several lines of evidence suggest that nucleosome phasing plays a role in modulation of transcription initiation (FELSENFELD 1992; WOLFFE 1994). Thus, interruption of chromosome structure by triple helix formation may alter transcription. Several studies by our team (AKIYAMA and HOGAN 1997) and by others (MAHER et al. 1992; SHIN and KOO 1996) have suggested that the triple helix formation can change the helical pitch (SHIN and KOO 1996) or the flexibility of the helix (MAHER et al. 1992; AKIYAMA and HOGAN 1997). These properties of the triple helix may attenuate the communication between activator and general transcription factors, which is known to occur in the initiation of the activated level of transcription, by altering the 3-D structure of the chromosome fold or by making the site of triple helix formation too stiff to be correctly folded into a transcriptionally active state.

II. Inhibition of Transcription Factor Binding by Triple Helix Formation

One promising approach to inhibition of transcription is to prevent the binding of transcription factors which regulate the expression of the targeted gene. The 5'-flanking region of a gene, which usually contains the enhancer or promoter activity, is an obvious target for inhibitory triple helix formation. As a first step to address such a possibility, several biochemical studies on the inhibition of protein binding by triple helix formation have been carried out in a cell-free system (MAHER et al. 1989; GEE et al. 1992; GRIGORIEV et al. 1992; EBBINGHAUS et al. 1993; ING et al. 1993; MAYFIELD et al. 1994; MAYFIELD and MILLAR 1994; NOONBERG et al. 1994a; KIM et al. 1995; NEURATH et al. 1995; REDDOCH et al. 1995; KOCHETKOVA and SHANNON 1996; KOVACS et al. 1996). In these studies, the gel shift method has been used to show that stable triple helix formation can inhibit transcription factor binding. All reported data suggest that triple helix formation at a site which has overlapped or is close to the recognition site of a transcription factor can interfere with binding of transcription factors. The transcription factors which have been studied thus far, the corresponding DNA binding site elements, and the TFO motif used to obtain binding inhibition are summarized in Table 1.

III. Inhibition of Transcription by Triple Helix Formation in a Cell-free System

In vitro transcription analysis using a bacteriophage promoter has been used as a model system to analyze the possibility of triple-helix-mediated inhibition

Table 1. A summary of transcription factors studied, DNA binding site elements, and the TFO motif used to obtain binding inhibition. Studies on binding inhibition of transcription factors *in vitro* are summarized. Transcription factor investigated (column 1), binding element for the transcription factor (column 2), and the TFO motif (column 3; see also Figs. 1, 2) are shown

Transcription factor	Gene	Motif	Reference
Sp1	Human Ha-*ras* promoter	Antiparallel	Mayfield et al. 1994
Sp1	Dihydrofolate reductase promoter	Antiparallel	Gee et al. 1992
Sp1	Metallothionein promoter	Parallel	Maher 1989
Progesterone receptor	Progesterone responsive gene	Antiparallel	Ing 1993
NF-κB	Interleukin 2-receptor α promoter	Parallel	Grigoriev 1992
NF-κB	Human granulocyte macrophage colony stimulating factor promoter	Antiparallel	Kochekova 1996
Myc-associated zinc finger proteins	Human c-*myc* P2 promoter	Antiparallel	Kim 1995
PU1	HER-2/neu/c-*erb* B2 promoter	Antiparallel	Noonberg 1994
B-cell specific activator protein	Immunoglobulin 3'-α enhancer	Antiparallel	Neurath 1995
Nuclear proteins from HeLa	HER-2/neu promoter	Antiparallel	Ebbinghaus 1993
Nuclear proteins from murine T lymphocyte	Murine K-*ras* promoter	Antiparallel	Mayfield 1994
Nuclear proteins from Murin YC8	Murine c-myc promoter	Antiparallel	Reddoch 1995
Nuclear proteins from rat cardiac fibroblast	Rat α1(I) collagen promoter	Antiparallel	Kovacs 1996

of the RNA polymerase reaction. The formation of triple helix at a site close to or overlapping the initiation site of T7 or *Escherichia coli* RNA polymerase appeared to inhibit the initiation of the transcription (Duval-Valentin et al. 1992; Maher 1992; Skoog and Maher 1993a; Alunni-Fabbroni 1994, 1996; Xodo et al. 1994). Surprisingly, triple helix formation at a site 40 bp downstream from the initiation site still seemed to inhibit the *initiation* of RNA synthesis (Alunni-Fabbroni 1994; Xodo et al. 1994).

Inhibition of transcription initiation by the binding of TFO to a eukaryotic promoter region was also studied in a cell-free system. In this type of experiment, transcription from a plasmid construct containing a promoter region is initiated and elongated by a nuclear extract, which includes the transcription factors required for the initiation process. Inhibition of eukaryotic transcription by triple helix formation in a cell-free system was originally described by Hogan and co-workers in 1988 (COONEY et al. 1988; POSTEL 1992). Miller's group has demonstrated the inhibition of transcription with a TFO targeted to the HER-1/neu promoter (EBBINGHAUS et al. 1993), the human Ha-*ras* promoter (MAYFIED et al. 1994), or the human c-*myc* promoter (KIM and MILLER 1995) by using transcription in a cell-free system. Transcription initiation activated by NF-κB in a hybrid promoter consisting of the interleukin II receptor α regulatory promoter and *fos* promoter was suppressed by the binding of a TFO conjugated with psoralen to the sequence overlapping the NF-κB site (GRIGORIEV et al. 1993). Interestingly, in this study the TFO conjugated with psolaren blocked transcription initiation only after photo cross-linking of the psoralen, which generates the covalently attached triple helix. Triple-helix mediated inhibition of transcription initiation activated by progesterone receptor was also studied via a G-free cassette in vitro (ING et al. 1993), which allows simplified analysis of the results (SAWADOGO and RODER 1985). A TFO targeted to $\alpha1(I)$ collagen promoter (KOVACS et al. 1996) was also found to inhibit the transcriptional activity of a promoter in a cell-free system.

Maher et al. have reported the inhibition of transcription initiation by parallel triple helix formation from a synthetic construct containing a Sp1 transcription factor binding site overlapping the site of triple helix formation (MAHER et al. 1992). Interestingly, they concluded that inhibition was not due to occlusion of Sp1 binding. Based on circular permutation analysis, they proposed that parallel triple helix formation may stiffen the duplex, which could suppress the duplex bending required to mediate contacts between Sp1 and other general transcription factors. Recent data from our lab, based upon ring closure analysis (SHORE et al. 1981), have also suggested that DNA is stiffened by antiparallel triple helix formation (AKIYAMA and HOGAN 1997), adding credence to the idea that transcription inhibition could be obtained at a distance due to local stiffening at sites of TFO binding.

As mentioned above, inhibition of transcription elongation by blocking of RNA polymerase reaction is another possible mechanism of triple-helix-mediated transcription inhibition. In vitro analysis using T7 and T3 promoter was exploited to address this possibility (MAHER 1992; SKOOG and MAHER 1993b; RANDO et al. 1994). There are two inconsistent results on the attenuation of transcription elongation by triple helix formation in a prokaryote system (MAHER 1992; RANDO et al. 1994). The reason for this inconsistency remains unclear, but it might be due to the condition and/or property of the TFO used for the assays (RANDO et al. 1994). In a eukaryotic transcription system, the inhibition of transcription elongation was observed *in vitro* (YOUNG et al. 1991). However, the inhibition was transitory because the TFO was forced to be dissociated by the RNA polymerase II complex stalled at the triple helix site. This dissociation effect was overcome by the covalent linkage of the TFO to the targeted duplex (YOUNG et al. 1991).

IV. Change of Nucleosome Positioning by Triple Helix Formation

Analysis of cleavage patterns in reconstituted histone-DNA complexes has suggested that the binding of a parallel TFO altered the positioning of the histone complex on a mouse mammary tumor virus (MMTV) promoter during in vitro nucleosome reconstitution (WESTIN et al. 1995). We have also observed the inhibition of nucleosome reconstitution with a 171 bp DNA fragment when an antiparallel TFO binds to the center of the fragment (T. AKIYAMA and M. HOGAN, unpublished data). These results suggest that the triple helix formation in genomic DNA might alter chromosome structure. This effect could cause a modulation of mRNA expression because nucleosomes are known to be important for transcription gene regulation as described above.

V. Transcription Inhibition by Triple Helix Formation in the Cell

Since the first report by Hogan and co-workers (POSTEL et al. 1991), many studies on the inhibition of mRNA expression by triple helix in living cells were carried out (Table 2). In several studies, chloramphenicol acetyltransferase (CAT) or luciferase activity expressed transiently in the cell by plasmid transfection was used to monitor the efficiency of the suppression of promoter function by triple helix formation (Table 2). However, the suppression of endogenous mRNA production is more important for the final goal, which is the rational design of an artificial gene regulator functioning in living cells. Both antiparallel and parallel motifs have been exploited; however, the examples of antiparallel motif are more numerous to date (Table 2).

Unmodified oligonucleotides are known to be degraded by serum or cellular nucleases. In some cases, unmodified TFOs exhibited statistically significant inhibition activity of mRNA expression in the cell (POSTEL et al. 1991; ROY 1994; SCAGGIANTE et al. 1994; THOMAS et al. 1995; KOVACS et al. 1996; PORUMB et al. 1996). However, chemical modification of TFOs may be required for maximum activity. A simple and effective modification is to block the free 3' hydroxy group of TFO (ORSON et al. 1991; MCSHAN et al. 1992; ING et al. 1993; FEDOSEYEVA et al. 1994; NEURATH et al. 1995; KOCHETKOVA and SHANNON 1996; LAVROVSKY et al. 1996). This modification provides increased stability against exonuclease degradation in the serum or cells (SHAW et al. 1991a; GAMPER et al. 1993). Phosphorothioate modification may also be effective to prevent nuclease-mediated degradation of TFOs (SONG et al. 1995; TU et al. 1995).

Uptake of TFOs into cells is another important factor in their application as gene-specific inhibitors. Many studies on the cellular uptake of oligonucleotides have been reported in the context of antisense technology. The difference relative to these antisense-oriented studies is that, in order to function in the cell, a TFO must enter into the nucleus rather than the cytoplasma (the site

Table 2. A summary of the inhibition of mRNA expression by triple helix formation in living cells. Studies on inhibition of gene expression in living cells are summarized. The name of the gene suppressed (column 1), triple helix formation site in the gene (column 2), cell type (column 3), and the TFO motif (column 4; see also Figs. 1, 2) are indicated

Gene	Site	Cell type	Motif	Assay	Reference
c-myc	Promoter	HeLa	Antiparallel?	Northern (endogenous RNA)	POSTEL 1991
GM-CSF	Promoter	Jurkat T cell	Antiparallel	Luciferase, RNase protection (endogenous RNA)	KOCHETKOVA 1996
ALDH2	Promoter (Sp1)	Hepatoma cell	Antiparallel	RT-PCR (endogenous RNA)	TU 1995
HER2	Promoter	Human epithelial cell	Antiparallel	RT-PCR (endogenous RNA), ELISA (HER2 protein)	PORUMB 1996
Rat α1(I) collagen	Promoter	Rat cardiac fibroblast	Antiparallel	CAT (transient transfection)	KOVACS 1996
Progesterone responsive gene	Promoter (progesterone receptor)	Monkey kidney	Antiparallel	CAT (transient transfection)	ING 1993
Interleukin 2 R	Promoter	Peripheral blood mononuclear cell	Parallel?	Northern (endogenous RNA)	ORSON 1991
Interleukin 2 R	Promoter (NF-κB)	Tumor T cell	Parallel	CAT (transient transfection), Northern (endogenous)	GRIGORIEV 1993
					GRIGORIEV 1992

Table 2. (*Continued*)

Gene	Site	Cell type	Motif	Assay	Reference
c-*myc*	Promoter	HeLa	Antiparallel	Viable cell counting	Helm 1993
Murine immunoglobulin	B-cell specific activator protein	Murine B-cell lymphoma	Antiparallel	Northern (endogenous), Western (IgA), Luciferase (transient transfection)	Neurath 1995
Synthetic Promoter	Interferon responsible element	HeLa	Antiparallel	CAT (transient transfection)	Roy 1994
c-*myc*	Promoter	Breast cancer cell	Antiparallel	Northern (endogenous RNA)	Thomas 1995
c-*fos*	Promoter (FBS2/AP2)	Endothelial cell	Parallel	CAT-ELISA (transient transfection)	Lavrovsky 1996
HIV proviral	Promoter	HIV infected T lymphocyte	GT-Parallel?	Northern (viral transcripts), p24 antigen	McShan 1992
Human MHC class II	Promoter (X/X2 box)	Peripheral blood T cell	Antiparallel	ICAM-1 and Fc receptor	Fedoseyeva 1994
Human multidrug-resistance gene	Exon 3	MDR cancer cell	GT-parallel	Northern (endogenous RNA)	Scaggiante 1994
Androgen receptor	Promoter	COS-1	Antiparallel	Luciferase (transient transfection)	Song 1995

of antisense oligonucleotide action). 3'-Modified TFOs appear to enter the nucleus; however, there is variability among published data, depending on the method of analysis (POSTEL et al. 1991; MCSHAN et al. 1992; FISHER et al. 1993). When antisense oligonucleotides are transfected with cationic lipid, enhanced accumulation in the nucleus is generally observed using fluorescently labeled oligonucleotide (BENNETT et al. 1992). This represents a reason why cationic delivery systems should be investigated more thoroughly as a vehicle for enhancing triple helix activity in the cell.

J. Other Pharmaceutically Interesting Effects of Triple Helix Formation

I. Inhibition of DNA Replication

The effect of TFO binding on DNA replication has been less studied than its effect on RNA transcription. Biochemical studies using a cell-free system have shown that the elongation reaction of T7 DNA polymerase from a primer was arrested at the triple helix formation site in single-stranded substrate DNA (SAMADASHWILY et al. 1993). This effect was also observed for DNA synthesis by Klenow fragment (HACIA et al. 1994), DNA polymerase I (SAMADASHWILY and MIRKIN 1994), or Taq polymerase (SAMADASHWILY and MIRKIN 1994). A "fold back" TFO, described below, was applied to the inhibition of DNA synthesis from a single-stranded substrate (GIOVANNANGELI et al. 1993).

Inhibition of the initiation of DNA synthesis from a primer by the binding of a TFO was observed when the triple helix was formed on the primer (GUIEYSSE et al. 1995). However, the inhibitory effect was relieved when the duplex region of the primer protruded beyond 3 bp from the boundary of the triple helix (GUIEYSSE et al. 1995). In a culture cell assay system, a parallel TFO linked to acridine as a stabilizer inhibited the DNA synthesis of simian virus 40 (SV40) infected in CV-1 cells (BIRG et al. 1990). The details of the mechanism remain to be confirmed but could have been due to direct inhibition of viral DNA replication.

II. Inhibition of HIV Integration by Triple Helix Formation

Integration of duplex DNA, which is reverse-transcribed from viral RNA, into host cell is one of the key steps in the infection of retroviruses such as HIV-1. This event is catalyzed by an integrase, which is a product of the HIV *pol* gene. The reaction is initiated by integrase binding to LTR regions at both ends of proviral DNA. Triple helix formation at a synthetic LTR sequence was found to inhibit the reaction of the integrase in a cell-free system (MOUSCADET et al. 1994a), which is probably due to the blockage of integrase binding to the LTR sequence. The inhibition of the integrase reaction by the TFO was specific to the substrate bearing the site of the triple helix formation and was dependent on the concentration of the TFO, with an EC_{50} of approximately 100 nM (MOUSCADET et al. 1994a).

K. Targeting Single-Stranded DNA or RNA by Triple Helix Formation

Triple helix formation can be used for the recognition of single-stranded DNA or RNA. One purine and one pyrimidine strand bearing methylphosphonate internucleotide linkages, which are designed to form intermolecular parallel triple helix at a polypyrimidine RNA sequence, were shown to arrest protein synthesis from that RNA template in a cell-free system (REYNOLDS et al. 1994).

Extending that idea one step further, "fold-back" oligonucleotides have been designed that have one domain for Watson-Crick base pairing and one for Hoogsteen base pairing linked together so as to form a hairpin-like fold back structure (GIOVANNANGELI et al. 1991; BROSSALINA and TOULME 1993a; BROSSALINA et al. 1993b; BOOHER et al. 1994; KANDIMALLA and AGRAWAL 1994; WANG and KOOL 1994b; AZHAYEVA et al. 1995; KANDIMALLA et al. 1995; RUMNEY and KOOL 1995). Related circular oligonucleotides have also been synthesized by chemical or biological ligation. Such circular oligonucleotides also formed a triple helix with single-stranded nucleic acid (KOOL 1991; PRAKASH and KOOL 1991, 1992; WANG and KOOL 1994a,b, 1995; CHAUDHURI and KOOL 1995; KANDIMALLA et al. 1995b; VO et al. 1995; WANG et al. 1995). UV melting analysis of the fold-back (GIOVANNANGELI et al. 1991) or circular parallel triple helix (PRAKASH and KOOL 1991) showed enhanced stability of the Watson-Crick and Hoogsteen base pairing, which was ascribed to the forced cooperativity of the dual form of base pairing (PRAKASH and KOOL 1991). In addition, the circular oligonucleotide displayed higher mismatch discrimination for its complement as compared to normal DNA duplex (KOOL 1991; WANG et al. 1995). A study of the structure of fold-back (BOOHER et al. 1994; RUMNEY and KOOL 1995) and circular molecules (PRAKASH and KOOL 1992) was executed, suggesting that length of the linker connecting two domains was very important to the stability of the triple helix. A detailed kinetic study indicated that the rates for triple helix formation by circular oligonucleotides were approximately 100 times faster than that for normal triple helix formation (WANG et al. 1995).

A fold-back TFO conjugated to psoralen was shown to arrest DNA synthesis upon a single-stranded DNA target (GIOVANNANGELI et al. 1992a) and was also shown, as expected, to have resulted in stoppage of the synthesis reaction at the site of fold-back TFO binding. The effect was accentuated by photo cross-linking of a psolaren-conjugated homologue.

L. DNA Modification Mediated by TFO Binding

I. Sequence-Specific DNA Cleavage

Sequence-specific cleavage of double-strand DNA has been achieved by attaching various cleaving agents to TFOs. In an early study by Dervan's group, EDTA-Fe, which can produce active oxygen species by the addition of a reducing agent, was conjugated to TFO (MOSER and DERVAN 1987; STROBEL et al. 1988; STROBEL and DERVAN 1990). Selective double-strand cleavage of 48.5 kb genomic DNA at a single site was reported with that TFO-EDTA conjugate (STROBEL et al. 1988). Subsequently, various oxidative damaging agents such as 1,10-phenanthroline (FRANCOIS et al. 1989a,b; SHIMIZU et al. 1994, 1996), Cu(II) desferal (JOSHI and GANESH 1994) or methaloporphyrin (BIGEY et al. 1995) have been linked to TFOs to achieve site-selective damage. They also showed sequence-specific DNA cleavage; however, the yield of the cleavage reaction was usually low. Since the active species of the reaction mediated by these molecules are believed to be diffusible active oxygen, analysis of the products by sequencing reactions has shown multiple cleavage products around the site where the cleaving moiety was positioned in the triple helix.

Schultz and co-workers has reported a conjugate of a semisynthetic nuclease with a TFO (PEI et al. 1990). This conjugate would produce a hydrolyzed DNA product by the cleavage reaction; however, multiple cleavage bands were observed even if the active species was not diffusible. This observation might be due to conformational flexibility of the cleaving reagent or the superimposition of several overlapping binding sites.

II. Sequence-Specific Alkylation

Alkylating agents such as the N-bromoacetyl group (POVSIC and DERVAN 1990; Povsic et al. 1992; GRAND and DERVAN 1996), ethan-5-methycytidine (SHAW et al. 1991b), or aromatic chloroethylamine (FEDOROVA et al. 1988; VLASSOV et al. 1988) have been incorporated in TFOs so as to cause sequence specific alkylation of the underlying duplex. The alkylation at N^7 of guanine, which is most susceptible for reaction by such a nucleophile, has been targeted in this strategy. The reaction by N-bromoacetyl TFO caused almost quantitative cleavage of substrate (POVSIC et al. 1992). In addition, after the treatment of this alkylated product with piperidine, which generated a free 3'-hydroxy group, the cleaved product was ligated to a DNA fragment bearing a 5'-phosphate end (POVSIC et al. 1992).

III. Single-Site Enzymatic Cleavage

Triple helix formation can prevent the binding of various kinds of proteins, including restriction enzymes or methylases. Using this effect, single-site enzymatic cleavage of yeast genome or human chromosome DNA was demonstrated (STROBEL and DERVAN 1991; STROBEL et al. 1991) according to so-called "Achilles heel cleavage" techniques originally developed by Szybalski's group in protein technology (KOOB et al. 1988; KOOB and SZYBALSKI 1990).

IV. Photo-Induced Cross-Linking

A photo cross-linking reagent such as the p-azidophenacyl group (PRASEUTH et al. 1988b), azidoproflavine (LEDOAN et al. 1987), or proflavine (PRASEUTH et al. 1988a) has been covalently attached to TFOs. After photo-irradiation, the photosensitive group conjugated with TFO can form a covalent linkage with the targeted double strand. The cross-linked product is detected as slowly migrating bands in gel electrophoresis, which can be converted to a fast migrating band by alkaline treatment.

The attachment of an ellipticine derivative to a TFO was shown to induce site-specific photodamage (PERROUAULT et al. 1990). The cleavage reaction proceeded without the treatment of piperidine in this reaction. The detailed mechanism of this reaction remains to be determined.

The photosensitized reaction of psoralen to DNA has been well-characterized (CIMINO et al. 1985). Under irradiation with UVA light (350–400nm), either the 3,4-double bond of the pyrone ring or the 4′,5′-double bond of the furane ring in psoralen reacts with the 5,6-double bond of the thymine base by the mechanism of [2+2] photosensitized reaction of two double bonds. Psoralen is known to intercalate into duplex DNA. When psoralen intercalates at the TpA sequence, the reaction of both double bonds of psoralen is possible, which would generate a covalent linkage between two strands in a DNA duplex. The standard alkaline treatment of the reaction products allows one to determine the position of the modification. The first sequence-specific reaction by a psoralen-conjugated TFO was reported by Helene's group (TAKASUGI et al. 1991). Besides the bis adduct, comprising two strands of duplex and TFO, the mono adduct was observed as an intermediate product in the reaction (TAKASUGI et al. 1991; GIOVANNANGELI et al. 1992b) Recently, several groups have employed this modified psoralen chemistry to obtain targeted DNA damage in living cells (GASPARRO et al. 1994; VASQUEZ et al. 1996).

V. Targeted Mutagenesis by a Psoralen-Linked TFO

Based on site-specific cross-linking with a psoralen-TFO conjugate, Glazer's group showed the method of introducing mutation at a specific gene of λ DNA in bacteria (HAVRE and GLAZER 1993b). The mutation was probably induced by nucleotide misincorporation during the repair of the psoralen adduct. Subsequently, targeted mutagenesis has been shown to occur on a SV40 shuttle vector in monkey COS cells with a higher efficiency than seen in bacteria (HAVRE and GLAZER 1993a). In addition, efficient mutagenesis by a psoralen-

TFO was observed in a *Xeroderma pigmentosum* variant, which has revealed hypermutability but almost normal levels of nucleotide excision repair (RAHA et al. 1996). Several other examples of TFO-targeted mutagenesis have also been reported (BREDBERG et al. 1995; GUNTHER et al. 1996). In mammalian cells, psoralen adducts are known to be repaired by the nucleotide excision repair pathway (REARDON et al. 1991). The time course of repair for a triple-helix-mediated psoralen adduct has been reported (DEGOLOS et al. 1994; SANDOR and BREDBERG 1994; WANG and GLAZER 1995). The length of the triple helix moiety appeared to influence the repair reaction (DEGOLOS et al. 1994; WANG and GLAZER 1995), suggesting that it is related to the stability of complex formation. Interestingly, a recent report demonstrates that triple helix itself induced mutations within living cells (WANG et al. 1996b). This mutation seemed to be related to excision repair and to transcription-coupled repair.

M. Recent Topics and Prospects

I. Artificial DNA Bending by Triple Helix Formation

DNA bending might be important for transcription (MARTIN and EPISONA 1993; REES et al. 1993), DNA recombination (GOODMAN and NASH 1989), and DNA translation (BEESE et al. 1993). A number of DNA binding protein are known to induce bending of the DNA duplex. Recently, both we (AKIYAMA and HOGAN 1996a,b, 1997) and others (LIEBERES and DERVAN 1996) have independently developed an artificial agent to bend DNA via the triple helix motif. The idea is based on a previous study in our laboratory (KESSLER et al. 1993) in which a pair of TFOs connected by a linker moiety proved able to bind to two noncontiguous sites of triple helix formation separated by one helical turn of the intervening duplex. In that study, the linker passed over the minor helix groove of the intervening duplex. We reasoned that shortening the linker in that structure would induce bending in the minor helix groove of the intervening duplex. The idea of that artificial bending was verified (AKIYAMA and HOGAN 1997) by phasing analysis (ZINKEL and CROTHERS 1987), circular permutation analysis (WU and CROTHERS 1984), and ring closure analysis (SHORE et al. 1981). More importantly, DNA flexibility of the intervening duplex region was successfully monitored by this artificial bending system. Detailed gel titration experiments indicated that, surprisingly, DNA bending occurred without significant expenditure of the free energy (AKIYAMA and HOGAN 1996a,b), which is inconsistent with the elastic coil model (BARKLEY and ZIMM 1979). As a result, we have hypothesized that DNA flexibility might be asymmetric and minor groove compression of DNA is easier than that predicted by the elastic coil model (AKIYAMA and HOGAN 1996a,b). Furthermore, we have found that, by using this system, DNA bending can be induced until the minor helix groove was almost completely collapsed (T. AKIYAMA et al., unpublished results) with a remarkably small expenditure of binding free energy.

Since the artificial DNA bending system that we have developed was found to be stable under physiological temperature and pH, it appears that it can be applied to pharmacological or biotechnological purposes as a new "twist" to the triple helix approach.

N. Conclusion

A review of the current literature suggests that remarkable progress has been made in the development of oligonucleotides as gene-specific agents. Based upon the fact that this effort has required that the physical chemistry of triple helix formation be understood (information which was already well known in the field of antisense therapeutics), the pharmaceutical application of triple helix forming (antigene) oligonucleotides has lagged behind that of the antisense field.

However, recent advances made in the use of triple helix compounds as agents for site-selective DNA modification and advancements made in the use of these compounds to block gene expression at the transcriptional level suggest that the field is probably on the threshold of a major advance.

This advance will probably be based upon application of several technological improvements. First, the success obtained with cationic lipid delivery agents in the antisense arena now provides a set of technologies which can be used to greatly enhance uptake of triple helix compounds. Secondly, recent advances in the use of modified bases and backbones suggest that high affinity triple helix formation can now be obtained at physiological potassium ion concentration and pH. And, finally, although substantial work remains to be done, it now appears that several approaches have been defined which allow the target site range of triple helix formation to be extended to include many sequences which are not of the simple homopurine–homopyrimidine class.

Perhaps the greatest long-term pharmaceutical value to be obtained from the field of triple helix forming oligonucleotide design is the recognition that this may be but one example of the use of rationally designed, major groove specific polymers for the purpose of sequence-specific DNA binding. Although nucleic acid or modified nucleic acid building blocks have served as an excellent first step towards that goal, it is more likely that nonnucleic acid motifs, still based upon the general concept of Hoogsteen H-bonding, may prove to be the final implementation of the technology discovery process which has been set in motion.

Acknowledgements. We are grateful to all our collaborators in M. Hogan's laboratory at the Baylor College of Medicine for general assistance and to Ms. Nobuko Akiyama and Mr. Sean R. Smith for editorial assistance. T. Akiyama is a Research Fellow of the Japan Society for the Promotion of Science (JSPS) and is grateful for the Fellowships of JSPS for Young Scientists he has received.

References

Akiyama T, Hogan ME (1996a) The design of an agent to bend DNA. Proc Natl Acad Sci U S A 93:12122–12127

Akiyama T, Hogan ME (1996b) Microscopic DNA flexibility analysis. J Biol Chem 271:29126–29135

Akiyama T, Hogan ME (1997) Structural analysis of DNA bending induced by tethered triple helix forming oligonucleotides biochemistry. Biochemistry 36:2307–2315

Alunni-Fabbroni M, Manfioletti GM, Manzini G, Xodo LE (1994) Inhibition of T7 RNA polymerase transcription by phosphate and phosphorothioate triplex-forming oligonucleotides targeted to a R.Y site downstream from the promoter. Eur J Biochem 226:831–839

Alunni-Fabbroni M, Manzini G, Quadrifoglio F, Xodo LE (1996) Guanine-rich oligonucleotides targeted to a critical R.Y site located in the Ki-ras promoter. The effect of competing self-structures on triplex formation. Eur J Biochem 238:143–151

Arnott S, Bond PJ (1973). Triple-stranded polynucleotide helix containing only purine bases. Science 181:68–69

Arnott S, Selsing E (1974) Structures for the polynucleotide complexes poly d(A) with poly d(T), and poly d(T) with poly d(A) with poly d(T). J Mol Biol 88:68–69

Arnott S, Hukins DWL, Dover SD (1972) Optimized parameters for RNA double-helices. Biochem Biophys Res Commun 48:1392–1399

Azhayeva E, Azhayev A, Guzaev A, Hovinen J, Lonnberg H (1995) Looped oligonucleotides form stable hybrid complexes with a single-stranded DNA. Nucleic Acids Res 23:1170–1176

Balatskaya SV, Belotserkovskii BP, Johnston BH (1996) Alternate-strand triplex formation: modulation of binding to matched and mismatched duplexes by sequence choice in the Pu.Pu.Py block. Biochemistry 35:13328–13337

Barawkar DA, Rajeev KG, Kumar VA, Ganesh KN (1996) Triple helix formation at physiological pH by 5-Me-dC-N4-(spermine) [X] oligodeoxynucleosides: non protonation of N3 in X of X*G:C triad and effect of base mismatch/ionic strength on triplex stabilities. Nucleic Acids Res 24:1229–1237

Barkley MD, Zimm BH (1979) Theory of twisting and bending of chain macromolecules analysis of the fluorescence depolarization of DNA. J Chem Phys 70:2991–3007

Bates PJ, Macaulay VM, McLean, MJ, Jenkins TC, Reszka AP, Laughton CA, Neidle S (1995) Characterization of triplex-directed photoadduct formation by psoralen-linked oligonucleotides. Nucleic Acids Res 23:3627–3632

Bazile D, Gautier C, Rayner, B, Imbach JL, Paoletti C, Paoletti, J (1989) alpha-DNA X: alpha and beta tetrathymidilates covalently linked to oxazolopyrido-carbazolium (OPC): comparative stabilization of oligo beta[dT]:oligo beta [dA] and oligo alpha [dT]:oligo beta [dA] duplexes by the intercalating agent. Nucleic Acids Res 17:7749–7759

Beal PA, Dervan PB (1991) Second structural motif for recognition of DNA by oligonucleotide-directed triple-helix formation. Science 251:1360–1363

Beal PA, Dervan PB (1992a) The influence of single base triplet changes on the stability of a pur.pur.pyr triple helix determined by affinity cleaving. Nucleic Acids Res 20:2773–2776

Beal PA, Dervan PB (1992b) Recognition of double helical DNA by alternate strand triple helix formation. J Am Chem Soc 114:4976–4982

Beese LS, Derbyshire V, Steitz TA (1993) Structure of DNA polymerase I Klenow fragment bound to duplex DNA. Science 260:352–355

Bennett CF, Chiang, MY, Chan H, Shoemaker JEE, Mirabelli CK (1992) Cationic lipids enhance cellular uptake and activity of phosphorothioate antisense oligonucleotides. Mol Pharmacol 41:1023–1033

Berressem R, Engels JW (1995) 6-Oxocytidine a novel protonated C-base analogue for stable triple helix formation. Nucleic Acids Res 23:3465–3472

Best GC, Dervan PB (1995) Energetics of formation of sixteen triple helical complexes which vary at a single position within a pyrimidine motif. J Am Chem Soc 117:1187–1193

Bigey P, Pratviel G, Meunier B (1995) Cleavage of double-stranded DNA by "metalloporphyrin-linker-oligonucleotide" molecules: influence of the linker. Nucleic Acids Res 23:3894–3900

Birg F, Praseuth D, Zerial A, Thuong NT, Asseline U, Doan TL, Helene C (1990) Inhibition of simian virus 40 DNA replication in CV-1 cells by an oligodeoxynucleotide covalently linked to an intercalating agents. Nucleic Acids Res 18:2901–2908

Booher MA, Wang S, Kool ET (1994) Base pairing and steric interactions between pyrimidine strand bridging loops and the purine strand in DNA pyrimidine.purine.pyrimdine triplexes. Biochemistry 33:4645–4671

Bouziane M, Cherny DI, Mouscadet JF, Auclair C (1996) Alternate strand DNA triple helix-mediated inhibition of HIV-1 U5 long terminal repeat integration in vitro. J Biol Chem 271:10359–10364

Bredberg A, Sandor Z, Brant M (1995) Mutational response of Fanconi anaemia cells to shuttle vector site-specific psoralen cross-link. Carcinogenesis 16:555–561

Broitman SL, Im DD, Fresco JR (1987) Formation of the triple-stranded polynucleotide helix, poly (A.A.U). Proc Natl Acad Sci U S A 84:5120–5124

Brossalina E, Toulme JJ (1993a) A DNA hairpin as a target for antisense oligonucleotides. J Am Chem Soc. 115:796–797

Brossalina E, Pascolo E, Toulme JJ (1993b) The binding of an antisense oligonucleotide to a hairpin structure via triplex formation inhibits chemical and biological reactions. Nucleic Acids Res 21:5616–5622

Brown PM, Drabble A, Fox KR (1996) Effect of a triplex-binding ligand on triple helix formation at a site within a natural DNA fragment. Biochem J 314:427–432

Brunar H, Dervan PB (1996) Sequence composition effect on the stabilities of triple helix formation by oligonucleotides containing N-7 deoxyguanosine. Nucleic Acids Res 24:1987–1991

Callahan DE, Trapane TL, Miller PS, T'so POP, Kan LS (1991) Comparative circular dichroism and fluorescence studies of oligodeoxyribonucleotide and oligodeoxyribonucleoside methylphosphonate pyrimidine strands in duplex and triplex formation. Biochemistry 30:1650–1655

Cassidy SA, Strekowski L, Wilson D, Fox, KR (1994) Effect of a triplex-binding ligand on parallel and antiparallel DNA triple helices using short unmodified and acridine-linked oligonucleotides. Biochemistry 33:15338–15347

Chalikian TV, Plum GE, Sarvazyan AP, Breslauer KJ (1994) Influence of drug binding on DNA hydration: acoustic and densimetric characterizations of netropsin binding to the poly(dAdT).poly(dAdT) and poly(dA).poly(dT) duplexes and the poly(dT).poly(dA).poly(dT) triplex at 25 degrees C. Biochemistry 33:8629–8640

Chandler SP, Fox KR (1993) Triple helix formation at A8XA8.T8YT8. FEBS Lett 332:189–192

Chandler SP, Fox KR (1995) Extension of DNA triple helix formation to a neighbouring (AT)n site. FEBS Lett 360:21–25

Chandler SP, Strekowski L, Wilson D, Fox KR (1995) Footprinting studies on ligands which stabilize DNA triplexes: effects on stringency within a parallel triple helix. Biochemistry 34:7234–7242

Chaudhuri NC, Kool ET (1995) Very high affinity DNA recognition by bicyclic and cross-linked oligonucleotides. J Am Chem Soc 117:10434–10442

Cheng AJ, Van Dyke MW (1993) Monovalent cation effects on intermolecular purine-purine-pyrimidine triple-helix formation. Nucleic Acids Res 21:5630–5635

Cheng AJ, Van Dyke MW (1994) Oligodeoxyribonucleotide length and sequence effects on intermolecular purine-purine-pyrimidine triple-helix formation. Nucleic Acids Res 22:4742–4747

Cheng YK, Pettitt BM (1992) Hoogsteen versus reversed-Hoogsteen base pairing: DNA triple helixes. J Am Chem Soc 114:4465–4474
Cheng YK, Pettitt BM (1995) Solvent effects on model d(CG.G)7 and d(TA.T)7 DNA triple helices. Biopolymers 35:457–473
Cherny DI, Malkov VA, Volodin AA, Frank-Kamenetskii MD (1993) Electron microscopy visualization of oligonucleotide binding to duplex DNA via triplex formation. J Mol Biol 230:379–383
Cho BP, Evans FE (1991) Structure of oxidatively damaged nucleic acid adducts: Tautomerism, ionization, and protonation of 8-hydroxyadenosine studied by 15N NMR. Nucleic Acids Res 19:1041–1047
Cimino GD, Gamper HB, Isaacs ST, Hearst JE (1985) Psoralen as photoactive probes of nucleic acid structure and functions: organic chemistry, photochemistry, and biochemistry. Annu Rev Biochem 54:1151–1193
Collier DA, Mergny JL, Thuong NT, Helene C (1991a) Site-specific intercalation at the triplex-duplex junction induces a conformational change which is detectable by hypersensitivity to diethylpyrocarbonate. Nucleic Acids Res 19:4219–4224
Collier DA, Thuong NT, Helene C (1991b) Sequence-specific bifunctional DNA ligands based on triple-helix-forming oligonucleotides inhibit restriction enzyme cleavage under physiological condition. J Am Chem Soc 113:1457–1458
Colocci N, Dervan PB (1995) Cooperative triple helix formation at adjacent DNA sites: Sequence composition dependence at the junction. J Am Chem Soc 117:4781–4787
Cooney M, Czernuszewicz G, Postel EH, Flint SJ, Hogan ME (1988) Site-specific oligonucleotide binding represses transcription of the human c-myc gene in vitro. Science 241:456–459
Dagle JM, Weeks DL (1996) Positively charged oligonucleotides overcome potassium-mediated inhibition of triplex DNA formation. Nucleic Acids Res 24:2143–2149
Dagneaux C, Liquier J, Taillandier E (1995) Sugar conformations in DNA and RNA-DNA triple helices determined by FTIR spectroscopy: role of backbone composition. Biochemistry 34:16618–16623
de Bizemont T, Duval-Valentin G, Sun JS, Bisagni E, Garestier T, Helene C (1996) Alternate strand recognition of double-helical DNA by (T,G)-containing oligonucleotides in the presence of a triple helix-specific ligand. Nucleic Acids Res 24:1136–1143
de los Santos C, Rosen M, Patel D (1989) NMR studies of DNA (R+)n.(Y-)n.(Y+)n triple helices in solution: imino and amino proton markers of T.A.T and C.G.C+ base-triple formation. Biochemistry 28:7282–7289
Degols G, Clarenc JP, Lebleu B, Leonetti JP (1994) Reversible inhibition of gene expression by a psoralen functionalized triple helix forming oligonucleotide in intact cells. J Biol Chem 269:16933–16937
Distefano MD, Dervan PB (1993) Energetics of cooperative binding of oligonucleotides with discrete dimerization domains to DNA by triple helix formation. Proc Natl Acad Sci U S A 90:1179–1183
Distefano MD, Shin JA, Dervan PB (1991) Cooperative binding of oligonucleotides to DNA by triple helix formation: dimerization via Watson-Crick hydrogen bonds. J Am Chem Soc 113:5901–5902
Dittrich K, Gu J, Tinder R, Hogan M, Gao X (1994) T.C.G triplet in an antiparallel purine.purine.pyrimidine DNA triplex. Conformational studies by NMR. Biochemistry 33:4111–4120
Durand M, Maurizot JC (1996) Distamycin A complexation with a nucleic acid triple helix. Biochemistry 35:9133–9139
Durand M, Thuong NT, Maurizot JC (1992a) Binding of netropsin to a DNA triple helix. J Biol Chem 267:24394–24399
Durand M, Peloille S, Thuong NT, Maurizot JC (1992b) Triple-helix formation by an oligonucleotide containing one (dA)12 and two (dT)12 sequences bridged by two hexaethylene glycol chains. Biochemistry 31:9197–9204

Durland RH, Kessler DJ, Gunnell S, Duvic M, Pettitt BM, Hogan ME (1991) Binding of triple helix forming oligonucleotides to sites in gene promoters. Biochemistry 30:9246–9255

Durland RH, Rao TS, Revankar GR, Tinsley JH, Myrick MA, Seth DM, Rayford J, Singh P, Jayaraman K (1994) Binding of T and T analogues to CG base pairs in antiparallel triplexes. Nucleic Acids Res 22:3233–3240

Durland RH, Rao TS, Bodepudi V, Seth DM, Jayaraman K, Revankar GR (1995) Azole substituted oligonucleotides promote antiparallel triplex formation at non-homopurine duplex targets. Nucleic Acids Res 23:647–653

Duval-Valentin G, Thuong NT, Helene C (1992) Specific inhibition of transcription by triple helix-forming oligonucleotides. Proc Natl Acad Sci U S A 89:504–508

Duval-Valentin G, de Bizemont T, Takasugi M, Mergny JL, Bisagni E, Helene C (1995) Triple-helix specific ligands stabilize H-DNA conformation. J Mol Biol 247:847–858

Ebbinghaus SW, Gee JE, Rodu B, Mayfield CA, Sanders G, Miller DM (1993) Triplex formation inhibits HER-2/neu transcription in vitro. J Clin Invest 92:2433–2439

Egli M, Williams LD, Gao Q, Rich A (1991) Structure of the pure-spermine form of Z-DNA (magnesium free) at 1-A resolution. Biochemistry 30:11388–11402

Escude C, Nguyen CH, Mergny JL, Sun JS, Bisagni E, Garestier T, Helene C (1995) Selective stabilization of DNA triple helixes by benzopyridoindole derivatives. J Am Chem Soc 117:10212–10219

Escude C, Giovannangeli C, Sun JS, Lloyd DH, Chen J-K, Gryaznov SM, Garestier T, Helene C (1996a) Stable triple helices formed by oligonucleotide N3'–>N5' phosphoramidates inhibit transcription elongation. Proc Natl Acad Sci U S A 93:4365–4369

Escude C, Sun JS, Nguyen CH, Bisagni E, Garestier T, Helene C (1996b) Ligand-induced formation of triple helices with antiparallel third strands containing G and T. Biochemistry 35:5735–5740

Fedorova OS, Knorre DG, Podust LM, Zarytova VF (1988) Complementary addressed modification of double-stranded DNA within a ternary complex. FEBS Lett 228:273–276

Fedoseyeva EV, Li Y, Huey B, Tam S, Hum A, Benichou G, Garovoy MR (1994) Inhibition of interferon-gamma-mediated immune functions by oligonucleotides. Suppression of human T cell proliferation by downregulation of IFN-gamma-induced ICAM-1 and Fc-receptor on accessory cells. Transplantation 57:606–612

Felsenfeld G (1992) Chromatin as an essential part of the transcription mechanism. Nature 355:219–223

Felsenfeld G, Davies DR, Rich A (1957) Formation of a three-stranded polynucleotide molecule. J Am Chem Soc 79:2023–2024

Fisher TL, Terhorst T, Cao X, Wangner RW (1993) Intracellular disposition and metabolism of fluorescently-labeled unmodified and modified oligonucleotides microinjected into mammalian cells. Nucleic Acids Res 21:3857–3865

Fossella JA, Kim YJ, Shih H, Richards EG, Fresco JR (1993) Relative specificities in binding of Watson-Crick base pairs by third strand residues in a DNA pyrimidine triplex motif. Nucleic Acids Res 21:4511–4515

Fox KR (1994) Formation of DNA triple helices incorporating blocks of G.GC and T.AT triplets using short acridine-linked oligonucleotides. Nucleic Acids Res 22:2016–2021

Fox KR (1995b) Kinetic studies on the formation of acridine-linked DNA triple helices. FEBS Lett 357:312–316

Fox KR, Polucci P, Jenkins TC, Neidle S (1995a) A molecular anchor for stabilizing triple-helical DNA. Proc Natl Acad Sci U S A 92:7887–7891

Francois JC, Saison-Behmoaras T, Chassignol M, Thuong NT, Helene C (1989a) Sequence-targeted cleavage of single- and double-stranded DNA by oligo-thymidilates covalently linked to 1,10-phenanthroline. J Biol Chem 264:5891–5898

Francois JC, Saison-Behmoaras T, Barbier C, Chassignol M, Thuong N T, Helene C (1989b) Sequence-specific recognition and cleavage of duplex DNA via triple-helix formation by oligonucleotides covalently linked to a phenanthroline-copper chelate. Proc Natl Acad Sci U S A 86:9702–9706

Gaffney BL, Kung PP, Wang C, Jones RA (1995) Nitrogen-15-labeled oligodeoxynucleotides 8. Use of ^{15}N-NMR to probe Hoogsteen hydrogen bonding at guanine and adenine N7 atoms of a DNA triplex. J Am Chem Soc 117:12281–12283

Gamper HB, Reed MW, Cox T, Virosco JS, Adams AD, Gall AA, Scholler JK, Meyer RB Jr (1993) Facile preparation of nuclease resistant 3'-modified oligodeoxynucleotides. Nucleic Acids Res 21:145–150

Gasparro FP, Havre PA, Olack GA, Gunther EJ, Glazer PM (1994) Site-specific targeting of psoralen monoadduct and crosslink formation. Nucleic Acids Res 22:2845–2852

Gee JE, Blume Scott, Snyder RC, Ray R, Miller DM (1992) Triplex formation prevents Sp1 binding to the dihydrofolate reductase promoter. J Biol Chem 267:11163–11167

Gee JE, Yen RL, Hung MC, Hogan ME (1994) Triplex formation at the rat neu oncogene promoter. Gene 149:109–114

Gee JE, Revankar GR, Rao TS, Hogan ME (1995) Triplex formation at the rat neu gene utilizing imidazole and 2'-deoxy-6-thioguanine base substitution. Biochemistry 34:2042–2048

Giovannangeli C, Montenay-Garestier T, Rougee M, Chassignol M, Thuong NT, Helene C (1991) Single-stranded DNA as a target for triple helix formation. J Am Chem Soc 113:7775–7777

Giovannangeli C, Rougee M, Garestier T, Thuong NT, Helene C (1992a) Triple-helix formation by oligonucleotides containing the three bases thymine, cytosine, and guanine. Proc Natl Acad Sci U S A 89:8631–8635

Giovannangeli C, Thuong NT, Helene C (1992b) Oligodeoxynucleotide-directed photo-induced cross-linking of HIV proviral DNA via triple-helix formation. Nucleic Acids Res 20:4275–4281

Giovannangeli C, Thuong NT, Helene C (1993) Oligonucleotide clamps arrest DNA synthesis on a single-stranded DNA target. Proc Natl Acad Sci U S A 90:10013–10017

Goodman SD, Nash HA (1989) Functional replacement of a protein-induced bend in a DNA recombination. Nature 341:251–254

Grant KB, Dervan PB (1996) Sequence-specific alkylation and cleavage of DNA mediated by purine motif triple helix. Biochemistry 35:12313–12319

Greenberg WA, Dervan PB (1995) Energetics of formation of sixteen triple helical complexes which vary at a single position within a purine motif. J Am Chem Soc 117:5016–5022

Griffin LC, Dervan PB (1989) Recognition of thymine adenine base pairs by guanine in a pyrimidine triple helix motif. Science 245:967–970

Griffin LC, Kiessling LL, Beal PA, Gillespie P, Dervan PB (1992) Recognition of all four base pairs of double-helical DNA by triple-helix formation: design of nonnatural deoxyribonucleosides for pyrimidine:purine base pair binding. J Am Chem Soc 21:7976–7982

Grigoreiv M, Praseuth D, Robin P, Hemar A, Saison-Behmoaras T, Dautry-Varsat A, Thuong NT, Helene C, Harel-Bellan A (1992) A triple helix-forming oligonucleotide-intercalator conjugate acts as a transcriptional repressor via inhibition of NF kappa B binding to interleukin-2 receptor alpha-regulatory sequence. J Biol Chem 267:3389–3395

Grigoreiv M, Praseuth D, Guieysse AL, Robin P, Thuong NT, Helene C, Harel-Bellan A (1993) Inhibition of gene expression by triple helix-directed DNA cross-linking at specific sites. Proc Natl Acad Sci U S A 90:3501–3505

Gryaznov SM, Lloyd DH, Chen JK, Schultz RG, DeDionisio LA, Ratmeryer L, Wilson WD (1995) Oligonucleotide N3'–>P5' phosphoramidates. Proc Natl Acad Sci U S A 92:5798–5802

Guieysse AL, Paseuth D, Francois JC, Helene C (1995) Inhibition of replication initiation by triple helix-forming oligonucleotides. Biochem Biophys Res Commun 217:186–194

Gunther EJ, Havre PA, Gasparro FP, Glazer PM (1996) Triple-mediated, in vitro targeting of psoralen photoadducts within the genome of a transgenic mouse. Eur J Biochem 63:207–212

Guschlbauer W, Duplaa AM, Guy A, Teoule R, Fazakerley GV (1991) Structure and in vitro replication of DNA templates containing 7,8-dihydro-8-oxoadenine. Nucleic Acids Res 19:1753–1758

Hacia JG, Dervan PB, Wold BJ (1994) Inhibition of Klenow fragment DNA polymerase on double-helical templates by oligonucleotide-directed triple-helix formation. Biochemistry 33:6192–6200

Hampel KJ, Crosson P, Lee JS (1991) Polyamines favor DNA triplex formation at neutral pH. Biochemistry 30:4455–4459

Han H, Dervan PB (1993) Sequence-specific recognition of double helical RNA and RNA.DNA by triple helix formation. Proc Natl Acad Sci U S A 90:3806–3818

Hardenbol P, van Dyke MW (1996) Sequence specificity of triplex DNA formation: analysis by a combinational approach, restriction endonuclease protection selection and amplification. Proc Natl Acad Sci U S A 93:2811–2816

Hausheer FH, Singh UC, Saxe JD, Colvin OM, T'so POP (1990) Can oligonucleotide methylphosphonates form a stable triplet with a double DNA helix? Anticancer Drug Des 5:159–167

Hausheer FH, Singh UC, Saxe JD, Flory JP, Tufto KB (1992) Thermodynamic and conformational characterization of 5-methylcytosine-versus cytosine-substituted oligomers in DNA triple helices: ab initio quantum mechanical and free energy perturbation studies. J Am Chem Soc 114:5356–5362

Havre PA, Glazer PM (1993a) Targeted mutagenesis of simian virus 40 DNA mediated by a triple helix-forming oligonucleotide. J Virol 67:7324–7331

Havre PA, Glazer PM (1993b) Targeted mutagenesis of DNA using triple helix-forming oligonucleotides linked to psoralen. Proc Natl Acad Sci U S A 90:7879–7883

Helm CW, Shresta K, Thomas S, Shingleton HM, Miller DM (1993) A unique c-myc-targeted triplex-forming oligonucleotide inhibits the growth of ovarian and cervical carcinomas in vitro. Gynecol Oncol 49:339–343

Higuchi S, Tsuboi M, Iitaka Y (1969) Infrared spectrum of a DNA-RNA hybrid. Biopolymers 7:909–916

Holland JA, Hoffman DW (1996) Structural features and stability of an RNA triple helix in solution. Nucleic Acids Res 24:2841–2848

Hopkins HP, Hamilton DD, Wilson WD, Zon G (1993) Duplex and triple helix formation with dA19 and dT19, thermodynamic parameters from calorimetric, NMR, and circular dichroism. J Phys Chem 97:6553–6563

Horne DA, Dervan PB (1990) Recognition of mixed-sequence duplex DNA by alternate-strand triple-helix formation. J Am Chem Soc 112:2435–2437

Horne DA, Dervan PB (1991) Effects of an abasic site on triple helix formation characterized by affinity cleaving. Nucleic Acids Res 19:4963–4965

Howard FB, Miles HT, Liu K, Frazier J, Raghunathan G, Sasisekharan V (1992) Structure of d(T)n.d(A)n.d(T)n: the DNA triple helix has B-form geometry with C2'-endo sugar pucker. Biochemistry 31:10671–10677

Howard FB, Miles HT, Ross PD (1995) The poly(dT).2poly(dA) triple helix. Biochemistry 34:7135–7144

Huang CY, Miller PS (1993) Triple helix formation by an oligodeoxyribonucleotide containing N4-(6-aminopyridinyl)-2'-deoxyctyditide. J Am Chem Soc 115:10456–10457

Huang CY, Bi G, Miller PS (1996) Triplex formation by oligonucleotides containing novel deoxycytidine derivatives. Nucleic Acids Res 24:2606–2613

Hunziker J, Priestley ES, Brunar H, Dervan PB (1995) Design of an N7-glycosylated purine nucleoside for recognition of GC base pairs by triple helix formation. J Am Chem Soc 117:2661–2662

Husler PL, Klump HH (1995) Prediction of pH-dependent properties of DNA triple helices Arch Biochem Biophys 317:46–56

Ing NH, Beekman JM, Kessler DJ, Murphy M, Jayaraman K, Zendegui JG, Hogan ME, O'Malley BW, Tsai MJ (1993) In vivo transcription of a progesterone-responsive gene is specifically inhibited by a triple helix-forming oligonucleotide. Nucleic Acid Res 21:2789–2796

Jayasena S, Johnston BH (1992a) Intramolecular triple-helix formation at (PunPyn).(PunPyn) tracts: recognition of alternate strands via Pu.PuPy and Py.PuPy base triplets. Nucleic Acids Res 20:5279–5288

Jayasena S, Johnston BH (1992b) Oligonucleotide-directed triple helix formation at adjacent oligopurine and oligopyrimidine DNA tracts by alternate strand recognition. Biochemistry 31:320–327

Jayasena S, Johnston BH (1993) Sequence limitations of triple helix formation by alternate-strand recognition. Biochemistry 32:2800–2807

Jetter MC, Hobbs FW (1993) 7,8-Dihydro-8-oxoadenine as a replacement for cytosine in the third strand of triple helices. Triplex formation without hypochromicity. Biochemistry 32:3249–3254

Jin R, Breslauer KJ, Jones RA, Gaffney BL (1990) Tetraplex formation of a guanine-containing nanomeric DNA fragment. Science 250:543–544

Jin R, Gaffney BL, Wang, C, Jones RA, Breslauer KJ (1992) Thermodynamics and structure of a DNA tetraplex: spectroscopic and calorimetric study of the tetramer complexes of d(TG3T) and d(TG3T2G3T). Proc Natl Acad Sci U S A 89:8832–8836

Jin R, Chapman WH Jr, Srinivasan AR, Olson WK, Breslow R, Breslauer KJ (1993) Comparative spectroscopic, calorimetric, and computational studies of nucleic acid complexes with 2',5"-versus 3',5"-phosphodiester linkage. Proc Natl Acad Sci U S A 90:10568–10572

Johnson KH, Durland RH, Hogan ME (1992) The vacuum UV CD spectra of G.G.C triplexes. Nucleic Acids Res 20:3859–3864

Jones RJ, Swaminathan S, Milligan JF, Wadwani S, Froehler BC, Matteucci MD (1993) Oligonucleotides containing a covalent conformationally restricted phosphodiester analog for high-affinity triple helix formation: the riboacetal internucleotide linkage. J Am Chem Soc 115:9816–9817

Joshi RR, Ganesh KN (1994) Duplex and triplex directed DNA cleavage by oligonucleotide-Cu(II)/Co(III) metallodesferal conjugates. Biochem Biophys Acta 1201:454–460

Kamiya M, Trigoe H, Shindo H, Sarai A (1996) Temperature dependence and sequence specificity of DNA triplex formation: an analysis using isothermal titration calorimetry. J Am Chem Soc 118:4532–4538

Kandimalla ER, Agrawal S (1994) Single-strand-targeted triplex formation: stability, specificity and RNase H activation properties. Gene 149:115–121

Kandimalla ER, Agrawal S (1995a) Single strand targeted triplex formation: parallel-stranded DNA hairpin duplexes for targeting pyrimidine strands. J Am Chem Soc 117:6416–6417

Kandimalla ER, Agrawal S (1995b) Single strand targeted triplex-formation. Destabilization of guanine quadruplex structures by foldback triplex-forming oligonucleotides. Nucleic Acids Res 23:1068–1074

Kandimalla ER, Manning AN, Venkataraman G, Sasisekharan V, Agrawal S (1995) Single strand targeted triple helix formation: targeting purine-pyrimidine mixed sequences using abasic linkers. Nucleic Acids Res 23:4510–4517

Kessler DJ, Pettitt BM., Cheng YK, Smith SR, Jayaraman K, Vu HM, Hogan ME (1993) Triple helix formation at distant sites: hybrid oligonucleotides containing a polymeric linker. Nucleic Acids Res 21:4810–4815

Kibler-Herzog L, Kell B, Zon G, Shinozuka K, Mizan S, Wilson WD (1990) Sequence dependent effects in methylphosphonate deoxyribonucleotide double and triple helical complexes. Nucleic Acids Res 18:3545–3555

Kiessling LL, Griffin LC, Dervan PB (1992) Flanking sequence effects within the pyrimidine triple-helix motif characterized by affinity cleaving. Biochemistry 31:2829–2834

Kim HG, Miller DM (1995) Inhibition of in vitro transcription by a triplex-forming oligonucleotide targeted to human c-myc P2 promoter. Biochemistry 34:8165–8171

Kim SG, Tsukahara S, Yokohama S, Takaku H (1992) The influence of oligodeoxyribonucleotide phosphorothioate pyrimidine strands on triplex stability. FEBS Lett 314:29–32

Klink R, Guittet E, Liquer J, Taillandier E, Gouyette C, Huynh-Dinh T (1994) Spectroscopic evidence for an intramolecular RNA triple helix. FEBS Lett 355:297–300

Klink R, Liquer J, Taillandier E, Gouyette C, Huynh-Dinh T, Guittet E (1995) Structural characterization of an intramolecular RNA triple helix by NMR spectroscopy. Eur J Biochem 233:544–533

Kochetkova M, Shannon MF (1996) DNA triplex formation selectively inhibits granulocyte-macrophage colony-stimulating factor gene expression in human T cells. J Biol Chem 271:14438–14444

Koh JS, Dervan PB (1992) Design of a nonnatural deoxyribonucleoside for recognition of GC base pairs by oligonucleotide-directed triple helix formation. J Am Chem Soc 114:1470–1478

Koob M, Szybalski W (1990) Cleaving yeast and Escherichia coli genomes at a single site. Science 250:271–273

Koob M, Grimes E, Szybalski W (1988) Conferring operator specificity on restriction endonuclease. Science 241:1084–1086

Kool ET (1991) Molecular recognition by circular oligonucleotides: increasing the selectivity of DNA binding. J Am Chem Soc 113:6265–6266

Kopel V, Pozener A, Baran N, Manor H (1996) Unwinding of the third strand of a DNA triple helix, a novel activity of the SV40 large T-antigen helicase. Nucleic Acids Res 24:330–335

Koshlap KM, Gillespie P, Dervan PB, Feigon J (1993) Nonnatural deoxyribonucleoside D3 incorporated in an intramolecular DNA triplex binds sequence-specifically by intercalation. J Am Chem Soc 115:7908–7909

Kovacs A, Kandala JC, Weber KT, Guntaka RV (1996) Triple helix-forming oligonucleotide corresponding to the polypyrimidine sequence in the rat alpha 1(I) collagen promoter specifically inhibits factor binding and transcription. J Biol Chem 271:1805–1812

Krawczyk SH, Milligan JF, Wadwani S, Moulds C, Froehler BC, Matteucci MD (1992) Oligonucleotide-mediated triple helix formation using an N3-protonated deoxycytidine analog exhibiting pH-independent binding within the physiological range. Proc Natl Acad Sci U S A 89:3761–3764

Laughton CA, Neidle S (1992a) Molecular dynamics simulation of the DNA triple helix d(TC)5.d(GA)5.d(C+T)5. Nucleic Acids Res 20:6535–6541

Laughton CA, Neidle S (1992b) Prediction of the structure of the Y+.R–.R(+)-type DNA triple helix by molecular modelling. J Mol Biol 223:519–529

Lavrovsky Y, Stoltz RA, Vlassov VV, Abraham NG (1996) c-fos protooncogene transcription can be modulated by oligonucleotide-mediated formation of triplex structures in vitro. Eur J Biochem 238:582–590

Le Doan T, Perrouault L, Praseuth D, Habhoub N, Decoudt JL, Thuong NT, Lhomme J, Helene C (1987) Sequence-specific recognition, photo-crosslinking, and cleavage of the DNA double helix by an oligo-[alpha]-thymidilate covalently linked to an azidoproflavine. Nucleic Acids Res 15:7749–7760

Lee JS, Johnson DA, Morgan AR (1979) Complexes formed by (pyrimidine)n.(purine)n DNAs on lowering the pH are three-stranded. Nucleic Acids Res 6:3073–3091

Lee JS, Woodsworth ML, Latimer LJP, Morgan AR (1984) Poly(pyrimidine).poly(purine) synthetic DNAs containing 5-methylcytosine form stable triplexes at neutral pH. Nucleic Acids Res 12:6603–6614

Lee JS, Latimer L, Hampel KJ (1993) Coralyne binds tightly to both T.A.T- and C.G.C(+)-containing DNA triplexes. Biochemistry 32:5591–5597

Lehrman EA, Crothers DM (1977) An ethidium-induced double helix of poly (dA)-poly (rU). Nucleic Acids Res 4:1382–1392

Liberles LS, Dervan PB (1996) Design of artificial sequence-specific DNA bending ligands. Proc Natl Acad Sci U S A 93:9510–9514

Lipsett MN (1963) The interactions of poly C and guanine trinucleotide. Biochem Biophys Res Commun 11:224–228

Lipsett MN (1964) Complex formation between polycytidylic acid and guanine oligonucleotides. J Biol Chem 239:1256–1260

Live DH., Radhakrishnan I, Mirsa V, Patel DJ (1991) Characterization of protonated cytidine in oligonucleotides by nitrogen-15 NMR studies at natural abundance. J Am Chem Soc 113:4687–4688

Lu M, Guo Q, Kallenbach NR (1993) Thermodynamics of G-tetraplex formation by telomeric DNAs. Biochemistry 32:598–601

Macaya RF, Gilbert DE, Malek S, Sinsheimer JS, Feigon J (1991) Structure and stability of X.G.C mismatches in the third strand of intramolecular triplexes. Science 254:270–274

Macaya RF, Schultze P, Feigon J (1992a) Sugar conformations in intramolecular DNA triplexes determined by coupling constants obtained by automated simulation of P.COSY cross peaks. J Am Chem Soc 114:781–783

Macaya RF, Wang E, Schultze P, Sklenar V, Feigon J (1992b) Proton nuclear magnetic resonance assignments and structural characterization of an intramolecular DNA triplex. J Mol Biol 225:755–773

Maher LJ III (1992) Inhibition of T7 RNA polymerase initiation by triple-helical DNA complexes: a model for artificial gene repression. Biochemistry 31:7587–75894

Maher LJ III, Wold B, Dervan PB (1989) Inhibition of DNA binding proteins by oligonucleotide-directed triple helix formation. Science 245:725–730

Maher LJ III, Dervan PB, Wold BJ (1990) Kinetic analysis of oligodeoxyribonucleotide-directed triple-helix formation on DNA. Biochemistry 29:8820–8826

Maher LJ III, Dervan PB, Wold BJ (1992) Analysis of promoter-specific repression by triple-helical DNA complexes in a eukaryotic cell-free transcription. Biochemistry 31:70–81

Malkov VA, Voloshin ON, Soyfer VN, Frank-Kamnenetskii MD (1993) Cation and sequence effects on stability of intermolecular pyrimidine-purine-purine triplex. Nucleic Acids Res 21:585–591

Manzini G, Xodo LE, Gasparotto D, Quadrifoglio F, van der Marel GA, van Boom JH (1990) Triple helix formation by oligopurine-oligopyrimidine DNA fragments. Electrophoretic and thermodynamic behavior. J Mol Biol 213:833–843

Marchand C, Bailly C, Nguyen CH, Bisagni E, Garestier T, Helene C, Waring MJ (1996) Stabilization of triple helical DNA by a benzopyridoquinoxaline intercalator. Biochemistry 35:5022–5032

Martin JP, Episona M (1993) Protein-induced bending as a transcriptional switch. Science 260:805–807

Matteucci M, Lin KY, Butcher S, Moulds C (1991) Deoxyoligonucleotides bearing neutral analogs of phosphodiester linkages recognize duplex DNA via triple-helix formation. J Am Chem Soc 113:7767–7768

Mayfield C, Millar D (1994) Effect of abasic linker substitution on triple helix formation, Sp1 binding and specificity in an oligonucleotide targeted to the human Ha-ras promoter. Nucleic Acids Res 22:1909–1916

Mayfield C, Ebbinghaus S, Gee JE, Jones D, Rodu B, Squibb M, Miller D (1994) Triplex formation by the human Ha-ras promoter inhibits Sp1 binding and in vitro transcription. J Biol Chem 269:18232–18238

McShan WM, Rossen RD, Laughter AH, Trial J, Kessler DJ, Zendegui JG, Hogan ME, Orson FM (1992) Inhibition of transcription of HIV-1 in infected human cells by oligodeoxynucleotides designed to form DNA triple helices. J Biol Chem 267:5712–5721

Mergny JL, Sun JS, Rougee M, Montenay-Garestier T, Barcelo F, Chomilier J, Helene C (1991a) Sequence specificity in triple-helix formation: experimental and theoretical studies of the effect of mismatches on triplex stability. Biochemistry 30:9791–9798

Mergny JL, Sun JS, Rougee M, Montenay-Garestier T, Helene C (1991b) Intercalation of ethidium bromide into a triple-stranded oligonucleotide. Nucleic Acids Res 19:1521–1526

Mergny JL, Duval-Valentin G, Nguyen CH, Perrouault L, Faucon B, Rougee M, Montenay-Garestier T, Bisagni C, Helene C (1992) Triple helix-specific ligands. Science 256:1681–1684

Mestre B, Jakobs A, Pratviel G, Meunier B (1996) Structure/nuclease activity relationships of DNA cleavers based on cationic metalloporphyrin-oligonucleotide conjugates. Biochemistry 35:9140–9149

Michel J, Toulme JJ, Vercauteren J, Moreau S (1996) Quinazoline-2,4(1H,3H)-dione as a substitute for thymine in triple-helix forming oligonucleotides: a reassessment. Nucleic Acids Res 24:1127–1135

Miller PS, Cushman CD (1993) Triplex formation by oligodeoxyribonucleotides involving the formation of X.U.A triads. Biochemistry 32:2999–3004

Miller PS, Dreon N, Pulford SM, McParland KB (1980) Oligothymidylate analogues having stereoregular, alternating methylphosphonate/phosphodiester backbones. Synthesis and physical studies. J Biol Chem 255:9659–9665

Miller PS, McParland KB, Jayaraman K, Ts'o POP (1981) Biochemical and biological effects of nonionic nucleic acid methylphosphonate. Biochemistry 20:1874–1880

Miller PS, Bhan P, Cushman CD, Trapane TL (1992) Recognition of a guanine-cytosine base pair by 8-oxoadenine. Biochemistry 31:6788–6793

Miller PS, Bi G, Kipp SA, Fok V, DeLong RK (1996) Triplex formation by a psoralen-conjugated oligodeoxyribonucleotide containing the base analog 8-oxo-adenine. Nucleic Acids Res 24:730–736

Milligan JF, Krawczyk SH, Wadwani S, Matteucci MD (1993) An anti-parallel triple helix motif with oligodeoxynucleotides containing 2′-deoxyguanosine and 7-deaza-2′-deoxyxanthosine. Nucleic Acids Res 21:327–333

Mohan V, Smith PE, Pettitt BM (1993) Evidence for a new spine of hydration: solvation of DNA triple helixes. J Am Chem Soc 115:9297–9298

Mooren MM, Pulleyblank DE, Wijmenga SS, Blommers MJ, Hilbers CW (1990) Polypurine/polypyrimidine hairpins form a triple helix structure at low pH. Nucleic Acids Res 18:6523–6539

Moser HE, Dervan PB (1987) Sequence-specific cleavage of double helical DNA by triplex formation. Science 238:645–650

Mouscadet JF, Carteau S, Goulaouic H, Subra F, Auclair C (1994a) Triplex-mediated inhibition of HIV DNA integration in vitro. J Biol Chem 269:21635–21638

Mouscadet JF, Ketterle C, Goulaouic H, Carteau S, Subra F, Bret ML, Auclair C (1994b) Triple helix formation with short oligonucleotide-intercalator conjugates matching the HIV-1 U3 LTR promoter end sequence. Biochemistry 33:4187–4196

Musso M, Van Dyke MW (1995) Polyamine effects on purine-purine-pyrimidine triple helix formation by phosphodiester and phosphorothioate oligodeoxyribonucleotides. Nucleic Acids Res 23:2320–2327

Neurath MF, Max EE, Strober W (1995) Pax5 (BSAP) regulates the murine immunoglobulin 3′ alpha enhancer by suppressing binding of NF-alpha P, a protein that controls heavy chain transcription. Proc Natl Acad Sci U S A 92:5336–5340

Noonberg SB, Scott GK, Hunt A, Hogan ME, Benz CC (1994a) Inhibition of transcription factor binding to the HER2 promoter by triplex-forming oligonucleotide. Gene 149:123–126

Noonberg SB, Scott GK, Hunt CA, Benz CC (1994b) Detection of triplex-forming RNA oligonucleotides by triplex blotting. Biotechniques 16:1070–1072

Noonberg SB, Francois JC, Garestier T, Helene C (1995) Effect of competing self-structure on triplex formation with purine-rich oligodeoxynucleotides containing GA repeats. Nucleic Acids Res 23:1956–1963

Olivas WM, Maher LJ III (1994) DNA recognition by alternate strand triple helix formation: affinities of oligonucleotides for a site in the human p53 gene. Biochemistry 33:983–991

Olivas WM, Maher LJ III (1995a) Overcoming potassium-mediated triplex inhibition. Nucleic Acids Res 23:1936–1941

Olivas WM, Maher LJ III (1995b) Competitive triplex/quadruplex equilibria involving guanine-rich oligonucleotides. Biochemistry 34:278–284

Olivas WM, Maher II LJ (1996) Binding of DNA oligonucleotides to sequences in the promoter of the human bcl-2 gene. Nucleic Acids Res 24:1758–1764

Ono A, Chen CH, Kan LS (1991a) DNA triplex formation of oligonucleotide analogue consisting of linker groups and octamer segments that have opposite sugar-phosphate backbone polarities. Biochemistry 30:9914–9921

Ono A, Ts'o POP, Kan LS (1991b) Triplex formation of oligonucleotides containing 2'-O-methylpseudoisocytidine in substitution for 2'-deoxycytidine. J Am Chem Soc 113:4032–4033

Ono A, Ts'o POP, Kan LS (1992) Triplex formation of an oligonucleotide containing 2'-O-methylpseudoisocytidine with a DNA duplex at neutral pH. J Org Chem 57:3225–3231

Orson FM, Thomas DW, McShan WM, Kessler DJ, Hogan ME (1991) Oligonucleotide inhibition of IL2R alpha mRNA transcription by promoter region collinear triple helix formation in lymphocytes. Nucleic Acid Res 19:3435–3441

Orson FM, Kinsey BM, McShan WM (1994) Linkage structures strongly influence the binding cooperativity of DNA intercalators conjugated to triplex forming oligonucleotides. Nucleic Acid Res 22:479–484

Ouali M, Letellier R, Sun JS, Akhebat A, Adnet F, Liquier J, Taillandier E (1993a) Determination of G*G:C triple-helix structure by molecular modeling and vibrational spectroscopy. J Am Chem Soc 115:4264–4270

Ouali M, Letellier R, Sun JS, Adnet F, Liquier J, Sun JS, Lavery R, Taillandier E (1993b) A possible family of B-like triple helix structures: comparison with the Arnott A-like triple helix. Biochemistry 32:2098–2103

Park YW, Breslauer KJ (1992) Drug binding to higher ordered DNA structures: netropsin complexation with a nucleic acid triple helix. Proc Natl Acad Sci U S A 89:6653–6657

Pegg AE (1988) Polyamine metabolism and its importance in neoplastic growth and a target for chemotherapy. Cancer Res 48:759–774

Pei D, Corey DR, Schultz PG (1990) Site-specific cleavage of duplex DNA by a semisynthetic nuclease via triple-helix formation. Proc Natl Acad Sci U S A 87:9858–9862

Perrouault L, Asseline U, Rivalle C, Thuong NT, Bisagni E, Giovannangeli C, Doan TL, Helene C (1990) Sequence-specific artificial photo-induced endonucleases based on triple helix-forming oligonucleotides. Nature 344:358–360.

Pilch DS, Breslauer KJ (1994) Ligand-induced formation of nucleic acid triple helices. Proc Natl Acad Sci U S A 91:9332–9336

Pilch DS, Levenson C, Shafer RH (1990a) Structural analysis of the (dA)10.2(dT)10 triple helix. Proc Natl Acad Sci U S A 87:1942–1946

Pilch DS, Brousseau R, Shafer RH (1990b) Thermodynamics of triple helix formation: spectrophotometric studies on the d(A)10.2d(T) and d(C+3T4C+3).d-(G3A4G3).d(C3T4C3) triple helices. Nucleic Acids Res 18:5743–5750

Pilch DS, Levenson C, Shafer RH (1991) Structure, stability, and thermodynamics of a short intermolecular purine-purine-pyrimidine triple helix. Biochemistry 30:6081–6087

Pilch DS, Waring MJ, Sun JS, Rougee M, Nguyen CH, Bisagni E, Garestier T, Helene C (1993) Characterization of a triple helix-specific ligand. BePI(3-methoxy-7H-8-methyl-11-[(3′-amino)propylamino]-benzo[e]pyrido[4,3-b]indole) intercalators into both double-helical and triple-helical DNA. J Mol Biol 232:926–946

Plum GE, Breslauer KJ (1995) Thermodynamics of an intramolecular DNA triple helix: a calorimetric and spectroscopic study of the pH and salt dependence of thermally induced structural transition. J Mol Biol 248:679–695

Plum GE, Park YW, Singleton SF, Dervan PB, Breslauer KJ (1990) Thermodynamic characterization of the stability and the melting behavior of a DNA triplex: a spectroscopic and calorimetric study. Proc Natl Acad Sci U S A 87:9436–9440

Porumb H, Gousset H, Letellier R, Salle V, Briane D, Vassa J, Amor-Gueret M, Israel L, Taillandier E (1996) Temporary ex vivo inhibition of the expression of the human oncogene HER2 (NEU) by a triple helix-forming oligonucleotide Cancer Res 56:515–522

Postel EH (1992) Modulation of c-myc transcription by triplex formation. Ann N Y Acad Sci 660:57–63

Postel EH, Flint SJ, Kessler DJ, Hogan ME (1991) Evidence that a triplex-forming oligodeoxynucleotide binds to the c-myc promoter in HeLa cells, thereby reducing c-myc mRNA levels. Proc Natl Acad Sci U S A 88:8227–8231

Potaman VN, Sinden RR (1995) Stabilization of triple-helical nucleic acids by basic oligopeptide. Biochemistry 34:14885–14892

Povsic TJ, Dervan PB (1989) Triple helix formation by oligonucleotides on DNBA extended to the physiological pH. J Am Chem Soc 111:3059–3061

Povsic TJ, Dervan PB (1990) Sequence specific alkylation of double-helical DNA by oligonucleotide-directed triple-helix formation. J Am Chem Soc 112:9428

Povsic TJ, Strobel SA, Dervan PB (1992) Sequence-specific double-strand alkylation and cleavage of DNA mediated by triple helix formation. J Am Chem Soc 114:5943–5941

Prakash G, Kool ET (1991) Molecular recognition by circular oligonucleotides. Strong binding of single-stranded DNA and RNA. J Chem Soc Chem Commun 1991:1161–1163

Prakash G, Kool ET (1992) Structural effects in the recognition of DNA by circular oligonucleotides. J Am Chem Soc 114:3523–3527

Praseuth D, Doan TL, Chassignol M, Decout JL, Habhoub N, Lhomme J, Thuong NT, Helene C (1988a) Sequence-targeted photosensitized reactions in nucleic acids by oligo-alpha-deoxynucleotides and oligo-beta-deoxynucleotided covalently linked to proflavin. Biochemistry 27:3031–3038

Praseuth D, Perrouault L, Doan TL, Chassignol M, Thuong NT, Helene C (1988b) Sequence-specific binding and photocrosslinking of alpha and beta oligodeoxynucleotides to the major groove of DNA via triple-helix formation. Proc Natl Acad Sci U S A 85:1349–1353

Priestley ES, Dervan PB (1995) Sequence composition effects on the energetics of triple helix formation by oligonucleotides containing a designed mimic of protonated cytosine. Am Chem Soc 117:4761–4765

Radhakrishnan I, Patel DJ (1992a) Solution conformation of a G:TA triple in intramolecular pyrimidine:purine:pyrimidine DNA triplexes. J Am Chem Soc 114:6913–6915

Radhakrishnan I, Patel DJ (1992b) Three dimensional homonuclear NOESY-TOCSY of an intramolecular pyrimidine.purine.pyrimidine DNA triplex containing a central G.TA triple: nonexchangeable proton assignments and structural implications. Biochemistry 31:2514–2523

Radhakrishnan I, Patel DJ (1993) NMR structural studies on a nonnatural deoxyribonucleoside which mediates recognition of GC base pairs in pyrimidine-purine-pyrimidine DNA triplexes. Biochemistry 32:11228–11234

Radhakrishnan I, Patel DJ (1994a) Solution structure and hydration patterns of a pyrimidine.purine.pyrimidine DNA triplex containing a novel T.CG base-triple. J Mol Biol 241:600–619

Radhakrishnan I, Patel DJ (1994b) DNA triplexes: solution structures, hydration sites, energetics, interactions, and function. Biochemistry 33:11406–11416

Radhakrishnan I, Gao X, de los Santos C, Live D, Patel DJ (1991a) NMR structural studies of intramolecular (Y+)n.(R+)n.(Y–)n DNA triples in solution: Imino and amino proton and nitrogen markers of G.TA base triple formation. Biochemistry 30:9022–9030

Radhakrishnan I, Patel DJ, Gao X (1991b) NMR assignment strategy for DNA protons through three-dimensional proton-proton connectivities. Application to an intramolecular DNA triplex. J Am Chem Soc 113:8542–8544

Radhakrishnan I, de los Santos C, Live D, Patel DJ (1991c) Nuclear magnetic resonance structural studies of intramolecular purine.purine.pyrimidine DNA triplexes in solution. Base triple pairing alignments and strand direction. J Mol Biol 221:1403–1418

Radhakrishnan I, de los Santos C, Patel DJ (1993) Nuclear magnetic resonance structural studies of A.AT base triple alignments in intramolecular purine.purine.pyrimidine DNA triplexes in solution. J Mol Biol 234:188–197

Raha M, Wang G, Seidman MM, Glazer PM (1996) Mutagenesis by third-strand-directed psoralen adducts in repair-deficient human cells: high frequency and altered spectrum in a xeroderma pigmentosum variant. Proc Natl Acad Sci U S A 93:2941–2946

Rajagopal P, Feigon J (1989a) Triple-strand formation in the homopurine:homopyrimidine DNA oligonucleotides d(G-A)4 and d(T-C)4. Nature 339:637–640

Rajagopal P, Feigon J (1989b) NMR studies of triple-strand formation from the homopurine-homopyrimidine deoxyribonucleotides d(GA)4 and d(TC)4. Biochemistry 28:7859–7870

Rando RL, DePaolis L, Durland RH, Jayaraman K, Kessler DJ, Hogan ME (1994) Inhibition of T7 and T3 RNA polymerase directed transcription elongation in vitro. Nucleic Acids Res 22:678–685

Rao TS, Durland RH, Sethe DM, Myrick MA, Bodepudi V, Revankar GR (1995) Incorporation of 2'-deoxy-6-thioguanosine into G-rich oligodeoxyribonucleotides inhibits G-tetrad formation and facilitates triplex formation. Biochemistry 34:765–772

Reardon JT, Spielman P, Huang JC, Sastry S, Sancar A, Hearst JE (1991) Removal of psoralen monoadducts and crosslinks by human cell free extracts. Nucleic Acids Res 19:4623–4629

Reddoch JF, Miller DM (1995) Inhibition of nuclear protein binding to two sites in the murine c-myc promoter by intermolecular triplex formation. Biochemistry 34:7659–7667

Rees WA, Keller RW, Vesenka JD, Yang G, Bustamante C (1993) Evidence of DNA bending in transcriptional complexes imaged by scanning force microscopy. Science 260:1646–1649

Reynolds MA, Arnold LJ Jr, Almazan MT, Beck TA, Hogrefe RI, Metzeler MD, Stoughton SR, Tseng BY, Trapane TL, Ts'o POP, Woolf TM (1994) Triple strand forming methylphosphonate oligodeoxynucleotides targeted to mRNA efficiently block protein synthesis. Proc Natl Acad Sci U S A 91:12433–12437

Roberts RW, Crothers DM (1991) Specificity and stringency in DNA triplex formation. Proc Natl Acad Sci U S A 88:9397–9401

Roberts RW, Crothers DM (1992) Stability and properties of double and triple helices: dramatic effects of RNA or DNA backbone composition. Science 258:1463–1466

Roberts RW, Crothers DM (1996) Prediction of the stability of DNA triplexes. Proc Natl Acad Sci U S A 93:4320–4325

Robles J, Rajur SB, McLaughlin LW (1996) A parallel-stranded DNA triplex tethering a Hoechst 33258 results in complex stabilization by simultaneous major groove binding and minor groove binding. J Am Chem Soc 118:5820–5821

Rougee M, Faucon B, Mergny JL, Barcelo F, Giovannangeli C, Garestier T, Helene C (1992) Kinetics and thermodynamics of triple-helix formation: effects of ionic strength and mismatches. Biochemistry 31:9269–9278

Roy O (1994) Triple helix formation interferes with the transcription and hinged DNA structure of the interferon-inducible 6-616 gene promoter. Eur J Biochem 220:493–503

Rumney S IV, Kool ET (1995) Structural optimization of non-nucleotide loop replacement for duplex and triplex DNA. J Am Chem Soc 117:5635–5646

Samadashwily GM, Mirkin SM (1994) Trapping DNA polymerases using triplex-forming oligonucleotides Gene 149:127–136

Samadashwily GM, Dayn A, Mirkin SM (1993) Suicidal nucleotide sequence for DNA polymerization. EMBO J 12:4975–4983

Sandor Z, Bredberg A (1994) Repair of triple helix directed psoralen adducts in human cells. Nucleic Acids Res 22:2051–2056

Sanger (1984) Principles of Nucleic Acids Structures. Springer, Berlin Heidelberg New York

Sawadogo M, Roder RG (1985) Factors involved in specific transcription by human polymerase II: analysis by a rapid and quantitative in vitro assay. Proc Natl Acad Sci U S A 82:4394–4398

Scaggiante B, Morassuti C, Tolazzi G, Michelutti A, Baccarani M, Quadrifoglio F (1994) Effect of unmodified triple helix-forming oligodeoxyribonucleotide targeted to human multidrug-resistance gene mdr1 in MDR cancer cells. FEBS Lett 352:380–384

Scaria PV, Shafer RH (1991) Binding of ethidium bromide to a DNA triple helix. Evidence for intercalation. J Biol Chem 266:5417–5423

Scaria PV, Shafer RH (1996) Calorimetric analysis of triple helices targeted to the d(G3A4G3).d(C3T4C3) duplex. Biochemistry 35:10985–10994

Scaria PV, Shire SJ, Shafer RH (1992) Quadruplex structure of d(G3T4G3) stabilized by K^+ or Na^+ is an asymmetric hairpin dimer. Proc Natl Acad Sci U S A 89:10336–10340

Scaria PV, Will S, Levenson C, Shafer RH (1995) Physicochemical studies of the d(G3T4G3).d(G3A4G3).d(C3T4C3) triple helix. J Biol Chem 270:7295–7303

Semerad CL, Maher LJ III (1994) Exclusion of RNA strands from a purine motif triple helix. Nucleic Acids Res 22:5321–5325

Sen D, Gilbert W (1990) A sodium-potassium switch in the formation of four stranded G4 DNA. Nature 344:410–414

Shaw JP, Kent K, Bird J, Fishback J, Froehler B (1991a) Modified deoxyoligonucleotide stable to exonuclease degradation in serum. Nucleic Acids Res 19:747–750

Shaw JP, Milligan JF, Krawczyk SH, Matteucci M (1991b) Specific, high efficiency, triple-helix-mediated cross-linking to duplex DNA. J Am Chem Soc 113:7765

Shimizu M, Inoue H, Ohtsuka E (1994) Detailed study of sequence-specific DNA cleavage of triplex-forming oligonucleotides linked to 1,10-phenanthroline. Biochemistry 33:606–613

Shimizu M, Morioka H, Inoue H, Ohtsuka E (1996) Triplex-mediated cleavage of DNA by 1,10-phenanthroline-linked 2'-O-methy RNA. FEBS Lett 384:207–210

Shin C, Koo HS (1996) Helical periodicity of GA-alternating triple-stranded DNA. Biochemistry 35:968–972

Shindo H, Trigoe H, Sarai A (1993) Thermodynamic and kinetic studies of DNA triplex formation of an oligopyrimidine and a matched duplex by filter binding assay. Biochemistry 32:8963–8969

Shore D, Langowski J, Baldwin R (1981) DNA flexibility studied by covalent closure of short fragments into circles. Proc Natl Acad Sci U S A 78:4833–4837

Singleton SF, Dervan PB (1992a) Influence of pH on the equilibrium association constants for oligodeoxyribonucleotide-directed triple helix formation at single DNA site. Biochemistry 31:10995–11003

Singleton SF, Dervan PB (1992b) Thermodynamics of oligodeoxyribonucleotide-directed triple helix formation: an analysis using quantitative affinity cleavage titration. J Am Chem Soc 114:6957–6965

Singleton SF, Dervan PB (1993) Influence of pH on the equilibrium association constants for oligodeoxyribonucleotide-directed triple helix formation at single DNA sites. Biochemistry 32:13171–13179

Singleton SF, Dervan PB (1994) Temperature dependence of the energetics of oligonucleotide-directed triple-helix formation at a single DNA site. J Am Chem Soc 116:10376–10382

Sklenar V, Feigon J (1990) Formation of a stable triplex from a single DNA strand. Nature 345:836–838

Skoog JU, Maher LJ III (1993a) Repression of bacteriophage promoters by DNA and RNA oligonucleotides. Nucleic Acids Res 21:2131–2138

Skoog JU, Maher LJ III (1993b) Relief of triple-helix-mediated promoter inhibition by elongating RNA polymerases. Nucleic Acids Res 21:4055–4058

Song CS, Jung MH, Supakar PC, Chen S, Vellanoweth RL, Chatterjee B, Roy AK (1995) Regulation of androgen action by receptor gene inhibition. Ann N Y Acad Sci 761:97–108

Stilz HU, Dervan PB (1993) Specific recognition of CG base pairs by 2-deoxynebularine within the purine.purine.pyrimidine triple-helix motif. Biochemistry 32:2177–2185

Stonehouse TJ, Fox KR (1994) DNase I footprinting of triple helix formation at polypurine tracts by acridine-linked oligopyrimidines: stringency, structural changes and interaction with minor groove binding ligands. Biochem Biophys Acta 1218:322–330

Strobel SA, Dervan PB (1990) Site-specific cleavage of a yeast chromosome by oligonucleotide-directed triple-helix formation. Science 249:73–75

Strobel SA, Dervan PB (1991) Single-site enzymatic cleavage of yeast genomic DNA mediated by triple helix formation. Nature 350:172–174

Strobel SA, Moser HE, Dervan PB (1988) Double strand cleavage of genomic DNA at a single site by triple helix formation. J Am Chem Soc 110:7927–7929

Strobel SA, Doucette-Stamm LA, Riba L, Housman DE, Dervan PB (1991) Site-specific cleavage of human chromosome 4 mediated by triple-helix formation. Science 254:1639–1642

Sun JS, Lavery R (1992) Strand orientation of [alpha]-oligodeoxynucleotides in triple helix structures: dependence on nucleotide sequence. J Mol Recognit 5:93–98

Sun JS, Francois JC, Montenay-Garestier T, Saison-Behmoaras T, Roig V, Thuong NT, Helene C (1989) Sequence specific intercalating agents: intercalation at specific sequences on duplex DNA via major groove recognition by oligonucleotide-intercalator conjugates. Proc Natl Acad Sci U S A 86:9198–9202

Sun JS, Giovannangeli C, Francois JC, Kurfurst R, Montenay-Garestier T, Asseline U, Saison-Behmoaras T, Thuong NT, Helene C (1991) Triple-helix formation by alpha oligodeoxynucleotides and alpha oligodeoxynucleotide-intercalator conjugates. Proc Natl Acad Sci U S A 88:6023–6027

Svinarchuk F, Monnot M, Merle A, Malvy C, Fermandjian S. (1995a) Investigation of the intracellular stability and formation of a triple helix formed with a short purine oligonucleotide targeted to the murine c-pim-1 proto-oncogene. Nucleic Acid Res 23:3831–3836

Svinarchuk F, Paoletti J, Malvy C (1995b) An unusually stable purine(purine-pyrimidine) short triplex. The third strand stabilized double-stranded. DNA J Biol Chem 270:14068–14071

Szewczyk JW, Baird EE, Dervan PB (1996) Cooperative triple-helix formation via sequence specific minor groove dimerization domain. J Am Chem Soc 118:6778–6779

Tabor CW, Tabor H (1984) Polyamines. Annu Rev Biochem 53:749–791

Takasugi M, Guendouz A, Chassignol M, Decout JL, Lhomme J, Thuong NT, Helene C (1991) Sequence-specific photo-induced cross-linking of the two strands of double-helical DNA by a psoralen covalently linked to a triple helix-forming oligonucleotide. Proc Natl Acad Sci U S A 88:5602–5606

Thomas T, Thomas TJ (1993) Selectivity of polyamines in triplex DNA stabilization. Biochemistry 32:14068–14074

Thomas TJ, Faaland CA, Gallo MA, Thomas T (1995) Suppression of c-myc oncogene expression by a polyamine-complexed triplex forming oligonucleotide in MCF-7 breast cancer cells. Nucleic Acid Res 23:3594–3599

Trapane TL, Hogrefe RI, Reynolds MA, Kan LS, Ts'o POP (1996) Interstrand complex formation of purine oligonucleotides and their nonionic analogs: the model system of d(AG)8 and its complement d(CT)8. Biochemistry 35:5495–5508

Tsuboi M (1969) Application of infrared spectroscopy to structure studies of nucleic acids II. Assignments of the absorption bands. Appl Spectrosc Rev 3:54–55

Tu GC, Cao QN, Israel Y (1995) Inhibition of gene expression by triple helix formation in hepatoma cells. J Biol Chem 270:28402–28407

Tuite E, Norden B (1995) Intercalative interactions of ethidium dyes with triplex structures. Bioorg Med Chem 3:701–711

Tung CH, Breslauer KJ, Stein S (1993) Polyamine-linked oligonucleotides for DNA triple helix formation. Nucleic Acids Res 21:5489–5494

Tung CH, Breslauer KJ, Stein S (1996) Stabilization of DNA triple-helix formation by appended cationic peptides. Bioconjug Chem 7:529–531

Uhlmann E, Peyman A (1990). Chem Rev 90:544

Umemoto K, Sarma MH, Gupta G, Luo J, Sarma RH (1990) Structure and stability of a DNA triple helix in solution: NMR studies on d(T)6:d(A)6:d(T)6 and its complex with a minor groove binding drug. J Am Chem Soc 112:4539–4545

van Dongen MJP, Heus HA, Wymenga SS, van der Marel GA, van Boom JH, Hilbers CW (1996) Unambiguous structure of characterization of a DNA-RNA triple helix by 15N- and 13C-filtered NOESY spectroscopy. Biochemistry 35:1733–1739

Van Meervelt L, Vlieghe D, Dautant A, Gallois B, Precigoux G, Akennard D (1995) High-resolution structure of a DNA helix forming (C.G)*G base triplets. Nature 374:742–744

Vasquez KM, Wensel, TG, Hogan ME, Wilson JH (1995) High affinity triple helix formation by synthetic oligonucleotides at a site within a selectable mammalian gene. Biochemistry 34:7243–7351

Vasquez KM, Wensel TG, Hogan ME, Wilson JH (1996) High efficiency triple-helix mediated photo-cross-linking at a target site within a selectable mammalian gene. Biochemistry 35:10712–10719

Vigneswaran N, Mayfield CA, Rodu B, James R, Kim HG, Miller DM (1996) Influence of GC and AT specific DNA minor groove binding drugs on intermolecular triple helix formation in the human c-Ki-ras promoter. Biochemistry 35:1106–1114

Vlassov VV, Gaidamakov SA, Zarytova VF, Knorre DG, Levina AS, Nekona AA, Podust LM, Fedorova OS (1988) Sequence-specific chemical modification of double-stranded DNA with alkylating oligodeoxyribonucleotide derivatives. Gene 72:313–322

Vlieghe D, Meervelt LV, Dautant A, Gallois B, Precigoux G, Kennard O (1996) Parallel and antiparallel (G:GC)2 triple helix fragment in a crystal structure. Science 273:1702–1705

Vo T, Wang S, Kool ET (1995) Targeting pyrimidine single strands by triple helix formation: structural optimization of binding. Nucleic Acids Res 23:2937–2944

Volker J, Botes DP, Lindsey GG, Klump HH (1993) Energetics of a stable intramolecular DNA triple helix formation. J Mol Biol 230:1278–1290

von Krosigk U, Benner SA (1995) pH-independent triple helix formation by an oligonucleotide containing a pyrazine donor-donor-acceptor base. J Am Chem Soc 117:5361–5362
Wang G, Glazer PM (1995) Altered repair of targeted psoralen photoadducts in the context of an oligonucleotide-mediated triple helix. J Biol Chem 270:22595–22601
Wang S, Kool ET (1994a) Circular RNA oligonucleotides. Synthesis, nucleic acid binding properties, and a comparison with circular DNAs. Nucleic Acids Res 22:2326–2333
Wang S, Kool ET (1994b) Recognition of single-stranded nucleic acids by triplex formation. The binding of a pyrimidine-rich sequence. J Am Chem Soc 116:8857–8858
Wang S, Kool ET (1995) Relative stabilities of triple helices composed of combinations of DNA, RNA, and 2'-O-methyl-RNA backbones: chimeric circular oligonucleotides as probes. Nucleic Acids Res 23:1157–1164
Wang E, Malek S, Feigon J (1992) Structure of a G.T.A triplet in an intramolecular DNA triplex. Biochemistry 31:4838–4846
Wang Q, Tsukahara S, Yamakawa H, Takai K, Takaku H (1994) pH-independent inhibition of restriction endonuclease cleavage via triple helix formation by oligonucleotides containing 8-oxo-2'-deoxyadenosine. FEBS Lett 355:11–14
Wang S, Friedman A, Kool ET (1995) Origins of high sequence selectivity: a stopped-flow kinetics study of DNA/RNA hybridization by duplex- and triplex-forming oligonucleotides. Biochemistry 34:9774–9784
Wang E, Koshlap KM, Gillespie P, Dervan PB, Feigon J (1996a) Solution structure of a pyrimidine-purine-pyrimidine triplex containing the sequence-specific intercalating no-natural base D3. J Mol Biol 257:1052–1069
Wang G, Seidman MM, Glazer PM (1996b) Mutagenesis in mammalian cells induced by triple helix formation and transcription-coupled repair. Science 271:802–805
Ward B (1996) Type IIS restriction enzyme footprinting I. Measurement of a triple helix dissociation constant with Eco57I at 25†C. Nucleic Acids Res 24:2435–2440
Waring MJ (1974a) Stabilization of two-stranded ribohomopolymer helices and destabilization of a three-stranded helix by ethidium bromide. Biochem J 143:484–486
Waring MJ (1974b). J Phys Chem 143:483–486
Washbrook E, Fox KR (1994) Comparison of antiparallel A.AT and T.AT triplets within an alternate strand DNA triple helix. Nucleic Acids Res 22:3977–3982
Weerasinghe S, Smith PE, Mohan V, Cheng YK, Pettitt BM (1995) Nanosecond dynamics and structure of a model DNA triple helix in saltwater solution. J Am Chem Soc 117:2147–2158
Westin L, Blomquist P, Milligan JF, Wrange R (1995) Triple helix DNA alters nucleosomal histone-DNA interactions and acts as a nucleosome barrier. Nucleic Acids Res 23:2184–2191
White AP, Powell JW (1995) Observation of the hydration-dependent conformation of the (dG)20(dG)20(dC)20 oligonucleotide triplex using FTIR spectroscopy. Biochemistry 34:1137–1142
Wilson WD, Tanious FA, Mizan S, Yao S, Kiselyov AS, Zon G, Strkowski L (1993) DNA triple-helix specific intercalators as antigene enhancers: unfused aromatic cations. Biochemistry 32:10614–10621
Wilson WD, Hopkins HP, Mizan S, Hamilton DD, Zon G (1994a) Thermodynamics of DNA triplex formation in oligomers with and without cytosine bases: influence of buffer species, pH, and sequence. J Am Chem Soc 116:3607–3608
Wilson WD, Mizan S, Tanius FA, Yao S, Zon G (1994b) The interaction of intercalators and groove-binding agents with DNA triple helical structures: the influence of ligand structure, DNA backbone modification, and sequence. J Mol Recognit 7:89–98
Wolffe AP (1994) Transcription: in tune with the histones. Cell 77:13–16
Wu HM, Crothers DM (1984) The locus of sequence-directed and protein-induced DNA bending. Nature 308:509–513

Xiang G, Soussou W, McLaughlin LW (1994) A new pyrimidine nucleoside (m5oxC) for the pH-independent recognition of G-C base pairs by oligonucleotide-directed triplex formation. J Am Chem Soc 116:11155–11156

Xiang G, Bogacki R, McLaughlin LW (1996) Use of a pyrimidine nucleoside that functions as a bidenate hydrogen bond donor for the recognition of isolated or contiguous G-C rich base pairs by oligonucleotide-directed triplex formation. Nucleic Acids Res 24:1963–1970

Xodo LE (1995) Kinetic analysis of triple-helix formation by pyrimidine oligodeoxynucleotides and duplex DNA. Eur J Biochem 228:918–926

Xodo LE, Manzini G, Quadrifoglio F (1990) Spectroscopic and calorimetric investigation on the DNA triplex formed by d(CTCTTCTTTCTTTTCTTTCTTCTC) and d(GAGAGAAAGA) at acidic pH. Nucleic Acids Res 18:3557–3564

Xodo LE, Alunni-Fabbroni M, Manzini G, Quadrifoglio F (1993) Sequence specific DNA triplex formation at imperfect homopurine-homopyrimidine sequences within a DNA plasmid. Eur J Biochem 212:395–401

Xodo LE, Alunni-Fabbroni M, Manzini G, Quadrifoglio F. (1994) . Nucleic Acids Res 22:3322–3330

Yang M, Ghosh SS, Millar DP (1994) Direct measurement of thermodynamics and kinetic parameters of DNA triple helix formation by fluorescence spectroscopy. Biochemistry 33:15329–15337

Yoon K, Hobbs CA, Koch J, Sardaro M, Kutny R, Weis AL (1992) Elucidation of the sequence-specific third-strand recognition of four Watson-Crick base pairs in a pyrimidine triple-helix motif: T.AT, C.GC, T.CG, and G.TA. Proc Natl Acad Sci U S A 89:3840–3844

Young SL, Krawczyk SH, Matteucci MD, Toole JJ (1991) Triple helix formation inhibits transcription elongation in vitro. Proc Natl Acad Sci U S A 88:10023–10026

Zhou BW, Puga E, Sun JS, Garestier T, Helene C (1995) Stable triple helixes formed by acridine-containing oligonucleotides with oligopurine tracts of DNA interrupted by one or two pyrimidines. J Am Chem Soc 117:10423–10428

Zinkel SS, Crothers DM (1987) DNA bend direction by phase sensitive detection. Nature 328:178–181

Subject Index

A
AACGTT 245
absorption, phosphorothioate pharmacokinetic properties 17, 160–165, 219, 220, 538–540
– local administration 164
– non-parenteral administration 164
– parenteral administration 161–163
– radioactivity, absorbed oligonucleotide-derived 538
accumulation, high accumulation 150
acetylcholin receptors 281, 282
ACh 267
– m1-m3, muscarinic 267, 269
– nicotinic ($\alpha3$, $\alpha4$, $\alpha7$, subunits) 267
adenine
– $C2$-carbon 61
– 7-deaza-7-substituted 63
– $N2$-adenine 62
adenosine derivates, triple helix 572
A1-adenosine receptor 429, 439
A3-adenosine receptor, G-proteins by antisense ODNs 351
adrenergic
– G-proteins by antisense ODNs 351, 361
– – α1A-adrenergic 351
– – α2-adrenergic 361
– receptors, 267, 269, 282, 283, 351
– – $\alpha2$ 267, 269, 351
– – G-proteins by antisense ODNs 351
aerosol administration 25
affinity constants 13
AIDS 205, 207, 445, 450
albumin 13, 106
ALDH2 585
2'-alkoxy 200
alkylating agents 35
– of $N7$ 589
– sequence-specific, triple helix 589
– 2'-O-alkyls 70
– 2'-O-alkyl chain 70
$\alpha4\beta7$ 375

αd$\beta2$ 375
alternative pathway 29, 182, 183
– alternative complement cascade 186
3'-amidate
– bis-fluoro 83
– modification 84, 85
amide-3
– bis-methoxy 83
– modification 82, 84
5'-C18 amine conjugate 34
5'-C18 amino 88
2'-O-aminopropoxy modification 70, 72
2'-O-aminopropyl (AP) modification 70, 72, 91
amphetamine-induced, antisense ODNs 317–322
amygdala, antisense ODNs 322–328
– kindling 326
androgen receptor 586
anemia 236
angiotensin AT1A, G-proteins by antisense ODNs 351
angiotensinogen 429, 435
– mRNA 435
anti-apoptotic effects, oligonucleotides 250
antibodies
– anti-ODNs 198
– ISIS 2302, antibody formation 511
anticancer activities, antisense oligonucleotides 395–425
anticoagulation activation/effects 161, 467
anti-HCMV 209
antisense
– antiviral, clinical (see there) 445–462
– – future development 455, 456
– cancer, antisense-based approaches 396
– documentation 490
– drugs

antisense
- – G proteins   341–369
- – molecular mechanisms of   5
- effects, specificities   356, 357
- mechanism   4, 103, 490
- medicinal chemistry   51–92
- molecules   217
- ODNs   169, 263–302
- oligonucleotides   129, 463–476
- oral and colorectal administrations, in animals   525–543
- specifity   4
- therapeutics   1, 169 ff.
- treatment stratigies, antisense-based   417
anti-TNF Ab (chimeric)   516
antitumor activities   395–425
- CPG 69846A in stealth liposomes   411, 415
- phase II   75
antiviral
- activities, clinical   445–462
- assays   3
- clinical research   450–455
- drugs, antiviral (approved in the US)   446
- future antisense antiviral development   455, 456
- pharmacokinetic assessment strategies   448, 449
- pharmacology, antisens oligo-nucleotide   447
- scal-up and development, chemical   450
- toxicology for antisense oligo-nucleotide antivirals   449, 450
AP-1 transcription   436
AP-c-*raf* protein   77
apoptosis   251
- anti-apoptotic effects, oligo-nucleotides   250
aPTT (activated partial thromboplastin time)   29, 188–193, 235, 239, 467–471, 500, 502, 510–516
- dose-related increase   502
- includet 2'-fluoro substituents and substitution   191
- prolongation   191, 467, 510
- transient increased   516
a-*raf*   465
arthritis, rheumatoid   500, 513, 517
asialglycoprotein-poly-L-lysine   434
association
- with polycations   128
- rates   13
asthma   429

atherosclerosis   429, 432–434
AUC (area under the plasma concentration curve)   186, 222, 223, 507, 513
AUG start codon   76
availability, phosphorothioates   20, 56, 162, 165, 173, 527–533
- mixed-backbone oligonucleotides (MBO)-1 and -2   527–538
- oral bioavailability   20, 56, 173, 527–540
- systemic   165
6-azathymine   64

B
B cells   193, 194, 197, 372
- human   197
- polyclonal   372
- proliferation   194
- stimulation   194, 197
backbone
- DNA and RNA   577
- mixed-backbone
- – oligonucleotide-1   529, 535, 537
- – oligonucleotide-2   532, 533, 536
- modifications   10, 37, 55, 577
- – chemical   577
- phosphodiester-phosphorothioate   187
- phosphorothioate   230, 254
- – immune effects   254
- research   86
- triple helix, backbone modification and strand switching   577–579
basic oligopeptides, triple helix   576, 577
Bb   185, 187
bcl-2   218, 400, 408, 487
bcr-abl   25, 400, 407, 485
- RNA   25
- transcripts   485
behavioral responses with antisense ODNs   331
benzopyridonindole derivated, triple helix   575
bFGF   401
bicyclonucleosides   69
binding
- affinity   74
- protein (*see there*)   3, 14, 17
bioavailability of phosphorothioate ODNs   20, 56, 162, 173, 527–543
- absolut   527–530
- antisense oligonucleotides   527–540
- oral/oral absorption   20, 56, 173, 527–533

Subject Index

blood
- clearance 145
- kinetics, distribution 145
- pressure 29

blood-brain barrier 17

B-lymphocyte proliferation (B cells) 27, 246–248, 250
- B-cell activation 247
- B-cell cytokine and Ig secretion by CpG DNA 251
- B-cell differentiation 253

bone marrow 205
- purging agents 484, 491, 492

bradycardia 29

bradykinin B2 439

b-*raf* 408, 465, 466

brain, antisense ODNs 276
- distribution in brain 276
- CNS (*see there*) 309–339
- immediate-early genes in brain function 309–311

C

C2′
- endo 564
- carbon of adenine 61
- position of hypoxanthine 62

C3′-endo 564

C3′a complement 185, 467, 510

C5′
- complement 29
- position 32

C5′a complement 185, 467, 503, 511
- split products 237, 503, 511

calcium
- intracellular calcium concentration, G-protein-mediated 354
- phosphate precipitation 128
- voltage-gated calcium channels in neuroendocrine cells, G-proteins 352–354

calf serum, fetal (FCS) 104

cancer targets for pharmaceutical intervention, antisense oligonucleotides 395, 396

capillary gel
- electropherogram of ISIS 1082 127
- electrophoresis 155–157, 449

5′-capping 8

carbohydrate modifications 55, 66

C2′-carbon of adenine 61

cardiac allograft 379, 380
- heterotopic 379
- rejection 380
- survival 379

cardiovascular
- activation 161
- collapse 180, 193
- effects, phosphorothioate oligonucleotides 187, 239
- intolerance 234
- safety 233, 234

cationic
- lipids 16, 105, 128
- lipid formulation 494
- liposomes 105, 129

CCK (cholecystokinin) 314

CDAI (*Crohn's* disease activity index) 515, 516

cdc-2/cdk-2 (cyclin-dependent kinases) 22, 429–433

cell(s)
- adhesion molecules 373–385
- association 107
- B cells (*see there*) 193, 194, 197, 372
- cellular uptake
- – internationalization and distribution 132, 436
- – in vitro 2, 103–133
- – in vivo 147–150
- HUVEC cells 77
- kupffer cells 176, 200, 202
- stability 2

central vein occlusion 435

CETP (cholesteryl ester transfer protein) 429, 434

c-*fos* 310, 312, 320, 324, 332, 333, 397, 406, 436, 437, 586
- antisense technology 332, 333
- expression 320, 324, 437
- family 312
- *fos*B 436
- striatum 315, 316

CG (cytosine-guanosine) 196, 197, 249
- combinations 196

CGE 224, 228, 230
- analysis 230

CGP
- 64128A 141, 409
- – antitumor activity 409
- – [3H]CGP 64128A 164
- 69845A 413, 414
- – antitumor activity 413, 414
- 69846A 141, 409, 414, 415
- – antitumor activity 409
- – stealth liposomes 411, 415
- 71849A 191, 414
- – antitumor activity 414
- – elimination 160

chamber infalmmation, anterior 29

chemotherapie, cytotoxic 465

chimera 2′-O-modification 72
chimeric
– oligonucleotides containing 2′-deoxyoligonucleotide 36
– strategy (gapmer technology) 72, 78
chiral 83, 106
– methylphosphonate bis-methoxy 83
cholecystokinin (CCK) 314
cholesterol
– conjugate phosphorothioate oligonucleotide 416, 438
– modification 89, 90
– – 5′-cholesterol modification 34, 90
– – PSs report 90
cholesteryl, CETP (cholesterylester transfer protein) 429, 434
chromosome structure by triple helix 581
c-jun 398, 436
c-junB 436
c-kit 487
clearance (CL) 145, 161, 178, 222
– blood clearance 145
– of unlabelled CGP 69846A 161
3′-cleavage 158
clinical antiviral activities (see antiviral activities) 445–462
clotting 29, 180, 189
– abnormalities 29
– cascade 180
– polyanions 189
CML (chronic myelogenous leukemia) 407, 408, 482, 487, 491
– cells 487, 491
– patients 484
CMV (cytomegalovirus) 29, 206, 446, 451, 452
– human (see HCMV) 206
– mutates 447
– ocular side effects 454
– retinitis 29, 207, 209, 451, 455
CMV-DNA 447
c-myb 22, 397, 400, 402, 408, 428–430, 477–492
– antisense ODNs 479–482, 488
– AS ODN 486, 487, 490, 492
– c-myb mRNA 428
– efficacy of c-myb ODNs in vivo, animal models 486, 487
– expression 408
– function 483
– human hematopoiesis, c-myb-encoded protein 479
– leukemia, clinical experience with ODNs 477–483
– myb-killing leukemic cells 487

– patient material 481, 482
– pharmacodynamic studies 488
– proto-oncogene 477
– sequence-specific 483
– targeting c-myb gene 479, 488
c-myc 397, 400, 402, 408, 429, 430, 585, 586
– c-myc mRNA 430
– triple helix, c-myc promotor 583
CNS, antisense oligonucleotides 309–339
– alternative modifications 328–330
– amygdala 322–328
– c-fos 315, 317–322
– – basal ganglia function 317
– – in the striatum 315
– – amphetamine-induced 317–322
– immediate-early genes in brain function 309–311
coagulation, savety, phosphorothioates 234, 235
$\alpha 1(1)$ collagen 585
colorectal administrations in animals, antisense oligonucleotides 525, 534–543
complement activation 29, 55, 161, 180–187, 193, 237, 467–471, 503
– alternative pathway 237, 467–471
– – cascade 186
– – in monkeys 237
– – in patients 467, 468
– C3′a 185, 467, 510
– C5′ 29
– C5′a 185, 467, 503, 511
– pathway 55
– plasma thresholds 185
– safety 237
– split products B′b, C3′a, C5′a 181, 185
– toxicity
– – acute 180–187
– – dose-limiting 471
Cmax (maximum concentration) 178, 186, 222, 223, 469
conjugation
– of intercalators 579
– of small compounds enhance triple helix stability 579, 580
corneal neovascularization 435
corticosteroid requirements, daily 516
CpG
– CpG DNA 251–253
– CpG motiv 27, 199, 245–257
– CpG ODN 254
– CpG PS-ODN 252
– dinucleotide 248, 249

Subject Index

- immune activation 247–250
- immune effects 250–254
- leukocyte activation by CpG 254
- oligonucleotides 252, 254
- palindromes, CpG-containing 246
- sequence-specific immune effects in ODNs 245–257
- transcriptional activation by CpG 256

c-*raf* kinase 77, 218, 399, 403, 408, 409, 465, 466
- AP-c-*raf* protein 77
- antisense oligonucleotides 465, 466
- c-*raf* mRNA
- – expression 466
- – inhibition 410, 412
- as cancer targets for antisense 408
- expression 409
- sequence 77
- toxicities, target-specific 466
- in vivo activity 410

Crohn's disease 222, 381, 513–520
- CDAI (*Crohn's* disease activity index) 515, 516
- ISIS 2302 513, 514
- steroid-dependent 516, 520

cyclin
- B1 429–433
- cyclin-dependent kinase activity 22, 429–433
- – cdc-2 22, 429–433
- – cdk-2 22, 429–433
- D1 402

cyclodextrins 130
cyclonucleosides 69
cyclosporin 380
cynomolgus monkey 382
CYP2B1 90
cytidine
- 5′-ME-cytidine 81
- 5′-position pyrimidine- and tricyclic cytidine-modified oligonucleotides 59
- triple helix, cytidine derivates 572

cytochrome P450 3A2 438
cytofectin 128
cytokine(s) 196, 243, 251–253, 427
- B-cell cytokine and Ig secretion by CpG DNA 251
- IL-12 252
- monocyte cytokine secretion by CpG DNA 252
- production 243
- PS-ODN, cytokine secretion 243

cytomegalovirus (*see* CMV / HCMV)
cytoplasmic targets 133

cytotoxic
- chemotherapie 465
- dermatitis 279

D

3-deaza purines 61
7-deaza
- guanine 33, 63
- inosine 33
- 7-Me-guanine 63

degradation 122
- 3′-degradation 158

delivery systems 273
5′-deoxy-5′-octadecylamine 89
3-(2′-deoxy-B-D-ribofuranosyl)pyrrolo 65

deoxyribonucleotides 10
- chimeric oligonucleotides containing 2′-deoxyoligonucleotide 36
- ribonucleotide 10

dermatitis
- contact dermatitis 500
- cytotoxic 379

development
- consequences, developmental 206
- embyonic 205

dextran sulfate, antitumor activity of GGP 69846A 411, 416

2-6-diaminopurine 33

distribution
- antisense ODNs 276
- blood kinetics 145
- cellular distribution, in vitro 2, 103–133
- distribution phase 145
- dose dependence 150
- – multiple dosing 153
- extravascular 223
- half-life 221
- intracellular (*see there*) 106, 109–112, 116–119
- methylphosphonate 112
- pharmacokinetic properties 145–147
- phase 511
- phosphodiesters 109, 110
- phosphorothioate (*see there*) 17, 116–119, 145–147
- sequence dependence 150, 151
- tissue distribution 145, 147
- volume of distribution 145, 511

DNA
- antisense 63
- backbone 577
- binding 592
- CMV-DNA 447

- CpG DNA 251
- DNA backbone, sequence-independent immune effects 243, 244
- flexibility 591
- gap 11
- PNA
- - PNA-DNA chimeras 553–557
- - strand invasion in double-stranded DNA 555, 556
- polymerases 14
- - α-polymerase 14
- - β-polymerase 14
- - χ-polymerase 14
- - δ-polymerase 14
- - inhibitors 447
- replication 10
- 4'-thia-DNA 67
- triple helix DNA 561–593
- - artificial DNA 591
- - DNA cleavage 589
- - DNA replication 587–589
dopamine receptors, antisense ODNs 268–271, 280, 283–295
- D1 269, 286
- - receptor agonists 315
- D1a 267
- D2 268–271, 280, 288, 290–294, 300, 351
- - G-proteins by antisense ODNs 351
- - intracerebroventricular treatment 290, 291
- - intrastriatal injection 291
- - intraventricular administration 288, 292
- D3 271, 294, 295
- - intraventricular infusion 294
- design and synthesis 286
- in vitro studies 289, 293, 294
- in vivo models/-studies 285, 286, 290
dopaminergic receptors, pharmacological inhibition 263–276
dose/dosing, phosphorothioate
- distribution, dose dependence 150
- dose-dependency of human plasma pharmacokinetics 222
- dose-limiting 470
- intra-tracheal dosing 165
- multiple dosing 153
- response curves 4
- subcutaneous dosing of ISIS 2302 519, 520
DOTAP 105
DOTMA 105

drug
- accumulation 471
- properties 55
- - multidrug resistance 405, 406

E
edema 519
efflux 106–113
- methylphosphonate oligonucleotides 112, 113
- phosphodieseter oligonucleotides 106, 110
- phosphorothioate oligonucleotides 119, 120
EGFR 398, 401
EIS 515
electrocardiographic (ECG) monitoring 501
electron microscopy, triple helix 565, 566
electropherogram
- of ISIS 1082, capillary gel 127
- of ISIS 2302 505
electrophoresis
- capillary gel 155, 449
- PAGE-(polyacrylamide gel electrophoresis)-autoradiography 144, 224
- phase-sensitive, triple helix 565
elimination, phosphorothioate 159, 160, 228–230
- CGP 160
- dependence 230
- pharmacokinetics 228–230
embryonic development 205
3'-endo
- conformation 67, 84
- sugar conformation 70
endocytosis/endocytotic process 3, 108, 112, 118, 133, 489, 490
- fluid-phase 112, 118
- receptor-mediated 3, 133, 489, 490
endonuclease / endonucleolytic activity 16, 75, 104, 123
endothel cell-leukocyte adhesion molecules 384
extrinsic pathways 188
enzymatic cleavage, single-site, triple helix 589
erythema 519
escherichia coli RNase H 11
ethidium bromide, triple helix 574, 575
EWS-erg 402
EWS-fli 402
exonuclease 16
- 3'-exonucleases 104, 153, 158
extracellular milieu, stability 104

F

factor H 182
FCS (fetal calf serum) 104
fluid phase 112, 118
– endocytosis 112, 118
– pinocytosis 112
5'-fluorescein-labled poly A PSs 79
2'-fluoro modification 10
fomivirsen 53, 446, 450–453
– intraocular 455
– intravitreal injection 450, 451
– response to 453
– sodium 446, 450
formacetal
– modified oligomers 85
– TFO 578
fos (*see* c-*fos*)
fos/jun TF 314
fra-1 312

G

G0 355
G3139 218, 220
– G3139e 218
GABA (y-aminobutyric acid) receptors, antisense ODNs 268, 271, 295–298, 319, 327
– α subunit 268, 271
– α6 subunit 268
– c-*fos* decreases 319
– τ2 subunit 268, 271, 297
GABAA 297
GABAB 268
– G-proteins by antisense ODNs 351
galanin, G-proteins by antisense ODNs 351
gapmer technology (chimeric strategy) 72, 78
gastrointestinal tract 530, 539, 540
GEM-91 18, 191, 200, 202, 205, 218–221, 224, 229, 233–239
– GEM-91B 200
– GEM-91C 200
– GEM-91H 200
– GEM-91 per kg 220
– clinical studies 234
– metabolized 224
GEM-132 164, 237
gene(s)/genetic
– gene walks 5, 409, 548
– genomic knockout, G-protein functions 345–350
– HIV rev gene 246
– immediate-early genes (IEGs) in brain function 309–311
– mechanisms of inhibition of gene expression 276, 277
– morphogenesis 205
– transcription 205
– toxicity, genetic 179
– triple helix, gene regulation 580
genital warts 29
genotoxicity 27
mGluR, G-proteins by antisense ODNs 351
– mGluR1 351
– mGluR5 351
N7-glycosylated guanosine 573
GM-CSF 585
G-proteins by antisense drugs 341–369
– antisense ODNs 350–369
– cell culture medium 358
– coupled receptor 342, 352
– expression 344, 362
– functions
– – by antisense ODNs 350–360
– – by antisense RNA 345
– – genomic knockout 345, 349
– Gα11 346, 362
– Gα12 346, 362
– inhibition, antisense ODNs 350, 360
– – in animals 360
– involvements detected by antisense ODNs (*table*) 351, 361
– – in animals (*table*) 360
– nuclear microinjection 350, 354
– opiod-receptors (*see there*) 361
– patch-clamp pipette, antisense ODNs 357
– techniques 344
– transcription of antisense RNA 347
– transduction pathways 344
– – signal transduction, G-protein--mediated inhibition 352–354
– voltage-gated calcium channels in neuroendocrine cells 352–354
grooves, major or minor 75, 576, 578, 579, 591
growth factors and growth factor receptors 404, 405, 427
– NGF (nerve growth factor) 311, 314
– VEGF (vascular endothelial growth factor) 401, 429, 436
GS2888 128
guanine 61
– 7-deaza 33, 63
– 7-deaza-7-Me-guanine 63
– 7-deaza-7-substituted 63
– 7-iodo-7-deaza-guanine 33, 63
– tetraplex formation by guanine-rich oligonucleotides, triple helix 570

G (guanosine) 2, 573
- triple helix, guanosine derivates 573
- - N7-glycosylated guanosine 573
- quartet 15, 121, 430
- - phosphodiesters, structured 121

H
3'-hairpin loop 34
Ha-*ras* 401
Harvey-ras sequence 463
HCMV (human cytomegalovirus) 206–209
- anti-HCMV 209
- IE2 206
- retinitis 29, 207, 209, 451, 455
- second generation 209
hematological
- malignancies 407, 408
- savety, phosphorothioates 234–237
- - coagulation 234, 235
- - platelets 235, 236
- - red blood cells 236, 237
hematopoiesis, ODNs 105, 193, 205, 479
- c-*myb*-encoded protein in normal human hematopoiesis 479
- extramedullary 193
- in vitro experience in hematopoietic cell system 479
hemodynamic
- activation 161
- changes 511
hemostasis, acute toxicity 187–191
hepatic/hepatocytes 200–202
- dysfunction 200
- effects, ODNs, toxicology treatment 200–202
- metabolism 158
hepatitis C virus 76
hepatocellular toxicities 200
hepatocytomegaly 193
HER2 585
heterocycle-modified oligonucleotides 32, 55, 56, 59, 64, 65
- nucleobases 59
heteroduplex 75
2'-hexylaminocarbonyloxy-cholesteryl 88
3'-hexylaminocarbonyloxy-cholesteryl 88
HIV (human immune deficiency virus) 7, 14, 21, 218, 246, 445, 586
- *gag/pol* 218
- proviral 586
- rev gene 246

- transcriptase, HIV-reverse 14
- triple helix, HIV integration 587
HIV-1 219
homopolymers 107, 122
Hoogsteen hydrogen 561, 564
- reverse *Hoogsteen* hydrogen 561, 564
HPLC 224
HPV (human papilloma virus) 29, 218, 219
- HPV-6/11 218, 219
hsp 70 402
5-HT 271, 299, 351
- 5-HT1c, G-proteins by antisense ODNs 351
- 5-HT2c, G-proteins by antisense ODNs 351
- 5-HT6, antisense ODNs, serotonin receptor 271, 299
humans 177, 510
- ISIS 2302 499–524
- ODNs 217–232, 233–258
- oligonucleotides, pharmacokinetic properties 217–232, 384
HUVEC cells 77
hybridise 141
hybridization 170
- PNA 545, 552
- specifity of 13
- *Watson-Crick* 20
hypertension 429, 434, 435
hypoxanthine, $C2$-position 62

I
ICAM (intercellular adhesion molecule) 9, 206
- ICAM-1 53, 62, 77, 79, 89, 129, 206, 218, 219, 235, 258, 372–387, 499, 514
- - antiflammary activity 387
- - antisense oligonucleotide 377, 382
- - contribution 378
- - cytokine-induced 500
- - expression 381, 499
- - function 500
- - inflammatory diseases 514
- - mechanisms 381
- - message 500
- - mRNA 89, 377, 378
- - monoclonal antibodies 499
- - toxicology 382
- ICAM-2 374
- ICAM-3 374
- ICAM-1-LFA-1 373

- pharmacology, antisense oligonucleotides 377
- - colitis 381, 382
- - human xenografts 378, 379
- - proof of mechanism 377
- - renal ischemia 382
- - rodent allografts 379
IE2 of HCMV 206
IEGs (immediate-early genes) 309, 312–314, 324
- inductible transcription factor 312, 313
- origin 312
- stimulus-transcription coupling 311, 312
- targets 313, 314
IFN (interferon)
- IFN-y 253, 258, 378
- production 245
Ig secretion, B-cell cytokine and Ig secretion by CpG DNA 251
IgA 198
IgE receptor 250
IGF-IR 398, 401
IgG 198, 199
IgM 198, 199
IgM secretion 253
$2'$-O-imidazolbutyl modification 72
$N2$-imidazolylpropyl-2-aminoadenines 62
immune
- mediated cellular infiltrates 200
- stimulation 28, 55, 192, 196, 197, 243–262
- - activation by antisense and control 246, 247
- - activation by CpG motivs 247–250, 254
- - immune effects of CpG motivs 250, 256, 257
- - immune effects of PS-ODN 244
- - immunostimulatory sequence motiv 196
- - leukocyte activation by CpG 254–256
- - non-specific 500
- - by oligonucleotides 243–258
- - phosphorothioate backbone 254
- - sequence-independent immune effects, DNA backbone 243, 244
- - sequence-specific immune effects, CpG motivs 245
- - sequence-specific immune effects, Poly(G) motifs 258
immunoglobulins
- $\alpha2$-macroglobulin 106

- $\beta2$-microglobulin 204
- murine 586
- production 197
- in vitro studies 197
- in vivo studies 197
in vitro
- antisense ODNs, in vitro effects 281–295
- cellular uptake, distribution and metabolism 2, 103–133
- metabolism 123
- pharmacokinetics 103, 144
- toxicological proterties 26
- uptake, intracellular distribution and efflux 106–120
in vivo
- antisense ODNs, in vivo effects 281–295
- cellular uptake 147–150
- pharmacological activities 22
- pharmakokinetics 16, 17
inflammatory diseases/process
- antisense oligonucleotides 371–393
- ICAM-1 514
- responses, local 193
infrared spectroscopy, triple helix 565
infusion, continuous intravenous 220
- OL(1)p53c 229
injection
- antisense ODNs 353
- dopamin D2 intrastriatal injection 291
- intravitreal injection, fomivirsen 450, 451
- microinjection (see there) 128, 350, 352, 354
INOS (inducible nitric oxide synthase) 429, 439
inosine 33
- 7-deaza 33
inteferon (see IFN) 245
$\beta2$-integrins 499
interactions with proteins 13
intercellular adhesion molecule-1, in humans 499–524
interleukin (IL) receptors
- IL-1 386
- IL-1α 27
- IL-2 194
- IL-2R 585
- IL-6 194, 252
- IL-10 516
- IL-12 252, 253
- - cytokine 252
- - secretion 252

internationalization and distribution, cellular uptake 132, 436
intracellular
– distribution 109, 110
– – methylphosphonate oligonucleotides 112
– – phosphodieseter oligonucleotides
   106, 109, 110
– – phosphorothioate oligonucleotides
   116–119
– metabolism 122
– stability 120–128
– – methylphosphonate oligonucleotides 122
– – phosphodiester oligonucleotides
   121
– – phosphothioate oligonucleotides
   122–128
intracerebroventricular treatment,
   D2 dopamine 290, 291
intradermally 29
intraocular
– fomivirsen 455
– inflammation 455
– pressure (IOP) 454
intrastriatal injection, D2 dopamine
   291
intra-tracheal dosing 165
intravenous administration 162
intraventricular administration
– D2 dopamine 288, 292
– D3 dopamine 294
intravitreally injection, fomivirsen 29,
   450, 451
7-iodo-7-deaza-guanine 33, 63
IOP (intraocular pressure) 454
ISIS
– 1047 107
– 1082 119, 125, 127
– – capillary gel electropherogram
   of ISIS 1082 127
– – 35S ISIS 1082 119
– 1939 376
– 2105 27, 123, 125, 164, 198, 201,
   203, 205, 218–221, 228, 229
– – 2105^a 218
– 2302 18, 26, 53, 117, 186, 190,
   197–199, 205, 206, 220–230,
   235–237, 376–382, 499–520
– – 2302^d 218
– – 2302 per kg 220, 221, 228
– – antibody formation 511
– – 20-base phosphorothioate ODNs
   499
– – clinical studies 383
– – Crohn's disease 514–520

– – cytokine-induced ICAM-1
   expression 500
– – future plans 520
– – intercellular adhesion molecule-1,
   in humans 499–524
– – and metabolism 230, 508, 509,
   513
– – in normal volunteers 237, 383
– – peak plasma concentration 513
– – pharmacokinetics 504, 506
– – phase I intravenous trial 501
– – phase II trial 513–519
– – placebo 501, 515
– – plasma concentration-time curves
   for subjects 504
– – plasma concentration-time profiles
   506, 511
– – remitters 516
– – responders 516
– – representative electropherogram
   505
– – safety 502
– – subcutaneous dosing 519, 520
– – tolaration 516
– – toxicity 519
– 2922 27, 52, 53, 125, 206–209, 446
– – 3H ISIS 2922 115
– – intravitreal 209
– 3082 206, 225, 379, 381, 500
– 3521 141, 198, 202, 218, 464, 465,
   468, 469
– – 3521/CGP 64128A 219, 223
– – clinical studies 468, 469
– 5132 75, 77, 141, 178, 191, 198,
   202, 218, 466, 471, 472
– – 5132/CGP 69846A 219, 223, 225
– – clinical studies 471, 472
– – pharmacokinetics 471
– 9125 379
– 2'-methoxy analog of 9
isomers, mixture of Rp and Sp 106

K
K562 489
Kd 13
keratinocytes 436
kidney 158, 203, 439
– ischemia 429
– proximal tubules 176
– toxicity 237
killer cells, natural (*see* NK) 253, 372
kinase
– cyclin-dependent kinase activity 22
– mitogen-activated protein kinase
  signaling 466

Subject Index

- protein kinase C 14
- *raf* kinases (*see also* c-*raf*) 77, 218, 399, 403, 408, 409, 465, 466, 468
- tyrosine kinase receptor 487
Ki-*ras* 401
kupffer cells 176, 200, 202

L
LD50 28
leukemia
- chronic myelogenous (CML) 407, 408, 482, 487
- c-*myb*, clinical experience with ODNs (*see also* c-*myb*) 477–484
- hematopoiesis, c-*myb* protein requirement 480
- leukemic cells 483
- *myb*-killing leukemic cells 487
- nucleic acid therapeutics for human leukemia 477–497
leukocyte(s)
- activation by CpG oligonucleotides 254
- circulating 196
LFA-1 (lymphocyte function associated antigen) 373, 375, 379
lipids, cationic 16, 105, 128
lipofectamine 128
lipofectase 128
lipofectin 128
5'-lipophilic PS 89
lipopolysaccharide (LPS) 243, 253
lipoproteins
- HDL 89
- LDL 89
liposomes 105, 115, 128, 431
- antitumor activity of GGP 69846A in stealth liposomes 411, 415
- cationic 105, 129
liver 237, 438, 439
- toxicity 237
local administration 164
LPS (lipopolysaccharide) 243, 253
- LPS-activated secretion 253
lung 439
lymph node 193
lymphadenopathy 244, 520
lymphocyte
- LFA-1 (lymphocyte function associated antigen) 373, 375, 379
- proliferation 27, 246–248
- - B-lymphocyte proliferation 27, 246–248
- - by ODNs 246
lymphoid

- hyperplasia 28, 192
- organs 193

M
M6G receptor, G-proteins by antisense ODNs 361
MAC-1 (macrophage 1) 373, 375
machrophages (MAC-1) 373, 375, 385
α2-macroglobulin 106
macular degeneration 435
MadCAM-1 (mucosal addressin cell adhesion molecule) 374
marrow-purging agents 484, 491, 492
matrilysin 402
MBO-1 and -2, phosphorothioate 527–533
- oral bioavailability/oral absorption 530, 533
- two end-modified MBO 533
mechanism
- action of antisense ODNs 276–281
- - inhibition of gene expression 276, 277
- proof of 1
- terminating mechanisms 3
medicinal chemistry
- antisense 51–92
- of oligonucleotides 31–38
melting
- lower duplex melting 143
- transition (*see* Tm) 13, 57
mephosphonates 73
metabolism 122, 123, 153–159, 175, 223–228
- GEM-91 224
- hepatic 158
- intracellular 122
- ODNs 175–179
- - degradation, metabolic 177
- pharmacokinetic properties 153–159, 223–228
- plasma 155
- tissues 155–159
- in vitro 123
metabolites 16
5-(Me-thiazolyl)-2-deoxyuridine 59
methotrexate 516
2'-methoxy
- analog of ISIS 9
- modification 10, 20
- - phosphorothioate oligonucleotides 20
- substitutions 200
2'-*O*-methoxy ribofuranosyl groups 84

2'-modified methoxyethoxy 74, 80, 209, 410
2'-O-Me-(methoxyethyl)-modification 58, 70, 71, 75, 76, 78, 81, 85, 92, 384
– with an RNase H-independent mode of action 76
methoxytriethoxy 78
2'-O-methyl (ME) 54, 164
– modification 164
5'-methyl C residues 191
5'-methylene group 82
methylphosphonate 83, 103
– chiral methylphosphonate bis-methoxy 83
– oligonucleotides 10, 105, 111–113, 122
– – efflux 112, 113
– – intracellular distribution 112
– – uptake 111, 112
– stability 122
methyl-phosphonate-modified ODNs 329
2'-O-(methyoxethyl)-modification 58
2'-methyoxyethyoxy 37, 191
– thiated (CGP 71849A) 191
MHC class II, human 586
mice, SCID (severly compromised immunodeficient) 486
micorinjection (see also injection) 128, 350, 352, 354
– antisense ODNs 352
– nuclear microinjection, G-protein antisense ODNs 350, 354
micro globulin, β2 204
β2-microglobulin 204
minor groove binding ligands, triple helix 576
miscellaneous, antisense-based approaches 406, 407
mismatching, triple helix 569, 570
mitogen-activated protein kinase signaling 466
mitogenic effects, oligonucleotides 250–252
MMI 82–84, 91
– bis-methoxy 83, 91
– modification 82–84
M-muscarinic, G-proteins by antisense ODNs 351
– M1-muscarinic 351
– M2-muscarinic 351
– M3-muscarinic 351
– M2/4-muscarinic 351
– M4-muscarinic 351
modifications / modified
– ODN modifications

– – structural modified 191
– oligonucleotides/oligonucleotide modifications
– – 2' modified 2, 69, 72, 79
– – – Rp-Me-phosphonates 85
– – 3' modified 34
– – 4' modified 67
– – 3'-amidate modified (see there) 84, 85
– – amide-3 modified 82
– – 2'-O-aminopropoxy modified 70, B2
– – 2'-O-aminopropyl (AP) modified 70, 72, 91
– – backbone modified 10, 37, 55, 75, 82
– – carbohydrates 55, 66
– – chimera 2'-O modified 72
– – cholesterol modified (see there) 89, 90
– – 7'-deaza modified 63
– – 2'-fluoro modified 70, 75
– – heterocycle (nucleobases) 32, 55, 56, 59, 64, 65
– – 2'-O-imidazolbutyl modified 72
– – linkage modified 55, 82
– – 2'-O-Me-(methoxyethyl)-modified (see there) 58, 70, 71, 75, 76, 78, 85, 91
– – 2'-O-methyl (ME) 54, 164
– – MMI-modified (see there) 82–84, 91
– – morpholino modified 82
– – N2-purine modified 61
– – phosphorothioate modified 143
– – 2-propoxy modified 64
– – 5-propynyl-desoxyuridine modified 64
– – 5-propynyl modified 61
– – purine 33
– – pyrimidine 32
– – sugar modified (see there) 36, 67, 81
– – 3'-thioformacetal modified 85
molecular mechanisms of antisense drugs 5
monkeys 28, 177, 180, 184, 187, 189, 204, 207, 225, 467, 510
– cynomolgus monkey 382
– monkey plasma 225
monocellular infiltrate, multiorgan 28
monoclonal antibodies, ICAM-1 499
monocyte cytokine secretion by CpG DNA 252
mononuclear
– cell infiltrates 192–195, 197

Subject Index

– – diffuse multiorgan mixed   192, 195
– peripheral blood cells (see PBMC)
  244
5′-monophosphates   66
5′-monothiophosphates   66
morphogenesis   205
morpholino
– bis-amidate   76
– diamidate-modified oligomers   69
– modification   82
– RNA heteroduplexes   69
MRT (h)   222
multidrug resistance   405, 406
– gene, human   586
multiorgan monocellular infiltrate   28
myelogenous leukemia, chronic   407, 408

N

N2-adenine   62
N2-imidazolylpropyl-2-aminoadenines
  62, 64
N2-purine-modified oligonucleotides
  61
N7 alkylating agents   589
N7-glycosylated guanosine   573
nanoparticles   128
naphtyl-quinoline derivate, triple helix
  575
NDMA-(N-methyl-D-aspartate)-R1   22
necrosis   193
neovascularization
– corneal   435
– ocular   429, 435, 436
nerve growth factor (NGF)   311, 314
nervous tissue, antisense ODNs   331
neuromedin B, G-proteins by antisense
  ODNs   351
neurotransmitter receptors, antisense
  ODNs   263–308
– dopamine receptors (see there)
  268–271, 280, 283–295
– GABA receptors (see there)   268,
  271, 295–298, 327
– NMDA receptors   268, 271, 298,
  299
– serotonin receptor   299, 300
neutrophil count and haemodynamic
  changes   511
NF (nuclear factor)
– κB   385, 386, 429, 432
– κB (p65)   246, 256, 398, 402
NGF (nerve growth factor)   311, 314
NK (natural killer) cells   253, 372
– IFN-y-secretion   253

NMDA receptors, antisense ODNs   268,
  271, 298, 299
– NMDA-R1   268, 271, 298
non-antisense effects   3
non-lymphoid-derived cells   196
non-mutagenetic concentrations   27
non-parenteral administration   164
northern blot   22
nuclear
– factor (see NF)   246, 256
– microinjection, G-protein antisense
  ODNs   350, 354
– nucleic acid therapeutics for human
  leukemia   477–497
– targets   133
nuclease   14, 122
– degradation   177
– inhibit nucleases   123
– resistance   57
– stability   15
nucleic acid binding proteins   14
– single-strand   14
nucleosome, triple helix   584
3′-O of the 3′-nucleotide   90

O

ocular
– intraocular (see there)   454, 455
– neovascularization   429, 435, 436
– side effects, CMV   454
– toxicity, antisense therapeutic agents
  206–209
– – inflammation, ocular   209
– – kinetics, ocular   207
– – pharmacokinetics   207, 208
– – profile, ocular   208, 209
– – toxicity profile   208–210
ODNs (oligodeoxynucleotide)
– antibodies, anti-ODNs   198
– antisense (AS-ODNs)   263–303, 350,
  405, 406
– – advantages   264
– – bone marrow-purging agents   484
– – c-myb antisense ODNs   479–484
– – CNS (see there)   315–332
– – decreased   405
– – delivery systems   273
– – distribution   273, 276
– – G-protein functions   350–360
– – injecting/microinjection   352, 353
– – mechanism   276–281
– – neurotransmitter receptors
  263–339
– – on Pgp   406
– – stability   274

ODNs
- – structure 266
- – transcript 266
- – uptake 273
- – in vitro and in vivo effects 281–295
- chemistry effects 199
- CpG ODN 254
- CpG PS-ODN 252
- G-rich 189, 258
- interactions, ODN-protein 171
- lymphocyte proliferation by ODNs 246
- metabolism (see there) 175–179
- methyl-phosphonate-modified ODNs 329
- phosphorothioate (see there) 147–168, 171–183, 184–225, 325, 466
- – OL(1)p53 512
- PS-ODN, immune effects 243, 244
- reproductive effects, antisense ODNs 205, 206
- synthetic ODNs 252, 254, 258
- toxicity 169–215
- treatment 200

OL(1)p53c 218, 220, 221, 229, 230, 234, 236, 238, 239, 512
- intravenous infusion 229
- oligonucleotide 230
- per kg 220

oligodeoxynucleotides (see ODNs)
oligomers 38, 69, 85, 122
- formacetal modified 85
- methylphosphonate oligonucleotides 122
- morpholino diamidate-modified 69
- PNA 38

oligonucleotides
- α-oligonucleotides 10
- anti-apoptotic effects 250
- anticancer activities 395–425
- antisense pharmacology 129, 141, 371–393, 428, 491, 525–543
- – adhesion molecules 373–385
- – bioavailability 525–543
- – c-raf kinase 465, 466
- – clinical setting 491, 492
- – efficacy 428
- – inflammatory process 371–393
- – interleukin-1 (IL-1) receptors 386
- – nuclear factor-κB 385, 386
- – protein kinase C-α 463–465
- – second and third generation 383
- chemical class 2
- chimeric (see there) 11, 36

- conjugated/pendants (conjugates) 33, 87, 89
- control 4
- CpG oligonucleotides 252
- detection and analysis 144
- immune stimulation 243–258
- medicinal chemistry 31
- methylphosphonate 105
- mitogenic effects 250–252
- mixed-backbone
- – oligonucleotide-1 529, 535, 537
- – oligonucleotide-2 532, 533, 536
- modifications (see there)
- pharmacodynamics 230, 231
- pharmacokinetic properties
- – in animals 525–543
- – in humans 217–232, 384
- phosphorothioate (see there) 3, 13, 17, 20, 31, 73, 103, 106, 113–120, 141–168, 169–215, 223, 466
- purity 1
- radioactivity, oligonucleotide-derived absorbed 538
- [35S]-labeled oligonucleotides 526
- structure 105
- TFO (triple helix oligonucleotide) 568–570, 578, 584

opioids, G-proteins by antisense ODNs 351, 361
- δ-opioid 351, 361
- μ-opioid 361
- κ1-opioid 361
- κ3-opioid 361

oral
- absorption 531
- administrations in animals, antisense oligonucleotides 525–534
- bioavailability/oral absorption 20, 56, 173, 527–543

organ toxicity 237–239
osteopontin 403

P

p53 218
p120 398, 403
P150/95 375
PAGE-(polyacrylamide gel electrophoresis)-autoradiography 144, 224
palindromic
- elements containing CG 197
- sequences 196, 245

parenteral
- administration 162, 173
- routes 173

patients 190, 222, 468, 471, 472, 514

Subject Index

- c-*myb* antisense ODNs 481
- ISIS 3521 468
- ISIS 5132 471, 472
- refractory 514
PBMC (peripheral blood monocuclear cells) 244
PCNA 398, 429, 431, 432
PCNA mRNA 431
PCR (polymerase chain reaction), reverse-ligation 11
PDGF-A 399
PECAM-1 (platelet/endothelial cell adhesion molecule) 373, 375, 384
pentofuranose ring 36
peptide nucleic acid (*see* PNA) 38, 83, 85, 545–559
peripheral organs, pharmacokinetics of oligonucleotides 437
Pgp 406
pharmacodynamic properties 31, 141, 230, 231
- c-*myb*-targeting
- limitations 53
pharmacokinetic properties 5, 15–17, 31, 103, 141–168, 217–232, 471
- antisense oligonucleotides, oral and colorectal administrations, in animals 525–543
- antiviral activities 448, 449
- dose-dependency of human plasma pharmacokinetics 222
- ISIS 5132 471
- limitations 54
- ocular pharmacokinetics 207, 208
- phosphorothioate 31, 141–168, 197, 217–232
- – absorption 161–165, 219, 220
- – distribution (*see there*) 145–150
- – elimination 159–161, 228–230
- – in humans 217–232
- – metabolism (*see there*) 153–159
- – modification 143, 144
- – multiple dosing 153
- – non-linear 150
- – ODNs, toxicity 173–179
- – oligonucleotide detection and analysis 144, 145
- – plasma 177
- in vitro 103
- in vivo 16, 17
pharmacology/pharmacological
- antisense oligonucleotide activities
- – animal models of disease 427–443
- – antiviral acitivities 447
- – in vivo 22

- effects in CNS 309–339
- inhibition of dopaminergic and other neurotransmitter receptors 263–308
pharmacology, antisense oligonucleotides 377–382
colitis, ulcerative
- ISIS 2302 513, 514, 517
- antisense oligonucleotides 381, 382
phase II
- antitumor 75
- trials, ISIS 2302 513–519
phophoramidates 76
phophoryl group 82
phosphodiesterase, stability 105
phosphodiesters (PO) 13, 37, 57, 103–105, 106–111, 119, 121, 372
- α-anomeric 105
- backbone, phosphodiester-phosphorothioate 187
- end-capped 121
- G-quartet, structured 121
- guanin-rich 105
- oligonucleotides
- – efflux 110, 111
- – intracellular distribution 109, 110
- – uptake, cellular 103–105
- – uptake, intracellular 106–111
- – uptake, mechanisms of 107
- 2'-propoxy 37
- stability 121
- SVPD (snake venom phosphorodiesterase) 57
- wings, phosphodiester 187
ME-phosphonates 76
- *Rp* 2'-*O*-Me 85
Rp-Me-phosphonates, 2'-modified 85
5'-phosphoramidate 105
N3'-5'-phosphoramidate 578
phosphoramidates 73
2,3'-phosphorothioate 105
phosphorothioate, oligonucleotide/ODNs 2, 3, 13, 17, 20, 31, 52, 73, 103, 106, 113–120, 141–168, 171–215, 223, 233–258, 266, 387, 465, 466, 486
- antisense
- – drug candidates 52
- – inhibitor 465
- congeners 107
- degradation 125
- distribution 17, 116–119, 145–147
- – blood kinetics 145
- – dose dependence 150
- – extravascular 223
- – intracellular 116–119

– – sequence dependence 150, 151
– – tissue distribution 145, 147
– end-capped 121
– ICAM-1, antiflammary activity 387
– immune effects 254
– limitations 53
– linkage, phosphorothioate 187
– MBO-1 and -2 527–531
– metabolism/metabolization 123, 223
– 2′-methoxy 20
– modification 143, 266
– ODNs 147–168, 171, 285, 289, 191, 197, 465, 466
– – 20-base 466
– – c-*myb* 486
– – complement activation 183
– – hepatic effects 200
– – local administration 164
– – non-parenteral administration 164
– – parenteral administration 162
– – pharmacokinetic 197
– – plasma 185
– – polyanions 189
– – structural modifications 191
– – substituted 325
– pharmacodynamics 31, 141
– pharmacokinetic properties (see also there) 31, 141–168, 217–232
– – absorption 161–165, 538, 539
– – distribution (see there) 145–150
– – elimination 159–161
– – in humans 217–232
– – metabolism (see there) 153–159
– – modification 143, 144
– – multiple dosing 153
– – non-linear 150
– – oligonucleotide detection and analysis 144, 145
– safety and tolerance in humans 233–258
– stability 122, 143
– toxicologic 31
– uptake 107, 113–116
phosphotriester 76
photoinduced azides 35
pinocytosis, fluid-phase 112
PKA (RIα) 403
PKC-α 218, 399, 403, 409, 463, 465, 466
– in abnormal proliferation 464
– in cell lines 464
– expression 409
– isoenzymes 464
– mRNA expression 465
– protein kinase 438

– toxicities, target-specific 466
PKC-β1 464
PKC-γ 464
placebo ratio, ISIS 2302 501, 515
plasma thresholds 185
– AUC (area under the plasma concentration curve) 186, 222, 223
– complement activation 185
– concentrations 188, 220, 471
– maximum concentration (Cmax) 178, 186, 222, 223
– monkey plasma 225
– pharmacokinetics 177, 220–223
– – calculated parameters 221–223
– – peak plasma concentrations 220, 221
– phosphorothioates metabolism 155
platelets
– counts, circulating 205, 236
– savety, phosphorothioates 235, 236
PLC-β 354
PNA (peptide nucleic acid) 38, 83, 85, 545–559
– antigene activity 555–557
– antisense activity 545–559
– cellular biology 549
– chemical structures 548
– complexes 552
– duplex 547
– hybridization 545, 552
– mechanism of action 548, 549
– oligomers 38
– PNA-DNA chimeras 553–557
– PNA-RNA 547, 548
– PNA2-RNA triplexes (homopurine targets) 548
– strand invasion in double-stranded DNA 555
– structure 547
– structure-activity relationship 549
poly (G) 258
poly (IC) 245
poly (L-lysine) 130
3′-polyadenylation 9
polyamines and basic oligopeptides, triple helix 576, 577
polyanions 27, 189
– polyanionic nature 55, 180
polycations 128
polynucleotides 27
polysaccharides, cyclic 130
porphyrin 35
2′position 36, 187
progesterone responsive gene 585
proof of mechanism 1

proof of mechanism, antisense oligonucleotides 377
2'-propoxy 37
– phosphodiester 37
5-propynyl
– deoxycytidine 59, 64
– deoxyuridine 59, 64
– modified oligonucleotides 61
– pyrimidine 32
proteases 405
protein
– AP-c-*raf* protein 77
– binding/binding protein 3, 14, 17
– – serum protein binding 17
– interactions with proteins 13, 171
– – ODN-protein 171
– mitogen-activated protein kinase signaling 466
– G proteins (*see* G) 341–369
– kinase C 14
– – C-α 408, 463–465
– – c-*raf* 408, 465, 466
– nucleic acid binding proteins 14
prothrombin times (PT) 188
PS-ODN
– cytokine secretion 243
– immune effects 244
psoralene 35
psoriasis 513, 517, 518
PT (prothrombin times) 188
ptt (partial thromboplastin time), activated (*see* aPTT) 29, 188, 190, 191, 193, 235, 239
pulmonary administration 165
purine modifications 33
– 3-deaza 61
PyPuPy formation, triple helix motif 567, 570
– antiparallel 567, 570
– parallel 567, 570
pyrimidine 32, 33, 59, 61
– 6-aza 33
– modifications 32
– 5'-position pyrimidine- and tricyclic cytidine-modified oligonucleotides 59
– 4-thyopyrimidines 32

R

radioactivity, absorbed oligonucleotide-derived 538
raf kinases
– a-*raf* 465
– b-*raf* 408, 465, 466

– c-*raf* (*see there*) 77, 218, 396, 399, 403, 408, 409, 465, 466
– clinical development 468
rank order 4, 80
ras kinases 396
RAS (renin-angiotensin system) 434
rats 189, 204
– rat restenosis 431
receptor
– receptor-mediated endocytosis 3, 133, 489, 490
– Y-Y1 receptor 22
rectal administrations in animals, antisense oligonucleotides 525, 534–543
red blood cells, savety, phosphorothioates 236, 237
remitters, ISIS 2302 516
renal
– acute allograft rejection/transplantation 513, 518
– – prophylaxis 513
– changes 204
– dysfunction 204
– effects, ODNs, toxicology treatment 202–205
– functions 470
– ischemia, antisense oligonucleotides 382
– ISIS 2302 518
– phase II segment 518
– toxicity 202
renin-angiotensin system (RAS) 434
reproductive effects, antisense ODNs 205, 206
– embryos 206
responders, ISIS 2302 516
restenosis 427–432
– efficacy of antisense oligonucleotides 428
– rat 431
retina/retinal/retinopathy/retinitis 435
– CMV/HCMV retinitis 29, 207, 209, 451, 455
– concentrations 207
– pigment epithelium (RPE) 209
– retinopathy of prematurity 435
reverse-ligation PCR 11
rheumatoid arhtritis 500, 513, 517
ribofuranosyl moieties 67
– β-D-ribofuranosyl moiety 69
ribonuclease H (*see* RNase H) 3, 10–12, 54, 58, 69, 76, 91, 143, 330
ribonucleotide 10
– deoxyribonucleotides 10
ribosome 8

RNA 2
- antisense RNA, G-protein functions 345–350
- – transcription 347
- backbone 577
- *bcr-abl* 25, 485
- cleavers/cleaving groups 35, 62
- gene walks 548
- mimics (2'-*O* modifications) 91
- PNA-RNA 547
- PNA2-RNA triplexes (homopurine targets) 548
- polymerase, triple helix 581–583
- structure 2
RNase 12, 91
- double-strand RNase 12
RNase H (ribonuclease H) 3, 10–12, 54, 58, 69, 76, 91, 143, 330, 383, 438, 548
- escherichia coli 11
- independent antisense 76
- mechanism 58, 91
- mode 58
- RNA 548
rodent allografts, antisense oligonucleotides 379
RPE (retinal pigment epithelium) 209

S
safety
- and tolerance of phosphorothioates in humans 233–258
- – cardiovascular safety 233, 234
- – complement activation 237
- – hematological savety (*see there*) 234–237
- – specific organ toxicity 237–239
- profile 191
SAR (structure-activity relationship) 53
scale-up and development, chemical, antiviral activities 450
scarring/wounding 429
scavenger receptor 176
SCID (severly compromised immunodeficient) mice 486, 487
- survival curves 487
secondary structures 2
selectin
- E-selectin 374
- L-selectin 374
- P-selectin 374
sequence 2
serotonin receptor
- antisense ODNs 271, 299, 300
- – 5-HT6 271, 299

- G-protein, serotonin-induced stimulation 346
signal transduction 352–354, 436
- G-protein-mediated inhibition 352–354
skin 436, 437
[35S]-labeled oligonucleotides 526
snake venom phosphorodiesterase (SVPD) 57, 75, 84
somatostatin, G-proteins by antisense ODNs 351
species differences 197
spleen 192–196, 253
- cells 253
- weights 192, 194, 196
splenomegaly 28, 192, 244
splicing, inhibition 5
stereochemistry 106
steroids 29
streptolyosin O 128
striatum, c-*fos*, antisense ODNs 315, 316
structure-activity relationship (SAR) 53, 194
structures 2
- secondary 2
- tertiary 2
subcutaneous administration/-regimen 17, 162
- dosing of ISIS 2302 519, 520
sugar
- 3'-endo sugar conformation 70
- modifications 36, 67, 81
- – 2'-sugar position 67
- – 4'-sugar position 67
SVPD (snake venom phosphorodiesterase) 57, 75, 84
synthetic promotor, triple helix 586

T
TAR (transactivator response element) 7
terminating mechanisms 3
tertiary structures 2
tetrameric structures 15
tetraplex formation by guanine-rich oligonucleotides, triple helix 570
TFO (triple helix oligonucleotide) 568, 569, 578, 584, 590
- antiparallel 578
- formacetal 578
- psoralen-linked 590
- unmodified 584
TGF-α 401
TGF-β, antisense oligonucleotide 399

- TGF-β1   401, 429, 433, 476
- TGF-β2   401
therapeutic index   29
thioformacetal   83
3'-thioformacetal modification   85
thiono triester   34
thrombocytopenia   205, 235, 236
thromboplastin time, partial, activated (see aPTT)   29, 188, 190, 191, 193, 235, 239
4'-thyopyrimidines   32
thyrotropin-releasing hormone (TRH)   314
tissue, phosphorothioates ODNs   174
- distribution   145, 147, 173, 174
- metabolism   155
Tm   567
Tm (melting transition)   13, 57
- increase of Tm greater than 1,5C   57
TNF (tumor necrosis factor)   244
- anti-TNF Ab (chimeric)   516
- TNF-α   244, 376, 378
topoisomerases   28
toxicological properties/toxic effects/ toxicities   26, 27, 31, 55, 169–179
- c-raf kinase, target-specific toxicities   466
- ICAM-1 antisense oligonucleotides   382
- implications of 2'-modified   80
- ISIS 2302, subcutaneous toxicity   519
- liability   65
- ODNs, therapeutic agents   169 ff.
- – acute   179
- – antisense-mediated   171
- – class-related   171
- – complement activation   180–187
- – dose-limiting toxicities   180
- – genetic   179
- – hematopoiesis   105, 193, 205
- – hemostasis   187–191
- – hepatic effects   200–202
- – immune stimulation   192–200
- – non-antisense-mediated pathways   171, 172
- – ocular toxicity (see there)   206–209
- – pharmacokinetics   173–179
- – profile, toxicologic   175
- – renal effects   202–205
- – response, toxicologic   171
- – studies in rodents, lagomorphs and primates   171
- – systemic   179
- – toxicologic effects   192, 204
- oligooxinucleotide   26 ff., 54

- – antivirals   449, 450
- – genotoxicity   27
- – limitations   54
- – organ toxicities, specific   237–239
- – in vitro   26
- PKC-α, target-specific toxicities   466
transaminase activity, serum   238
transcription(al)
- activation by CpG   256
- factors   14, 312, 313
- – IEGs   312, 313
- triple helix   580–584
- – factor   581
- – inhibition   581
- – initiation   583
- – vitro transcription   581
transfectam   128
translation/translational
- arrest   6
- initiation codon   6
TRH (thyrotropin-releasing hormone)   314, 351
- G-proteins by antisense ODNs   351
tricyclic cytidine-modified oligo- nucleotides, 5'-position   59
triple helix DNA   561–593
- adenosine derivates   572
- alternate strand recognition   578, 579
- backbone modification and strand switching   577–579
- benzopyridonindole derivated   575
- binding ligands   574
- c-myc promotor   583
- chromosome structure
- conjugation of small compounds enhance triple helix stability   579, 580
- cross-linking   590
- cytidine derivates   572
- detailed structure   562–566
- DNA replication   587–589
- – DNA cleavage   589
- effect of pH   571
- electron microscopy   565, 566
- electrophoresis, phase-sensitive   565
- enzymatic cleavage, single-site   589
- ethidium bromide   574, 575
- formation   583, 584
- gene regulation   580
- guanosin derivates   573
- HIV integration   587
- infrared spectroscopy   565
- kinetics of triple helix formation   571
- minor helix groove binding ligands   576

- minor helix groove bridging 578, 579
- mismatching 569
- mutagensis 590
- naphtyl-quinoline derivate 575
- nucleoside modification 571–573
- – antiparallel motif binding 573
- – parallel motif binding 573
- nucleosome 584
- oligonucleotide (*see* TFO) 568, 569, 578, 584, 590
- other intercalators 575, 576
- other pharmaceutically interesting 587
- polyamines and basic oligopeptides 576, 577
- PyPuPy formation, triple helix motif 567, 570
- – antiparallel 567, 570
- – parallel 567, 570
- RNA polymerase 581–583
- stability 566–571
- strategies and progress 561–593
- synthetic promotor 586
- tetraplex formation by guanine-rich oligonucleotides 570
- transcription (*see there*) 580–584
- *Watson-Crick* douplex 572
- X-ray crystallography 566

tubules/tubular (in kidney)
- degeneration 193, 204
- proximal 147, 176
- – epithelial cells 202
- – tubular epithelium 194

tumor
- growth 465
- necrosis factor (*see* TNF) 244

tyrosine kinase receptor 487

U

ulcerative colitis 381, 382, 513, 514, 517
3'-untranslated region 499
uptake
- of 21-mer 109
- enhancement, methods 128–130
- mechanisms 112, 118, 488
- methylphosphonate oligonucleotides 111, 112
- ODNs 273
- phosphodiester oligonucleotides 2, 103–133
- – cellular (*see there*) 2, 103–133
- – intracellular 106–120
- phosphorothioate
- – ODNs 147–150, 436, 494
- – oligonucleotides 113–116

urinary excretion 512
urine 230
urokinase-type PA 402

V

vascular system 427–436
- cardiovascular (*see there*) 161, 180, 187, 193, 233, 234, 239
- restenosis 427–432

VCAM-1 (vascular cell adhesion molecule 1) 373, 375, 384, 385
VEGF (vascular endothelial growth factor) 401, 429, 436
vein occlusion, central veins 435

virus/viral
- antiviral (*see there*)
- cytomegalovirus (*see* CMV/HCMV) 29, 206–209, 446
- hepatitis C virus 76
- human immune deficiency virus (*see* HIV) 7, 14, 21, 218, 219, 246, 445
- human papilloma virus (*see* HPV) 29, 218, 219

visual acuity 455
VLA-4 (very late activation, antigen) 373
volunteers, normal 190
v-*ras*-Ha 464
VSS 222

W

Watson-Crick duplex, triple helix 572
Watson-Crick
- base 64
- hybridzation 20
3' wings 11
5' wings 11
wounding 429

X

xenografts, human, antisense oligonucleotides 378, 379
X-ray crystallography, triple helix 566

Printing: Saladruck, Berlin
Binding: Buchbinderei Lüderitz & Bauer, Berlin